Communications
in Computer and Information Science 215

Song Lin Xiong Huang (Eds.)

# Advances in Computer Science, Environment, Ecoinformatics, and Education

International Conference, CSEE 2011
Wuhan, China, August 21-22, 2011
Proceedings, Part II

Springer

Volume Editors

Song Lin
International Science & Education Researcher Association
Wuhan Branch, No.1, Jiangxia Road, Wuhan, China
E-mail: 1652952307@qq.com

Xiong Huang
International Science & Education Researcher Association
Wuhan Branch, No.1, Jiangxia Road, Wuhan, China
E-mail: 499780828@qq.com

ISSN 1865-0929 e-ISSN 1865-0937
ISBN 978-3-642-23323-4 ISBN 978-3-642-23324-1 (eBook)
DOI 10.1007/978-3-642-23324-1
Springer Heidelberg Dordrecht London New York

Library of Congress Control Number: Applied for

CR Subject Classification (1998): I.2, C.2, H.4, H.3, D.2, H.5



*Typesetting:* Camera-ready by author, data conversion by Scientific Publishing Services, Chennai, India

Printed on acid-free paper

Springer is part of Springer Science+Business Media (www.springer.com)

# Preface

The International Science & Education Researcher Association (ISER) puts its focus on the study and exchange of academic achievements of international teaching and research staff. It also promotes educational reform in the world. In addition, it serves as an academic discussion and communication platform, which is beneficial for education and scientific research, aiming to stimulate the interest of all researchers.

The CSEE-TMEI conference is an integrated event concentrating on the field of computer science, environment, ecoinformatics, and education. The goal of the conference is to provide researchers working in this field with a forum to share new ideas, innovations, and solutions. CSEE 2011-TMEI 2011 was held during August 21–22, in Wuhan, China, and was co-sponsored by the International Science & Education Researcher Association, Beijing Gireida Education Co. Ltd, and Wuhan University of Science and Technology, China. Renowned keynote speakers were invited to deliver talks, giving all participants a chance to discuss their work with the speakers face to face.

In these proceeding, you can learn more about the field of computer science, environment, ecoinformatics, and education from the contributions of several researchers from around the world. The main role of the proceeding is to be used as means of exchange of information for those working in this area.

The Organizing Committee made a great effort to meet the high standards of Springer's *Communications in Computer and Information Science* (CCIS) series. Firstly, poor-quality papers were rejected after being reviewed by anonymous referees. Secondly, meetings were held periodically for reviewers to exchange opinions and suggestions. Finally, the organizing team held several preliminary sessions before the conference. Through the efforts of numerous people and departments, the conference was very successful.

During the organization, we received help from different people, departments, and institutions. Here, we would like to extend our sincere thanks to the publishers of CCIS, Springer, for their kind and enthusiastic help and support of our conference. Secondly, the authors should also be thanked for their submissions. Thirdly, the hard work of the Program Committee, the Program Chairs, and the reviewers is greatly appreciated.

In conclusion, it was the team effort of all these people that made our conference such a success. We welcome any suggestions that may help improve the conference and look forward to seeing all of you at CSEE 2012-TMEI 2012.

June 2011 Song Lin

# Organization

## Honorary Chairs

| | |
|---|---|
| Chen Bin | Beijing Normal University, China |
| Hu Chen | Peking University, China |
| Chunhua Tan | Beijing Normal University, China |
| Helen Zhang | University of Munich, Germany |

## Program Committee Chairs

| | |
|---|---|
| Xiong Huang | International Science & Education Researcher Association, China |
| Li Ding | International Science & Education Researcher Association, China |
| Zhihua Xu | International Science & Education Researcher Association, China |

## Organizing Chairs

| | |
|---|---|
| ZongMing Tu | Beijing Gireida Education Co. Ltd, China |
| Jijun Wang | Beijing Spon Technology Research Institution, China |
| Quan Xiang | Beijing Prophet Science and Education Research Center, China |

## Publication Chairs

| | |
|---|---|
| Song Lin | International Science & Education Researcher Association, China |
| Xiong Huang | International Science & Education Researcher Association, China |

## International Program Committee

| | |
|---|---|
| Sally Wang | Beijing Normal University, China |
| Li Li | Dongguan University of Technology, China |
| Bing Xiao | Anhui University, China |
| Z.L. Wang | Wuhan University, China |
| Moon Seho | Hoseo University, Korea |
| Kongel Arearak | Suranaree University of Technology, Thailand |
| Zhihua Xu | International Science & Education Researcher Association, China |

## Co-sponsored by

International Science & Education Researcher Association, China
VIP Information Conference Center, China

## Reviewers

| | |
|---|---|
| Chunlin Xie | Wuhan University of Science and Technology, China |
| Lin Qi | Hubei University of Technology, China |
| Xiong Huang | International Science & Education Researcher Association, China |
| Gang Shen | International Science & Education Researcher Association, China |
| Xiangrong Jiang | Wuhan University of Technology, China |
| Li Hu | Linguistic and Linguidtic Education Association, China |
| Moon Hyan | Sungkyunkwan University, Korea |
| Guang Wen | South China University of Technology, China |
| Jack H. Li | George Mason University, USA |
| Marry. Y. Feng | University of Technology Sydney, Australia |
| Feng Quan | Zhongnan University of Finance and Economics, China |
| Peng Ding | Hubei University, China |
| Song Lin | International Science & Education Researcher Association, China |
| XiaoLie Nan | International Science & Education Researcher Association, China |
| Zhi Yu | International Science & Education Researcher Association, China |
| Xue Jin | International Science & Education Researcher Association, China |
| Zhihua Xu | International Science & Education Researcher Association, China |
| Wu Yang | International Science & Education Researcher Association, China |
| Qin Xiao | International Science & Education Researcher Association, China |
| Weifeng Guo | International Science & Education Researcher Association, China |
| Li Hu | Wuhan University of Science and Technology, China, |
| Zhong Yan | Wuhan University of Science and Technology, China |
| Haiquan Huang | Hubei University of Technology, China |
| Xiao Bing | Wuhan University, China |
| Brown Wu | Sun Yat-Sen University, China |

# Table of Contents – Part II

# The Implementation of NPKI System Based LDAP

DaWei Xu[1] and XiaoHong Hu[2]

[1] Computer Science and Technology, Shandong University of Finance, Jinan, China
xudw76@163.com
[2] Department of Finance of Shandong ProvinceJinan, China
haohuhu@163.com

**Abstract.** This paper presents a NPKI model based LDAP directory service, describes the implementation of certificate and CRL management system in detail. The model uses three-tier structure of web browser-web server - the directory server, describes details of the implementation of the various functional components of certificate and CRL storage system, and describes certificate and CRL extensions, treatment of the recommendation of directory server and client, and several other basic services in detail. Meanwhile, it describes the details of the certificate and CRL in the inserted and asynchronous query implementation of directory information. The system is open, scalable, security, ease to maintain and so on, and provides users with a positive, intelligent, dynamic certificate services.

**Keywords:** LDAP; NPKI; Certificate; CRL.

## 1 Introduction

NPKI, which is based on nested certificate PKI, it use ordinary certificates and nested certificate, change the certificate verification path and the verification algorithm of the subject certificate and nested certificate in the path, reaching efficiency, flexibility and user-centered design purposes[1]. As the issuance of the nested certificate in the NPKI system, the certificate number of the entire system has increased significantly. Therefore, author introduces the LDAP directory into NPKI system, when the number of the certificate is substantial increased, it uses LDAP directory to store the certificate distributed, by using the division of the directory information and recommendations, whose network will be set up into a manageable, easy to control and orderly world. Through the organic integration of NPKI and LDAP directory services, it can provid fast, secure certificate services[2].

## 2 Implementation of System

As a core component of NPKI model, to realize a management system based LDAP directory of certificate and CRL, it uses three-tier B / S structure, directory partition and recommendations, to realize the distributed query of the certificate and CRL.

S. Lin and X. Huang (Eds.): CSEE 2011, Part II, CCIS 215, pp. 1–6, 2011.

### 2.1 Directory Information Structure

Design a directory service structure should be considered:

- the number of users need to use the network
- the location of network users (preferably to achieve centralized management of the account)
- types of enterprises and organization
- resources of users to access (type, quantity, location and management)[3]

According to the above demand, directory information is to be partitioned on geographical location, and be partitioned on provinces and municipalities as the unit of origin, with the lower-level directory server established, the server store all the local certificate and some other parts of the certificate.it can not find the certificate in local server, to provide services through the recommendation mechanism. As shown in fig1, the performance of this division, including: dc is a Domain Component, and ou is organise unit.

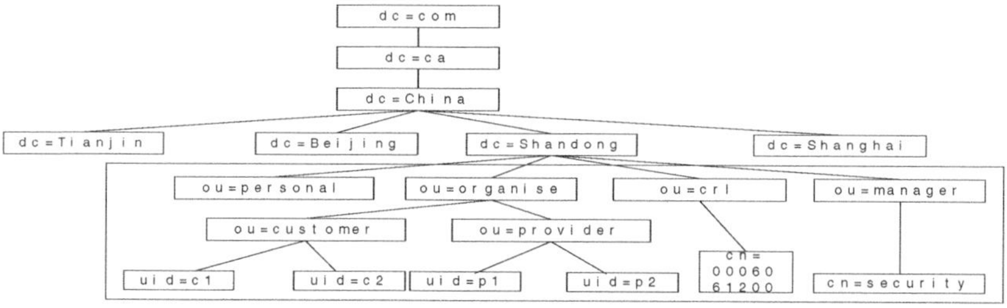


**Fig. 1.** Components of directory tree

### 2.2 Certificate and CRL Storage Process

**Extension of certificate class and attribute**

This model has two kinds of certificates: X.509 certificate and nested certificates. So it is necessary to extend the standard object class and attribute provided by LDAP. In the storage of X.509 certificate, LDAP provides an standard attribute usercertificate, but did not provide a standard attribute for the storage of nested certificate, so the model defines an extension attribute for nested certificate is nestedcertificate. In order to store the certificates in the directory server, records of any one user or server must have the corresponding attribute in the certificate, which also need to extend the standard object class to design a new object class certificateobject used to store information of users or unit and certificate. The object class attribute must be included: surname (sn), name (cn), ID (uid), it can contain attributes for the user e-mail (e-mail), X.509 certificates (usercertificate), nested certificates ( nestedcertificate) and so on.

**Extension of CRL class and attribute**

In order to meet the needs of storage and issue of CRL, the system design a new object class certificaterl based on the standard class provided by the LDAP. The object class attribute must be included: CRL attribute (sn), CRL name (cn), certificate revocation list (certificaterevocationlist), etc., and can contain attributes: whether

Delta-CRL (delta), the serial number (serialnumber), issue time(createtime), the next issue time (nexttime) and so on. Certificaterevocationlist is used to store CRL, createtime, nexttime and the delta are extended attributes, being respectively used to store the issued time of CRL, the issued time of the new CRL and whether the identity of Delta-CRL, that is the different revocation record with the base CRL.

**Storage Process**

Based on specifications of LDAP[4], the basic process of store model certificate as follows:

- receives and generate certificate file from the CA, obtain the certificate content;
- initialize connection with the directory server;
- bind the certificate to the directory server as an administrator;
- search directory, obtain the corresponding full DN by account of the user corresponding to the certificate;
- according to the certificate type, decide the certificate attribute should be stored. Write, write the contents of the certificate to the corresponding attribute of corresponding DN records;
- Disconnect from the directory server.

The basic process of store CRL as follows:

- CRL file generated from the CA obtain the latest CRL content;
- initialize connection with the directory server;
- bind to the directory server as a CRL administrator;
- using filters sn = new search directory, find and record the DN of the latest CRL before the update;
- Add a CRL record, write the latest contents of the CRL and some additional information to the record and set sn is new;
- If the 5th operation is successful, update the sn attribute of the latest CRL ewritten as old; disconnected from the directory server.

## 2.3 Implementation of LDAP Directory Recommendation

In order to achieve the directory recommendation, the model designs a referral class that has a attribute ref, ref attribute has the following purposes:

Name recommendation: name a reference server by ref attribute, the attribute value identify the appropriate directory items on the referenced server. Identifies name of ref attribute can be a naming context managed by the directory server, it can be some directory item identifies name of this server.

Higher recommendation: When the server receives a request, which does not contain the basic items or the name context in the request, the server will return a higher level recommendation, the value is the ref attribute of the root DSE.

No name recommendation: It is not only a concept of distributed queries and cross-index in multi-server, in using the recommendation server, recommendation is used for the name directory items, in the process of query, the directory items named by recommendation can contain a number of other attributes.

**Process of server treats with recommendation**

When a client send a directory request to the server, the server itself does not have the requested directory information, it should return a recommendation to the client, but when the client try to read a directory item may exist but does not actually exist or can not be returned due to licensing reasons, it returns an error message or no response. When directory server returns a recommendation, it allows map a directory item or a directory information tree to ldapURL, as follow situations:

- generates a pointer point to a different directory server containing the same namespace;
- generates a pointer of a different name space on the local server;
- generates a pointer of a different namespace on different server (can be a remote server). As shown in fig,2,

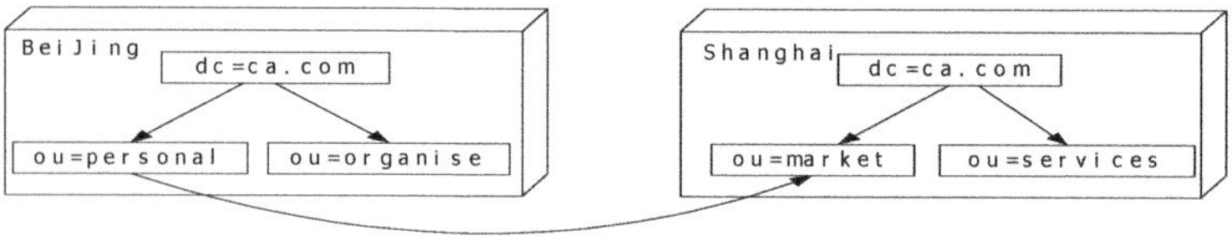


**Fig. 2.** Recommendation point to a different namespace of different server

If all of requests of "o = personal.beijing.ca.com, c = cn" on Beijing server are redirected to "o = market.shanghai.ca.com, c = cn", only set the following recommendations: " ldap: / / o = shanghai.com, c = cn" in the directory item "o = beijin.ca.com, c = cn", which can generate a pointer as shown in Figure3.If in a query request the immediate subordinates directory item of the basic item contains rel attribute, the server returning recommendation must contain base range in returning of corresponding SearchResultReferenceLdapURL, attribute ref of multi-valued can be used for showing different locations of the same resources, as shown,

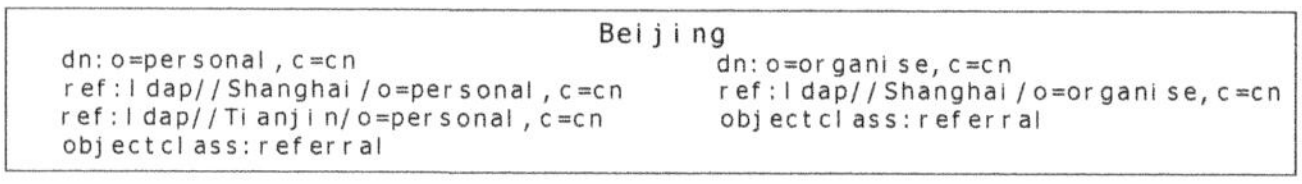


**Fig. 3.** Multiple-valued recommendation

If a client sends an queries to subtree "c = cn", this time, any directory item of naming context containing "c = cn" matchs to the filters, the server in Beijing will return two continued recommendation, one is "o = personal, c = cn", the other is "o = organize, c = cn". As following form,

```
...SearchResultEntry response...
searchResultReference
    {ldap://Shanghai/o=personal,c=cn
     ldap://Tianjin/o=personal,c=cn}
SearchResultReference
    {Ldap://Shanghai/o=organize,c=cn}
```

SearchResultDone"success"

If a client sends an single query request to "c = cn", the server will return two continued recommendation, as following form,

...SearchResultEntry response...
searchResultReference
{ldap://Shanghai/o=personal,c=cn?? base
ldap://Tianjin/o=personal,c=cn?? base}
SearchResultReference
{Ldap://Shanghai/o=organize,c=cn?? base}
SearchResultDone"success"

At this point, continued recommendation of the returned URL include base range, which indicates single query semantics.

**Process of client treats with recommendation**

When the client receive the recommendation, it should sends a same request to the server referred by the recommendation automatically. If the client provide connection authentication in the initial directory requests, the same authentication information should be used to connect new server after it receive the recommendation. Client has a track recommendation limitation, which can reduce query time of directory network service by limiting the number of tracking recommendations, also help to reduce process hang as a result of recommendations ring.

When the client receive the recommendation, the client will format the original request automatically to fit collection range of recommendation, then the client will re-process the request. The client will automatically query the directory and the information tree represented by the URL. In this case, the query basic items will be modified to fit the identification name of recommendation, the new operation's success depends on whether the remote directory system allows this operation. In the same case, if you try to modify a directory item, and this directory item is recommendation pointing to another server, the client will reformat the request to fit collection range of recommendation.

**Implementation process of recommendation**

In this model, the recommendation mechanism of the directory completed by a few pointers, they are: parent pointer, lower pointer, cross-pointer.

Parent pointer: In addition to the server storing the highest node of directory information tree, the other each server has only one parent pointer to point to higher server its own, if the directory item of a client request operation is not in the root node of directory information tree of local server, the server returned to the client a parent pointer;

Lower pointer: When parts of sub-tree of directory item stored in the server are in the other server, it needs to include a lower pointer, the number of pointers is the number of lower level server. Lower pointer is stored in the directory as a special directory items, whose object class is referral. when the server find the directory items of the class referral, it will return a lower pointer;

Cross-Pointer, this pointer is optional. The server can complete all of the distributed operations with the lower pointer and the parent pointer, but the cross-pointer can increase the efficiency of distributed operation.

### 2.4 Implementation of LDAP Directory Function

In the beginning of creating a directory, the most standard way to add data to the LDAP server is creating a LDIF (LDAP Directory Interchange Format) files.

The basic process of the conversation:

- The initialization of the LDAP conversation is implemented by calling ldap_open (), ldap_open() returns an handle, allowing open multiple conversation in the same time;
- Client authentication is implemented by Ldap_simple_bind_s () and related functions. Multiple authentication methods are supported, ranging from simple authentication to the Simple Authentication Security Layer (SASL) of the LDAP V3.
- Directory operations. Once a successful connection is established, a variety of LDAP operations can be carried on directory, such as search, add, delete, modify, and so on.
- End of the conversation is implemented by calling ldap_unbind ()[5].

**The choice of synchronization and asynchronization**

LDAP protocol allows multiple sessions processed at the same time, indicating that the server can handle multiple queries, however, their processing order depends on the server. LDAP protocol itself is an asynchronous protocol[6].

In the asynchronous communication mode, ldap_search sends "search" requests only, not returns the results directly, but returns a message ID. The message ID identifies the "search" request which the server should process, the same time, the server will set this message ID as parameters of ldap_result, so that you can retrieve results of "search".

## 3 Conclusion

LDAP provides a good directory (data) access in a heterogeneous network environment, which has been widely used with the development of Internet. NPKI model based LDAP, its core idea is the use of LDAP technology to store and access data, it is so greatly improved the quality of service in reliability, access speed, open and cross-platform, which provides strong technical support for modern open network.

## References

1. Levi, A., Ufuk, M.:Nested Certificates and its Application in PKI (May 1999)
2. Xu, D., Hu, X.: A PKI System Based on Nested Certificate. In: ICEIM 2011, vol. 4, pp. 264–268. ICEIM (2011)
3. Xu, J., Li, R., Jiang, Y.: Analysis and Practice of Directory Service based LDAP. Journal Of Xiangtan Mining Institue (March 2002)
4. RFC1777 Lightweight Directory Access Protocol
5. Zhangjun, Yan, K.: The Analysis Of The Light Weight Directory Access Protocol. Computer Applications 19(10), 40–44 (1999)
6. Guan, Y., Liu, J., Chen, B.: Study and System Realization of LDAP Protocol. Journal of Computer Research and Development (October 1998)

# Research on Backstepping Integral Control of Two Order Nonlinear Systems with Input Nonlinearity

Guoqiang Liang*, Jianhong Shi, and Junwei Lei

Department of Control Engineering, Naval Aeronautical and Astronautical University, 264001 Yantai, China
{liangguoqiang1024,leijunwei}@126.com, zhld2002@163.com

**Abstract.** A kind of backstepping integral control strategy is designed for a large family of nonlinear systems with both input nonlinearity and uncertain functions. With a assumption that the input nonlinearity satisfies the so called linear field nonlinearity condition, the backstepping technology is used to control high order systems and a Lyapunov function is constructed to guarantee the stability of the whole system. And the integral action is used to cope with general kinds of uncertainties. The situation of input nonlinearity of two order system is researched and a backstepping integral controller is designed to cope with the system uncertainties included both input nonlinearity and unknown nonlinear functions.

**Keywords:** Backstepping; Integral; Nonlinear; Control.

## 1 Introduction

Backstepping technology is a well-known method which is widely used in the design of high order systems[1-5]. The advantage of backstepping technology is that it is a universal design technolgy for any high order system. Especially, it can integrated with Lyapunov funciton and the stability of whole system can be analyzed thoroughly. Integral control is a old and traditional strategy. It has been widely used in all kinds of control systems for nearly 100 years. But it is still useful and also it is necessary in most control systems because of its strong ability to solve the robustbility problem.

In this paper, the backstepping technology is integrated with integral action perfectly. So it composes a universal adaptive and robust control strategy. It can be widely used in any high order systems under assumptions proposed in the below text. To explain this theory clearly, a general two order system is used to analyze the stability with this backstepping integral method.

## 2 Model Description

A simple general two order system is used as an example to illustrate the main theory of this paper as follows:

---

* Corresponding author.

S. Lin and X. Huang (Eds.): CSEE 2011, Part II, CCIS 215, pp. 7–10, 2011.

$$\begin{aligned}\dot{x}_1 &= f_1(x_1)+\Delta f_1(x_1)+x_2 \\ \dot{x}_2 &= f_2(x)+\Delta f_2(x)+g(u)\end{aligned}, \tag{1}$$

Where $x=(x_1,x_2)$, $x_1$, $x_2$ are states of the system, $\Delta f_1(x_1)$ and $\Delta f_2(x)$ are system uncertainties, $g(u)$ is input nonlinearity. And $\Delta f_1(x_1)$, $\Delta f_2(x)$ and $g(u)$ satisfies the follows assumptions respectively.

Assumption 1: $\Delta f_1(x_1)$, $\Delta f_2(x)$ satisfies the below condition:

$$\begin{aligned}|\Delta f_1(x_1)| &\le T_{11}g_1(x_1)+T_{12} \\ |\Delta f_2(x)| &\le T_{21}g_2(x_1)+T_{22}\end{aligned}, \tag{2}$$

Where $g_1(x_1)$ and $g_2(x)$ are known functions.

Assumption 2: $g(u)$ is a continues nonlinear function with $g(0)=0$, and there exist $\varepsilon_1$ and $\varepsilon_2$ such that

$$\varepsilon_1 u^2 \le ug(u) \le \varepsilon_2 u^2 . \tag{3}$$

The control objective is to design a control law $u$ such that the state $x_1$ can track to the desired value $x_1^d$.

## 3 Design of Backstepping Integral Controller

Considering the above subsystem, a backstepping integral controller is designed as follows.

First, define a new variable as $e_1 = x_1 - x_1^d$, then the first order subsystem can be written as

$$\dot{e}_1 = f_1(x_1)+\Delta f_1(x_1)+x_2-\dot{x}_1^d . \tag{4}$$

Design a virtual control $x_2^d$ as follows

$$x_2^d = -f_1(x_1)+\dot{x}_2^d - k_{11}e_1 - k_{12}\int e_1 dt - \eta_1(x_1) . \tag{5}$$

Define a new variable as $e_2 = x_2 - x_2^d$, then

$$e_1\dot{e}_1 = e_1e_2 - k_{11}e_1^2 - k_{12}e_1\int e_1 dt + e_1\Delta f_1(x_1) - e_1\eta_1(x_1) . \tag{6}$$

And it is easy to prove that

$$e_1\Delta f_1(x_1) - e_1\eta_1(x_1) \le |e_1T_{11}g_1(x_1)| + |e_1T_{12}| - e_1\eta_1(x_1) . \tag{7}$$

Design

$$\eta_1(x_1) = \hat{T}_{11} g_1(x_1) sign(e_1) + \hat{T}_{12} sign(e_1). \tag{8}$$

Define a new variable as $\tilde{T}_{11} = T_{11} - \hat{T}_{11}$, then

$$T_{11} g_1(x_1)|e_1| - \hat{T}_{11} e_1 g_1(x_1) sign(e_1) = \tilde{T}_{11} g_1(x_1)|e_1|. \tag{9}$$

Design the adaptive turning law as

$$\dot{\hat{T}}_{11} = g_1(x_1)|e_1|, \dot{\hat{T}}_{12} = |e_1|. \tag{10}$$

Considering the second order subsystem, it can be written as follows

$$\dot{e}_2 = f_2(x) + \Delta f_2(x) + g(u) - \dot{x}_2^d. \tag{11}$$

Design the control law as

$$u = -k_{21} e_2. \tag{12}$$

Then

$$e_2 \dot{e}_2 = e_2 f_2(x) + e_2 \Delta f_2(x) + e_2 g(u) - e_2 \dot{x}_2^d. \tag{13}$$

So

$$e_2 g(u) = e_2 g(-k_{21} e_2) = -\frac{1}{k_{21}} u g(u) \le -\frac{\varepsilon_1}{k_{21}} u^2. \tag{14}$$

It is easy to get

$$e_2 g(u) \le -\varepsilon_1 k_{21} e_2^2. \tag{15}$$

Assume that there exists $C_i (i = 1, 2, 3)$ and big enough $k_{21}$ such that

$$e_2 f_2(x) + e_2 \Delta f_2(x) - e_2 \dot{x}_2^d \le C_1 e_2^2 + C_2 |e_2| + C_3. \tag{16}$$

Then it is easy to prove that there exists constant $C$ such that if $|e_2| > C$, then

$$e_2 \dot{e}_2 \le 0. \tag{17}$$

Choose a Lyapunov function as

$$V = \frac{1}{2} e_1^2 + \frac{1}{2} e_2^2 + \frac{1}{2} \tilde{T}_{11}^2 + \frac{1}{2} \tilde{T}_{12}^2. \tag{18}$$

It can be proved that if $|e_2| > C$ , then $\dot{V} \le 0$. So the signal $e_2$ is bounded. Then it is easy to get $e_1$ is also bounded. Then all signal of the closed-loop system are proved to be bounded.

## 4 Conclusions

The situation of input nonlinearity of two order system is researched and a backstepping integral controller is designed to cope with the system uncertainties included both input nonlinearity and unknown nonlinear functions. The whole system signals are proved to be bounded with the help of the Lyapunov function. So a better control law, which can make the state $e_i$ converged to zero thoroughly, will be studied in our future work.

## Reference


1. Kim, S.-H., Kim, Y.-S., Song, C.: Contr. Eng. Prac. 12, pp. 149–154 (2004)
2. Hull, R.A., Qu, Z.: Design and evaluation of robust nonlinear missile autopilot from a performance perspective. In: Proce. of the ACC, pp. 189–193 (1995)
3. Lei, J., Wang, X., Lei, Y.: How many parameters can be identified by adaptive synchronization in chaotic systems? Phys. Lett. A. 373, 1249–1256 (2009)
4. Lei, J., Wang, X., Lei, Y.: A Nussbaum gain adaptive synchronization of a new hyperchaotic system with input uncertainties and unknown parameters. Commun. Nonlinear Sci. Numer. Simul. 14, 3439–3448 (2009)
5. Wang, X., Lei, J., Lei, Y.: Trigonometric RBF Neural Robust Controller Design for a Class of Nonlinear System with Linear Input Unmodeled Dynamics. Appl. Math. Comput. 185, 989–1002 (2007)

# Research on High Order Sliding Mode Control of Three Order Systems with Input Nonlinearity

Qiang Ma and Yuqiang Jin*

Department of Training, Naval Aeronautical and Astronautical University,
264001 Yantai, China
{maqiang1024,naau301yqj}@126.com

**Abstract.** A kind of high order sliding mode control strategy is designed for a three order nonlinear systems with both input nonlinearity and uncertain functions. With a assumption that the uncertain function satisfies the bounded uncertainty condition and the input nonlinearity satisfies the linear field nonlinearity condition, a novel kind of sliding mode is constructed with integral action which is useful to cope with the uncertainties of the system. Most important of all, the input nonlinearity problem is solved by the new kinds of sliding mode control method. And this method can also be applied in a large family of nonlinear systems with the same input nonlinearities situation.

**Keywords:** Sliding Mode; Control; Nonlinearity; Uncertainty.

## 1 Introduction

Sliding mode control method is widely used in the design of high order systems[1-5]. The advantage of sliding mode control is that it can reduce the system order and make the design of high order system to be very simple. Input nonlinearity is caused by many practical factors and it exists in many kinds of real systems. It is difficult to solve input nonlinear problem with traditional classic control methods.

In this paper, the input nonlinearity problem is solved with a high order integral sliding mode control strategy. To make it simple, a third order uncertain system is chosen as an example to explain the main theory of this paper.

## 2 Model Description

A simple general two order system is used as an example to illustrate the main theory of this paper as follows:

$$\begin{aligned} \dot{x}_1 &= f_1(x_1) + \Delta f_1(x_1) + x_2 \\ \dot{x}_2 &= f_2(x_1, x_2) + \Delta f_2(x_1, x_2) + x_3 \\ \dot{x}_3 &= f_3(x_1, x_2, x_3) + \Delta f_3(x_1, x_2, x_3) + g(u) \end{aligned} , \tag{1}$$

* Corresponding author.

S. Lin and X. Huang (Eds.): CSEE 2011, Part II, CCIS 215, pp. 11–14, 2011.

Where $x_1$ , $x_2$ , $x_3$ are states of the system, $\Delta f_1(x_1)$ , $\Delta f_2(x)$ and $\Delta f_3(x)$ are system uncertainties, $g(u)$ is input nonlinearity. And $\Delta f_1(x_1)$, $\Delta f_2(x)$, $\Delta f_3(x)$ and $g(u)$ satisfies the follows assumptions respectively.

Assumption 1: $\Delta f_1(x_1)$, $\Delta f_2(x)$ and $\Delta f_3(x)$ satisfies the below condition:

$$\begin{aligned} |\Delta f_1(x_1)| &\le T_{11} g_1(x_1) + T_{12} \\ |\Delta f_2(x)| &\le T_{21} g_2(x_1) + T_{22}, \\ |\Delta f_3(x)| &\le T_{31} g_2(x_1) + T_{32} \end{aligned} \tag{2}$$

Where $g_1(x_1)$, $g_2(x)$ and $g_3(x)$ are known functions.

*Assumption 2: $g(u)$ is a continues nonlinear function with $g(0)=0$, and there exist $\varepsilon_1$ and $\varepsilon_2$ such that*

$$\varepsilon_1 u^2 \le u g(u) \le \varepsilon_2 u^2 . \tag{3}$$

The control objective is to design a control law $u$ such that the state $x_1$ can track to the desired value $x_1^d$ .

## 3 Design of Sliding Mode Controller

Considering the above subsystem, a sliding mode controller is designed as follows.

First, define a new variable as $e_1 = x_1 - x_1^d$ , then the first order subsystem can be written as

$$\dot{e}_1 = f_1(x_1) + \Delta f_1(x_1) + x_2 - \dot{x}_1^d . \tag{4}$$

Design a sliding mode surface as follows

$$s_0 = e_1 + \int e_1 dt . \tag{5}$$

Solve its derivative and define a new variable as

$$s_1 = \dot{s}_0 = \dot{e}_1 + e_1 = f_1(x_1) + \Delta f_1(x_1) + x_2 + e_1 . \tag{6}$$

Solve its second order derivative and define a new variable as

$$\begin{aligned} s_2 = \ddot{s}_0 &= \dot{f}_1(x_1) + \Delta\dot{f}_1(x_1) + \dot{x}_2 + \dot{e}_1 \\ &= \dot{f}_1(x_1) + \Delta\dot{f}_1(x_1) + f_2(x_1, x_2) + \Delta f_2(x_1, x_2) . \\ &+ f_1(x_1) + \Delta f_1(x_1) + x_2 - \dot{x}_1^d + x_3 \end{aligned} \tag{7}$$

Solve its third order derivative and define a new variable as

$$\begin{aligned} s_3 = \dddot{s}_0 &= \ddot{f}_1(x_1) + \Delta\ddot{f}_1(x_1) + \dot{f}_2(x_1, x_2) \\ &+ \Delta\dot{f}_2(x_1, x_2) + \dot{f}_1(x_1) + \Delta\dot{f}_1(x_1) \\ &+ \dot{x}_2 - \ddot{x}_1^d + f_3(x_1, x_2, x_3) + \Delta f_3(x_1, x_2, x_3) + g(u) \end{aligned} . \tag{8}$$

Define a sliding mode surface for the whole system as

$$S = \sum_{i=0}^{2} c_i s_i , \tag{9}$$

Where $c_i (i = 0, 1, 2)$ are coefficients of a hurwitzs polynominal.

Solve its derivative, it satisfies

$$\dot{S} = \sum_{i-0}^{2} c_i \dot{s}_i = c_0 s_1 + c_1 s_2 + c_2 s_3 = c_2 g(u) + L , \tag{10}$$

Where $L$ constains both known information and unknown information of the system. Assume that there exists a constant $k_1$ such that

$$|L| \le k_1 |S| . \tag{11}$$

Design the control law as

$$u = -k_3 S . \tag{12}$$

Then

$$S\dot{S} = S c_2 g(u) + SL . \tag{13}$$

So

$$k_3 S c_2 g(u) = -u g(u) \le -c_2 \varepsilon_1 u^2 \le -c_2 \varepsilon_1 (k_3 S)^2 . \tag{14}$$

Then

$$S\dot{S} = S c_2 g(u) + SL \le -c_2 \varepsilon_1 k_3 S^2 + k_1 S^2 . \tag{15}$$

It is obvious that, choose proper parameters such that $c_2\varepsilon_1 k_3 > k_1$, then

$$S\dot{S} \leq 0. \tag{16}$$

So the system is stable.

## 4 Conclusions

The situation of input nonlinearity of three order system is solved by the constructing of a novel kind of sliding mode control strategy. With the help of a main assumption that all uncertianties are bounded by the sliding surface, the whole system is guaranteed to be bounded and stable. And the control objective is fulfilled completedly. The main defect of this paper is that the assumption is too strong and it will be relaxed in our furture works.

## References

1. Gao, W., Hung, J.C.: Variable Structure Control of Nonlinear systems: A New Approach. IEEE Trans. Indus. Electro. 40(1) (February 1993)
2. Tang, F., Wang, L.: An adaptive active control for modified Chua's circuit. Phys. Lett. A. 346, 342–346 (2005)
3. Lei, J., Wang, X., Lei, Y.: How many parameters can be identified by adaptive synchronization in chaotic systems? Phys. Lett. A. 373, 1249–1256 (2009)
4. Lei, J., Wang, X., Lei, Y.: A Nussbaum gain adaptive synchronization of a new hyperchaotic system with input uncertainties and unknown parameters. Commun. Nonlinear Sci. Numer. Simul. 14, 3439–3448 (2009)
5. Wang, X., Lei, J., Lei, Y.: Trigonometric RBF Neural Robust Controller Design for a Class of Nonlinear System with Linear Input Unmodeled Dynamics. Appl. Math. Comput. 185, 989–1002 (2007)

# PS-SIM: An Execution-Driven Performance Simulation Technology Based on Process-Switch

Xiaowei Guo, Yufei Lin, Xinhai Xu, and Xin Zhang

National Laboratory for Parallel and Distributed Processing
National University of Defense Technology, China
xiaow_g@126.com, {linyufei,xunxinhai,zhangxin}@nudt.edu.cn

**Abstract.** Nowadays, the performance of large-scale parallel computer system improves continuously, and the system scale becomes extremely large. Performance prediction has become an important approach to guide system design, implementation and optimization. Simulation method is the most widely used performance prediction technology for large-scale parallel computer system. In this paper, after analyzing the extant problems, we proposed a novel execution-driven performance simulation technology based on process-switch. We designed a simulation framework named PS-SIM, and implemented a prototype system based on MPICH2. Finally, we verified the proposed approach by experiments. Experimental results show that the approach has high accuracy and simulation performance.

**Keywords:** Performance Prediction, Large-Scale Parallel Computer System, Execution-Driven Simulation, MPICH.

## 1 Introduction

With advances in architecture and semiconductor technology, the performance of large-scale parallel computer system improves continuously. In recent years, the computing speed of parallel computer system has reached Petascale, and forwarding to Exascale[1]. The extremely large scale system will take years to design and implement. In order to reduce cost and shorten the development cycle, researchers often need to conduct a comprehensive performance analysis before the system developed, using performance prediction technologies to evaluate whether the application can achieve the expected performance, will further guide the hardware and software design and provide reference for the improvement and optimization.

Simulation is the most widely used performance prediction technology. Trace-driven simulation and execution-driven simulation are two well-known approaches for performance simulation. The basic idea of trace-driven is to trace the program execution, and take the tracing information as the simulator's input to simulate the program performance in target architecture. The simulation process includes trace generation and trace simulation. Execution-driven simulation is another current mainstream performance simulation technology. Because of its high accuracy, it becomes a research hotspot in the field of performance prediction since 20th century 90s. The basic idea is to simulate only the part which needs to be analyzed, and the

S. Lin and X. Huang (Eds.): CSEE 2011, Part II, CCIS 215, pp. 15–22, 2011.

remaining part actually executes. Combined with software and hardware, it can improve the simulation accuracy and shorten simulation run time.

Execution-driven simulator MPI-SIM[4] developed at University of California and trace-driven simulator BigSim[3] developed at University of Illinois are two well-known performance simulators. Trace-driven simulation needs to obtain prior execution traces, which makes the program have to execute many times in the simulation process. As the scale of systems and the complexity of applications continue to increase, the cost of performance simulation increases rapidly, trace-driven simulation will be further limited by the scalability. Meanwhile, the existing implementation of execution-driven simulator simulates process through the thread, which reduces accuracy.

In this paper, the scope of the research lies in high-scalability and high accurate performance simulation for large-scale parallel computer system. We proposed a novel execution-driven performance simulation technology based on process-switch. The main contribution of the paper is as follows:

— Proposed a novel execution-driven performance simulation technology based on process-switch, and designed the simulation framework PS-SIM.
— Based on MPICH2, studied the key technologies in PS-SIM and implemented a prototype system of PS-SIM.
— Verified the proposed approach and key technologies by Experiments. Experimental results show that this approach has high accuracy and simulation performance.

This paper is organized as follows. Section 2 gives our basic idea and describes the general simulation framework. Section 3 presents the key technologies and the implementation of PS-SIM. The experiments and experimental result analysis are showed in Section 4. Some related work in performance prediction is reviewed in Section 5. We present our conclusions in Section 6.

## 2 PS-SIM Framework

### 2.1 Basic Ideas

In this paper, we proposed a novel execution-driven performance simulation technology based on process-switch, which brings two advantages: First, simulating process by process, simulation is more accurate; and second is the high simulation efficiency since the program only needs to execute once.

The basic ideas of execution-driven performance simulation technology based on process-switch can be summarized as follows:

— Simulating process through process. Since the number of processor cores on target system is usually much larger than the host, multiple physical processes need to be mapped to one processor core in simulation environment. Traditional execution-driven simulation simulates multi-process on the target machine by running multiple threads on the host; we will simulate process by process, mapping a physical process in target machine to a logical process in the simulator, which means each processor core on the host will execute multiple logical process.

— Adopting execution-driven simulation method. The execution time of process includes the sequential computation time and the communication time. In the simulation, we obtain the sequential computation time through actual running, and the communication time is calculated using accurate communication model. Each process maintains local virtual time, sequential computation execution time and the calculated communication time will be accumulated to the virtual time for each logical process. When the simulation ends, we will get each process's execution time, in which the maximum is the predicted execution time.

— Scheduling process based on minimum virtual time. In order to obtain the exact computation time and communication time, simulator will take over scheduling for all the logical processes. To ensure the correctness of program execution, we schedule process based on minimum virtual time, this means that the simulator always switches to the process which has the smallest virtual time. Virtual time update algorithms ensured the correct order of message send and receive.

## 2.2 Overall Block Diagram

For MPI applications, we designed the PS-SIM simulation framework based on MPICH2. The overall block diagram is shown in Fig. 1.

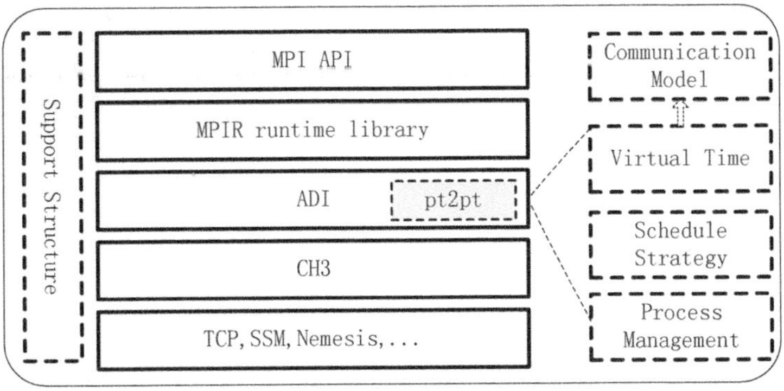


**Fig. 1.** PS-SIM Overall Block Diagram

The boxes drawn by thick dotted line are modules added based on the original library, the main changes is in the ADI layer of MPICH2 library and API layer completely retained. This ensures the application without modification, and we just need to use the framework to re-build before the program can be simulated.

PS-SIM modified all point to point communication interfaces in order to obtain the communication time through the communication model and update the virtual time. Once the process blocked, switching process through the Process Management interface, the Schedule Strategy ensures that switching does not affect the correctness of the program.

As shown above, Support Structure module is the additional structure added in the MPI library for other function module. Such as local virtual time, the process state variables, global process list pointer, etc. Process Management module provides process-switch relevant interfaces, when communication blocked, the process management module will select the next executable process through the schedule Strategy, and suspend the current process. Virtual time module provides local virtual time initialization, updating and other functions. The virtual time update for communication operations need to call the communication model to calculate the communication time.

## 3 Key Technologies and Implementation

### 3.1 Process Scheduling Strategy

The basic idea of the process scheduling includes two aspects: First, only one process is running at the same moment, each process updates the virtual time and then suspends it when switching, so that the virtual time can accurately record the sequential execution time; second, we always select the process which has the smallest virtual time to ensure the correctness of procedures.

There are two types of process in Linux system: real-time process and general process. The priority of real-time process is higher than the general process. The scheduling strategy for real-time process includes first come first served strategy (SCHED_FIFO) and time-slice round robin strategy (SCHED_RR). When using the first come first served scheduling policy, the process will occupy the processor unless the process yields or a higher priority real-time process comes into an executable state. Linux provides related system calls to set process scheduling strategy, such as *sched_setscheduler*, etc.

We set the scheduling policy of current process to SCHED_FIFO; other processes are general processes, so that the current process will not be preempted by other processes before yielding or switching.

The process of switching process includes two steps:

Step1: set the scheduling policy of target process to SCHED_FIFO
Step2: set the scheduling policy of current process to SCHED_OTHER

After Step1, the current and target process have the same priority, the target process cannot preempt CPU, then after Step2, the current process changes into a general process, CPU will immediately be preempted by the target process.

### 3.2 Virtual Time Updating Algorithm

Normal sequential execution time will be recorded to the local virtual time; the communication time of each message sent and received can be calculated by the communication model. Blocking event occurs only in the communication process. There are many MPI communication patterns, but most applications can be done by calling the following five functions: MPI_RECV, MPI_SEND, MPI_ISEND, MPI_IRECV, and MPI_WAIT.

MPI_SEND and MPI_RECV complete blocking message send and receive. According to different message length, the library uses two different transmission protocols. In which the eager protocol is used by short messages, and rendezvous protocol used by long messages. When using the eager way, the message body is immediately sent after the header, and the message body will be copied to the receive buffer once the receiving process found a matched request; when using the rendezvous way, the message header is sent first, and the receiver will return an ACK message to the sender if found an match, the sender will begin to send the message body after the response received. The virtual time updating process for blocking send and receive is shown in Fig. 2.

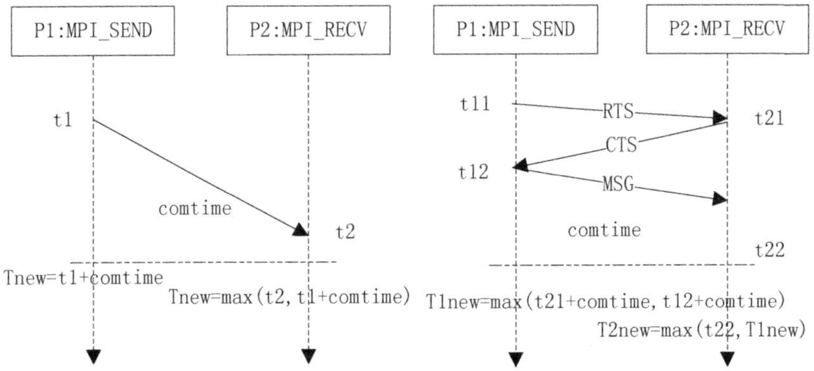


**Fig. 2.** Blocking Send and Receive

With eager protocol, the sender will send its virtual time with the message header to the receiver process, the receiver will update its virtual time when receiving operation completed, and the virtual time will be set to the maximum value of the sender virtual time and the current virtual time, this ensures the order of sending and receiving.

With rendezvous protocol, the send request packet RTS and the receive response packet CTS carries the virtual time of the source process; process will update their own virtual time according to local virtual time, communication time and the partner virtual time when the message transmission is complete.

Similarly, non-blocking communication protocol is also divided into eager and rendezvous. The difference is that the virtual time is updated when the MPI_WAIT completed.

### 3.3 Virtual Time Update and MPI Interface Modification

We add two virtual time related variables for each process: vTime and timeStall. In which vTime is the process local virtual time and timeStall saves the system time when the process suspends in the last time. In order to send the virtual time to target process in communication, we modified the message header data structure, adding the time variable, so that the virtual time could be sent contained in the message header.

In mpich2-1.3.1, the default communication interface is Nemesis. Considering these most widely used interfaces, we need to modify the implementation of MPI point to point communication interfaces, including MPI_Send, MPI_Recv, MPI_Isend, MPI_Irecv and MPI_Wait, the basic group communication interfaces MPI_Barrier, MPI_Reduce, MPI_Allreduce and MPI_Alltoall are also included. Most MPI applications can be implemented using these above interfaces.

## 4 Experimental Results and Analysis

The CPU used in experiment is Intel Xeon E5405, 2.00 GHz, 8 cores, 2.0G memory, the operating system is SUSE LINUX 11.2, kernel version of Linux 2.6.31.5. MPI library version is mpich2-1.3.1. We use the NPB 3.3-MPI as test program.

Four NPB test cases are selected: CG, EP, FT, LU. Simulator will bind multiple processes running on a single core. The test program scale is set to Class A, the size of target system are 1, 2, 4 and 8 processors. The real execution time and predicted time are shown in Fig. 3.

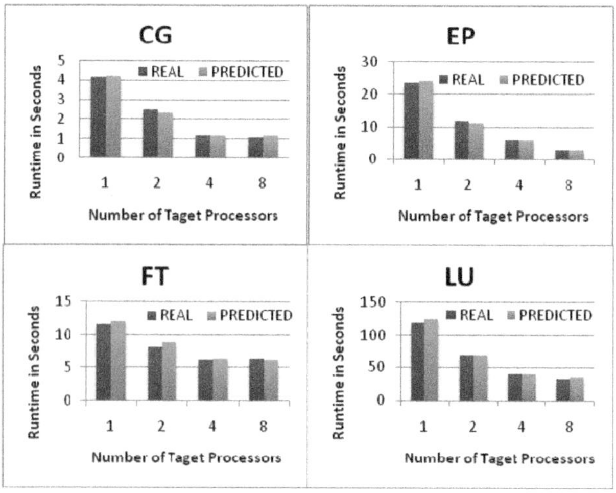


**Fig. 3.** Real execution time VS. Predicted execution time

Experimental results show that the simulation result is accurate, and all test cases have maintained a relative lower error; the average relative error for performance prediction is less than 3%.

In order to evaluate the performance of the simulator, we introduce the widely used metric slowdown, defined *slowdown(H,T)=SimTime(H)/RealTime(T).*

SimTime(H) is the application simulation time running in the H-cores host. RealTime(T) is the real execution time for T-cores target machine. This experiment simulated 1, 2, 4 and 8 processor cores target using single-core, so H=1, and T= 1, 2, 4, 8. The results are shown in Fig. 4.

It can be seen from the figure, with increase in the size of simulated target, slowdown continues to grow. When H=T=1, the average slowdown of four tests is 1.02, indicating that the modified MPI library has little effect to the computing performance. When H=T/2, the slowdown is only 2.13. When H=T/8, the slowdown for EP and FT is less than 8.3, CG and LU are respectively 8.89 and 9.03. Experimental results show that the simulator has high simulation performance.

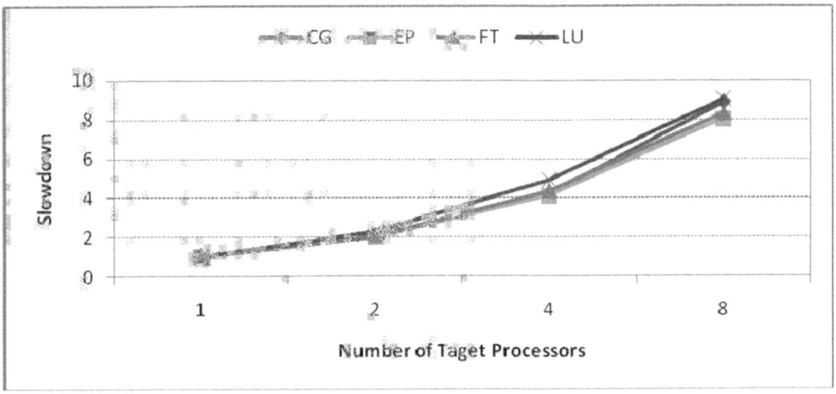


**Fig. 4.** Performance of the simulator

## 5 Related Work

Performance prediction for large-scale parallel system is a widely researched area. The main performance prediction method can be divided into two categories: modeling analysis and simulation [5][6]. The main advantages for modeling method is low-cost, but the model building requires a better understanding for program algorithm and implementation, most of models are constructed manually by domain experts, which limits the applications of modeling analysis method.

Simulation method is performed by constructing a system simulator to execute applications on it for performance prediction. Simulation technique can capture detailed performance behavior at all levels, and can balance the accuracy, cost and flexibility, therefore it is used more and more widely in the performance prediction fields. Execution-driven simulator MPI-SIM[4] developed at University of California and trace-driven simulator BigSim[3] developed at University of Illinois are two well-known performance simulators. They have a better simulation of parallel network, but the problem how to get more accurate serial execution time is still not solved well. In PPOPP 2010, JiDong Zhai[2] proposed an accurate method to obtain execution time: Phantom, and implement this method in the simulator SIM-MPI.

Our approach is inspired by these previous works. We expect this paper will motivate more progress in performance prediction for large-scale parallel systems.

## 6 Conclusion

In this paper, we proposed a novel execution-driven performance simulation technology based on process-switch, and designed a simulation framework named

PS-SIM. This approach brings two advantages: First, using process to simulate process, simulation is more accurate; and second is the high simulation efficiency since the program only need to execute once.

We also implemented a prototype system based on MPICH2. Finally, we verified the proposed approach by Experiments. Experimental results show that the approach has high accuracy and simulation performance.

The prototype system simulator is implemented by modifying the MPI library, it is simple and scalable. Our next step could be extending existing simulation system to support multi-core and multi-node platform for large-scale parallel computers.

**Acknowledgment.** We would like to thank the anonymous reviewers for their insightful comments. This work is supported by the National Natural Science Foundation of China (NSFC) No.60921062.

## References

1. Amarasinghe, S., Campbell, D., Carlson, W.: Exascale software study: Software challenges in extreme scale systems. DARPA IPTO, Air Force Research Labs, Tech Report (2009)
2. Zhai, J.D., Chen, W., Zheng, W.: PHANTOM: Predicting Performance of Parallel Applications on Large-Scale Parallel Machines Using a Single Node. In: PPoPP 2010, Bangalore, India (2010)
3. Zheng, G., Kakulapati, G., Kale, L.V.: BigSim: A parallel simulator for performance prediction of extremely large parallel machines. In: Proceedings of the International Parallel and Distributed Processing Symposium (2004)
4. Prakash, S., Bagrodia, R.L.: MPI-SIM: using parallel simulation to evaluate MPI programs. In: Proceedings of the 30th Conference on Winter Simulation, pp. 467–474 (1998)
5. Barnes, B.J., Rountree, B., Lowenthal, D.K., Reeves, J., de Supinski, B., Schulz, M.: A regression-based approach to scalability prediction. In: ICS 2008, pp. 368–377 (2008)
6. Yang, L.T., Ma, X., Mueller, F.: Cross-platform performance prediction of parallel applications using partial execution. In: SC 2005, p. 40 (2005)
7. Bourgeois, J., Spies, F.: Performance prediction of an NAS benchmark program with ChronosMix environment. In: Proceedings of Euro-Par 2000, Parallel Processing, 6th International Euro-Par Conference, pp.208–216 (2000)

# A Small System of Program-Controlled Amplification Amplifier Based on AD603

XuFang Zhu[1], XiaoFang Zhu[2], XiaoMin Li[3], and FengBo Zhu[3]

[1] Naval University of Engineering, Electronic engineering college, Wuhan, China
zhuxufang_love@163.com
[2] Wuhan University technology huaxia college, Wuhan, China
[3] Naval University of Engineering, Electronic engineering college, Wuhan, China

**Abstract.** This article provides a kind of programmable amplifiers AD603 as the main design of small systems. Input signal generated by the DDS integrated chip, the two integrated broadband amplifiers, low noise variable gain amplifier cascade from AD603, the maximum voltage gain up to 60dB, 10dB step adjustable gain control to achieve through the microcontroller。 Debugging results show that the system can meet the basic needs indicators.

**Keywords:** DDS, variable gain, program-controlled, amplifier.

## 1 Introduction

In the complex marine environment, the sonar will detect acoustic signals of different frequencies, objectives, environment, reverb and more. In this case, if the goal is low-frequency signals to detect the low frequency response to a more significant signal amplification, while the other band signal amplification or attenuation of small amplitude, which involves a single amplifier should be designed to gain Variable problem. According to actual requirements, a program-controlled amplifier is designed, which can amplify 10mv amplitude sine wave signals of different frequencies, the maximum gain of 60dB, 10dB adjustable gain step, the pass band is 100Hz ~ 40kHz. This design mainly includes three parts: the signal part; controllable gain amplifier part; MCU control part. The following specific analysis of these sections.

## 2 Signal Generator

We design a signal generator based on DDS(Direct Digital Synthesis) technology, which Can output frequency 100Hz ~ 200kHz, 10kHz sine wave signal step.

DDS chip AD9851 has a high-speed frequency switching time, high frequency resolution and low phase noise, when frequency is changed, DDS can maintain a continuous phase, it is easy to realize the frequency, Phase and amplitude modulation. In addition, DDS technology is based on most digital circuit technology with programmable control of the outstanding advantages. Frequency step by SCM to achieve simple and convenient, AD9851 DDS chip with the input signal frequency

S. Lin and X. Huang (Eds.): CSEE 2011, Part II, CCIS 215, pp. 23–27, 2011.

range is 100Hz ~ 200kHz, the microcontroller controls the frequency step, the main circuit shown in Figure 1. To meet the requirements of the input signal amplitude of 10mV, in the following additional emitter follower allows the signal amplitude are attenuated.

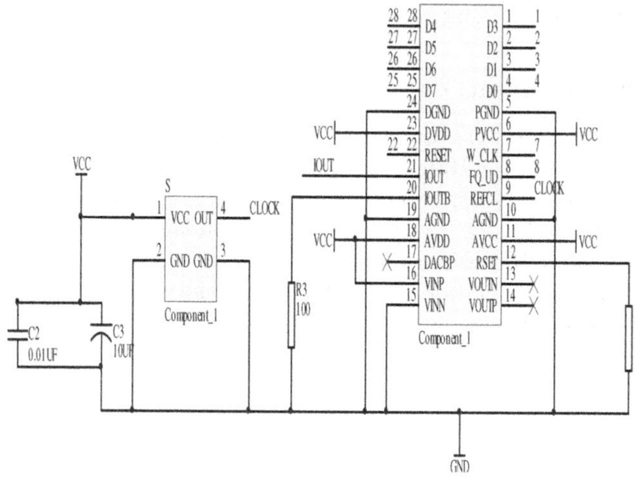


**Fig. 1.** Signal Generator

## 3 Controlled Gain Amplifier

Now, the controllable gain amplifier to achieve the following ways:

(1) The realization of the traditional discrete components. Multi-stage amplifier circuit using cascaded in order to meet the requirements of 60dB gain. Diode detector circuit produces an output voltage feedback circuit automatically adjust the preamp gain adjustment. However, due to a large number of discrete components, such as transistors and other, more complex circuits is difficult to adjust the operating point, in particular, it is very difficult to gain a quantitative regulation. And, circuit stability is poor, prone to self-excitation phenomenon becaue of using multi-level amplification.

(2) analog switch with integrated programmable gain amplifier form, this is a common way, that is, respectively, with a number of op-amp inverting input connected in a deep negative feedback after the series into a multi-stage amplifier inverting amplifier, and then control the analog switch inverting input of op amp unit external resistor and the public to side pass, off state, then control the amplifier gain. The main disadvantage of this approach is the analog switch on-resistance of the amplifier gain, affect the signal transmission accuracy.

(3) select op-amp, such as AD603.AD603 programmable gain adjustment is a dedicated chip produced by ADI Corporation in the United States, which is a low-noise, 90MHz bandwidth, integrated operational amplifier with adjustable gain, internally by the R-2R resistor ladder network and the fixed gain amplifiers, plus a ladder network in its input signal by the attenuation from a fixed gain amplifier

output, attenuation is added to the reference voltage gain control interface decisions; and this reference voltage can be controlled by the microcontroller operation and D/A chip output control voltage obtained in order to achieve more accurate numerical control. In addition, AD603 can provide more than the work of DC to 30MHz bandwidth, single-stage when the actual work of more than 30dB of gain, two or more cascaded gain of 60dB can be obtained. Advantage of this approach is that the circuit is highly integrated, coherent clearer, easy to control, easy-to-digital processing with a single chip. Therefore, Two pieces of AD603 can be cascaded in a way that constitutes amplifier. The program can meet the pass band 100Hz ~ 300kHz, and even wider, the voltage gain of 60dB, the amplifier output voltage without significant distortion. Specific circuit shown in Figure 2 Gain and control voltage relationship:

AG (dB) = 40 × V +10, level control is only 30dB, Tandem two-stage amplifier, the gain expression is

AG (dB) = 40 × V1 +40 × V2 +20, Gain range is -20 ~ +60 dB, up to system requirements.

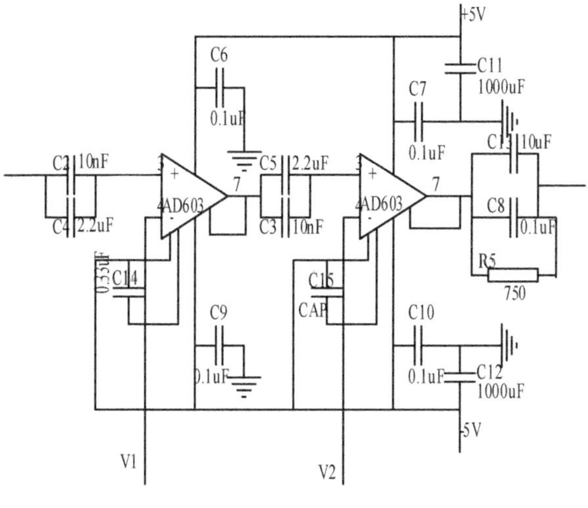


**Fig. 2.** Controlled Gain Amplifier

Two-stage amplifier circuit as the frequency response curve of the same, so when two AD603 series, its pass band up to 30MHz. Step by the microcontroller to control the gain. If the input voltage amplitude of 10mV, after the voltage by the amplifier should be able to reach 10V or so, but because the maximum output voltage amplitude of the AD603 in the 3V or so. It amplifies the signal to be amplified through the higher level after the output voltage amplitude. To improve the control precision of AD603 in its dual op amp plus control side LM358, the control voltage low-pass filter and voltage follower.9M typical first-level connection with AD603, single-stage gain-10dB ~ 30dB, second-level connection with 90M, single-stage gain of 10 dB ~ 50 dB. The actual gain of two cascaded up to 0dB ~ 80 dB, 10 dB gain step adjustable pass band up to 30Hz ~ 1MHz. Specific circuit connection shown in Figure 3:

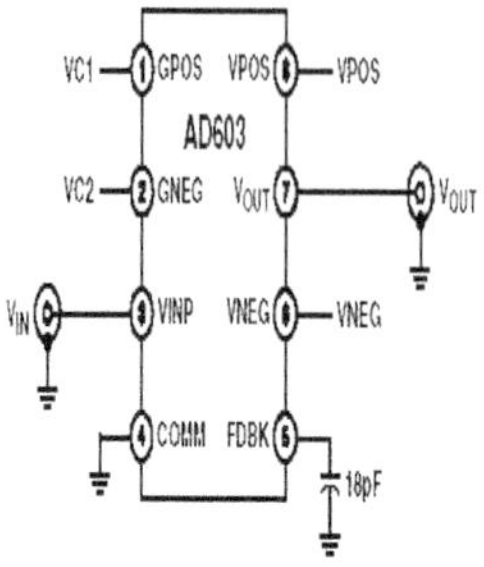


(a)9M

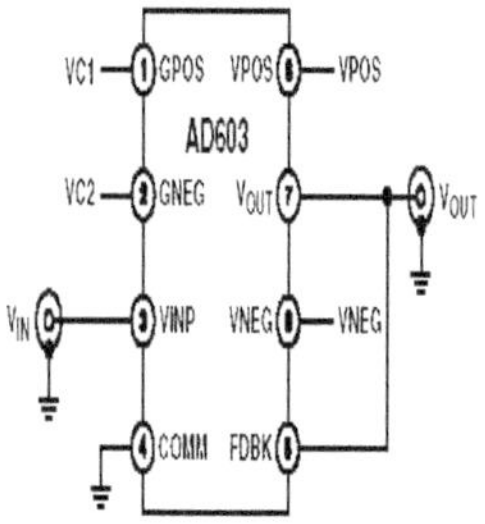


(b)90M

**Fig. 3.** AD603Typical connection

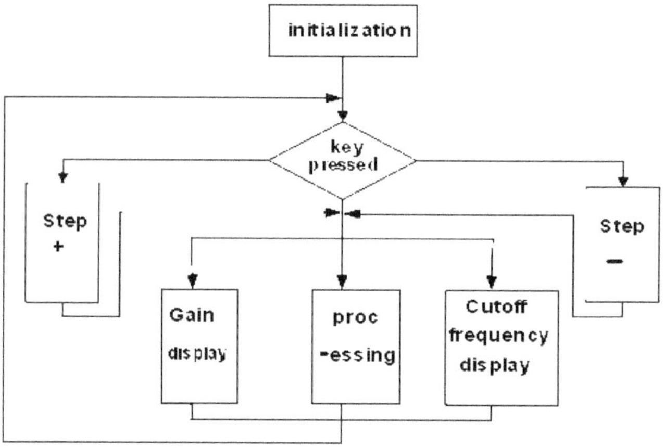


**Fig. 4.** Software flow chart

## 4 Software Design

In the small system, we select microcontroller AT89C51 to realize 10dB step adjustable gain, this design to application is very mature. Figure 4 shows the amplifier SCM software flow chart, the cutoff frequency and voltage gain step can be Presetde through the keyboard.

## 5 Results and Analysis

We connect various parts of the circuit after all the units are tested well. the main instrument include the oscilloscope and multimeter. Then we test the system In the following two cases:

(1) Input signal is a sine wave signal f = 1KHZ (2)Input signal is a sine wave signal f= 4KHZ, The results are shown in Table 1

**Table 1.** Preset gain compared with the actual gain

| Preset gain (dB) | 10 | 20 | 30 | 40 | 50 | 60 |
|---|---|---|---|---|---|---|
| actual gain (dB) (f=1kH) | 10 | 19 | 28 | 39 | 52 | 60 |
| actual gain (dB) (f=4kHz) | 11 | 19 | 29 | 41 | 51 | 60 |

We can see from the test data, the amplifier voltage gain up to 60dB, 10dB gain step adjustable voltage gain error is less than 4%. Basically meet system design requirements. To reduce the error, in the actual design should be noted: ① Digital part and analog part of their power should be isolated from each other, ② gain controling part should be installed in the shielded box to avoid interference and high-level self-excited. ③ In order to avoid high-frequency gain decreased. All electrolytic capacitors at both ends of the signal coupling high-frequency ceramic capacitors in parallel

## 6 Conclusion

The basic design to system requirements, but also by a few monolithic control further expansion of its functions, and programmable filter circuit filter phase can be achieved. Can be widely used in other communication electronic circuits.

## References

1. Huang, Z.W.: The Grid: National Undergraduate Electronic Design Competition circuit design. Beijing Electronic Industry Press, Beijing (2005)
2. Xie, Z.M.: The Grid: Electronic circuit design experiments testing. Huazhong University Press, Wuhan (2000)
3. Xu, Z.W.: Variable gain amplifier AD603 and the application. J. Journal of Hubei Institute 15(3), 27–29 (2009)

# Fractal Characteristics of Contact Surface of Needle Punched Nonwovens

XuHong Yang[1,*] and Masayuki Takatera[2]

[1] National Engineering Laboratory for Modern Silk, College of Textile and Clothing Engineering, Soochow University, Suzhou, 215123, China
[2] Shinshu University, Japan
yang_xuh@163.com

**Abstract.** In this work, we analyzed the fractal characteristics of the contact surface of needle punched nonwovens with different contents of polyurethane compressed under different pressures. Fractal dimensions, $Da$ and $Dg$, were used to evaluate the fiber area distribution and grayscale distribution of the contact surfaces respectively. The results show that both $Da$ and $Dg$ increase with the increasing compressing stress. The more polyurethane the nonwoven contains, the larger $Dg$ and smaller $Da$ it has.

**Keywords:** needle punched nonwovens, contact surface, Fractal dimensions.

## 1 Introduction

The wetting, adhesion, friction, wear and lubrication of materials, are, in most cases, closely associated with the materials' surface characteristics [1]. In many cases, fabric is subjected to normal compressive loads, and the physical, hydraulic properties and so on, change with the surface state depending on the compressive load of the fabric. It is important to understand the contact characteristics when they are subjected to certain pressures.

However, nonwovens have rough and irregular surfaces, and it is difficult to depict their surface structures by conventional geometric parameters. Fractal analysis is a method for assessing the randomness and irregularity of natural structures, and has been applied successfully for the texture analysis, especially for rough surface description [2,3]. The fractal dimension gives a feature of the roughness of a surface and can be used to quantify the texture information. The present work focuses on the fractal characteristics of the rough surface textures of needle punched nonwoven felts compressed under different pressures.

## 2 Experimental Methods

### 2.1 Specimens and Image Capturing

4 nonwoven felts containing polyurethane in content of 0, 3, 6 and 9%, named sample 1, 2, 3 and 4, respectively, were used in this study. The surface textures of the

* Corresponding author.

S. Lin and X. Huang (Eds.): CSEE 2011, Part II, CCIS 215, pp. 28–33, 2011.

specimens compressed at different pressures were observed by a polarized microscopy, and the images were captured by a computer through CCD camera. The apparatus is illustrated in Fig. 1.

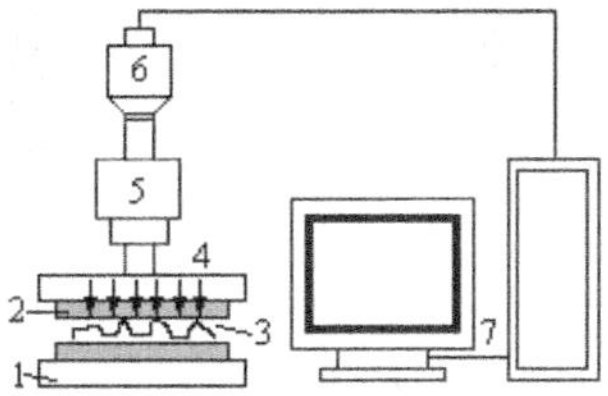


**Fig. 1.** Schematic diagram of apparatus for specimen compressing and image capturing. 1. compressing board 2. acrylic polarized board 3. sample 4. light source 5. reflecting microscope 6. CCD camera 7. Computer.

There is a polarized board which contacts the surface fibers of the felt at a certain pressure forced by an air pump system. The lights are irradiated to the polarized board as shown in Fig. 1, and they are adsorbed in the areas where there are fibers contacting the board, while reflected in other areas. Therefore, the fibers contacting the board are displayed in low grayscale level and the background in high grayscale level. The grayscale images and the corresponding binary ones are shown in Fig. 2 and Fig. 3 respectively (take sample 1 as an example).

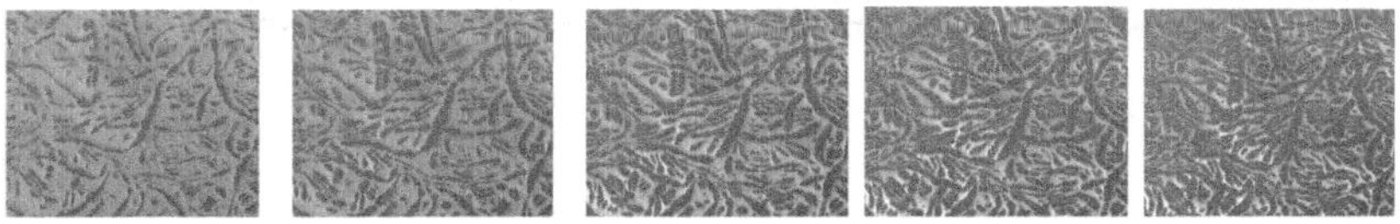

**Fig. 2.** Grayscale images of sample 1 under different compress stress $P$ (kgf·cm$^{-2}$)

**Fig. 3.** Binary images of sample 1, with fibers in black pixels and background in white pixels

## 2.2 Surface Characterization by Using Fractal Geometry

The fractal dimension can be used to quantify the irregular texture information. When a self-similar structure is covered with regular unit cell, it is known that the unit cell length ($\varepsilon$), and the number of occupied unit cells ($N$) have a linear relationship in the double logarithm plot.

In this work, the box-counting method was used to calculate the fractal dimensions. To get the box-counting dimension, a square with side length $\varepsilon$ is used as the unit cell, and regard the image as a set $A$ which is covered by the boxes (unit cells). For various side length $\varepsilon$, the number of coverings $N(\varepsilon)$ is calculated, and a log-log plot of $N(\varepsilon)$ versus $1/\varepsilon$ is made. The linear part of the curve of log $N(\varepsilon)$-log $(1/\varepsilon)$ is the scale invariance section and the slope of the least square linear fit line is the box-counting dimension. Here, we use two kinds of methods to calculate $N(\varepsilon)$. One is, to segment the grayscale image into a binary one displayed in black and white pixels, representing fiber and background area respectively (as shown in Fig. 3), and then cover it by boxes with side length $\varepsilon$. When a box contains any part of the fibers, it is counted. The total number of such boxes is $N_a(\varepsilon)$ [4]. We call the dimension obtained by this mean the fractal dimension of fiber area distribution, $D_a$.

The other method is, to take the grayscale image as the set A, and calculate $N_g(\varepsilon)$ by the following equation [5]:

$$N_g(\varepsilon) = \sum_{k=1}^{n} (g_{k\max} - g_{k\min})/\varepsilon \tag{1}$$

where $g_{kmax}$, $g_{kmin}$ are the maximum and minimum gray level in the kth box, respectively, and $n$ is the total number of boxes with side length $\varepsilon$. The dimension obtained in this way is called the fractal dimension of grayscale distribution, $D_g$.

## 3 Results and Discussion

### 3.1 Fractal Characteristic of the Fiber Area Distributions in the Images

Fig. 4 is the illustration of calculating $D_a$ under a certain pressure, taking sample 1 compressed at the stress of 100 kgf·cm$^{-2}$ as an example. It is obvious that all the data points at the range of the first 10 points is in the linear section which means the curve has a scale invariance section. By the least-squares method, the linear correlation coefficients of the data can be obtained. For all the binary sample images, with the compressive stress $P$ being 100~1000 kgf·cm$^{-2}$, respectively, the square of linear correlation coefficients of the data are larger than 0.997. The high degree of linear correlation of the data proves the existence of fractal structures in the fiber area distribution. That means the fiber area distributions of the sample images under different compressive stresses are fractals. The slope of the best fit straight line through the data yields the fractal dimension of fiber area distribution, $D_a$.

### 3.2 Fractal Characteristic of the Grayscale Distributions in the Images

Fig. 5 is the illustration of calculating $D_g$ under a certain pressure, also taking sample 1 compressed at the stress of 100 kgf·cm$^{-2}$ as an example. The data points of each series in log-log plot are fitted by the least-squares method. For all the grayscale sample images, with the compressive stress $P$ being 100~1000 kgf·cm$^{-2}$, respectively, the square of linear correlation coefficients of all the images are larger than 0.999. That means the grayscale distributions of the sample images under different compressive stresses are also fractals. The fractal dimension of grayscale distribution, $D_g$, is determined by the slope of the best fit straight line through the data.

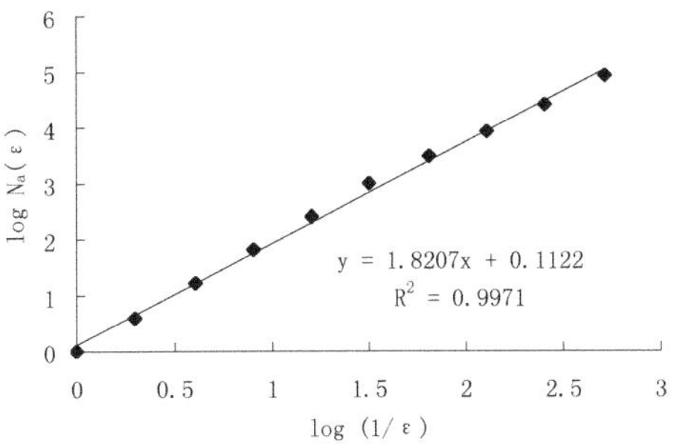


Fig. 4. Plot of the number of boxes $N_a(\varepsilon)$ as a function of box side length ε. The slope of the best fit straight line through the data yields the fractal dimension of fiber area distribution, $D_a$.

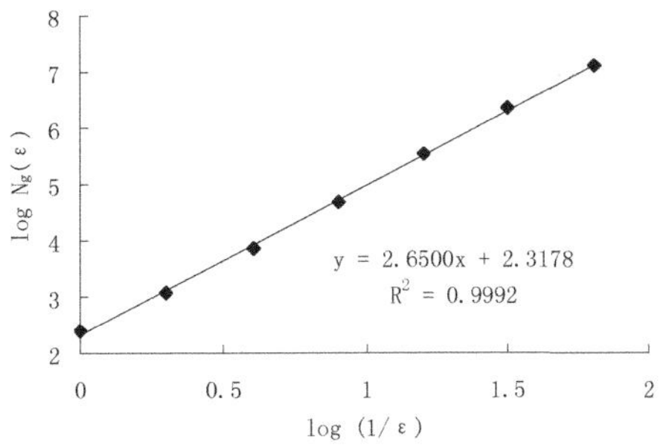


Fig. 5. Plot of the number of boxes $N_g(\varepsilon)$ as a function of box side length ε. The slope of the best fit straight line through the data yields the fractal dimension of grayscale distribution, $D_g$.

### 3.3 The Effect of the Compressive Stress on $D_a$

The effect of the compressive stress $P$ on the fractal dimension of fiber area distribution is illustrated in Fig. 6. It shows that $D_a$ becomes larger with the increase of $P$. It can be explained that with the increase of P, the fibers are compressed to be flatter and the felt to be denser, thus the area of surface fibers contacting the polarized board increases. Suppose $R_f$ is the ratio of the actual contact area to the nominal surface area which can be calculated as:

$$R_f = \frac{\text{number of black pixels in binary image}}{\text{number of total pixels in binary image}} \times 100(\%) \tag{2}$$

From the binary images of different samples under different compress stress (as shown in Fig. 3), we can obtain the $R_f$s. There obviously exists a function relationship between $D_f$s and Rfs as shown in Fig. 7. Gathering the data points of all the samples, the obtained regression equation is y = $0.0005e^{6.0235x,}$ and the square of correlation coefficient is 0.9979

It also can be seen from Fig. 6 that more polyurethane contained in the felt leads to the smaller $D_a$ and $R_f$. It is possibly because that, the polyurethane makes the felt harder and more difficult to be deformed at a certain pressure.

### 3.4 The Effect of the Compressive Stress on $D_g$

Fig. 8 shows the effect of compressive stress $P$ on the fractal dimension of grayscale distribution. For all the samples, $D_g$ increases with the increase of $P$ at first, and then inclines to be stable or a slight decrease. It can be explained that with the increase of $P$, the area of surface fibers contacting the polarized board increases, which enlarges the difference of gray level between fiber pixels and background pixels. When $P$ reaches to a certain degree, it hardly deforms the fibers and the felt any longer, which results the little change of $D_g$. It also can be seen from Fig. 8 that more polyurethane contained in the felt leads to a larger $D_g$. The larger $D_g$ indicates that more polyurethane makes rough surface more difficult to be deformed.

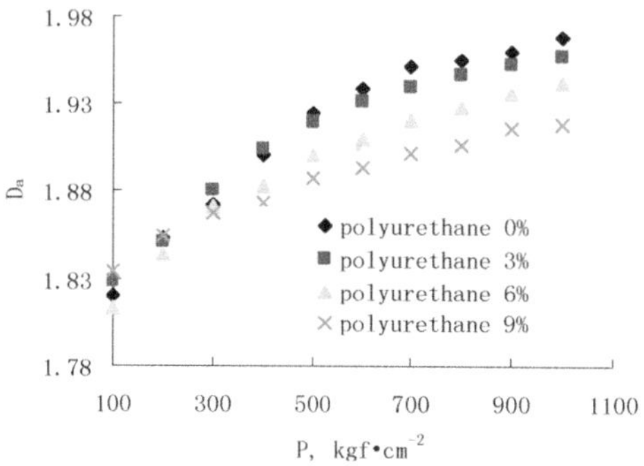


**Fig. 6.** Effect of compressive stress on $D_a$

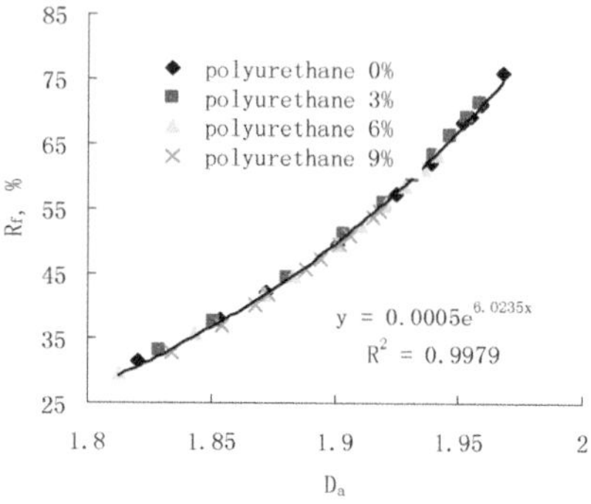


**Fig. 7.** $R_f$ as a function of $D_a$

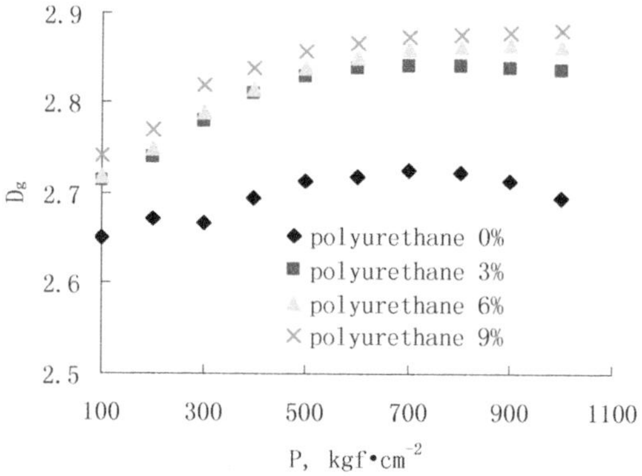


**Fig. 8.** Effect of compressive stress on $D_g$

## 4 Conclusion

Nonwoven felts have fractal surfaces when they are compressed under different pressures. Two kinds of fractal dimensions, $D_a$ and $D_g$, reflecting the fiber area distribution and grayscale distribution respectively, have been used to evaluate the surface characteristics. $D_a$ is closely associated with the actual contact area of surface fibers, and $D_g$ is related to the roughness of surfaces. The results show that both $D_a$ and $D_g$ increase with the increase of compressing stress. The more polyurethane the nonwoven contains, the larger $D_g$ and smaller $D_a$ it has.

**Acknowledgments.** This is a project funded by Natural Science Foundation of Jiangsu Province under Grant No. BK2009148; by China National Textile and Apparel Council under Grant No. 2009088; by Opening Project of Silk Engineering Key Laboratory of Jiangsu Province; and by the Priority Academic Program Development of Jiangsu Higher Education Institutions.

## References

1. Buzio, R., Boragno, C., Biscarini, F.: The Contact Mechanics of Fractal Surfaces. Nature Materials 2(4), 233–236 (2003)
2. Toshio, M., Hiroyuki, K., Hirokazu, K., et al.: Prediction Method for Wrinkle Grade by Image Analysis. Journal of Textile Engineering 54(5), 135–141 (2008)
3. Militky, J., Bajzfk, V.: Surface Roughness and Fractal Dimension. Journal of Textile Institute 3, 91–113 (2001)
4. Bisoi, A.K., Mishra, J.: On Calculation of Fractal Dimension of Images. Pattern Recogn. Lett. 22, 631–637 (2001)
5. Hou, Z.D., Qin, Y.W.: The Study of Fractal Correlation Method in the Displacement Measurement and Its Application. Optics and Lasers in Engineering 39(4), 465–472 (2003)

# A Cepstral PDF Normalization Method for Noise Robust Speech Recognition

Yong Ho Suk[1] and Seung Ho Choi[2,*]

[1] Creatz Inc., #101, Business Center, KICET,
233-5, GaSan-dong, Geumcheon-gu, Seoul 153-801, Korea
yhsuk@mycreatz.com
[2] Dept. of Electronic and Information Engineering,
Seoul National University of Science and Technology, Nowon-Gu, Seoul 139-743, Korea
shchoi@seoultech.ac.kr

**Abstract.** In this paper, we propose a novel cepstrum normalization method based on the scoring procedure of order statistics for speech recognition in additive noise environments. The conventional methods normalize the mean and/or variance of the cepstrum, which results in an incomplete normalization of the probability density function (PDF). The proposed method fully normalizes the PDF of the cepstrum, providing an identical PDF between clean and noisy cepstrum. For the target PDF, the generalized Gaussian distribution is selected to consider various densities. In recognition phase, a table lookup method is devised in order to save computational costs. From the speaker-independent isolated-word recognition experiments, we show that the proposed method gives improved performance compared with that of the conventional methods, especially in heavy noise environments.

**Keywords:** Cepstrum normalization, noisy speech recognition, order statistics, generalized Gaussian distribution.

## 1 Introduction

When speech signals are contaminated by additive noise, the statistical properties of a speech feature vector vary according to the types of noise and signal-to-noise ratio (SNR) levels. The statistical differences between clean and noisy feature vectors result in performance degradation of noisy speech recognition. Some methods are devised to decrease the statistical differences by normalizing cepstral moments [1][2][3]. However, these methods partly normalize the probability density functions (PDFs).

In this work, we introduce a cepstrum PDF normalization (CPN) method, which fully normalizes the statistical properties. The CPN normalizes the PDFs of the clean and the noisy cepstrum to a target PDF so that the statistical properties are identical. In order to consider various densities, the generalized Gaussian distribution (GGD) is used as the target PDF. Moreover, a table lookup method is also devised to alleviate the computational load of the CPN.

---

* Corresponding author.

S. Lin and X. Huang (Eds.): CSEE 2011, Part II, CCIS 215, pp. 34–39, 2011.

## 2 Derivation of CPN

For a sequence of cepstrum vectors $\{\mathbf{x}(t), t=1,2,\cdots,N\}$, we obtain the sequence of normalized vectors $\{\mathbf{x}_{CPN}(t), t=1,2,\cdots,N\}$ by the scoring procedure [4] that uses expected values of the order statistics of a target PDF. With the independence assumption between elements, the cepstrum vector is normalized element-wise, and we use scalar notation $x(t)$ and $x_{CPN}(t)$ for each vector element. The $x_{CPN}(t)$ is obtained by

$$x_{CPN}(t) = E[z_{(r(t))}] = \int_{-\infty}^{\infty} y f_{r(t)}(y)dy \tag{1}$$

where $z_{(r)}$ is the $r$-th order statistic of the target PDF and $r(t)$ is the rank of cepstrum $x(t)$ and $f_r(\cdot)$ is the PDF of $z_{(r)}$. Although exact solution of the expected value of the order statistics (1) is available for some PDF, it is hard to find the exact solution as the rank $r$ and/or the sample size $N$ increase. David and Johnson [5] presented approximations with an error of $O(N^{-3})$ for (1) as

$$\begin{aligned} E[z_{(r)}] = Q(p_r) + \frac{p_r q_r}{2(N+2)}\frac{d^2Q(p_r)}{dp^2} \\ + \frac{p_r q_r}{(N+2)^2}\{\frac{1}{3}(q_r - p_r)\frac{d^3Q(p_r)}{dp^3} + \frac{1}{8}p_r q_r \frac{d^4Q(p_r)}{dp^4}\}, \end{aligned} \tag{2}$$

where $Q(\cdot)$ is the inverse function of the cumulative distribution function of target random variable $z$ and $p_r = \frac{r}{N+1}$ and $q_r = 1 - p_r$.

The GGD selected as the target PDF is a symmetric, unimodal density function parameterized by two constants, a variance $\sigma^2$ and a decay parameter $k > 0$ defined as

$$f(z;k) = \frac{k}{2A(k)\Gamma(1/k)} e^{-\{|z|/A(k)\}^k}, \tag{3}$$

where $A(k) = \{\sigma^2 \frac{\Gamma(1/k)}{\Gamma(3/k)}\}^{\frac{1}{2}}$ and $\Gamma(\cdot)$ is the gamma function [6].

The GGD is capable of varying its shape by changing the decay parameter $k$. With $k=1$ and $k=2$, the GGDs are Gaussian and Laplacian, respectively [5]. Let $Q = Q(p;k)$ be the inverse function of CDF of the GGD with the decay parameter $k$, and it is obtained using the numerical integration. The derivatives of $Q = Q(p;k)$ in (2) are derived as follows:

$$\frac{d^2Q(p;k)}{dp^2} = \frac{\eta_k \, sign(Q) \, |Q|^{k-1}}{\{f(Q;k)\}^2} \tag{4}$$

$$n\frac{d^3Q(p;k)}{dp^3} = \frac{\eta_k \, |Q|^{k-2}}{\{f(Q;k)\}^3}\{(k-1)+2\eta_k \, |Q|^k\} \tag{5}$$

$$\frac{d^4Q(p;k)}{dp^4} = \frac{\eta_k \, sign(Q) \, |Q|^{k-3}}{\{f(Q;k)\}^4}\{(k-1)(k-2)+7\eta_k \, |Q|^k + 6\eta_k^2 \, |Q|^{2k}\} \tag{6}$$

where $\eta_k = \frac{k}{\{A(k)\}^k}$. Substituting (4), (5), and (6) into (2), we have the approximations of expected values with the decay parameter $k$.

In addition to the above numerical method, we also use a table lookup method in the recognition phase. Let $S_R(r)$ and $S_C(r)$ be the expected values of $r$-th order statistic with sample size $N_R$ and $N_C$, respectively. First, to reduce the computation load, we make a reference table $T_R = \{S_R(r), r = 1,2,\cdots,N_R\}$ with the frame length $N_R$. Then, using the reference table $T_R$, we approximate the table $T_C = \{S_C(r), r = 1,2,\cdots,N_C\}$ of the current utterance as

$$S_C(r) = S_R(1 + \left\lfloor \frac{r-1}{N_C - 1}(N_R - 1) \right\rfloor), r = 1,2,\cdots,N_C \tag{7}$$

where $N_C$ is the frame length of current utterance and $N_C\lfloor \cdot \rfloor$ is the rounding off operator. Using the current table $T_C$, the normalized cepstrum sequence $\{x_{CPN}(t)\}$ is obtained as

$$x_{CPN}(t) = S_C(r(t)), t = 1,2,\cdots,N_C \tag{8}$$

## 3 Experimental Results

We performed isolated-word recognition experiments in the environments of additive white Gaussian noise and car noise at various SNR levels. The database used for the recognition experiments consists of 100 isolated-words. The training data consist of 2,400 utterances from 12 male speakers, while 1,600 utterances from 8 male speakers are used for testing. Each utterance was recorded in an ordinary office environment, and then sampled at 8 kHz with a resolution of 16 bits. We used 12-th order mel-frequency cepstral coefficients as a feature vector and constructed the recognition system based on the simple left-to-right semi-continuous hidden Markov model. We use the GGDs as the target PDF with the unit variance $\sigma^2 = 1$ and various values of the decay parameter $(k = 1, 2, 3)$. The reference table with frame length $N_R = 100$ is used for the table lookup method in the recognition phase.

As shown in Fig. 1 and Fig. 2, the proposed CPN methods outperform the conventional cepstrum normalization methods, cepstrum mean normalization (CMN) and cepstral mean-variance normalization (CMVN). When the table lookup method is used, the recognition results are similar to those of the numerical integration method as shown in Fig. 3 and Fig. 4. We can see that the CPN method is useful especially in heavy noise environments.

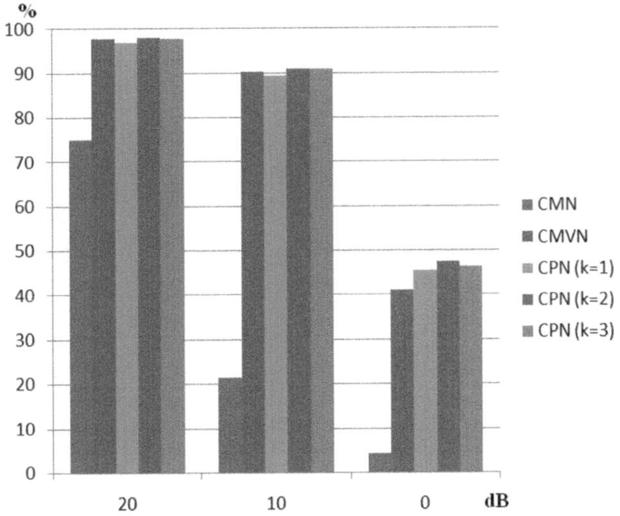


**Fig. 1.** Recognition result under additive white Gaussian noise environments (numerical integration method)

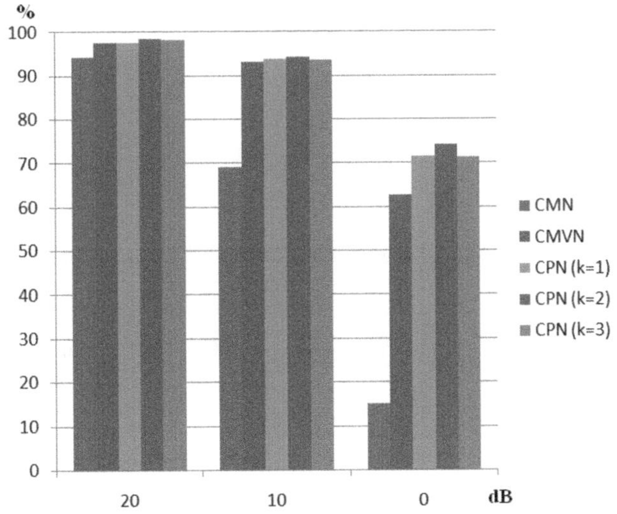


**Fig. 2.** Recognition result under 100 km/h car noise environments (numerical integration method)

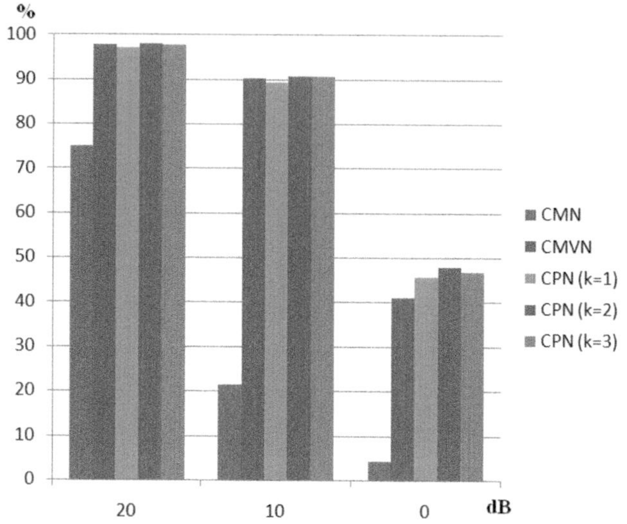


**Fig. 3.** Recognition result under additive white Gaussian noise environments (table lookup method)

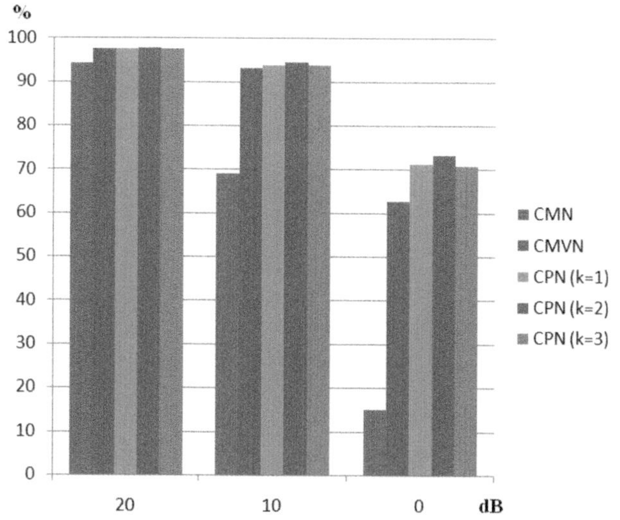


**Fig. 4.** Recognition result under 100 km/h car noise environments (table lookup method)

## 4 Conclusion

In this work, we introduced a cepstrum PDF normalization (CPN) method. In contrast with the conventional methods of cepstrum mean and/or variance normalization, the CPN fully normalizes the PDF of cepstrum vectors. The isolated-word recognition results show that the CPN yields an improved performance, especially, in heavy noise conditions. Furthermore, the CPN can give the best performance by appropriately selecting the decay parameter of the generalized Gaussian density.

## References

1. Furui, S.: Cepstral analysis technique for automatic speaker verification. IEEE Tr. on Acoust., Speech, Signal Process., ASSP 29, 254–272 (1981)
2. Viiki, O., Bye, D., Laurila, K.: A recursive feature vector normalization approach for robust speech recognition in noise. In: Proc. ICASSP, pp. 733–736 (1998)
3. Hsu, C.W., Lee, L.S.: Higher order cepstral moment normalization (HOCMN) for robust speech recognition. In: Proc. ICASSP, pp. 197–200 (2004)
4. David, H.A.: Order Statistics. John Wiley & Sons, NY (1981)
5. David, F.N., Johnson, N.L.: Statistical treatment of censored data, Part I. fundamental Formulae. Biometrica 41, 228–240 (1956)
6. Kassam, S.A.: Signal Detection in Non-Gaussian Noise. Springer, Heidelberg (1988)

# On the Best Condition of Electrolysis Coking Wastewater by Orthogonal Method

Xuwen He, Lu Xin, Chunhui Zhang, Hui Lin, Liyuan Liu, Jingjing Zhang, and Xiong He

School of Chemical & Environmental Engineering, China University of Mining & Technology, Beijing, 100083 Beijing, China
hexuwen@sina.com.cn

**Abstract.** Coking wastewater was advanced treated by electrolytic experiment. Through the orthogonal method, we found out the optimum experimental condition as current density 20 mA/cm2, electrode plate span 1cm, electrolytic time 60 min, PH 7 and electrical conductivity 8000 μs/cm. The removal rates of COD, NH3-N and phenol were 58.62%, 89.9% and 96.8% respectively and the corresponding concentrations were 97.78mg/L, 14.3mg/L and 0.43mg/L respectively, which can meet first discharge standard of GB 13456-1992.

**Keywords:** The orthogonal method; electrolysis; coking wastewater; advanced treatment.

## 1 Introduction

It is difficult for coking wastewater to be treated because of its complex component, high consistency and nonbiodegradability [1]. It contains complex inorganic and organic pollutants, such as ammonium, cyanide, phenol and polynuclear aromatic hydrocarbons [2], most of which are refractory, toxic, mutagenic and carcinogenic [3]. This means that the coking wastewater if not treated, in the environment will become a permanent presence of pollutants and continue to accumulate, will cause great environmental harm [4]. At present, the biochemical treated coking wastewater generally have high concentrations of COD and $NH_3$-N [5]. Selecting proper advanced treatment process of coking wastewater is very important.

Coking wastewater was advanced treated by electrolytic experiment [6]. It was studied that the impact of these parameters, current density, electrode plate span, electrolytic time, PH and electrical conductivity, on treatment effects of COD, $NH_3$-N and phenol based on an orthogonal test. We got the optimum experimental condition and the variation rule of the removal rate with change of parameters levels. It can guide the future experiment.

## 2 Experimental Part

### 2.1 Materials and Apparatus

The treated water from a secondary clarifier of one coking plant was employed in the electrolytic experiment. The quality of raw water was shown in table 1.

S. Lin and X. Huang (Eds.): CSEE 2011, Part II, CCIS 215, pp. 40–46, 2011.

**Table 1.** Raw water quality and GB 13456-1992 indexs

| Characteristics | COD (mg/L) | $NH_3N$ (mg/L) | Phenol. (mg/L) | PH | Conductivity (μs/cm) |
|---|---|---|---|---|---|
| Raw water quality | 200~300 | 100~150 | 10~15 | 6~9 | 6250 |
| GB 13456-1992 | ≤100 | ≤15 | ≤0.5 | 7~8 | |

Coking wastewater was pumped into electrolysis bath, which was made by organic glass with a dimension of 20cm × 10cm× 15cm. Both the specification of the stainless steel cathode plates and the titanium anode plates coated with noble $RuO_2$ and $TiO_2$, were 20cm × 10cm ×1cm [7]. The reaction system was shown in Figure1.

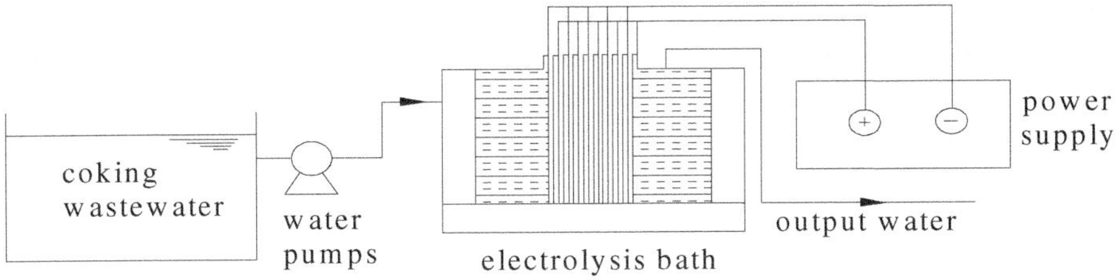


**Fig. 1.** Experimental apparatus of electrolytic experiment

## 2.2 Detecting Instrument and Method

DC electrical source (MPS702); Ultraviolet and visible spectrophotometer (UNIC 2100); PH meter ( PHS-3C); Conductivity meter(DDS-11A). COD was detected by dichromate titration. $NH_3$-N was detected by Leonard's reagent spectrophotometric method. Phenol was detected by directive light intensity method of 4-anino-antipyrine [8].

There were many parameters which have effect on the treatment effects. The experiment chose current density, electrode plate span, electrolytic time, PH and electrical conductivity as the influence factor, and marked them with A, B, C, D and E respectively. Table 2 was the parameters and levels table.

**Table 2.** Parameters and levels in orthogonal test

| Level | Parameter<br>A Current density ($mA/cm^2$) | B plate span (cm) | C Electrolytictime (min) | D PH | E Conductivity (μs/cm) |
|---|---|---|---|---|---|
| 1 | 10 | 0.75 | 40 | 6 | 6500 |
| 2 | 15 | 1 | 50 | 7 | 7000 |
| 3 | 20 | 1.25 | 60 | 8 | 7500 |
| 4 | 25 | 1.5 | 70 | 9 | 8000 |

# 3 Result and Analysis

## 3.1 The Result of Orthogonal Test

The test had 5 parameters, and each of them had 4 levels. Don't consider interaction between the parameters. The L16 ($4^5$) orthogonal table was selected [9]. Arrangement and results of the orthogonal test were shown in table 3.

**Table 3.** Arrangement and results of the orthogonal test

| Mumber | A | B | C | D | E | COD (%) | $NH_3$-N (%) | Phenol (%) |
|---|---|---|---|---|---|---|---|---|
| 1 | 10 | 0.75 | 40 | 6 | 6500 | 12.22 | 23.08 | 89.44 |
| 2 | 10 | 1 | 50 | 7 | 7000 | 39.34 | 20.42 | 87.55 |
| 3 | 10 | 1.25 | 60 | 8 | 7500 | 23.61 | 12.87 | 71.28 |
| 4 | 10 | 1.5 | 70 | 9 | 8000 | 25.25 | 10.47 | 64.18 |
| 5 | 15 | 0.75 | 50 | 8 | 8000 | 38.64 | 30.18 | 99.73 |
| 6 | 15 | 1 | 40 | 9 | 7500 | 44.08 | 20.88 | 85.48 |
| 7 | 15 | 1.25 | 70 | 6 | 7000 | 38.55 | 5.76 | 71.20 |
| 8 | 15 | 1.5 | 60 | 7 | 6500 | 36.84 | 18.22 | 78.18 |
| 9 | 20 | 0.75 | 60 | 9 | 7000 | 40.23 | 51.88 | 94.29 |
| 10 | 20 | 1 | 70 | 8 | 6500 | 57.62 | 80.09 | 99.63 |
| 11 | 20 | 1.25 | 40 | 7 | 8000 | 33.19 | 21.59 | 89.40 |
| 12 | 20 | 1.5 | 50 | 6 | 7500 | 25.97 | 10.48 | 60.86 |
| 13 | 25 | 0.75 | 70 | 7 | 7500 | 56.25 | 94.03 | 99.15 |
| 14 | 25 | 1 | 60 | 6 | 8000 | 56.52 | 77.23 | 99.42 |
| 15 | 25 | 1.25 | 50 | 9 | 6500 | 20.25 | 14.48 | 80.00 |
| 16 | 25 | 1.5 | 40 | 8 | 7000 | 23.29 | 7.29 | 57.00 |

With reference to table 3, advanced treating coking wastewater by electrolytic experiment, the treatment effects of $NH_3$-N and phenol were well. But it was difficult to removal COD. Results of the orthogonal test could be analyzed in detail using extreme difference analysis.

## 3.2 The Optimum Experimental Condition

The extreme difference analysis is used widely in analyzing the results of the orthogonal test. Analysis results were shown in table 4.

As shown in table 4,the current density, plate span, electrolytic time had a great effect on the removal rate of pollutants. The removal rate of COD could reach quickly with increasing of current density when it was less than 15mA/cm$^2$, but it was decreased slightly if it exceed 15mA/cm$^2$. This was because there existed removal and energy input requirement extreme for pollutants [10]. For the input energy was under the extreme circumstance, there was a higher removal efficiency and better redox reaction with the input energy increases [11]. On the contrary, with the increasing of input energy, the removal efficiency might decrease. The proper current density for the removal of COD was 15mA/cm$^2$, $NH_3$-N was 25mA/cm$^2$, phenol was 20 mA/cm$^2$.

**Table 4.** Analysis results of the extreme difference analysis

| Result | Analysis | A | B | C | D | E |
|---|---|---|---|---|---|---|
| COD | K1 | 25.11 | 36.84 | 28.20 | 35.82 | 31.73 |
| | K2 | 39.53 | 40.39 | 31.05 | 41.41 | 35.35 |
| | K3 | 39.25 | 28.90 | 39.30 | 35.79 | 37.48 |
| | K4 | 39.08 | 27.84 | 44.42 | 32.45 | 38.45 |
| | range | 14.42 | 12.55 | 16.22 | 8.96 | 6.72 |
| | level influence | A2>A3> A4>A1 | B2>B1> B3>B4 | C4>C3> C2>C1 | D2>D1> D3>D4 | E4>E3> E2>E1 |
| | parameter influence | C>A>B>D>E | | | | |
| $NH_3$-N | K1 | 16.71 | 49.79 | 18.21 | 29.14 | 33.87 |
| | K2 | 18.76 | 49.66 | 18.89 | 38.57 | 21.34 |
| | K3 | 41.01 | 10.82 | 40.05 | 32.61 | 34.57 |
| | K4 | 48.26 | 11.62 | 47.59 | 24.43 | 34.87 |
| | range | 31.55 | 38.97 | 29.38 | 14.14 | 13.53 |
| | level influence | A4>A3> A2>A1 | B1>B2> B4>B3 | C4>C3> C2>C1 | D2>D3> D1>D4 | E4>E3> E1>E2 |
| | parameter influence | B>A>C>D>E | | | | |
| Phenol | K1 | 55.50 | 95.93 | 80.33 | 80.38 | 86.90 |
| | K2 | 83.72 | 93.26 | 82.10 | 88.94 | 77.51 |
| | K3 | 86.14 | 77.97 | 85.94 | 82.07 | 79.41 |
| | K4 | 84.25 | 65.06 | 83.85 | 81.99 | 88.40 |
| | range | 30.64 | 30.87 | 5.61 | 8.56 | 10.89 |
| | level influence | A3>A4> A2>A1 | B1>B2> B3>B4 | C3>C4> C2>C1 | D2>D3> D4>D1 | E4>E1> E3>E2 |
| | parameter influence | B>A>E>D>C | | | | |
| the optimum conditions | | A3B2C3D2E4 | | | | |

The removal rate of COD reached a maximum when the electrode plate span was 1cm, and it decreased conspicuously with the increase of the plate span. It might because when the plate span was small, there was much current around the electrode, and it had a higher concentration of $H_2O_2$ and·OH which influenced the degradation of organic pollutants. With the plate span increasing, there would be higher system resistance [12]. With the plate span became too small, the current on the electrode would decreased as the generation of spark in the reaction process and increasing electrode resistance, the removal rate was reduced. So we should choose a optimum electrode plate span to get an extreme input energy. Table 4 indicated that the proper plate span for the removal of COD was 1cm, $NH_3$-N was 0.75cm, phenol was 0.75cm.

The removal rates of COD, $NH_3$-N kept a rapid growth with the length of electrolytic time in the first 60min and became slowly in the following 10min, and The removal rates of phenol decreased after 60min.

PH and electrical conductivity had a slight effect on the removal rate of the pollutant. The optimum PH was 7 and the optimum electrical conductivity was 8000μs/cm. The removal rate decreased probably because some substances produced in the acidic or alkaline liquor had a influence on the actinity of $H_2O_2$ and·OH.

In the applications, considering the cost of power consumption, The current density and electrolytic time should be controled as low as possible when meeting the demand. Through a comprehensive comparision and analysis of the treatment effect of COD, $NH_3$-N and phenol in table 4, the optimum technological conditions were determined as current density 20 mA/cm$^2$, plate span 1cm, electrolytic time 60 min, PH 7 and electrical conductivity 8000μs/cm.

### 3.3 Removal Rate in the Optimum Condition

The optimum level of each parameter was elected through orthogonal test. Electrolyzing coking wastewater in the optimum condition (current density 20 mA/cm$^2$, plate span 1cm, PH 7 and electrical conductivity 8000μs/cm, the treatment effects were shown in Figure2.

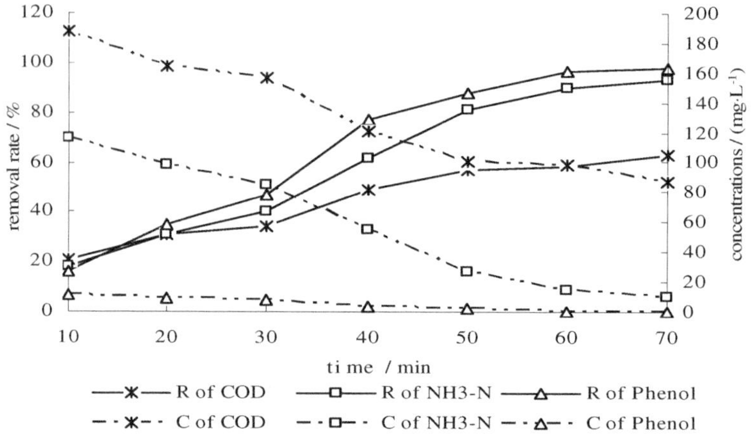


**Fig. 2.** The removal efficiency of COD, $NH_3$-N and phenol in the optimum condition

As could be seen from Figure 2, the removal efficiency was better in 70 min compared with it in 60 min. After 60 min, the removal rates of COD, $NH_3$-N and phenol were 58.62%, 89.9% and 96.8% respectively and the corresponding residual concentrations were 97.78mg/L, 14.3mg/L and 0.43mg/L respectively, which can meet first discharge standard of GB 13456-1992. For reducing cost, the optimum electrolytic time was 60 min.

### 3.4 The Experimental Mechanism

Electrolyzing coking wastewater could produce·OH, which has a strong oxidation ability to remove pollutants in wastewater [13]. Electrode reaction were shown as follows.:

$$2\,H_2O \rightarrow 2\cdot OH + 2\,H^+ + 2e^- . \tag{1}$$

$$\text{Organic pollutants} + \cdot OH \rightarrow CO_2\uparrow + H_2O\,. \tag{2}$$

$$2\,NH_3 + 6\cdot OH \rightarrow N_2\uparrow + 6\,H_2O\,. \tag{3}$$

$$2\cdot OH \rightarrow H_2O + 1/2\;O_2\uparrow. \tag{4}$$

Much chloride ions in coking wastewater could emit electron at the anode to turn to $Cl_2$ [14]. It can produce $ClO^-$ to destroy pollutants in wastewater by the strong oxidation. The reaction happened as follows:

$$2\,Cl^- \rightarrow Cl_2\uparrow + 2\,e^-\,. \tag{5}$$

$$Cl_2 + H_2O \rightarrow ClO^- + 2H^+ + Cl^-\,. \tag{6}$$

$$\text{organic pollutants} + ClO^- \rightarrow CO_2\uparrow + H_2O + Cl^-\,. \tag{7}$$

## 4 Conclusion

a) The current density, plate span, electrolytic time had a great effect on the removal rate of pollutant. PH and electrical conductivity had a slight effect on the removal rate of the pollutant.
b) The optimum technological conditions were determined as current density 20 $mA/cm^2$, plate span 1cm, electrolytic time 60 min, PH 7 and electrical conductivity 8000μs/cm.
c) The removal rates of COD, $NH_3$-N and phenol were 58.62%, 89.9% and 96.8% respectively and the corresponding residual concentrations were 97.78mg/L, 14.3mg/L and 0.43mg/L in the optimum consition.

**Acknowledgments.** The financial support of this research by National High Technology Research and Development Program (2009AA05Z306) was deeply appreciated.

## References

1. Cao, W.P., Xu, S.R., Chen, H.L.: Advanced Treatment of bio-treated coking effluent by Fenton Oxidation. J. Environmental Science & Technology 33, 305–308 (2010) (in Chinese)
2. Lai, P., Zhao, H.Z., Zeng, M.: Study on treatment of coking wastewater by biofilm reactors combined with zero-valent iron process. J. Hazardous. Materials 162, 1423–1429 (2009) (in Chinese)
3. Sag, Y., Kutsal, T.: The selective biosorption of chromium and copper( II )ions from binary metal mixtures by arrhizus. J. Process Biochemistry 31, 561–572 (1996)
4. Hosseini, S.H., Borghei, S.M.: The treatment of phenolic wastewater using a moving bed bio-reactor. J. Process Biochemistry 40, 1027–1031 (2005)
5. Lai, P., Zhao, H.Z., Wang, C.: Advanced treatment of coking wastewater by coagulation and zero-valent iron processes. J. Journal of Hazardous Materials 147, 1–2 (2007) (in Chinese)

6. Huang, Y., Huang, G., Lee, S.N.: Method of wastewater treatment by electrolysis and oxidization. P. US 6126838 (2000)
7. Yusuf , Yavuz., Savas, Koparal.: Electrochemical oxidation of phenol in a paralled plate reactor using ruthenium mixed metal oxide electrode. J. Hazardous Mater. B136, 296 (2006)
8. China SEPA.: The Water and Wastewater Monitoring Method (editor4). China Environmental Science Press, Beijing (2002) (in Chinese)
9. Hao, L.D., Yu, H.D.: On use of orthogonal experimental design. J. Actaeditologica. 17, 334–335 (2005) (in Chinese)
10. Takeshi, H., Kimihito, S., Taro, T., Tsutomu, S.: Coke powder heat-treated with boron oxide using an Acheson furnace for lithium battery anodes. J. Carbon. 40, 2317–2322 (2002)
11. Zhang, X., Wen, Y.B.: Advanced treatment of coking wastewater with electrocoagulation. J. Shanxi Architecture 35, 9 (2009) (in Chinese)
12. Zhu, X.P., Ni, J.R., Lai, P.: Advanced treatment of biologically pretreated coking wastewater by electrochemical oxidation using boron-doped diamond electrodes. J. Water Research 43, 4347–4355 (2009) (in Chinese)
13. Chen, W., Li, F.X., Mei, P.: Mechanism of COD Degradation in Wastewater Using Three Dimension Electrode Method. J. Environmental Science and Technology 25, 11–12 (2005) (in Chinese)
14. Liu, Y., Liu, D., Zhao, S.L.: Impact of chloride ion on undegradable landfill leachate treatment by electrochemical oxidation. J. Journal of Chongqing University. 31, 1431–1435 (2008) (in Chinese)

# An Analysis on Modern Technology Applied to Education

Xiaojuan Liu

School of Foreign Languages, Shandong University of Technology,
Zibo, Shandong, China, 255049

**Abstract.** Modern technology, due to its powerful features, has been widely applied to education. However, as more and more modern techniques applied in education, some problems have occurred. This paper aims at studying the status, problems and countermeasures of applying modern technology to education.

**Keywords:** application, modern technology, education.

## 1 Introduction

In this information age, among various traditional ways of acquiring and spreading knowledge, such as books, newspapers, magazines, radio, television, and so on, people turn to computers and Internet for help more and more frequently. In the realm of education, as a result of the development of modern technology, a lot of advanced techniques, especially computerized equipments, have been employed to improve the efficiency and effect of teaching and learning. It's almost believed that the new educational technology can cause a revolution in education and studies.

## 2 Modern Teaching Technology Has Shown Great Power

In the late 1950s and early 1960s, developed countries and a few developing countries showed great interest in applying the new communications techniques to education. People hoped to improve the quality of education, to offer equal opportunities of education to all people, to overcome the lack of teachers, etc. People also hoped to lower the cost of education greatly, and meanwhile, improve the efficiency of teaching.

Some experts believed that the utilization of new technology in education can bring about the following results: 1) A cultural and educational environment in which knowledge and information can have more sources will be created. 2) Information can be better stored and more widely spread, thus offering larger space for life-long learning. 3) In formal education, new teaching technology can enable the students to study according to their own plan and progress, effectively avoiding the failure in their studies. 4) Cultivate students' ability in acquiring and processing information, and benefit them in their future jobs and integration with the social environment.

It has been proved that the modern educational technology has influenced education in three aspects: on the material facet, in the educational systematic structure and policies, and the concept of education.

S. Lin and X. Huang (Eds.): CSEE 2011, Part II, CCIS 215, pp. 47–52, 2011.

### 2.1 The New Technology Has Modernized the Facilities in Schools and Bettered the Surroundings of Teaching and Learning

It adds the interaction of the learners and the computers to the traditional methods of education, making the contexts of teaching more acceptable and easier to acquire for the students. Moreover, the vast range of information online enriches and develops the material used in education.

### 2.2 The New Technology Enables the Learners to Have Access to the Material beyond the Limitation of Time and Places

With the help of computer network, long-distance education is possible and convenient to be carried out, making classroom teaching only one of the choices in teaching and learning. New teaching method, such as personal consultation, differentiated teaching according to one's own capability and situation, and so on, will be more and more prevalent, and will ultimately realize the revolution and transformation of the whole education system and, especially school education. The tendency of education is surely to be more open and flexible.

### 2.3 The New Technology of Education Greatly Influences People's Concept in Education

It brings the new fruits and discoveries of international studies in education, providing people with the conditional elements of modernizing and transforming their concept in education. The new context, new methods and new policies in education brought by the new technology urge people to face a series of problems in education, giving people the inner motivation to reform the existing educational system.

## 3 Negative Effects Have Occurred in the Utilization of the New Educational Technology

In the latest years, new methods of teaching and learning, which are characterized by the using of computers and Internet, have been widely use in school education. People attitudes towards this phenomenon diversified from complete support to resolute opposition. Experts study the new educational technology by doing lots of experiments and researches of its utilization in teaching practice. Many of the experiments proved that in the traditional system of teaching, the new technology of education didn't bring revolutionary influence. As emphasized by Bagley, "Only when closely combined with the reform of schools and complete change of teaching process, can the new technology have ideal effect." Priscilla Norton also believed, "Offering computer to the teachers and students does not necessarily result in a much better effect in the teaching process. When using the new teaching technology, its influence on the different factors which forming the whole teaching and learning process must be taken into consideration."

According to a survey done by Japan's Education Labor Union, private computers are very popular in both urban and rural elementary schools. Even in one of the primary schools surveyed in the remote mountain area, only one of the twenty students in a class does not have private computer in his family. Parents buy computers for their children for the purpose of "creating" a companion for their children besides the aim of enabling the children to learn the operation of computers. This "companion" not only helps the children study, but also brings them lots of entertainments. The kids sit in front of the computers, pressing buttons, playing games, listening to sons…for many hours, and seem never be tired. Undoubtedly, this benefits the children in their learning. However, it causes some negative effects also. It's reported that more than half the students surveyed have shown some negative tendencies brought by computers. For example, some of them have bad eye-sight, some are easily enraged, some are very anxious, and some are emotionally far away from parents and teachers because they mainly make friends through the Internet. More seriously, some kids become very selfish and difficult to communicate with.

In addition, because of the dependence on computers, the students' show an apparent decline in their capabilities of dealing with mathematic symbols and words used in verbal language, only having some improvement in the emotional and modal reaction to the visual images. They are used to the attractive learning material displayed by computers, and gradually becoming bored at the issues and phenomena in real life. The reason is that the learning materials in computer have effectively combined the inner structure and external forms, yet in reality, things don't often have lively appearance. As a result, the learners feel it difficult to accept the normal educational contexts and patterns.

The similar situation also holds true for television, another frequently used learning media. Television, as a popular form of media, not only has educational power, but also brings people various entertainments. Some people even think that television is mainly for entertainments. As a form of media, the functions of television are difficult to define. Whether it is mainly for education or for entertainments depends on the users' differentiated understanding.

For most children, watching TV is mainly a way of entertaining themselves. They spend large amount of time in front of televisions. A survey, made by N. Postmain, an expert in television and a professor in media in New York University, showed that American children, from six to eighteen years old, spend the average of 15,000 to 16,000 hours watching TV, while they spend less than 13,000 hours studying their lessons. It's definitely true that no matter for studying or for entertaining, watching TV has become the "primary course" for American students. Professor Postmain believes that there are two devastating consequences caused by spending so much time watching TV: Firstly, when watching TV, the students focus their attention and have positive attitude and become emotionally excited, however, they get no practice in their abstract thinking. By and by, their capabilities of language, abstract thinking and dealing with abstract symbols in mathematics will inevitably decline. It's reported that some pupils in primary schools have a lot of difficulties in reading simple articles and doing some basic calculations. Moreover, TV programs have formed a peculiar cultural atmosphere, in which the differences of adults and children are totally neglected, for they have equal access to the same programs and information. This so-called "spiritual civilization" is almost a disaster for the children, for it deprives them of their childhood.

The negative influence of modern educational technology can be illustrated as the following:

Politically, the most powerful countries or companies mastering the core technology have deciding influence on culture and politics. The cultural industries and products, monopolized by few countries, spread their own values and attitudes to the whole world, putting some developing countries at a loss.

Socially, the ability to acquire and process information becomes more and more important, and it has become one of the deciding factors for a person's employment and integrating harmoniously with the society. Therefore, different levels of mastering the technology may bring imbalance of the social groups, and deepen the gaps of them.

In educational activities, the new technology has shown much deficiency. In moral education, because of the lack of face-to-face communication, the teachers' model behavior and influence as examples may be greatly weakened. In terms of the social intercourse, the students who are indulged in playing computers lack chances to communicate with the other people, hence hindering the improvement of their personality and the development of their individuality. And also, when the students spend a lot of time sitting in front of the computers, their physical and mental health will inevitably be harmed. For example, many may be near-sighted and ache in necks. Those who often surf Internet may be influenced by some false even harmful information for there is no effective way to prevent this kind of information from occurring online. Though many people may think that computers help a lot in the students' learning knowledge and widen their horizons greatly, the basic knowledge and skills should be taught by teachers, and the fact is that teachers can do much better in this factor than computers. In addition, the effect of computer learning is largely decided by the quality of the teaching software. Though there has been rapid development of all sorts of soft wares, whether these soft wares and Internet sources of learning are good for learning or not is still a question without a definite answer, and people have a long way to go to find the ideal way to make use of the computer aided instructions (CAI) and sources online.

## 4 There Are Many Difficulties in Popularizing the New Education Technology

The popularization of the new teaching and learning technology will be restricted by many elements. They can be listed as follows:

### 4.1 Limitation Comes from People's Concept

The fact is that people's so-called understanding of the new technology only stays on the surface. It has been clearly shown that the new technology can improve the students' interest in learning and can greatly enlarge the range of spreading knowledge. However, few people can see the potential meaning of utilizing the modern technology in teaching and learning, such as its influence on the development even revolution in the education concept. What's more, many people don't think it possible to combine the new technology and education theories effectively. The reason is that they think the new technology mainly involves machinery and equipment, while education theories are much more "human" and emotional, and the two opposites almost have polar difference.

### 4.2 There Exists the Economic Limitation

The popularization and effective utilization of the new education technology require comparatively good material conditions. In many schools in the developing countries, the basic teaching facilities may not be guaranteed, let alone the modern equipments for the new teaching and learning technology. Some schools have experimental classes and try to use the new technology, but since the students in these classes must have private computers at home to finish some after-class activities, the proportion of the students in the experimental classes is usually very small. Thus, only a small part of students can have access to the new teaching and learning technology.

### 4.3 The Quality of Teachers May Also Bring Some Trouble in Popularizing the Modern Teaching Technology

Though most teachers have the basic skills to operate computers and use the teaching soft wares in teaching, many of them lack and special training and have difficulty making best use of the modern technology in teaching and guiding the students in their learning. Only when most teachers have the ability to operate computers skillfully and can develop teaching soft wares according to their own teaching purposes, can the modern education technology show its best and greatest effect in education.

Some experts point that, the new technology can enable those educating and educated to improve the efficiency of their work and better accomplish their tasks of teaching and learning, however, the new technology can never replace the activities of "human" in the realm of education. In another word, the modern technology can only be a supplementary way in education and a good tool for all educators. The right attitude towards the modern technology in education is to combine it with the reforms of the schools and to make the best use of it in teaching and learning. When using the technology in classroom teaching, teachers should think more about the determination of teaching and learning aims, the management of teaching and learning process, the assessment of teaching and learning effects, and so on.

## 5 Modern Technology Can Never Replace School Education

American educator Bagley believes, "In this information age, reforms in all fields may have been influenced more or less by modern technology. When carrying out the reform of educational system, the utilization of modern education technology must be combined with the reforms of schools and teaching." It's true that the reform of education can have the best result when all aspects are taken into consideration.

It must be admitted that some serious problems exist in education: fossilization, monopolization and closeness. As a result, new context, disciplines and methods must be taken in to better the educational system. We cannot deny the achievements of the school education and ignore the problems existing in this field. A feasible way of improving the whole education system is to make better use of school education. In the near future, most students can enjoy both school education and home-learning aided by computerized technology. Self-education will be an important factor in one's learning process and contribute a lot in their acquirement of knowledge and skills.

Through all the factors discussed above, the conclusion can be drawn: Modern technology brings great changes in education; however, many problems still hinder its effectiveness and popularization. Only when the technology is combined harmoniously with the reforms of education concept and system, can it function well. In addition, school education will never be replaced completely by modern teaching technology. Its function in the whole educational system will surely be improved by the new way of education brought by modern technology.

## References

1. Snow, D., Barley, Z.A.: Classroom Strategies for Helping At-risk Students. Assoc. for Supervision and Curriculum Development, Alexandria VA (2005)
2. Son, J.-B.: Learner Experiences in Web-based Language Learning. Computer Assisted Learning (2007)
3. Chuang, Y.S.: Creating online multimedia courseware efficiently and interestingly. In: The Proceedings of the Fourth Conference on Interactive English Teaching at Chainmen Institute of Technology. Crane Publishing Company, Taipei (2004)
4. Lin, X.: Principles of Applying Multimedia in Teaching. Reading and Writing (2008)
5. Lei, X.: The Transformation of English Teacher's Role in the Multimedia Teaching. Research on the Tertiary Education (2001)

# Multimedia Technology in College English Teaching

Xiaojuan Liu

School of Foreign Languages, Shandong University of Technology,
Zibo, Shandong, China, 255049

**Abstract.** Multimedia Assisted Instruction (MAI) has great functions because of the powerful functions of the multimedia. Recently, MAI is widely used in the classroom teaching, especially in English classroom teaching in colleges and universities in China. However, in the application of the MAI, there exist some problems. This paper makes a survey to investigate the situation of the application of MAI in English classroom in colleges and universities.

**Keywords:** multimedia, modern technology, college English teaching.

## 1 Introduction

The twenty-first century is the era of information and network. Progress in science and technology changes people's lifestyle. The development of modern educational technology provides convenient conditions for all-round development of the students. Multimedia used in the teaching process is a good way to support teaching. Multimedia technology has been widely used by English teachers. This changes the traditional means and methods of English teaching. English classes are brought into a new realm. There are many advantages when multimedia technology is used in English teaching.

## 2 Multimedia Is Modern Technology

Multimedia technology refers to the technology composing text, graphics, sound, animation, video and other information with the media by computer processing to render the knowledge and information. It offers a variety of means in teaching and learning, so that students in class can get plenty of training in listening, speaking, reading and writing.

### 2.1 Audio-Visual Recording Media Shows Students the Correct Pronunciation and Intonation

Students hear the standard pronunciation from native English-speaking countries; together with high-density of spoken language training, students can independently correct errors in their pronunciation, and lay a good foundation of speech.

### 2.2 Audio-Visual Effects Makes Heuristic Introduction of New Courses

The aim of teaching is to introduce a new course, drawing the students' attention to the new knowledge, and establishing substantive contact with the previous knowledge. In

S. Lin and X. Huang (Eds.): CSEE 2011, Part II, CCIS 215, pp. 53–57, 2011.

the conventional teaching, it is not easy to realize. Multimedia provides colorful, clear images on the screen, allowing students to quickly focus on learning objects.

### 2.3 Multi-media Create Scenarios, Presenting the Important and Difficult Points Vividly

In the traditional teaching process, teachers take lots of wall charts and write a lot on the blackboards, which make them exhausted. Multimedia can use animation, graphics and image conversion features to reveal the rich content, to help students understand the nature of things in depth. With the help of multimedia, teaching points can be conveyed smoothly. For the complicated and difficult text, the use of multimedia technology, coupled with intuitive image of the sound and picture, the students can have profound impression.

### 2.4 Multimedia Increases Density of Teaching and Improves the Learning Efficiency

Multimedia-aided teaching can cover large amount of information, greatly improving the efficiency of teaching. Graphic is not language, but more lively, more concrete than language. Students in the classroom do high-quality practice in listening, speaking, reading and writing, whose basic skills are developed efficiently. Multimedia uses sound, shape, and vivid images to make the students understand, digest, and finally absorb the learning content.

## 3 Multimedia Technology in Teaching Effectively Improves the Students' Practical Use of English

In the twenty-first century, the education pays special attention to developing people's overall quality. Every teacher is facing new challenges. How to make full use of multimedia technology in English teaching and to enable the students to acquire various abilities is a problem that every English teacher should think over. Computer aided instruction (CAI) show advantages in English teaching.

### 3.1 It Stimulates Students' Interest in Learning So That They Actively Participate in Classroom Activities

Interest plays a vital role in learning. It helps the learners overcome the inherent psychological difficulties. The more interested they are in a course, the more time and energy they will invest in learning it. Multimedia technology, by showing real life, choosing proper examples, illustrations, fully mobilizes the students' senses, and further expands their thinking, so that students can be actively involved in the content.

### 3.2 It Fosters the Students' Ability in Innovation and Creation

In the traditional teaching, some teaching materials can not be displayed to the students. The use of multimedia caters to the student's cognitive and physiological characteristics, providing students with an adjustable environment of hearing, speaking, reading, writing and doing, stimulating their learning motivation, so that their imagination, creativity can be developed effectively.

### 3.3 It Brings Timely Feedback, Inspiring and Encouraging Students to Discover Rules and Knowledge, Improving Their Self-learning Ability

The involvement of CAI provides a variety of choice, providing the results of study, feedback of information, and proper selection of learning methods for students. There are many soft wares for students to choose according to their own pace and interest, by which, students can always detect a problem with their timely correction of errors, and their ability of finding rules and knowledge will be developed gradually.

The advantages of multimedia in English teaching are undeniable. But in the actual teaching process, if it is not applied reasonably, some undesirable consequences may occur.

## 4 Some Problems Have Appeared in Current Multimedia Teaching

### 4.1 Teachers Are Not Good Enough at Utilizing Multimedia Technology, and Courseware Is Unsatisfactory

Although most English teachers get training in the use of computer-based training, many still feel it difficult to make out satisfactory teaching courseware. Some teachers download some teaching PPT from the Internet, and then apply them to classroom teaching without modification. However, downloaded courseware is difficult to achieve the desired effect in teaching. Some course wares pursue beautiful pictures, appealing music and stunning video, so the students are distracted by too many fancy things. Some English teachers just move the contents from the blackboard to the screen, simply combining the audio and video and slide, and thus greatly reduced the advantages of multimedia teaching.

### 4.2 Multimedia Occupies the Dominant Position of Teachers and Hinders Their Personal Development

The term “multimedia English teaching” suggests that multimedia is a tool to help teachers complete the task of English teaching, that the teacher still plays the dominant role at class. However, in the implementation process, an English class is often made a class to display multimedia courseware; teachers function as a "projectionist" and "narrator". Teaching is a complex process of human thoughts, and teachers should reasonably adjust their teaching methods to accumulate a wealth of teaching experience, improving their teaching.

### 4.3 Multimedia Reduced the Students’ Enthusiasm of Participation in the Classroom, Splitting the Teacher-Student Interaction

Many colleges and universities adopt large class teaching, and teaching contents are rich, time is limited, so some teachers take the advantage of the multimedia, using a brief pause in showing the slides, demonstrating the content quickly, ignoring whether the students have taken in the information or not. Some students hardly have time to understand and take notes. In order to keep up with the teacher, the students have to concentrate on the screen, passively following the slide, and there is no time to think. That not only greatly reduces the initiative and enthusiasm, hindering students' ability of innovation and development of thinking, but also split the interactive relationship between teachers and students.

## 5 Some Countermeasures Have Been Worked Out as Follows

### 5.1 Improve the Quality of Teachers; Promote Rational Use of Multimedia Technology

Teachers' awareness and operating skills of multimedia technology bring direct impact on the production of courseware and implementation of multimedia teaching. Teachers should continue to learn, reflect, and improve the skills in using modern educational technology, constantly improving teaching means and methods to optimize classroom teaching, and actively exploring new ways of CAI, making the technology a good helper for English teaching. In addition, schools should offer opportunities for the teachers to study multimedia operation, creating conditions to encourage teachers to observe and learn through short-term training.

### 5.2 Teachers Demonstrate Their Leading Roles and Explore Modern Teaching Methods

In English class, the teacher plays the leading role that the computer can not replace. Teaching content, teaching methods, classroom activities are designed and implemented by the teachers; more importantly, interaction between teachers and students is the most direct and most effective teaching and learning methods. Teachers convey information to students through posture, facial expressions, eye contact, tone and intonation. These can not be realized by computers. Therefore, in the teaching process, on the one hand, teachers should improve all-round quality, continuing to explore innovative teaching model; on the other hand, the "people" and "machine" function together.

### 5.3 Focus on Students' Position to Promote Effective Interaction between Teachers and Students

Students are the ones who are learning the subject, so, no matter how good the teachers are, there is no substitute for students' learning. Teachers should respect the subjectivity and initiative of students, recognizing the leading role of teachers and the main role of students in the process, guiding students to actively participate in the learning process. Therefore, the teachers should organize the teaching content and classroom activities according to the students' specific conditions, taking advantage of multimedia, so that students may actively participate in classroom activities.

In short, multimedia technology in English teaching reform has been a hot topic. It significantly enhanced the effect of teaching, effectively presenting the important and difficult points in the teaching content. But the media is only a supplementary teaching tool, not a substitute for teachers. Only the rational selection and use of modern teaching media, together with its organic combination with traditional teaching methods can contribute to a reasonable mode of teaching, and achieve the most optimal teaching.

## References

1. Li, X.: English Teaching Disadvantages Analysis and Countermeasures . Teaching and Management (2008)
2. Hou, Y.: Multimedia Network of Meta Cognitive Theory in the Teaching of English Writing. Liaoning Provincial College of Communications (2007)
3. Qin, S.: Factors Influencing English Writing Ability. Liling University (2003)
4. Son, J.-B.: Learner Experiences in Web-based Language Learning. Computer Assisted Learning (2007)

# Research on Chinese People's Physical Education under the Concept of Quality Education

Chengbao Zhang

Library, Shandong University of Technology, Zibo, Shandong, China (255049)

**Abstract.** Today's education is guided by the concept of quality education, but the concept of quality education in sports has not been formed. In fact, the quality education of the national sports can promote national stability, economic prosperity and social progress. However, the quality of China's National Sports is not satisfactory. This article focuses on how to promote people's quality by bettering physical education.

**Keywords:** quality physical education, physical education, sports.

## 1 Introduction

In the 21st century, China's economy, technology and civilization have been improved; the overall quality of people has also been improved and strengthened. However, the quality of national sports is lagging far behind in comparison with developed countries. There are many problems: people do not exercise, lack awareness and guidance of exercise; there are few large-scale sports industry and brokers.

The concept of physical quality has been elaborated by the famous scholars Gong Zhengwei and Lai Tiande. The meaning of physical quality refines the study of sports, providing a theoretical basis for understanding and thinking.

Unlike quality of the physical education, general physical fitness is the ability of human activities, which includes the motion, strength, speed, stamina, agility and flexibility and other functional capabilities of the body. The physical quality is the integrated system of a person's capability in sports, awareness of fitness and good exercise habits, knowledge on sports and sports shows, sports management capabilities. It is a person's overall acknowledgement and implement of sports.

## 2 Problems of Chinese People's Physical Quality

### 2.1 Sports Skills Are Poor

In the physical quality, technical ability is a major external factor in sports. In general, the better one masters the skills, the better body quality he will have, and the more comprehensive technical movements he will do. Therefore, he will have a sense of pride and accomplishment, which inspires him to participate in sports more frequently. Instead, a person who doesn't master the skills cannot be happy and satisfied, and can hardly experience the fun of sports.

S. Lin and X. Huang (Eds.): CSEE 2011, Part II, CCIS 215, pp. 58–62, 2011.

### 2.2 Awareness of Sports Mainly Includes

People should improve learning ability and to master sports skills, to organize and to participate in sports activities, to develop and to operate sports industries. Social development shows that the right concept of sports is the core of the development of physical quality. Only when the national sports consciousness is increased, can people's physical quality be strengthened. In the current physical education, the concept of sports quality is seldom mentioned in textbooks.

### 2.3 Sports Expertise and the Relevant Knowledge Are Not Enough

Now, knowledge of sports has developed from sports activities, exercise physiology, biochemical, to exercise care, management. What's more, a three-dimensional structure of biological, psychological and social sports knowledge has been formed. But most people still lack the basic sports knowledge. For example, college students don't know how to deal with closed soft tissue injury which often appears at P.E. class.

### 2.4 The Cultural and Technological Content of Physical Education Is Not High

Physical culture is one of the indispensable qualities, but physical education involves little content about culture. In physical education, there is not much technological content. For example, different students have different physical features, yet most teachers only use subjective feelings to judge each student's situation.

### 2.5 The Developers of the Sports Industry Do Not Have High Quality at Sports

Sports industry developers should have both the wisdom and high quality at sports, and this can maintain the continuous improvement of the sports industry. Therefore, improving the quality the sports developers is a deciding factor to improve sports industry. Meanwhile, we should learn more advanced foreign experience in the sports industry to identify a suitable system for China's sports industry development.

### 2.6 Most People Lack Ability of Appreciating Sports

A person's sports quality involves many factors. If one is not good at doing sports, he can enjoy participating in or appreciating sports. If interested in watching the games or sports, one will also enjoy a lot of fun. One can experience emotions such as the pain, frustration, excitement, and so on when participating or appreciating sports. But today, many people don't take or watch any sports at all; some watch the game only to vent their anger.

## 3 Cause of the Problem

### 3.1 In School Education

**1) There is no systematic Physical Education in schools**

There are many problems in the setup of the P.E. course. As a course in curriculum, P.E class should have the teaching content which is from the simple to the complex, from the easy to the difficult. The current situation is that physical education is often repeated. For example, in primary schools, it is not necessary for students to study and master the technology. It's the period to develop the main physical qualities of students such as sensitivity and coordination, and to develop their exercise habits, so they can be happy when doing sports. Junior middle-schools should focus on developing the students' basic physical fitness and basic technology. Comprehensive technical teaching can be done in high schools and in universities.

**2) Physical education is not taken seriously**

School is the main place of Physical Education, where a person can have a variety of physical quality. Physical Education is one of the longest courses at school in one's lifetime. Schools should have systematic quality education of sports for students. From the present situation of education in China, although more importance is attached to students' performance in sports, the overall view of physical education is still at a disadvantageous stage. Many school leaders and teachers, including parents (especially in high school), don't have enough awareness of the importance of P.E. Students make little use of P.E. textbooks: I have surveyed more than 600 high school sophomores. Among them, only 10% had ever read books of sports, most students did not read their P.E. textbook, even some of them didn't know they got P.E textbooks. Most students are passive in doing sports and lack proactive sports consciousness.

Currently, several serious problems exist in school physical education: one is that the goal of physical education is unreasonable, too much emphasis being put on technical education and too little on cultivating interests. The second is the teaching content is old and materials are not fundamentally improved. Third, the evaluation of physical education is not proper. Teachers focus only on the form, while the physical and mental pleasure, which can reduce stress and develop intelligence, may be lost. Fourth, sports economy, consumption, and sports industry haven't been adopted as the teaching materials in schools.

## 3.2 Seen from the Social Environment

Social environment doesn't provide proper understanding and favorable support for the physical quality education. Since the National Fitness Program was issued and implemented, awareness of health and participation in sports increased and people enjoy watching the games. But some problems still exist:

1) Bias in understanding sports interferes with fostering physical quality. Among the many qualities of people, cultural quality tends to be attached more importance, while the other qualities tend to be ignored. The physical quality is not taken seriously.

2) Lack of places to exercise is another important factor. In our country, the facilities still can not meet the needs of people involved in physical exercise. With the urbanization process in recent years, green space and trees increasingly are replaced by "concrete jungle", forcing people to pay for exercise indoors. But with the improvement of the community sports, this problem is expected to be resolved.

3) Pace of sports socialization process is slow. Physical exercise does not form a climate. Serious differences exist between urban and rural areas. Sports socialization is linked with the economic development. In the vast rural areas, due to the lack of awareness and some economic reasons, people rarely do exercise and they need to vigorously promote the implementation of Physical Education.

# 4 Solution

## 4.1 Innovate School Education

1) The schools should attach importance to the quality education of sports, improving ideological and moral quality of training, cultivating the psychological maturity and developing the intellectual abilities of students, so that all-round quality improvement and development can be realized.

2) Reform the physical education curriculum, forming a complete system of physical education from primary schools to universities, from the easy tasks to the difficult, from the simple to the complex, so that the course has a vertical linkage. Specify stages of the teaching tasks, choosing appropriate teaching content and teaching methods. Break the traditional technology teaching style in primary schools and junior middle-schools, trying to make students interested in improving the physical fitness.

3) Increase the technological content of physical education. The 21st century is the era when people believe in science. The facts speak for themselves. Using science to guide physical education is a major principle in the new century.

4) School leaders and physical education teachers should pay attention to physical education for all students, putting the teaching, training and scientific research on the same position. Physical education teachers must value their work and be physically fit and strong, civilized in behavior, rich in wisdom.

5) Schools should create a good atmosphere, encouraging the students to participate in sports activities and competitions on a regular basis. For excellent performance, schools should treat the participants in sports the same as treating outstanding students in intellectual performance, giving them recognition and reward.

6) The school should absorb some of the non-sports events as a teaching content, such as folk games for the region to not only develop the physical fitness of students, but also a combination of local features. The form of some events an also be changed to be more suitable for students.

## 4.2 Seen from the Social Environment

1) Try to create a good atmosphere, so that the physical exercise may become an important part of life. Sports promote health; health is the basic guarantee of life and work. Only when understanding this truth, the whole community can have a good public opinion, and people may have sense of physical training.

2) Increase investment in construction and facilities, protecting of our country's economy by people's physical fitness. Investment in the construction of sports venues can not be measured by the economic value. If not for sports and exercise, human life span will be shortened; physical condition will decline, thereby causing negative effects to the entire social and economic development.

3) Use extensive publicity to praise prominent figures in sports activities, teaching people good ways of physical exercise, telling people the importance of good health conditions. Individuals achieving a lot in sports industries must also be praised, so that people know about the sports industry, thus increasing people's confidence in investment and consumption.

In short, the concept of Physical Education, more in line with the concept of the new century demands of sports. Quality education of sports is not only a goal for schools; it is the goal of the whole society. If people's understanding of sports is as deep as people's understanding of culture, then the quality of sport in our country will be greatly improved. Consequently, it will accelerate the pace of China's socialist construction and economic development.

## References

1. Liu, H.: Vocational Education. Guangdong Tertiary Education Press (2004)
2. Plan of Mass Sports (1995)
3. Geng, P., Cao, W.: Acknowledgements of Developing Textbooks of P.E. China's School P.E. (2005)
4. Sun, S.: Theory of Educational Information. Shanghai Education Press (2000)
5. Sun, X.: P.E. and Psychological Health. Journal of Cangzhou Normal College (2005)

# The Origin and Development of Library

Chengbao Zhang

Library, Shandong University of Technology, Zibo, Shandong, China, 255049

**Abstract.** From the buildings used to store books to the modern library, the term "library" contains a long history. This paper focuses on the transformation of library from ancient China to the modern form, and demonstrating the transformation of library in China.

**Keywords:** origin, development, library.

## 1 The Ancient Library

According to research, the word "library" originated in Latin, Libraria, with the meaning of "place storing books". So, the ancient libraries and libraries of today are essentially different.

### 1.1 The Foreign Ancient Library

Early libraries appear in Babylon, Greece, Rome, Egypt and other countries. In fact, these ancient Libraries were dominated by the state. According to archaeological discoveries, the world's first library was that of the kingdom of Babylon, which appeared in the 10th century BC, followed by Nineveh Palace Library of 7th century BC, in the Assyrian empire and the Library in the Kingdom of the 6th century BC Greek city of Athens. The most noteworthy is Alexandria Library (now Alexandria), which was built in the 4th century BC. King Ptolemy I had a dream: to convert the entire world's literature and to write down all the things worth remembering. Ptolemaic dynasty took a very big effort to recruit the famous scholars to work for the library, to take all means to collect books, expand collections, hoping to achieve this goal. Indeed, the Alexandria Library was truly the center of the era of Greek literature and, in up to 200 years, playing a unique role in Greek culture. After Roman Empire conquered Greece, all books were moved to Rome, and in the 2nd century BC, a new library was established in Rome. In addition, in the early Middle Ages, religious rules hanging over the West, many monastic libraries appeared in this period. During 11th to 13th century AD, with the rise of universities, university libraries also developed quickly. Therefore, the National Library, the monastery and university libraries together constituted the foreign Ancient Library.

### 1.2 Ancient Library in China

Before the term "library" was introduced into China, "places storing books" were referred to as the "building storing books." According to historical records, our

S. Lin and X. Huang (Eds.): CSEE 2011, Part II, CCIS 215, pp. 63–67, 2011.

official collection was formed in the Western Zhou Dynasty to the Warring States period; the Zhou Dynasty had a collection agency called "storage room." According to "Historical Records", Laos had been in charge of this agency and can be regarded as the earliest chief of the National Library.

The colorful Chinese characters led to the fact that a variety of names for "libraries" appeared. For example, there was "Gu fu（府）" in the Western Zhou Dynasty, " Epang gong（宫）", such in Qin Dynasty, "Tianlu ge（阁）" in Han Dynasty, "Guanwen dian（殿）" in Sui Dynasty, "Chongwen yuan（院）" in Song Dynasty, "Dansheng tang（堂）" in Ming Dynasty, "Buzu zhai（斋）", "Tieqin tongjian lou（楼）" in Qing Dynasty, and so on.

Collection of Books in Ancient China consists of four branches:

1) Official collections: In China, in the Yin Period of the 13th century BC, the royal family have a place to collect books, which can be seen as the origin of library in China. It also showed that he beginning of book collection was state-run business.

2) Private Collections: The rise of private collections appeared in the Warring States

Period to the Han Dynasty, for the reason that the government attached importance to the academic culture, emphasized education, tried efforts to promote private libraries, and so on.

3) College collections: Starting from the Song Dynasty, College library became more popular, among which there were the famous "Bailudong Academy", "Yuelu Academy", "Yingtian Academy" and "Songyang Academy". These are called "the four Academies in Song Dynasty", having a large number of books.

4) Temple collections: Taoism began in China and generated a lot of Taoist books, which resulted in the collection of "Taoist books". Buddhism is a foreign religion. With the introduction of Buddhist scriptures, the books were gradually spread to more people, composing another important branch of Temple Books. It can be said that temple library is the product of religious culture.

No doubt, the form and functions of the ancient libraries had a profound social background. Ancient Libraries focused on collection and preservation of books. The name "Building Storing Books" showed that these places were built mainly for the collection of literature, and because there was very few or even only one copy, these books were seldom seen outside the ancient libraries.

## 2 The Recent Epoch Library—The Modern Library

The occurrence of recent Epoch Library is accompanied by the formation of the capitalist system, evolved from the ancient library.

The industrial revolution in the 18th century resulted in the sprouting of capitalism in the West, creating new conditions for the development of libraries. Production requires large machines operated by knowledgeable workers in the premise of process control technology. Therefore, the bourgeois government not only has to promote school education, but to set up new libraries for an open society, to improve the workers' scientific and cultural knowledge. In this case, some libraries in European countries were separated from the palaces and the churches, becoming the Modern Library for the open society.

China's modern library is a product of Western ideology and culture. After the Opium War in 1840, shocked by the Western capitalist culture, feudal culture declined greatly, and there was the increasing trend of learning Western culture and science. In this context, the feudal library building was no longer meeting the needs of social development. With the introduction of Western models of modern library, the old buildings storing books gradually disintegrated. Xu Shulan, in 1902, imitated western modern library to build up a new library, the Guyue Book-storing Building, with his private efforts. Although the name was in the old term, it is truly a new library—to make articles of association, public borrowing, with the prototype of the modern library. At the same time, a group of government-run new libraries were built and completed in 1904, such as Library of Hunan Province, Library of Hubei Province (China's first collection unit using the name "library"), Library of Fujian Province, and so on. In 1910, the Qing government issued" Charter of libraries in capital city and provinces" and announced the establishment of Imperial Library (now National Library), which was completed and opened in 1912. Then the provinces set up public libraries on after another. From the Revolution to the Sino-Japanese War, the transition from the ancient libraries to the modern libraries was completed, and it is the most important revolution in the history of China's literature. The most essential difference between the ancient and modern libraries books lies in whether they are open to public. Modern libraries attach importance to both collection and application. Therefore, the Modern Library is more than just a place to store books, it is a place to spread knowledge culture, is a social organization.

## 3 The Contemporary Library

After World War II, the new information technology brought the library into the modern period of development. The difference between modern Library and the traditional library lies in the management philosophy—to be the service center or to be the center of storing books. Modern Library owns service-oriented operation mechanism, and its business activated with of every aspect of services for the readers. Books in the library system and services are often based on social needs, the needs of the readers. Characteristics of the Contemporary Library are as follows:

Focus of work: transferring from books to people—readers.

Business focus: from the second line to the first line, which is a major adjustment of business structure, through the application of new technology and business outsourcing, etc., to reduce the cost of information processing, paying more attention to the service focused on the first line.

Service focus: from general services to consulting services, librarians should become consultants, librarians should be a knowledge navigator.

Collection: from a focus on "ownership" to focus on "access", online networks, interlibrary loan, document delivery and other means to make the library resource shared, greatly increasing service capacity.

Services: extending to the community and family. The modern library is becoming an integral part of community culture, being family teachers.

In addition, the library is a center of life-long education and entertainment center, and the center of information dissemination and exchange. Therefore, the modern library's characteristic as a social and cultural system becomes more significant.

## 4 Definition of Library

What is a library? The question seems very simple, and some people even blurt out: "Library is the place to borrow books."

This answer is not the scientific definition of a library, because it did not explain the nature of the Library. To accurately and scientifically answer this question, we must point out the core of the definition of the meaning and to find the real nature of a library—a comprehensive system of the quality of a library, creating the proper concept of it. The scientific definitions of library, due to the different angles, are expressed in different words.

So far, domestic and foreign scholars and experts of libraries gave different definitions for libraries, as the following:

France's "Great Larousse Encyclopedia" defines: the task of the library is to preserve the human mind information written in different words, expressed in various ways, ... ... library collects books and materials of various categories, organized according to certain methods. These data can be used for study, research or serve as general information.

"Soviet Encyclopedia" defines: The library is the subsidiary organs in the form of cultural organization of social education through publications and science. Library is engaged in the systematic collection, preservation, and promotion of publications as well as lending them to the readers; meanwhile, it carries out the task of book and information processing.

"British encyclopedia" defines: Library is a facility of collecting books for people's reading, studies or reference.

"American Encyclopedia" defines: Since its appearance, the library has gone through many centuries, carrying three main functions: to collect, to preserve and to provide information. The library is an important tool for the books and their predecessors to implement their inherent potentials.

In Japan, "Dictionary of Library Terms" defines: Library is a public service organization to collect, organize, preserve all books and other materials and information, and to provide them in accordance with the requirements of users.

In China, since 1930s, scholars have been exploring what libraries are.

Mr. Liu Guojun thought that the library was to collect all records of human thoughts and activities, using the most scientific and most economical way to save them, organize them so that all people in society can make use of them.

Mr. Huang Zongzhong pointed out in his "Introduction to Library Science" that a library is to collect, process, sort, store, select, control, transform and deliver all the documents containing information, knowledge, and scientific literature as the content, providing readers with a information system. In short, the library is the center of storing and transferring information.

Mr. Wu Weici pointed out in his "Introduction to the Library" that the library is the cultural and educational institution that can collect, collate, store books and information, serving a certain kind of economic society.

"Library Encyclopedia" defined library in this way: the library is a scientific, cultural, educational institution to collect, collate and save the document information, providing benefits to the readers.

Despite divergent views on the definition of the library, these explanations include two things in common: First, the library is a place of collection of books and materials; Second, the collection of books and reference materials are for people to use.

To sum up, the question "what is a Library" can be replied in this way: Library is the cultural and educational institution to collect, organize, store and use literature information, thus to serve the politics and economy of a society. This explanation has three meanings: the library is a scientific, educational and cultural institution; Library is the institution to collect, collate, store and utilize the documents and literature; Library is for the community's political, economic services.

## References

1. Zhu, L.: On the Origin of Book Storage. Journal of Jiangxi Library (2009)
2. Zhang, R.: On the Origin of Modern Libraries in China. Frontline (2007)
3. Lifeng, L.: From Book-storing House to Liarbry. Journal of Shaoxing Art and Science College (2005)
4. Chen, H.: A Reflection on the Histiry of Libraries in China. Research on Books and Intelligence in Shanghai (2003)

# A Vehicle Tracking System Overcoming Occlusion and an Accident Detection Monitoring Using Tracking Trace

In Jeong Lee

Computer Engineering Part, Hoseo University,
336-795, Sechuli 165, Baebangmyun, Asan, Chungnam, South Korea
Tel.: 82-41-540-5713, Fax : 82-41-540-5713
leeij@hoseo.edu

**Abstract.** We consider the video image detector systems using tracking techniques which can be handling of the all kind of problems in the real world, such as shadow, occlusion, and vehicle detection by nighttime. Also we have derived the traffic information, volume count, speed, and occupancy time, under kaleidoscopic environments. In this system we propose a shadow cast algorithm and this system was tested under typical outdoor field environments at a test site. We evaluated the performance of traffic information, volume counts, speed, and occupancy time, with 4 lanes in which 2 lanes are upstream and the rests are downstream. And the performance of our video-based image detector system is qualified by comparing with laser detector installed on testing place. And we propose an accident detection monitoring system through this tracking trace.

**Keywords:** vehicle detector systems, mean square error, accident detection system.

## 1 Introduction

Vision-based traffic monitoring systems are widely used in intelligent transportation system. The goal of a traffic monitoring system is to extract traffic information, such as the vehicle volume count, traffic events, and traffic flow, which plays an important role for traffic analysis and traffic management [1]. Because video-based image detector systems are more efficient than other systems such as the loop detector, spatial traffic information based on video image can be more useful than spot information at a single point. In fact, by averaging trajectories over space and time, the traditional traffic parameters are more stable than corresponding measurements from point detectors, which can only be averaged over time. Additional information from the vehicle trajectories could lead to improved incident detection, both by detecting stopped vehicles within the camera's field of view and by identifying lane change maneuvers and acceleration/deceleration patterns that are indicative of incidents beyond the camera's field of view [2, 3].

The occlusions have taken place when multiple vehicles are regarded as one vehicle because of overlapping of objects from the viewpoint of camera. In the explicit occlusion, multiple vehicles that enter a scene separately merge into a moving

S. Lin and X. Huang (Eds.): CSEE 2011, Part II, CCIS 215, pp. 68–76, 2011.

object region in the scene. And the other case, implicit occlusion, multiple vehicles enter the scene of being overlapped with each other from beginning. The explicit occlusion can be detected easier than implicit occlusion because of identifying individual vehicles before occlusion. In implicit occlusion case, we cannot obtain information of individual vehicles in the occluded region beforehand.

The existence of the cast shadow in the moving object degrades extraction accuracy. For example, one solution to identify the vehicle segmentation is based on model matching, however this is a difficult task by itself, and it may not always work property.

This may misclassify shadows as foreground objects, which can cause various unwanted behavior such as object shape distortion and merging, affecting surveillance capability like target counting and identification [4]. The lighting condition changes by the weather, light sources, and other factors such as daytime and nighttime. It is difficult to solve the all kinds of situations with a single approach. The vehicle segmentation at nighttime is a complicated issue due to the lack of valuable visual information. Especially disturbances caused by headlights and their corresponding reflections into road have been badly affected vehicle detection.

Generally, image processing and object tracking techniques have been mostly applied to traffic video analysis to address queue detection, vehicle classification, and volume counting [5, 6]. Model-based tracking is highly accurate for a small number of vehicles [7]. The most serious weakness of this approach, however, is its reliance on detailed geometric object models. It is unrealistic to expect it to have detailed models for all vehicles on the roadway. In region-based tracking, the process is typically initialized with the background subtraction technique. In this approach, the VIPS identifies a connected region in the image, a 'blob' associated with each vehicle, and then tracks it over time using a cross-correlation measure. The Kalman-filter-based adaptive background model allows the background estimate to evolve as the weather and time of day affect the lighting conditions. If a separate contour could be initialized for each vehicle, then each one could be tracked even in the presence of partial occlusion [2, 8].

Objects in the world exhibit complex intersections. When captured in a video sequence, some intersections manifest themselves as occlusions. A visual tracking must be able to track objects which are partially or even fully occluded [9].

Occlusions are classified as track occlusions due to other moving objects, background object occlusions due to fixed objects, and apparent or false occlusions due to sudden shape changes [10].

The mainly adaptive field of occlusion is tracking people in public places, such as airport, subway station, bank, and park. The most cited paper [11] for occlusion is a real time visual surveillance system for detecting and tracking multiple people and monitoring their activities in an outdoor environment.

To resolve more complex structures in the track lattice produced by the bounding box tracking, it used by appearance based modeling [9]. The appearance model is an RGB color model with probability mask similar to that used by Ismail Haritaoglu et. al [11]. In methods to solve the implicit occlusion problem in moving objects, the fusions of multiple camera inputs are used to overcome occlusion in multiple object tracking [1,12,13,14].

The Predictive Trajectory Merge-and-Split (PTMS) proposed to uses a multi stage approach to determining the vehicle motion trajectories and eventually the lane geometry [15]. Approaches based on spatial reasoning use more complex object representations such as templates or trained shape models, or appearance models. However this is dependent on image resolution and only works under partial occlusion. Some shadow elimination techniques have been classified in the literature into two groups, model-based and property-based technique [16]. Model-based techniques are based on matching sets of geometric features such as edges, lines, or corners to 3D object models, and rely on models representing the a prior knowledge of geometry of the scene, the objects, and illumination. However they are generally only applicable to the specific application and designed for specific applications. The shadow removal approaches are based on an assumption that the shadow pixels have the same chrominance as the background but are of lower luminance [4]. In generally, shadow pixels are classified as shaded background if they have similar chromaticity but lower brightness than the same background pixel [4, 17]. The earliest investigations in shadow removal proposed by Scanlan et. al., the image was split into square blocks and produced an image based on the mean intensity of each block [18]. The drawback of this method: the reduction of intensity contrast between image pixels, and the existence of blocking effect [19]. A. Bevilacqua proposed the gradient based technique, similar to SEBG algorithm, this method needed not prior knowledge in order to detect and remove shadows, represented one of the most general purpose systems to date for detecting outdoor shadows [20].

Successful video surveillance systems for real traffic monitoring must be adaptive to different illumination conditions, including heavy shadows under strong sunlight, dim illumination in the morning or evening, vehicle headlight at night and reflections in a rainy day as well. Detecting and tracking vehicles accurately under such situations, especially the nighttime images happened everyday, is always an important but challenging work to whole system [21]. We know the vision-based with nighttime images was RACCOON system[22] which has been integrated into a car experiment on the CMU Navlab II, tracks car taillights. Another pattern classifier algorithm, Support Vector Tracking (SVT) integrates the SVM classifier into optic-flow based on tracker [23]. Ilkwang Lee et. al.,[24] applied the Retinex algorithm as preprocessing to reduce the illumination effects at nighttime images. Samyong Kim et. al.,[25] proposed that the vehicle detection method in the nighttime can extract bright regions by the heuristic classified light sources and verify by vehicle width in several continuous frames. The bright regions in the nighttime generated by headlights, tail lights, break lights, and reflected lights around light sources are used as the vehicle feature.

In this paper, we propose VIPS measurement system and we shall show that our video-based image detector system is more appropriate by comparing with laser detector installed on testing place. And we propose an accident detection monitoring system through this tracking trace.

## 2 Methodology for Vehicle Detection

This system was tested under typical outdoor field environments at the test site, shown in Fig. 1. We have gathered for traffic data, volume count, speed, and occupancy time, with 4 lanes as 2 lanes are upstream and the rests are downstream for three days. And we have evaluated the performance of our system at four different time periods, i.e., daytime, sunrise, sunset, and nighttime, for 30 minutes each.

And the performance of our video-based image detector system is qualified by comparing with laser detector installed on testing place as illustrated in Fig. 1. It would be more appropriate to establish the verification of the proposed system by comparing to a non VIPS measurement system. The CCTV camera has a 640*480 resolution with 30fps, and the analog image sequences are transformed to digital 256 gray-levels via frame grabber board. The weather conditions during the test are clear two days, and one is rainy day. In the rainy day, we could not gather the baseline data from laser detector because the laser detector cannot operate in rainy day.

**Fig. 1.** The outdoor test site

In this section, we explain the basic idea behind the tracking algorithm developed in this research. Vehicle tracking has been based on the region-based tracking approach. For individual vehicle tracking the first step, acquisition image sequences and predetermining the detection zones at each lane. The second, we have conducted the background subtraction, deciding threshold for binary images. The background subtraction algorithm requires a relatively small computation time and shows the robust detection in good illumination conditions [26]. The third step, morphology for small particles removal as noise, sets in mathematical morphology represent the shapes of objects in an image, for example, the set of all white pixels in a binary image is a complete description of the image. The next step, it is important to remove cast shadows due to extract the vehicle area exactly, we developed the new algorithm in this paper using by edge detection and vertical projections within the vehicle particles. And the fifth step generates the vehicle ID and labeling to each vehicle, and individual vehicle's bounding rectangle data, i.e., left, top, right, bottom coordinates. These particle data are saved into reference table which can be referred to next sequence frames.

In this system, the occlusion detection is easy relatively because of short length of detection zones, less than 15m. And we have considered only limited to explicit occlusion. The explicit occlusion cases have taken place several times during the field test, that is, multiple vehicles enter a scene separately into detection zone, and merge into a moving object region in the scene. In this case, we have maintained each vehicle ID continuously as referred to previous frame.

In the nighttime, diffused reflections on the road due to vehicle headlights pose a serious concern. Thus we need to pre-processing by reflection elimination to adjust the light parameters such as luminosity or brightness, contrast, intensity.

And finally, generated the traffic information and saved these data into table. The performance evaluation of our system is computed by mean square error (MSE) comparing with baseline data, generated by laser detectors. For the vehicle extraction exactly, we have to consider about background estimation, occlusion, cast shadow detection and elimination, and light condition processing at night.

We have considered only limited to explicit occlusion. Our system has covered the four lanes with single camera. As the more system has to be processed, the less performance of system has been. The reason for that if we have included the implicit occlusion process, the performance evaluation marked low grade especially the calculation of velocity. The occlusion detection of this system is easy relatively because of short length of detection zones, less than 15m.

So many algorithms of cast shadow elimination are proposed, the various cases are occurred in the real traffic flows, for example, dark or light gray shadows, shadow from trees or clouds. The proposed algorithms as mentioned before, have been applied to our experiment, the shadows cannot be extracted exactly as a result. Thus we have developed the appropriate algorithm in our test site. The basic concept is that the shadow area has less edge because of no variance within shadow. On the other side hand, vehicle area has more edges relatively. Let B be a binary image plane and $B_x$ be a set of number of vertical pixels which value is 1 at $x$. We define a function $Verti : B/x \to B_x$

$$by\ Verti\ (x) = \sum_y B_1(x, y)\ ,$$

where $B_1(x, y)$ is a pixel of which value is 1 at $(x, y)$ and $B/x$ is a projection of B into $x$. The graph of $y = Verti(x)$ is shown in Fig. 2c. after smoothing. In Fig. 2b., the distribution of edges from moving object area can be discriminated between vehicle and cast shadow and Fig. 2c. shows the density of edges via vertical projection. And then discard under 25%, that is cast shadow area, Fig. 2d.

(a) Original image

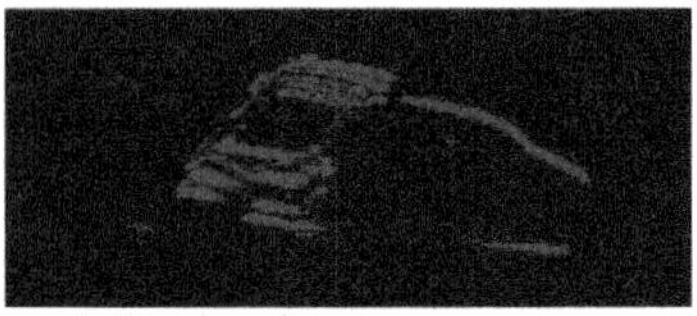

(b) Edge of moving object

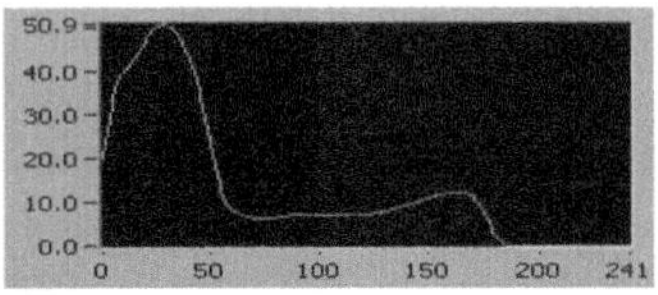


(c) Vertical Projection

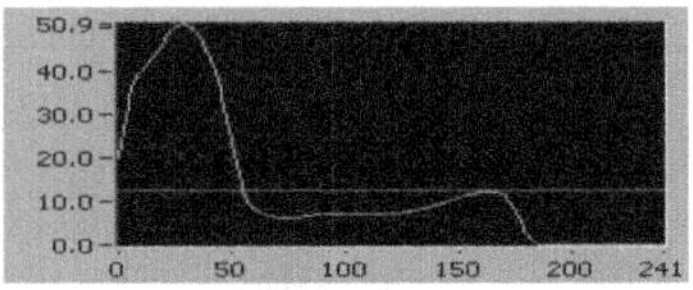


(d) Shadow area under blue line, 25%

**Fig. 2.** Cast shadow detection process in our test site

The light condition is a lot different from whether streetlamps are on the road or not. The around road with streetlamps has a bright gray under night, whereas around the vehicle headlight can be classified distinctly without streetlamp. In the test site has no streetlamp. Diffused reflections on the road due to vehicle headlights pose a serious concern during the night processing even though no streetlamp. Thus we need reflection elimination to adjust the light properties such as luminosity or brightness, contrast, intensity. We have adjusted the luminosity in this test in order to eliminate the diffused reflections on the road. Then, vehicle detection can be achieved through the detection of headlight of 255 gray alone as shown in Fig. 3.

(a) Diffused reflections on the road (b) Adjust the luminosity of headlights

**Fig. 3.** Diffused reflection elimination during the night

As an accident detection system, we propose a system of tracking vehicle trace Fig. 4., in this trace we can find an accident from change of trace image.

**Fig. 4.** Vehicle trace has no change

# 3 Experimental Results

We have tested under typical outdoor field environments at the test site. Since, as in Fig. 5. , we can find some change of trace image, and we present warning to system manager like as in Fig. 6.

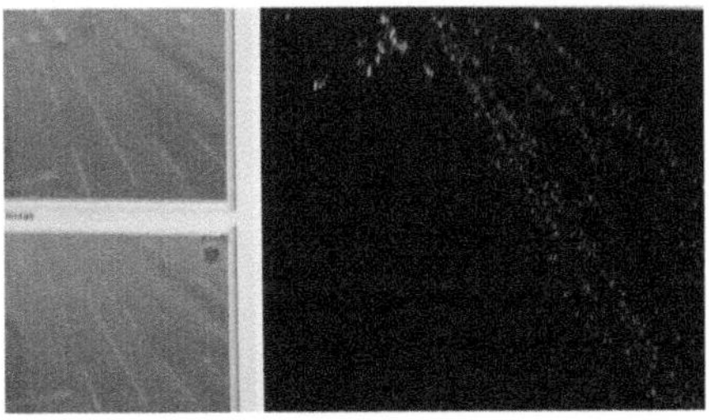

**Fig. 5.** Vehicle trace is changed

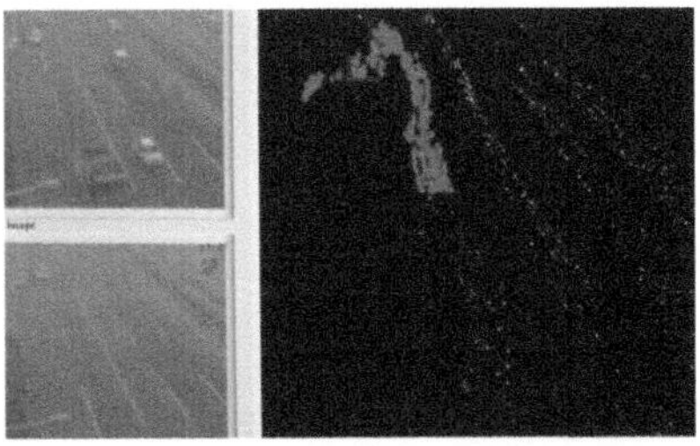

**Fig. 6.** Warning is delivered to manager

And have evaluated the performance of traffic information, volume count, speed, and occupancy time. We cannot obtain the baseline data from the radar detection on second day, because of rainy day, for the reason that radar detector have generated the diffused reflections from raindrops under rainy day, therefore the baseline data are incorrect. The traffic information can be obtained by aggregating count vehicles passing through the detection zones for one minute.

The evaluation basis, refer to mean square error (MSE), was calculated by following equation.

$$\mathrm{MSE}=\frac{\sum_{t=1}^{T}\frac{(B_t-M_t)^2}{B_t}}{n}\times 100$$

Where, $B_t$ , $M_t$ is baseline data generated by laser detectors and measured mean data at time $t$ respectively. $T$ is total measuring time unit. Generally, the time unit $T$ can be divided into four different time periods, i.e., daytime, sunrise time, sunset time, and night, the traffic information for time unit $T$ is measured for 30 minutes each. The MSE of measuring for each time period are shown as Table 1.

The more detailed measuring results, which are compared with aggregating one minute of baseline data and measuring volume counts for 30 minutes within each time period, are illustrated in Fig. 7. as followings.

**Table 1.** The MSE of measuring traffic information on the test site

| Time | Volume Error Rate | Speed Error Rate | Occupancy Time Error Rate |
|---|---|---|---|
| First Day 15:00~15:30(daytime) | 3.32 | 1.69 | 3.18 |
| 2nd Day 07:25~07:55(sunrise) | 5.73 | 4.01 | 7.84 |
| First Day 20:00~20:30(night) | 4.64 | 3.30 | 5.56 |

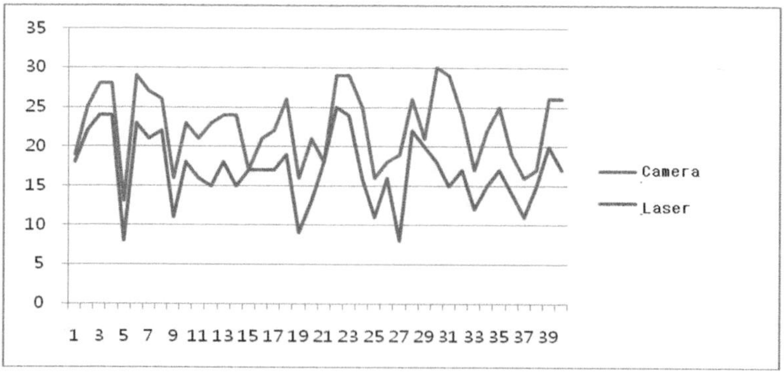


**Fig. 7.** Volume counts for 30 minutes within each time period

## 4 Conclusions

In this paper, we use the measuring traffic information which is computed by mean square error (MSE) comparing with baseline data, generated by laser detectors. These systems can be handling of the all kind of problems in the real world, such as shadow, occlusion, and vehicle detection by nighttime. Also we have derived the traffic information, volume count, speed, and occupancy time, under kaleidoscopic environments. For the cast shadow elimination of our system, we have developed the new algorithm using analysis of edge distribution of vehicle and shadows. And the performance of our video-based image detector system is qualified by comparing with laser detector installed on testing place. In future works, applications of the tracking system developed in this study should be tested based on various environmental conditions for the any other sites. We have developed the accident detection system using tracking trace image.

## References


1. Yoneyama, A., Yeh, C.-H., Kuo, C.-C.J.: Robust Vehicle and Traffic Information Extraction for Highway Surveillance. EURASIP Journal on Applied Signal Processing, 2305–2321 (2005)
2. Coifman, B., Beymer, D., McLauchlan, P., Malik, J.: A Real-Time Computer Vision System for Vehicle Tracking and Traffic Surveillance. Transportation Research Part C 6, 271–288 (1998)
3. Oh, J., Min, J.: Development of a Real Time Video Image Processing System for Vehicle Tracking. Journal of Korean Society of Road Engineers 10(3), 19–31 (2008)
4. Liu, H., Li, J., Liu, Q., Qian, Y.: Shadow Elimination in Traffic Video Segmentation. In: MVA 2007 IAPR Conference on Machine Vision Applications, Tokyo Japan, May 16-18 (2007)
5. Chen, S.-C., Shyu, M.-L., Peeta, S., Zhang, C.: Learning-based Spatio-temporal Vehicle Tracking and Indexing for a Transportation Multimedia Database System. IEEE Trans. on Intelligent Transportation Systems 4(3), 154–167 (2003)
6. Oh, J., Min, J., Kim, M., Cho, H.: Development of an Automatic Traffic Conflict Detection System based on Image Tracking Technology, submitted to TRB (2008)
7. Koller, D., Daniilidis, K., Nagel, H.: Model-based Object Tracking in Monocular Image Sequences of Road Traffic Scenes. International Journal of Computer Vision 10, 257–281 (1993)
8. Koller, D., Weber, J., Huang, T., Malik, J., Ogasawara, G., Rao, B., Russell, S.: Towards Robust Automatic Traffic Scene Analysis in Real Time. In: ICPR, Israel, vol. 1, pp. 126–131 (1994b)
9. Senior, A., Hampapur, A., Tian, Y.-L., Brown, L., Pankanti, S., Bolle, R.: Appearance Models for Occlusion Handling. Journal of Image and Vision Computing 24(11), 1233–1243 (2006)
10. Cucchiara, R., Grana, C., Tardini, G., Vezzani, R.: Probabilistic people tracking for occlusion handling. In: Proceedings of the 17th International Conference on ICPR 2004, August 23-26, vol. 1, pp. 132–135 (2004)
11. Haritaoglu, I., Harwood, D., Davis, L.S.: $W^4$: Real-Time Surveillance of People and Their Activities. IEEE Trans. on Pattern Analysis and Machine Intelligence(PAMI) 22(8) (August 2000)

12. Chang, T.-H., Gong, S., Eng-Jong.: Tracking multiple people under occlusion using multiple cameras. In: Proc. 11th British Machine Vision Conference (2000)
13. Dockstader, et al.: Multiple camera tracking of interacting and occluded human motion. Proceedings of the IEEE 89(10), 1441–1455 (2001)
14. Dockstader, S.L., Tekalp, A.M.: Multiple camera fusion for multi-object tracking. In: Proc. IEEE Workshop on Multi-Object Tracking, pp. 95–102 (2001)
15. Melo, J., Naftel, A., Bernardino, A., Santos-Victor, J.: Viewpoint Independent Detection of Vehicle Trajectories and Lane Geometry from Uncalibrated Traffic Surveillance Cameras. In: International Conference on Image Analysis and Recognition, Porto, Portugal, September 29-October 1 (2004)
16. Xiao, M., Han, C.-Z., Zhang, L.: Moving Shadow Detection and Removal for Traffic Sequences. International Journal of Automation and Computing, 38–46 (January 2007)
17. Horprasert, T., Harwood, D., Davis, L.: A Statistical Approach for Real-time Robust Background Subtraction and Shadow Detection. In: Proc. 7th IEEE ICCV Frame-rate Workshop Corfu., pp. 1–19 (1999)
18. Avery, Zhang, Wang, Nihan.: An Investigation into Shadow Removal from Traffic Images. TRB 2007 Annual Meeting CD-ROM (2007)
19. Wang, J.M., Chung, Y.C., Chang, C.L., Chen, S.W.: Shadow detection and Removal for Traffic Images. In: Proceedings of the IEEE International Conference on Networking, Sensing & Control, Taipei Taiwan, March 21-23, pp. 649–654 (2004)
20. Bevilacqua, A.: Effective Shadow Detection in Traffic Monitoring Applications. Journal of International Conference in Central Europe on Computer Graphics, Visualization, and Computer Vision 11(1), February 3-7 (2003) ISSN 1213-6972
21. Gao, D., Zhou, J., Xin, L.: SVM-based Detection of Moving Vehicles for Automatic Traffic Monitoring. In: Proceedings of IEEE Intelligent Transportation Systems Conference, Oakland (CA) USA, August 25-29 (2001)
22. Sukthankar, R.: RACCOON: A Real-time Autonomous Car Chaser Operating Optimally at Night. Proceedings of IEEE Intelligent Vehicles (1993)
23. Avidan, S.: Support Vector Tracking. IEEE Transactions on Pattern Analysis and Machine Intelligence 26(8), 1064–1072 (2004)
24. Lee, I., Ko, H., Han, D.K.: Multiple Vehicle Tracking based on Regional Estimation in Nighttime CCD images. In: Proc. IEEE Int. Conf. Acoustic, Speech, Signal Processing (ICASSP 2002), Orlando, Fla., USA, vol. 4, pp. 3712–3715 (May 2002)
25. Kim, S., Oh, S.-y., Kim, K., Park, S.-c., Park, K.: Front and Rear Vehicle Detection and Tracking in the Day and Night Time using Vision and Sonar Sensors. In: Proceedings of 12th World Congress of ITS, November 6-10 (2005)
26. Kim, Z.: Real Time Object Tracking based on Dynamic Feature Grouping with Background Subtraction. In: Proc. IEEE Conf. Computer Vision and Pattern Recognition (2008)

# A Study on the Application of the Streaming-Media Technology in English Teaching Based on Network Circumstance

JinBao Cai[1] and Ying Lin[2]

[1] Jiangxi University of Science and Technology, Ganzhou City, Jiangxi Province, P.R. China, 34100
[2] Gannan Normal University, Ganzhou City, Jiangxi Province, P.R. China, 34100
caijinbao@yahoo.com.cn

**Abstract.** The streaming-media technology has been widely used in language teaching in classroom. This paper has discussed how to transform English language teaching from traditional classroom-oriented teaching method to the method combination of multi-media network and classroom teaching by adopting advanced streaming-media technology, and further apply personalized teaching. Meanwhile, it has demonstrated the application of the streaming-media technology in English teaching based on network circumstance. Finally, this paper has studied the means of developing common audio-visual materials in foreign language teaching.

**Keywords:** Streaming-Media Technology, Net-based English Teaching, Interactive Learning, Personalized Learning.

## 1 Introduction

With the rapid development of computer-based network multi-media technology, streaming media technology, which is a kind of upgraded technology based on the traditional multi-media broadcasting technology, has been further exploitured. The traditional broadcasting technology means that the completed file is downloaded by the users and played. On the otherhand, the streaming-media technology indicates that the streaming transmissiong method is adopted, which means that the whole multi-media file is compressed into several compression packing bag, and then broadcasted in order through network. Thus the file can be being received and played by the users, which means the storing space of computer disk and time can be saved by the users. So the development of streaming-media technology has further broken through the limitation of the time and place of information broadcasting, which means that the streaming-media technology can be technologically adopted in English language teaching. The traditional teaching model that teacher-blackboard-chalk is oriented has been greatly challenged. The new stereocopic teaching system is being established in English language teaching. The appropriateness of vision effect, the learning content, the learning method and the means of interactive practice is being reviewed by the teachers and students. This paper, based the author's two-decade

S. Lin and X. Huang (Eds.): CSEE 2011, Part II, CCIS 215, pp. 77–80, 2011.

teaching experience in the language lab, has dicussed the practical application and development of streaming-media technology under the network-based circumstance in English language teaching.

## 2 The Application of Streaming-Media Technology in English Teaching

*(1) Video Broadcasting and Live Network Teaching*

The streaming-media technology can continuously transmit multi-media files including audio-frequency, video-frequency,document and photos ect.. The key point of the streaming-media technology is that the transmission of network database and the broadcasting of the users' end can be done at the same time. Based on the network, the multi-media technology system organically combines various modern teaching equipment into integrated one, which can effectively collect, process and broadcast the information of text, photo, voice and picture in time. Thus it can dig out the potential and whole capability of the users and the equipments, and eventually hit the purpose of sharing the resources of teaching information and equipments together.

The video live broadcasting server can capture the various different signals from TV, vidicon and video cassette recorder. It can broadcast through internet or intranet after being encoded, which can be done within 15 seconds. The video live broadcasting server has one strong function of video compression encoding, which can easily combine voice and image into multi-media format and transfer it to Web point. Thus it can carry out 5-30 frame internet live broadcasting per second. Based on this kind of technology, we can set several roads of video into the same video live broadcasting server. The live English teaching scene is transferred to video live broadcasting server through the vidicon equiped in the centered classroom. Then through the function of video compression encoding of video live broadcasting server, the voice and image are mixed into multi-media format and transferred to web point which delivers them to different classroom by the internet or intranet. The students an other learners can study at home, dorm or classroom through moden, PSTN, ADSL, cable modem which is connected with internet. This demonstrates the "student-centered" teaching idea, which can effectively improve the student's language ability.

*(2) Interactive Learning*

That the interactive learning environment can be provided in the process of classroom teaching is the prominent feature that multi streaming-media technology teaching network is better than other media technologies. The information symbol of multi-media include text, graphics, diagram, audio frequency, videl frequency and so on. It is divided into quiet & live forms, scattered & mixed forms, and the forms of vision & audition. Thus a integrated body of multi-media information is formed after these multi-media information is dealt with by computer. The English teaching way based on text, recording, PPT, electronic teaching plan and internet PPT is helpful to creat real language learning scene and motivate the students to be interested in studying.

The language is a kind of social intercourse behavior generated in the specific environment but not only the unipol symbol system used for social intercourse. The good learning environment is the precondition and foundation for obtaining language knowledge. Because the network teaching gathers text, ICON, sound and animation as one integrated body, the English teachers and studens can make use of its diversified advantage and overcome the features of language studying such as simple, repeat and mode singleness. On the contrary, the teachers and students can create real diversified simulation language learning environment, and make student acquire vivid, abundant and visualized materials and have the motivation to keep high interests in studying. The current language teaching environment and teaching method cannot meet with the teaching requirements, and we are subjected to many subjective or objective condition in the language teaching. The development of network multi-media technology based on the computer help us find out one way of improving language teaching environment. The approval makes use of the multi-media technique as fundamental streaming medium technique, we can easily convert classroom into classroom into a live simulation scene by streaming media technique based on multi-media technology. Thus, being shown with vivid picture, music, the forms like animation,etc., the students seems to obtain a real language learning environment, which can motivate them be interested in the study, stir up their knowledge desire and participating desire.

The application of streaming-media technique in the process of English teaching under the network environment is helpful for the teachers and students to establish one new interactive teaching method in the process of English teaching. The stereopic teaching system in English language teaching is set up, which is completely different from traditional English teaching system.

*(3) Personalized Learning*

The teaching model based on the streaming-media technique is a student-centered and personalized studying model that the students teach themselves by using internet. The good environment for personalized studying is provided through multi-media language lab, VOD system and internet. The students can log in the campus intranet at any time and at any place, according to their own requirements for the contents, method, means and strategies, to do speaking and listening practice. The resources provided for personalized studying are more rich under the condition of streaming-media technology. The teachers can dub for animation by the multi-media streaming format, and make PPT by the software of Powerpoint and Flash. Also, the teachers can adopt the internet resources, original English movies and parts of animation into the classroom. In addition, the teachers can re-organize the information resources according to the requirement for teaching and structure of the students' major to create the language studying atmosphere. Thus,the purposes of intensive teaching&more practice, creative teaching and improving learning efficiency can be carried out by repeated interactive practice between the teachers and the students. The environment based on the network teaching also provide the students with relaxing and harmonious atmosphere to develop their personality, which means that the students would like to be interested in studying more actively under this teaching model.

## 3 The Development and Utilization of Common Video&Audio Resources under the Condition of Streaming-Media Technology

*(1) To Select DVD Film and Original Video&Audio Resources as Listening&Speaking Textbook*

Most DVD films in the market have different subtitles when they are played. If these films are input into the controlling machine or server in the controlling desk by using streaming-media technique, original video&audio teaching materials are obtained. The students can be attracted to participate in the interactice teaching and learning actively to improve the listening and speaking ability when the above mentioned teaching materials are used by the teachers in the classroom.

*(2) The Utilization of Video&Audio Resources Stored in the Library*

There are a lot of video&audio teaching resources stored in the universities library, but it is not used effectively. The library only provide service by lending them to students. Service by this way means that it only benefits very few of students and the DVD is easily damaged. If these materials are input into the controlling machine or server in the controlling desk by the making them into streaming-media format, and provide service through internet or intranet, the utilization ratio of these materials is greatly raised.

*(3)To Build up Foreign Language Broadcasting Station*

The English programmes of CCTV, BBC and VOA can be broadcasted by establishing multi-media teaching system. The students can obtain these resources by loging in the campus internet at any time. Thus the students can expand the capacity of words and expressions and improve their listening and speaking ability.

## 4 Conclusions

The application of network-based streaming-media technology in English teaching can develop the traditional English teaching method toward personalized teaching and learning method. Meanwhile, it indicates that listening&speaking English teaching is more practical, more cultural and more interesting. It can motivate both the teachers and the students. More importantly, the student-centered position is established in the process of teaching. Technically, it shows interactive, practical and operable features. In the process of English language teaching, the teachers should adopt new teaching ideas and take full advantage of computer and internet technologies to create better language learning environment for the students.

## References

[1] Skrinda, A.: The challenge of language teaching: Shifts of paradigms (2004)
[2] Warschauer, M., Meskill.: Technology and Second Language Teaching (2001)
[3] Knmhen, S.: Second Language Acquisition and Second Language Learning. Pergamon, Oxford (1981)

# Robust Image Matching Method Based on Complex Wavelet Structural Similarity*

Jianwei An[1] and Xiaochen Zhang[2]

[1] Department of Communication Engineering, School of Computer and Communication, University of Science & Technology Beijing, 100083, Beijing, China
[2] School of Automation, Beijing Institute of Technology, 100081, Beijing, China

**Abstract.** We apply the complex wavelet structural similarity index to image matching system and propose an image matching method which has strong robustness to image transform in spatial domain. Experimental results show that the structural similarity index in complex wavelet domain reflects to a large extent structural similarity of the images compared, which is more similar to human visual cognitive system; in the meanwhile, because of approximate shift invariance of complex wavelet, this index shows good robustness to such disturbance as contrast ratio change and illumination change to template image, so it is more suitable to be used as similarity index for image matching under complex imaging conditions. Moreover, matching simulation experiment shows that this method has higher correct matching rate in complicated disturbance environment.

**Keywords:** navigation technology; image matching; structural similarity; complex wavelet transform.

## 1 Introduction

Image matching1 is in such a matter that first use two different sensors to obtain image from the same image and then align the two images in space to determine the relative translation. It is widely used in such fields as pattern recognition, computer vision, navigation and guidance, remote sensing, etc [2][3].

When shooting template images, due to various changes in shooting conditions such as cloud and mist, cloudy weather or sunshine intensity, it is unavoidable that for the real-time images there are such disturbances as partial occlusion, contrast ratio change, illumination change, hazy image, noise disturbance and so on. Hence, seeking image matching algorithm with robustness to such disturbances is an urgent matter.

As for partial occlusion, conventional method is that: first detect the region occluded, and then adopt the same occlusion technology to occlude the corresponding region in the other image before matching. However, the complexity of occlusion makes it difficult toaccurately detect the region occluded, so this method is not practical. The method based on Hausdorff distance can resolve partial occlusion problem for image matching5. However, this method is often in failure in case of some

* Sponsored by the National Natural Science Foundation (60873192).

complex occlusion and noise disturbance. As for contrast ratio change, illumination change, hazy image and noise disturbance, some algorithms adopt spatial domain image enhancement and restoration or other preprocessing technology. In the course of preprocessing, other image features will change to some extent, this will also influence the matching accuracy.

The key to seeking image matching method, which has robustness to complex disturbance to template images, lies in the similarity index, that is, the similarity index to be adopted should show sufficient robustness to the foresaid disturbances. Basing on the cognitive principle of human visual system, Wang 6 et al propose a novel image similarity evaluation model, which is a index model based on structural similarity. On this basis, Zhao7 and Mehul 8 et al respectively present structural similarity index model in complex wavelet domain. For the template image problems such as partial occlusion, contrast ratio change, illumination change and hazy image, as well as noise disturbance to image matching, this paper presents image matching method adopting the structural similarity index model in complex wavelet domain as similarity index, which satisfies image matching robustness requirement due to the feature that it is insensitive to image spatial domain change.

## 2 Complex Wavelet Transform

With good space and frequency domain localization property as well as a higher computational efficiency, the first generation and the second generation real discrete wavelet transform become an important tool for image multi-scale analysis and make great success in image compression field. However, it still has the faults as below: (1) lack of shift invariance; (2) poor directional selectivity, there is only horizontal, vertical and diagonal directional selectivity, and for the detail in diagonal direction, ±45° cannot be distinguished. To make up this deficiency of conventional real discrete wavelet, many scholars extend structural space of wavelet to complex field, such complex wavelet transform not only keeps the time frequency localization property of conventional wavelet transform but also has good directivity. Among these, dual-tree complex wavelet transform based on Q-shift filter proposed by Kings-bury is widely used 12. Fig. 1 shows dual-tree complex wavelet decomposition diagram for one-dimensional signal based on Q-shift filter.

The key properties required for the Q-shift filters are that they should provide a group delay of either 1/4 or 3/4of a sample period, while also satisfying the usual 2-hand filterbank constraints of no aliasing and perfect reconstruction. Since the filters are even-length, the time reverse of a 1/4 sample delay filter is a 3/4 sample delay filter, giving the required delay difference of 1/2 sample if the filter and its time reverse are used in trees *a* and *b* respectively.

Dual-tree complex decomposition of two-dimensional image can be realized through separation, that is, first conduct one-dimensional transform from the row direction of image, and then carry out one-dimensional transform from the column direction of image 7, which is as shown in Fig. 1. As for the directivity, six sub-bands under each scale for two-dimensional image dual-tree complex wavelet decomposition can output the details in six directions, in such case, the directional selectivity is better than real discrete wavelet transform and a better effect for image processing can be obtained.

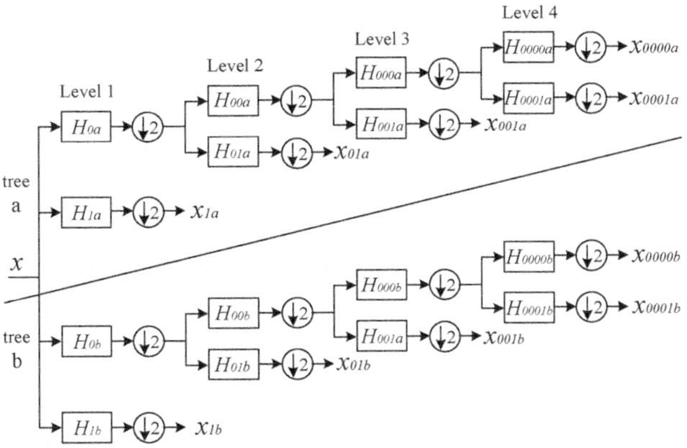


**Fig. 1.** Dual-tree complex decomposition of one-dimensional signal

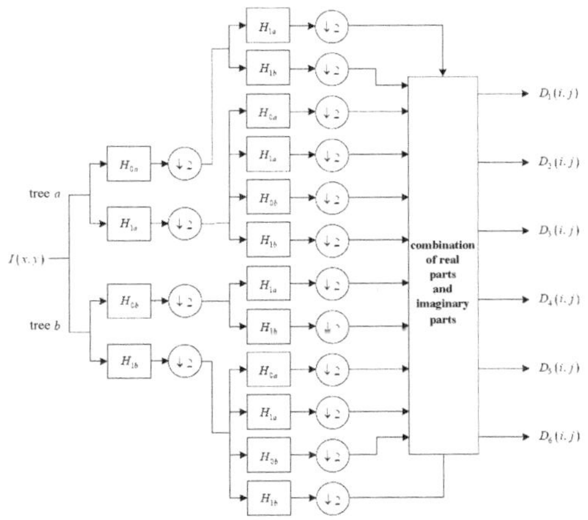


**Fig. 2.** Fliters of dual-tree complex for two-dimensional signal

# 3 Complex Wavelet Structural Similarity Index

## 3.1 Structural Similarity Index Measure

Under the assumption that human visual perception is highly adapted for extracting structural information from a image, as in [6], Wang introduce an alternative complementary framework for quality assessment based on the degradation of structural information. As a specific example of this concept, we develop a Structural Similarity Index and demonstrate its promise through a set of intuitive examples. The SSIM provides surprisingly good image quality prediction performance for a wide variety of image distortions.

In the spatial domain, the SSIM index between two image patches $X = \{x_i | i = 1, 2, \cdots M\}$ and $Y = \{y_i | i = 1, 2, \cdots M\}$ is defined as

$$\mathrm{SSIM}(X,Y)=l(X,Y)^{\alpha}c(X,Y)^{\beta}s(X,Y)^{\gamma} \quad (1)$$

Where $\alpha>0$, $\beta>0$, $\gamma>0$ are parameters used to adjust the relative importance of the three components. $l(X,Y)$, $c(X,Y)$ and $s(X,Y)$ are luminance comparison function, contrast comparison function and structure comparison function, which are defined respectively as

$$l(X,Y)=\frac{2\mu_x\mu_y+C_1}{\mu_x^2+\mu_x^2+C_1},\quad c(X,Y)=\frac{2\sigma_x\sigma_y+C_2}{\sigma_x^2+\sigma_y^2+C_2},\quad s(X,Y)=\frac{\sigma_{xy}+C_3}{\sigma_x\sigma_y+C_3} \quad (2)$$

Where $C_1$, $C_2$, $C_3$ are three small positive constants. $\mu_x$ and $\mu_y$ are mean intensity of two image patches, which are defined respectively as:

$$\mu_x=\frac{1}{M}\sum_{i=1}^{M}x_i,\quad \sigma_x^2=\frac{1}{M}\sum_{i=1}^{M}(x_i-\mu_x)^2,\quad \sigma_{xy}=\frac{1}{M}\sum_{i=1}^{M}(x_i-\mu_x)(y_i-\mu_y) \quad (3)$$

In order to simplify the expression, we set $\alpha=\beta=\gamma=1$ and $C_3=\frac{C_2}{2}$. This results in a specific form of the SSIM index

$$\mathrm{SSIM}(X,Y)=\frac{(2\mu_x\mu_y+C_1)(2\sigma_{xy}+C_2)}{(\mu_x^2+\mu_x^2+C_1)(\sigma_x^2+\sigma_y^2+C_2)} \quad (4)$$

### 3.2 Complex Wavelet Structural Similarity Index

In the complex wavelet transform domain, suppose $c_x=\left\{c_{x,i}\middle|i=1,2,\cdots N\right\}$ and $c_y=\left\{c_{y,i}\middle|i=1,2,\cdots N\right\}$ are two sets of coefficients extracted at the same spatial location in the same wavelet sub-bands of the two images being compared, respectively. Based on (3) and (4), the CW-SSIM index is defined as [8]:

$$\text{CW-SSIM}=\frac{2\left|\sum_{i=1}^{N}c_{x,i}c_{y,i}^{*}\right|+K}{\sum_{i=1}^{N}\left|c_{x,i}\right|^2+\sum_{i=1}^{N}\left|c_{y,i}\right|^2+K} \quad (5)$$

Here $c^*$ denotes the complex conjugate of and $K$ is a small positive constant. The purpose of the small constant K is mainly to improve the robustness of the CW-SSIM measure when the local signal to noise ratios is low.

CW-SSIM id considered as a useful measure of image structural similarity based on the believes that [13]:

- The structural information of local image features is mainly contained in the relative phase patterns of the wavelet coefficients.
- Consistent phase shift of all coefficients does not change the structure of the local image feature.

## 4 Comparison Test

To test CW-SSIM robustness, we select Fig. 3(a) as original reference map and conduct various complex disturbances that might happen during imaging to simulate template image. As shown in Fig. 3(b)-(h) respectively represent the seven disturbances as partial occlusion, contrast ratio change, hazy image, illumination change (average brightness transform and point light illumination) and Gaussian noise, salt-and-pepper noise. After three-level decomposition of dual-tree complex wavelet for reference map and all simulated real-time images, conduct structural similarity computation and make comparison with the indexes widely used in image matching such as MSE, HDM, and SSIM defined in (4). The definition of MSE and HDM is:

$$\mathrm{MSE}(X,Y)=\frac{1}{M}\sum_{i=1}^{M}(x_i-y_i)^2 \tag{6}$$

$$\mathrm{HDM}(X,Y)=\max\left[h(X,Y),\,h(Y,X)\right] \tag{7}$$

Where

$$h(X,Y)=\max_{x_i}\min_{y_j}\left\|x_i-y_j\right\| \tag{8}$$

$$h(Y,X)=\max_{y_j}\min_{x_i}\left\|y_j-x_i\right\| \tag{9}$$

(a) reference image (b) occlusion image (c) contrast stretch (d) blurring

(e) mean luminance shift (f) partial luminance shift (g) Gaussian noise (h) salt noise

**Fig. 3.** Test images

Table 1. shows reference map, template images in all sorts of circumstances and four similarity index values. The results of comparison test for the template images under various complex disturbance conditions show that although from human visual point the two images compared have large similarity, they show a greater difference in

MSE than the other three indexes, and the difference is especially obvious in the event of a large image contrast ratio change and uneven illumination; HDM and SSIM have similar realization, but the former's calculated amount is obviously higher than that of the latter; according to the whole sample data, due to approximate shift invariance of complex wavelet, CW-SSIM to the maximum extent reflects the structural similarity of the images compared, which is more similar to human visual system, so it has good robustness to image spatial domain change and is more suitable to be used as image matching similarity index under complex imaging conditions.

**Table 1.** Results of comparison test

| Number of image | Similarity index values | | | |
|---|---|---|---|---|
| | MSE | HDM | SSIM | CW-SSIM |
| a | 0 | 0 | 1.000 | 1.000 |
| b | 503 | 62 | 0.978 | 0.948 |
| c | 1566 | 37 | 0.851 | 0.873 |
| d | 279 | 34 | 0.802 | 0.946 |
| e | 443 | 30 | 0.929 | 0.961 |
| f | 1365 | 21 | 0.814 | 0.883 |
| g | 2217 | 96 | 0.742 | 0. 864 |
| h | 512 | 36 | 0.875 | 0.890 |

## 5 Image Matching Experiment

To test the actual result of image matching with CW-SSIM as similarity index, we adopt another image as reference image 15and clip 50×50 image blocks at 50 random positions in the map as template images, as shown in Fig. 4. We respectively conduct the seven disturbances as shown in (b)-(h) in Fig.3 (disturbance 1 to disturbance 7) and respectively carry out matching experiments based on MSE, HDM, SSIM and CW-SSIM. With the distance between two pixels as allowable error, the times of correct matching and correct matching probability are given in Table 2.

**Fig. 4.** Reference image

According to the experiment results, average correct matching times of the four calculation methods is 253, 284, 259 and 304 respectively, and average matching probability is 72.3%, 81.1%, 74.0% and 86.9% respectively. The algorithm based on CW-SSIM is better than other algorithms, especially in the event of illumination change and contrast ratio change of real-time image, it has better robustness. It is necessary to note that because of the complexity of complex wavelet decomposition, the algorithm based on CW-SSIM is time-consuming and the time consumed is just less than that by the algorithm based on Hausdorff distance index. Therefore, how to simplify the algorithm and satisfy the real-time requirements of project is a matter worthy of concern.

**Table 2.** Results of image matching experiment

| Disturbance | MSE | | HDM | | SSIM | | CW-SSIM | |
|---|---|---|---|---|---|---|---|---|
| | correct matching | matching probability | correct matching | matching probability | correct matching | matching probability | correct matching | matching probability |
| 1 | 37 | 74% | 35 | 70% | 39 | 78% | 39 | 78% |
| 2 | 33 | 66% | 43 | 86% | 32 | 64% | 47 | 94% |
| 3 | 42 | 84% | 41 | 80% | 37 | 74% | 42 | 84% |
| 4 | 34 | 68% | 50 | 100% | 38 | 76% | 45 | 90% |
| 5 | 35 | 70% | 37 | 74% | 45 | 90% | 48 | 96% |
| 6 | 32 | 64% | 36 | 72% | 32 | 64% | 43 | 86% |
| 7 | 40 | 80% | 42 | 84% | 36 | 72% | 40 | 80% |

## 6 Conclusion

For spatial domain disturbance to template imaging due to weather, illumination and other shooting conditions, this paper presents an image matching method based on complex wavelet structural similarity index. This method makes full use of the insensitive property to image spatial domain change and improves robustness under such interference conditions to template images as partial occlusion, contrast ratio change, illumination change, hazy image and noise interference. Simulation experiment results show that this method has a higher correct matching rate under complex conditions. And at the same time, this paper proposes that the real-time requirement for this method in project practice is a matter worthy of concern.

## References

1. Sun, Z.-k., Shen, Z.-k.: Digital image processing and its application. National Defence Industry Press, Beijing (1985)
2. Zitova, B., Flusser, J.: Image registration methodspp. A survey. Image and Vision Computing 21, 977–1000 (2003)
3. Brown, L.G.: A survey of Image registration techniques. ACM Computing Surveys 24, 325–376 (1992)
4. Li, H.-s., Xu, D.-x., Xu, G.-s.: A image matching algorithm based on complex–valued wavelet transform. Systerms Engineering and Electronics 24, 83–86 (2002)

5. Huttenlocher, D.O., Klanderman, G.A., Rucklidge, W.J.: Comparing images using the Hausdorff Distance. IEEE Trans. on Pattern Analysis and Machine Intelligence 15, 850–863 (1993)
6. Wang, Z., Bovik, A.C., Sheikh, H.R., et al.: Image quality assessment: from error visibility to structural similarity. IEEE Transactions on Image Processing 13, 600–612 (2004)
7. Zhao, W.-f., Shen, H.-b., Yan, X.-l.: Structural similarity measurement based on dual-tree complex wavelet. Journal of Zhejiang University (Engineering Science) 42, 1385–1388 (2008)
8. Mehul, P.S., Wang, Z., Shalini, G., et al.: Complex wavelet structural similarity: a new image similarity index. IEEE Transactions on Image Processing 18, 2385–2401 (2009)
9. Kingsbury, N.G.: The dual-tree complex wavelet transformpp. a new technique for shift invariance and directional filters. In: Proceedings of 8th IEEE DSP Workshop, pp. 86–89. IEEE, Bryce Canyon (1998)
10. Kingsbury, N.G.: Complex wavelets for shift invariant analysis and filtering of signals. Applied Computational Harmonic Analysis 10, 234–253 (2001)
11. Kingsbury, N.G.: A dual-tree complex wavelet trans-form with improved orthogonality and symmetry properties. In: Proceedings of IEEE Conference on Image Processing, pp. 375–378. IEEE, Los Alamitos (2000)
12. Kingsbury, N.G.: Design of q-shift complex wavelets for image processing using frequency domain energy minimization. In: Proceedings of IEEE International Conference on Image Processing, pp. 1013–1016. IEEE, Barcelona (2003)
13. Wang, Z., Simoncelli, E.P.: Translation insensitive image similarity in complex wavelet domain. In: IEEE International Conference on Acoustics, Speech and Signal Processing, pp. 573–576. IEEE, Philadelphia (2005)
14. Quan, W., Wang, X.-d., Gan, J.: Comparison and analysis of similar measure in image matching. Aeronautical Computing Technique 38, 18–20 (2008)
15. Gan, J., Wang, X., Quan, W.: A fast image matching algorithm based on characteristic points. Electronics Optics & Control 16 (2009)

# Optimized Process for Efficiently Extraction Protein from Excess Activated Sludge by Alcalase Hydrolysis

Ruijing Su, Wenfeng Zhang, Enqiang Wang,
Mincong Zhu, Penghui Shi, and Dengxin Li*

College of Environmental Science and Engineering, Donghua University,
Shanghai 201620, China
lidengxin@dhu.edu.cn

**Abstract.** In this paper, we developed the utilization of excess activated sludge (EAS) to extract protein. Firstly, activated sludge was hydrolyzed, using six different commercial proteolytic enzymes by the Kjeldahl method, at their optimal conditions respectively, and Alcalase was chosen as the ideal for further analysis. Secondly, important parameters on protein extraction efficiency(PEE) of EAS by hydrolysis were investigated. Experimental results have been shown that 52.5 % of protein was extracted from EAS (accounting for 30.5%of initial dry sludge (DS) weight) and reducing the quality of activated sludge about 35% under optimum hydrolysis conditions. This novel technology is more environmentally friendly than the conventional sludge treatments.

**Keywords:** Excess Activated Sludge (EAS), Protein Extraction Efficiency(PEE), Enzymatic Hydrolysis, Alcalase.

## 1 Introduction

Excess sewage sludge, produced in sedimentation after wastewater biotreatment, is an inevitable by-product of wastewater treatment processes[1] .McCarty[2] ancticipated a quasi-exponential growth of excess sludge production in USA. In1984, the excessive sludge to be treated in the European Union countries reached 5.56 million dry materials[3].With the expansion of population and industry, the increased excess sludge production is generating a real challenge in the field of environmental engineering technology. The USEPA report estimates an annual 2% increase in the quantity of sewage sludge produced[4], while the annual increase rate in China is over 10% although more the four million tons(dry weight) of sewage sludge have been produced annually in China. The treatment of waste sludge in an economically and environmentally acceptable manner is one of the critical issues facing modern society today.

In this paper, we have developed a novel technology to make use of the bacterial protein in the sewage sludge. Sewage sludge can be used as the protein source because its main constituents are protein and carbohydrate. The objective of the paper

* Corresponding author.

S. Lin and X. Huang (Eds.): CSEE 2011, Part II, CCIS 215, pp. 89–95, 2011.

work (i) investigate the effects of six wide-spectrum proteaseson the PEE; (ii) optimize the enzymatic hydrolysis reaction parameters(temperature, pH and enzyme to substrate(E:S) ratio)for one optimized enzyme.

## 2 Materials and Methods

### 2.1 Materials and Chemicals

The excess sewage sludge samples used in this study were obtained from Qingpu Wastewater Treatment Plant with a capacity of $1.5\times10^4 m^3/d$, located at Qingpu District of Shanghai. This plant uses the typical anaerobic-anoxic-oxic process and the sludge is mechanically dewatered by a belt press filter. The excess sewage sludge samples were kept in a refrigerator at a temperature of 4±0.5 °C before experiments.

### 2.2 Methods and Laboratory Instruments

Moisture of sewage sludge was determined by the weight method. pH value was measured by pH monitor (METTLER TOLEDO(PE20)). The total nitrogen content of the non-hydrolyzed raw material and the aqueous phase generated by hydrolysis was determined using the Kjeldahl method. Crude protein was estimated by multiplying the total nitrogen content (%N) by the factor 6.25. (GB 5009.5-2010)[5,6] using KDN-1000 Kjeldahl apparatus(Tianwei Instrument Co., Ltd., Suzhou) and Spectrophotometry[5]. Absorbance measurements were carried out on a UV-7504 ultraviolet-visual spectrophotometer from SHANGHAI XINMAO INSTRUMENT CO., LTD. Heavy metals concentration of sludge oxidized solution were determined by Inductively coupled plasma emission spectrometer ( Prodigy, Leeman Co., Ltd., USA).

### 2.3 Extraction of Protein from Sewage Sludge by Enzymatic Hydrolysis

5.0 g of activated sewage sludge (water content 84.3%, DS 0.785 g) were mixed with 20 ml of de-ionized water in a water bath shaker and heated at a required temperature and pH for moderate time. Then 0.1 g enzyme was added to initiate the hydrolysis for a certain time, respectively. The enzyme was inactivated by incubating in boiled water for 15 min, then centrifuged at 4000 rpm for 10 min. The protein solution (i.e., the hydrolyzate) was collected. The effects of the ratio of liquid to solid, reaction temperature, pH value and enzyme to substrate (E:S) ratio, reaction time, and heating temperature on the protein extraction efficiency(PEE) were investigated. The residual solid (RS) was weighted after oven drying for 24h at 105◦C. The protein content is given by Equation (1) :

$$C_P = \frac{cv\times0.014\times6.25}{m}\times 100 \tag{1}$$

where $C_P$ is the protein content of the sludge sample, g/100g; c is the concentration of HCl, i..e 0.005mol/l; v is the volume of HCl, ml; m is the weight of the sludge sample, g.

The PEEs at different hydrolysis parameters were calculated as following:

$$\mathrm{PEE}(\%) = \frac{C_P\times V_P}{M_B\times V_B}\times 100\% \tag{2}$$

where $C_P$ is the content of protein in the protein solution, g/100g;$V_P$ the volume of the protein solution, mL; $M_B$ the content of protein in the blank sludge, g/100g; $W_B$ the weight of the blank sludge, g.

# 3 Results and Discussion

## 3.1 Blank Sludge Characterization

The blank sludge contained 84.3% water, 58.7% protein(DS). These results were in accordance with those reported by several researchers. Tanaka et al. [7] reported that protein and carbohydrate were the main constituents of domestic sludge. Chen et al.[8] demonstrated that the waste sludge consisted of 61% protein.

## 3.2 Enzymes Hydrolysis of the Blank Sludge for Protein Extraction

The protein-extraction mechanism is usually destroyed the cell walls leading to the solubilization of extracellular and intracellular materials into the aqueous phase. Hence, different enzymes obviously influence the protein extractive rate. In this paper, six different commercial proteolytic enzymes were investigated at their optimal conditions respectively. The results were showed in Table 1.

**Table 1.** Effects of different enzymes on protein extraction efficiency (PEE,%)

| Enzyme | pH value | E:S(%) | T (℃) | time(h) | PEE(%) |
|---|---|---|---|---|---|
| Papain | 7.0 | 3 | 55 | 4 | 48.0 |
| Alcalase | 8.0 | 3 | 55 | 4 | 52.4 |
| Trypsin | 7.0 | 3 | 35 | 4 | 21.3 |
| Acid proleslytic enzyme | 3.5 | 3 | 50 | 4 | 31.5 |
| Neutral protease | 6.0 | 3 | 50 | 4 | 51.6 |
| Pepsin | 2.0 | 3 | 25 | 4 | 36.8 |

Table 1 demonstrated that Alcalase has the strengths of high activity, highest efficiency, followed by Neutral protease and Papain. Through the six protein obtained from the sewage sludge testes were compared, and alkaline protease was selected as the ideal enzyme ultimately. Further study was proceeded in the next experiment to the protein extraction of sewage sludge and optimize the best extraction conditions.

## 3.3 Optimization of Parameters

### 3.3.1 Effect of Reaction Time (h) and the pH Value on the PEE (%)

Experiments were selected at pH7.5, pH8, pH9, pH10 respectively. Other conditions were: E:S 2%, reaction temperature 55 °C, the ratio of liquid to solid 4:1.

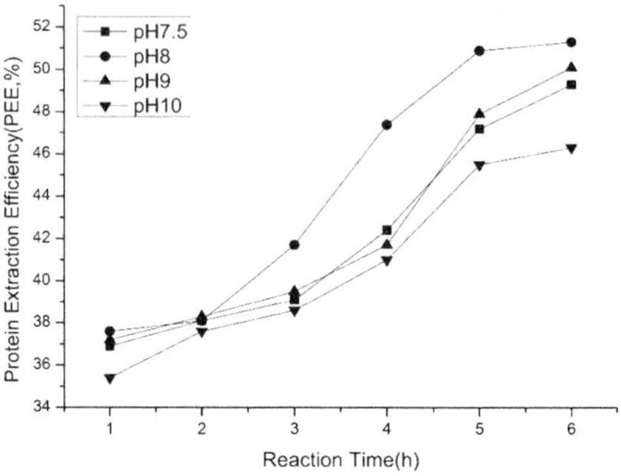


**Fig. 1.** Effect of reaction time (h) and the pH on the PEE (%)

Fig.1 illustrates the effects of the reaction time and pH on the PEE. It seemed that pH 8 was beneficial for the protein extraction, and the PEE reached the highest level at pH 8 at any reaction time. At a set pH value (e.g. pH 1.0), the PEE increased with an increase of the reaction time, and seems to continued to increased. This is because, the enzymatic hydrolysis process is gently compared to conventional methods(such as acidic hydrolysis), and part of the extracted protein won't decomposed into amino acids until about 24h later[9,10]. Therefore, a hydrolysis pH 8 was used as standard in all further investigations.

### 3.3.2 Effect of Reaction Time (h) and Enzyme to Substrate (E:S) (%) on Protein Extraction Efficiency (PEE,%)

Experiments were selected at enzymatic concentration3%, 2%, 1%, 0.5% respectively. Other conditions were: pH 8, reaction temperature 55 °C, the ratio of liquid to solid 4:1.

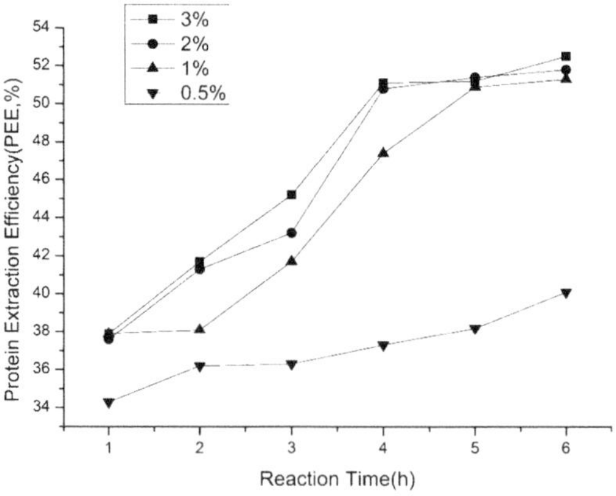


**Fig. 2.** Effect of reaction time (h) and enzyme to substrate (E:S) (%)) on PEE (%)

Fig. 2 shows the effect of enzyme to substrate (E:S) (%) on the PEE. The PEE increased gradually with an increase of the E:S and then rised moderately when the E:S was over 2%. The maximum PEE value was around 52.7% obtained at E:S 3%, which was close to E:S 2%. At the set of E:S 0.5%, PEE increased gently with the time go on. There was a sharp jump of PEE between E:S 0.5% and E:S 1%, while the change between E:S 1% and E:S 2% was little. Considered from the commercial, a hydrolysis E:S of 2% was therefore used as standard in all further investigations.

### 3.3.3 Effect of Reaction Time (h) and Reaction Temperature (°C) on PEE (%)

Experiments were selected at temperature 50 °C, 55 °C, 60 °C and 65°C respectively. Other conditions were: pH 8, E:S 2%, the ratio of liquid to solid 4:1.

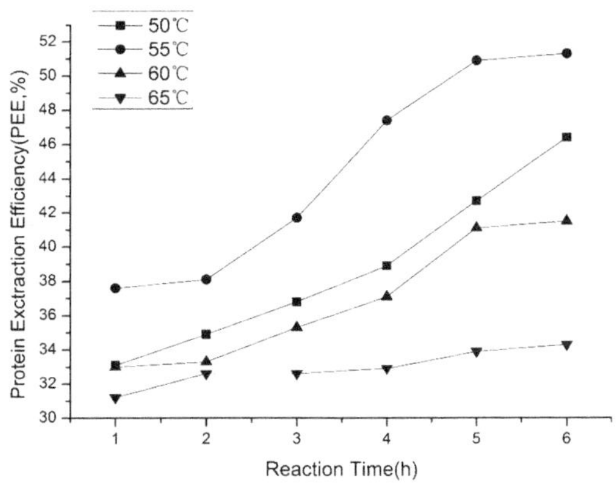


**Fig. 3.** Effect of reaction time (h) and temperature (°C) on PEE (%)

Fig. 1 shows the effect of hydrolysis temperature on the PEE. The PEE increased gradually with an increase of the temperature and then fell quickly when the temperature was over 65 °C. The maximum PEE value was around 52.4% obtained at temperatures of 55 °C. With the increase of the hydrolysis temperature, the sludge floc was disintegrated and the bacterial cell-wall was ruptured, so the protein in the cell was gradually released [11], resulting in the increased protein content in the hydrolytic solution. However, when the temperature was above 60 °C, the PEE was rapidly decreased due to the enzyme activity was inhibited. Therefore, a hydrolysis temperature of 55 °C was used as standard in all further investigations.

### 3.3.4 Effect of Reaction Time (h) and the Ratio of Liquid to Solid on PEE (%)

Experiments were selected the ratio of liquid to solid1:3, 1:4, 1:5,1:6 respectively. Other conditions were: pH 8, E:S  2%, temperature 55 °C.

The ratio of liquid to solid is also a factor to protein extraction rate, plus the amount of water directly affect the probability of enzyme and the size of the protein interactions, but also affect the interaction of the mass transfer rate. Fig.4.demonstrate that when the ratio is up to 6:1, the appropriate enhance of  the ratio, extraction rate will increased although in a small apart, but also the extract reduced the total solids content which is not conducive to industrial production.

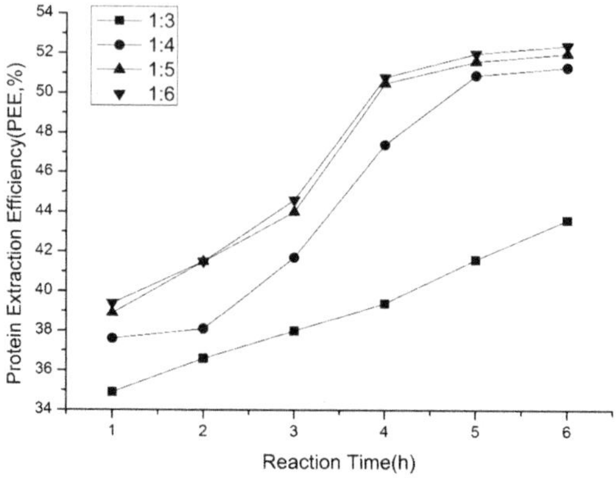


**Fig. 4.** Effect of reaction time (h) and the ratio of liquid to solid on PEE (%)

Based on the above results, the optimal condition for protein extraction from the blank sludge was as follows: E:S of 2%, reaction time of 6 h, the ratio of liquid to solid of 6:1, pH of 8 and hydrolysis temperature of 55 °C. Under these parameters, the highest PEE was 52.7%.

### 3.3.5 Heavy Metal in the Protein Extraction

The concentration of heavy metals in the liquid was measured. The table 2 summarized the experiment results of heavy metals concentration.

**Table 2.** Heavy metals concentration of the extraction liquid

| Experiment number | Cu | Pb | Cr | Cd | Hg | As | Ni | Zn |
|---|---|---|---|---|---|---|---|---|
| 1 | 0.0951 | ≤0.01 | ≤0.01 | ≤0.01 | ≤0.01 | ≤0.01 | 0.0825 | 0.0198 |
| 2 | 0.1028 | ≤0.01 | ≤0.01 | ≤0.01 | ≤0.01 | ≤0.01 | 0.0743 | 0.0256 |
| 3 | 0.0989 | ≤0.01 | ≤0.01 | ≤0.01 | ≤0.01 | ≤0.01 | 0.0901 | 0.0320 |
| mean | 0.0989 | ≤0.01 | ≤0.01 | ≤0.01 | ≤0.01 | ≤0.01 | 0.0823 | 0.0258 |

Table 2 indicates that the leaching contents of Cu and Zn were far below their limit values required by the China Environmental Quality Standards for Surface Water [12], while those of Pb, As, Ni, Cd, Cr and Hg were below their detection limits.

## 4 Conclusions

This study developed a novel technology for the sewage sludge utilization. The optimal condition for protein extraction from the blank sludge was as follows: E:S of 2%, reaction time of 6 h, the ratio of liquid to solid of 6:1, pH of 8 and hydrolysis temperature of 55 °C. Under these parameters, the highest PEE was 52.7%. The extraction protein solution produced in this study was in accordance with China Standard for feed protein standard. Accordingly, it could a very effective hydrolysis protocols for excess sludge protein extraction.

**Acknowledgements.** This work was supported by Innovation Foundation of Donghua University for Doctoral Candidates (BC201132) and Shanghai Leading Academic Discipline Project (B604). In addition, the authors appreciate the constructive suggestions of the anonymous reviewers, which are invaluable in improving the quality of the manuscript.

## References

1. Wei, Y.J., Liu, Y.S.: Effects of sewage sludge compost application on crops and cropland in a 3-year field study. J. Chemosphere 59, 1257–1265 (2005)
2. McCarty, P.: Sludge concentration-needs, accomplishments and future goals. J. Water Pollut Control Fed. 38, 493 (1966)
3. Chudoba, P., Capdeville, B.: A possible way towards reduction of waste sludge production. In: Sixth IAWPRC Conference on Design and Operation of Large Wastewater Treatment Plants, Prague (1991)
4. Bagreev, A., Bashkova, S., Locke, D.C., Bandosz, T.J.: Sewage sludge-derived materials as efficient adsorbents for removal of hydrogen sulfide. J. Environ. Sci. Technol. 35, 1537–1543 (2001)
5. Ministry of Health of PRC, GB 5009.5-2010. Determination of protein in foods. Standards Press of China, Beijing (2010)
6. Crooke, W.M., Simpson, W.E.: Determination of ammonium in Kjeldahl digests of crops by an automated procedure. J. Agric. Food Chem. 27, 1256–1262 (1971)
7. Tanaka, S., Kobayashi, T., Kamiyama, K., Bildan, M.N.: Effects of thermochemical pretreatment on the anaerobic digestion of waste activated sludge. J. Water Sci.Technol. 35, 209–215 (1997)
8. Chen, Y.G., Jiang, S., Zhou, H.Y.Q., Gu, G.W.: Hydrolysis and acidification of waste activated sludge at different pHs. J. Water Res. 41, 683–689 (2007)
9. Church, F.C., Swaisgood, H.E., Catignani, G.L.: J. Appl. Biochem. 6, 205 (1984)
10. Aniello, A.D., Petrucelli, L., Gardner, C., Fisher, G.: Anal. Biochem. 213, 290 (1993)
11. Wang, Z.J., Wang, W., Zhang, X.H., Zhang, G.M.: Digestion of thermally hydrolyzed sewage sludge by anaerobic sequencing batch reactor. J. Hazard. Mater. 162, 799–803 (2009)
12. State Environment Protection Administration, China Environmental quality standards for surface water, GB 5086.2–1997. China Environmental Science Press, Beijing (1997)

# Harmonic Detection for Hybrid Filter

Yunjing Liu and Fengwen Wang

Department of Automation Engineering, Northeastern University at Qinhuangdao
Qinhuangdao, China

**Abstract.** Harmonic distortion can be caused by both active and passive non-linear devices in a power system. These harmonic currents can cause voltage distortion and excessive power losses in the system. To maintain high quality power supply, the system harmonic levels must be measured and kept under control. Passive, active and hybrid filters can be used in a power system to eliminate harmonic currents at the load sides. In order to compensate harmonic in real time, each harmonic component is detected. Due to this complicated process, it is difficult to improve the processing speed and simplify harmonic detection process. A method of describing the total harmonic distortion in current detection is suggested in order to simplify harmonic detection. The results show that the approach of harmonic detection is applicable to detect each harmonic effectively.

**Keywords:** Harmonic detection, power systems, hybrid filters, distortion.

## 1 Introduction

Modern power industries are using more and more power-electronics equipment, such as frequency changers, motor-drive systems, etc. Such equipment presents nonlinear impedance to the utility, generating large harmonic currents with well known adverse effects, such as low power factor, low efficiency and destruction of other equipment. Also, some precision instruments and communication equipment will be interfered.

To improve the power quality, some methods for harmonic suppression have been used in power systems. Compensation techniques based on active and passive filters to eliminate current harmonics and to compensate reactive power have already been presented and published in the technical literature.

Though traditional passive filter (PF) is cheap and simple, the effect of harmonic suppression is greatly influenced by impedance and it's own parameters, and resonance may occur between power impedance and passive filter. The PF can't satisfy the needs of modern industry and power system.

An active filter (APF) basically consists of a power electronic inverter circuit and a control circuit. The common operation of the active filter is as follows. First the harmonic detecting unit in the control circuit samples the instantaneous waveform of the power-line voltage and current and then computes the required compensating current. Second, based on the computed amount of current, the control unit generates gating signals to drive the inverter power circuit. Third, according to the gating signals, the inverter generates the compensating current and injects it into the power line to cancel its harmonic distortion.

S. Lin and X. Huang (Eds.): CSEE 2011, Part II, CCIS 215, pp. 96–101, 2011.

Active power filter has good harmonic suppression effect, but is very expensive. So hybrid active power filter which combines low-capacity active power filter with high-capacity passive filter becomes a "must-be" choice in application.

An effective way for harmonic elimination is the harmonic compensation by using active power filter. An active power filter detects harmonic current from the distorted wave in power line, then generates a negative phase current same as the detected harmonic to cancel out the harmonic in power system. However, active power filter has complicated structure such that this causes time delay for harmonic compensation. In order to compensate harmonic in real time, each harmonic component is detected. Harmonic current detection method used in active power filter is very important, which determines the harmonic current detection accuracy and tracking speed and the effect of harmonic current compensation.

This paper presents a method of harmonic detection for hybrid active filter. A system is modeled in Matlab Simulink to consist of the non-linear loads. Harmonic detection is developed to demonstrate the performance of the proposed method. The simulation results show that the method has the expected performance.

## 2 Harmonic Detection Technology

Generally, power electronic devices introduce mainly characteristic harmonics whose frequencies are integer multiple of fundamental frequency. In practice, the fundamental frequency of power signal always deviates slightly from its nominal value, and therefore its initial value used for iterative optimization can be chosen as 50 Hz.

Harmonic current detection method used in active power filter is very important, which determines the harmonic current detection accuracy and tracking speed and the effect of harmonic current compensation. The harmonic parameters, except for the fundamental frequency, used to generate the voltage waveform are obtained from input current of an actual three-phase AC. Several detection algorithms are as following: low-pass filter, adaptive detection, and so on.

In general, harmonic detection does not need to calculate the various aspects of harmonic components, but need to detect the harmonic current or reactive current or their sum. This paper mainly discusses the harmonic detecting and processing signal based on instantaneous reactive power theory.

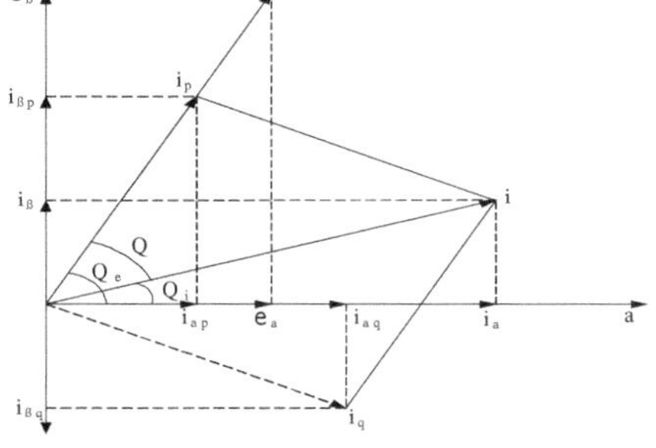


**Fig. 1.** α-β coordinate system voltage and current vector

According to instantaneous reactive power theory, we obtain instantaneous active power $p$ and instantaneous reactive power $q$. The principle is shown in Figure 1.
The process can be defined as follows:

$$\begin{aligned} e_a &= E_1 \sin \omega t \\ e_b &= E_1 \sin(\omega t - \frac{2\pi}{3}) \\ e_c &= E_1 \sin(\omega t + \frac{2\pi}{3}) \end{aligned} \tag{1}$$

Where : $e_a$ _ instantaneous voltage of phase $a$
$e_b$ _ instantaneous voltage of phase $b$
$e_c$ _ instantaneous voltage of phase $c$

$$\begin{bmatrix} e_\alpha \\ e_\beta \end{bmatrix} = \sqrt{\frac{2}{3}} E_1 \begin{bmatrix} \sin \omega t \\ -\cos \omega t \end{bmatrix} \tag{2}$$

$$\begin{bmatrix} p \\ q \end{bmatrix} = \sqrt{\frac{2}{3}} E_1 \begin{bmatrix} \sin \omega t & -\cos \omega t \\ -\cos \omega t & -\sin \omega t \end{bmatrix} \tag{3}$$

Where $p$ is instantaneous active power, and $q$ is instantaneous reactive power

$$\begin{bmatrix} p \\ q \end{bmatrix} == |e| \begin{bmatrix} i_p \\ i_q \end{bmatrix} = \sqrt{\frac{2}{3}} E_1 \begin{bmatrix} i_p \\ i_q \end{bmatrix} \tag{4}$$

$i_p$ _ instantaneous active current
$i_q$ _ instantaneous reactive current

$$\begin{bmatrix} i_p \\ i_q \end{bmatrix} = \begin{bmatrix} \sin \omega t & -\cos \omega t \\ -\cos \omega t & -\sin \omega t \end{bmatrix} \begin{bmatrix} i_\alpha \\ i_\beta \end{bmatrix} \tag{5}$$

$$\begin{bmatrix} p \\ q \end{bmatrix} = E_1 \begin{bmatrix} \sin \omega t & -\cos \omega t \\ -\cos \omega t & -\sin \omega t \end{bmatrix} \begin{bmatrix} 1 & -\frac{1}{2} & -\frac{1}{2} \\ 0 & \frac{\sqrt{3}}{2} & -\frac{\sqrt{3}}{2} \end{bmatrix} \begin{bmatrix} i_a \\ i_b \\ i_c \end{bmatrix} \tag{6}$$

$$\begin{bmatrix} i_p \\ iq \end{bmatrix} = \sqrt{\frac{2}{3}} \begin{bmatrix} \sin \omega t & -\cos \omega t \\ -\cos \omega t & -\sin \omega t \end{bmatrix} \begin{bmatrix} 1 & -\frac{1}{2} & -\frac{1}{2} \\ 0 & \frac{\sqrt{3}}{2} & -\frac{\sqrt{3}}{2} \end{bmatrix} \begin{bmatrix} i_a \\ i_b \\ i_c \end{bmatrix} \tag{7}$$

$$\begin{bmatrix} i_p \\ i_q \end{bmatrix} = \begin{bmatrix} i_{pf} \\ i_{qf} \end{bmatrix} + \begin{bmatrix} i_{ph} \\ i_{qh} \end{bmatrix} \tag{8}$$

By combining and rearranging (7) and (8), Eq. (9) can be written as

$$\begin{bmatrix} i_{\alpha f} \\ i_{\beta f} \end{bmatrix} = \begin{bmatrix} \sin \omega t & -\cos \omega t \\ -\cos \omega t & -\sin \omega t \end{bmatrix} \begin{bmatrix} i_{pf} \\ i_{qf} \end{bmatrix} \tag{9}$$

Clearly, a low pass filter (LPF) can also get three-phase current transformation and separate the fundamental frequency current and harmonic current.

Harmonic detection has been carried out as shown in Fig2. One goal of these detections was finding the distribution system background harmonic levels. The harmonics were analyzed. The data were exported to MATLAB, where a detailed analysis of the data was performed.

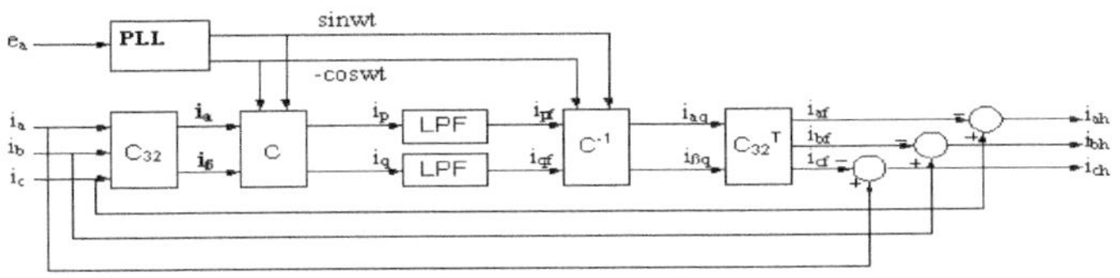


**Fig. 2.** Block diagram of harmonic detection circuit of ip-iq method

## 3 Simulation Model and Results

To analyze the performance of the proposed method of harmonic detection for hybrid active power filter, simulation model as shown in Fig3 and Fig4 has been carried out in the simulation environment MATLAB.

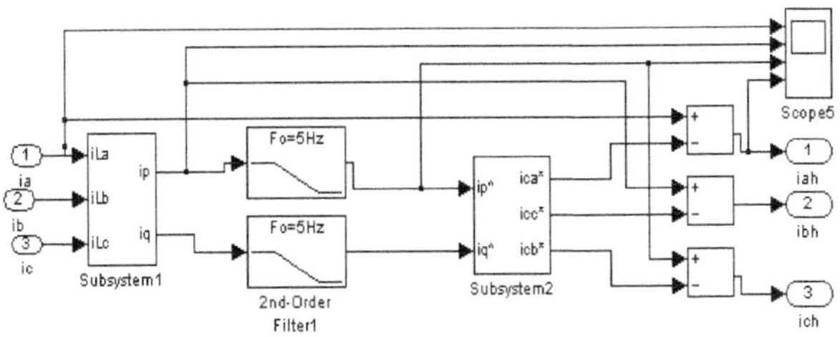


**Fig. 3.** Simulation model of harnomic detection

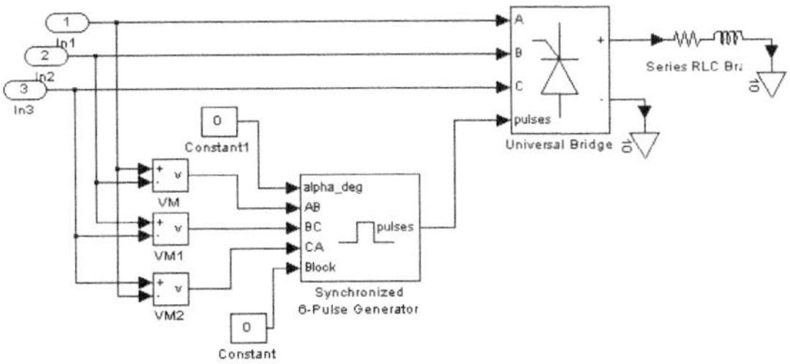


**Fig. 4.** Fully controlled three-phase bridge rectifier circuit

Because of non-sinusoidal load currents consisting primarily of lower-order 5th, 7th, 11th, and 13th harmonics that distort the system power quality, we considered the 5th, 7th, 11th, and 13th harmonics detection. distortion current waveforms are shown in Fig.5.

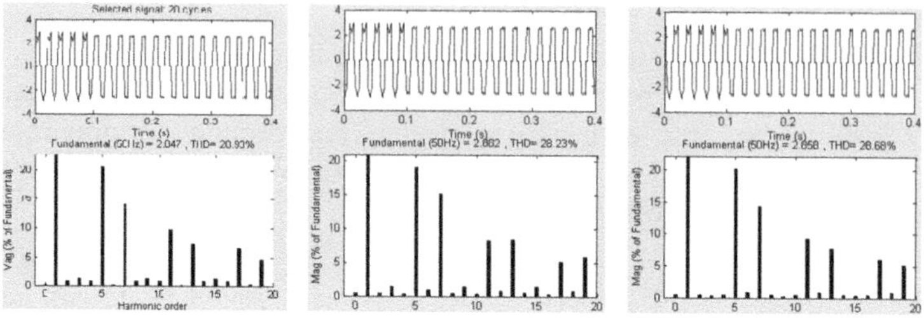

**Fig. 5.** Three-phase ( A,B and C ) distortion current of FFT

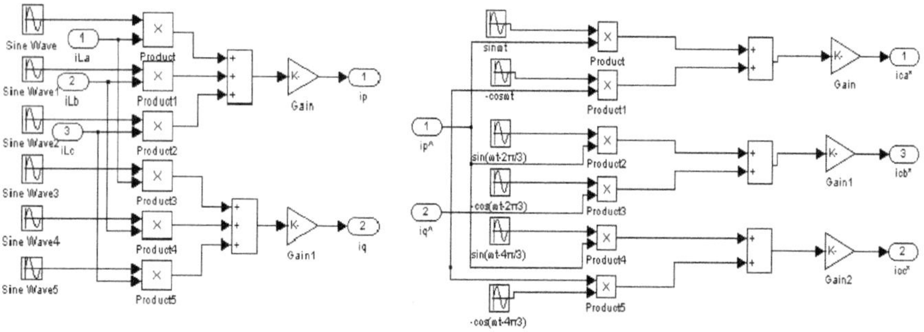


**Fig. 6.** Simulation model of hybrid filter

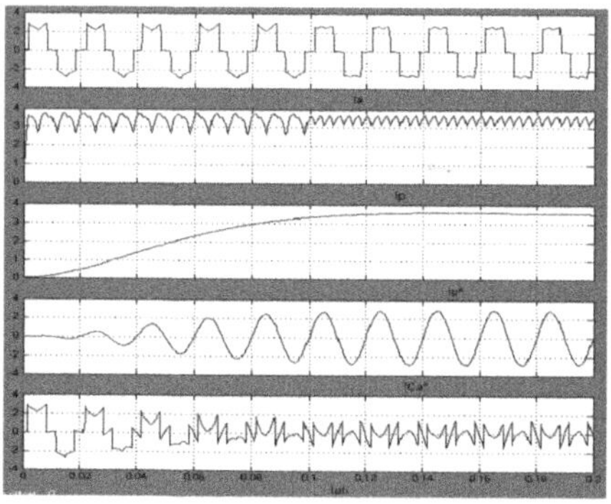

**Fig. 7.** Simulation results

In hybrid filters, passive power filters are tuned to suppress part of harmonic currents and improve the power factor. The active power filter generates harmonic currents opposite to the harmonic currents caused by non-linear loads and improves the performance of passive power filters. The compensating current is injected to the power grid through power feeder. The active power filter focuses on suppressing harmonic currents and improving the performance of passive power filters, it is not necessary to compensate reactive power. To verify the performance of the proposed method of harmonic detection for hybrid active power filter, simulation model as shown in Fig.6 has been carried out in the simulation environment MATLAB.

The comparison between before-compensation and after-compensation of the current waveforms are shown in Fig.7. Simulation results show the effectiveness of the proposed method of harmonic detection for hybrid active power filter.

## 4 Conclusions

This paper presents harmonic current detection and regulation for the hybrid active filter to enhance the harmonic suppression capability in industrial power system. Harmonic current distortion in FFT analysis can be induced that there are more serious harmonic interference when THD was 28.93%. From the harmonic order, these harmonics are mainly 5, 7th, 11th, 13th, 17th and 19th harmonics. Simulation results indicate that we can achieve satisfactory effectiveness by adopting hybrid active power filter to achieve harmonic suppression and improve the power quality. From simulation results we can get the following conclusions: the detecting method is much better in stability and detection accuracy.

## References

1. Rastegar, S.M.R., Jewell, W.T.: A new approach for suppressing harmonic disturbances in distribution system based on regression analysis. Electric Power Syst. Res. 59, 165–184 (2001)
2. Jintakosonwit, P., Fujita, H., Akagi, H.: Control and Performance of a Fully-Digital-Controlled Shunt Active Filter for Installation on a Power Distribution System. IEEE Trans. Power Electron 17, 132–140 (2002)
3. Hirofumi, A.: Active Harmonic Filters. Proceedings of the IEEE 93(12), 2128–2141 (2005)
4. Nunez-Zuniga, T.E., Pomilio, J.A.: Shunt Active Power Filter Synthesizing Resistive Loads. IEEE Trans. Power Electron 17, 273–278 (2002)
5. Zhou, L., Li, Z.: A Novel Active Filter Based on the Least compensation Current control Method. IEEE Trans. Power Electronics 15(4), 655–659 (2000)
6. Li, D., Chen, Q., Jia, Z.: A High-Power Active Filtering System With Fundamental Magnetic Flux Compensation. IEEE Trans. Power Delivery 21, 823–830 (2006)
7. Jintakosonwit, P., Akagi, H., Fujita, H., Ogasawara, S.: Implementation and performance of automatic gain adjustment in a shunt active filter for harmonic damping throughout a power distribution system. IEEE Trans. Power Electron 17(3), 438–447 (2002)
8. Bhende, C.N., Mishra, S., Jain, S.K.: TS-Fuzzy Controlled Active Power Filter for Load Compensation. IEEE Trans. Power Delivery 21, 1459–1465 (2006)
9. Wu, T.-F., Nien, H.-S., Shen, C.-L.: A single-Phase Inverter System for PV Power Injection and Active Power Filtering With Nonlinear Inductor Consideration. IEEE Trans. Ind. Applicant 41, 1075–1083 (2005)

# The Application of Delay Estimation in Binary Control of Sequence Control in Power Plant Based on ZigBee

Renshu Wang, Yan Bai, and Fengzhu Wang

School of Control and Computer Engineering
North China Electric Power University, 102206, Beijing, P.R. China
{wangrenshu_1,god_soup,wangfengzhu1}@126.com

**Abstract.** This paper introduces the application of wireless sensor networks (WSNs) based on ZigBee in the binary control of sequence control in power plant. And the process of the chemical water treatment is taken as an example to explain the system architecture and the problem. As the signals of binary control are transmitted through wireless channel and the process is implemented according to the logic sequence, the problem caused by delay is described in detail which isn't encountered in cable networks and a solution based on delay estimation is proposed to settle it. Then on condition that the network is constrained in single-hop, the analysis of the method is carried out and the merits are also presented.

**Keywords:** sequence control, ZigBee, binary control, delay estimation.

## 1 Introduction

In recent years, with the rapid development of microprocessor and wireless technology, the application of wireless sensor networks(WSNs) in industrial field will inevitably become a direction of research. And there have been wireless technologies or standards that we are familiar with, such as Bluetooth, WiFi, ZigBee and so on. There are many merits of ZigBee such as low energy consumption, ease of deployment, scalability, reliable data transmission and low cost [1]. Combined with the practical demands, considering the situation of the field and cost of the device, ZigBee is competent for our goal to realize the binary control in the process control. Chemical water treatment, which is chosen for the implementation of our research in binary control, is one part of the process of the treatment of boiler feedwater. There are several valves and pumps with different functions. The system architecture and the control streams are presented in Fig. 1. Obviously it is a closed-loop sensor network with binary control signals. The sink node works as the controller in the sequence control with the designed configuration program resided in, receiving DI and generating DO. Except the control packets (DI and DO) transmitted to realize the sequence control, other information, such as the energy state of wireless module and so on, is not presented in Fig. 1.

Most of papers studying the control system with WSNs focus on the processing of analog signals and the controllers are designed [2,3]. Although the main problems of the wireless transmission, such as packet loss, delay and so on, are taken into

S. Lin and X. Huang (Eds.): CSEE 2011, Part II, CCIS 215, pp. 102–106, 2011.

consideration in these papers and the effective models are presented to settle the problem, there are few papers about the problem of the binary control in WSNs. Problems introduced while using wireless networks to transmit the digital signals, especially the delay of the signals, will result in serious consequence. So the main purpose of this paper is analyzing the problem caused by delay and proposing a feasible and effective method to solve the problem.

The rest of the paper is organized as follows. In Section 2, the problem caused by delay is described in detail while introducing the wireless technology into the binary control and the shortcomings of the method of using time stamps are stated. Section 3 presents the delay estimation of the wireless sensor networks, based on which the problem described in Section 2 is settled. Section 4 gives the conclusion.

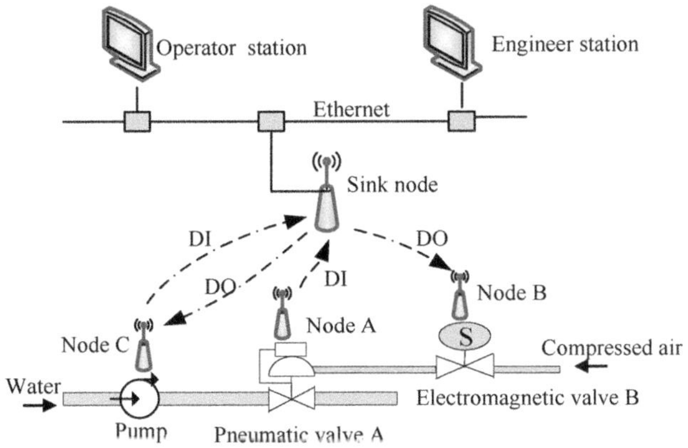


**Fig. 1.** The architecture of control system with wireless networks introduced

## 2 Problem

In the continuous control system, the delay may lead to oscillating. And the things will be worse in discrete control system that controller may send out an instruction totally contrary to the actual need. At the same time, in traditional cable control system, because the binary signals are transmitted through the lead directly, the delay of transmission is so small that it won't cause much trouble. But things are quite different in the wireless sensor network. There are lots of factors resulting in the delay of signal transmission or even the packet disorder while the signals are transmitted through the wireless channels. Firstly, because the CSMA/CA is adopted in ZigBee, each node has to compete for the access of media. Secondly, there may be a congestion happened in sink node, so that the arrival of the data will be discarded, requiring the retransmission of the data. Thirdly, in multi-hop networks, the packets have to be received and transmitted through the route nodes. All these will contribute to the delay of the transmission of the data.

The delay of the signals transmitted through wireless channel would cause troubles in the sequence control. Fig 2 illustrates the situation. In the sink node with configuration program resided in, there is an output C derived from the logic operation C=A&B. In Fig. 2, A and B are the signals from two different end nodes. The signals of dashed line stand for the sampled signals and the solid ones are the delayed signals. It shows clearly that there is no overlap of the actual signals in time domain and the output should be zero all the time. However, because of the delay, there comes an undesirable result that the signal is "1" between $t_3$ and $t'_{A2}$.

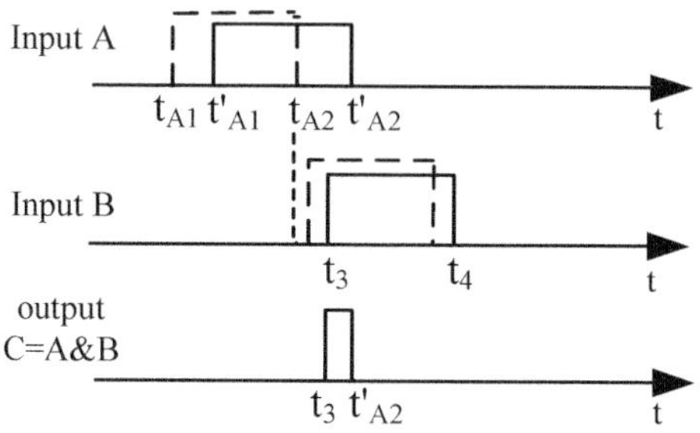


**Fig. 2.** The unexpected result caused by delay of transmission

One solution to this problem is using the time stamps while sampling the signals and packing the time stamps into the packets with the sampled signals in the end nodes. Then the packets are transmitted to the sink node. After the sink node receives the packets, the signals can be reverted according to their time stamps and then be used as the inputs of the logic operation. So the output will coordinate with the actual situation. However the implementation of this method must base on rigid time synchronization, otherwise the outcome will be wrong. In fact, the precisions of different time synchronization algorithms rank from microseconds to seconds [4]. Even if the rigid time synchronization is implemented, it has to consume lots of extra energy, which will affect the lifetime of the whole networks since the nodes are all power constrained [5]. What is more, with the time information carried in the frame, it increases the probability of the errors of the transmitted frames. So we introduce a method of using delay estimation to solve the problem.

## 3 Delay Estimation

There are lots of metal pipes in the field where the wireless end nodes are installed, which will cause distortions of the signals. So we locate the sink node, which works as a controller and gateway, close to the end nodes as possible to alleviate the interference to the transmitted signals through long distance transmission. All the end nodes are the neighbors of the sink node within single hop. And we model the wireless network as directed graph G=(V,E), where V (|V|=n)is the set of wireless nodes and E is set of the delays of wireless links. The weight of the link is noted by the delay.

For the further research, we have to make some assumptions. Firstly, the node mean-time-to-fail(MTTF) is relatively large compared to the message transmission time. Secondly, the whole network is working at a stable state, the sink node being in the state of activity all the time and the end nodes sampling and transmitting signals at a fixed interval. Thirdly, the time of end nodes sampling the valid device information to the buffer of the transceiver can be neglected. At last, the process speed of the sink node is greater than the arrivals of signals, so there is no congestion happened in sink node. We propose the solutions of the problem in two situations.

*Situation 1*: the delay of the same wireless link keeps unchanged

In [6,7], it presents that the expected normalized end-to-end packet delay using CSMA/CA is $D^i$ and the deduction of it is independent of time. For the end nodes sample signals at a fixed rate and the delay of the same wireless link keeps unchanged, the received signals in sink node are shifted $D^i$ back off compared to the sampled signals. Thus, for each end node linked to the sink node, there is $D^i$ (i=1,2,...,n-1) . The function of the logic operation is g. The output in sink node between $t_0$ and $t_1$ is $f(k)|_{t0-t1}$, and k is the number of the input signals. So we have

$$f(k)|_{t0\to t1} = g(S^1(t_0+D^1,t_1+D^1),\cdots,S^k(t_0+D^k,t_1+D^k)) \qquad (1)$$

$S_i(t_0+D^i,t_1+D^i)$ is the signal sent by the ith end node and received by sink node between $t_0+D^i$ and $t_1+D^i$.

*Situation 2*: the delay of the wireless link changes with time.

Because ZigBee adopts CSMA/CA as the MAC protocol, the delay of each node accesses wireless channel is stochastic. And we suppose that the delays at $t_0,t_1,\dots,t_{n-1}$ are all independent. The delay evaluation model can be derived as

$$\begin{cases} D^i(t_0) = d^i_0 & k=0 \\ D^i(t_k) = d^i_{k-1} + \delta^i(t_k) & k \ge 1 \end{cases} \qquad (2)$$

$d^i_0$ is the value of delay at $t_0$. The $\delta^i(t_k)$ is sampled from gaussian white noise, with the mean value being 0 and the variance $\zeta_x=2\zeta$. And the $\delta^i(t_k)$ is independent. Thus, the unbiased estimation of $D^i$ with minimum variance is given by

$$d^i_{n-1} = \frac{1}{n}\sum_{m=0}^{n-1} D^i(t_m) \qquad n \ge 1 \qquad (3)$$

The estimated variance $\zeta^2/n$ decreases linearly as the n increases. In (2), replacing dik-1 with (3), we have

$$\begin{cases} D^i(t_0) = d^i_0 & k=0 \\ D^i(t_k) = \frac{1}{k}\sum_{m=0}^{k-1} D^i(t_m) + \delta^i(t_k) & k \ge 1 \end{cases} \qquad (4)$$

The delay estimation with minimum deviation can be derived. Using (1) and (4), we can derive an output closer to the practical condition with minimum deviation. The result of the application is show in Fig. 3.The delayed signals can be transformed according to (4), thus we have $t_0= t_1-D^1(t_0)$, $t_n=t_{n+1}-D^1(t_n)$, $t_m=t_{m+1}-D^2(t_m)$, $t_p=t_{p+1}-D^2(t_p)$. The signals after transformation eliminate the overlap in time domain, so the output is always zero which coordinates with the actual situation. It can be seen clearly that the process of signals bases on the clock of the sink node. So it will alleviate the reliance on time synchronization greatly.

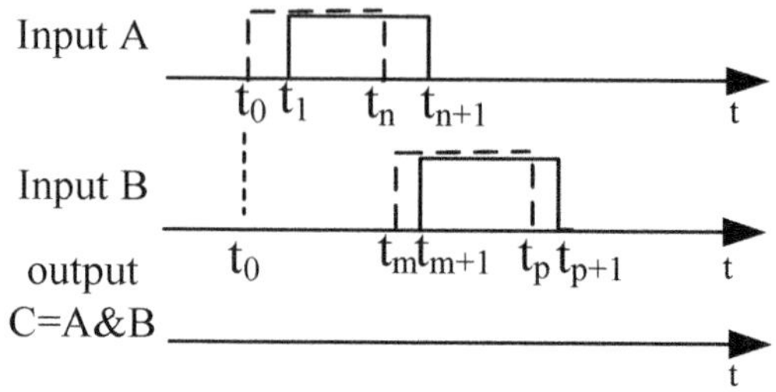


**Fig. 3.** The result derived with the method of delay estimation which accords with the actual situation

## 4 Conclusion

To settle the problem caused by delay while introducing wireless sensor networks into the application of the industrial control field, a method of using delay estimation is proposed. With the delay estimation of minimum deviation derived, the transformation of the delayed signals can be implemented and the transformed signals are taken as the inputs of the logic operation. And the output coordinates with the actual situation. Thus, this method not only works effectively, but also reduces the reliance on time synchronization.

## References

1. Egan, D.: The emergence of ZigBee in building automation and industrial control. Computing & Control Engineering Journal, 6–9 (2005)
2. Hu, L., Chen, X., Chen, H., Chen, Y., Li, Q.: Fuzzy adaptive power control based on PID algorithm for clusterheads of wireless sensor networks. In: IET International Conference on Wireless, Mobile and Multimedia Networks, pp. 1–4 (2006)
3. Swartz, R.A., Chin-Hsiung, L., Lynch, J.P.: Near real-time system identification in a wireless sensor network for adaptive feedback control. In: American Control Conference, ACC 2009, pp. 3914–3919 (2009)
4. Sivrikaya, F., Yener, B.: Time synchronization in sensor networks: a survey. Network IEEE 18(4), 45–50 (2004)
5. Chen, J., Yu, Q., Zhang, Y., Chen, H.-H., Sun, Y.: Feedback-Based Clock Synchronization in Wireless Sensor Networks: A Control Theoretic Approach. IEEE Transactions on Vehicular Technology 59(6), 2963–2973 (2010)
6. Takagi, H., Kleinrock, L.: Throughput Analysis for Persistent CSMA Systems. IEEE Transactions on Communications COM-33(7), 12 (1985)
7. Gkelias, A., et al.: Average Packet Delay of CSMA/CA with Finite User Population. IEEE Communications Letters 9(3), 3 (2005)

# Financing Methods of MBO and Relative Problems

Xin Luo

Department of Accounting
Zhengzhou Institute of Aeronautical Industry Management
Zhengzhou, Henan Province, China
luo_xin33@yahoo.com.cn

**Abstract.** MBO plays a significant role in solving Chinese corporate property structure defects as well as promoting improvements of corporate governance. This is just the focus of this paper that is to discuss and evaluate the MBO financing methods and study some important derivative problems about MBO financing. Firstly, this paper introduced the features of MBO financing. Secondly, some significant issues about MBO were discussed and the paper pointed out some measures to solve these issues as well. The last but not least, one important derivative problem was studied. That was the method of pricing. The conclusion of this paper is that for China, there are some issues about MBO and some measures must be taken to solve them.

**Keywords:** MBO (management buyout), financing obstacles, financing methods, pricing methods.

## 1 Introduction

There are two main financing methods of management buyout (MBO) in current China.

### 1.1 Equity Funds

In the process of management buyout, the equity funds supplied by management itself only take 10% to 20% of all needed funds that is essential for completing the whole management buyout. Compared with the all needs of funds, this number of equity funds is far more to be enough. As a result, in order to finish one management buyout the investors or purchasers often use some external funds to complete the whole process.

### 1.2 External Funds

In the general, the financing methods of external funds often include the following three approaches: loan from financial institutions, issuing bonds and issuing stocks.

The last two financing methods are more easy to illustrate, which just mean that the purchaser of MBO go into the capital market to issue relevant securities to get money, including bonds, preferred stocks and common stocks.

The first financing method refers to that the relevant people apply for loans from some certain financial institutions through taking some assets as the pledges and these

S. Lin and X. Huang (Eds.): CSEE 2011, Part II, CCIS 215, pp. 107–112, 2011.

assets often include inventories and accounts receivable or other current assets. For one thing, the risk of this financing method is lower and commercial banks always intend to be the lenders. For another, the term of this financing method is shorter (1to 3 years) and the interest rate is lower. In general, for all funds of MBO, there are portions of 15%--35% that are from this kind of financing method.

## 2 Issues in the Process of MBO in China

### 2.1 General Problems in the Process of MBO

Firstly, it is difficult for Chinese enterprises to get the loan from commercial banks and other financial institutions when they plan to do MBO. The reason is that both Chinese commercial banks and other financial institutions have some serious restrictions about loans. Some regulations that are constituted by People's Bank of China forbid commercial banks and other financial institutions to lend money to enterprises to use for equity investments. Moreover, some relevant legal clauses also have the same rules. In addition, the current laws forbid enterprises to take some assets or equities as pledges to get loans from financial institutions and then transfer this money to individual persons. Also, in the goal of equity investments, individual persons are forbidden to get money from financial institutions just like enterprises. Finally, other relevant laws have the rules that security companies can not supply money to enterprises for equity investments directly.

Secondly, the conditions of one enterprise to issue securities (such as common stock, preferred stock, bond and convertible bond) are very rigorous. Take bonds for one example. In some relevant Chinese laws, only some qualified enterprises can issue bonds. The first is joint-stock companies whose net assets should not lower than 30,000,000 RMB. The second is limited liability companies whose net assets should not lower than 60,000,000 RMB. In addition, two important conditions must be satisfied. The first is that the cumulative amount of bonds outstanding should be lower than 40% of the issuer's net assets. Another one is that the coupon rate of bonds should be lower than national interest rate. All of these restrictions about issuing scale of bonds and issuing conditions of bonds make it difficult for most enterprises to get money through the financing method of issuing bonds. Other marketable securities have the same characteristics such as rigorous issuing conditions and exorbitant issuing thresholds. All the above mentioned situations make it unrealistic for most enterprises to get money through the financing method of issuing bonds.

The last but not least, the number of active institutional investors is too small. The institutional investor refers to one entity or one institution that holds a large number of bonds or stocks and participates in the process of strategic decisions actively. It is one normal phenomenon in the process of MBO in western developed countries to finance money from certain institutional investors. Nowadays, in Chinese financial market, several institutional investors construct special funds to help enterprises to do MBO, including New China Trust, Shanghai Rongzheng and so on. However, the scale of these funds is still very small compared with an increasingly large capital market.

## 2.2 Measures to Solve These Problems

### 2.2.1 Vigorous Supports of National Regulations and Laws

As one special method of capital operating, MBO needs vigorous supports of government and financial system.

Firstly, many present laws have two obvious characteristics. The first is about "shunt" that means one economic system is changing to another system. Another is about "immature" that means Chinese market economy is still in the primary phase. With the construction and growth of Chinese market economy, two things must be done as soon as possible. The first is to do anything should be according to the requirements of market economy and constrict the range of legal functions of government. Another is to correct current relevant laws according to emerging situations.

Secondly, it is necessary to have a developed financial system. This system should strengthen the construction of bank system and relax some restrictions about the range of commercial banks' operations. At the mean time, the government should relax the control to the bond market and permit enterprises to issue junk securities based on the security rating, which can help one enterprise to get more money to use in the process of MBO.

### 2.2.2 Reform of Financing Methods

Due to the existing of narrow financing channels, management often misses the best opportunity of MBO because they have not the enough money to finish the acquisition, which will bring about some losses. As a resuly, it is necessary to expand financing channels and create some new financing methods.

Firstly, some trust companies should interpose the financing methods of MBO. According to some relevant regulations of People's Bank of China, one trust company can reform some new trust products due to the needs of capital market. Moreover, it is no any legal barriers for one trust company to lend money to one enterprise's management to do MBO. As a result, trust companies can strut their stuff in the field of MBO because the funds of them can be gotten from many sources and their operating activities are very flexible and variable. There are many operating methods to do MBO with the form of trust. For one thing, the entity of MBO is one trust company and the management will not do it directly. Another, the management does the direct financing through using trust loan. In addition, the management can use the combination of equity funds and trust load to do MBO.

Secondly, private-issued funds can play a role in the process of MBO. Funds for MBO are one kind of venture capital. At present, the funds that supply money to the process of MBO generally are private-issued funds. Private-issued funds refer to one type of funds that are issued only to some institutional investors and some rich individuals in the informal form. The sales and redemptions of private-issued funds are completed through negotiations between management of funds and the investors. On the other hand, one indirect financing method can be used by management of one process of MBO, which is private-issued funds can supply money to one trust company. This approach is a lawful activity. In a word, private-issued funds have the stronger flexibility, which is suitable for any projects and fields. In addition, according to its good secrecy, its wide investing fields and its higher annual return rate, private-issued funds are very appropriate for the financing of MBO.

Thirdly, institutional investors should play more important role in the process of MBO, which can optimize the financing structure of MBO in Chinese enterprises. Qualified institutional investors are the key factor and core that can help MBO of China to step in benign growth. The following contents are the major effects of institutional investors in the process of MBO. For one thing, based on the long-term investment, they can impel management to focus on the strategic reconstructions about enterprises' growth in the long-run. Another, based on the equity capital, they can participate some post-MBO operating activities efficiently, including reorganization, supervising, restraint and controlling. This point can help one enterprise to complete structure optimization. In addition, based on the characteristics of specialization, management, human resource and investing experience, they can reduce the blindness of investing and promote the reasonable and healthy growth of MBO market. Finally, based on the large amount of funds and diversified investments, they can reduce the investing risks and defuse financial risks resulted in the individual financing.

# 3 Pricing Problem in MBO

## 3.1 Main Pricing Methods of MBO in China

### 3.1.1 The Method of NAPS

In China, the earliest pricing policies of MBO existed in the relevant regulations taken by the Committee of National Asset Management in 2003. According to these regulations, the transferring price of national shares in listed companies should be taken based on the profitability and the market value. On the mean time, the price can not be lower than one important financial ration that is net assets per share (total net assets divided by numbers of common stock), or NAPS.

In the process of MBO, the pricing method that is based on net assets per share is named "the method of NAPS". However, in practice, it is unrealistic that one enterprise only uses "the method of NAPS" as the pricing method of MBO. The reason is this method has several shortcomings.

Firstly, the real value of national share is not equal to the value of net assets. On the other words, net assets can not be looked on as the proof of the value of national assets. It is very simple to take the value of net assets as the only basis of transferring price of national shares but it is short of truthfulness and effectiveness.

Secondly, the premium of controlling is neglected when the value of net assets is taken as the pricing mechanism of MBO. Based on this method, management will get the control right by use one low price or even free. In China, in the process of MBO, the price paid by management often does not include the factor of control right, which is no difference with the trading of general minority shares.

The last but not least, the factor of risk is neglected when the value of net assets is taken as the pricing mechanism of MBO. In any types of trading, all of political risks, legal risks and financial risks should not be ignored. However, this method does not take the factors of risk into the price.

### 3.1.2 The Method of Income

In addition to the method of NAPS, the method of income is often used in the process of MBO. This method refers that when management value the value of the target company, the incomes that are based on assets are capitalized. On the other words, the normal future incomes of the target company are discounted to get the present value based on one appropriate capitalized rate. However, there still exist some shortcomings about this method of income. Firstly, it is difficult to confirm the appropriate capitalized rate. Secondly, in the analysis of the income, the different figures of income will affect the price directly. Nevertheless, many factors of accounting principles, such as changes of relevant accounting rules, can affect the level of the income, which means there exist uncertainties to the future incomes.

### 3.1.3 The Method of Discounted Cash Flows

The method of discounted cash flows means when management want to know the value of one enterprise, they can discount expected future cash flows to get the present value based on one appropriate discount rate. Compared with the above mentioned two methods, this method has some advantages.

Firstly, MBO belongs to one kind of leverage acquisition, which means that the management or managers of the target company go to purchase the shares by using the debt funds. This activity requires that the target company can get enough cash in the future to meet the debt funds, which can remit the financial pressures.

Secondly, all the factors that can affect the value of the target company, such as expected future cash flows, discount rate and the term, are able to be confirmed based on some sufficient information. As a result, the subjectivity and uncertainty are reduced sharply.

## 3.2 Measures to Solve the Pricing Problems of MBO

### 3.2.1 To Formulate Complete and Compulsory System of Information Disclosure

There are two main reasons for many enterprises to evade open market pricing. Firstly, it is absence of guidelines from explicit policies. Secondly, it is absence of restraints from compulsive laws. Some measures should be taken, such as open of acquisitions and open of information, which can increase the costs of private trading and decrease the occurrence of private trading. Moreover, national assets can be protected.

### 3.2.2 To Develop Intermediaries

To develop intermediaries is one significant direction of the marketization. Management know some factors of one enterprise very well, including financial position, operating income and firm's value. As a result, they can be the dominant position in the process of negotiation. Therefore, in order to avoid the losses caused by asymmetric information, the sellers of MBO should employ professional intermediaries to value the firm's value and confirm the internal value of shares, which can reduce the disadvantages in the process of negotiation.

### 3.2.3 To Complete the Market of Ownerships Trading

Relevant departments should complete the market of ownerships trading as soon as possible and change the situation that only one buyer in the process. Through increasing the numbers of buyer, management will take part on the bid auction of shares as one of several buyers, which is helpful to many aspects.

### 3.2.4 To Complete the Legal Environment

The higher level legislation should be taken because MBO is relevant to the management mechanism of national assets and pricing principles. As a result, the specialized laws and regulations should be formulated. The sufficiency of information disclosure, the transparency of negotiation and the fairness of trading can be ensured through the formulation of relevant laws and stronger monitors.

## References

1. Liu, Y.: Our management buy-out financing difficulties and countermeasures. Science & Technology Industry in China (August 2008) (in Chinese)
2. Liao, Y. P.: Issues in the process of Chinese enterprise MBO. Co-operative Economy & Science (January 2009) (in Chinese)
3. Liu, F.: The discussion of issues in the process of MBO. Caikuai Tongxun (June 2009) (in Chinese)
4. Yu, L.: MBO and loss of state-owned assets. China Business & Trade (June 2009) (in Chinese)

# Synchronous Measuring Device and Its Operating Method for Air Fuel Ratio of CNG Engines

Xiqin Li[1], Ruijin Wang[1], and Bing Liu[2]

[1] School of Mechanical & Automotive Engineering,
Zhejiang University of Science & Technology, Hangzhou, 310023, China
[2] Library, Zhejiang University of Science & Technology, Hangzhou, 310023, China
{lixiqin,liubinglee}@yahoo.com.cn, wrj5188@163.com

**Abstract.** Air fuel ratio has a great influence on emissions, power and economy of a CNG engine. When the CNG engine is at transient conditions, the air fuel ratio can not be precisely controlled at theoretical value by traditional means, which impairs the efficiency of catalytic converter and increases the emissions of CNG engines. To overcome the shortcomings of existing technology, a device and its operating method measuring air fuel ratio of CNG engines synchronously are developed. The moments which intake and exhaust valves open are used as the trigger signals for ECU to collect the test data of CNG engine simultaneously by sensors. Thus the measured air, intaking and than exhausting the cylinder, is at the same working cycle. And the throttle angle, the intake air, the speed and air fuel ratio of the CNG engine are measured simultaneously in transient conditions.

**Keywords:** CNG Engine, Transient Conditions, Air Fuel Ratio, Synchronous Measuring.

## 1 Introduction

Compressed natural gas (CNG) is a cleaner burning alternative to conventional petroleum automobile fuels. Combustion of CNG produces the least amount of CO2 of all fossil fuels. CNG vehicles now generally equip with three-way catalytic converters. According to its characteristics, three-way catalytic converter can simultaneously convert HC, CO and NOX into harmless gases efficiently when air fuel ratio of the CNG engines nears the theoretical value. When CNG engines run at stable condition, the electronic control unit (ECU) can accurately force the air-fuel ratio to near the theoretical air-fuel ratio by detecting the intake air mass and applying closed loop control technology. However when the engines work in transient condition (acceleration or deceleration), the throttle angle changes so suddenly that there is a dynamic air fill-empty phenomena in intake pipe, which leads to the air fuel ratio measured by sensors is no longer the actual value [1][2]. Moreover, the oxygen sensor has response lag. The closed-loop control or the adaptive control methods commonly used can not be adopted to control the air fuel ratio precisely and can not meet the requirements in transient condition[3]. Therefore, the open-loop control method is generally used to control the air fuel ratio in transitional conditions at present. However, due to the restrictions on road conditions, most cars, especially the CNG bus

S. Lin and X. Huang (Eds.): CSEE 2011, Part II, CCIS 215, pp. 113–118, 2011.

or taxi, run in transition conditions in urban areas for most of the time. It is necessary to control the air fuel ratio of CNG engine effectively in transition conditions.

When CNG engines run in transition conditions, the signals exported by throttle position sensor, speed sensor, intake air pressure sensor and air flow sensor only represent engine state parameters at the moment, these parameters will change in a second. Therefore, controlling air fuel ratio by detecting the intake air flow has a certain degree of error. In addition, controlling air fuel ratio by feedbacking the oxygen sensor signals can not meet the requirements of real time in transition conditions. Because the air fuel ratio from oxygen sensor belongs to the air-fuel mixture having burned in this working cycle. The air fuel ratio will change in the next cycle.

This paper takes the moments which intake and exhaust valves open as the trigger signals for ECU to collect the test data simultaneously. The throttle angle, the intake air flow, the speed and air fuel ratio of the CNG engine are measured simultaneously in transient conditions.

## 2 The Principle of Measuring Air Fuel Ratio of CNG Engine

The intake air and the fuel supply system of CNG engine mainly include intake air pipe and fuel pipe. Air fuel ratio $R$ is defined as the ratio of the air mass $Mac$ inhaled into the cylinder to the fuel mass $Mfc$ injected into the cylinder in per cycle, that is to say, $R=Mac/Mfc$. The fuel mass $Mfc$ injected into the cylinder in each cycle can precisely be controlled by ECU. Measuring the air mass $Mac$ inhaled into cylinder accurately in each cycle is the key to achieve simultaneous measurement of air fuel ratio. According to the average value model of engine's intake air, the air mass $Mac$ passing through the throttle is a nonlinear function of throttle angle $\alpha$ and intake air pressure $Pm$ [4], which can be described as:

$$\frac{dMat}{dt} = f_1(\alpha, Pm) \tag{1}$$

The air mass $Mac$ inhaled into the cylinder can be described as a nonlinear function of engine speed $n$ and intake air pressure $Pm$, which can be expressed as:

$$\frac{dMac}{dt} = f_2(n, Pm) \tag{2}$$

When the engine is at transient conditions, there is a dynamic air fill-empty phenomenon in intake air system. So the air mass inhaled into cylinder is not equal the air mass flowing out the cylinder. According to the ideal gas state function, the relationship can be indicated as the following:

$$M_{am} = \frac{PmVm}{RTm} \tag{3}$$

Where: $Mam$ represents the air mass inside the intake pipe, $Vm$ is the volume of intake pipe; $R$ is the gas constant; $Tm$ is the temperature of intake air.

Derivative of the formula (3) on both sides gets:

$$\frac{dPm}{dt} = \frac{RTm}{Vm}\frac{dMam}{dt} + \frac{Pm}{Tm}\frac{dTm}{dt} \tag{4}$$

Taking into account the facts that heat conduction is very small in intake pipe, the pressure changes with time more greatly than the temperature does, the right second term of the formula (4) is ignored. The intake pipe system is taken as a pneumatic pump, the law of mass conservation is applied to the intake pipe system.

$$\frac{dMam}{dt} = \frac{dMat}{dt} - \frac{dMac}{dt} \tag{5}$$

Combining formulas (4) and (5).

$$\frac{dPm}{dt} = \frac{RTm}{Vm}\left(\frac{dMat}{dt} - \frac{dMac}{dt}\right) \tag{6}$$

From formulas (1), (2) and (6), the air mass entering the cylinder through intake pipe in transition condition can be described as:

$$\frac{dMac}{dt} = f_3(n, Pm, \alpha) \tag{7}$$

Formula (7) shows that the air mass *Mac* entering the cylinder in each cycle has a relationship with the engine speed $n$, the intake air pressure $Pm$ and the throttle angle $\alpha$. The inlet air pressure is influenced by atmospheric pressure, the adiabatic coefficient, volumetric efficiency, engine speed and throttle angle. Therefore equation (7) is expressed in complex function.

Analysing the intake process shows the air pressure is mainly affected by engine speed and throttle angle in transition condition [5]. The parameters, such as engine speed, air pressure, intake air temperature and throttle angle, are measured by installing related sensors in intake pipe. Then the air mass entering the cylinder can be calculated according to the parameters.

## 3 Synchronous Measuring Device for Air Fuel Ratio of CNG Engines

Figure 1 shows the composition of the device measuring air fuel ratio of CNG synchronously. An air flow meter, a throttle position sensor, an intake air pressure sensor, an intake air temperature sensor, an intake valve displacement sensor, a exhaust valve displacement sensor, a three-way catalytic converter, a wide range oxygen sensor, a speed sensor and a ECU are installed on a multi-point injecting CNG engine.

The first cylinder of CNG engine is chosen to collect test data. An intake valve displacement sensor and an exhaust valve displacement sensor are mounted on the intake valve and exhaust valve respectively. These sensors detect both valves' actions and provide both valve's opening signals. The signals are a trigger for ECU to collect test data. So the CNG engine state parameters, such as throttle angle, intake air mass, speed and air fuel ratio, are measured simultaneously in transient condition.

ECU detects the signals of throttle angle with throttle position sensor and calculates the derivative of throttle angle. The value of derivative is a criterion to determine whether the engine is at transient condition. If the derivative of throttle angle is larger than the preset value, the engine is at transient condition; otherwise, the engine is at stable condition. A set of sensors, such as air flow meter, air pressure sensor and intake air temperature sensor, are installed in the intake pipe. ECU gets the signals with the sensors and calculates the intake air mass.

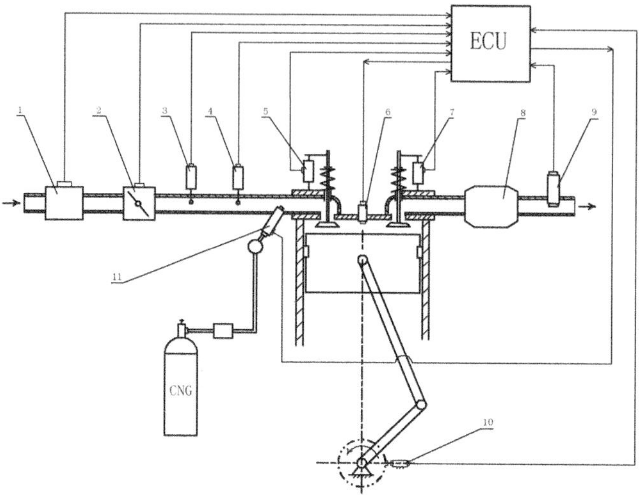


1. air flow meter; 2. throttle position sensor; 3. intake air pressure sensor;
4. intake air temperature sensor; 5.intake valve displacement sensor; 6.spark plug;
7. exhaust valve displacement sensors; 8.three-way catalytic converter;
9.wide range oxygen sensor; 10.speed sensor; 11.injector

**Fig. 1.** Synchronous measuring device for air fuel ratio of CNG engines engine

A wide-range oxygen sensor is installed in the exhaust pipe to test the air fuel ratio of CNG engine. An engine speed sensor is installed on one end of crankshaft.

# 4 Simultaneous Measuring Methods for Air Fuel Ratio of CNG Engine

Figure 2 shows the simultaneous measuring flow process diagram for air fuel ratio of CNG engine. The CNG engine's throttle is changed programmatically at different rates during tests. ECU constantly receives the signals from throttle position sensor and calculates the derivative of throttle angle. When the derivative is larger than the preset value stored in ECU, the engine is at transient condition. At this moment, ECU begins working in preparation for detecting the signals from intake valve position sensor and waiting the signal when the intake valve opens. As soon as the intake valve opens, the ECU sends the commands collecting the signals exported by air flow meter, throttle position sensor, intake air pressure sensor, intake air temperature sensor and engine speed sensor simultaneously. After these signals are converts into digital data by A / D converter, ECU reads and stores away these data in files; When the throttle derivative is less than the preset value, the engine is at stable condition, ECU stops collecting data.

The measured gas, intaking and exhausting first cylinder, come from same working cycle. That is, the collected intake-air dada comes from the air being inhaled the cylinder during intake process, the collected exhaust-air dada comes from the air after

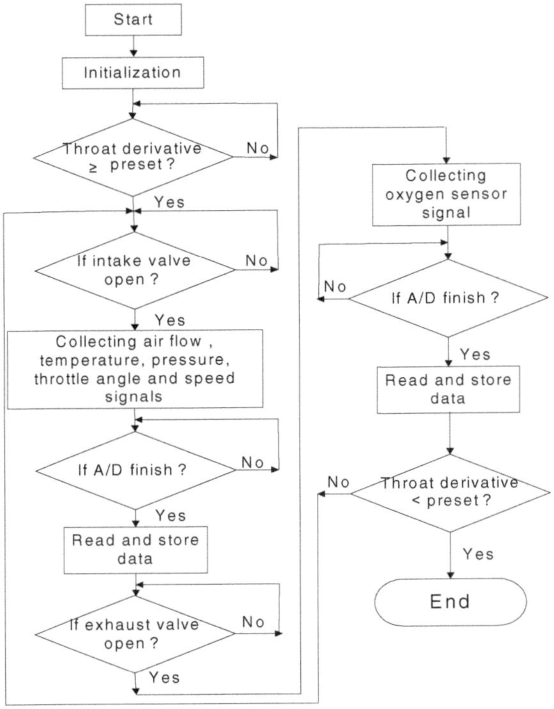


**Fig. 2.** Simultaneous measuring flow process diagram for air fuel ratio of CNG engine

inhalation, compression, combustion, and discharge from the cylinder during exhaust process. The CNG engine working parameters, such as throttle angle, intake air mass, speed and air fuel ratio, are measured simultaneously in transient condition.

## 5 Conclusions

When CNG engine works in transition condition, there is a dynamic air fill-empty phenomenon in intake system, which leads to the air fuel ratio measured by sensors is no longer the actual value. Due to the response lag of oxygen sensor, the closed-loop control or the adaptive control methods commonly used can not control the air fuel ratio precisely and can not meet the requirements in the condition.

This paper takes the derivative of throttle angle as a criterion to determine whether CNG engine is at transition condition. When the derivative is larger than the preset value, the engine is at transient condition; otherwise, the engine is at stable condition. This method is simple and efficient.

The first cylinder of CNG engine is chosen to collect test data. Two displacement sensors are installed on the intake valve and exhaust valve respectively. The sensors detect both valves' actions and provide both valve's opening signals. The signals are a trigger for ECU to collect test data. The measured air mass, entering and then discharging the first cylinder, come from the same working cycle. So the simultaneous measurement of air fuel ratio is achieved.

**Acknowledgements.** This work is sponsored by the Natural Science Foundation of Zhejiang Province (No.Y1110071).

## References


1. Li, D.-g., Shu, Y.-q., Li, X.-z.: Gasoline Engine Transient Air-fuel Ratio Control Based on Modified MVEM. Vehicle Engine 1, 20–23 (2010) (in Chinese)
2. Yao, J., Wu, B., Zhou, D.: Analysis on the Transportation Characteristics of Air Fuel Ratio of Electronic Controlled LPG Engine. Small Internal Combustion Engine and Motorcycle 2, 78–80 (2009) (in Chinese)
3. Arsie, I., Pianese, C., Rizzo, G.: An Adaptive Estimator of Fuel Film Dynamics in the Intake Port of a Spark Ignition Engine. Control Engineering Practice (11), 303–309 (2008)
4. Wu, Y.-h., Hou, Z.-x., Song, D.-d.: Air Fuel Ratio Control of Gasoline Engine Based on Neural Network Multi-Step Predictive Model during Transient Conditions. Journal of Combustion Science and Technology 14(1), 11–15 (2008) (in Chinese)
5. Zhang, J.: Fuzzy Neural Network Control Theory and Application. Harbin Polytechnic University Press, Harbin (2004) (in Chinese)

# Research and Simulation of Linear Instantaneous Blind Signal Separation Algorithm

Xinling Wen

Zhengzhou Institute of Aeronautical Industry Management, 450015, China
wenxinling@zzia.edu.cn

**Abstract.** The process of separating initial source information from mixing signal is called blind signal separation (BSS). This paper mainly introduces the original, development, and style, etc of BSS, as well as analyzes model types of BSS algorithm. We simulate two signal of a sine and a triangle wave, then mixing them by non-linear matrix. Through simulation, based on BSS technology of typical Fast ICA or JADE algorithm, we concluded that the BSS algorithm can get the accuracy separation effect. However, because the observed signals in actual project may be non-linear mixing signal, and hard to determine, BSS algorithm of further research is very significant.

**Keywords:** Blind signal separation (BSS), Signal process, ICA, JADE, Linear instantaneous model.

## 1 Introduction

In real life, we observed signals are often unknown, multi-input and multi-output linear system signal. If have not other priori information, only according to the source signal statistically independent of each other, use the observed signals to realize the mixed stack together separation of source signals, which is called the blind source separation (BSS). BSS [1] is a process that returned from the mixed signal to recover cannot directly observation initial signal. Here, the 'blind' means two aspects, the source signal can not be observed and mixing system is unknown. Obviously, when we are hard to build mathematic model from source to sensor based on the transmission properties, or hard to establish a priori knowledge about the transmission or unable to obtain, BSS is a very natural selection. The core of BSS is matrix learning algorithm of separation (or solution mixed), the basic idea is act statistically independent feature extraction as input, and the representation of information is not lost. [2]

## 2 The Development of BSS Technology

BSS is a powerful signal processing method, which is early researched in late 1980s of 20th century. Due to the BSS has wide application foreground in the past decade in wireless communication, medical analysis, speech recognition, image processing [3] fields, the relevant theories and algorithms are rapidly development.

S. Lin and X. Huang (Eds.): CSEE 2011, Part II, CCIS 215, pp. 119–124, 2011.

The international first researcher of study BSS method is Treichler and Larimore, they put forward with CMA can finish the source signal blind source separation and channel equilibrium in the same time, but this algorithm is unstable in much more signal mixing. [4] In the late 1980s, the French scholars Herault and Jutten put forward the basic concept of independent component analysis as well as independent components extraction nonlinear irrelevant law [5]. This method is based on feedback neural network by choosing odd nonlinear function constitute, which achieves the Hebb training, and the purpose of signal separation. But this method cannot accomplish more than two aliasing source signal separation, non-linear function selection is arbitrary, and lack of theory explanation. Linskesr also put forward greatly standards unsupervised learning thought based on the information transmission of neural network in 1988. [6] In 1991, Herault and Jutten published separation algorithm [7] [8] [9], which is considered the start of the study in the world. They pioneered the artificial neural network algorithm used to blind separation problem, which ushering a new field. In 1994, Comon puts forward independent component analysis (ICA) algorithm, provide a new thought for BSS, defined the basic assumptions problem of solving method of ICA, clearly defines the concept of independent component analysis, and presents an effective independent component analysis algorithm [10]. Tong present the perfect separation mathematical framework and analyzes separation problem reparability and uncertainty [11], and proposed decomposition methods based on higher order statistics matrix algebra features, which can solve transmission of linear algebra method. In 1995, Bell and Sejnowski using neural network, non-linear characteristics to eliminate the statistical correlation of high-order observed signals, and make the problem of BSS into information framework, let information maximization criterion to construct the goal function (Infomax algorithm). This paper gives the neural network optimal weight matrix iterative learning algorithm, which becomes the follow various algorithm basic. [12] In 1997, Cichocki improved Herault-Jutten algorithm network, the source signal mixed in worse scale and ill-posed problems are got better solution. [13] In 1999, Hyvarinen provides a quick nod algorithm based on the source signal gaussian measure, [14] this algorithm can extract single or multiple source signals according to the size order of non-gaussian character, which became the most representative batch algorithm, and has a very fast convergence speed, so, it is also called Fast ICA. In addition, the other scholars proposed method in different angles without using the perspective of the information theory to solve the separation question, Pham and Garat, etc first proposed maximum likelihood method [15]. Girolami proposed negative entropy maximization method [16]. Amari, Kung and Cichocki given respectively the source signal extracted algorithm in order based on kurtosis method, which has good effect. However, the algorithm has problems of stability. At the same time, non-linear PCA algorithm and maximum likelihood ICA algorithm is presented.

## 3 The Application of BSS Linear Instantaneous Model

According to the difference of mixing process, the mathematical model of BSS can be divided into two kinds, namely linear blind separation that source signal passing through linear mixing and source signal passing through the non-linear mixing. Linear blind separation can divide again into signal delay instantaneous linear separation and the signal delay convolution blind source separation. Here, we only introduce the linear model without signal delay instantaneous researches and simulation.

Linear instantaneous model is the simplest and most basic model and its principle is shown as Fig.1.[17]

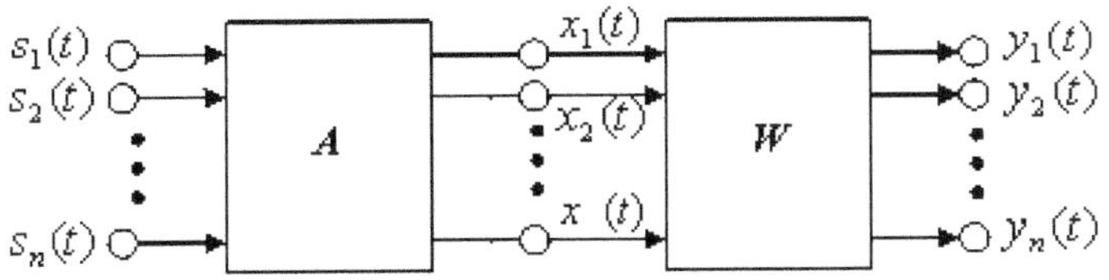


**Fig. 1.** BSS linear instantaneous model

If we suppose $n$ number of unknown source signal of $s_1(t), s_2(t), \cdots, s_n(t)$ $(m \geq n)$, during mixed stack matrix $\boldsymbol{A}$ and got $m$ number mixed observation signal (or sensor signal) of $x_1(t), x_2(t), \cdots, x_m(t)$. Usually, observation signal can express as formula (1).

$$x_i(t) = \sum_{j=1}^{n} a_{ij} s_j(t) \tag{1}$$

Among formula (1), $a_{ij}$ is mixing coefficient, and we can present it in matrix style as formula (2).

$$\boldsymbol{x}(t) = \boldsymbol{A}\boldsymbol{s}(t) \tag{2}$$

$\boldsymbol{s}(t) = [s_1(t), s_2(t), \cdots, s_n(t)]^T$ is a $n$ dimension source signal vector, $\boldsymbol{x}(t) = [x_1(t), x_2(t), \cdots, x_m(t)]^T$ is a $m$ dimension mixed signal vector, and $\boldsymbol{A}$ is *a* $m \times n$ dimension matrix. The basic idea of BSS is that in the premise of source signal $\boldsymbol{s}(t)$ and mixed matrix $\boldsymbol{A}$ unknown, only based on observation signal $\boldsymbol{x}(t)$ to look for a separable matrix $\boldsymbol{W}$, and through constant adjustment separable matrix $\boldsymbol{W}$, at last get a source signals as accurate estimate.

$$\boldsymbol{y}(t) = \boldsymbol{W}\boldsymbol{x}(t) \tag{3}$$

$y(t)$ is the best estimation of source signal $\boldsymbol{s}(t)$. From Figure1, the front part is unknown source signal mixing process, this part is blind to signal processing port, and the back part is blind signal separation process, we can get separated signal through the processing observed signals.

Linear instantaneous mixture model algorithm is the most simple and basic algorithm of BSS, the other model algorithm is obtained based on linear instantaneous model. According to the different ways of signal extracted, BSS algorithm can be divided into batch processing algorithm, adaptive algorithm and detect projection chasing method, etc. [17].

Batch processing method is based on higher order statistics joint approximate diagonalization of eigenmatrices (JADE) method, JADE is the method of fourth-order accumulation learning algorithm, which is a kind of typical offline algorithm, to the number of various situations, all have certain function, but with the increases of source signal interrelation, JADE algorithm increases, the separation effect becomes poor. FOBI algorithm uses four order statistics to decompose independent source and identify mixed matrix method, it is essentially JADE predecessor.

Adaptive algorithm handle observation signal with increasing input signal, which makes results to approach expected output gradually, is a kind of online processing method. Currently combining with neural network adaptive algorithm is the research focus. Adaptive includes maximum informatization method (Informax) and the expansion of maximum informatization method (EnInformax) [18].

In addition, projection chasing method separates one group of source signal every times, and cannot separate all source signals in one time. The ICA anchors algorithm is a kind of typical order extracts independent source algorithms. After fourth-order cumulate is improved, brings forward algorithm based on negative primogeniture fixed algorithm. The algorithm adopts fixed-point iterative optimization algorithm and the convergence speed is faster than the adaptive algorithm, also known as 'Fast ICA algorithm'. Later, negative maximum Fast ICA algorithm primogeniture was expanded to complex field, which can solve the complex signal mixed separation problem. At present, the algorithm is the most widely used blind processing algorithm. Fast ICA algorithm is based on non-gaussian maximization principle, using the theory of anchors iteration for non-gaussian maximum [19].

## 3 Simulations and Analysis

In order to understand the separation algorithm's role in separated signal, we using two signals and blending, by using Fast ICA algorithm and JADE algorithm to observe accuracy of BSS separation algorithm. We assume a sine signals and a triangle wave respectively, which is shown a Fig.2.

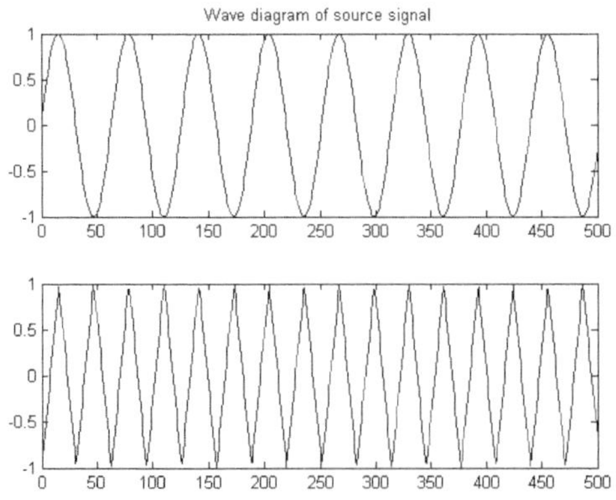


**Fig. 2.** Wave diagram of source signal

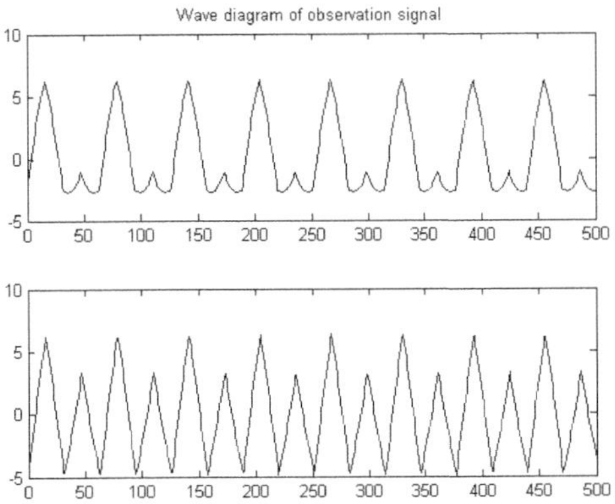


**Fig. 3.** Wave diagram of observation signal

We mix the two signals in a 2×2 matrix $\boldsymbol{A}$ to get observation signal, observation signal is shown as Fig.3. Through Fast ICA separation algorithm and JADE separation algorithm BSS application, the result of separated signal is shown as Fig.4. From the Figure 4 we can see, BSS algorithm can get better identification effect.

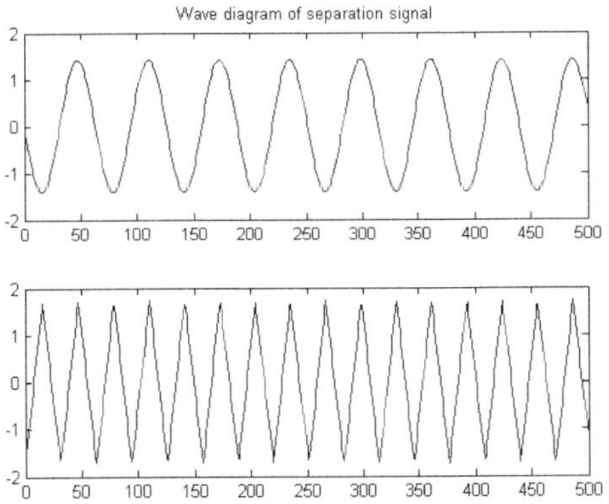


**Fig. 4.** Wave diagram of BSS signal

## 5 Conclusions

Although development of BSS technology has more mature, but because most of the algorithm is based on assumptions, the blind source separation algorithm is hard to realize in some special situation. Therefore, the study also should be in the more common non-linear mixed circumstances in quickly and effective separation algorithm. In addition, mixing matrix is full rank, or source signal is unknown, and in all kinds of noise circumstance, research of BSS algorithm is future research and focus.

**Acknowledgment.** This paper is supported by the Aeronautical Science Foundation in China. (No.2010ZD55006)

## References

1. Jutten, C., Herault, J.: Space or time adaptive signal processing by neural network models. In: Intern. Conf. on Neural Network for Computing. Snowbird (Utah,USA), pp. 206–211 (1998)
2. Li, G., Zhang, J., Mao, Y.: Development and Research Condition of BSS. Aeronautical Electronic Countermeasure 13 (2006)
3. Gao, Y., Xie, S.: An Algorithm for Blind Signal Extraction of Linear Mixture. Journal of Electronics & Information Technology 999 (2006)
4. Hu, J., Zhang, G., Xiong, G.: BSS Separation Technology. Digital Communication World 67 (2010)
5. Herault, J., Jutten, C.: Space or time adaptive signal processing by neural network models. In: AIP Conference Proceedings, p. 151 (1986)
6. Linskesr, R.: Self-organization in a perceptual network. Computer, 105–117 (1988)
7. Jutten, C., Herault, J.: Blind separation of sources, Part I: An adaptive algorithm base on neuromimatic architecture. Signal Processing, 1–10 (1991)
8. Common, P., Jutten, C., Herault, J.: Blind separation of sources, Part II: Problems statement. Signal Processing, 11–20 (1991)
9. Jutten, C., Herault, J.: Blind separation of sources, Part I: An adaptive algorithm based on neuromimetic architecture. Signal Processing, 21–30 (1991)
10. Comon, P.: Independent component analysis, a new concept. Signal Processing, 287–314 (1994)
11. Tong, L., Li, R.W., Soon, V.C.: Indeterminacy and identifiability of blind identification. IEEE Trans. On Circuits and Systems, 499–509 (1991)
12. Bell, A.J., Sejnowski, T.J.: An information-maximization approach to blind separation and blind deconvolution. Neural Computation, 1004–1034 (1995)
13. Cichocki, A., Bogner, R.E., Moszczynski, L.: Modified Herault-Jutten Algorithms for Blind Separation of Sources. Digital Signal Processing, 80–93 (1997)
14. Hyvarinen, A.: Fast and Robust Fixed-Point Algorithms for Independent Component Analysis. IEEE Trans. On Neural Networks, 626–634 (1999)
15. Pham, D.T., Garat, P.: Blind separation of mixture of independent sources through a quasi-maximum likelihood approach. IEEE Trans. Signal Processing, 1712–1725 (1997)
16. Girolami, M., Fyfe, C.: Kurtosis extrema and identification of Independent components: a neural network approach. In: Proc. ICASSP 1997, pp. 3329–3333 (1997)
17. Zhang, X., Du, Z., Zhou, J.: Blind Source Separation Technique and Its Application. Elementary Electroacoustics 4 (2004)
18. Zhou, Z., Chen, H.: Research and Survey on Algorithms of Blind Signal Separation Technology. Computer Science, 16–19 (2009)
19. Ximing, C., Shuoyi, H.: Blind Source Separation :Problem, Principle and Method, Electronic information against technology, 1–4 (2008)

# The CAT Bézier Curves

Jin Xie[1], XiaoYan Liu[2], and LiXiang Xu[1]

[1] Department of Mathematics and Physics, Hefei University, 230601Hefei, China
[2] Department of Mathematics and Physics, University of La Verne, 91750 La Verne, USA

**Abstract.** A class of cubic Bézier-type curves based on the blending of algebraic and trigonometric polynomials, briefly CAT Bézier curves, is presented in this paper. The CAT Bézier curves retain the main superiority of cubic Bézier curves. The CAT Bézier curves can approximate the Bézier curves from the both sides, and the shapes of the curves can be adjusted totally or locally. With the shape parameters chosen properly, the introduced curves can represent some transcendental curves exactly.

**Keywords:** C-Bézier curve, CAT-Bézier curve, shape parameter, local and total control, transcendental curve.

## 1 Introduction

In CAGD, the cubic Bézier curves, have gained wide-spread applications, see [1]. However, the positions of these curves are fixed relative to their control polygons. Although the weights in the non-uniform rational B-spline (NURBS) curves, have an influence on adjustment of the shapes of the curves, see [2]. For given control points, changing the weights to adjust the shape of a curve is quite opaque to the user. On the other hand, it is also noticed that those curves based on algebraic polynomials have many shortcomings, especially in representing transcendental curves such as the cycloid and helix, etc. Hence, some author proposed new methods in the space of mixed algebraic and non-algebraic polynomials, see [3-10]. Pottman and Wagner [4] investigated helix splines. GB-splines were constructed in [5] for tension generalized splines allowing the tension parameters to vary from interval to interval and the main results were extended to GB-splines of arbitrary order in [6]. Han, Ma and Huang [7] studied the cubic trigonometric Bézier curve with two shape parameters, the shape of which can be adjusted locally by shape parameters. Zhang [8, 9] constructed the C-B curves based on the space span{1, t, sint, cost}, which can precisely represent some transcendental curves such as the helix and cycloid, but can only approximate the curves from the single side. Wang, Chen and Zhou studied non-uniform algebraic-trigonometric B-splines ($k \geq 3$) for space span{$1, t, t^2 \ldots t^{k-2}$, sint, cost } in [10]. Note that these existing methods can deal with some polynomial curves and transcendental curves precisely. But the forms of these curves are adjusted by shape parameters with complicated procedures.

In this paper, we present a class of cubic Bézier-type curves with two shape parameters based on the blending of algebraic and trigonometric polynomials. This approach has the following feature: 1.The introduced curves can approximate the

S. Lin and X. Huang (Eds.): CSEE 2011, Part II, CCIS 215, pp. 125–130, 2011.

Bézier curve from the both sides, and the change range of the curves is wider than that of C-Bézier curves. 2. The shape of the curves can be adjusted totally or locally. 3. With the shape parameters and control points chosen properly, the CAT Bézier curves can be used to represent some transcendental curves such as helix and cycloid.

## 2 The CAT Bézier Base Functions and Their Related Propositions

**Definition 2.1.** For two arbitrarily selected real values $\lambda$ and $\mu$ ,where $\lambda,\mu \le \frac{\pi^2}{\pi^2-8} =$ 5.27898, the following four functions in variable t with are called CAT-Bézier base functions:

$$\begin{cases} b_{0,3}(t) = \dfrac{(3-2\lambda)\pi - 3\pi t + 3\pi\lambda t^2 - \pi\lambda t^3 + 6(\lambda-1)\cos\left(\frac{\pi}{2}t\right)}{3\pi - 6 + (6-2\pi)\lambda}, \\ b_{1,3}(t) = \dfrac{2-\pi+\pi t - 2\left(\sin\left(\frac{\pi}{2}t\right) - \cos\left(\frac{\pi}{2}t\right)\right)}{4-\pi} - b_{0,3}(t), \\ b_{2,3}(t) = \dfrac{2-\pi t + 2\left(\sin\left(\frac{\pi}{2}t\right) - \cos\left(\frac{\pi}{2}t\right)\right)}{4-\pi} - b_{3,3}(t), \\ b_{3,3}(t) = \dfrac{3(1-\mu)\pi t + \pi\mu t^3 + 6(\mu-1)\sin\left(\frac{\pi}{2}t\right)}{3\pi - 6 + (6-2\pi)\mu}. \end{cases} \tag{1}$$

It can be verified by straight calculations that these CAT Bézier base functions possess the properties similar to those of the cubic Bernstein base functions.

(a) Properties at the endpoints:

$$\begin{cases} b_{0,3}(0) = 1 \\ b_{3,3}(1) = 1 \end{cases}, \quad \begin{cases} b_{i,3}^{(j)}(0) = 0 \\ b_{3-i,3}^{(j)}(1) = 0 \end{cases}, \tag{2}$$

where $j = 0,1,\cdots,i-1, i = 1,2,3$ ,and $b_{i,3}^{(0)}(t) = b_{i,3}(t)$ .

(b) Symmetry:

$$b_{i,3}(t;\lambda,\mu) = b_{3-i,3}(1\text{-}t;\lambda,\mu) \text{ ,for } i = 0,1,2,3. \tag{3}$$

(c) Partition of unity:

$$\sum_{i=0}^{3} b_{i,3}(t) = 1. \tag{4}$$

**Proposition 2.1.** These CAT Bézier base functions are nonnegative for $\lambda,\mu \le \frac{\pi^2}{\pi^2-8}$, namely,

$$b_{i,3}(t) \ge 0, \; i = 0,1,2,3 \tag{5}$$

**Proof:** By simple computation, for $t \in [0,1]$ and $\lambda \le \frac{\pi^2}{\pi^2-8}$, we have $b_{0,3}'(t) \le 0$ , so the non-negativity of $b_{0,3}(t)$ can be easily proved.

For $t \in [0,1]$ and $\lambda < \frac{3\pi-6}{2\pi-6}$, since $b_{1,3}(0) = b_{1,3}(1) = b'_{1,3}(1) = 0$, $b'_{1,3}(0) = \frac{3\pi}{(3\pi-6)-(2\pi-6)\lambda} > 0$, and $b''_{1,3}(1) = \frac{\pi^2}{8-2\pi} > 0$, if $b_{1,3}(t)$ fails to retain the same sign in the interval (0,1), then $b'_{1,3}(t)$ has at least four zeros in [0,1]. So

$$b'''_{1,3}(t) = \left(\frac{\pi^3}{16-4\pi} - \frac{6\pi\lambda}{(3\pi-6)-(2\pi-6)\lambda}\right) cos\frac{\pi}{2}t + \frac{6(1-\lambda)+\pi\lambda}{(16-4\pi)((3\pi-6)-(2\pi-6)\lambda)} sin\frac{\pi}{2}t$$

has at least two zeros in (0,1), which contradicts the properties of trigonometric functions. Since $b_{1,3}(t)$ retains the same sign and $b_{1,3}(0.5) = \frac{24\sqrt{2}(\lambda-1)+(12-11\lambda)\pi}{8((3\pi-6)-(2\pi-6)\lambda)} > 0$ for $\lambda < \frac{3\pi-6}{2\pi-6}$, the property of non-negativity of $b_{1,3}(t)$ is proved. By symmetry, the non-negativity of $b_{3,3}(t)$ and $b_{2,3}(t)$ can be proved similarly.

So, for $t \in [0,1]$ and $\lambda, \mu \le \frac{\pi^2}{\pi^2-8}$, $b_{i,3}(t) \ge 0$, $i = 0,1,2,3$.

Fig.1 shows the figures of the four CAT-Bézier base functions and C-Bézier base functions.

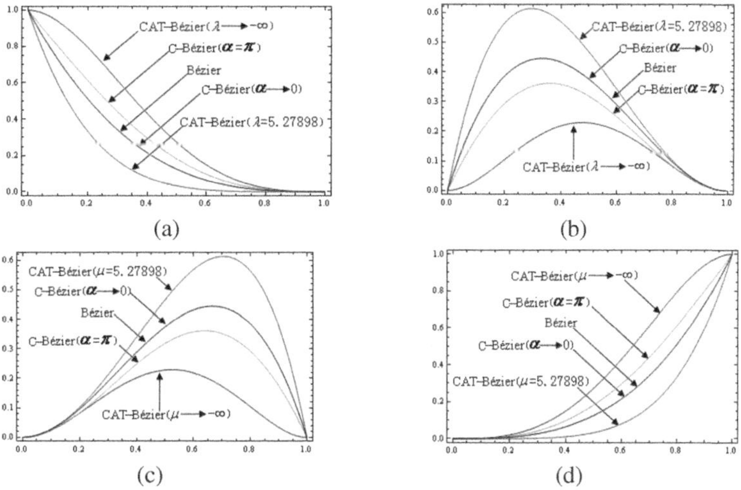


**Fig. 1.** The figures of four CAT-Bézier base functions

# 3 The CAT Bézier Curves and Their Related Properties

## 3.1 Construction of the CAT-Bézier Curves

**Definition 3.1.** Given points $\boldsymbol{P}_i\,(i = 0,1,2,3)$ in $R^2$ or $R^3$, then

$$\boldsymbol{r}(t) = \sum_{i=0}^{3} \boldsymbol{P}_i b_{i,3}(t) \tag{6}$$

is called a CAT Bézier curve, where $t \in [0,1]$, $\lambda, \mu \le 5.27898$, $b_{i,3}(t)$, $i=0,1,2,3$ are the CAT Bézier base functions.

From the definition of the base functions, some properties of the CAT-Bézier curves can be obtained as follows:

**Theorem 3.1.** The CAT-Bézier curve (6) has the following properties:
(a) Terminal properties :

$$\begin{cases} \boldsymbol{r}(0)=\boldsymbol{P}_0, \boldsymbol{r}(1)=\boldsymbol{P}_3, \\ \boldsymbol{r}'(0)=\dfrac{3\pi}{3\pi-6+(6-2\pi)\lambda}(\boldsymbol{P}_1-\boldsymbol{P}_0), \boldsymbol{r}'(1)=\dfrac{3\pi}{3\pi-6+(6-2\pi)\mu}(\boldsymbol{P}_3-\boldsymbol{P}_2). \end{cases} \tag{7}$$

(b) Symmetry:
$\boldsymbol{P}_0, \boldsymbol{P}_1, \boldsymbol{P}_2, \boldsymbol{P}_3$ and $\boldsymbol{P}_3, \boldsymbol{P}_2, \boldsymbol{P}_1, \boldsymbol{P}_0$ determine the same CAT-Bézier curve in different parameterizations, i.e.

$$\boldsymbol{r}(t,\lambda,\mu;\boldsymbol{P}_0,\boldsymbol{P}_1,\boldsymbol{P}_2,\boldsymbol{P}_3)=\boldsymbol{r}(1-t,\lambda,\mu;\boldsymbol{P}_3,\boldsymbol{P}_2,\boldsymbol{P}_1,\boldsymbol{P}_0). \tag{8}$$

(c) Geometric invariance:
The shape of a CAT Bézier curve is independent of the choice of coordinates, namely, (3.1) satisfies the following two equations:

$$\boldsymbol{r}(t,\lambda,\mu;\boldsymbol{P}_0+\boldsymbol{q},\boldsymbol{P}_1+\boldsymbol{q},\boldsymbol{P}_2+\boldsymbol{q},\boldsymbol{P}_3+\boldsymbol{q}) = \boldsymbol{r}(t,\lambda,\mu;\boldsymbol{P}_0,\boldsymbol{P}_1,\boldsymbol{P}_2,\boldsymbol{P}_3)+\boldsymbol{q}, \tag{9}$$

$$\boldsymbol{r}(t,\lambda,\mu;\boldsymbol{P}_0*\boldsymbol{T},\boldsymbol{P}_1*\boldsymbol{T},\boldsymbol{P}_2*\boldsymbol{T},\boldsymbol{P}_3*\boldsymbol{T})=\boldsymbol{r}(t,\lambda,\mu;\boldsymbol{P}_0,\boldsymbol{P}_1,\boldsymbol{P}_2,\boldsymbol{P}_3)*\boldsymbol{T}, \tag{10}$$

where $\boldsymbol{q}$ is an arbitrary vector in $\boldsymbol{R}^2$ or $\boldsymbol{R}^3$, and $\boldsymbol{T}$ is an arbitrary $\boldsymbol{d}\times\boldsymbol{d}$ matrix, $\boldsymbol{d}$ = 2 or 3.
(d) Convex hull property:
The entire CAT-Bézier curve segment must lie inside its control polygon spanned by $\boldsymbol{P}_0, \boldsymbol{P}_1, \boldsymbol{P}_2, \boldsymbol{P}_3$.

## 3.2 Shape Control of the CAT-Bézier Curves

By introducing the shape parameters $\lambda$ and $\mu$, the CAT-Bézier curves possess the better representation ability than C-Bézier curves.

As shown in Fig.2, when the control polygon is fixed, by letting $\lambda$ be equal to $\mu$ and adjusting the shape parameter from -∞ to 5.27898, the CAT-Bézier curves can range from below the C-Bézier curves to above the cubic Bézier curves. And, the shape parameters are of the property that the larger the shape parameter is, the more closely the curves approximate the control polygon.

For $t \in [0,1]$, we rewrite (6) as follows:

$$\begin{aligned} \boldsymbol{r}(t)=&\frac{2-\pi(1-t)+2\left(cos\left(\frac{\pi}{2}t\right)-sin\left(\frac{\pi}{2}t\right)\right)}{4-\pi}\boldsymbol{P}_1+\frac{2-\pi t+2\left(sin\left(\frac{\pi}{2}t\right)-cos\left(\frac{\pi}{2}t\right)\right)}{4-\pi}\boldsymbol{P}_2 \\ &+b_{0,3}(t;\lambda)(\boldsymbol{P}_0-\boldsymbol{P}_1)+b_{3,3}(t;\mu)(\boldsymbol{P}_3-\boldsymbol{P}_2). \end{aligned} \tag{11}$$

Obviously, shape parameters $\lambda$ and $\mu$ only affect curves on the control edge $\boldsymbol{P}_0\boldsymbol{P}_1$ and $\boldsymbol{P}_2\boldsymbol{P}_3$ respectively. In fact, from (11), we can adjust the shape of the CAT-Bézier curves locally: as $\mu$ increases and $\lambda$ fixes, the curve moves in the direction of the edge $\boldsymbol{P}_2\boldsymbol{P}_3$; as $\mu$ decreases and $\lambda$ fixes, the curve moves in the opposite direction to the edge $\boldsymbol{P}_2\boldsymbol{P}_3$, as shown in Fig.3. The same effects on the edge $\boldsymbol{P}_0\boldsymbol{P}_1$ are produced by the shape parameter $\lambda$.

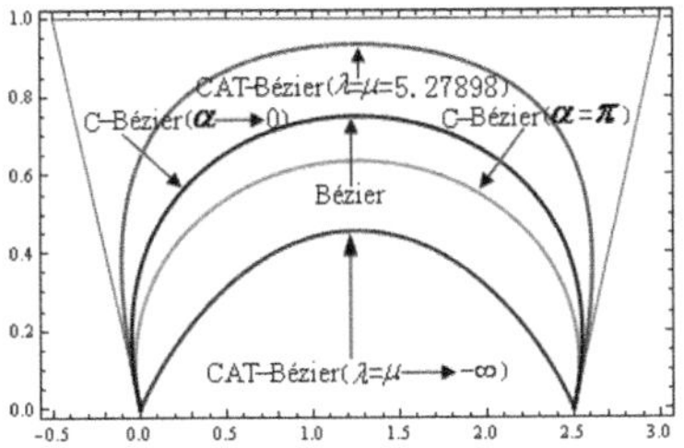


**Fig. 2.** Adjusting the curves totally

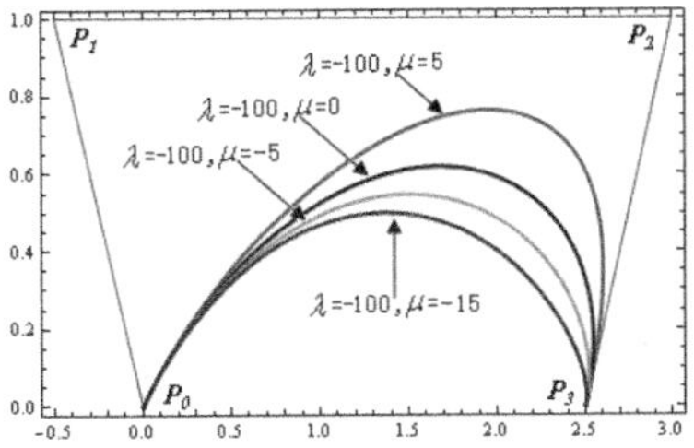


**Fig. 3.** Adjusting the curves locally

# 4 The Applications of the CAT-Bézier Curves

In this section, we can represent some transcendental curves by means of CAT-Bézier curves precisely.

**Proposition 4.1.** Let $P_0$, $P_1$, $P_2$ and $P_3$ be four control points as follows,

$$\boldsymbol{P}_0=(0,0),\boldsymbol{P}_1=\left(-\frac{(2-\pi)^2}{2\pi}a,0\right),\boldsymbol{P}_2=\left(\frac{2-\pi}{\pi}a,\frac{4-\pi}{2}a\right),\boldsymbol{P}_3=(0,a)(a\neq 0).$$

Then the corresponding CAT-Bézier curve with the shape parameters $\lambda=\mu=0$ and $t\in[0,1]$ represents an arc of cycloid.

**Proof.** Substituting $\boldsymbol{P}_0=(0,0),\boldsymbol{P}_1=\left(-\frac{(2-\pi)^2}{2\pi}a,0\right),\boldsymbol{P}_2=\left(\frac{2-\pi}{\pi}a,\frac{4-\pi}{2}a\right),\boldsymbol{P}_3=(0,a)$ into (6) yields the following coordinates of the CAT-Bézier curve,

$$\begin{cases} x(t)=a\left(t-sin\frac{\pi}{2}t\right), \\ y(t)=a\left(1-cos\frac{\pi}{2}t\right), \end{cases} \tag{12}$$

which is a parametric equation of a cycloid.

**Proposition 4.2.** Let $P_0$, $P_1$, $P_2$ and $P_3$ be four properly chosen control points such that

$$\boldsymbol{P}_0=(0,a,0),\boldsymbol{P}_1=\left(\frac{\pi-2}{2}a,a,\frac{\pi-2}{\pi}b\right),\boldsymbol{P}_2=\left(a,\frac{\pi-2}{2}a,\frac{2b}{\pi}\right),\boldsymbol{P}_3=(a,0,b)(a\neq 0,b\neq 0).$$

Then the corresponding CAT-Bézier curve with the shape parameters $\lambda=\mu=0$ and $t\in[0,1]$ represents an arc of a helix.

**Proof.** Substituting $\boldsymbol{P}_0=(0,a,0),\boldsymbol{P}_1=\left(\frac{\pi-2}{2}a,a,\frac{\pi-2}{\pi}b\right),\boldsymbol{P}_2=\left(a,\frac{\pi-2}{2}a,\frac{2b}{\pi}\right),\boldsymbol{P}_3=(a,b)$ into (6) yields the coordinates of the CAT-Bézier curve

$$\begin{cases} x(t)=a\,cos\frac{\pi}{2}t, \\ y(t)=a\,sin\frac{\pi}{2}t, \\ z(t)=bt, \end{cases} \tag{13}$$

which is the parametric equation of a helix.

**Remark:** By selecting proper control points and shape parameters, one can show that the segments of sine, cosine curve and ellipse can also be represented via CAT-Bézier curves.

## 5 Conclusions

As mentioned above, the CAT-Bézier curves have all the properties of C-Bézier curves. What is more, the introducing curves can precisely represent some transcendental curves. Also, we can adjust the shapes of the CAT-Bézier curves totally or locally. Because there is hardly any difference in structure between a CAT-Bézier curve and a cubic Bézier curve, it is not difficult to adapt a CAT-Bézier curve to a CAD/CAM system that already uses the cubic Bézier curves.

**Acknowledgments.** This work is supported by the National Nature Science Foundation of China (No.61070227), the Key Project Foundation of Teaching Research of Department of Anhui Province (No.20100935).

## References

1. Farin, G.: NURBS curves and surfaces. A.K.Peter, Wellesley (1995)
2. Shi, F.Z.: Computer aided geometric design & non-uniform rational B-spline. Higher Education Press, Beijing (2001)
3. Peña, J.M.: Shape Preserving Representations for Trigonometric Polynomial Curves. Comput. Aided Geom. Design 14, 5–11 (1997)
4. Pottman, H., Wagner, M.G.: Helix splines as an example of affine tchebycheffian splines. Adv. Comput. Math. 2, 123–142 (1994)
5. Kvasov, B.I.: GB-splines and their properties. Ann. Numer. Math. 3, 139–149 (1996)
6. Kvasov, B.I., Sattayatham, P.: GB-splines of arbitrary order. J.Comput.Appl. Math. 104, 63–88 (1999)
7. Han, X.L., Ma, Y., Huang, X.: The cubic trigonometric Bézier curve with two shape parameters. Appl. Math. Lett. 22, 226–231 (2009)
8. Zhang, J.W.: C-curves:An Extension of Cubic Curves. Comput. Aided Geom. Design 13, 199–217 (1996)
9. Zhang, J.W.: Two different forms of C-B-splines. Comput. Aided Geom. Design 14, 31–41 (1997)
10. Wang, G.Z., Chen, Q.Y., Zhou, M.H.: NUAT-spline curves. Comput. Aided Geom. Design 14, 31–41 (1997)

# A Secure Searchable Public Key Encryption Scheme with a Designated Tester against Keyword Guessing Attacks and Its Extension

Chengyu Hu[1] and Pengtao Liu[2]

[1] School of Computer Science and Technology, Shandong University, P.R. China
hcy@mail.sdu.edu.cn
[2] Institute of Information Science and Technology,
Shandong Institute of Political Science, P.R. China
ptwave@163.com

**Abstract.** Public key encryption with keyword search enables user to send a trapdoor to a server that will enable the server to locate all encrypted messages containing the keyword *W*, but learn nothing else. In a searchable public-key encryption scheme with a designated tester (dPEKS), only the designated server can test which dPEKS ciphertext is related with a given trapdoor by using his private key. PEKS/dPEKS scheme does not allow the user to decrypt the encrypted keyword or decrypt the encrypted massage which limits its applicability. Decryptable searchable encryption which enables decryption can resolve this problem. In this paper, we study the keyword guessing attack of dPEKS and propose an enhanced secure searchable public key encryption scheme. At last, we extend the dPEKS scheme to a decryptable searchable encryption scheme with designated tester.

**Keywords:** searchable encryption scheme, decryptable searchable encryption scheme, designated tester, keyword guessing attack.

## 1 Introduction

With the development of internet technologies, the amount of sensitive data to be stored and managed on networked servers rapidly increases. To ensure the privacy and confidentiality of sensitive data from even inside attackers such as a malicious system administrator of a server, a user may encrypt the sensitive data before storing the data into a database server. However, this renders a server unable to perform searches for retrieving the data upon a query from a user.

To resolve this problem, in 2004, Boneh et al. [1] proposed the concept of public key encryption with keyword search scheme(PEKS) to enable one to search encrypted keywords without compromising the security of the original data and proposed a universal transformation from anonymous identity-based encryption (IBE) [2, 3, 4] to PEKS. Abdalla et al. presented an improved universal transformation from anonymous IBE to PEKS and a novel expansion of PEKS that public-key encryption with temporary keyword search (PETKS) [5]. To achieve combinable multi-keyword search, two schemes with conjunctive keyword search (PECKS) [6,7] were respectively

S. Lin and X. Huang (Eds.): CSEE 2011, Part II, CCIS 215, pp. 131–136, 2011.

proposed. In [8], Baek et al. proposed a searchable public key encryption for a designated tester (dPEKS) in which only the server can test whether or not a given dPEKS ciphertext is related with a trapdoor by using its private key. Under the practical condition that the size of keyword space is not more than the polynomial level, Byun et al. firstly defined off-line keyword guessing(KG) attacks [9] and showed that the PEKS scheme in [1] is insecure against KG attacks. Jeong et al. proved that any PEKS scheme satisfying at least computationally indistinguishable consistency implies successful keyword guessing attacks [10]. However, it is possible to construct a secure dPEKS scheme against KG attacks [11,12].

All of the schemes do not allow to retrieve the keyword, also they do not guarantee any relation between message and keyword. However, in some situation, users may want to sort the received encrypted emails using keywords without decrypting them. The PEKS/dPEKS is not applicable in this scenario. Decryptable searchable encryption which extends the notion of PEKS and enables decryption of keyword can resolve this problem. Fuhr and Paillier [13] put forward a construction for PEKSD scheme. However, it is improper that its test method can decrypt the ciphertext and get the associated keyword. Fang et al. present a decryptable searchable encryption scheme without random oracle [14]. But it is not the scheme with designated tester.

This paper suggests that the schemes in [11,12] are insecure against KG attacks and construct an enhanced dPEKS scheme. We prove that the scheme is secure against KG attacks. Then we extend our scheme to a secure decryptable searchable encryption with designated tester. The rest of this paper is organized as follows. In Section 2, we review some preliminaries. In Section 3, we analyze two schemes in [11,12]. In Section 4, we present our new dPEKS scheme against keyword guessing attacks. The security analysis is given in Section 5. In Section 6, we extend the dPEKS scheme to a secure decryptable searchable encryption scheme with designated tester. Finally, we draw our conclusions in Section 7.

## 2 Preliminaries

### 2.1 Bilinear Pairings

Let $G_1$ be a cyclic additive group generated by $P$, with a prime order $p$, and $G_2$ be a cyclic multiplicative group with the same prime order $p$. Let $e : G_1 \times G_1 \rightarrow G_2$ be a map with the following properties [15]:

1) *Bilinearity*: $e(aP, bQ) = e(P, Q)^{ab}$ for all $P, Q \in G_1$, $a, b \in Z^*_p$ ;
2) *Non-degeneracy*: There exists $P, Q \in G_1$ such that $e(P, Q) \neq 1$;
3) *Computability*: There is an efficient algorithm to compute $e(P,Q)$ for all $P,Q \in G_1$;

### 2.2 Searchable Public Key Encryption for a Designated Tester(dPKES)

A searchable public-key encryption scheme for a designated tester consists of the following polynomial time algorithms where *gp* denotes a set of global parameters.

1) GlobalSetup($\lambda$):takes a security parameter $\lambda$ as input, and generates a global parameter *gp*.

2) KeyGen$_{Server}$($gp$):takes input gp, outputs a pair of public and secret keys ($pk_S$, $sk_S$), of server $S$.

3) $KeyGen_{Receiver}(gp)$:takes as input gp and generates a pair of public and secret keys, $(pk_R, sk_R)$, of the receiver $R$.

4) dTrapdoor($gp$, $pk_S$, $sk_R$, $w$):takes as input, gp, the server's public key, $pk_S$, the receiver's secret key, $sk_R$, and a keyword, w. It then generates a trapdoor, $T_w$.

5) dPEKS($gp$, $pk_R$, $pk_S$, $w$):takes as input, gp, the receiver's public key, $pk_R$, the server's public key, $pk_S$, and a keyword, $w$. It returns a dPEKS ciphertext, $C$ of $w$.

6) dTest($gp$, $C$, $sk_S$, $T_w$):takes as input, gp, a dPEKS ciphertext, $C$, the server's secret key, $sk_S$, and a trapdoor, $T_w$. It outputs 'yes' if $w$=$w$' and 'no' otherwise, where $C$= dPEKS($gp$, $pk_R$, $pk_S$, $w$').

Consistency requires that for any keyword $w$, $(pk, sk)$= KeyGen($1^\lambda$), $T_w$ = dTrapdoor($sk$,$w$), we have dTest($pk$, dPEKS($pk$,$w$), $T_w$) = 1.

# 3 Analysis of Two dPEKS Schemes against KG Attacks

## 3.1 Attack on Schme in [11]

We review Rhee et al.'s dPEKS scheme [11] and show that the scheme is insecure against KG attacks. Rhee et al.'s dPEKS scheme works as follows:

1) GlobalSetup($\lambda$): Given a security parameter $\lambda$, it returns a global parameter gp=( $G_1$, $G_2$, $e$, $H_1$, $H_2$, $g$, $KS$), where $KS$ is a keyword space.

2) $KeyGen_{Server}(gp)$: It randomly chooses $\alpha \in_R Z^*_p$ , and $Q \in_R G_1$, and returns $sk_S=\alpha$ and $pk_S=(gp, Q, y_s)=(gp, Q,g^\alpha)$ as a server's pair of secret and public keys.

3) $KeyGen_{Receiver}(gp)$:This algorithm randomly chooses $x \in_R Z^*_p$, and returns $sk_R=x$ and $pk_R=g^x$ as a receiver's pair of secret and public keys, respectively.

4) dPEKS($gp$, $pk_R$, $pk_S$, $w$)**:** This algorithm randomly picks a rand value $r \in_R Z^*_p$, and outputs $C=[A,B]=[(pk_R)^r, H_2(e(y_s, H_1(w)^r))]$, where $w \in KS$.

5) dTrapdoor($gp$, $pk_S$, $sk_R$, $w$)**:** This algorithm randomly picks a rand value $r'$ $\in_R Z^*_p$, and outputs $T_w=[T_1, T_2]=[y_s^{r'}, H_1(w)^{1/x} \cdot g^{r'}]$, where $w \in KS$.

6) dTest(gp, $C$, $sk_S$, $Tw$)**:** This algorithm compute $T=(T_2)^\alpha/T_1$ and checks if $B=H_2(e(A,T))$. If the equality is satisfied, then output '1'; otherwise, output '0'.

We show that the scheme in [11] is not secure against KG attacks by the server. Suppose that a server is given a ciphertext $C$ and a trapdoor $T_w$ for a keyword $w$ such that dTest(gp, $C$, $sk_S$, $T_w$) =1. The server can determine which keyword is used in generating $C$ and $T_w$ as follows:

(1) The server can get $g^{r'}$ from $T_1$ using its secret key $sk_S=\alpha$.

(2) The server guess a keyword $w$' in KS and compute $H_1(w')$.

(3) The server checks if $e(pk_R, T_2)=e(pk_R, g^{r'})e(g, H_1(w'))$. If so, the guessed keyword $w$' is a valid keyword. Otherwise, go to (1).

## 3.2 Attack on Schme in [12]

Rhee et al.'s dPEKS scheme works as follows:

1) GlobalSetup($\lambda$): Given a security parameter $\lambda$, it returns a global parameter $gp$=( $G_1$, $G_2$, $e$, $H_1$, $H_2$, $g$, $h$, $u$, $t$,$KS$), where $KS$ is a keyword space and $h$, $u$, $t \in_R G_1$.

2) KeyGen$_{Server}$($gp$): It randomly chooses $\alpha \in_R Z^*_p$, and returns $sk_S=\alpha$ and $pk_S=(gp, y_{s1}, y_{s2}, y_{s3})=(\text{gp}, g^{\alpha}, h^{1/\alpha}, u^{1/\alpha})$ as a server's pair of secret and public keys, respectively.

3) KeyGen$_{Receiver}$($gp$):It randomly chooses $x \in_R Z^*_p$, and returns $sk_R=x$ and $pk_R=(y_{R1}, y_{R2}, y_{R3})=(g^x, h^{1/x}, t^x)$ as a receiver's pair of secret and public keys.

4) dPEKS($gp$, $pk_R$, $pk_S$, $w$): This algorithm randomly picks a rand value $r \in_R Z^*_p$, and outputs $C=[A,B]=[(y_{R1})^r, H_2(e(y_{s1}, H_1(w)^r))]$, where $w \in KS$.

5) dTrapdoor($gp$, $pk_S$, $sk_R$, $w$): This algorithm outputs $Tw = H_1(w)^{1/x}$.

6) dTest($gp$, $C$, $sk_S$, $Tw$): This algorithm checks if $B = H_2(e(A, T_w^{sk_S}))$. If the equality is satisfied, then output '1'; otherwise, output '0'.

We show that the scheme in [12] is not secure against KG attacks as follows:

(1) The attacker guess a keyword $w$' in KS and compute $H_1(w')$.

(2) The attacker checks if $e(y_{R1}, T_w) = e(g, H_1(w'))$. If so, the guessed keyword $w$' is a valid keyword. Otherwise, go to (1).

## 4 Our Enhanced dPEKS Scheme

Our dPEKS scheme works as follows:

1) GlobalSetup($\lambda$): Given a security parameter $\lambda$, it returns a global parameter $gp=(G_1, G_2, e, H_1, H_2, g)$, $H_1$:{0,1}*→$G_1$, $H_2$: $G_2$→{0,1}*.

2) KeyGen$_{Server}$($gp$): It randomly chooses $\alpha \in_R Z^*_p$, and $Q \in_R G_1$, and returns $sk_S=\alpha$ and $pk_S=(gp, Q, y_s)=(gp, Q, g^{\alpha})$ as a server's pair of secret and public keys.

3) KeyGen$_{Receiver}$($gp$):It randomly chooses $x,t \in_R Z^*_p$, and returns $sk_R=x,t$ and $pk_R=(y_{R1}, y_{R2}, y_{R3}, y_{R4})=(g^x, g^{tx2}, g^{xt}, y_s^t)$ as a receiver's pair of secret and public keys.

4) dPEKS($gp$, $pk_R$, $pk_S$, $w$): This algorithm randomly picks a rand value $r \in_R Z^*_p$, and checks if $e(y_{R1}, y_{R4}) = e(y_{R3}, y_s)$. If the equality is satisfied, then outputs $C=[A,B]=[(y_{R2})^r, H_2(e(y_{R4}, H_1(w)^r))] = [g^{x^2 tr}, H_2(e(y_{R4}, H_1(w)^r))]$.

5) dTrapdoor($gp$, $pk_S$, $sk_R$, $w$): This algorithm randomly picks a rand value $r$' $\in_R Z^*_p$, and outputs $T_w=[T_1, T_2]=[y_s^{r'}, H_1(w)^{1/x^2} \cdot g^{r'}]$.

6) dTest($gp$, $C$, $\text{sk}_S$, $T_w$): This algorithm computes $T=(T_2)^{\alpha}/T_1$ and checks if $B=H_2(e(A,T))$. If the equality is satisfied, then output '1'; otherwise, output '0'.

## 5 Analysis

Since our scheme is similar to one in [11], we can show the security in same manner in [11]. We omit the proof of security of a dPEKS ciphertext. We show that our scheme is computationally consistent and is secure against KG attacks on the base that a discrete logarithm problem is hard.

Firstly, similar to [11],our dPEKS scheme satisfies computationally consistency.

Secondly, our scheme is secure against KG attacks. Suppose $A$ is a attacker with advantage $\varepsilon$. Assume that $T_w=[T_1, T_2]$ is a trapdoor. To obtain a correct keyword $w$ from the given $T_w$, it should be possible that $A$ get $H_1(w)^{1/x^2}$ or $H_1(w)$ from $T_w$.

Since a discrete logarithm problem is hard, $A$ cannot easily get the unknown $r'$ or $\alpha \in Z^*_p$ from $T_1 = y_s^{r'}$ where $y_s = g^{\alpha}$. Furthermore, even though $A$ can compute

$$e(y_s, T_2) / e(g, T_1) = e(y_s, H_1(w)^{1/x^2} \cdot g^{r'}) / e(g, y_s^{r'}) = e(y_s, H_1(w)^{1/x^2})$$

A cannot guess $w$ such that $e(y_s, H_1(w)^{1/x^2})$ with out a knowledge of a receiver's secret key $x$ or a server's secret key $\alpha$. Therefore, it is hard that A guesses $H_1(w)^{1/x^2}$ or $H_1(w)$ from $T_w$.

Even the server can get $g^{r'}$ from $T_1$ using its secret key $sk_S=\alpha$. The server cannot guess a keyword $w'$ by checking if $e(y_{R4}, T_2)=e(y_{R4}, g^{r'})e(g, H_1(w'))$.

## 6 Extension to Decryptable Searchable Encryption with Designated Tester

We extend the dPEKS scheme to a secure decryptable searchable encryption scheme with designated tester as follows:

1) GlobalSetup($\lambda$): Given a security parameter $\lambda$, it returns a global parameter gp=( $G_1, G_2, e, H_1, H_2, H_3, g$), $H_1$:{0,1}*→$G_1$, $H_2$: $G_2$→{0,1}*, $H_3$:{0,1}*→$Z_p$.

2) KeyGen$_{Server}$($gp$):It randomly chooses $\alpha \in_R Z^*_p$, and $Q \in_R G_1$, and returns $sk_S=\alpha$ and $pk_S$=(gp, $Q$, $y_s$)=(gp, $Q$,$g^{\alpha}$) as a server's pair of secret and public keys.

3) KeyGen$_{Receiver}$(gp):It randomly chooses $x,t \in_R Z^*_p$, and returns $sk_R=x,t$ and $pk_R=(y_{R1}, y_{R2}, y_{R3}, y_{R4})=(g^x, g^{tx2}, g^{xt}, y_s^t)$ as a receiver's pair of secret and public keys.

4) dPEKS(gp, pk$_R$, $pk_S$, $w$): It randomly picks a rand value $r \in_R Z^*_p$, and checks if $e(y_{R1},y_{R4})=e(y_{R3},y_s)$. If the equality is satisfied, then outputs $C=[A,B,D,E]=[(y_{R2})^r, H_2(e(y_{R4},H_1(w)^r)), w \oplus H_2(g^r), e(g,g)^{rh}]= [g^{x^2tr}, H_2(e(y_{R4},H_1(w)^r))$, $w \oplus H_2(g^r), e(g,g)^{rh}]$, where $h=H_3(A||B||D||g^r)$.

5) dTrapdoor(gp, $pk_S$, $sk_R$, $w$): This algorithm randomly picks a rand value $r'$ $\in_R Z^*_p$, and outputs $Tw=[T_1, T_2]=[y_s^{r'}, H_1(w)^{1/x^2} \cdot g^{r'}]$.

6) dTest(gp, $C$, $sk_S$, $Tw$): This algorithm computes $T=(T_2)^{\alpha}/T_1$ and checks if $B=H_2(e(A,T))$. If the equality is satisfied, then output '1'; otherwise, output '0'.

7) KeywordDec($gp$, $C$, sk$_R$): This algorithm computes $g^r$ from $A= g^{x^2tr}$ using $sk_R=\{x,t\}$ and outputs $w=D \oplus H_2(g^r)$ if $E= e(g,g^r)^h$ where $h=H_3(A||B||D||g^r)$.

## 7 Conclusion

The public key encryption with keyword search enables one to search encrypted data without compromising the security of the original data. In this paper, we analyze two dPEKS schemes and suggest that they are insecure against keyword guessing attack and construct an enhanced dPEKS scheme. We prove that the scheme satisfies the consistency and is secure against KG attacks. At last, we extend our dPEKS scheme to a secure decryptable searchable encryption scheme with designated tester.

## References

1. Boneh, D., Di Crescenzo, G., Ostrovsky, R., Persiano, G.: Public key encryption with keyword search. In: Cachin, C., Camenisch, J.L. (eds.) EUROCRYPT 2004. LNCS, vol. 3027, pp. 506–522. Springer, Heidelberg (2004)
2. Boneh, D., Franklin, M.: Identity-based encryption from the weil pairing. In: Kilian, J. (ed.) CRYPTO 2001. LNCS, vol. 2139, pp. 213–239. Springer, Heidelberg (2001)
3. Boyen, X., Waters, B.: Anonymous hierarchical identity-based encryption (Without random oracles). In: Dwork, C. (ed.) CRYPTO 2006. LNCS, vol. 4117, pp. 290–307. Springer, Heidelberg (2006)
4. Ducas, L.: Anonymity from asymmetry: New constructions for anonymous HIBE. In: Pieprzyk, J. (ed.) CT-RSA 2010. LNCS, vol. 5985, pp. 148–164. Springer, Heidelberg (2010)
5. Abdalla, M., Bellare, M., Catalano, D., Kiltz, E., Kohno, T., Lange, T., Malone-Lee, J., Neven, G., Paillier, P., Shi, H.: Searchable encryption revisited: Consistency properties, relation to anonymous IBE, and extensions. In: Shoup, V. (ed.) CRYPTO 2005. LNCS, vol. 3621, pp. 205–222. Springer, Heidelberg (2005)
6. Park, D.J., Kim, K., Lee, P.J.: Public key encryption with conjunctive field keyword search. In: Lim, C.H., Yung, M. (eds.) WISA 2004. LNCS, vol. 3325, pp. 73–86. Springer, Heidelberg (2005)
7. Hwang, Y.-H., Lee, P.J.: Public key encryption with conjunctive keyword search and its extension to a multi-user system. In: Takagi, T., Okamoto, T., Okamoto, E., Okamoto, T. (eds.) Pairing 2007. LNCS, vol. 4575, pp. 2–22. Springer, Heidelberg (2007)
8. Baek, J., Safavi-Naini, R., Susilo, W.: Public key encryption with keyword search revisited. In: Gervasi, O., Murgante, B., Laganà, A., Taniar, D., Mun, Y., Gavrilova, M.L. (eds.) ICCSA 2008, Part I. LNCS, vol. 5072, pp. 1249–1259. Springer, Heidelberg (2008)
9. Byun, J.W., Rhee, H.S., Park, H.-A., Lee, D.-H.: Off-line keyword guessing attacks on recent keyword search schemes over encrypted data. In: Jonker, W., Petković, M. (eds.) SDM 2006. LNCS, vol. 4165, pp. 75–83. Springer, Heidelberg (2006)
10. Jeong, I.R., Kwon, J.K., Hong, D., Lee, D.H.: Constructing PEKS schemes secure against keyword guessing attacks is possible? Computer Communications 32(2), 394–396 (2009)
11. Rhee, H.S., Susilo, W., Kim, H.J.: Secure searchable public key encryption scheme against keyword guessing attacks. IEICE Electronics Express 6(5), 237–243 (2009)
12. Rhee, H.S., Park, J.H., Susilo, W., Lee, D.H.: Improved Searchable Public Key Encryption with Designated Tester. In: ASIACCS, pp. 376–379 (2009)
13. Fuhr, T., Paillier, P.: Decryptable searchable encryption. In: Susilo, W., Liu, J.K., Mu, Y. (eds.) ProvSec 2007. LNCS, vol. 4784, pp. 228–236. Springer, Heidelberg (2007)
14. Fang, L., Wang, J., Ge, C., et al.: Decryptable Public Key Encryption with Keyword Search Schemes. International Journal of Digital Content Technology and its Applications 4(9), 141–150 (2010)
15. Menezes, A.J., Okamoto, T., Vanstone, S.A.: Reducing elliptic curve logarithms to a finite field. IEEE Transactions on Information Theory 39(5), 1636–1646 (1993)

# The Modeling of Radar Electromagnetic Propagation by Parabolic Equation

Bole Ma, Xiushan Zhang, and Zifei Zhang

Dept. of Computer Technology, Naval Univ. of Engineering, Wuhan Hubei, 430033

**Abstract.** In modern battlefield with the utilization of radar frequently, the detection-range of the radar has become an important factor for commandant. Because of the invisibility of electromagnetic wave, there are much more value to simulate and display it. The simulation of electromagnetic wave has two steps, including simulation and visualization. In this paper, we studied parabolic equation and introduced some new theories to modify the factor of DMFT, otherwise compute the impedance by a new rough surface reflection coefficient. At last, we construct a model about electromagnetic radar, while we apply it on the clutter of sea to simulate radar detection-range. The result of simulation is the same as some papers.

**Keywords:** parabolic equation, DMFT, reflection coefficient, propagation loss.

## 1 Introduction

The electromagnetic atmosphere of the battlefield is becoming more and more complex because the electric devices are used frequently in modern battlefield. The acquisition of the electromagnetic wave is the key to win a war. Radar is the most used electronic equipment in various battlefields. The simulation and the visualization of radar have high value in military. The simulation of the complex electromagnetic environment and constitution of three-dimensional states can increase senior commandant's ability of processing the information in battlefield by the modern suppositional technique.

Because of the invisibility of the radar, two steps are needed to appear to commander: modeling and visualization. Until now many models of propagation of the electronic wave have been created, normally by three methods: 1. method of moment, 2.FDTD, 3.Parabolic Equation. Method of moment applies to low frequency. FDTD needs enormous calculation and is confined by the choice of step[1]. Parabolic equitation is proposed by Leontovich and Fock for research of propagation of the electromagnetic wave. PE, applied to the independent reflected structure in atmosphere in which every distance and every height are independent, is the only effective method to calculate the propagation loss of the long distance between the visual zones to the shadow zone. PE synthesizes the influence affected by the atmosphere, terrain and so on[3]. By the research of the parabolic equitation, this paper is going to introduce the latest achievement in related field to construct the model of propagation of the radar electromagnetic wave.

S. Lin and X. Huang (Eds.): CSEE 2011, Part II, CCIS 215, pp. 137–149, 2011.

## 2 Brief Introduction of Parabolic Equation

As reference [4].Parabolic Equation (PE), derived from wave equation is a determined propagation model. When radar electromagnetic wave is being calculated, concerning the forward wave, ignoring the backward wave, second-order wave equation simplifies to one-order wave equation. This is PE equation.

We set $e^{-i\omega t}$ as time harmonic factor; $\psi$ (x,y) field quantity. In rectangular coordinate system meets Helhorthz's wave equation.

$$\frac{\partial^2\psi}{\partial x^2}+\frac{\partial^2\psi}{\partial z^2}+k_0 n^2\psi=0 \tag{2.1}$$

$k_0=2\pi/\lambda$: spatial wave number;

$\lambda$: length of wave;

N: reflection coefficient

Equation (2.1) is explained by: $\psi(x,z)=\mu(x,z)e^{ik_0 x}$ (2.2)

Substitute (2.2) to (2.1) equals： $\frac{\partial^2\mu}{\partial x^2}+\frac{\partial^2\mu}{\partial z^2}+2ik_0\frac{\partial\mu}{\partial z}+k_0^2(n^2-1)\mu=0$ (2.3)

Factorization: $\left[\frac{\partial}{\partial x}+ik_0(1-Q)\right]\left[\frac{\partial}{\partial x}+ik_0(1+Q)\right]u=0$ (2.4)

Q as pseudo-differential operator equals $\sqrt{\frac{\partial^2}{k_0^2\partial z^2}+n^2}$

From above equation, the propagation of the electronic wave can be factorized of forward wave and backward wave. In actual spread, concerning the forward component of wave, the final PE is:

$$\frac{\partial\mu}{\partial x}=-ik_0(1-Q)\mu \tag{2.5}$$

Q's different approximation will get different types of PE

$$Q\approx\sqrt{1+\upsilon}+\sqrt{1+\varepsilon}-1 \tag{2.6}$$

Inter alia: $\upsilon=\sqrt{\frac{\partial^2}{k_0\partial z^2}}, \varepsilon=n^2-1$

Substitute above (2.5) to get: $\frac{\partial\mu}{\partial x}=ik_0\left[\sqrt{1+\frac{\partial^2}{k_0^2\partial z^2}}-1\right]\mu+ik_0(n-1)\mu$ (2.7)

## 3 Boundary Condition

PE's upper boundary is an infinite area. To avoid overflow and the reflection of wave's energy due to the limitation of the calculation, we can add absorbing layer to the upper boundary to filter the field intensity, as add Tuky's window function [5].

$$w_{(z)}=\begin{cases}1, & 0\le z\le 3/4z_{max}\\ 1/2+1/2\cos\left[\frac{4\pi(z-3/4z_{max})}{z_{max}}\right], & \text{others}\end{cases} \tag{3.1}$$

In research of the surface boundary, at the moment the most effective method is to research surface boundary by impedance boundary condition, In which electronic field intensity will satisfy the boundary condition[13]:

$$\left.\frac{\partial u}{\partial z}\right|_{z=0} + \alpha u\big|_{z=0} = 0 \tag{3.2}$$

Impedance index $\alpha$ reflects the impedance of the surface boundary. On smooth surface according to different polarization $\alpha$ shows below form [5]:

$$\alpha = \sqrt{\varepsilon_r' - \cos^2 \theta_r'} \,/\, \varepsilon\text{; vertical polarization} \tag{3.3}$$

$$\alpha = \sqrt{\varepsilon_r' - \cos^2 \theta_r'}\text{ ; Horizontal Polarization} \tag{3.4}$$

$\theta_r'$ : g razing angle in step,

$\varepsilon_r'$ :Surface's complex permittivity. The expression is [14];

$$\varepsilon_r' = \varepsilon_r + i60\lambda\sigma \tag{3.5}$$

$\varepsilon_r$ : Surface's relative dielectric constant;

$\sigma$ : Surface's conductivity

## 4 Definition of Initial Field

The solution of PE is the research of initial value. Only showing the distribution of initial field by radar's radiation pattern, the distribution of other field can be calculated precisely. Traditional definition adopts Fourier's transition through antenna's radiation pattern. This paper will adopt Green's function in reference [4,5] to solve initial field.

$$u(0,z) = \sqrt{\frac{k_0}{2\pi}} \exp(i\pi/4) \int_{-\infty}^{+\infty} \frac{A(p)}{(k_0^2 - p^2)^{1/4}} \exp(ipz)dp \tag{4.1}$$

$A(p)$ : Antenna's radiation pattern $p = k_0 \sin\theta$ ,

$\theta$ : Propagation angle in initial field,

By the symmetry and anti-symmetry of initial field's distribution in Space P, we can get(Vertical polarization):

$$u(0,z) = e^{i\pi/4}\sqrt{\frac{2k_0}{\pi}} \int_0^{\infty} \frac{[A(p)e^{-ipa} - RA(-p)e^{ipa}]\cos(pz)}{(k_0^2 - p^2)^{1/4}} dp \tag{4.2}$$

R: Reflection coefficient of surface wave in vertical and horizontal polarization.

$a$ : height of antenna.

$R = \dfrac{\varepsilon_r' \sin\theta - \sqrt{\varepsilon_r' - \cos^2\theta}}{\varepsilon_r' \sin\theta + \sqrt{\varepsilon_r' - \cos^2\theta}}$ [4] Vertical polarization （4.3） （4.2） can be realized by DST or DCT.

# 5 Parameter of Models

Previously I have introduced the characteristic of parabolic equation and the definition of initial value. How to unite them with natural environment to form integrated model is main content in this section.

## 5.1 Model of Atmospheric Reflection Coefficient

Radar electronic wave is mainly reflected and absorbed by atmosphere. Radar wave is distorted because of the reflection effect. The loss of the electronic wave changes by absorbing of atmosphere.

Relative index of refraction N is used as it is not convenient to use standard atmospheric refraction index. As reference [3]. $N=(n-1)\times 10^6=\frac{77.6p}{T}+\frac{d_s\times 3.73\times 10^5}{T^2}$ (5.1.1)

$d_s$ : Steam partial pressure

$$d_s=\frac{6.105\rho e^x}{100} \quad (5.1.2) \qquad x=25.22\frac{T-273.2}{T}-5.31\ln(\frac{T}{273.2}) \quad (5.1.3)$$

P: Atmospheric Pressure,

T: Absolute Temperature,

$\rho$ : Percentage of Relative Humidity.

To inspect the gradient of reflection coefficient and the influence to the propagation of the electric wave the equation of reflection coefficient concerning the change of the height should be modified.

$$M=N+0.157h \quad (5.1.4)$$

h: height.

Sea wave guide effect includes: evaporation duct, surface duct and elevating duct. Evaporation duct changes reflection coefficient in height and direction most commonly and affects radar detection greatly. This paper will simulate evaporation duct to acquire the height division of atmospheric refraction index in evaporation duct. The paper adopts PJ model[15]. Atmospheric absorption includes Oxygen attenuation and steam attenuation. Reference [3] shows two kinds of numerical calculation formula and shows numerical value calculation equation in two situations.

Oxygen absorption loss

$$\gamma_a=[1+0.01(t-15)](a_1+a_2+0.00719)(\frac{F}{1000})^2 10^{-3} \quad (5.1.5)$$

T: bottom surface temperature,

F: electromagnetic wave frequency

$$a_1=\frac{6.09}{(\frac{F}{1000})^2+0.227} \qquad a_2=\frac{4.81}{(\frac{F}{1000}-57)^2+1.5}$$

water-vapour absorption loss

$$g_w=(0.05+0.0021C+b_1+b_2+b_3)(\frac{F}{1000})^2 C*10^{-4} \quad (5.1.6)$$

C: Surface absolute humidity

$b_1 = \frac{3.6}{(\frac{F}{1000} - 22.2)^2 + 8.5}$ $b_2 = \frac{10.6}{(\frac{F}{1000} - 183.3)^2 + 9}$ $b_3 = \frac{8.9}{(\frac{F}{1000} - 325.4)^2 + 26.3}$

Total absorption loss:

$$l_a = (\gamma_a + \gamma_w)H \tag{5.1.7}$$

H: relative ground level

## 5.2 Terrain Model

In dealing with irregular topography, reference [3] adopts boundary shift conversion algorithm by model APM. This paper will quote piecewise linear terrain translation frequency-Spectrum-transformation from reference [6] and [7]. As reference[6,7] ,the basic idea is boundary conditions processing of each point in irregular terrain by coordinate conversion. Traditional translational converter technique proposed by Beilis-Tappert, relying on terrain curvature, is not applicable to the simulation of the electronic map constituted subsection. From reference [7] piecewise linear parabolic equation can solve the above problems. Until now it is the most precise method which applies to Fourier's fraction stepping method[6]. We will adopt the method of reference[6,7].

Coordinate conversion:

$$x = u, z = v - T(u)$$

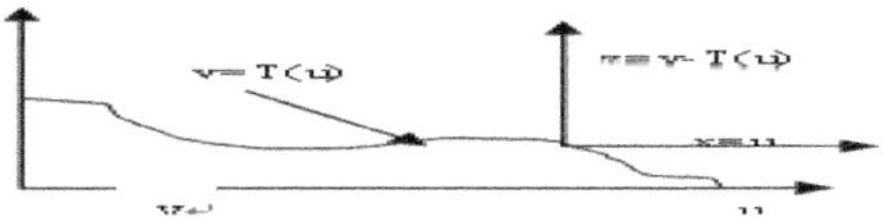

**Fig. 1.** Irregular terrain shift conversion

Piecewise linearity wide angle PE:

$$\frac{\partial u}{\partial x} = i\sqrt{\frac{k_0^2}{1+T'^2} + \frac{\partial^2}{\partial z^2}}u + ik_0\sqrt{n^2 - \frac{T'^2}{1+T'^2}}u \tag{5.2.1}$$

$T'$: Gradient of subsection topography.

In branch points, forced multiplied by phase position factor the boundary shift conversion is continuous in branch points. As in the picture, the gradient of A and B are different. They are $T_A'$ and $T_B'$. The upper field of A and B is $u_A, u_B$ in branch point AB, the field is $u_B(x_{AB}, z) = u_A(x_{AB}, z)e^{ik_0 z(T_A' - T_B')}$ (5.2.2)

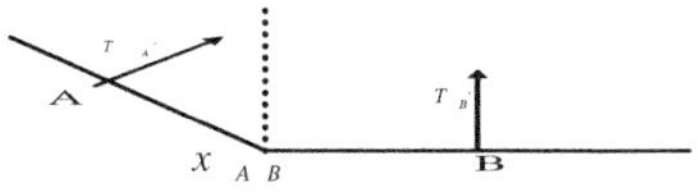


**Fig. 2.** Inflexion Calculation

Define new impedance boundary condition by subsection linearity shift conversion:

$$\alpha_{new} = ik_0 \cos a\left[\sin\theta\left(\frac{1-R}{1+R}\right)+\frac{\sin a}{\cos a}(1-\cos\theta)\right] \quad (5.2.3)$$

$\theta$ : calculated grazing angle in surface.

a : intersection angle between surface and horizontal plane, $R$ is Fresnel reflection coefficient

## 5.3 Ocean Model

Ocean can be considered as horizontal terrain as give the value 0 of $T'$ in above equation. The processing of impedance boundary condition of rough ocean surface and terrain are different. Now frequently when we deal with the influence of roughness, the smooth sea surface reflecction index is multiplied by revision factor of roughness, which we use MB model usually. We should consider shadowing effect. So in this paper revision factor of roughness will substitute MB model quoted from reference [8,9,15]. The traditional expression of the revision factor of roughness is:

$$\int_{-\infty}^{+\infty} \exp(-2ik_0 \sin\theta\varepsilon)p(\varepsilon)d\varepsilon \quad (5.3.1)$$

$p(\varepsilon)$: probability density function obeyed rough sea height. This paper will use Ament's model to suppose rough sea height obey Gaussian function.

$$p(\varepsilon)=\frac{1}{\sqrt{2\pi}h}\exp(-\frac{\varepsilon^2}{2h^2}) \quad (5.3.2)$$

According to reference [12] Standard deviation h is related to wind speed. $h = 6.28*10^{-3}*v_{10}^{2.02}$ [15], $v_{10}$ is wind speed of ten meters on sea surface. Reference [8] points out due to the shadowing effect when the rough sea surface is irradiated by plane wave, the average height of sea surface will raise. So as reference [8], we adopt new revision factor of roughness concerning to the shadowing effect. This paper will substitute (5.3.2) by revision factor of roughness from reference [15]:

$$p_{new}(\theta,\varepsilon) = p(\varepsilon)(1+2\wedge)[F(\varepsilon)]^{2\wedge} \quad (5.3.3)$$

$$\wedge(\theta,h) = \wedge(v) = \frac{\exp(-v^2)-v\sqrt{\pi}\,erfc(v)}{2v\sqrt{\pi}},$$

$$v=\frac{\tan\theta}{\sqrt{2}\sigma_r}$$

$$F(\varepsilon) = 1-1/2*erfc\left[\frac{\varepsilon}{\sqrt{2}h}\right]$$

$\theta$: calculated grazing angle in surface. $\sigma_r$: variance of slope of wave height $\sigma_r = 5.62*10^{-2} v_{10}^{0.5}$ [15]. Finally, we can get the reflection index of rough sea surface [8]:

$$R = R_S \int_{-\infty}^{+\infty} \exp(2ik_0 \sin\theta\varepsilon) p_{new}(\varepsilon)\ d\varepsilon \tag{5.3.4}$$

As reference [15], we simplify that to:

$$R = R_s \exp(-2ik_0 \sin\theta m_\varepsilon - \frac{4k_0^2 \sin^2\theta\sigma_\varepsilon^2}{2}) \tag{5.3.5}$$

$m_\varepsilon$, $\sigma_\varepsilon$ are mean and standard deviation of (5.3.3) .

We can get $m_\varepsilon$, $\sigma_\varepsilon$ by equation (19), (21) from reference[12] processed by Almet quadrature formulation. Finally, Impedance index of rough sea surface $\alpha$ [8] :

$$\alpha = ik_0 \cos\theta(\frac{1-R}{1+R}) \tag{5.3.6}$$

From this we can get that the solution of the impedance factor of rough earth surface and sea surface relies on the determination of the grazing angle. Now the definition of grazing angle can be divided in two theories: spectrum estimation and geometrical optics. Stated by reference [8], Optics can't solve the angle of irregular terrain because of the presence of shadow zone. Computational complexity of spectrum is huge. So in this paper the two theories will be combined. Optics applies to sea condition, meanwhile, spectrum estimation applies to irregular terrain

## 6 Improvement of DMFT

By the setting above, quoted Fourier's mixed conversion can meet boundary condition. We can solve parabolic equation by DMFT. The theory of DMFT is using second-order difference to fit the differential form of boundary condition (3.2). By the auxilliary function we change Fourier's mixed conversion to unilateral sine transformation.

Because this kind of transformation exists oscillation of numerical value and huge computational complexity, this paper will quote the backward difference from reference [10] and improve the relative factor to solve the field intensity of piecewise linearity wide-angle parabolic equation and unification of the solution of terrain and ocean.

The auxilliary function by backward difference fits the boundary condition.

$$w(x,t\triangle z) = \frac{u(x,t\triangle z) - u[x,(t-1)\triangle z]}{\triangle z} + \alpha u(x,t\triangle z) \tag{6.1}$$

We set $r = (1+\alpha*\triangle z)^{-1}$ ( 6.2) to solve difference equation.

So (6.1) is simplified to:

$$w(x,t\triangle z) = u(x,t\triangle z) - ru[x,(t-1)\triangle z] \tag{6.3}$$

$$w(x,0) = w(N\Delta z) = 0\text{, t=1:N-1} \tag{6.4}$$

DST conversion is made to equation (6.3):

$$W(m\Delta p) = \sum_{t=1}^{N-1} w(x,t\Delta z)\sin(\frac{mt\pi}{N}) \tag{6.5}$$

$m$ =0: N

N: point number of division.

Equation (6.5) multiply propagator in spectral domain $e^{i\Delta x\sqrt{\left(k_0/\sqrt{(1+T\cdot^2)}\right)^2-p^2}}$ to get $W(x+\Delta x, m\Delta p)$

By IDSF conversion of $W(x+\Delta x, m\Delta p)$ we can get:

$$w(x+\Delta x, t\Delta z) = \frac{2}{N}\sum_{m=1}^{N-1} W(x+\Delta x, m\Delta p)\sin(\frac{mt\pi}{N}) \tag{6.6}$$

T=0:N. (6.3) is a first-order difference equation. Characteristic root: $r$ , the solution:

$$u(x,t\Delta z) = u_p(x,t\Delta z) + A(x)r^t \tag{6.7}$$

By setting $u(x,0)$, we can get:

$$u_p(x,t\Delta z) = w(x,t\Delta z) + ru_p\left[x,(t-1)\Delta z\right] \tag{6.8}$$

Above is the solution of each step. From (6.7) we can get $w(x+\Delta x, t\Delta z)$, then we can get $u_p(x+\Delta x, t\Delta z)$ by the substitution of above equation.

First we define $B(x)$ [10], we can get solution of A: $B(x) = \sum_{t=0}^{N} r^t u(x,t\Delta z)$ (6.9)

So $$A(x) = B(x) - \sum_{t=0}^{N} u_p(x,t\Delta z)r^t \tag{6.10}$$

$B(x+\Delta x)$ can be solved by stepping of $B(x)$:

$$B(x+\Delta x) = B(x)e^{i\Delta x\sqrt{\left(k_0/\sqrt{(1+T\cdot^2)}\right)^2-\left(\frac{\ln r}{\Delta z}\right)^2}} \tag{6.11}$$

From (6.9) , (6.10) , (6.11) we can get $A(x+\Delta x)$

$$A(x+\Delta x) = B(x+\Delta x) - \sum_{t=0}^{N'} u_p(x+\Delta x, t\Delta z)r^t \tag{6.12}$$

$N'$ is the point number of field value on stepping of $x+\Delta x$ . Finally, we substitute $u_p(x+\Delta x, t\Delta z)$ and $A(x+\Delta x)$ into (6.7) to get field intensive distribution $u(x+\Delta x, t\Delta z)$ on the stepping of $x+\Delta x$. Concerning the reflection of atmosphere we multiply $u(x+\Delta x, t\Delta z)$ to medium refractive factor $e^{ik_0\sqrt{n^2-\frac{T'^2}{1+T'^2}}\Delta x}$, to get the field intensity which meet the impedance boundary condition of the stepping $\Delta x$. We should pay attention to spectral domain stepping factor. To solve $B(x+\Delta x)$ multiplied stepping factor and medium refractive factor are the factors corresponding to piecewise linear wide-angle equation.

## 7 Simulation of Models

Based on the above constructed model, the piecewise linearity wide-angle PE solved by DMFT can get different positions of field value in space. The propagation loss of each point solved by these values can get the electromagnetic situation in relative terrain and natural environment of certain radar.

The basic step as below Fig. 3.

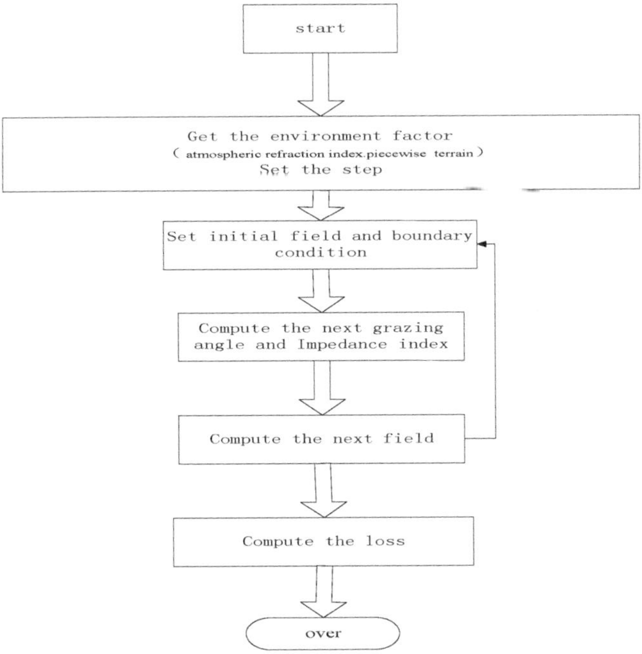


**Fig. 3.** Simulation Process

In the opening stage the basic parameter of radar should be set. In this paper we use settings: Gaussian antenna; vertical antenna 3dB width is 3°, vertical polarization, transmitting frequency: 7e9HZ, height of antenna: 12m, launching elevation of antenna0°. Natural model parameter: (1) sea condition, set gradient $T'$ as 0， $\varepsilon_r$ =58.6027 ， $\sigma$ =13.9266s/m; Temperature T=15°C, sea surface temperature: 10.2°C; Air pressure P=1013hpa, atmospheric absolute humidity: 11. (2)Irregular terrain adopts half wedge-shape simulation mountain, as Fig. 3-1.

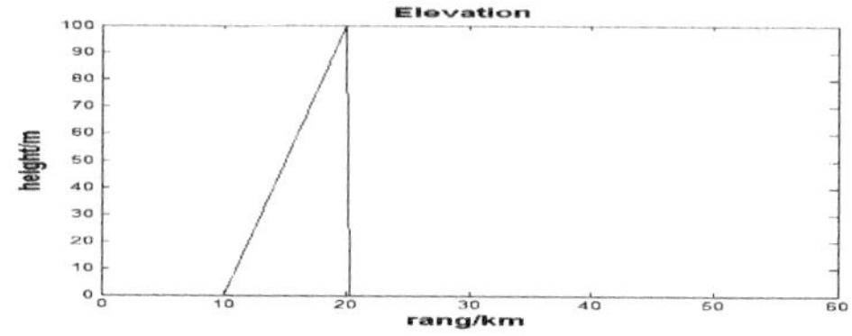


**Fig. 3-1.** Elevation Map

(3) Fig. 3-2 shows the distribution of atmospheric Refraction index solved by Model PJ

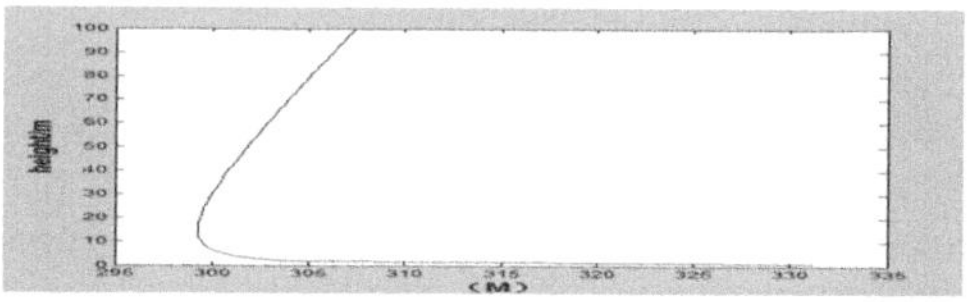

**Fig. 3-2.** Atmospheric Height Distribution of Modified Refraction Index

Firstly, we use smooth surface of low glancing angle. Calculation parameter: horizontal maximum range of 60,000 meters, maximum height: 512meters. Propagation loss of each point of field value in space can be solved by the method of reference [11]. $L = 20\log 4\pi + 10\log x - 20\log\lambda - 20\log|u(x,z)| + l_a$,

Fig. 7 to 8 shows propagation loss of ocean and irregular terrain.

Fig. 4 shows propagation loss of ocean

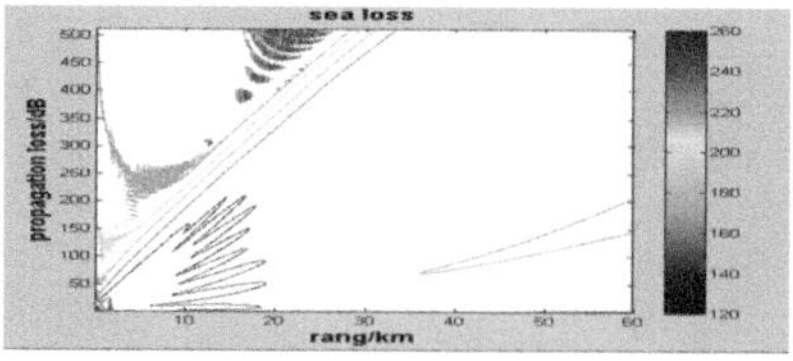


**Fig. 4.** Propagation loss of Ocean

Fig. 5 shows propagation of irregular terrain (half wedge shape)

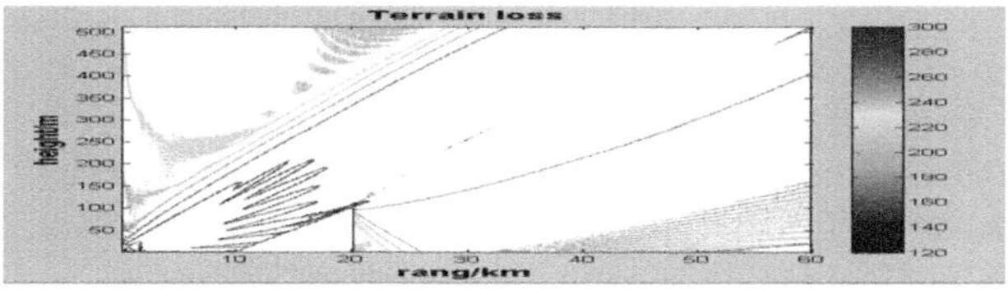


**Fig. 5.** Propagation loss of irregular terrain

Fig. 6 and Fig. 7 show propagation loss of height of 50m.

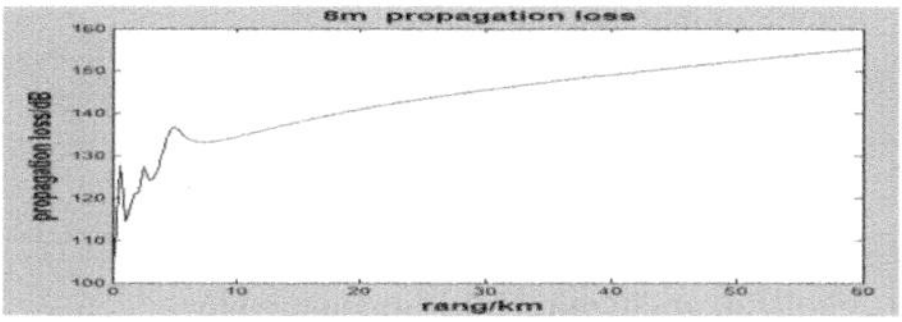


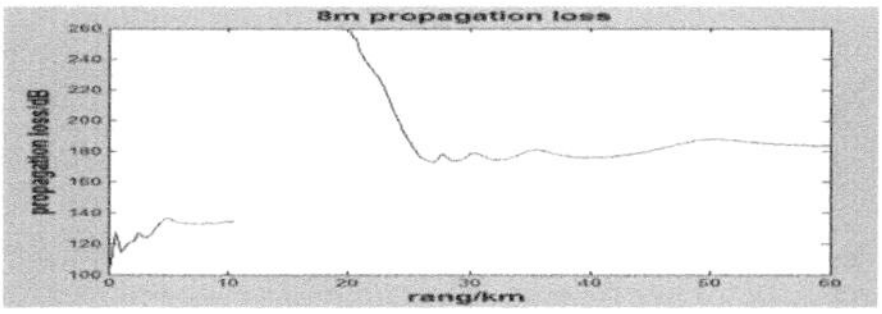


**Fig. 6.** Ocean Condition

**Fig. 7.** Irregular Terrain

Rough sea:
Here we set the wind in the height of 10m is 12m/s.

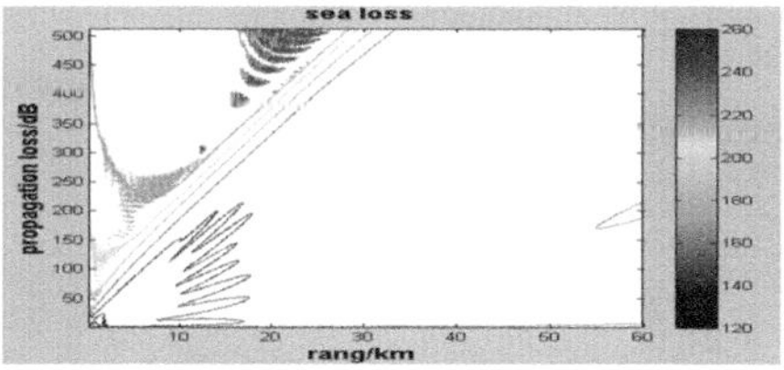


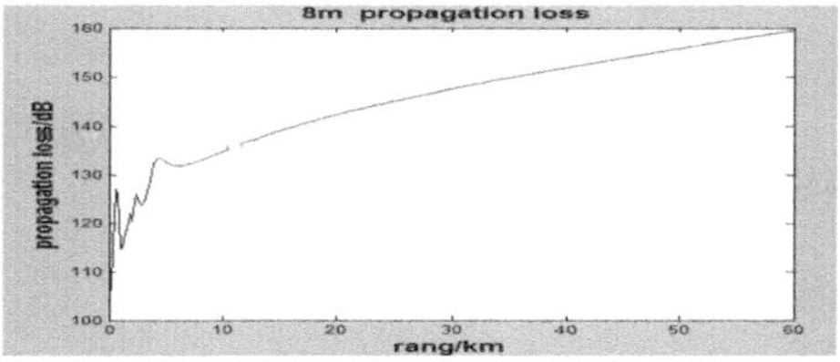


**Fig. 8.** Rough sea

**Fig. 9.** Rough sea condition

This result shows that the propagation of the radar electric wave under the synthetic effect of atmosphere and earth surface is not straight but curved. Radar's radiation lope formatted by periphery boundary outline of certain loss blank map, under the situation of irregular terrain, electromagnetic wave reflects (burr of the third lobe contour) and refracts. This is corresponded to the actual radar conduct operation. From the above simulation result we can prove that the result of radar campaigning affected by environment by PE is more comprehensive than other methods.

## 8 Conclusion

This paper based on piecewise linearity wide-angle PE, quotes relative new theories of this field and improves the solution of the reflection coefficient of relative factor and rough surface in DMFT to apply to various sea surface and mountain. Theoretically we improve the calculation model of reference [4]. Finally, on this basis we model the propagation loss uniformly. This sets the numerical source as the stage of constitution of three-dimension radar electromagnetic situation in later period. In the final experiment, the result is obvious and meets the conclusion of relative papers. Concerning the simplicity, we only simulate the smooth surface in low glancing angle. Impedance factor in rough surface can be solved by solving the equation (5.3.5). We should focus on this method in next step.

**Acknowledgements.** I acknowledge Marine Electromagnetic Environment Institute of Naval Univ. of Engineering for all of the help with the experiment and all the authors of references.

## References

1. Wu, Y., Zhang, L., Zhang, L., Zhang, W.: The Research of Simulation and Visualization of Electromagnetic Environment. System Simulation Paper 20(21), 6332–6338 (2009) (in Chinese)
2. Wu, L., Song, H.: The Technique of Constitution of Three-Dimensional Battlefield. System Simulation Paper 1(21), 91–94 (2009) (in Chinese)
3. Yang, C.: Three-Dimension Modeling of Maximum Detection Range of Rader and The Research and Realization of Visualization. Defence Science University, 1–50 (2006) (in Chinese)
4. Hu, H.: Forecasting the Algorithm of Electric Wave's Characteristic of Propagation in Complex Environment. Defence Science University, 1–106 (2006) (in Chinese)
5. Li, H.: Propagation Characteristic of Electric Wave in Atmospheric Waveguide Environment and Inversion Based on Genetic Algorithm. Xi'an Electronic And Science Unversity, 1–61 (2009) (in Chinese)
6. Guo, J.: Parabolic Equation of Complex Electric Wave's Propagation. Sun Yat-sen University, 1–127 (2008) (in Chinese)
7. Donohue, D.J., Kuttler, J.R.: Propagation Modeling Over Terrain Using the Parabolic Wave Equation. IEEE, 0018-926x(00)02620-x 260-277 (2000)
8. Guillet, N., Fabbro, V., Bourlier, C., Combes, P.F.: Low Grazing Angle Propagation Above Rough Surface By the Parabolic Wave Equation. IEEE, 0-7803-7929-2/03 4186-4188 (2003)
9. Pan, Y., Yang, K., Ma, Y.: Microwave's Waveguide over-the-horizon Propagation Influence by Rough Sea Surface. Simulation of Computer 5(25), 324–428 (2008) (in Chinese)
10. Guo, J., Wang, J., Long, Y.: Analysis of Electric Wave's Propagation on Rough Sea Surface by Parabolic Equation. Communication Paper 30(6), 48–51 (2009) (in Chinese)
11. Li, F., Cai, H.: Electric Wave's Propagation Affected by Irregular Terrain by Parabolic Equation. Fire Control And Command Control 35(8), 115–116 (2010) (in Chinese)
12. Fabbro, V., Bourlier, C., Combes, P.F.: Forward Propagation Modeling Above Guassian Rough Surfaces By The Parabolic Equation: Introduction Of The Shadowing Effect. Progress in Electromagnetic Research, PIER 58, 243–269 (2006)
13. Dockery, G.D., Kuttler, J.R.: An Improved Impedence-Boundary Algorithm for Fourier Split-Step Solution of the Parabolic Wave Equation. IEEE Transantions on Antennas and Propagation 44(12), 1592–1599 (1996)

14. Barrios, A.E., Patterson, W.L.: Advanced Propagation model(APM) Ver.1.3.1computer software configuration item(CSCI)documents2002(EB/OL) 2010-4-15:50-54,130, http://sunspot.spawar.navy.mil
15. Fabbro, V., Bourlier, C., Combes, P.F.: Forward propagation Modeling above Gaussian rough surfaces by the parapoloc shadowing effect progress. Electromagnetics Research, PIER 58, 243–269 (2006)
16. Yao, J.S., Yang, S.X., Xin, M.: The PJ model of Evaporation duct and the extent of the detection of radar. Modern Radar 30(8), 32–36 (2008) (in Chinese)

# Sharing Computation Resources in Image and Speech Recognition for Embedded Systems

Seung Eun Lee

Dept. of Electronic and Information Engineering
Seoul National University of Science and Technology, Seoul, Korea
seung.lee@seoultech.ac.kr

**Abstract.** An important class of features centered on recognition (i.e. the ability to reconize images, speech, gestures, etc) is rapidly becoming available on embedded systems. This requires high performance computing to support recognition with low-latency, low power and high throughput. In this paper, we investigate the image and speech recognition algorithms and find opportunity to share the computation resources, reducing the cost without losing any system performance.

**Keywords:** Embedded System, Image Recognition, Speech Recognition, Mobile Augmented Reality, Design Optimization.

## 1 Introduction

Over the last decade, smart interface becomes a major issue that is how to support more natural and immersive user interface. The user interface has evolved from a simple controller such as a pointing device or a button, to a tangible input device. The tangible interface could be able to interact with the virtual space through a physical behavior. Emerging smart interface applications such as gesture recognition, motion tracking, and speech recognition are quickly entering the mobile domain and realistic game systems. As user interacts more with devices such as game platforms, smart phones, smart TVs, entertainment equipments in automobile etc., the enhancing personalized experience become challenging to enable more natural modes of input-output as compared to traditional devices and new usage models in order to provide interface customized to the user context. Because natural interface is more convenient to users than using existing methods of keyboard or mouse, recognition based interfaces are emerging. An important class of features centered on recognition (i.e. the ability to reconize images, speech, gestures, etc) is rapidly becoming available on embedded systems.

In this paper, we investigated image and speech recognition algorithms and presented an opportunity to sharing the computation and storage resources on embedded systems by proposing the unified computation unit. We first describe the characteristics of image and speech recognitions in section 2 including the algorithms in details and then propose the way to share the computing resources in designing such recognition systems in section 3. Finally, we conclude in section 4 by outlining the direction for future works on this topic.

S. Lin and X. Huang (Eds.): CSEE 2011, Part II, CCIS 215, pp. 150–156, 2011.

## 2 Backgrounds

Smart phones and mobile internet devices (MIDs) have gained widespread popularity by placing compute power and novel applications conveniently in the hand of end users. Emerging smart interfaces based on image recognition, motion/gaze tracking are quickly entering the mobile domain. For instance, Mobile Augmented Reality is an upcoming application [1] that enables users to point their handheld cameras to an object and the device recognizes the objects and overlays relevant metadata on the object. Automatic speech recognition is particularly attractive as an input method on embedded systems enabling users to make voice calls, dictate notes and initiate searches. In this paper, we focus on mobile augmented reality and speech recognition and present the way to share computing resources in these emerging recognition applications.

### 2.1 Mobile Augmented Reality

The MAR workload of interest is best described with an example as follows. Consider a tourist walking the streets of a foreign city and scanning the surroundings using the camera in their smart phone. The smart phone should recognize the objects in the camera image and provide contextual data overlaid on the object in the display [2]. In order to achieve the usage model, it is required to compare query image against a set of pre-existing images in a database for a potential match, performing following three major steps.

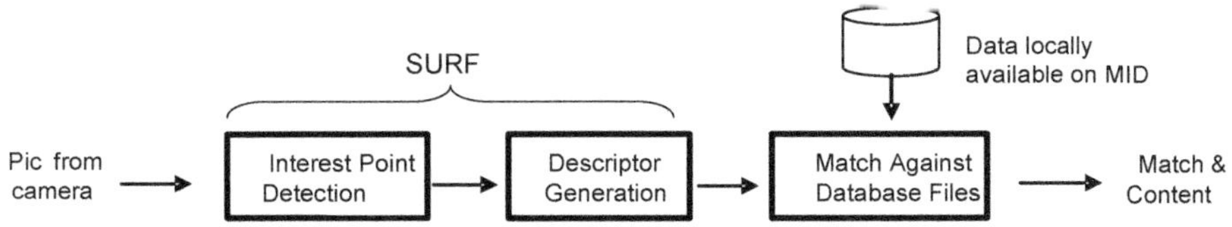


**Fig. 1.** Image recognition flow

- **Interest-point detection:** identify interest points in the query image
- **Descriptor generation:** create descriptor vectors for these interest points
- **Match:** compare descriptor vectors of the query image against descriptor vectors of database images

Figure 1 illustrates the MAR system flow. There are several algorithms that have been proposed to detect interest points and generate descriptors. The most popular algorithms amongst these are variants of SIFT (Scale-Invariant Feature Transform) [3] and SURF (Speeded up Robust Features) [4]. In this paper, we chose the SURF algorithm for our MAR application because it is known to be faster and has sufficient accuracy for the usage model of interest. In addition, researchers have also used SURF successfully for mobile phones for MAR [5]. Below we provide a brief explanation of the Match step, which we are going the share with speech recognition, although we refer the readers to [2] for a more detailed description.

In order to match two images (query and database), we use a brute force match algorithm that exhaustively compares a pair of interest point descriptor vectors from each image based on the Euclidean (a.k.a. L2) distance. Manhattan (a.k.a. L1) distance is another option for the descriptor comparison.

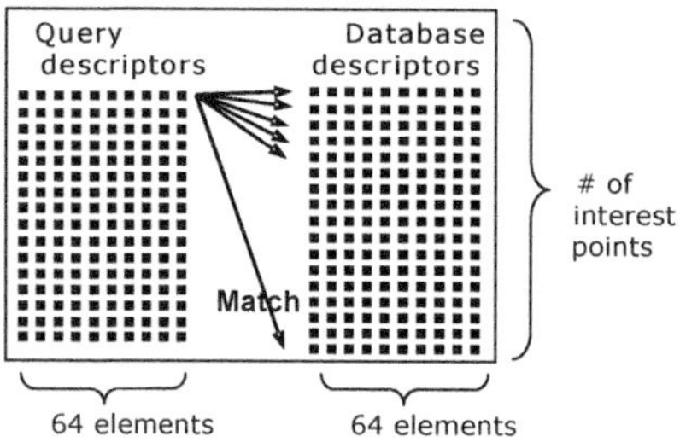


(a) overall match process

```
// Q and D are vectors from the query and db descriptors,
respectively
Distance(Q, D)
{ Loop from i =1 to 64 in groups of 4
      sum += (Q[i] – D[i])^2   // (for L2 distance)
      if (sum > 2nd min)  break; //check every 4
}
This function is performed on all pairs of query and database
descriptors whose laplacian values match.
```

(b) Brute force matching algorithm

**Fig. 2.** Brute force match between a query image and a database image

Figure 2 describes how a query image from the camera is matched against a candidate image from the data-base. One descriptor represents one interest point. The key function is a simple loop, assuming 64 elements per descriptor. For each descriptor of the query image, performing the Distance function on all descriptors of a database image gives us the minimum and second minimum ($2^{nd}$min) values of sum. A match for a query descriptor is found in the database image if min < $0.5 \times 2^{nd}$min. Database images are then ranked based on how many matches they have for the query image, and the highest ranked candidate is selected as the winner. This requires computation of the Euclidean distance square for every pair of descriptor vectors given by equation (1)

$$sum = \sum_{i=1}^{64} (Q_i - D_i)^2 \tag{1}$$

## 2.2 Speech Recognition

Automatic speech recognition (ASR) is particularly attractive as an input method on handheld platforms due to their small form factors. Future usage case projections for ASR on embedded systems include natural language navigation, meeting transcription, and web search, just to name a few. These applications are built on large vocabulary, continuous speech recognition (LVCSR) systems. The most widely researched LVCSR software is CMU Sphinx 3.0 [6], which requires three steps: (1) acoustic front-end, (2) GMM scoring, and (3) back-end search. Sphinx3 uses hidden Markov models (HMMs) as statistical models to represent sub-phonemic speech sounds. One key step of an HMM based ASR system is a Gaussian mixture probability density function evaluation which computes Gaussian mixture model (GMM) scores. GMM scoring consumes 82% of total execution time in Sphinx3 on Intel®Atom™ based platform [7]. Figure 3 contains a drastically stripped-down description of the GMM score computation in Sphinx3. This requires computation of the sum of weighted squares of distances between audio feature vectors and the mean values in the Gaussian table, followed by score generation. The computation intensive one is the sum of weighted squares of distances given by equation (2)

$$sum = \sum_{i=1}^{39} (X_i - M_i)^2 \times V_i \tag{2}$$

```
For every audio frame x
    For senone from 0 to S   // e.g., S = 8000.
        For mix from 0 to N_M    // e.g., N_M = 16. Number of Gaussian mixtures
            lrd = LRD[mix];
            For dim from 0 to D  // e.g., D=39.  Feature vector dimensions
                lrd = lrd - (x[dim] – M[senone][mix][dim])^2×V[senone][mix][dim];
            gauscr = (f × lrd) + weight[senone][mix];   // f is a constant
        score = LogAdd(score, gauscr)
```

**Fig. 3.** Sphinx3 GMM scoring code flow

## 3 Sharing Computation Resources for the Recognitions

In designing embedded systems, area and power cost should be reduced while providing the desired functionality and performance. Sharing resources for the different applications or tasks is the simplest and efficient way to reduce the cost.

Sharing resources among different applications is required to keep the following thins in key focus: (1) need to specify the common operation, (2) need to ensure the same data types in the common operation, (3) need to provide a control flow for shared resources, and (4) need to ensure that it is extremely efficient in terms of low power, low cost and high performance. In particular, the application space such as accuracy, execution time, and throughput should be satisfied for the recognition applications.

In order to improve energy efficiency and execution time for the recognition based embedded system, a recognition server (CogniServe) based on heterogeneous architecture with hardware accelerators for the two recognition algorithms described in Section 2, image and speech recognition, were proposed and implemented [7]. For the speech recognition, a GMM accelerator was proposed, calculating sum of weighted square of distances between audio feature vectors and the mean values in the Gaussian table, followed by score generation. For the image recognition, they designed two hardware accelerators: (1) a Match accelerator which computes distance calculations for matching descriptors from a query image to descriptors from a set of database image and (2) a Hessian accelerator that identifies the interest points in an input image. In this paper we focus on the GMM accelerator and the Match accelerator since they have the similar operations as shown in equation (1) and (2). Figure 4 shows the microarchitecture of the two accelerators.

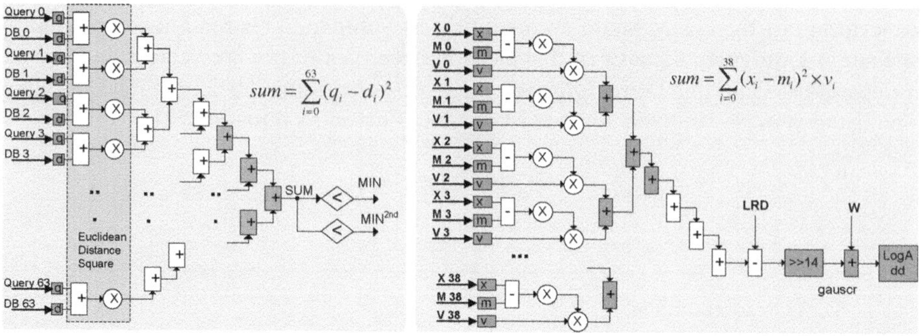

(a) Match (b) GMM

**Fig. 4.** Microarchitecture of the (a) Match accelerator and (b) GMM accelerator [7]

The differences between two accelerators are (1) the GMM accelerator requires one more multiplication in addition to the Euclidean distance calculation between two variables, (2) the number of calculation units is 64 and 39 in the Mach and GMM accelerators, respectively, and (3) types of input data (the Match accelerator takes 8bit data and the GMM accelerator takes 32bit data). Based on these observations, we propose a computation unit for both GMM and Match accelerators as shown in Figure 5. Four 8bit adders in the Match are used to implement one 32bit adder for the GMM. Initiating 16 elements in parallel (see Figure 5) completes the computation of the Match process. In case of GMM scoring, only 16 calculations are processed in parallel, requiring three iterations to finish the calculations of 39 elements. Our initial analysis reveals that iterating three times for GMM calculation does not reduce the recognition time by adopting the grouping factor (N). In GMM calculation, instead of iterating over all senones for the current feature vector, the control unit loads and processes N feature vectors before loading the next senones. The N score's are written to system memory before the M, V, LRD and W data in the SRAM are reloaded. Since the Gaussian table for the thousands of senones accounts for the vast majority of memory traffic, this audio frame grouping technique can reduce memory bandwidth to 1/N of the original algorithm. Complexity of the control logic for this optimized processing order is the same as for the original algorithm. The only downside is extra initial delay of N audio frames. In our default sampling granularity of 10ms per audio frame, N=8 will cause an initial delay of 80ms, which is hardly perceivable to the human being. This grouping technique was proposed by Mathew [8] and used in a number of GMM software optimizations and hardware accelerators [9, 10, 11].

The Mach and GMM accelerators employ SRAM as a staging buffer in the address space, without which all subsequent values would have to be read from DRAM one at a time. One approach is to support a SRAM size that can accommodate the entire size of data. However, since an SRAM of this size would consume an unrealistically large area, smaller SRAMs that support a partial data buffering were embedded and the loop in Algorithms are rearranged using loop block technique. However, two accelerators have dedicated SRAMS in the size of 8KB and 5.33KB for the Match and GMM, respectively. Because the control flows of the two accelerators are similar, we can further reduce the size of SRAM by sharing between two accelerators, embedding only 8KB memory.

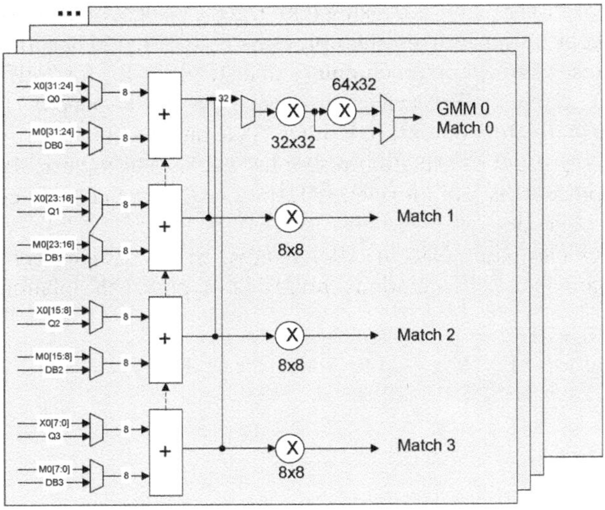


**Fig. 5.** Microarchitecture of the proposed accelerator

## 4 Conclusions

In this paper, we investigated image and speech recognition algorithms and presented an opportunity to sharing the computation and storage resources on embedded systems by proposing the unified computation unit. We plan to continue studying additional recognition workload and further refining the computation unit in order to reduce the hardware cost in embedded systems. We also plan to integrate the proposed computation unit on FGPA and emulate the unified accelerator proposed along with RISC cores to study the interaction between hardware and software components. Last but not least, we also plan to implement our recognition processor on silicon. We expect that sharing resources among different applications on embedded systems will bring forth a new spectrum of optimizations for emerging digital systems.

## References

1. Takacs, G., et al.: Outdoors Augmented Reality on Mobile Phone using Loxel-Based Visual Feature Organization. In: ACM ICMR (2008)
2. Lee, S.E., et al.: Accelerating Mobile Augmented Reality on a Handheld Platform. In: IEEE Int'l. Conf. on Computer Design (ICCD), pp. 419–426 (2009)
3. Lowe, D.G.: Distinctive image features from scale-invariant keypoints. Int. J. Comput. Vision 60(2), 91–110 (2004)

4. Bay, H., Tuytelaars, T., Van Gool, L.: SURF: Speeded up robust features. In: Leonardis, A., Bischof, H., Pinz, A. (eds.) ECCV 2006. LNCS, vol. 3951, pp. 427–434. Springer, Heidelberg (2006)
5. Takacs, G., Chandrasekhar, V., Gelfand, N., Xiong, Y., Chen, W.-C., Bismpigiannis, T., Grzeszczuk, R., Pulli, K., Girod, B.: Outdoors augmented reality on mobile phone using loxel-based visual feature organization. In: MIR 2008, pp. 427–434 (2008)
6. Carnegie Mellon University, the Sphinx Project, `http://www.speech.cs.cmu.edu/`
7. Iyer, R., et al.: CogniServe: Heterogeneous Server Architecture for Large-Scale Recognition. In: IEEE Micro (2011)
8. Mathew, B., Davis, A., Fang, Z.: A Low Power Accelerator for the Sphinx3 Speech Recognition System. In: CASES 2003 (2003)
9. Choi, Y.K., You, K., Choi, J., Sung, W.: VLSI for 5000-word Continuous Speech Recognition. In: ICASSP 2009 (2009)
10. Lin, E.C., Yu, K., Rutenbar, R.A., Chen, T.: A 1000-word Vocabulary, Speaker-Independent, Continuous Live-mode Speech Recognizer Implemented in a Single FPGA. In: FPGA 2007 (2007)
11. Ma, T., Deisher, M.: Novel C.I-backoff Scheme For Real-Time Embedded Speech Recognition. In: ICASSP 2010 (2010)

# Differential Detection Techniques for Spectrally Efficient FQPSK Signals

Hyung Chul Park

Dept. of Electronic and Information Engineering,
Seoul National University of Science and Technology, Seoul, 139-743, Korea
hcpark@seoultech.ac.kr

**Abstract.** A new class of differential detection techniques for standardized Feher patented quadrature phase-shift keying (FQPSK) systems is proposed and studied by computer aided design/simulations. It is shown that significant BER performance improvements can be obtained by increasing the received signal's observation time over multi-symbol as well as by adopting trellis-demodulation. While the FQPSK transmitted signal is not trellis coded, its trellis coded interpretation leads to trellis demodulation. For example, our simulation results show that a BER=10-4 can be obtained for an Eb/N0=14.1 dB.

**Keywords:** FQPSK, Differential detection, Multi-symbol observation, Viterbi algorithm.

## 1 Introduction

Multiyear studies by the U.S. Department of Defense (DoD), NASA, AIAA, and the International Committee Consultative on Space Data Systems (CCSDS) confirmed that Feher patented quadrature phase-shift keying (FQPSK) [1], [2] technologies and Commercial-Off The-Shelf (COTS) FQPSK products offer the most spectrally efficient and robust (smallest degradation from ideal theory) bit error rate (BER) performance of non-linear amplification (NLA) RF power efficient systems. Based on numerous FQPSK airplane-to-ground, ground-to-ground and satellite tests in the 1Mb/s to 600Mb/s range, FQPSK has been specified in the new spectrally efficient telemetry standard known as IRIG 106-00 and recommended in the international CCSDS for use in high speed space communications-data systems.

The coherent detection methods have been used for FQPSK systems. However, since phase noise caused by oscillators and frequency synthesizers, and relatively large Doppler spread may degrade the performance of relatively low bit rate coherent demodulators and may increase the synchronization time, non-coherent detection is preferable for certain mobile applications. In [3], the limiter-discriminator based non-coherent detection techniques for FQPSK-B have been introduced.

In this paper, differential detector based non-coherent detection techniques for FQPSK-B signal are proposed and their BER performance in a Gaussian channel is compared using a simulation study.

S. Lin and X. Huang (Eds.): CSEE 2011, Part II, CCIS 215, pp. 157–161, 2011.

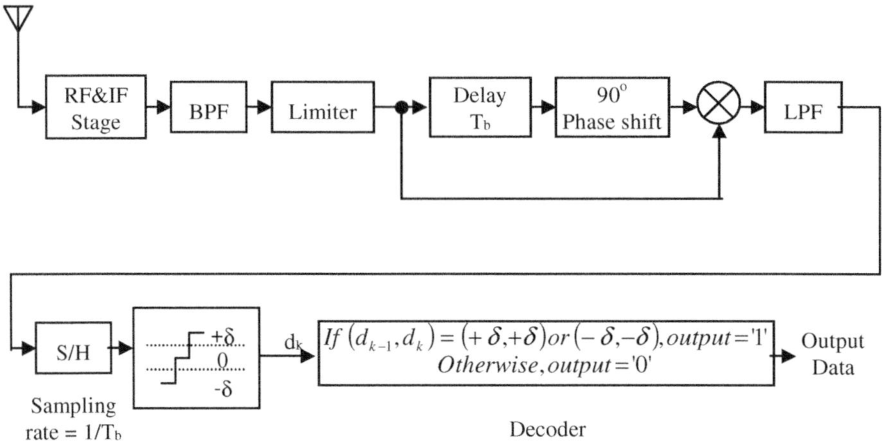


**Fig. 1.** Receiver structure based on one-bit differential detector followed by symbol-by-symbol decision for QDE FQPSK-B signal

## 2 One-Bit Differential Detection Followed by Symbol-by-Symbol Decision

It was shown that FQPSK-B modulated signals with amplitude parameter $A=1/\sqrt{2}$ is quasi-constant envelope signal and is interpreted as a continuous phase modulation (CPM), which can be demodulated by frequency discrimination and/or by phase detection [3-4].

The one of the NLA-FQPSK signals characteristics is the following [3]:

*The total allowed phase changes between $n{*}T_b$ and $(n+2)T_b$ are* $\Delta\varphi = \pm\pi, \pm\frac{3}{4}\pi, \pm\frac{\pi}{2}, \pm\frac{\pi}{4}, 0$. *Among these,* $|\Delta\varphi| > \frac{\pi}{4}$ *indicates the data at $(n+2)T_b$ has been changed from that at $n{*}T_b$. Otherwise, there is no change.*

It means that the FQPSK-B signals can be demodulated with differential detector based non-coherent detection techniques. Likewise in limiter discriminator based non-coherent detection for FQPSK-B signals, FQPSK-B transmitter employs *quadrature differential encoder (QDE)*, which has been presented in [3], for differential detection.

In the one-bit differential detector for FQPSK-B, intermediate frequency (IF) band pass filtered signal is multiplied by a variation of itself that is delayed by one bit duration ($T_b$) and phase-shifted by 90°. A limiter is used to normalize the signal envelope. The output of one-bit differential detector represents the *sine* of the phase change over $1T_b$ interval, i.e., $-\frac{\pi}{2}, -\frac{\pi}{4}, 0, +\frac{\pi}{4}, +\frac{\pi}{2}$ [3]. If the consecutive two-outputs of one-bit differential detector are (+δ, +δ), in which +δ is corresponding to $\sin\left(+\frac{\pi}{4}\right), \sin\left(+\frac{\pi}{2}\right)$, or (-δ, -δ), in which -δ is corresponding to $\sin\left(-\frac{\pi}{4}\right), \sin\left(-\frac{\pi}{2}\right)$, '1' is transmitted. Otherwise, '0' is transmitted. The conceptual block diagram of one-bit differential detection followed by symbol-by-symbol decision for FQPSK-B signals is presented in Fig. 1.

## 3 One-Bit Differential Detection Followed by Multi-Symbol Observation and Decision

It is noticed that only certain combinations of phase changes of FQPSK-B modulated signals are allowed, i.e., FQPSK signal has memory [3]. In this case, it is well known that the detection based on multi-symbol observation performs better than symbol-by-symbol detection [5]. Note that the concept of multi-symbol observation is similar to coherent detection using trellis decoding of FQPSK signals, described in [6], which traces allowed I and Q amplitude signals while we trace the phase trellis in this work.

As described in [1] and [3], 5 data sequence defines two phase change for $1T_b$ interval at two symbol points. So, N+3 data sequence is required to get a phase change vector that is composed of "N" phase change components. This means that many of the input vectors in L-space are mapped into the identical vector in m-space. Thus, the ratio for this is shown as the L/m in Table. 1. As the symbol observation interval increases, the ratio, m/M, of number of the allowed phase change vector to the total number of random combinations of symbol-by-symbol data (3 level in our case of Section II) reduces significantly, as shown in Table. 1. This means that the correction ability for the erroneous one-bit differential detector output value improves significantly.

Three detection techniques with multi-symbol observation are studied in our work, i.e., multi-symbol observation and middle bit decision, multi-symbol observation and majority voting, and maximum likelihood sequence detection (MLSD) with multi-symbol observation.

**Table 1.** Relationship between symbol observation interval and number of the allowed phase change, for $T_b$ interval, vector for the N-step process during $(N+3)T_b$ interval

| N | $L=2^{N+3}$ | $M = 3^N$ | M | m/M | L/m |
|---|---|---|---|---|---|
| 3 | 64 | 27 vectors | 31 vectors | 115% | 2.06 |
| 5 | 256 | 243 vectors | 127 vectors | 52.3% | 2.02 |
| 7 | 1024 | 2,187 vectors | 511 vectors | 23.4% | 2.00 |

N : Number of symbols observed.
L : Number of possible vectors tupled with (N+3) binary data sequence
M : Total number of vectors formed by random combination of symbol-by-symbol data
m : Number of allowed N-tupled vector

In the proposed MLSD, the states of trellis are the allowed phase change vectors in m-space. The branch metric is defined as the distance between N-tupled vector, formed by consecutive one-bit differential detector output data, and that by the allowed phase change vectors. The survival path is the path that has the smallest accumulated branch metric, i.e., state metric. MLSD output data are the middle bit of states, which are the allowed phase change vectors in m-space, on the survival path. The decoder uses them for final output data.

## 4 Two-Bit Differential Detection for FQPSK Signals

In the two-bit differential detector for FQPSK-B signals, IF band pass filtered signal is multiplied by a version of itself that is delayed by $2*T_b$. In the two-bit differential detector followed by symbol-by-symbol decision shown in Fig. 2, if the output of two-bit differential detector is corresponding to $\cos(0), \cos\left(\pm\frac{\pi}{4}\right)$, '0' is transmitted. Otherwise, '1' is transmitted. The decoder, which is presented in Fig. 1, isn't required.

Likewise in Section III, the multi-symbol observation method can improve the performance of two-bit differential detection. However, the allowed phase change vectors and element values differ from that of one-bit differential detection.

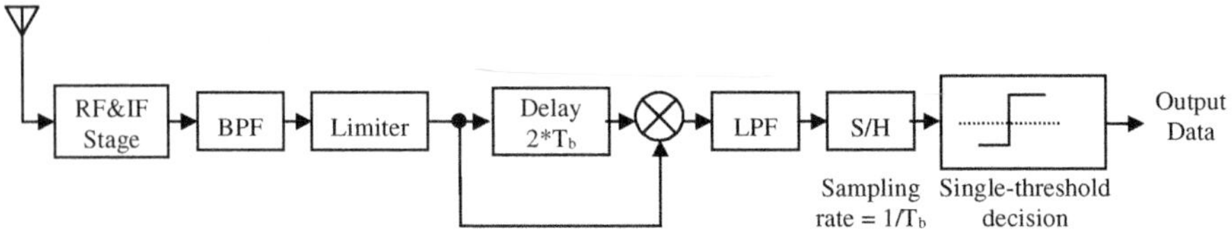


**Fig. 2.** Receiver structure based on two-bit differential detector followed by symbol-by-symbol decision for QDE FQPSK-B signal

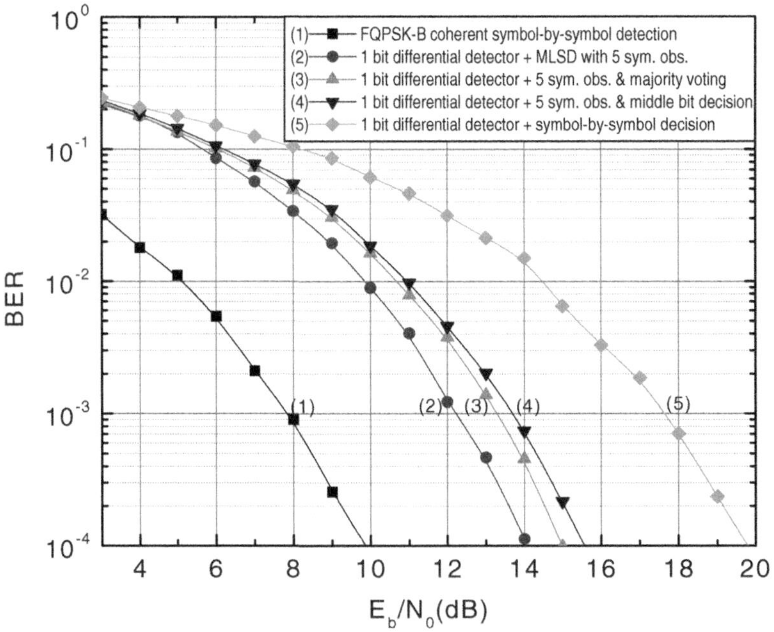


**Fig. 3.** BER performance of one-bit differential detector-based non-coherent detection techniques

## 5 Simulation Results and Discussion

In the simulation, hard limiter is assumed to approximate the Non-Linear Amplifier (NLA) in the transmitter. Fig. 3 shows the BER performance of one-bit differential detector-based non-coherent detection techniques in the presence of AWGN. The one-bit differential detector followed by symbol-by-symbol decision scheme suffers as large as 10.0dB degradation at BER=$10^{-4}$ from the best symbol-by-symbol coherent detection of FQPSK-B. But this degradation decreases significantly as we increase the observation time. Middle bit decision, majority voting, and MLSD based on 5-symbol observation lead to 5.8dB, 5.2dB, 4.3dB degradation at BER=$10^{-4}$ compared with best symbol-by-symbol coherent detection of FQPSK-B performance.

In addition, the two-bit differential detector followed by symbol-by-symbol decision scheme suffers 11.0dB degradation at BER=$10^{-4}$ from the best symbol-by-symbol coherent detection of FQPSK-B performance. But this degradation decreases significantly as we increase the observation time. Middle bit decision, majority voting, and MLSD based on 5-symbol observation lead to 6.3dB, 6.2dB, 5.6dB degradation at BER=$10^{-4}$ compared with best symbol-by-symbol coherent detection of FQPSK-B performance.

## 6 Conclusions

Based on the CPM based interpretation, we have proposed differential detection techniques for FQPSK-B. It has been shown that the BER performance of the one-bit/two-bit differential detector-based non-coherent detection techniques improves significantly using the inherent memory in the FQPSK-B signal phase, i.e., middle bit decision, majority voting, and MLSD with multi-symbol observation.

Simulation result shows that one-bit differential detector followed by MLSD with 5-symbol observation performs BER=$10^{-4}$ at $E_b/N_0$ =14.1dB. It suffers 4.3dB degradation at BER=$10^{-4}$ from the best symbol-by-symbol coherent detection of FQPSK-B performance.

## References


1. Feher, K., et al.: U.S. patents: 4,567,602; 4,644,565; 5,491,457; 5,784,402; 6,198,777, International Patent Cooperation Treaty (PCT) No: PCT US99/17995; International No: WO 00/10272 (2000)
2. Feher, K.: Spectrally Efficiency of Commercial Wireless and Telemetry Systems Is Doubled with IRIG 106-00 Standardized FQPSK and Is Quadrupled with FQAM. In: Proc. of ITC 2000, pp. 225–237 (2000)
3. Park, H.C., Lee, K., Feherk, K.: Non-coherent Detection of FQPSK Spectrally Efficient Systems. In: Proc. of European Test and Telemetry Conf., pp. 133–138 (2001)
4. Park, H.C., Lee, K., Feherk, K.: Continuous Phase Modulation of F-QPSK-B Sginals. IEEE Trans. Veh. Technol. 56(1), 157–172 (2007)
5. Anderson, J.B., Aulin, T., Sundberg, C.E.: Digital Phase Modulation. Plenum Press, New York (1986)
6. Lee, D., Simon, M., Yan, T.: Enhanced Performance of FQPSK-B Receiver Based on Trellis-Coded Viterbi Demodulation. In: Proc. of ITC 2000, pp. 631–640 (2000)

# Improved Car-Following Model for Traffic Flow and Its Numerical Simulation on Highway with Gradients

Wen-Xing Zhu[1,2,*] and Zhi-Ping Jia[1]

[1] School of Computer Science and Technology Shandong University Jinan, Shandong 250061, China
[2] School of Control Science and Engineering Shandong University Jinan, Shandong 250101, China
{zhuwenxing,jzp}@sdu.edu.cn

**Abstract.** We investigated the effect of the slope upon traffic flow on a single lane highway with an uphill gradient and a downhill gradient. The model was improved by introducing the variable brake distance on different gradient. A simulation is carried out to examine the validity and reasonability of the improved model. The result of the simulation shows that the amplitude of the density waves decreases with the decrease of the slope of the gradient on highway. Moreover the density waves propagate backward. The results indicated that the new model was reasonable and valid in describing the motion of the vehicles on highway with some gradients.

**Keywords:** Car-Following Model, Numerical Simulation, Traffic flow, Gradients.

## 1 Introduction

Recent decades, various traffic models including car-following models, cellular automaton models, hydrodynamics models, gas kinetics models *et al* are developed to disclose the nature of the traffic jams by the physicists with different background [1- 21]. In 1953, Pipes [1] proposed a classical car-following model to describe the motion of the two successive cars. After a long time about forty years Bando *et al* [2] improved the classical car-following model by proposing an optimal velocity model (OVM) to overcome the shortcoming of the Pipes' model in 1995. The OVM may describe the dynamical behaviors of the vehicles on a highway under a high density condition more realistically. Since then, many physicist and scholars improved the car-following models based on the OVM. In these modified models various effects were taken into account including the interactions [7, 17, 19], forward-looking [12- 14, 18], backward-looking [12], velocity difference [15], reaction-time delay [20] and gravitational force effect [21] so on. Z. P. Li [16] and W. X. Zhu [17, 18] *et al* discussed the density waves of the traffic flow and its performance. K. Komada *et al* took into account the effect of gravitational force upon traffic flow on highway with sags, uphill and downhill. The gravitational force was considered as an external force which

* Corresponding author.

S. Lin and X. Huang (Eds.): CSEE 2011, Part II, CCIS 215, pp. 162–168, 2011.

acts on vehicles. Author studied the traffic states and jamming transitions induced by the slope of the highway and derived the fundamental diagram for the traffic flow on the sag. But they did not investigate the effect of the different slope of the sags which can show different traffic states and jamming transitions.

## 2 Model

We consider a situation as K. Komada *et al* [21] had done such that vehicles move ahead on a single lane highway with uphill and downhill gradient. Traffic flow is under a periodic boundary condition. The gravitational force acts on the vehicles on the slope of the gradient. Fig. 1 shows the illustration of the gravitational force working upon a vehicle moving on a downhill. The slope of the gradient is represented by $\theta$, the gravity is g and the mass of the vehicle is m. Then, a horizontal force mg sin$\theta$ acts on the vehicle when the driver does not operate the brake. If a driver operates the brake, the external force would be reduced by the brake control.

The extended model with an effect of the slope was formulated as follows:

$$\frac{d^2x_i(t)}{dt^2} = a\left\{V(\Delta x_i) - \frac{dx_i(t)}{dt}\right\} \tag{1}$$

with $V(\Delta x_i) = \frac{v_{f,max}}{2}[\tanh(\Delta x_i - x_c) + \tanh(x_c)] + \frac{v_{g,up,max}}{2}\left[\tanh\left(\Delta x_i - x_{up,b}\right) + \tanh\left(x_{up,b}\right)\right]$ (2)

for a highway with an uphill gradient, and

$$V(\Delta x_i) = \frac{v_{f,max}}{2}[\tanh(\Delta x_i - x_c) + \tanh(x_c)] + \frac{v_{g,down,max}}{2}\left[\tanh\left(\Delta x_i - x_{down,b}\right) + \tanh\left(x_{down,b}\right)\right] \tag{3}$$

for a highway with a downhill gradient, where $x_i(t)$ is the position of vehicle i at time *t*, $\Delta x_i(t) = x_{i+1}(t) - x_i(t)$is the headway of vehicle *i* at time *t*, $V(\Delta x_i)$ is the optimal velocity function of vehicle *i* at time *t*, $v_{f,max}$ is the maximal velocity on the highway without any slope, $v_{g,up,max}$ and $v_{g,down,max}$ are the maximal reduced and enhanced velocity on uphill and downhill gradient which are formulated as $v_{g,up,max} = -\frac{mgsin\theta}{\gamma}$, $v_{g,down,max} = \frac{mgsin\theta}{\gamma}$ respectively where $\gamma$ is the friction coefficient, $x_c$ is the safety distance, $x_{up,b}$ and $x_{down,b}$ are the brake distance for the uphill and downhill gradient respectively.

As well known the brake distance of a vehicle is not a constant on the uphill or downhill gradient. It should vary with the slope of gradient. In the above mathematical equations the brake distance model is not reflect the real situation by taking a constant.

We proposed a new brake distance model which is a function of the slope $\theta$. The brake distance is related to the slope of the uphill and downhill gradient. We give two brake distance models respectively.

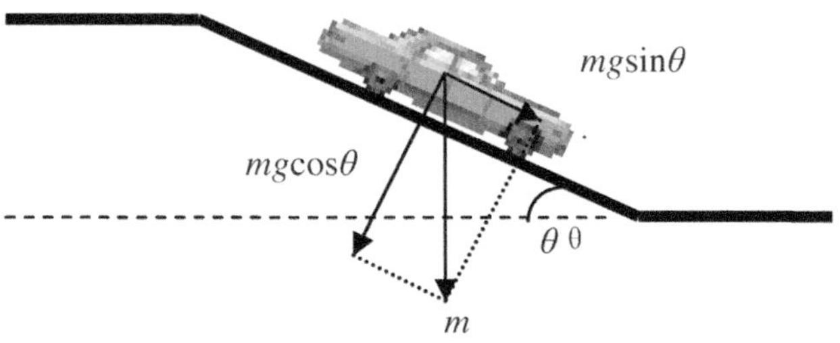


**Fig. 1.** Illustration of the gravitational force working upon a vehicle on a downhill slope

$$x_{up,b} = x_c(1 - \sin\theta) \tag{4}$$

$$x_{down,b} = x_c(1 + \sin\theta) \tag{5}$$

These two models can reflect the relationship between the slope and the brake distance. With the increase of the slope the brake distance of the vehicle increases on the downhill gradient and decreases on the uphill gradient. They can be used to examine the effect of the slope upon traffic flow. When the slope $\theta$ takes zero then $x_{up,b} = x_{down,b} = x_c$ and the whole model has the same form as Bando's optimal velocity model.

## 3 Simulation and Result

We perform a computer simulation to check out the performance of the new model. In order to reflect the real traffic situations we choose the slope $0° \leq \theta \leq 30°$ and there are an uphill gradient and a downhill gradient on a highway with a total length $1600m$. Without loss of generality length of the uphill gradient, the downhill gradient and the link between the uphill and downhill gradient is taken 100m respectively. We put 200 vehicles homogeneously on the highway and they move according to the motion equation. The 0 point is set at the position before the uphill gradient 100m. For simplicity we take $v_{g,up,max} = -\sin\theta$ and $v_{g,down,max} = \sin\theta$ . When $= 0°$ $v_{g,up,max} = v_{g,down,max} = 0\ m/s$ . When $\theta = 30°$ $v_{g,up,max} = -0.5\ m/s$ $v_{g,down,max} = 0.5\ m/s$.

We conduct simulations under four conditions $\theta = 0°, 10°, 20°, 30°$. The simulation duration is taken as long as $200000s$ which is enough for the vehicles to reach the steady state and the time step are taken as $\Delta t = 0.1s$. The result of the simulation is plotted in Fig.2- Fig.7. Fig.2 and Fig.3 are headways profile and velocities profile for 200 vehicles on the highway with uphill and downhill slope varying from $30°$ *to* $0°$.

Patterns (a) to (d) in Fig.2 are headway profiles at $t$=19800$s$ are corresponding to $\theta = 30°, \theta = 20°, \theta = 10°, \theta = 0°$, respectively. From these patterns we can observe that the amplitude of the headway decrease with the decrease of the slope $\theta$ of the uphill and downhill. When the slope takes zero the amplitude of the headways is also equal to zero. In Fig. 4 patterns (a) to (d) are velocities profiles which have the same law as that of the headways profiles. From the above analysis we can conclude that the slope of the gradient on highway plays an important role on the traffic flow. So the road should be built with no gradient or low slope of gradient on the highway. Fig. 4 to Fig. 7 are space-time evolution of the headways and velocities after $t$=19800$s$ corresponding to the slope $\theta = 30°, \theta = 20°, \theta = 10°, \theta = 0°$, respectively. Patterns (a) and (b) represent space-time evolution of headways and velocities in each figure. From these patterns we can observe that density waves of the traffic flow take place and propagate backward when the vehicles move on the highway with uphill and downhill gradients. Moreover we can also draw the same conclusions the above with increase of the slope of gradients the amplitude of the density waves increases. When the slope takes $\theta = 0°$ the density waves disappear.

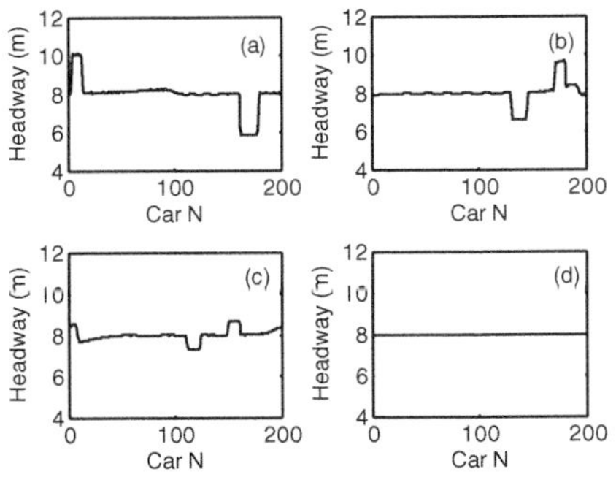


**Fig. 2.** Profiles of the headways at $t$=19800$s$ are corresponding to the slope (a)30°, (b)20°, (c)10 °, (d) 0°

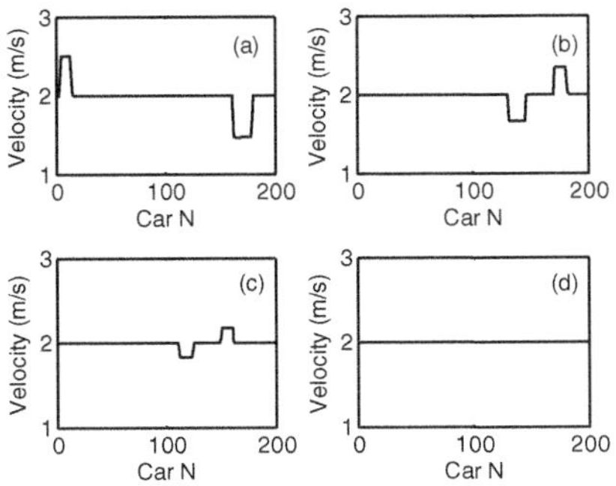


**Fig. 3.** Profiles of the velocities at $t$=19800$s$ are corresponding to the slope (a)30°, (b)20°, (c)10 °, (d) 0°

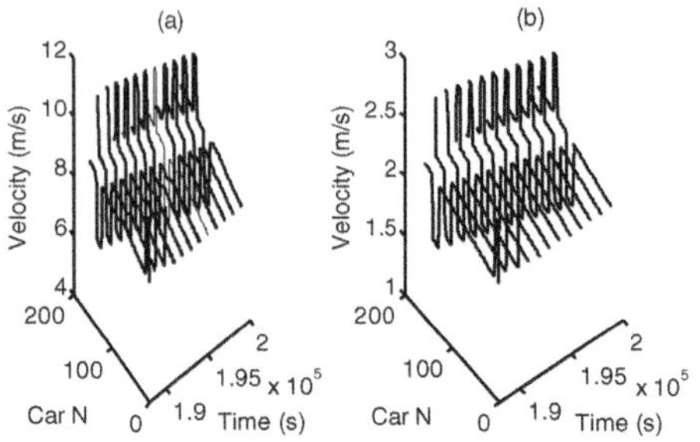


**Fig. 4.** Space-time evolution of the headways and velocities profiles after $t$=19800$s$ are corresponding to the slope $\theta = 30°$

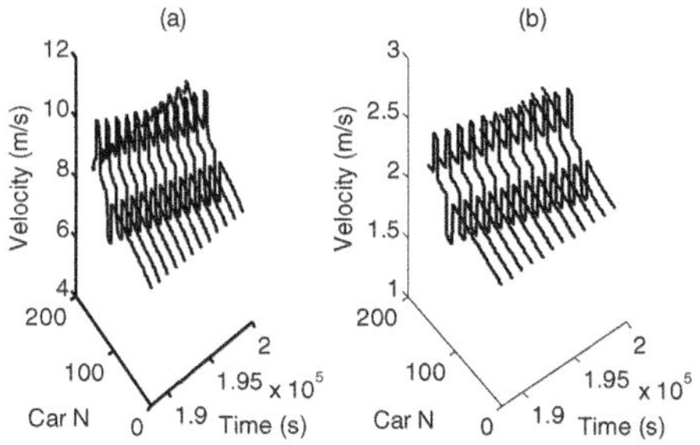


**Fig. 5.** Space-time evolution of the headways and velocities profiles after $t$=19800$s$ are corresponding to the slope $\theta = 20°$

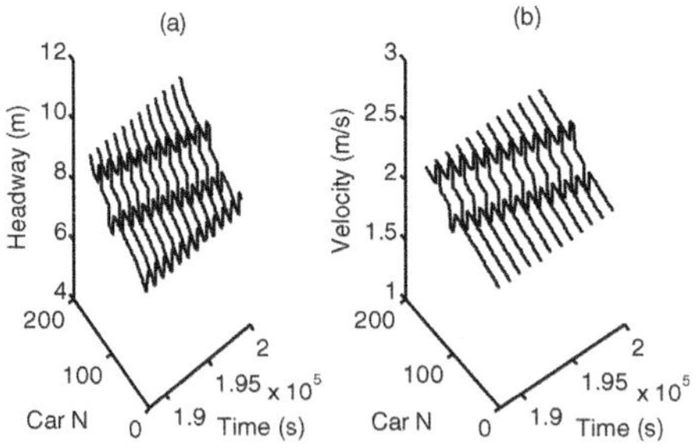


**Fig. 6.** Space-time evolution of the headways and velocities profiles after $t$=19800$s$ are corresponding to the slope $\theta = 10°$

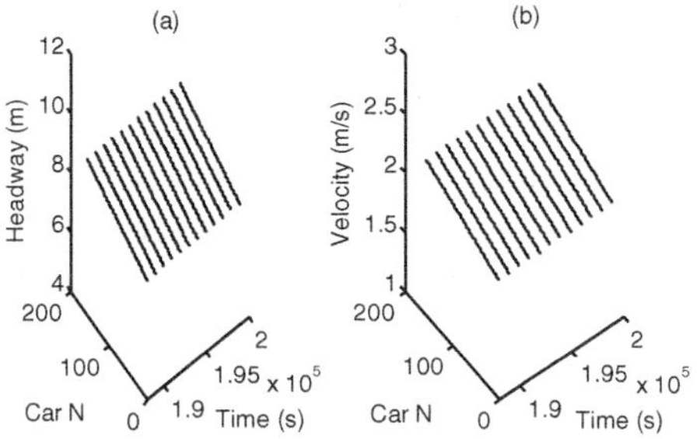


**Fig. 7.** Space-time evolution of the headways and velocities profiles after $t$=19800$s$ are corresponding to the slope $\theta = 0°$

# 4 Summary

We improved the Komada's model by introducing the variable slope effect of the gradient on highway. The mathematical models were formulated. A simulation was carried out to examine the effect of the variable slope of the gradients. The result of the simulation was in good agreement with the real traffic situations. The density waves were also got through the simulation. The new model is reasonable and valid in describing the motion of the vehicles on the highway with some gradients.

**Acknowledgements.** This work is partially supported by Middle-Aged and Young Scientists Research Incentive Fund of Shandong Province under Grant No. 2007BS01013, Shandong Province Natural Science Foundation under Grant No.ZR2009GM032, ZR2010FM036, Independent Innovation Foundation of Shandong University, GIIFSDU under Grant No. 2009TS046 and China Postdoctoral Science Foundation under Grant No. 20100481265.

# References

[1] Pipes, L.A.: An operational analysis of traffic dynamics. J. Appl. Phys. 24, 274–281 (1953)

[2] Bando, M., Hasebe, K.: Dynamical model of traffic congestion and numerical simulation. Phys. Rev. E 51, 1035–1042 (1995)

[3] Kerner, B.S., Konhauser, P.: Structure and parameters of clusters in traffic flow. Phys. Rev. E 48, 2335–2367 (1993)

[4] Sawada, S.: Nonlinear analysis of a differential-difference equation with next-nearest-neighbour interaction for traffic flow. J. Phys. A: Math. Gen. 34, 11253–11259 (2001)

[5] Hasebe, K., Nakayama, A., Sugiyama, Y.: Dynamical model of a cooperative driving system for freeway traffic. Phys. Rev. E 68, 26102 (2003)

[6] Komatsu, T., Sasa, S.: Kink soliton characterizing traffic congestion. Phys. Rev. E 52, 5574–5582 (1995)
[7] Nagatani, T.: Stabilization and enhancement of traffic flow by the next-nearest-neighbor interaction. Phys. Rev. E 60, 6395–6401 (1999)
[8] Lighthill, M.J., Whitham, G.B.: On Kinematic Waves. I. Flood Movement in Long Rivers. In: Proc. Roy. Soc. A, vol. 229, pp. 281–316 (1995)
[9] Richards, P.I.: Shock Waves on the Highway. Oper. Res. 4, 42–51 (1960)
[10] Tang, T.Q., Huang, H.J., Xu, X.Y., Xue, Y.: Analysis of density wave in two-lane traffic. Chin. Phys. Letters 24, 1410–1413 (2007)
[11] Hasebe, K., Nakayama, A., Sugiyama, Y.: Dynamical model of a cooperative driving system for freeway traffic. Phys. Rev. E 68, 26–102 (2003)
[12] Hasebe, K., Nakayama, A., Sugiyama, Y.: Equivalence of linear response among extended optimal velocity models. Phys. Rev. E 69, 17103 (2004)
[13] Lenz, H., Wagner, C.K., Sollacher, R.: Multi-anticipative car-following model. Eur. Phys. J. B 7, 331–340 (1998)
[14] Ge, H.X., Dai, S.Q., Dong, L.Y., Xue, Y.: Stabilization effect of traffic flow in an extended car-following model based on an intelligent transportation system application. Phys. Rev. E 70, 66134 (2004)
[15] Jiang, R., Wu, Q.S., Zhu, Z.J.: Full velocity difference model for a car-following theory. Phys. Rev. E 64, 17–101 (2001)
[16] Li, Z.P., Liu, Y.C.: Analysis of stability and density waves of traffic flow model in an ITS environment. Euro. Phys. J. B 53, 367–372 (2006)
[17] Zhu, W.X., Jia, L.: Stability and kink-antikink soliton solutions for total generalized optimal velocity model. International Journal of Modern Physics C 19, 1321–1334 (2008)
[18] Zhu, W.X., Jia, L.: Nonlinear analysis of a synthesized optimal velocity model for traffic flow. Communications in Theoretical Physics 50, 505–509 (2008)
[19] Zhu, W.X., Liu, Y.C.: Total generalized optimal velocity model and its numerical test. Journal of Shanghai Jiaotong University (English Edition) 13, 166–170 (2008)
[20] Yu, L., Li, T., Shi, Z.K.: Density waves in a traffic flow model with a reaction time delay. Physica A 398, 2607–2616 (2010)
[21] Komada, K., Masakura, S., Nagatani, T.: Effect of gravitational force upon traffic flow with gradients. Physica A 388, 2880–2894 (2009)

# Cycle Accurate Power and Performance Simulator for Design Space Exploration on a Many-Core Platform

Seung Eun Lee

Dept. of Electronic and Information Engineering
Seoul National University of Science and Technology, Seoul, Korea
seung.lee@seoultech.ac.kr

**Abstract.** As the computation demands increases to meet the design requirements for computation-intensive applications, the pressure to develop high performance parallel processors on a chip is increasing. However, software supports that enable harnessing parallel computing power of this type of architecture have not followed suit and it has become an important stumbling block for future many core application developments. In this paper, we introduce a cycle-accurate simulator for parallel programming on many-core platform, enabling fast design space exploration.

**Keywords:** Multiprocessor, Network-on-Chip, simulator, power model, design space exploration.

## 1 Introduction

High performance processor design is rapidly moving towards many-core architectures that integrate tens or hundreds of processing cores on a single chip. For instance, Intel recently announced its research prototype of 48 cores on a single die and IBM will soon release the IBM Cyclops-64 chip that will support 160 hardware thread units per chip. This new generation of architectural technology will integrate a large number of tightly-coupled simple processor cores on a single chip. As the demand for network bandwidth increase for many-core architecture, on chip interconnection network, the idea of Network-on-Chip (NoC) has shown great promise in terms of performance, power, and scalability in many-core design. NoC conceptually consists of processors and memories connected together using a network of routers. In this organization, resources communicate with each other incorporating packets that are routed through the network, as a traditional packet switching network does.

Although hardware development is making great progress for the many-core technology, software supports that enable harnessing parallel computing power of this type of architecture have not followed suit and it has become an important stumbling block for future many-core application developments. More specifically, the limitation with the current many-core technology is due to little support for programming, limiting applications to fully benefit from the continued growth of processing capability. In this paper, we propose a software tool that provides cycle accurate power/performance simulation for many-core platform.

S. Lin and X. Huang (Eds.): CSEE 2011, Part II, CCIS 215, pp. 169–175, 2011.

The rest of this paper is organized as follows. We first introduce the related works in this area in section 2 and describe an overview of the many-core platform in section 3. Section 4 introduces the cycle accurate simulator for power and performance analysis and performs design space exploration with a benchmark in order to show the utility of the simulator. We conclude in section 5 by outlining the direction for future work on this topic.

## 2 Related Works

NoC architecture simulators have been developed for design space exploration support NoC architecture performance estimation. Noxim [1] was developed in SystemC and allows exploring the design space spanned by the different parameters of the architecture for the analysis and evaluation of a large set of performance metrics including delay, throughput, energy consumption, and others. [2] is a system-level tool-chain called Polaris for NoC that helps designers predict the most suitable interconnection network tailored for their performance needs and power/silicon area constraints. It is built upon three tools, Trident -a traffic synthesizer/analyzer, LUNA - link utilization tool for network power analysis and ORION -a library of power, area and router pipeline delay models. Polaris, like Noxim, also focuses on design space exploration of NoC architecture and does not incorporate programmable processing elements in the simulator. CAFES (Communication Analysis For Embedded Systems) [3] enables users to map application onto NoCs. The behavior of an application can be described with models that consider different aspects, with respect to computation and communication. In this system, the user can choose one of the six application models and describe some of the NoC parameters. Based on the configurations, it can estimate the NoC behavior and the energy consumption. Our simulator is a cycle-accurate NoC architecture simulator implemented in the SysetmC library. Unlike other simulators, the programmability support of embedded processors makes it a strong candidate tool for extending the software development for application mapping on a many-core platform.

## 3 Many-Core Platform

Networked Processor Array (NePA) [4] is a 2-dimensional many-core NoC platform with mesh-topology as shown in Figure 1. This platform interconnects processing elements (PE), each of which includes a programmable processor and memory modules. By virtue of scalability of NoC, the number of connected processors or IPs is not fixed in this platform. A scalable multi-processor architecture allows parallel processing for several applications. Our platform includes a complete multi-core NoC simulator written in SystemC and a synthesizable RTL model written in Verilog HDL where adaptive routers, Network Interface (NI), and processors (compact version of OpenRISC) are integrated. The compact OpenRISC is connected with our optimized router [5] through the Network Interface [6]. NePA supports programmability and some of digital signal processing such as ray tracing, encryption, fft, etc., have been implemented and tested, demonstrating its potential as an embedded many-core solution [7, 8, 9].

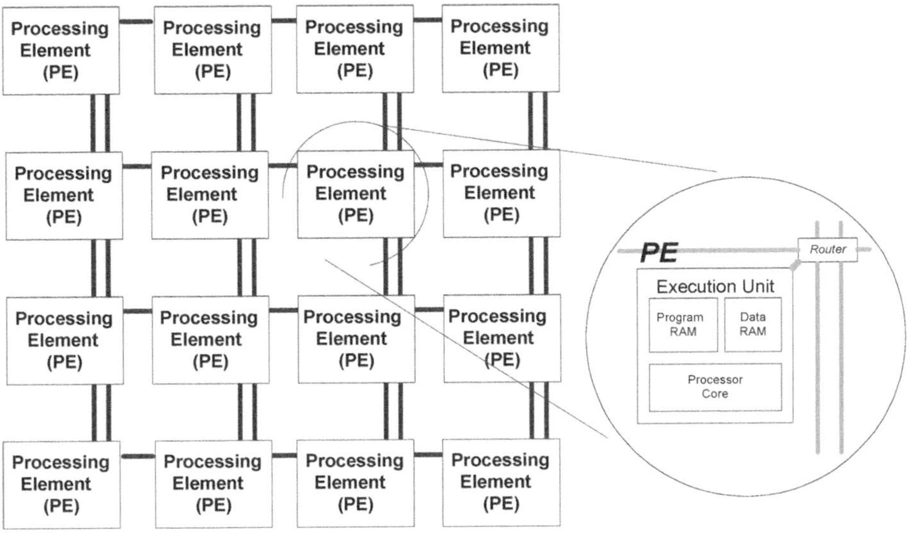


**Fig. 1.** Homogeneous many-core NoC platform

### 3.1 Fully Adaptive Router

The fully adaptive router uses a wormhole switching with a deadlock- and livelock-free algorithm for 2D-mesh topology. There is an input FIFO queue per each input channel and each output port has an associated arbiter to choose the proper packet among the given incoming packets from each candidate input port. A router with seven interfaces, suitable for a 2D-mesh, is considered which includes interface to an integrated Processing Element (PE). It is assumed that a packet coming through an input port does not loop back, thus, each input port is connected to four output ports. The separated routing paths for a vertical direction and the unidirectional path for a horizontal direction allow the network to avoid cycles in its channel-dependency-graph, resulting in a dead lock-free operation. Also, by choosing the shortest path in routing, a livelock free operation is guaranteed.

### 3.2 Processing Elements

Each of the PEs can perform kernel-level operations in parallel, depending on applications. OpenRISC core is adopted as a homogeneous PE, which is a 32-bit RISC processor with Harvard architecture. After extensive refinement, compact OpenRISC core is designed where some specific blocks such as I/D-cache, I/D-MMU (memory management unit), and debugging unit are omitted from the original OpenRISC core for simplicity and cost minimization. The compact OpenRISC core is interconnected throughout 2D meshed network where each of OpenRISC cores is encapsulated with network related components such as our generic NI and a router. The compact OpenRISC processing element has local program/data memory which is directly connected with OpenRISC core for fast access.

### 3.3 Network Interface

Network Interface (NI) translates packet-based communication into a higher-level protocol that is required by PEs resulting in packetizing and de-packetizing the requests and responses from the cores. Decoupling of communication from computation is a key concept in NoC design. This requires a well-defined NI that integrates IP cores onto the on-chip network in order to hide the implementation details of an interconnection. NI is located between a router and a PE, decoupling the communication from computation. It offers a memory-mapped view on all control registers in NI so that a PE can access the registers in NI by using read/write operations.

## 4 Performance and Power Simulator

Both SystemC model and synthesizable Verilog HDL description for NePA were developed. From the SystemC model and its simulation with traffic patterns, timing-accurate test vectors are generated for the RTL/gate-level simulation and verification. Throughout the synthesis with 90nm technology, physical characteristics including power model are extracted and applied to build the development environment.

### 4.1 High-Level Power Model

Although today's processors are much faster and far more versatile than their predecessors using high-speed operation and parallelism, they also consume a lot of power. Therefore, power awareness is another challenging issue in high performance many-core system. It is desirable to get detailed trade-offs in power and performance early in the design flow, preferably at the system level. The high level power macro model allows network power to be readily incorporated into simulation infrastructures, providing a fast and cycle accurate power profile, to enable power optimization such as power aware compiler, core mapping, and scheduling techniques for the many-core platform. Our power model allows extremely accurate power estimation (average absolute cycle error within 5%) with more than 1000x speed up over PrimeTime™ based gate level analysis. Below we provide a brief explanation of the router power model, although we refer the reader to [10] for a more detailed description.

Our power model consists of variables, that have a strong correlation to power consumption, and regression coefficients, that reflect the contributions of the variables for power consumption as follow:

$$P = \alpha_0 + \alpha_H \cdot \psi_H + \alpha_S \cdot \psi_S + \alpha_{\Delta S} \cdot \psi_{\Delta S} \tag{4}$$

where $\psi_H$ is Hamming distance of outgoing flits; $\psi_S$ is the number of outgoing ports passing body flits; and $\psi_{\Delta S}$ is the number of state transitions of outgoing ports. The coefficients for the variables are obtained by using multiple regression analysis. When validated against gate level simulation, our power model derives power profiles that match closely with that of gate level analysis.

## 4.2 Cycle Accurate Simulator

Customization of many-core platforms for an application requires the exploration of a large design space. Our simulator enables for designers to simulate various system level techniques on parallel programming, observing their impact on power and performance rapidly, which is infeasible with gate-level simulation due to the long simulation time and large generated file sizes. Our design environment also provides a way to program integrated PEs and evaluate the performance of the platform for application development. The simulator captures the utilization status and power consumption of the components of NoC so that an application designer intuitively would know which part has congestion or hot-spotted, enabling application level optimization. Our simulator provides following information for application development on the many-core platform.

**PE utilization level:** provides the utilization of each PE for given time frame or different functionality. Application programmer can use this information for scheduling on many-core.

**Latency information:** provides latency profile for each packet. If there's packet with high latency, programmer can determine the critical path in data-flow.

**Power for each routing node:** provide temporal and spatial power profiling with less than 5% average absolute cycle error and less than 1.5% average error against that of gate level analysis.

For the each supported feature, user can set constraint such as maximum latency, maximum peak power, or maximum average power for given time window. Our tool generates log file that contains detail information for the exceptional cases for the constraints. For instance, if maximum latency for PE0 is set to 100, whenever an arrived packet to PE0 has more latency than 100 cycles, the simulator generates corresponding message for user.

## 4.3 Design Space Exploration with Our Simulator

Application mapping on parallel processors becomes more complex as the number of processors increase. In order to fully utilize resources efficiently, rigorous performance and power analysis is necessary to compare several application mapping exploring design space. In order to show the feasibility of our simulator for design space exploration on many-core platform, we use benchmark from the SPLASH-2 [11] which provides the traffic among cores for each application.

Figure 2 shows the power profile of routers located on (0,0) and (2,1). During simulation, an application developer can zoom in the power profile in temporally. Fig. 2(a) provides cycle accurate power waveform and Fig. 2(b) shows the average power consumption in each 10 cycles. Our simulator also provides spatial power profile during the time windows as shown in Fig. 3. An application programmer can define the time window for power analysis.

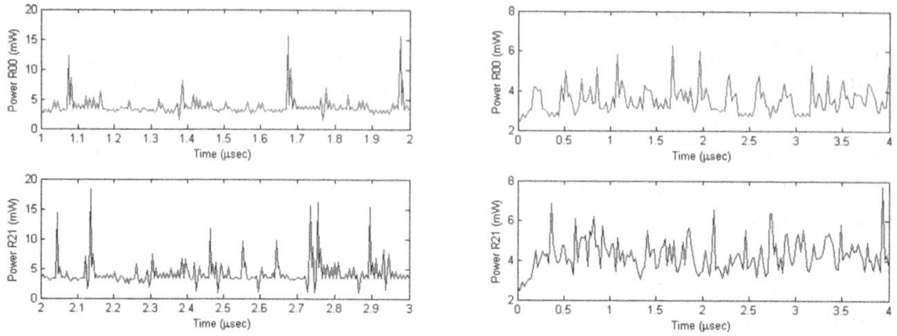


**Fig. 2.** Power waveforms for different time window setting: (a) cycle accurate power consumption during 1μsec to 2 μsec, (b) average power consumption during 0 μsec to 4.5 μsec

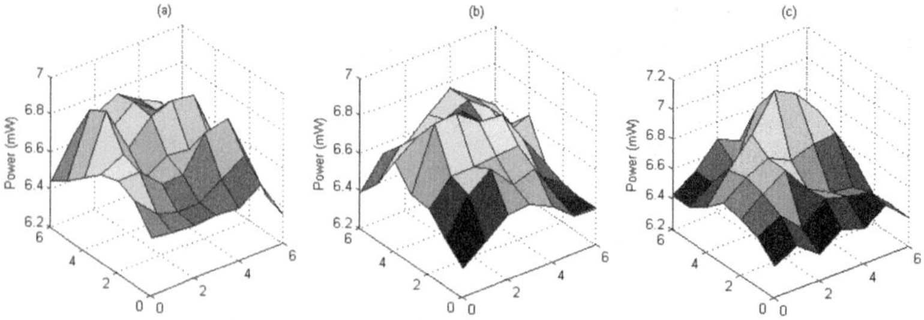


**Fig. 3.** Average power consumption for each router for the different time windows with ray-tracing benchmark

# 5 Conclusions

In this study, we presented a simulator for a many-core platform and demonstrated that this fairly fast simulator can help to explore design space in many core programming such that a parallel programming can be optimized in terms of performance and power. We plan to extend this work to make a complete IDE that enables application mapping and performance evaluation on our many-core platform.

# References


1. NOXIM, `http://noxim.sourceforge.net/`
2. Vassos, S., Noel, E., Hangsheng, W., Bin, L., Li-Shiuan, P.: Polaris: A System-Level Roadmapping Toolchain for On-Chip Interconnection Networks. IEEE Trans. VLSI Syst. 15(8), 855–868 (2007)
3. Csar, A.M., Jos, C.S.P., Ney, L.V.C., Fernando, G.M., Altamiro, A.S., Ricardo, R.: Modeling the Traffic Effect for the Application Cores Mapping Problem onto NoCs. In: VLSI-SoC, pp. 179–194 (2005)

4. Bahn, J.H., Lee, S.E., Yang, Y.S., Yang, J.S., Bagherzadeh, N.: On Design and Application Mapping of A Network-on-Chip (NoC) Architecture. Parallel Processing Letters (PPL) 18(2), 239–255 (2008)
5. Lee, S.E., Bahn, J.H., Bagherzadeh, N.: Design of a Feasible On-Chip Interconnection Network for a Chip Multiprocessor (CMP). In: International Symposium on Computer Architecture and High Performance Computing (SBAC-PAD 2007), pp. 211–218 (October 2007)
6. Lee, S.E., Bahn, J.H., Yang, Y.S., Bagherzadeh, N.: A Generic Network Interface Architecture for an NoC based Multiprocessor SoC. In: Brinkschulte, U., Ungerer, T., Hochberger, C., Spallek, R.G. (eds.) ARCS 2008. LNCS, vol. 4934, pp. 247–260. Springer, Heidelberg (2008)
7. Yang, Y.S., Bahn, J.H., Lee, S.E., Yang, J.S., Bagherzadeh, N.: Parallel Processing for Block Ciphers on a Fault Tolerant Networkd Processor Array. International Journal of High Performance Systems Architecture 2(3-4), 156–167 (2010)
8. Yang, J.S., Lee, S.E., Chen, C., Bagherzadeh, N.: Ray Tracing on a Networked Processor Array. International Journal of Electronics 97(10), 1193–1205 (2010)
9. Yang, J.S., Chen, C., Bagherzadeh, N., Lee, S.E.: Load Balancing for Data-parallel Applications on Network-on-Chip Enabled Multi-processor Platform. In: The 19th Euromicro International Conference on Parallel, Distributed and Network-Based Computing (PDP 2011), Ayia Napa, Cyprus (February 2011)
10. Lee, S.E., Bagherzadeh, N.: A High-level Power Model for Network-on-Chip (NoC) Router. Computer & Electrical Engineering 35(6), 837–845 (2009)
11. Woo, S.C., Ohara, M., Torrie, E., Singh, J.P., Gupta, A.: The splash-2 programs: characterization and methodological considerations. In: ISCA 1995: Proceedings of the 22nd Annual International Symposium on Computer Architecture, pp. 24–36 (1995)

# A New Hydrological Model (SEM-SHA): II. Mathematical Theory for Rainfall-runoff Module

Yu Tian[1,2], Genwei Cheng[1,*], Jihui Fan[1], and Yunchuan Yang[1,2]

[1] Institute of Mountain Hazards and Environment, Chinese Academy of Sciences and Ministry of Water Conservancy, Chengdu 610041, China
gwcheng@imde.ac.cn
[2] Graduate University of Chinese Academy of Sciences, Beijing 100049, China

**Abstract.** SEM-SHA is a new distributed hydrological model which stands for soil erosion model for small-watershed of hilly areas. Its rainfall-runoff module can simulate hydrological processes, such as canopy storage, root uptake, evapotranspiration, etc. In this paper, mathematical theory for rain-runoff module was introduced. There are six main themes: calculation of canopy interception, snow pack and snowmelt, calculation of soil water-holding capacity, calculation of evapotranspiration, calculation of unit runoff yield, and calculation of unit and watershed conflux.

**Keywords:** canopy interception, evapotranspiration, hydrological model, runoff formation and concentration.

SEM-SHA is a new distributed hydrological model, which stands for soil erosion model for small-watershed of hilly areas. The model is developed by Research Fellow Cheng GW and Doctor Fan JH in institute of mountain hazards and environment of Chinese Academy of Science (CAS), and consists of three main functional modules, rainfall-runoff module, slope erosion module and watershed sediment yield module [1, 2]. The rainfall-runoff module can simulate canopy interception, root uptake, evapotranspiration, runoff yield and runoff influence, respectively [1]. It first disperses a watershed into multiple grids, then calculates the water yield and confluence of each grid, and then computes the water flow of the whole basin [2]. In this paper, the mathematical theory for rainfall-runoff module was introduced.

## 1 Calculation of Canopy Interception

### 1.1 The Maximum Canopy Interception

The maximum canopy interception varies with the season and vegetation type. It is calculated by

$$Pm = PI \times Lm. \tag{1}$$

* Corresponding author.

S. Lin and X. Huang (Eds.): CSEE 2011, Part II, CCIS 215, pp. 176–185, 2011.

Where, *Pm* is the maximum canopy interception (mm/d), *PI* is phonological index, *Lm* is the forest type index (mm/d). SEM-SHA determines the PI with equation 14 [2].

$$PI = a + b \times \sin(\frac{Doy - 91}{366} \times 2\pi) \quad (2)$$

Where, *Doy* is a sequence day number of a year, *a* and *b* are tree species indexes.

### 1.2 The Actual Canopy Interception

The actual canopy interception for a given t period ($I(t)$) is determined by the minimum value among precipitation (P(t)), canopy interception ability ($P(t)$) and water deficiency of canopy ($Wcd(t)$):

$$I(t) = \text{Min}\{P(t), Pm(t), Wcd(t)\}. \quad (3)$$

$Wcd(t)$ is given by

$$Wcd(t) = Pm(t) - Wc_0(t-1). \quad (4)$$

Where, $Wcd(t)$ is water deficiency of canopy for a specific t period of time (mm), $Wc_0(t-1)$ is the canopy storage before the t period (mm).

The canopy storage in a t period of time is calculated with equation 5.

$$Wc_0(t) = Wc_0(t-1) + I(t) \quad (5)$$

## 2 Snow Pack and Snowmelt

When the air temperature ($T$) is less than the temperature at which snow forms ($Ts$), snow began to accumulate. The amount of accumulated snow is given by equation 6 [2].

$$Sw(t) = Sw(t-1) + P \quad (6)$$

Where, $P$ is precipitation (mm), $Sw(t)$ is the total amount of accumulated snow during a period of t (mm $H_2O$).

When the air temperature ($T$) is above the threshold at which snow forms ($Ts$), the snow began to melt. The amount of snowmelt is calculated by degree-day factor method [3, 4].

$$Sm = (T - Ts) \times M_s \quad (7)$$

Where, $Sm$ is snowmelt amount for a day(mm/d), $M_s$ is degree-day factor (mm/°C·d).

## 3 Calculation of Soil Water-holding Capacity

The Xin'anjiang model was introduced into the SEM-SHA to calculate the water-holding capacity of forest soil [2].

### 3.1 Absorption-water Storage Capacity of Upper Soil Layer (*WUM*)

*WUM* changes with the growth of forest (see Fig.1).

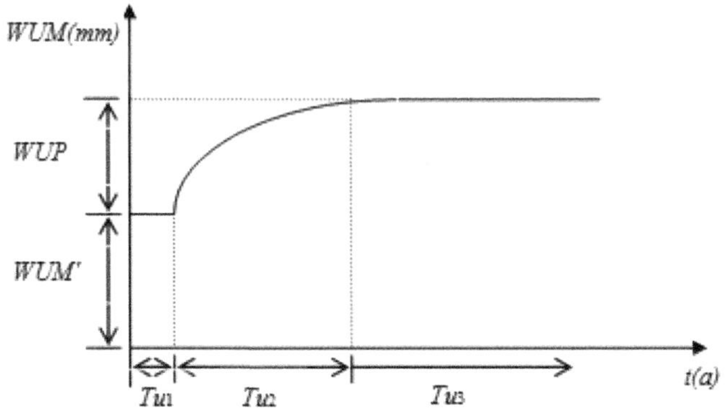


Note: *WUM'* is the water capacity of upper soil in non-forest region (mm $H_2O$), *WUP* is the increased absorption-water storage capacity resulting from forest (mm $H_2O$), $Tu_1$ is the young stage, $Tu_2$ is the middle stage, and $Tu_3$ is the maturity stage of forest growth. The same below.

**Fig. 1.** Absorption-water storage capacity of upper forest soil varies with plant growth

The *WUM* in different land types is calculated by

$$\begin{cases} \text{Non}-\text{forest lands: } WUM = WUM' \\ \text{Forest lands:} WUM = WUM' + WUP \end{cases} \tag{8}$$

*WUP* is the increased absorption-water storage of upper soil layer resulting from forest (mm $H_2O$). It is obtained by equation 9.

$$WUP = \begin{cases} 0 & T \le T_{u1} \\ WUPM \times (1 - e^{-a_1 T}) & T_{u1} < T \le T_{u2} \\ WUPM & T > T_{u2} \end{cases} \tag{9}$$

Where, *WUPM* is the highest absorption-water storage of upper forest soil (mm $H_2O$), $a_1$ is the distribution curve index, and $T$ is the time (a).

### 3.2 Absorption-water Storage Capacity of Lower Soil Layer (*WLM*)

The WLM in non-forest lands and forest lands is calculated respectively by

$$\begin{cases} \text{Non}-\text{forest lands: } WLM = WLM' \\ \text{Forest lands:} WLM = WLM' + WLP \end{cases} \tag{10}$$

Where, *WLM'* is the water capacity of lower soil layer in non-forest region (mm $H_2O$), *WLP* is the increased absorption-water storage of lower soil layer resulting from forest (mm $H_2O$), it is obtained by

$$WLP=\begin{cases}0 & T\le T_{l1}\\ WLPM\times(1-e^{-a_2T}) & T_{l1}<T\le T_{l2}\\ WLPM & T>T_{l2}\end{cases} \tag{11}$$

Where, *WLPM* the highest absorption-water storage of lower forest soil layer (mm $H_2O$), and $a_2$ is the distribution curve index.

### 3.3 Absorption-water Storage Capacity of Deep Soil Layer (*WDM*)

The forest has no significant effect on *WDM*, so the same value is adopted in forest lands and non-forest lands.

### 3.4 Free Water Storage Capacity (*SM*)

Forest also has an effect on soil free water capacity. Based on the assumption that free water has the same variation with *WUM* and *WLM*, SEM-SHA calculates the *SM* by

$$\begin{cases}\text{Non-forest lands: } SM=SM' \\ \text{Forest lands:} SM=SM'+SP\end{cases} \tag{12}$$

Where, *SM'* is the free-water storage (mm $H_2O$), SP is the increased absorption-water storage of soil layer resulting from forest (mm $H_2O$), and SP is expressed as

$$SP=\begin{cases}0 & T\le T_{s1}\\ SPM\times(1-e^{-a_3T}) & T_{s1}<T\le T_{s2}\\ SPM & T>T_{s2}\end{cases} \tag{13}$$

Where, *SPM* is the highest free-water storage of forest soil (mm $H_2O$), and $a_3$ is the distribution curve index.

## 4 Calculation of Evapotranspiration

The SEM-SHA first calculates the potential evapotranspiration, and then calculates the actual evapotranspiration.

### 4.1 Potential Evapotranspiration

The amount of potential evapotranspiration ($E_m$) can be calculated by Hamon's equation [5].

$$E_m=0.1651D_L\bullet\rho. \tag{14}$$

Where, *DL* is daylength (hr), $\rho$ the density of saturated vapor at daily mean air temperature, it is given by

$$\rho=216.7\frac{e_s}{T_a+273.3} \tag{15}$$

Where, $T_a$ is air temperature (°C), $e_s$ is saturated vapor pressure (h Pa), and it is given by equation 16.

$$e_s = 6.108\, e^{17.28 T_a /(T_a + 273.3)} \tag{16}$$

The equation 14 neglected the influences of plant on the evapotranspiration. Taking into consideration of plant, the potential evapotranspiration ($E_{mp}$) is calculated by

$$E_{mp} = E_m \times (1.0 + cp \times PI). \tag{17}$$

Where, *cp* is plant factor.

## 4.2 Actual Evapotranspiration

The actual evapotranspiration includes vegetation evapotranspiration, sublimation and evaporation from soil. The potential evapotranspiration is first deducted from vegetation storage, if the vegetation storage is not enough, the remaining evaporation capacity is deducted from sublimation, if the sublimation is also not enough, and it will be deducted from soil evaporation [2].

(1) vegetation evaportranspiration

If $Wc_0(t) \geq E_{mp}$, then

$$\begin{cases} Ev = E_{mp} \\ Wc_0(t-1) = Wc_0(t) - Ev \\ Ev' = 0 \end{cases} \tag{18}$$

If $Wc_0(t) < E_{mp}$, then

$$\begin{cases} Ev = Wc_0(t) \\ Wc_0(t) = 0 \\ Ev' = E_{mp} - Wc_0(t) \end{cases} \tag{19}$$

In formula 18 and 19, *Ev* is vegetation evapotranspiration, *Ev'* is the remaining evaporation capacity.

(2) sublimation

The sublimation calculation loop is as Fig. 2.

(3) evaporation from soil

Soil evaporation is calculated by tri-layer evaporation model [6]. The input parameters are *P*, *Emsoil*, *WUM*, *WLM*, *WDM* and *C*. The calculation process is as follows:

•to calculate the potential evaporation (*EP*)

$$EP = K \times Emsoil. \tag{20}$$

Where, *K* is conversion coefficient of water surface evaporation, *Emsoil* is the remaining evaporation capacity after deduction of vegetation storage and sublimation.

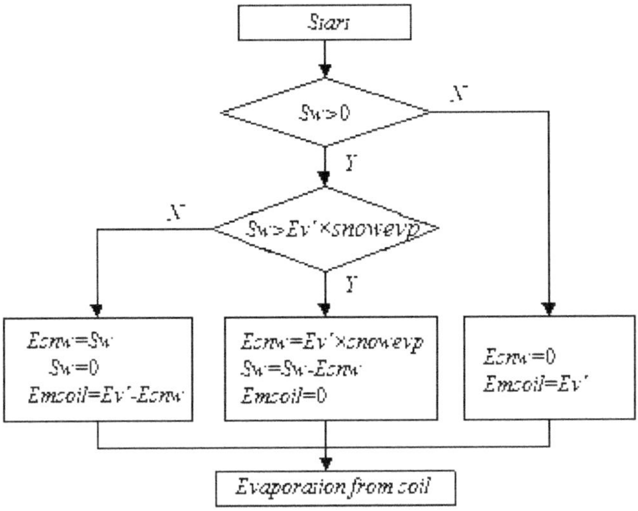


Note: *Sw* is the depth of snow cover (mm), *Ev'* is the remaining evaporation capacity (mm), snowevp is sublimation coefficient (mm/°C·d), *Esnw* is amount of sublimation, and *Emsoil* is remaining evaporation capacity from soil (mm).

**Fig. 2.** Sublimation command loop

•to calculate *EU, EL* and *ED*

If $WU+P \geq EP$, then

$$EU=EP,\ EL=0,\ ED=0. \tag{21}$$

If $WU+P < EP$ and $WL \geq C\times WLM$, then

$$EU=WU+P,\ EL=(EP-EU)\times WL/WLM,\ ED=0. \tag{22}$$

If $WU+P < EP$ and $C\times(EP-EU) \leq WL \leq C\times WLM$, then

$$EU=WU+P,\ EL=C\times(EP\quad EU),\ ED=0. \tag{23}$$

If $WU+P < EP$ and $WL < C\times(EP-EU)$, then

$$EU=WU+P,\ EL=WL,\ ED=C\times(EP\quad EU)\quad EL. \tag{24}$$

In equation 21 to equation 24, *P* is precipitation to land surface (mm), *C* is evaporation coefficient of deep soil layer, *EU* is the water content in upper soil layer (mm $H_2O$), *EL* is the water content in lower soil layer (mm $H_2O$), *ED* is the water content in deep soil layer (mm $H_2O$), *WU* is the absorption-water capacity of upper soil layer (mm $H_2O$), *WL* is the absorption water storage capacity of lower soil layer (mm $H_2O$), *WD* is the absorption water storage capacity of deep soil layer (mm H2O), *WM* is the total absorption water storage capacity of soil (mm $H_2O$), *W* is the absorption-water content of soil (mm $H_2O$).

•to calculate $E_s$

$$E_s=EU+EL+ED. \tag{25}$$

Where, $E_s$ is total soil evaporation (mm).

(4) the total evapotranspiration

The total evapotranspiration is the sum of vegetation evapotranspiration, sublimation and soil evaporation.

$$E = Ev + Esnw + E_s. \quad (26)$$

## 5 Calculation of Unit Runoff Yield

SEM-SHA adopts saturation excess (Dunne) runoff model to calculate water yield [2, 6]. The model introduced watershed absorption-water storage capacity curve to calculate yielded water (Fig. 3).

The input of the model is PE (PE=P–E), and the equation 27 to 29 are used in the model.

$$\frac{f}{F} = \left[1 - (1 - \frac{WM'}{WMM'})^{B}\right]. \quad (27)$$

$$WMM' = WM' \times (1+B). \quad (28)$$

$$A = WMM' \times \left[1 - \left(1 - \frac{W_0}{WM'}\right)^{\frac{1}{1+B}}\right]. \quad (29)$$

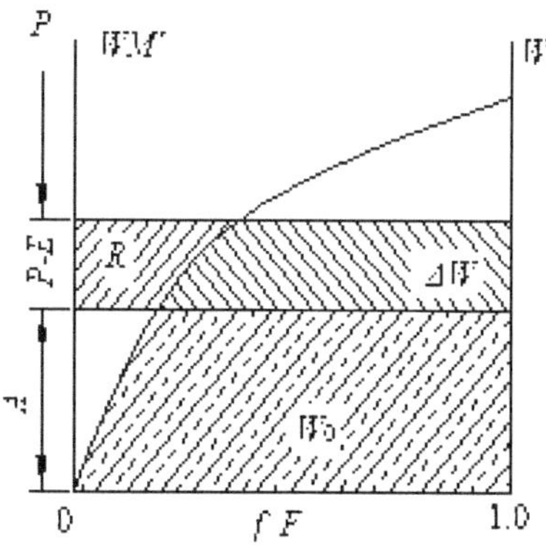


Note: $f$ is contributing area ($km^2$), $F$ is basin area ($km^2$), $P$ is precipitation (mm), $E$ is the actual evapotranspiration (mm), $B$ is the index of absorption-water storage curve, $WM'$ is absorption-water storage for a given place in a basin (mm $H_2O$), $WMM'$ is the maximum absorption-water storage capacity in a basin(mm $H_2O$) and $W_0$ is the final mean soil water content for a given time step (mm $H_2O$).The same below.

**Fig. 3.** Watershed absorption-water storage capacity curve

If $PE$ is greater than zero, runoff began to generate. The yielded runoff is calculated by equation 30 and 31.

If $PE+A<WMM'$, then

$$R = PE - WM + W_0 + WM \times \left[ 1 - \left( \frac{PE + A}{WMM'} \right)^{(1+B)} \right]. \tag{30}$$

If $PE+A \geq WMM'$, then

$$R = PE - WM + W_0. \tag{31}$$

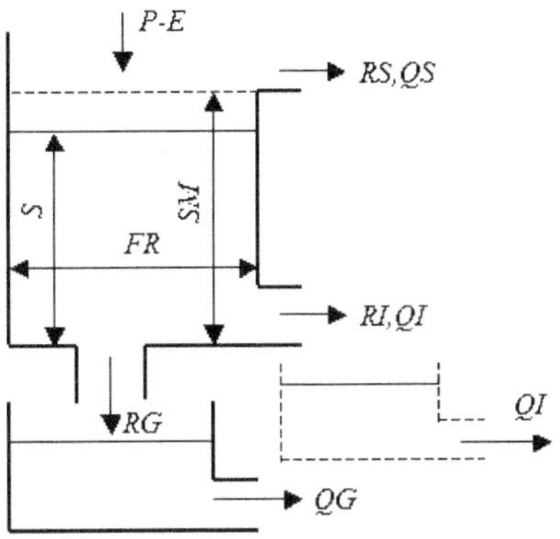


Fig. 4. The structure of free-water sluice reservoir

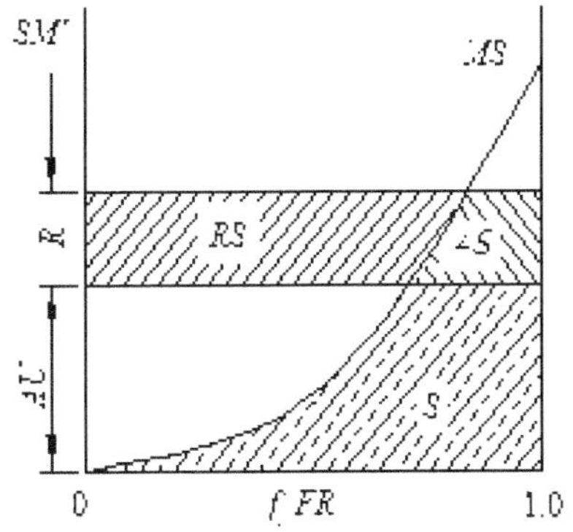


Fig. 5. Free-water storage capacity curve of a basin

The free-water sluice reservoir (Fig. 4) was introduced into SEM-SHA to divide the runoff types. The runoff ($R$) was divided into surface runoff ($RS$), interflow ($RI$) and underground runoff ($RG$). The soil free-water storage capacity curve (Fig. 5) was also introduced into the SEM-SHA.

The equations 32 to 35 are adopted in the SEM-SHA.

$$\frac{f}{FR} = 1 - \left( 1 - \frac{SM'}{MS} \right)^{EX}. \tag{32}$$

$$MS = SM \times (1+EX). \tag{33}$$

$$AU = MS \times \left[ 1 - (1 - \frac{S}{SM})^{\frac{1}{1+EX}} \right]. \tag{34}$$

$$FR = R/PE. \tag{35}$$

In the above equation 32 to 35, $SM$ is free-water storage of upper soil layer (mm $H_2O$), $EX$ is the index of free-water storage curve for upper soil layer, $KG$ is the constant of influx from upper free-water to underground water, and $KI$ is the constant of influx from upper free-water to interflow.

If $PE \leq 0$, then

$$\begin{cases} RS = 0 \\ RI = KI \times S \times FR \\ RG = KG \times S \times FR \end{cases} \tag{36}$$

If $PE > 0$, and $PE + AU < MS$, then

$$\begin{cases} RS = \left\{PE - SM + S + SM \times \left[1 - (PE + AU)/MS\right]^{1+EX}\right\} \times FR \\ RI = \left\{SM - SM \times \left[1 - (PE + AU)/MS\right]^{1+EX}\right\} \times KI \times FR \\ RG = \left\{SM - SM \times \left[1 - (PE + AU)/MS\right]^{1+EX}\right\} \times KG \times FR \end{cases} \tag{37}$$

If $PE > 0$, and $PE + AU \geq MS$, then

$$\begin{cases} RS = (PE - SM + S) \times FR \\ RI = KI \times SM \times FR \\ RG = KG \times SM \times FR \end{cases} \tag{38}$$

In the above mentioned formula 36 to 38, *MS* is the maximum free-water storage capacity (mm $H_2O$), *FR* is contributing area ($km^2$), *S* is mean free-water storage of a basin (mm $H_2O$), and *AU* is the corresponding y-coordinate of *S* on the free-water storage capacity curve (see Fig. 5).

## 6 Calculation of Unit and Watershed Conflux

SEM-SHA divides runoff into 3 types, surface runoff (*RS*), interflow (*RI*) and underground runoff (*RG*), the 3 types of runoff has different confluence path and flow rate. The surface runoff flows into the river network and form *QS*(*I*). The interflow flows into the interflow reservoir and forms *QI*(*I*) after regulating. Similarly, the underground runoff flows into the underground water reservoir and forms QG(*I*) after regulating *QG*(*I*). Surface runoff (*RS*), interflow (*RI*) and underground runoff (*RG*) of each unit are calculated respectively by equation 39.

$$\begin{cases} QS(I) = RS(I) \\ QI(I) = CI \times QI(I-1) + (1 - CI) \times RI(I) \\ QG(I) = CG \times QG(I-1) + (1 - CG) \times RG(I) \end{cases} \tag{39}$$

Where, *QS*(*I*) is the influx from surface runoff to river network, *QI*(*I*) is the influx from interflow to river network, *QG*(*I*) is influx from underground water to river network, *CI* is the regulating coefficient of interflow, CG is the regulating coefficient of ground water.

The total confluence for each grid (*Q*(*I*))is calculated by equation 40.

$$Q(I) = QS(I) + QI(I) + QG(I). \tag{40}$$

The confluence of a basin is generally calculated by Muskingum method [6, 7, 8]. The SEM-SHA is mainly used for small watershed, it omits the river network confluence time. So, the daily water flow of a basin ($Q$) is expressed as the sum of daily water flow of each grid ($Q(I)$). It is calculated by

$$Q = \sum Q(I). \tag{41}$$

## References


1. Tian, Y., Fan, J.H., Xiao, F.P., et al.: A New Soil Erosion Model for Hilly Region Based on Information Technology. In: Shen, G., Huang, X. (eds.) CSIE 2011(Part II), pp. 473–483. Springer, Heidelberg (2011)
2. Xiao, F.P.: Study on Runoff and Sediment Models for Small Watershed in Hilly Areas of Jialing River. Institute of Mountain Hazards and Environment, CAS, Chengdu (2009) (in Chinese)
3. Braithwaite, R.J., Zhang, Y.: Sensitivity of Mass Balance of Five Glaciers to Temperature Changes Assessed by Tuning a Degree-day Model. J. Glaciology 46(152), 7–14 (2000)
4. Jhannesson, T., Sigurdsson, O., Laumann, T., et al.: Degree-day Glacier Mass-balance Modeling with Applications to Glaciers in Iceland, Norway and Greenland. J. Glaciology 41(151), 559–567 (1995)
5. Hamon, W.R., Walk, W.R.: Evapotranspiration under Depleting Soil Moisture. J. Irrigation Drainage Division, ASCE 105(IR4), 392–402 (1979)
6. Xu, Z.X.: Hydrological Models. Science press, Beijing (2009) (in Chinese)
7. Cunge, J.A.: On the Subject of a Flood Propagation Computation Method (Muskingum Method). J. Hydr. Res. 7(2), 205–230 (1969)
8. Das, A.: Parameter Estimation for Muskingum Models. J. Irrigation Drainage Division, ASCE 130(2), 140–147 (2004)

# Design and Practice Research of City Ad Titles in the Image Software

Guobin Peng and Youneng Wu

Design and Art College
GuiLin University of Electronic Technology
Guilin, P.R. China
pengguobin@guet.edu.cn

**Abstract.** In the process of rapid city society and economy development, the growing number of cities use advanced media technologies means to promote and build their own city image, on the other hand, along with the continuous technology development, types of image software continuously meet today's society advertising propaganda to variety of products, play a huge role in the economy and society, how to effectively use these advanced technology to design a good ad products, and better serve economy and society becomes a problem which the industry must pay attention to and explore, this paper takes the design of city Ad titles of ZhangShu as an example, launch a practical design activity of this case in video image technology practice, conclude and explore some technical design application rules, enrich practical experience in this field.

**Keywords:** Urban; Advertising Film; Image Software; Design Manufacturing.

## 1 Introduction

At present society economy grow rapidly, promotion of activities become an indispensable means of social development, which runs through all aspects of the area of social economy development, city advertising film is currently the more popular topics at domestic, using media technology for city propaganda is an important means, this propaganda is a product based on a combination of technology and art, it is necessary to grasp the artistic taste, performance in the technology is more important, as an advertising product only familiar with its technical requirements can we select appropriate technical means service for the advertising design, saving the necessary human and material resources to ensure the advertising picture quality, to achieve the unity of the two, based on this goal, this paper we take study of urban advertising to explore graphics software technology in the application of Ad titles practice, this paper starts on discussion mainly on the technical analysis and technical practices.

## 2 Technical Analysis and Selection of Prophase

City advertising film production technology can be divided into two categories: one is the film and television production technology, also known as video technology, the

S. Lin and X. Huang (Eds.): CSEE 2011, Part II, CCIS 215, pp. 186–191, 2011.

other is animation production techniques, they all have different production technical requirements of different advertising films, with different technical characteristics.

### 2.1 Animation Production Technology

Animation technology includes two-dimensional and three-dimensional animation technology. Two-dimensional animation technology mainly uses the human eye "visual retention effect ", when people watch object, the object residentes in the brain optic nerve about 1 / 24 seconds, if the replacement of 24 pictures per second or more Screen, then before the previous screen disappearing in the human brain, the next screen to enter the human brain, which form a continuous image. So many people use this principle form the content of a continuous motion picture together, and then continuously play by the rate of 24 frames per second, generate visual images of continuous activities in the eyes, which formed what we call the animation. from the points of production animation is divided into and semi-animation." full animation " is the animatiom made in animation process in order to obtain perfect images, smooth movements and delicate in accordance with the number of images at 24 frames per second animation. This method of animation because of good viewing, usually used to produce large-scale animation and commercial advertising, the United States Walt Disney Company produced a large number of works is the full animation. "Semi-animation" generally use 6 frames per second, often use to repeat the action, stop motion picture screen to extend the number of case to fill the equivalent of 24 frames per second the screen time to show the animation, so to reduce economic spending and reduce the huge workload. Animation production needs large amount of work, time and energy consuming, producing a huge funding, as for publicity trailer for the city is not city ad titles it is not necessarily to achieve the desired effect of publicity, not suitable for the real scene style of ZhangShu city propaganda film, therefore, to abandon this kind of technology means.

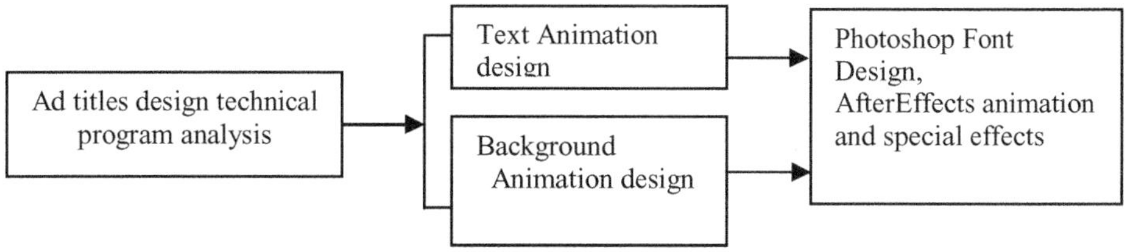


**Fig. 1.** Technical requirements analysis of ad titles design

### 2.2 Film Production Technology

Film production technology is the process to carry out material screen, collect, organize, and by disposing means of video production technology to form a complete promotional video based on pre-designed program according to preliminary design ideas. because of high operational efficiency, good production effect, television production techniques is the technical means often used by city image film, ZhangShu city propaganda film also applied the technology in the field to complete the ad titles design, and propose their needed technical requirements according to different content, so as to allow smooth development of design plans, such as figure 1 ad titles design technical analysis.

## 3 Software Technology Practice of Advertising Titles Design

Ad titles script says: morning glow of color rising slowly, accompanied by passionate music, red colored silk floating slowly from the air, the inscription of the four characters "Chinese medicine city" is a symbol of the medicine image sculptures - "medicine Grind ", demonstrated in front of people, while "medicine city ZhangShu " emergence with music and sculptures background, a fresh new day begins, music climax away, fade, lens end. According to the script and a shooting script analysis, ad titles design includes the followings:First, the color glow slowly rising, freeze at 5 seconds, beginning of the fade out at 11 seconds.Second, the red colored silk floating slowly from the air for 4 seconds, fade out at the 7 seconds.Third, the " medicine city ZhangShu " topic text effects and text animation.Fourth, the "Chinese medicine city" Sculpture appears at 7 seconds, beginning of fade out at 11 seconds 12 frames.

### 3.1 Design of Ad Titles Constitution Elements

1)"Color Glow" rise Background Design

Open AfterEffects software, first set the size of videos, now start the design of television size, size 720x576, frame rate of 25 frames / second, in the new project file, input the dynamic pink clouds, move the material onto the timeline for editing, since the original length of Dynamic Clouds is 23 seconds, clouds speed is too slow, so click the left mouse button select the pink clouds, use "effect command - time stretch command" to faster pink clouds, 16 seconds to complete play. Meanwhile, pink clouds is from dark to bright by the time of 5 seconds, then freeze, stay 7 seconds, 11 seconds 12 frames starts fading out, the action is complete, select the pink clouds, in 5 seconds use "effective command - time reproduce command " pink clouds at 5 seconds to change from dark to bright, and then freeze, stay 7 seconds to complete the action such as (Figure 2).

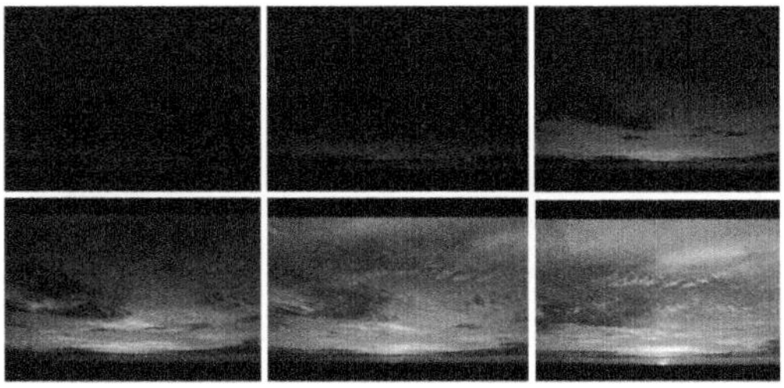

**Fig. 2.** ChoiHa rising background

2) The Animation of Red "Colored Silk " Slowly Drift

Material selection is a shooting scene material: blue background, yellow silk slowly falling down from the sky, but the design needs a red silk without blue background, you need to do the following to do adjustments.

(a) adjust color: select silk layer in AfterEffects software, execute the command "effect - color correction - color balance", the color change parameter is adjusted to "-31",thus the silk becomes red.

(b) deduction like this: because only needs the red silk, does not require a blue background, so then remove the blue background by keying, only to get a clean red silk. Select the silk layer, execute the command "effect - Keying - color key" to adjust the color tolerance parameters for the "104" edge feather "1.9", so to remove most of the blue, but, silk the edges and dark points still remained red shadow, in order to remove for cleaner, execute the command "effect - keying - spill suppression" to set suppress parameter of "200. " Such a clean, falling red silk is produced.

3) "Chinese Medicine City" Sculpture Design

**Fig. 3.** Theme text design of " medicine city ZhangShu "

"Chinese medicine city " sculptures - medicine grind is on behalf of the medicine culture of medicine city ZhangShu, where we need it to appear on the screen background image, to represent the medicine city image, but there is only one shot of original material, advertising titles only needs "chinese medicine city" sculptures - medicine grind image, we need to get rid of other background and trees branches. the following we expand production, photo material will be transferred into AfterEffects software, move onto the timeline, select the sculptures layer, use pen in the toolbar, tick down the "Chinese medicine city" sculptures –medicine grind to obtain a desired image, which is the application of the mask features in AfterEffects software, it can cover the range of other not selected, to get an independent medicine grind shape, while the production of pink clouds background in early phase shows out instead of the original sculptures background.

4)" Medicine City ZhangShu "Topic Text Effects and Text Animation Design

Ad titles " medicine city ZhangShu " written by script, reflects the long history and ancient medicine culture, in conjunction with the text " medicine city ZhangShu " changes from far to near, and then freeze, and then do a "sweep light " special effects, to complete ad titles theme text effects and animation production.

(a) text design. The " medicine city ZhangShu" theme text designed as a handwritten line calligraphy which is suitable for the theme (Figure3), after the font design we need enter into Photoshop software to change the font color, and in order to benefit optical scan" effect in AfterEffects software ", need to be designed to be independent and transparent layers, so transferred to the AfterEffects software background is transparent, or it will have colored background.

(b) change the font color and layer transparent processing. Enter into the Photoshop software, select the text image layer, then use the Magic Wand to select the red font filled with white, change the font to white, then execute "cut - paste"command, so the font as a separate layer, and name figure layer as " medicine city ZhangShu", save as psd format, so a separate transparent layers are preserved, into AfterEffects software, import "medicine city ZhangShu" file which is already produced, set the layer options: select layer - select "medicine city ZhangShu" layer, click on the OK button, the layer is imported, and then drag to the timeline window to adjust the font size, complete the production.

(c) special effects production of font "optical scan". "optical scan" effect for more fit for human visual aesthetic, separately sweep light "medicine city" and "zhangshu", "medicine city" swept light from left to right, "zhangshu" sweep from the bottom to top, so as not to appear so stiff. First, copy the font layer called layer 1 and layer 2, so that layer 1 is used for production of " medicine city " swept light, layer 2 is used for production of " zhangshu"swept optical effects.

Select Layer 1 implementation of "effect - trapcode - shine " by adding special effects, this effect is an external plug-in, set the beginning sweep light of 149.2 and end 352.2, the beginning of luminous intensity of "4" and the end of the intensity of "0", the length of light Start "1.7 " end "0 ", the key frame to be set to 3, from first key frame to key frame 2 is made parallel sweep of light, the thirfd key frame is to receive lighting. Therefore, in order to highlight the effect in the process of scanning light, to increase the length of light and light intensity of the 2nd key frame parameters, but also set the "start issuing light - continuing " the length of time is 1 second 16 frames, set the collection of light length time as 15 frames, using the same method to set the layer 2 "optical scan"parameter, the difference is selecting the direction of up and down adjusted parameters. After the completion of setting all layers we can achieve effect.

(d) font animation production. Make a deformation animation of theme text font of "medicine city zhangshu " from small to large according to the plan requirement, then to implement optical effects. Select the text layer and use layer properties command - the proportion deformation, adjust the deformation parameters to "0 ", the time length of 20 frames, the parameter is "102",so to complete a deformation animation from small to large.

## 4 Transitions and Summary Design of Ad Titles

After all constitution element and animation effect of ad titles finished, the next required by the visual rhythm and advertising interpretation process to design a transition, in the design we should pay attention to the following:

1, rhythm changes should be reasonable, smooth, to avoid too fast or too slow, action to comply with human visual characteristics and nature and humanity design needs, improve the apperance.

2, transition design is naturally interfaced, transition change way not only needs to meet the changes of the design elements but also as far as possible to avoid repetition, dull appearance and such discord factors.

3, in line with sound design, sound performance in a passionate and excited in the ad titles, with more relaxed traditional music after the end of ad titles, and the ad titles design needs to unify with later propaganda design, to avoid the mutual isolation, out of connection, ad titles rhythm and transitions design are as follows:

(a) within 5 seconds color glow finishes rising action, begin to freeze 7 seconds,

**Fig. 4.** Screenshot of final Effect

starting fading out at 11 seconds 12 frames, to complete fading out action at 12 seconds frame.
(b) red silk fade in at 4 seconds, fade duration of 12 frames, fade out at 7 seconds, at 7 seconds 12 frames to complete fading action.
(c) "Chinese medicine city" sculptures fade at the 7 seconds, action completion at 9 seconds, starting fading out at 11 seconds 12 frames, to complete fading out action at 12 seconds frame.
(d) Text of "medicine city zhangshu" beginning of fade in at 8 seconds 16 frames, execute light sweeping movements, completion of light sweep at 11 seconds, to complete fade at12 seconds. Ad titles within this short time use a sharp contrast from black to bright, heavy red and light blue background, slow pink clouds motion and fast red silk falling, with passionate, magnificent music, played most expression of various elements, attracted the audience attention, stimulating, highlight the theme. (Figure4. Ad titles screenshot of final Effect ).

## 5 Summary

In urban Ad titles design process, the first is to promote a make clear the propaganda position and design position of the city, according to the pre- positioning analysis of the technical requirements and implementation methods may be involved in the production process, make sure you are clear in the hearts, benefit to implementation of the program by the expected designed script in front phase, rather than aborted in the medium process, resulting in waste of manpower and material resources, but also facilitate the rational division of labor and personnel scheduling in later phase. Through the ad titles design and practice, lay a good foundation to master the specific application of video image softwar, and further define the practice relationships of technology and design art, lay a solid base for the provision of high quality of urban ad titles design.

**Acknowledgement.** Project Fund: 2010 scientific research established by Guangxi Education Department .Project number: 201010LX136.

## References

1. Luo, G., Chen, X.: Ad titles production experts, the application skills of After Effects. Qinghai Meteorological (March 2007)
2. Shi, L.: Using After Effects to achieve the scroll bar animation of film ad titles. 25 West China (2008)
3. Guan, Y.: After Effect comprehensive application of the three-dimensional features in the film ad titles application (first) - the "new dynamic shadow engine" section titles. TV Caption Effects Animation (May 2008)
4. Zhang, Z., Zhou, L.: Video compositing technical tools exploration based on Adobe. TV subtitles. Special Effects and Animation (June 2008)

# Three-Dimensional Game Modeling and Design Research Based on 3Dmax Software

Guobin Peng, Yueqing He, Yu Sun, and Kai xi Zhou

Design and Art College
GuiLin University of Electronic Technology
Guilin, P.R. China
pengguobin@guet.edu.cn

**Abstract.** Currently there is a number of application software on the market, of which the most representative is 3DS MAX software, its interface is simple and clear, powerful modeling function, widely applied in the game development and design. 3DS MAX software has its own unique design idea, the characteristics of the software requires quickly find workable rules of its own in three-dimensional modeling design, thus to speed up our design progress and improve the quality of design. The following is the exploration about 3DS MAX role in game modeling and related design according to three-dimensional game modeling design experience.

**Keywords:** 3DS MAX; Three-dimensional modeling; Design; Research.

## 1 Preface

The game currently has occupied the mainstream in the market of domestic animation industry, game design and development followed by attention, much developed three-dimensional design software in the current market, generally three-dimensional software can be divided into large and medium-sized and small three-dimensional design software, 3DS MAX software is a three-dimensional design software currently widely used in the design community, it is popular among the majority of the game designers by its feature of a strong design function, good interface, easy to use and several other features, especially in the three-dimensional scene modeling design it has its own unique places, The following is the exploration about 3DS MAX in game modeling and related design problems.

## 2 Modeling Approach and Model Analysis

In the process of game design and development, the game scenes, characters and its creation background all come from the real world, and the real world is constituted of different rules and irregular objects, the object of these realities propose different requirements for modeling. 3DS MAX software is actually to recreate these various rules and irregular objects in the virtual world of games, using computer and software to virtualize a three-dimensional world.

S. Lin and X. Huang (Eds.): CSEE 2011, Part II, CCIS 215, pp. 192–196, 2011.

### 2.1 Software Modeling Approach

3DS MAX now offers a variety of modeling methods: basic objects and extended objects, composite objects modeling, patch modeling, Surface Tools Modeling, Nurbs modeling, polygonal modeling, 3DS MAX software offers rich modeling capabilities according to different objects, and also provides a variety of powerful modeling function based on different model structure, can create complex and irregular models. Among these modeling ways software provides a variety of modeling methods for the real nature and man-made objects, first as for our daily tables and stools, can rapidly facilitate the application of basic objects to create model, then according to the object shape and Forming rules, the software also designes modify command to establish models, such as extrusion, rotation model, can also be combined in many ways the model can be used as a model, such as the Boolean operation command in modeling methods of compositing objects, two different model structures can be Interaction of be added or subtracted to produce a model result, this modeling approach is widely used in the establishment of architectural renderings, such as the windows button hole, usually use this way to carry out, sometimes there are dozens of building windows, and even first need to order well the model to be dug, and combine them into a model, finally do Boolean operations with building wall model, thus it is easy to produce dozens or even hundreds of needed holes. For irregular objects, we can use geometric objects deformation plus modify command to achieve the model formation, of course, a variety of modeling methods software provided, but also depending on what the object model to determine ways to resolve.

### 2.2 Analyzing Modeling Object

The first is modeling stage, first in the mind we must have a full picture of the object to be achieved, and understand the structure and complexity of the object. The basic idea of modeling is from whole to part, and gradually thinning, be aware of in the hands before starting, first to build the full picture the object in your mind, and then divided into different independent parts, then refinement of the part, finally divide into the basic geometry which constitutes the object, these basic geometry can be achieved through the simple steps. Such as Figure 1 housing modeling structure,first establish two levels of film here, and to draw a good map is set to front view, side view of two side pieces were attached to it, and then a BOX, click the right mouse button to the BOX into editable polygon, by squeezing come forward, dotted, image editing operations such as merge points, the final built model. Here the final model to be able to meet the late great textures, even taking into consideration the application of the final destination, such as game model needs to take into account the number of required model plane, model plane over more than the required number of the face number, and finally the final product may not

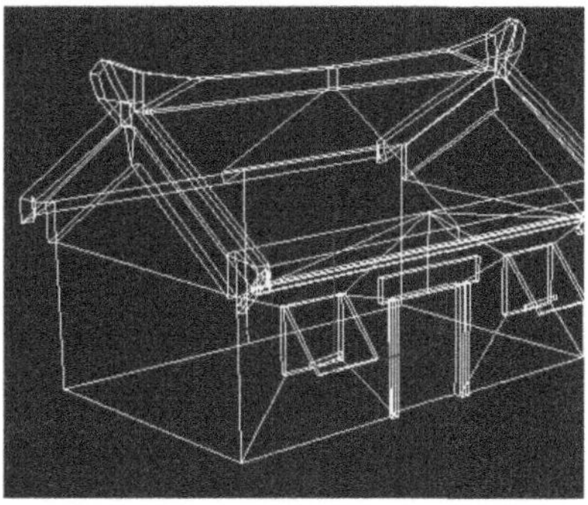

**Fig. 1.** Housing modeling structure

function properly. in using 3DS MAX mode to create model, need to analyze the object will be designed, have a certain judgement for the future movement of the establishd model, this affects the achievement of designating texture mapping and to achieve the effect of texture mapping in the later stage, to be able to understand thoroughly the key structure for the model, we must learn to break down complex models of the formation of a simple interconnect, analyzing these structures is to study use which ways and commands can conveniently, fastly and successfully build these models. For example, in creating a multi-storey houses, in addition to the top floor and the ground floor, other floors are the same basic structure, complete by replication after design a floor; one floor of the building design can be divided into walls, pillars, stairs, doors, windows, etc; walls, pillars and stairs can basically be attained through simple processed rectangular and cylinder, the doors and windows can be further refined. In addition, we should also take into account of the overall structure in refinement of the object, only by detailed local achieving and coordinated and appropriate overall co-ordination can we in order to obtain satisfactory scene effect.

## 3 The Material and Texture of the Model

a)Authenticity

Games material like people dressed, to set real texture for the scene, so as to distinguish a variety of things, but when we texture for all kinds of objects in the nature, different materials have different physical properties, such as stones, wood, glass, and plastic, because physical properties of these objects surfaces such as texture, transparency, color, and reflective properties are distinguished, which constitute the needs of material authenticity, the authenticity is closer to the game objectives that the design to be achieved, to enhance people's sense of identity and curiosity with virtual world and meet people in the real and virtual worlds, such we can achieve the goal of three-dimensional game design.

3DS MAX provides a variety of color patterns for the model objects, there are eight color types: First, the opposite sex (Anisotropic) commonly used in metal and glass.Second, reflex (Blinn) commonly used in rubber and plastic materials.Third, Metal (Metal) is mainly used for metal materials. Ford, multilayer high light (Multi-Layer) used in the glass rubber and plastic aterials. Five, shading (Oren-Bayar-Blinn) for non-smooth surfaces such as cloth and fabric materials. Six, multi-faceted (Phong) for hard and smooth as glass as the material. Seven,metal strengthening (Strauss) more easily performance of metal material. Eight, transparent (Translucent shader) for glass and some translucent material, each color chosen depends on the type of scene objects created by the demand. To

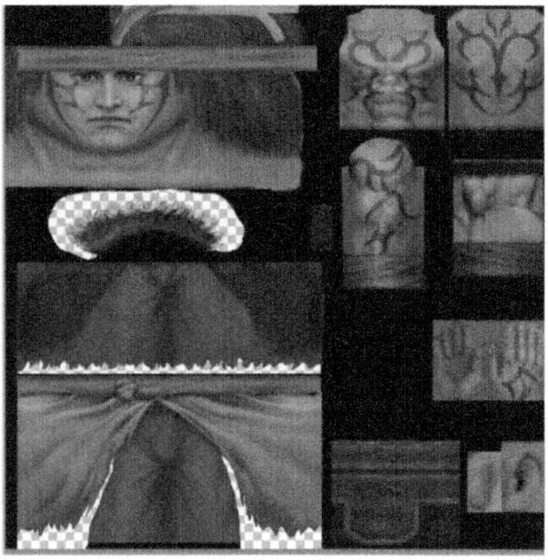

**Fig. 2.** Photoshop software to draw the map

create a choice of glass or plastic objects or reflective colored multi-faceted way, to make models with metal texture, then select the metal color mode, type in the complete color selection, basic parameters of colored paper under the column can show the color of the material components, diffuse color, reflection, opacity set, set a reasonable mediation. These features well maintained the authenticity of the material, while create more choices for design.

b) Artistic quality

Artistic quality mainly is demonstrated through the degree that performance the aesthetic mood such as lines, colors, lighting effects, layout and contrast level, the higher artistic the game works is, the stronger its appeal is, thus also better able to attract gamers. The use of 3DS MAX to texture is divided into two kinds, one is materials which 3DS MAX material library bring with itself, such as glass, wood, metal and other materials that can be directly given model object after the set, and can change the parameters to get satisfactory results. The other is affixed designed bitmap texture to the surface of the model, these images can be created by their own image processing software (Figure 2: Using Photoshop software to draw the map), with this function of creativing freedom module, initiative of material art and design handling to be greatly enhanced, given the game of man-made art atmosphere, and also save a lot of the texture time and performance of the game designers.

## 4 The Game Scene Lighting Design

Light plays a pivotal role in the game atmosphere. Through light polish to improve the expressiveness of the game scene, enhancing the realism of the scene and three-dimensional depth effect. Lighting is usually divided into three types: natural light, artificial light and natural light, artificial light combination of the two natural light sunlight, use of natural light, consider the time, place and the weather sunshine and artificial light can be almost any form, is light produced by the natural world, out of the artificial lighting, artificial light need to consider the direction of light, the intensity of light, several light sources, the main and auxiliary light source is how the distribution of radiation, and finally a combination of natural light and artificial light. The purpose of the first scene set the lights to emphasize the tone and atmosphere of the scene, it is very important. In some cases, the target may be only one or a few clearly see the object, but usually not the case, the actual goal is rather complex, which test the understanding of lighting and application capabilities of the designers, so as to create lighting for scenes, first we must know whether the tone position of the scene and lighting set promote the emotional expression of the game story and the atmosphere contrast, whethre respects the virtual under real. 3DS MAX can provide six light: ambient light, scattered light, the target directional light, free directional light, the target spotlight, free spotlight, sunlight, you can use these lights to simulate the three-dimensional scene that you like, generally lighting design steps are composed by the following steps:①in the early program based on the overall idea to design a good light targeting position, according to the nature and position of light source, projection direction and other factors in the scene to design the program of light layout, do best to remain the scene the same tone, using lighting to render the scene atmosphere, lay emphasis on the expression subject. ②determine scene composition and spatial arrangements, using the light position and intensity of main light to control primary light colors of the scene, to avoid production the effect of reducing modeling scene with other light. ③ set auxiliary light to make up for the main light, floodlights without a shadow and soft lighting as auxiliary light, generally auxiliary light should ease the rigid rough shadow which the main light projected under the condition of not affecting the main light, alleviate contrast of bright and dark, to enhance and enrich the scene lighting levels, so that the scene would not get stiff, while helping the main light highlight the main part of the scene and the sense of space. ④ environmental modified light should result in a special atmosphere according to the environment which the principal objects lie in and needs to form some sort of against the background and the comparison to the main scene, no matter what kind of lighting it is, they are mututally affected and constraint, these lights can influence the brightness of the whole scene by light brightness and light attenuation range of values, you can adjust the distance between the light and the exposure object to influence the object, the farther the light away from the objects, the greater the range of light is, more uniform the objects light is, the closer lighting away from objects, the smaller the range of light is, will produce a strong "light pool", objects light is not very uniform, Should try to avoid these phenomena occur.

## Summary

To sum up the above content, we describe from the three-dimensional game model estanlishment to the material and lighting design, we understand that in the application of 3DS MAX modeling software first need to analyze and understand model structure of the object which needs to be created, choose a reasonable modeling methods according to different structural features, make a clear direction for materials design and lighting, propose some solutions, 3DS MAX has a certain guidance for three-dimensional game design, we should on the basis of understanding the software design principles and design rules to constantly sum up, to observe and learn more in the arts aesthetic, so as to performance the power and performance of these software, make game design become truly unified between art of and technology.

**Acknowledgement.** Project Fund: innovative research of animation Writing and Chinese traditional artistic elements, the second phase of "National College of innovative experiment scheme" funded projects of Guilin University of Electronic Technology.

## References

1. Yu, D.: Three-Dimensional modeling examples and techniques. People Posts & Telecom Press Excellent Books, Beijing (2000)
2. Yan, F., Wu, Y.(translator): A special case of three-dimensional modeling design 3ds max5 completely real. Ocean Press, Beijing (2004)
3. Li, J.: Novice from zero learning 3ds Max. Shanghai Popular Science Press, Shanghai (2008)
4. Liu, K.: Creative design of three-dimensional video game scene. Beijing Hope Press, Beijing (2000)

# Teaching System Modeling Based on Topic Maps

Xilun Chen, Xia Hou, and Ning Li

Computer School, Beijing Information Science & Technology University
Beijing, China
cupidcxl@gmail.com, {houxia,3lining}@bistu.edu.cn

**Abstract.** In a teaching system, there are many confusing concepts and complicated relationships. It is difficult to analyze and organize all the information clearly and reasonably. By using Topic Maps, we are able to build a model for teaching system with lots of advantages, the Scope property of topics in Topic Maps is used to classify and retrieve information, which will largely shorten the information querying time and minimize the amount of data transmission. In this paper, a model is constructed to illustrate the modeling method from the perspective of teaching and demonstrate the use of Scope.

**Keywords:** Topic Maps, e-Teaching, Model.

## 1 Introduction

With the development of computer technology, the growing diversity of information presents in front of us. Information modeling and analyzing becomes more complex and complicated than ever. The huge information makes it difficult to quickly obtain the required information correctly. Tedious searching and querying wastes too much precious time, results in decreased efficiency and error. In the area of education, the same problems occur.

Curriculum is a huge repository, not only contains important information for all courses, such as course descriptions, teaching and studying plans and their relevant resources, but also covers the data for all teachers, teaching arrangements and other information. At the same time, among the curriculum, teachers, teaching resources and students there exists many complex relationships. Therefore, it is significant to fully understand and manage these information and relationships for constructing a reasonable curriculum system which is important to make educational research and to improve teaching and learning efficiency. This complex system can be analyzed and described by modeling.

Topic maps, known as a meta-data format which describes the structure of information resources, can largely improve information management and delivery. It has great value to study and worth of being researched. At present, many people use Topic Maps in the field of E-learning. Dicheva etc. discussed how to efficiently organize and classify learning content [1]. They noted the educational and research benefits of applying an innovative technology - Topic Maps - for the organization and retrieval of online information in the context of e-learning courseware [2]. Kuang etc. mentioned that in an e-learning system, topic maps can provide a good semantic

S. Lin and X. Huang (Eds.): CSEE 2011, Part II, CCIS 215, pp. 197–204, 2011.

model for the course document [3]. An algorithm is presented in [4] for automated building of a topic maps starting from a relational database to offer students in e-learning. These studies use Topic Maps resources for effective integration of teaching and studying, so that students can improve the efficiency of information retrieval and self-learning. But none of them have modeled in the perspective of teaching.

This paper will focus on modeling of curriculum in computer science major, using Topic Maps as a tool, from the perspective of teaching and learning, especially through the use of the Scope in Topic Maps to fully describe the resources of teaching and learning context. Modeling methods provided by this paper, offer an effective way for educational researchers to analyze the internal relationship between teaching and learning in curriculum resources, allocate work for teachers and make a reasonable teaching plan for teachers.

## 2 Topic Maps and Scope

### 2.1 Topic Maps

Topic Maps is a standard for the representation and interchange of knowledge. The origins of the topic maps paradigm itself date back to 1993, when it was first expressed as a working document in the context of the Davenport Group. The paradigm was more fully developed thereafter in the context of the GCA Research Institute (now known as IDE Alliance), in an activity called Conventions for the Application of HyTime, during and after which the paradigm was independently developed, implemented, and promulgated. Early in 2000, after several years of continuous effort by an international group of individuals, the topic maps paradigm was fully formalized for the first time as an ISO International Standard, ISO/IEC 13250:2000. Almost immediately thereafter, TopicMaps.Org was founded in order to develop the applicability of the paradigm to the World Wide Web, and to realize its enormous potential to improve the findability and manageability of information. In the year 2000 Topic Maps was defined in an XML syntax, XML Topic Maps (XTM). This is now commonly known as "XTM 1.0" and is still in fairly common use. The ISO standards committee published an updated XML syntax in 2006, XTM 2.0[5]. which is increasingly in use today. The most recent work standardizing Topic Maps (ISO/IEC 13250) is taking place under the umbrella of the ISO/IEC JTC1/SC34/WG3 committee (ISO/IEC Joint Technical Committee 1, Subcommittee 34, Working Group 3 - Document description and processing languages - Information Association) [6].

The key concepts in topic maps are topics, associations, and occurrences [7].

A topic is a resource within the computer that stands in for (or “reifies”) some real-world subject. Topics represent any concept, from people, countries, and organizations to software modules, individual files, and events.

Occurrences represent information resources relevant to a particular topic. Topics can have occurrences, that is, information resources that are considered to be relevant in some way to their subject.

Finally, topics can participate in relationships, called associations, in which they play roles as members. Associations represent hypergraph relationships between topics.

Thus, topics have three kinds of characteristics: names, occurrences, and roles played as members of associations. The assignment of such characteristics is considered to be valid within a certain scope, or context.

The purpose of a topic maps is to convey knowledge about resources through a superimposed layer, or map, of the resources. A topic map captures the subjects of which resources speak, and the relationships between subjects, in a way that is implementation-independent.

In a word, a topic maps is a kind of data format that can be used to describe the structure of information resources. In addition, it can also quickly locate one or more concepts on a "map", and represent the relationship between them.

## 2.2 Scope in Topic Maps

In Topic Maps, there is an important character, Scope, which is used to assign a given topic context.

In our daily life, the case of polysemy is widespread, and sometimes even a simple word has more than 3 meanings. Of course, in different contexts, people with smart brains can determine the words' meanings according to the kind of situation. For example, when people talk about the word "buck", if no context attached, we do not know what exactly the "buck" here means, because the word "buck" has too many meanings, it can be explained as male deer, male rabbits, one dollar, against and so on. When sometimes we cannot understand a word's meaning, the only thing we can rely on is tips from its contexts. Buck was apparently interpreted as meaning one dollar during a small business, while in a drastic Debate, buck means opposition. The human brain has already been aware of this way of interpretation by analysing context. However, computers are not that smart as human brains, computers cannot automatically analyse the context and explain the meaning of a topic. What computers can do is showing all the meanings of a particular topic to users, and users then select from which and determine whether the information is useful for them according to analyzing the context by themselves. Like consulting dictionaries, when you query the meaning of a word, you have to choose a specific meaning. But in this way, precious time is wasted during the query processing, because you have to make a comparison among all the meanings by the context and determine which the best is.

Different people have different views and angles, thus to a particular thing, it would be complicated to make some classifications. For instance, the study plan and teaching resources are quite different between the students who is major in computer science and non-computer science. A variety of resources are mixed together, although the users eventually are able to find the information they need, but this will waste time and space, causes confusion, and even may bring out errors when searching and understanding.

Fortunately, Topic Maps can offer a way to help us deal with these issues. Scope is the answer. Scope specifies the extent of the validity of a topic characteristic assignment. It establishes the context in which a name or an occurrence is assigned to a given topic, and the context in which topics are related through associations. Namely, a declaration of a topic characteristic is valid only within a scope. When a topic characteristic declaration does not specify a scope, it is valid in the unconstrained scope.

In fact, scope is not limited to eliminate ambiguity, it is also used for information navigation, dynamically adjust scope based on the users' profiles during the query process, to achieve the purposes of being more specific in searching information. For example, the online system can load and present the corresponding teaching and learning resources based on the profile of the students and course requirements, dynamically adjusted the topics and subjects that will be shown. In this way, the whole query process saves a plenty of data that are originally supposed to be transferred, making the searching more effective and efficient with less errors and misunderstanding. From this, we can see the powerful role that scope is playing, and we will discuss the problems scope can solve in the following sections.

# 3 Teaching System Modeling Based on Topic Maps

## 3.1 Topics, Associations, and Occurrences

Now, we use XTM to construct some basic models to illustrate the three significant concepts in Topic Maps—topics, associations and occurrences.

Topic Maps Document based on XTM is written in XML format. Let's see a topic that represents a C++ course, the xml codes are given in Fig.1. Topic id is the identifier of a topic and may be a random string. Element <baseName> is the name for a specific topic. Element <occurrence> contains the information resources that are relevant to this topic. In this example, there is a website, course.bistu.edu.cn, as its relevant resource.

```
<topic id="C++">
      <baseName>
          <baseNameString>C++</baseNameString>
        </baseName>
        <occurrence>
          <instanceOf>
            <topicRef xlink:href="#Book1" />
          </instanceOf>
        </occurrence>
        <occurrence>
          <instanceOf>
            <topicRef xlink:href="#Website" />
          </instanceOf>
          <resourceRef xlink:href="course.bistu.edu.cn " />
        </occurrence>
</topic>
```

**Fig. 1.** XML codes of a topic represent a C++ course

Thus, topics and occurrences are the basic concepts in constructing topics, and in the following we can see the relationship between topics.

XML codes of an association representing the relationship between Zhang San and the course C++ might are given in Fig. 2.

These lines are defining a relationship between Zhang san and C++. The relationship name is "taught-by" represented by the element <instanceOf> and this relationship has two members, each member demonstrates one role in this relationship. In this example, one is Zhang san which is a sub-topic of teacher, and the other is C++ which is a sub-topic of Course.

```
<association>
        <instanceOf>
          <topicRef xlink:href="#taught-by " />
        </instanceOf>
        <member>
          <roleSpec>
            <topicRef xlink:href="#Teacher " />
          </roleSpec>
          <topicRef xlink:href="#Zhang san" />
        </member>
        <member>
          <roleSpec>
            <topicRef xlink:href="#Course " />
          </roleSpec>
          <topicRef xlink:href="#C++ " />
        </member>
</association>
```

**Fig. 2.** XML codes of an association

### 3.2 Constructing the Model

Here, we build a simplified model of the teaching system to demonstrate the constructing method that uses the Topic Maps. Assuming the teaching system consists of several main types of teaching resources: courses, teachers, and teaching aids such as books, audio and video resources. How to organize these resources is the key in modelling. Teaching system involves a great number of specific resources, each resource not only contains the basic description of the specific and interrelated topics, but also there will be a series of related reference resources. The formation of the concept layer and information layer in Topic Maps Modelling helps to make a clear description of this architecture. By constructing topic associations between the different types in the concept layer, we can make this layer hierarchical, which will help improve the analysis and study of curriculum efficiently. As shown in Fig. 3. The model is divided to concept layer and information layer. In concept layer, there are topics and their associations. In information layer, there are the real resources and material.

Any object in Topic Maps is a topic. The first step in topic map modelling is to create a basic set of topics. Each kind of topic contains a collection of entities, which are also topics in the model. For example, in Fig.3, four kinds of topics are modelled: Teacher, Course, Resource and Group. Each of them has some entities.

In the teaching system, teachers teaching related courses can be divided into different groups, which will help us improve the quality of teaching and promote the development of teaching teams. Therefore, in this model, the topic--"Group" refers to Teacher group, and topic--"Group1" refers a specific series of teachers teaching certain related courses.

The topics are represented by polygons in a tree. The relationship between them in the same tree, represented by solid lines, is topic and sub-topic, just like the class and sub-class relations in programming language. This is a kind of association. Topic and sub-topic is a very common relation in topic map model. For example, "Teacher" and "Li Si" are both topics, but "Li Si" a specific teacher. In topic maps, this kind of relationship is much more easily to understand.

Furthermore, there are many kinds of relationships in teaching system. For example teach and taught-by, related and belong-to.

- teach and taught-by: the association is between a teacher entity and a course entity. For example "Zhang san" teaches "C++", and "C++" is taught by "Zhang san". The relationship is represented by line ··· .
- related: the association is between a entity and a resource entity. For example "C++" is related to "Book1". The relationship is represented by line —·—
- belong-to: the association is between a teacher entity and a group entity. The relationship is represented by line ---.

Through this network graph like Fig.3, we can easily recognize the relationship of each topic. For simplified, we only give three kinds of relationships here. Associations can be built according to the requirement of the modelled system. As long as there are some explanations near the legend, very complicated relationships can be easy to read and understand.

### 3.3 Application of Scope

In particular, we will focus on the use of "Scope", which would help efficiently organize and classify resources. In section 2 we have talk about the role that scope will play, when it is used in the teaching system, ambiguity can be effectively solved, and helpful for information navigation. In the following, we will illustrate this point by an example of using the course C++ to demonstrate how scope exactly works during modelling.

In the system, the course C++ is a topic and a sub-topic of course topic. Because many majors take it as a course, it is difficult to be classified and categorized. For instance, C++ is an obligatory course for students whose major is computer science. However, for students whose major is in different fields other than computer science, C++ may be an elective course. So teachers, curriculum plans and teaching resources are not exactly the same for different majors. The traditional model cannot clearly describe this system and its internal relationships. But, using Topic Maps, we can take advantage of scope to overcome this shortcoming.

Some XML codes in Fig. 4 are given to illustrate the application of scope. The topic C++ has more than one occurrence. According to its scope, teacher can retrieved the corresponding teaching resource for a certain major.

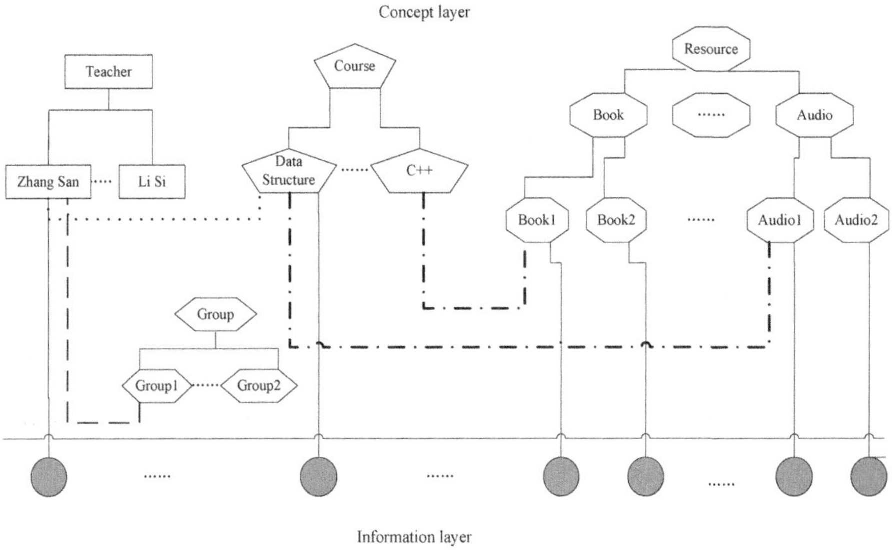


**Fig. 3.** Teaching System Model

```
<occurrence>
  <instanceOf>
    <topicRef xlink:href="# Video" />
  </instanceOf>
                              <scope>
    <topicRef xlink:href="#computer science " />
  </scope>
  <resourceRef xlink:href="www.bistu.edu.cn/video/major-cs" />
</occurrence>
<occurrence>
  <instanceOf>
    <topicRef xlink:href="#Video" />
  </instanceOf>
  <scope>
    <topicRef xlink:href="#Electrical Engineering" />
  </scope>
  <resourceRef xlink:href=" www.bistu.edu.cn/video/major-ee" />
</occurrence>
```

**Fig. 4.** XML codes of using Scope

## 4 Conclusion

This paper focuses on how to use Topic Maps to model the teaching system. We proposed a method of information classification by using scope as a core. Teaching system contains a large amount of information and complex resources. Each teacher has his own job and their corresponding responsibilities. If everyone in the teaching system tries to read the whole resource space to query information, he will waste much time. But through the Topic Maps model with scope, he would be easier to understand the concepts, their relationship between topics and their related resources because some unnecessary information will not be shown before him. This is particularly significant for teachers. Base on the different teaching works, each teacher may have his own teaching plans, and he must make his plans according to different majors' requirements. Scope can help teachers quickly locate and understand their specific major's work with less time spent. They can save time to study more about how to improve teaching efficiency.

Therefore, the Topic Maps with Scope is a good way to model a teaching system.

**Acknowledgment.** The authors thank Scientific Research Common Program of Beijing Municipal Commission of Education (No. KM201010772013) and Funding Project for Academic Human Resources Development in Institutions of Higher Learning Under the Jurisdiction of Beijing Municipality (No. PHR201108256) for funding this research.

## References

1. Dichev, C., Dichev, D., Aroyo, L.: Using Topic Maps for E-learning. In: Proc. of Computers and Advanced Technology in Education (2003)
2. Dichev, C., Dicheva, D.: TM4L: Creating and Browsing Educational Topic Maps. British Journal of Educational Technology 37(3), 391–404 (2006)
3. Kuang, W., Luo, N.L.: User Interests Mining based on Topic Map. In: Proc. of Seventh International Conference on Fuzzy Systems and Knowledge Discovery (2010)
4. Mihai, G., Stanescu, L., et al.: A Topic Map for Subject-Centric Learning. Intelligent Distributed Computing III 237, 141–150 (2009)
5. JTC1/ SC34. ISO 13250: Topic Maps - XML Syntax, http://www.isotopicmaps.org/sam/sam-xtm/
6. Home of SC34/WG3 - Information Association (June 03, 2008), http://isotopicmaps.org (retrieved May 5, 2011)
7. Steve Pepper. The TAO of Topic Maps, Senior Information Architect, Norway (April 11, 2000)

# The Reform of the Practice Teaching and the Training Mode of Creative Talents for Electronic-Information-Type Specialties

Liqun Huang, Ping Gong, and Jie Zhang

Department of Electronic Information, Northeastern University at Qinhuangdao, Qinhuangdao, China
persistent_hlq@sina.com

**Abstract.** In this paper, we take communications engineering specialty as an example, and investigate the practice teaching and the training mode of creative talents for electronic-information-type specialties. During formulating the training program, the course system needs to be optimized with emphasis on the teaching contents of practical skills. The experiments of fundamental courses and theoretical courses mainly fall within validation-type experiments, and the ones of professional platform courses and practical optional courses mainly belong to design-type experiments. Moreover, it's stressed that the subjects of course projects and graduation design should be comprehensive, practical and innovative; extracurricular activities in science and technology focus on training student's technical expertise. Finally, The construction scheme of professional laboratory is presented for further enhancing student's engineering practical ability and innovation ability.

**Keywords:** course system; laboratory planning; practice teaching; creative talents; training mode.

## 1 Introduction

Communications engineering specialty, as an important specialty of engineering colleges, aims to train senior professional technical personnels with the knowledges of communication technology, communication systems and communication networks, and graduates can be engaged in the research, design, development, manufacture, operations, and applications in the field of communications engineering.

In recent years, the number of universities with communications engineering specialty increase year by year in our country, and the enrollment scale is also growing[1]. Currently, there are 246 universities that have established communications engineering specialty. Because of broad social needs of communications engineering graduates, the admissions size per year is a great number for this specialty. The admissions number of each university averages about 100 to 200 each year, and the annual total admissions number of these universities has reached about 30,000 people.

In the coming years, many univeristies will pay more attention to the education training of communication engineering technical personnels, and need to significantly improve the teaching quality. Several important problems, about how to improve the

S. Lin and X. Huang (Eds.): CSEE 2011, Part II, CCIS 215, pp. 205–210, 2011.

teaching quality of communications engineering undergraduates, are presented and discussed in this paper, and focus on investigating the practice teaching course system and the training mode of creative talents.

## 2 Formulating the Reasonable Training Program with an Optimized Course System

Communications engineering specialty has the following features: this specialty involves three primary disciplines, namely information and communications engineering, electronic science and technology, and computer science and technology, respectively. All these three primary disciplines are the footstone of today's information society, the development of these disciplines is very rapid, and the science and technology achievements in these fields have profoundly changed people's production mode and life style.

The cutting-edge technologies of this specialty include information and signal processing, mobile communication systems, satellite communications systems, optical fiber communication systems, multimedia network communications, electromagneticfield and microwave technology, electromagnetic compatibility, VLSI design, and so on[2]. Obviously, communications engineering involves many disciplines, and these technologies are developing rapidly.

During formulating the training program, we need to have forward-looking teaching idea, and fully embody the guiding ideology of "solid theory foundations, wide professional knowledges and practical skills". In the optimization process of course system, we must always implement the principle of combining theory with practice, and make students grasp the solid theory foundation of mathematics and physics, systematic professional knowledges of information and communications engineering, and the strong practical skills of applied electronic technology and computer application technology.

The course system of communications engineering specialty is very rigorous, and there is a close relationship betweent preceding courses and follow-up courses. curriculum before and after the relationship between closely. For communications engineering undergraduates, the most basic courses include mathematics, physics, serial courses of electronic circuits, and serial courses of computer application technology. Only taking them as foundation course groups, students can further master higher-level professional knowledge and practical skills.

Professional platform courses include “signal and linear system”, “high-frequency electronic circuits”, “digital signal processing”, “electromagnetic field and microwave technology”, “microprocessor principle and interface technology”, “communications theory”, “computer networks”, “microcontroller principle and application”, “FPGA theory and applications”, “mobile communication systems”, and “modern switching technologies”. For professional foundation courses and professional platform courses, we should reserve enough teaching hours during establishing the training program, so that undergraduates have solid theory foundations and excellent basic practical skills, which lays a good foundation for further learning professional optional courses.

For optional course groups, we should particularly emphasize the professional skills training courses with strong practical applications. Through arranging serial courses of hardware and software design, students have chances to master the professional skills used in future job position. Thereinto, serial courses of hardware design include "fundamentals of analog electronic", "fundamentals of digital electronic", "microprocessor principle and interface technology", "microcontroller principle and application", "electronic circuit CAD", "FPGA theory and applications", "DSP technology and applications", and "electronic system design"; serial courses of software design include "the C/C++ programming language", "fundamentals of software technology", "database principles and applications", "the Jave programming language", "embedded systems and applications", and "communication software design and test". In addition, how can we bring the latest technical advances into the undergraduate teaching contents? this issue also needs further exploration and research.

## 3 The Organic Fusion of Systematic Practical Teaching Links and Multiform Teaching Methods

In order to meet social demands, the training program should focus on improving undergraduate's practical ability, follow the principle of combining theory with practice, and make teaching links systematized, which include theory teaching, course experiments, electronic technology practice, course projects, extracurricular activities in science and technology, and graduation design. These organic fusion teaching links provide the practical skills training from the elementary to the advanced.

For classroom teaching, we take advantage of multimedia courseware extensively, and strengthen the teaching of the important chapters. During compiling multimedia courseware, we need to pay attention to the visual effects. For example, the font size should be appropriate, and the line spacing should be moderate. In order to make the teaching more vivid and intuitive, we should adopt the figures and the tables as many as possible in multimedia coursewares. Meanwhile, using animations and hyperlinks makes the knowledge points to relate and confirm each other. Finally, the multimedia courseware has many advantages, such as clear structures, vivid demonstrations, and outstanding focal points, which is beneficial to enhancing the intuitive impression of students.

For course experiments, we pay attention to adjusting the ratio of validation-type experiments and design-type experiments. Thereinto, the experiments of fundamental courses and theory courses fall within validation-type experiments mainly, and the ones of professional platform courses and practical optional courses mostly belong to design-type experiments. On this basis, the open laboratory is constructed to facilitate students choosing their own time and their favorite experiments, which can enhance students' learning interest, widen the range of their knowledge, and improve the utilization efficiency of instruments and equipments.

After course experiments, course projects should further enhance the practical ability of undergraduate students, and it's stressed that the subjects of course projects should be comprehensive, practical and innovative, which is in favor of cultivating student's creative minds. Moreover, extracurricular activities in science and technology focus on training student's technical expertise according to the principle of individualized instruction, and lay a solid foundation for the creative personnel training system. We will consolidate the achievements obtained previously, make backbone teachers play a more important role, participate in the national electronic design competition actively, and create sound environments and beneficial conditions as much as possible for mobilizing the enthusiasm of instructors and students.

The subjects of graduation design should be closely related with student's future job position, follow along with the technical development direction of communications engineering specialty, and meet the social demand for student's practical capacity. Thus, student's initiative and creativity can be adequately mobilized. Students search relevant references by themselves, propose and demonstrate the design schemes, and do theoretical and experimental research independently. Finally, Students carry out the implementation of hardware and software, and compose the graduation thesis, all by themselves. The entire practice process can take advantage of student's initiative, independence and creativity, and enhance their engineering practical ability and innovation ability.

Perfect evaluation system should be established about instructor's responsibility during supervising students, reviewing their graduation thesis, and organizing graduation thesis defense. Morover, the following five parts must be realized to strengthen standardized management of graduation thesis: intensifying guidance to better topic selection of graduation thesis; strengthening inspection and building quality check system of graduation thesis; standardizing oral defense procedure; perfecting graduation thesis evaluation system; enhancing university-enterprise cooperation to seek social support[3].

## 4 The Professional Laboratory Planning of Communications Engineering Specialty

Modern communication technologies are developing rapidly, and have many research fields. In order to make the professional laboratory meet the demands of the practice teaching of communications engineering major, and follow along with the technical development direction, the laboratory construction needs the scientific planning. As shown in Fig. 1, the professional laboratory of communications engineering specialty can be divided into basic laboratories, applied technology laboratories, and modern communications technology laboratories[4].

Basic laboratories consist of electronic circuits lab, signal and linear system lab, high-frequency electronic circuits lab, communications theory lab, and computer networks lab. These basic laboratories are indispensable to communications engineering specialty, and need to be firstly established.

With the rapid development of modern electronic technology, digital communication systems has gradually occupied a leading place. Thereinto, electronic design automatic (EDA) technology and digital signal processing (DSP) technology are main development tools of digital communication systems. Therefore, after carrying out the construction of basic laboratories, we should focus on planning and establishing the applied technology laboratories, wich consist of microcontroller lab, FPGA lab, DSP lab, and embedded systems lab.

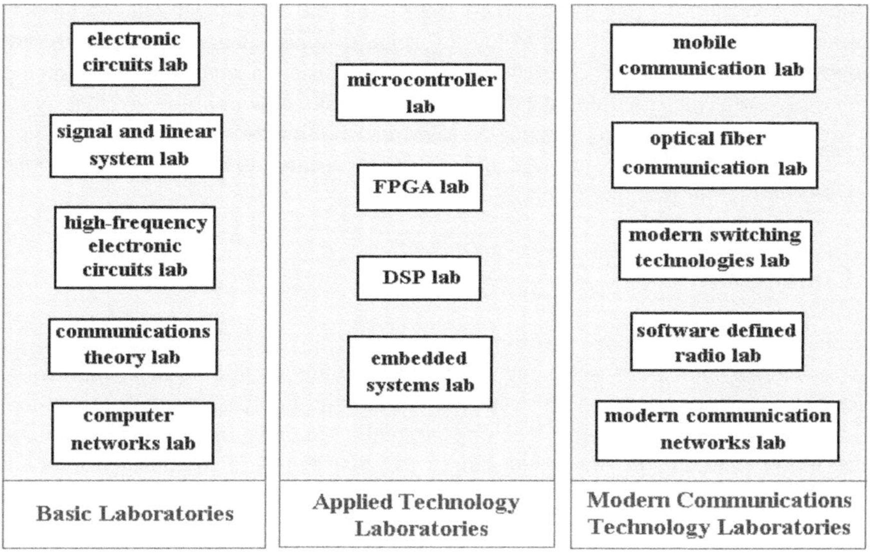


**Fig. 1.** The professional laboratory planning of communications engineering specialty

Combining with the knowledge of communications system learned in the classroom, students can further design and develop the function modules of digital communication systems by using the instruments and equipments of applied technology laboratories. In the integrative experiment platform of communication networks, students can also implement secondary development using EDA tools, and carry out comprehensive application of electronic technology and communication technology. These serial laboratories also give service to undergraduate's course projects, graduation design, and electronic design competition, and help students to improve their practical skills and innovation ability, especially for the design ability of electronic systems. Meanwhile, these laboratories can also provide good conditions for the research activities of teachers and graduate students .

Communications engineering major has many research fields, such as mobile communication systems, software defined radio, optical fiber communication systems, microwave communications, modern switching technologies, modern communication networks, and so on. The laboratory construction often has the shortage of funds, moreover, the experimental teaching hours are also restricted by the total teaching hours of the training program. Hence, the professional laboratories construction of modern communications technology can not cover everything. According to school specialty development planning, we should firstly choose the dominant academic fields of our school to establish partial professional laboratories of modern communications technology.

Normally, we give priority to establishing mobile communications lab, optical fiber communication lab, and modern switching technologies lab. The laboratory equipments related with communication networks are often very expensive, so that we don't construct these laboratories until school funds are sufficient[5].

Finally, we should pay attention to improving the utilization rate of laboratory equipments. If the related electronic-information-type specialties have the same equipments, the duplicate construction of laboratories needs to be avoided. We emphasize both on practice teaching and scientific researching, so that practice teaching equipments and scientific research equipments require coordination consideration as a whole. Moreover, the scientific research progress in our school will further promote our practice teaching.

## 5 Conclusion

Due to great social demand for communications engineering undergraduates, many universities expand the enrollment scale of communications engineering major. In order to keep up with the rapid development of modern communication technologies, we must adjust course system, teaching contents, teaching methods, and laboratory construction planning in time, and satisfy the diverse social demands for practical talents. In this process, we should grasp the direction of discipline development, carry out the periodic investigation of social demands, and actively participate in the discussion conferences on teaching experience among universities. Thus, we can shoot the arrow at the target during formulating the training program, and generally improve the teaching quality of communications engineering specialty.

**Acknowledgment.** This work was supported by the Eleventh Five-Year Planning Project of Education Science Research from the Education Department of Hebei Province under Grant No. O8020027. and also supported by the Scientific Research Program from the Education Department of Hebei Province under Grant No. z2005323.

## References

1. Sun, Y., Liu, T., Zhang, L.: Development and present situation of telecommunication engineering specialty in chinese universities. Journal of Taiyuan University of Technology (Social sciences edition) (12), 85–87 (2006)
2. Chen, W., Hu, J.: Exploration on formulating the training program of communications engineering major facing to the 21st Century. Higher Education Research in Areas of Communications (2), 21–22 (1996)
3. Wang, J.: Probe into standardized management of graduation thesis (design) of undergraduates. Journal of Higher Education Management 1(2), 85–88 (2007)
4. Zhao, S.: Laboratory construction and practice of communication and electronics specialties. Research and Exploration in Laboratory 24(5), 103–106 (2005)
5. Deng, Y., Wang, L., Zhou, Z.: Laboratory construction and teaching reform on practical training platform of modern communication. Research and Exploration in Laboratory 26(12), 122–125 (2007)

# Model of Fuzzy Comprehension Evaluation of Traffic Operations Safety on Urban Ice and Snow Road

Yulong Pei, Chuanyun Fu, Weiwei Qi, and Ting Peng

School of Transportation Science and Engineering, Harbin Institute of Technology, Harbin 150090, China

**Abstract.** In order to evaluate traffic operation safety of urban ice and snow road objectively, ice and snow pavement friction coefficient, visibility, following distance and snow-accumulated depth are chosen as evaluation indexes and their criteria are established. Evaluation indexes are weighted by using the improved entropy weight coefficient method, and then the model of fuzzy comprehension evaluation of traffic operation safety on urban ice and snow road is constructed. It is demonstrated through an example that the model has strong operability, and its result is relatively accurate. This model provides a theoretical basis for judging road traffic safety ranks objectively and formulating scientific policies and measures which can improve road traffic safety under ice and snow condition.

**Keywords:** Road traffic safety evaluation; Fuzzy comprehension evaluation; Urban ice and snow road; Road traffic safety ranks.

## 1 Introduction

In recent decades, many researches of road traffic safety evaluation have done. A macroscopic evaluation model based on fuzzy logic is established (Wang, et al., 2008). Criteria for dividing the rank of road traffic safety is confirmed on account of 85 percentages about value of single vehicle velocity difference (Zhou, et al., 2009). There is also another estimating way in view of technology method on monitoring road traffic accident (Ward, et al., 1982). These studies rarely refer to road traffic safety evaluation under ice and snow condition, and almost evaluation models are easily influenced by subjective factors. Therefore, the safety level of urban ice and snow road traffic operation is divided into three grades: safety, sub-safety, and danger. This paper adopts the method of fuzzy comprehension evaluation to estimate objectively the traffic operation safety on urban ice and snow road for there are no specific boundary among the three ranks.

## 2 Confirmation of Evaluation Indexes

It is found that the decrease of pavement friction coefficient dues to ice and snow is the inside reason of traffic accident, in addition, visibility affects greatly traffic accident of urban ice and snow road (Xiang, 2010). The braking distance of vehicles increases

S. Lin and X. Huang (Eds.): CSEE 2011, Part II, CCIS 215, pp. 211–215, 2011.

obviously for the decreasing of ice and snow pavement friction coefficient, and the following distance is too short to result in traffic accident. Besides, it is founded that the snow-accumulated depth is close related to traffic operation safety according to investigation and analysis. Therefore, ice and snow pavement friction coefficient, visibility, following distance and snow-accumulated depth are chosen as evaluation indexes to evaluate the traffic operation safety on urban ice and snow road.

# 3 Evaluating Criterion of Evaluation Indexes

## 3.1 Evaluating Criterion of Ice and Snow Pavement Friction Coefficient

Norwegian scholars have got by investigation the accident rates at different friction coefficient interval (Wallman, et al., 2001). Therefore, it can be concluded that traffic operation on urban ice and snow road is at risk when friction coefficient is < 0.15. When friction coefficient is 0.15~0.25, it is under sub-safety critical state. When friction coefficient is 0.25~0.30, it is under sub-safety state. When friction coefficient is 0.30~0.35, it is under safety critical state. And when friction coefficient is > 0.35, it is under safety state.

## 3.2 Evaluating Criterion of Visibility

When visibility is < 100 m, the danger of multi-vehicular accidents increases. A convoy may form also at visibilities of 100~200 m, at certain traffic volumes. In both cases, the danger of rear-end collision increases, as the traveling speed of the lead vehicle varies widely (Matsuzawa, et al., 2005). Under ice and snow condition, if two drivers heading toward each other are driving at 60 km/h, the distance at which they need to recognize each other in order to stop safely is approximately 350 m. It is found by investigation that the average speed of vehicles driving on the ice and snow road is 25~30 km/h. Based on this, it is relatively safe to travel on the urban ice and snow road while the visibility is 200~400m. Blowing snow will never happen for urban roads pavement timely and effective management in the winter, so it is safe to travel on the urban ice and snow road when the visibility is > 500m.

## 3.3 Evaluating Criterion of Following Distance

Following distance is the space between rear of vehicle ahead and head of following vehicle on the same lane. It is shown by investigation and analysis that speed on ice and snow pavement range from 15km/h to 40km/h.

Minimum safe following distance (MSFD) on the ice and snow road is analyzed and calculated according to theory of automobile driving (Luo, et al., 1999). Therefore, traffic operation on urban ice and snow road is at risk when the following distance is <= 8m. When following distance is 8~19m, it is under critical sub safety state. When following distance is 19~72m, it is under sub safety state. When following distance is 72~77m, it is under critical safety state. When following distance > 77m, it is under safety state.

### 3.4 Evaluating Criterion of Snow-Accumulated Depth

It is found that the frequency of traffic accidents is highest in Urumqi when snow-accumulated depth is 2~15cm (Ma, et al., 1985). The snow will be taken to roadside by vehicles when the depth is < 2cm, and traffic operation is under safety state. When snow-accumulated depth is 2~5cm, it is under critical safety state. When snow-accumulated depth is 5~10cm, is under critical sub safety state. When snow-accumulated depth is >30cm, vehicles cannot driving. So it is at risk when snow-accumulated depth is 15~30cm.

## 4 Model of Fuzzy Comprehension Evaluation

### 4.1 Determination of Membership Function

Membership functions of ice and snow pavement friction coefficient (ISPFC) are expressed as figure 1. Which of visibility (V) are shown as figure 2.

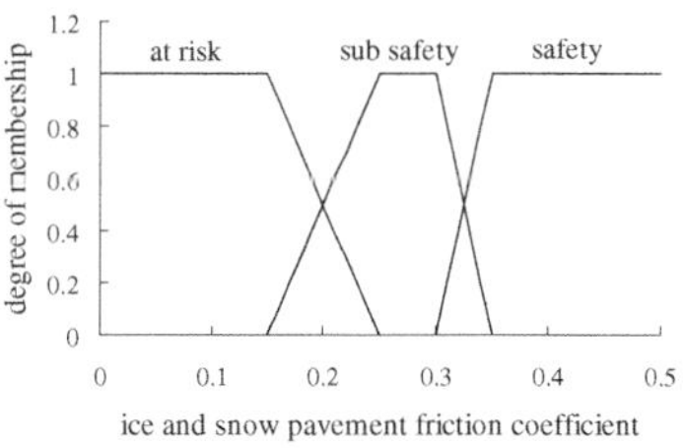


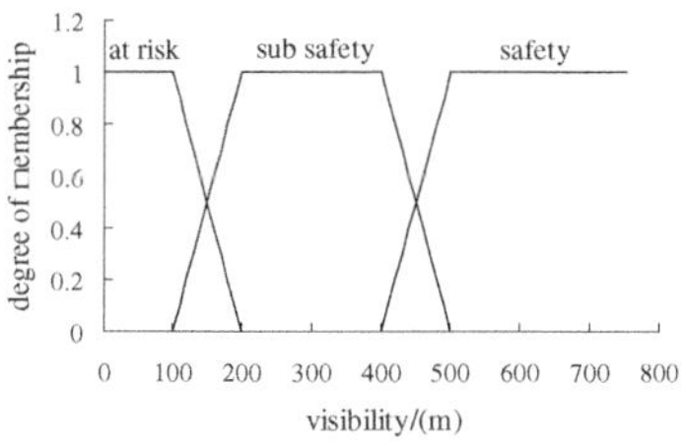


**Fig. 1.** Membership function of ISPFC **Fig. 2.** Membership function of V

Membership functions of following distance (FD) are shown as figure 3. Which of snow-accumulated depth (SAD) is shown as figure 4.

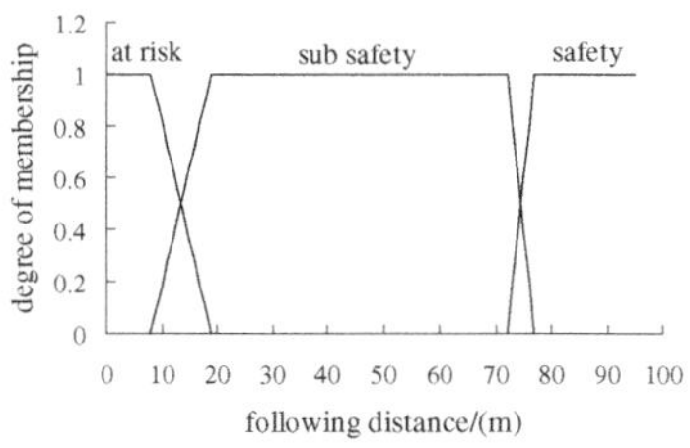


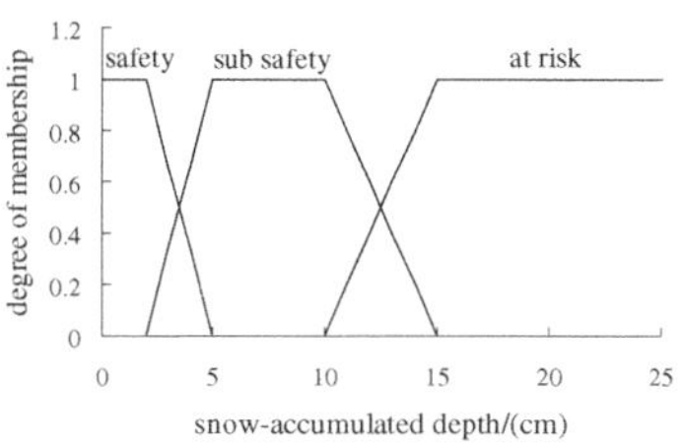


**Fig. 3.** Membership function of FD **Fig. 4.** Membership function of SAD

### 4.2 Entropy Coefficient

The weight of each evaluation index can be got by using fuzzy evaluation matrix, which is shown as follows (Fu, et al., 2010):

$$H_i = -\sum_{j=1}^{m} r_{ij} \ln r_{ij},\ j = 1, 2, \cdots, m \tag{1}$$

When $r_{ij}$ is equal, the maximal is $H_{\max} = \ln m$ . Supposing that $r_{ij} \ln r_{ij} = 0$ when $r_{ij} = 0$. The relative importance of each evaluation index $U_i$ can be calculated with the formula:

$$e_i = -\frac{1}{\ln m}\sum_{j=1}^{m} r_{ij} \ln r_{ij}, 0 \le e_i \le 1 \tag{2}$$

The weight of index $U_i$ can be measured with $1-e_i$, and then be normalized. Therefore, weight of $U_i$ is expressed as follows:

$$w_i = \frac{1}{n - \sum_{i=1}^{n} e_i}(1-e_i), 0 \le w_i \le 1, \sum_{i=1}^{n} w_i = 1 \tag{3}$$

### 4.3 Fuzzy Comprehension Evaluation

The evaluation muster $B$ can be calculated according to the membership functions and weight of each evaluation index:

$$B = W \cdot R = [w_1, w_2, w_3, w_4][r_{11}\ r_{12}\ r_{13}; r_{21}\ r_{22}\ r_{23}; r_{31}\ r_{32}\ r_{33}; r_{41}\ r_{42}\ r_{43}] = [b_1, b_2, b_3]$$

In this equation, $b_1$ , $b_2$ , $b_3$ indicates the safety state of traffic operations on urban ice and snow road is danger, sub safety, and safety respectively. The relative safety state of $\max(b_1, b_2, b_3)$ is the result of fuzzy comprehension evaluation.

## 5 Case Study

In a northeast city of China, The ice and snow pavement friction coefficient is 0.21; visibility is about 430m; average following distance is 20m; and snow-accumulated depth is about 13cm. The traffic operation on urban ice and snow road is under sub safety state.

## 6 Conclusions

The following conclusions can be drawn from this paper:

(1) It can be found that ice and snow pavement friction coefficient, visibility, following distance and snow-accumulated depth are the key factors for traffic operations safety of urban ice and snow road. It is suitable to select them as the evaluation indexes.

(2) The model of fuzzy comprehension evaluation of traffic operations safety on urban ice and snow road can reflect the actual condition, and it is practical.

(3) The method of entropy weight coefficient is a weighting way combining subjective and objective, which can make the weighting of indexes rationalization.

**Acknowledgement.** This paper was supported by National Natural Science Foundation of China (51078113).

## References

1. Wang, F., Nan, A.: Road Traffic Safety Evaluation Based on Fuzzy-set Theory. J. Communication Standardization (8), 56–60 (2008)
2. Zhou, Z., Wang, Y., Long, K.: Highway Traffic Safety Evaluation Based on Operating Speed. J. Journal of Transport Science and Engineering 25(1), 91–95 (2009)
3. Ward, Heather, Allsop, Richard: Area-wide Approach to Urban Road Safety-evaluation of Schemes by Monitoring of Traffic and Accidents. J. Traffic and Control 23(9), 424–428 (1982)
4. Xiang, W.: Research on Traffic Accidents Causes & Occurrence Mechanism on Urban Snow and Ice Road. D. A dissertation submitted to Harbin Institute of Technology for the Degree of Master of Engineering (2010)
5. Wallman, C.-G., Astrom, H.: Friction Measurement Methods and the Correlation between Road Friction and Traffic Safety. J. VTI Meddelande 911A, 40–42 (2001)
6. Matsuzawa, M., Kajiya, Y., Takeuchi, M.: The Development and Validation of a Method to Estimate Visibility during Dnowfall and Blowing Snow. J. Cold Regions Science and Technology (41), 91–109 (2005)
7. Luo, L., Gao, H., Pei, Y.: Determination the Minimum Safety Following Distance under Ice and Snow Condition. J. Hei Long Jiang Jiao Tong Ke Ji (1), 33–34 (1999)
8. Ma, S., Liu, S.: Urumqi Winter Climate and Highway Transportation. J. Weather (11), 15–17 (1985)
9. Fu, Y., Wu, X., Ye, Q., Peng, X.: An Approach for Information Systems Security Risk Assessment on Fuzzy Set and Entropy-weight. J. Acta Electronica Sinica 38(7), 1489–1494 (2010)
10. Guo, X.: Application of Improved Entropy Method in Evaluation of Economic Result. J. Systems Engineering-Theory & Practice (12), 98–102 (1998)

# Image Decomposition and Construction Based on Anisotropic Atom

Song Zhao[1], Hongliang Zhang[1], and XiaoFei Wang[2]

[1] ZhengZhou Institute of Aeronautical Industry Management,
Zhengzhou, 450015, China
[2] China Shipbuilding Industry Corporation NO.713 Reserch Institute,
Zhengzhou, 450052, China

**Abstract.** Image sparse decomposition is an efficient and key step for image compression, but hard problem. For sparse decomposition based on matching pursuit, this paper introduces a new anisotropy to construct over-complete dictionary, and makes use of an atom energy support to estimate the inner between atom and image. The results of image decomposition and construction show that image decomposition by the new atom has better performance and the computing speed is improved by 2 times.

**Keywords:** Sparse decomposition, matching pursuit, atom compact support property, anisotropy refinement.

## 1 Introduction

Image processing and analysis are needed as little as possible components to represent the image. The traditional image decomposition methods (such as discrete cosine transform and wavelet transform) used the base which limited by the orthogonal, the results obtained are not sparse image decomposition, and the image range of applications is limited. So the image extension is to be made from a complete orthogonal system to complete redundant system (known as the redundant dictionary.)

Signal decomposition based on the redundant dictionary is presented by Mallat. This concept has wide range of applications both on images and video processing [1-3]. Select the best way of approaching of an image in a redundant dictionary has been proved to be NP problem, Mallat gives a suboptimal algorithm - matching pursuit algorithm. Extensive research of fast algorithm for image sparse decomposition conducted on the basis of this algorithm has been carried out. However, even so, the sparse decomposition based on matching pursuit method (Matching pursuit of images, MP) need to calculate the image and redundant dictionary of atoms within each product, which led to enormous amount of computation. Domestic scholars conducted a study from the block perspective, but the reconstructed image quality is not high, and has blocking effect [4].

A new anisotropic atoms method is proposed in this paper, which has good time-frequency characteristics. Compared to Reference [5], the atoms have better ability of characterize the local regularity of images. According to the distribution of characteristics of atomic energy, the inner product of atomic and image is calculated to find the best match in the atom, then calculate the other parameters of atoms.

S. Lin and X. Huang (Eds.): CSEE 2011, Part II, CCIS 215, pp. 216–221, 2011.

## 2 Matching Pursuit Algorithm Based on Image Sparse Decomposition

Redundancy extend method of image sparse decomposition is not the only way, which is always looking for the practical application of the most sparse signal representation. This can be described as solving the following problems

$$\min \ \|c\|_0 \quad \text{s.t} \ \ f = \sum_{m=0}^{M-1} c_m g_m \tag{1}$$

Where $\|c\|_0$ is the number of non-zero entries in sequence $\{c_m\}$. Mallat uses decomposition algorithm with the matching pursuit to resolve the issue of the suboptimal solution [1]. Matching pursuit algorithm is adaptive algorithm, any function f the belonging to Hilbert space are decomposed in a redundant dictionary, the decomposition results are the sparse representation of images.

Suppose $D = \{g_\gamma\}_{\gamma\in\Gamma}$ is the redundant dictionary used for image sparse decomposition, which is a collection of parameter sets. Matching pursuit algorithm is iterative algorithm. In the course of the m iterations, select the largest atomic inner product as a result of the match, as follows:

$$\left|\left\langle g_{\gamma_m}, R_m \right\rangle\right| = \sup_{\gamma\in I}\left|\left\langle g_\gamma, R_m \right\rangle\right| \tag{2}$$

## 3 Fast Image Sparse Decomposition Algorithm Based on Anisotropic Atoms and Its Support Property

MP-based image sparse decomposition consists of two key components: the selection of generating functions to the redundant dictionary and the specific implementation of decomposition.

### 3.1 The Selection of Redundant Dictionary and the Anisotropic Generating Function

Edges of the image contour is the main components and features of image. Since the dictionary is not orthogonal redundant constraints, the selected generating function should be smooth low resolution function in the direction along the image contour, and the function perpendicular to the direction of the image edge has similar characteristics with wavelet, so that non-singular characteristics can be effectively approached to the image. In this paper uses the following atoms:

$$g(x,y) = (16x^4 - 48x^2 + 12)\exp[-(x^2 + y^2)] \tag{3}$$

It is obtained by the best time-frequency characteristics Gaussian function and its product of fourth-order derivative of one-dimensional Gaussian function at vertical direction.

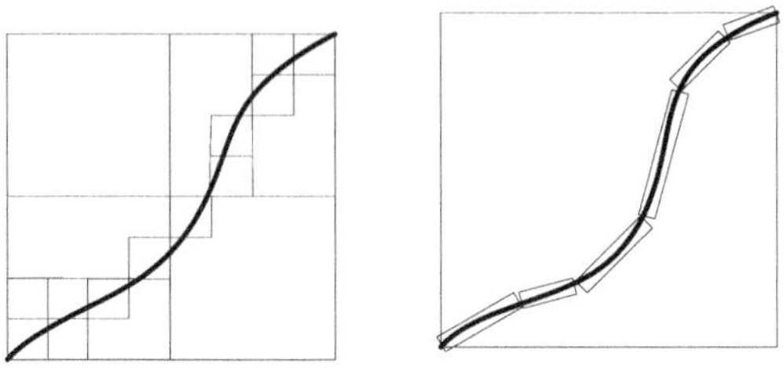

(a) Isotropic scale (b) Anisotropy scale

**Fig. 1.**

Redundant dictionary based on function $g(x, y)$ , $g(x, y)$ meet $g(x, y) \in L^2(R)$ , By the normalized operator $U_\gamma$ and the set of the given index $\Gamma$, it is obtained that:

$$D = \{U_\gamma, \gamma \in \Gamma\} \tag{4}$$

This set contains 3 basic types of operators: translation, anisotropic scaling and rotation stretching. In the actual calculations, all parameters of complete dictionary of must be discretized, according to the discrete theorem [1], the value of each parameter can be used as in the following form:

$$\begin{cases} 0 < u_x \le M \\ 0 < u_y \le N \\ s_x = 2^{\frac{stx}{\lambda}}, 0 < stx \le \lambda \times (\log_2 M - 1) \\ s_y = 2^{\frac{sty}{\lambda}}, 0 < sty \le \lambda \times (\log_2 N - 1) \\ \theta = \dfrac{k}{\max(M, N)} \times 2\pi, k \in [1, \max(M, N)] \end{cases} \tag{5}$$

If $P$ indicated the size of the complete dictionary, then

$$\mathrm{P} = M \times N \times \lambda \times (\log_2 M - 1) \times \lambda \times (\log_2 N - 1) \times \min(M, N) \tag{6}$$

### 3.2 Determination of the Atomic Support Scope

Signal decomposition process in a redundant dictionary used in the time-frequency atoms, which consists of a window function, so the amplitude of the atom from the center to the surrounding decay exponentially, as shown in Figure 2. From the energy point of view, atomic energy is concentrated in the main center of the atom, its distribution in a small area, a small number in other regions.

Support scope of atomic energy is determined by atomic-scale parameter. Energy of a large-scale support has a broader scope, more dispersed distribution, represents the background or smoothly change part in the image. Atoms of small-scale represent edges and details of the image, where energy is more concentrated.

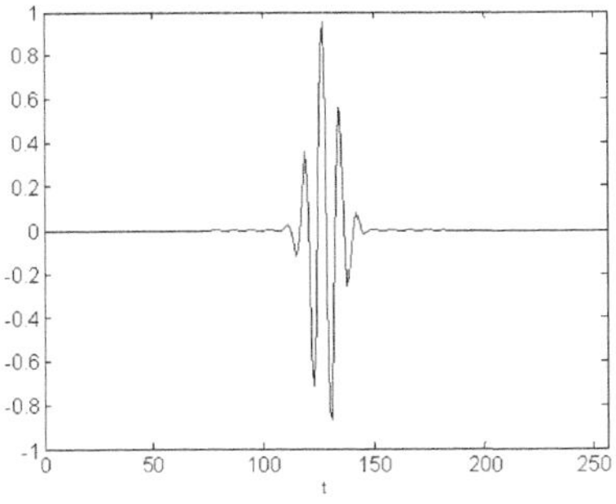


**Fig. 2.** Atom shape

$s_r$ represent atom compact support region. If $s_r$ selected properly and calculate atomic energy range from $s_r$, at this time the generation of atomic atom generation process can be converted into a small area. As the scope of the region is often less than the image size $M \times N$ , so this will greatly reduce the computational.

In the calculation of the atoms support area, generally calculated on the symmetric region of central atom. In order to identify the range of support sets, make the following change to atomic functions:

$$\begin{cases} g(x,y) = f(x)h(y) \\ f(x) = (16x^4 - 40x^2 + 12)\exp(-x^2) \\ h(y) = \exp(-y^2) \end{cases} \tag{7}$$

Integration on the absolute value of $f(x)$ and $h(y)$ respectively, to determine the matching pursuit algorithm calculations required for inner product areas. If the area covered by atomic is used to calculate the required inner product calculation area, we can follow (15) to calculate the boundary of integral area.

$$\begin{cases} \int_{-boundx}^{boundx} |f(x)|dx = a \times \int_{-\sup port}^{+\infty} |f(x)|dx \\ \int_{-boundy}^{boundy} |h(y)|dy = b \times \int_{-\infty}^{+\infty} |h(y)|dy \end{cases} \tag{8}$$

Generally, $a=b$; If the atoms in the range of coverage is 81%, $a = b = \sqrt{81\%} = 0.9$。 When $a_1 = b_1 = \sqrt{81\%} = 0.9$ $boundx_1 = 1.51$ , $boundy_1 = 1.65$ , $a_2 = b_2 = \sqrt{99\%} = 0.99$; $boundx_2 = 2.33$, $boundy_2 = 2.81$。

$$gain = \frac{boundx_2 \times boundy_2}{boundx_1 \times boundy_1} = 2.01 \tag{9}$$

Therefore, using 81% support area of atomic energy to calculate the image projected on a redundant dictionary to find the best matching atom, and then the number of inner product calculation is only half of the original.

## 4 Experiment Results

Experiment selected 256 × 256 × 8 standard test images of lena and pepper. Redundant dictionaries use the above-mentioned anisotropic atomic library. The reconstructed image is calculated with 500 atoms. Operating platform is PC with a Pentium processor at 2.8GHZ, 1G memory, and matlab of 7.0 version.

Table 1 shows that, for standard lena image, the peak signal to noise ratio (PSNR) of reconstructed image improved 0.43dB; for the standard image of peper, the reconstructed image PSNR increase 0.16dB. In this paper, the new atomic can represent the image edges and details better, so the reconstructed image by this new proposed atoms and Reference [5] both have better subjective quality and objective quality.

**Table 1.** Proposed method compared with Reference [5] on reconstruction quality

| Image | Reference [5] ( dB ) | Proposed method ( dB ) |
|---|---|---|
| Lena ( 256×256 ) | 27.48 | 27.71 |
| Peper ( 256×256 ) | 23.33 | 23.41 |

Figure 3 (b) is the image with the same number of wavelet coefficients of the largest reconstruction. Figure 3 (c) is the reconstruction image using atomic energy for 99% and 81% support area to calculate inner product. The above for the lena image, following the peppers image. For the lena image, the peak signal to noise ratio of reconstructed image (b), (c) and (d), are respectively 23.25dB, 27.83 dB, 28.05 dB; the reconstructed image PSNR decreased 0.21dB. For the peppers image, reconstructed image (b), (c) and (d) respectively, the peak signal to noise ratio is 21.98dB, 28.16 dB, 28.29 dB; the reconstructed image PSNR dropped 0. 13dB.

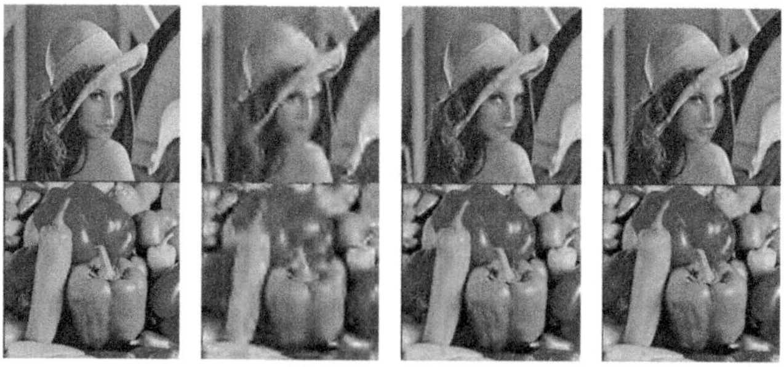

(a) Original image (b) wavelet method (c) fast algorithm (d) with new atom

**Fig. 3.** Comparison of reconstructed image quality with different atoms and algorithm

Theoretically, compared the matching time of atomic energy of 99% support regional and 81% support regional, calculation time is about 0.4975 times of the original, but the simulation calculation time were about 0.4651 and 0.4878 of the original, as total-region support at images updating and image reconstruction. In this paper we only use the central part of the atomic to calculate images, due to the matching pursuit algorithm, even if not obtained the best atom for a particular iteration, the useful information of image can also be compensated by later Iteration, so the objective and subjective quality of reconstructed image was not significant difference of the human accepted visual range.

## 5 Conclusion

In this paper, a new anisotropic atoms has been proposed to construct the redundant dictionary, using the atomic-scale anisotropy to adapt to the image line singularity; And the use the middle supporting part of atomic energy in to estimate the the inner product of atom and image, thus to achieve fast sparse decomposition of image. In addition, iterative algorithm of matching pursuit algorithms (such as Weak Matching Pursuit) is a notable new idea of sparse decomposition in improving the image quality and image reconstruction. In addition, the amount of extracted useful information in image sparse decomposition process still need further study with specific practical problems. This work is supported by the Natural science research project foundation of Education Department of Henan Province 2011A510024.

## Referencces


1. Mallat, S., Zhang, Z.: Matching pursuits with time-frequency dictionaries. IEEE Trans. Signal Processing 41(12), 3397–3415 (1993)
2. Ventura, R., Vanderheynst, P., Frossard, P.: Low-Rate and Flexible Image Coding With Redundant Representations. IEEE Trans. on Image Processing 15(3), 726–739 (2006)
3. Ebrahimi, A., Shirani, S.H.: Matching Pursuit-Based Region-of-Interest Image Coding. IEEE Trans. on Image Processing 16(2), 406–415 (2007)
4. Zhao, s.: Image denoising combined with blocks based on redundant Bandelet. Computer Engineering And Applications 44(15) (2008)
5. Vandergheynst, P., Frossard, P.: Efficient image representation by anisotropic refinement in matching pursuit. In: Proceedings of IEEE on ICASSP, Salt Lake City, UT, USA, vol. (3), pp. 1757–1760 (2001)
6. Yin, Z.K., Wang, J.Y., Vandergheynst, P.: New fast algorithm to sparsely decompose images. Journal Of Computer Applications 24(10), 92–96 (2004)

# Block Based Web Page Feature Selection with Neural Network

Yushan Jin, Ruikai Liu, Xingran He, and Yongping Huang

College of Computer Science and Technology, Jilin University
130012 Changchun, China
hyp@jlu.edu.cn

**Abstract.** To extract the feature of web page accurately is one of the basic topics of Web Data Mining. Considering the structure of web page, a block based feature selection method was imported in this article. A neural network could be used to recognize the priorities of different web page block and then the VPDom tree was built up. The experiment proves that Block Based Feature Selection could filter the "noisy" and enhance the main content of the web page.

**Keywords:** Web Page Block, VPDom Tree, Neural Network, SVM.

## 1 Introduction

Web pages have two notable characters: ruleless content and structured formation. The ruleless content is mainly because of the complexity of the source of the information contained in the web pages. There is also a large amount of noisy information contained in a web page such as navigation information, relevant hyperlinks. They could not show the real content of the web page which the author means and objectively enhance the ruleless of the web page. On the other hand, as the standard web pages are transferred according to the HTTP protocol and in the perspective the web pages are well-structured.

Web page feature selection is one of the basic works of Web Data Mining and Web Information Retrieval. Traditional web page feature selection usually focused on the formless plain text content. First it will filter all the HTML tags, and then extract the features such as IG, CHI, DF, ODDs and MI [1]. However this method will lose the structure information contained in the HTML tags. So a Block Based Web Page Feature Selection method is imported in this article. This method adds the web page structure feature selection to the traditional plain text feature selection and finally gets the features could represent the importance of HTML tags and Web Page Blocks so as to give a better representation of the structured character of web page. In experiment part, we use content features and structured features to large scale web page classifier and get an obvious promotion of the classification accuracy.

## 2 Web Page Block Feature and VPDom Tree

The structure feature of web page is extracted from HTML tags and could be used to modify the weight of plain text feature of a web page. There are two kinds of web

S. Lin and X. Huang (Eds.): CSEE 2011, Part II, CCIS 215, pp. 222–229, 2011.

page structure features: HTML tag feature and block feature. HTML tag feature is the feature contained in the HTML tags which could show the different importance of the web page content such as <title>, <meta>, Anchor Text and so on. During the feature extraction step, different terms from different HTML tags could be given different weights to modify the terms' frequency. So we could get modified TF*IDF to improve the accuracy of the web page classifier.

**Definition 1.** HTML web page could be divided into different areas and each of the areas is called Web Page Block [2]. Different block standards result in different block segmentations. The common web page block methods could be concluded as below: Heuristic-based segmentation [3], DOM-based segmentation [4], Location-based segmentation [5] and Vision-based Page Segmentation [6] (VIPS).

In this paper, we choose Document Object Model Tree as the standard block segmentation method. Based on Dom, this paper presents Visual Priority Model. While traversing the Dom tree, a Visual Priority property is added to the Dom segmentation block. The Visual Priority property could indicate the contribution of the block content to the whole web page. At last a web page block tree including block visual priority is created. We call it VPDom tree for short. VPDom tree does not change the content and the structure of the original web page. It is a reliable, automatic and common-used web information extraction method with both content and structure oriented.

According to the visual priority, the web page blocks could be divided into four sub-levels:

(1) Noisy Block: Embedded web page advertisement, version information and web navigation information.

(2) Relevant Block: Including the relevant content to the web page main content such as relevant hyperlinks and directory information.

(3) Main Block Dividable: Including Main Block (level 4) and blocks of other levels. It could be divided into sub blocks with different levels.

(4) Main Block: Including main content of the current web page. It could not be divided. Any of the sub blocks of main block could not indicate the main content of the web page.

The algorithm of creating a VPCom tree has four steps. Firstly traverse the blocks of the Dom tree by BFS (Breadth-first search) and get the position priorities of the block. Secondly judge the visual priority of the block by a function called 'CalcPri'. Thirdly add the visual priority information to the block. At last return to step 1 until all the blocks are given a visual priority. The pseudo code is just like below.

```
/*queue of the blocks of the Dom tree*/
Queue qBlock=Dom.root;
While (qBlock is not empty){
     block = qBlock.dequeue();
     /*calculate the priority of the block*/
     priority = CalcPri(block);
     /*add priority information to VPDom
     add priority to VPDom Tree; */
```

```
        /*Divide the block recursively only if the
        block is main block dividable*/
        if (priority==3){
          foreach(child in block.children)
          qBlock.enqueue(child);
        }
}
```

From the algorithm above we could find that the function ‘CalcPri’ which is used to calculate the block priority is the kernel of the VPDom algorithm. If we consider the visual and position priorities of a web page block as the input features and the visual priority of the block as the output result, the priority calculation problem could be considered as a function regression problem. A Supervised Artificial Neural Network could be used to solve this problem by using mass samples to train a web page block segmentation model.

The position properties of a web page will definitely add into the HTML document. We could get them form the ‘size’ attributes and the amount of the text from the HTML tags such as <table> <tr> <td> and <div>. Generally speaking the designer of a web page would like to put the main block in the center part of the web page with other blocks surrounded. So to determine the block visual priority by block’s position proprieties is reasonable.

This paper concluded the web page block visual priorities as ten priorities below:

(1) Width: ratio of current block width and the entire web page width

(2) Height: ratio of current block height and the entire web page height.

(3) X Position (xpos): ratio of abscissa of the block’s upper left Conner and the entire web page width.

(4) Y Position (ypos): ratio of ordinate of the block’s upper left Conner and the entire web page width.

(5) Relevant Width (rwidth): ratio of current block width and its parent block width.

(6) Relevant Height (rheight): ratio of current block height and its parent block height.

(7) Relevant X Position (rxpos): ratio of abscissa of the block’s upper left Conner and its parent block width.

(8) Relevant Y Position (rypos): ratio of ordinate of the block’s upper left Conner and its parent block height.

(9) Block Density (BD): ratio of the number of sub blocks in the current block and the number of all the blocks in the entire web page.

(10) Hypertext Density (HD): ratio of the number of letters contained in the hyperlink text and the number of all letters in the current block.

The property vector of the block could be descried like this:

$[Width, Height, Xpos, Ypos, Rwidth, Rheight, Rxpos, Rypos, BD, HD]$

If we consider the ten properties as input and the priority of the block as output, the web page block problem could be transformed into a regression problem with ten input parameters. This paper uses a Back Propagation Network [7] with Gradient descent method to solve the regression problem. The structure of the neural network is shown in Fig 1.

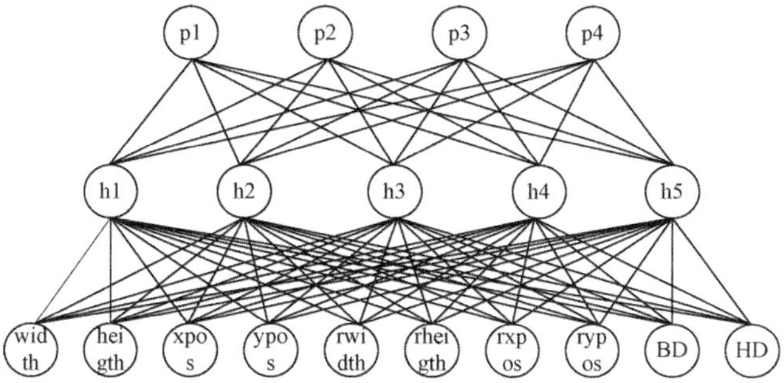


**Fig. 1.** Structure of BP Network to Web Page Block

Input layer contains ten nodes according to the ten properties of the block; hidden layer has five hidden nodes; output layer has four nodes according to the four visual priorities. Transfer function between input and hidden layer is the hyperbolic tangent sigmoid function 'tansig' (2.1). Transfer function between hidden and output layer is linear transfer function 'purelin'. The train method uses Gradient descent method with Memond Coefficients Modifying and learning rate dynamic adjusting.

$$\text{tansig}\left(x\right)=\frac{2}{1+e^{-2x}}-1 \tag{1}$$

We have crawled 1000 web pages from news.google.com which are from different web sites and then label the visual priority blocks of the web pages by hand. The ten-fold evaluation result of the Main Block could reach 88%.

## 3 Block Based Web Page Feature Selection

Term Frequency and Inverse Document Frequency [8] are used to describe the relationship between terms and the document in the Information Retrieval Scope. $TF(t_i)$ means the frequency when the term $t_i$ appears in the document. If a term $t_i$ appears in $n_i$ documents of all the $N$ documents, the formula for IDF is just like below.

$$IDF\left(t_i\right)=\log\frac{N}{n_i} \tag{2}$$

From the formula we could find that TF is just related to the current document, while IDF need to inspect all the documents in the training dataset. For one thing a dictionary should be set up to record the value $n_i$ of each term $t_i$. After tackling all the documents the value IDF for each word could be calculated. In text classification TF*IDF is usually used to evaluate the importance of term $t_i$ in a document.

Block Based Feature Selection will extract plain text feature, HTML structure feature and block feature of a web page. The kernel method is to modify the term frequency (TF) according to the HTML tag weight and web page block weight.

The modified term frequency $TF(t_i)$ for a term $t_i$ in a web page *doc* could be calculated as (3).

$$TF(i)=\sum_{j=1}^{m} w_j TF_j(i)+\sum_{k=1}^{n} w_k TF_k(i) \tag{3}$$

In this formula, $m$ is the total number of the HTML tag; $w_j$ is the term weight of HTML tag $tag_j$; $TF_j(i)$ is the frequency of $term_i$ appears in $tag_j$; $n$ is the total number of web page blocks; $w_k$ is the term weight of web page block $Block_k$; $TF_k(i)$ is the frequency of $term_i$ appears in $Block_k$. From what we have discussed above, in a web page, the modified term frequency $TF(i)$ of a term $term_i$ is not only related to its appearing times, but also related to the HTML tag that contains it and the Block that contains it.

According to the formula (2) and (3), the modified formula of TF*IDF is imported as (4).

$$Freq(i)=\frac{\sum_{j=1}^{m} w_j TF_j(i)+\sum_{k=1}^{n} w_k TF_k(i)}{\log\frac{N}{n_i}} \tag{4}$$

If the web page $doc$ includes $n$ different terms, the feature vector for the web page is $Feature(doc)=[Freq(1), Freq(2),..., Freq(i),...Freq(n)]$. Its normalized form is (5).

$$Normalized-Feature(doc)=\frac{1}{\sqrt{\sum_i^n Freq(i)^2}}[Freq(1), Freq(2),..., Freq(i),...Freq(n)] \tag{5}$$

# 4 Experiments

This paper will use Block Based Feature Selection in the problem of large-scale web page classification. The manual web page categories from Open Directory Project (http://dmoz.org) are selected as the testing dataset. Web page feature selection method is Block Based Web Page Feature Selection that is to say to a web page, feature selection method will extract its normalized frequency feature (3.4). A traditional plain text feature selection method will also be imported as a comparison. The classification algorithm in the paper is Hierarchical Support Vector Machine (HSVM). The kernel function of HSVM that we choose is Radial Basis Function (RBF). The off-line training will use 5-fold Cross Validation [9] to adjust the SVM threshold.

The definition of the measurement is described as follows: $C_i$ is the i$^{th}$ category

Precision: $$P(C_i)=\frac{\#\text{ of docs are correctly tagged as }C_i}{\#\text{ of docs are tagged as }C_i}$$

Recall: $$R(C_i)=\frac{\#\text{ of docs are correctly tagged as }C_i}{\#\text{ of docs whose category is }C_i}$$

F1-measure: $$F(C_i)=\frac{2\times P(C_i)\times R(C_i)}{P(C_i)+R(C_i)}$$

Micro Pre: $$microP = \frac{\sum_i \#\text{ of docs are correctly tagged as } C_i}{\sum_i \#\text{ of docs are tagged as } C_i}$$

Micro Recall: $$micrR = \frac{\sum_i \#\text{ of docs are correctly tagged as } C_i}{\sum_i \#\text{ of docs whose category is } C_i}$$

Micro F1: $$microF = \frac{2 \times microP \times microR}{microP + microR}$$

This paper uses ten-fold evaluation method which chooses 90% web page as training dataset and 10% web page as testing dataset.

Table1 shows the first four ODP levels' Micro-F1 of HSVM with block feature selection and traditional HSVM with plain text feature selection.

**Table 1.** First four levels' Micro-F1

| Micro-F1 | Level 1 | Level 2 | Level 3 | Level 4 |
|---|---|---|---|---|
| HSVM+Block | 80.90% | 74.60% | 66.40% | 59.10% |
| HSVM | 76.00% | 69.80% | 61.20% | 54.80% |

To the 15 categories of ODP level 1, the two methods detailed classification result is shown in Table 2.

**Table 2.** Detailed results for Level 1 classification

| Category | HSVM | | | HSVM+Block | | |
|---|---|---|---|---|---|---|
| | Precision | Recall | F1 | Precision | Recall | F1 |
| Arts | 75.60% | 86.30% | 80.60% | 84.30% | 85.70% | 85.00% |
| Business | 69.90% | 79.10% | 74.20% | 73.10% | 80.00% | 76.40% |
| Computers | 70.10% | 88.70% | 78.30% | 69.00% | 90.80% | 78.40% |
| Games | 70.40% | 82.20% | 76.40% | 79.10% | 86.90% | 82.80% |
| Health | 68.40% | 78.30% | 73.00% | 79.70% | 77.50% | 78.60% |
| Home | 62.00% | 94.60% | 74.90% | 72.20% | 93.00% | 81.30% |
| Kids and Teens | 69.10% | 79.00% | 73.70% | 96.60% | 78.30% | 86.50% |
| News | 77.40% | 77.00% | 77.20% | 86.30% | 78.70% | 82.30% |
| Recreation | 66.80% | 94.00% | 78.10% | 69.20% | 92.30% | 79.10% |
| Reference | 65.30% | 83.50% | 73.30% | 69.70% | 87.30% | 77.50% |
| Regional | 71.90% | 88.90% | 79.50% | 69.70% | 90.40% | 78.70% |
| Science | 62.80% | 82.50% | 71.30% | 85.80% | 85.00% | 85.40% |
| Shopping | 72.20% | 81.80% | 76.70% | 80.10% | 82.90% | 81.50% |
| Society | 69.10% | 86.90% | 77.00% | 71.00% | 90.10% | 79.40% |
| Sports | 66.70% | 87.50% | 75.70% | 74.00% | 88.50% | 80.60% |

Contrast curve of F1 value is shown in Fig 2.

Table 1 proves that HSVM with block based feature selection performance much better than traditional HSVM at the first four levels of ODP. The Micro-F1 gets a 5% promotion on the average. This is mainly because of the imbalance of the data distribution of ODP dataset. As the level increases, the number of training web pages will be smaller and the difference among different categories will be more evidence. Table 2 shows that usage of block based feature selection will both promote precision and recall, while the precision is more evidence. The average precision gets an 8% promotion, the average recall gets only a 1% promotion, and the average F1 promotes about 4.9%. This is mainly because of the labeling method of ODP. So far the ODP manual labeling job is done by volunteers and the pages collected are almost the homepages. The homepage always contains lots of noise such as pictures, embedded advertisements and navigation directories. Block Based Feature Selection takes great advantage of structure information in the web page and is able to filter noise and enhance the main content, so the precision could get an evidence promotion. On the other hand, to promote precision by enhancing a local feature will result in the decrease of the classifier's recall value.

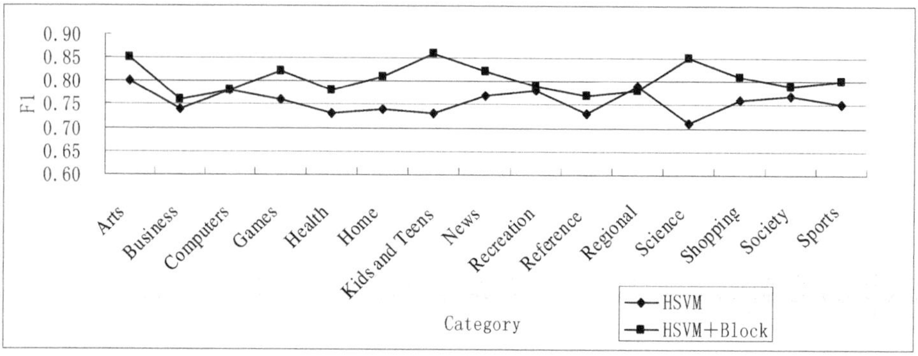


**Fig. 2.** Influence of Block Based Feature Selection to the F1 value in the ODP level 1

## 5 Conclusions

The experiments above prove that Block Based Feature Selection performs much better than traditional plain text feature selection in the scope of web page classification. It takes great advantage of the structure information of web page by adding in HTML tag features and web page block features. This could help to filter the noise and enhance main content. Next step we will focus on two points. One is to promote the accuracy of the kernel function 'CacPri'. Using more complex block segmentation algorithm such as VIPS and choose larger and more reliable training dataset will help to improve the accuracy. Another is to use VPDom Tree in other web data mining scopes such as web page key words extraction, sensitive web page detection and web browser in intelligence cell phone.

**Acknowledgments.** This work was supported by Project of Development Programs in Science and Technology, Jilin Province (20100504), and by Natural Science Fund of Jilin Province (20115025).

## References


1. Yang, Y., Pedersen, J.O.: A Comparative Study on Feature Selection in Text Categorization. In: Proc. of the 14th International Conference on Machine Learning, ICML 1997, pp. 412–420 (1997)
2. Wang, Q., Tang, S.W.: DOM-Based Automatic Extraction of Topical Information from Web Pages. Journal of Computer Research and Development 41, 1786–1792 (2004)
3. Embley, D., Jiang, S., Ng, Y.-K.: Record-boundary discovery in Web documents. In: Proc. 1999 ACM SIGMOD International Conference on Management of Data, pp. 467–478 (1999)
4. Chen, J., Zhou, B., Shi, J., Zhang, H.-J., Qiu, F.: Function-Based Object Model Towards Website Adaptation. In: The Proceedings of the 10th World Wide Web Conference (WWW 2010), Budapest, Hungary, pp. 587–596 (May 2001)
5. Kovacevic, M., Diligenti, M., Gori, M., Milutinovic, V.: Recognition of Common Areas in a Web Page Using Visual Information: a possible application in a page classification. In: The Proceedings of 2002 IEEE International Conference on Data Mining (ICDM 2002), Maebashi City, Japan, pp. 1345–1355 (2002)
6. Yu, S., Cai, D., Wen, J.-R., Ma, W.-Y.: Improving Pseudo-Relevance Feedback in Web Information retrieval Using Web Page Segmentation. In: The Proceedings of Twelfth World Wide Web Conference (WWW 2003), Budapest, Hungary, pp. 11–18 (2003)
7. Michael, T.M.: Machine Learning, pp. 60–72. McGraw-Hill, New York (1997)
8. Hiemstra, D.: A probabilistic justification for using tf.idf term weighting in information retrieval. International Journal on Digital Libraries 3(2), 131–139 (2000)
9. Liu, T.Y., Yang, Y., Wan, H., Zeng, H.J., Chen, Z., Ma, W.Y.: Support vector machines classification with a very large-scale taxonomy. SIGKDD Explor. Newsl (2005)

# Local Analytic Solutions of a Functional Differential Equation Near Resonance

LingXia Liu

Department of Mathematics, Weifang University,
Weifang, Shandong 261061, P.R. China
llxmath@126.com

**Abstract.** In this paper, a function differential equation is investigated in the complex field C. The existence of analytic solutions is discussed for some special cases of the above equation. Then, by reducing the equation with the *Schröder* transformation to the another functional equation with proportional delay, an existence theorem is established for analytic solutions of the original equation.

**Keywords:** iterative functional differential equation; *Schröder* transformation; resonance.

## 1 Introduction

Functional differential equation with state dependent delay have attracted the attentions of many authors in the last years ([1]- [6]). In [3], Petahov studied the existence of analytic solutions of the equation $x''(t) = ax(x(t))$. In [4]- [5], the authors studied local analytic solutions of the equation $x^{(n)}(z) = 1/x(az + bx'(z))$ and $\alpha z + \beta x'(z) = x(az + bx''(z))$. Nevertheless, there is still need for investigation of special equations, which may have interesting properties and provide insights into the more abstract theory. In this paper, we consider the equation of the form

$$x^{(n)}(z) = \frac{1}{x(az + bx''(z))}, \tag{1}$$

where a and b are complex numbers. The purpose of this paper is to discuss the existence of analytic solutions to (1) in the complex field. As in [4], we will seek explicit analytic solutions of Eq. (1) in the form of power functions in the case $a = 0,\ b \neq 0$ had been studied in [4]. However, the same idea cannot be applied for the general equation (1). So in section 3, we will considered the existence of analytic solutions of Eq. (1) in the case $a \neq 0,\ b \neq 0$ and $n = 2$ when $0 < |\lambda| < 1$.

## 2 Explicit analytic solutions

In case $a = 0, b \neq 0$, Eq. (1) changes into functional differential equation

S. Lin and X. Huang (Eds.): CSEE 2011, Part II, CCIS 215, pp. 230–235, 2011.

$$x^{(n)}(z)=\frac{1}{x(bx''(z))}. \tag{2}$$

For the equation, we have the following proposition.

**Proposition 2.1.** If $n\neq\frac{1}{4}(k^2\text{-}1)$, $k\in N^*=\{1,3,\cdots 2N-3\}$, $N\in N$ and $b\neq 0$, then Eq. (2) has two analytic solutions of the form

$$x_+(z)=2^{\frac{2n}{\sqrt{1+4n}+5}}(bn)^{\frac{-\sqrt{1+4n}+1}{\sqrt{1+4n}+5}}\left[(\sqrt{1+4n}+1)(\sqrt{1+4n}-1)\cdots(\sqrt{1+4n}-2n+3)\right]^{\frac{-2}{\sqrt{1+4n}+5}}z^{\frac{\sqrt{1+4n}+1}{2}}$$

$$x_-(z)=(-1)^{\frac{2n}{\sqrt{1+4n}-5}}2^{\frac{-2n}{\sqrt{1+4n}-5}}(bn)^{\frac{-\sqrt{1+4n}-1}{\sqrt{1+4n}-5}}\left[(\sqrt{1+4n}-1)(\sqrt{1+4n}+1)\right.$$

$$\left.\cdots(\sqrt{1+4n}+2n-3)\right]^{\frac{2}{\sqrt{1+4n}-5}}z^{\frac{-\sqrt{1+4n}+1}{2}}$$

in the region $|z-k_\pm|<k_\pm$, respectively, such that $k_\pm$ is the fixed point of $x_\pm(z)$, where

$$k_\pm=2^{-\frac{\pm\sqrt{1+4n}+1}{\pm\sqrt{1+4n}+5}}(bn)^{\frac{\pm\sqrt{1+4n}+2n+1}{n(\pm\sqrt{1+4n}+5)}}\left[(\pm\sqrt{1+4n}+1)(\pm\sqrt{1+4n}-1)\right.$$

$$\left.\cdots(\pm\sqrt{1+4n}-2n+3)\right]^{\frac{\pm\sqrt{1+4n}+1}{n(\pm\sqrt{1+4n}+5)}}.$$

*Proof.* We formally assume that

$$x(z)=\beta z^r.$$

Substituting it into (2), we see that

$$\beta^{r+2}r(r-1)\cdots(r-n+1)[br(r\quad 1)]-\beta z^{r^2-r-n}=1.$$

So we have $r^2-r-n=0$ and $\beta^{r+2}r(r-1)\cdots(r-n+1)[br(r-1)]^r=1$, that is $r_\pm=\dfrac{\pm\sqrt{1+4n}+1}{2}$, and

$$\beta_\pm=2^{\frac{2n}{\pm\sqrt{1+4n}+5}}(bn)^{\frac{\pm\sqrt{1+4n}+1}{\pm\sqrt{1+4n}+5}}\left[(\pm\sqrt{1+4n}+1)(\pm\sqrt{1+4n}-1)\cdots(\pm\sqrt{1+4n}-2n+3)\right]^{-\frac{2}{\pm\sqrt{1+4n}+5}}.$$

Thus, we have

$$x_\pm(z)=2^{\frac{2n}{\pm\sqrt{1+4n}+5}}(bn)^{\frac{\pm\sqrt{1+4n}+1}{\pm\sqrt{1+4n}+5}}\left[(\pm\sqrt{1+4n}+1)(\pm\sqrt{1+4n}-1)\cdots(\pm\sqrt{1+4n}-2n+3)\right]^{-\frac{2}{\pm\sqrt{1+4n}+5}}z^{\frac{\pm\sqrt{1+4n}+1}{2}}$$

and

$$x_\pm(k_\pm)=2^{\frac{2n}{\pm\sqrt{1+4n}+5}}(bn)^{\frac{\pm\sqrt{1+4n}+1}{\pm\sqrt{1+4n}+5}}\left[(\pm\sqrt{1+4n}+1)(\pm\sqrt{1+4n}-1)\cdots(\pm\sqrt{1+4n}-2n+3)\right]^{-\frac{2}{\pm\sqrt{1+4n}+5}}$$

$$\cdot\left\{2^{-\frac{\pm\sqrt{1+4n}+1}{\pm\sqrt{1+4n}+5}}(bn)^{\frac{\pm\sqrt{1+4n}+2n+1}{n(\pm\sqrt{1+4n}+5)}}\left[(\pm\sqrt{1+4n}+1)(\pm\sqrt{1+4n}-1)\cdots(\pm\sqrt{1+4n}-2n+3)\right]^{-\frac{\pm\sqrt{1+4n}+1}{n(\pm\sqrt{1+4n}+5)}}\right\}^{\frac{\pm\sqrt{1+4n}+1}{2}}=k_\pm,$$

where $n\neq\frac{1}{4}(k^2-1)$, $k\in N^*$, so $k_\pm$ is the fixed point of $x_\pm(z)$. On the other hand, in view of

$$(1+z)^k=1+\binom{k}{1}z+\binom{k}{2}z^2+\cdots+\binom{k}{n}z^n+\cdots,(|z|<1),$$

where $\binom{k}{n} = \frac{1}{n!}(k-1)(k-2)\cdots(k-n+1), k \in R$, we have

$$x_\pm(z) = \beta_\pm z^{r_\pm} = \beta_\pm(k_\pm + z - k_\pm)^{r_\pm} = \beta_\pm k_\pm^{r_\pm}(1+\frac{z-k_\pm}{k_\pm})^{r_\pm} = k_\pm(1+\frac{z-k_\pm}{k_\pm})^{r_\pm}$$

$$= k_\pm + r_\pm(z-k_\pm) + \frac{1}{k_\pm}\binom{r_\pm}{2}(z-k_\pm)^2 + \cdots + \frac{1}{(k_\pm)^{n-1}}\binom{r_\pm}{n}(z-k_\pm)^n + \cdots.$$

We see that (2) are analytic in the region $|z-k_\pm| < k_\pm$, respectively.

**Remark.** In case $n=1$, Eq. (2) changes into the functional differential equation $x'(z) = \frac{1}{x(bx''(z))}$, which has two solution of the form

$$x_1(z) = (\frac{\sqrt{5}-1}{2})^{\frac{\sqrt{5}-1}{2\sqrt{5}}} b^{-\frac{1}{\sqrt{5}}} z^{\frac{1+\sqrt{5}}{2}}, \quad x_2(z) = (-\frac{\sqrt{5}+1}{2})^{\frac{\sqrt{5}+1}{2\sqrt{5}}} b^{\frac{1}{\sqrt{5}}} z^{\frac{1-\sqrt{5}}{2}}.$$

In case $n=2$, Eq. (2) changes into the functional differential equation $x''(z) = \frac{1}{x(bx''(z))}$, which has two solutions of the form $x_1(z) = 2^{-\frac{3}{4}} b^{-\frac{1}{2}} z^2$, $x_2(z) = bz^{-1}$. In case $n=12$, Eq. (2) changes into the functional differential equation $x^{(12)}(z) = \frac{1}{x(bx''(z))}$, which has a solutions of form $x(z) = 143z^{-3}$, only if $b = \sqrt[3]{420^2}$. In case $n = \frac{1}{4}(k^2-1)$, $k = 3, \cdots, 2n-3$, Eq.(2) has only analytic solutions of the form (4).

## 3 Analytic Solutions of the Equation

A distinctive feature of the Eq. (1) when $b \neq 0$ is that the argument of the unknown function is dependent on the state derivative $x''(z)$, and this is the case we will emphasize in this paper. when $a \neq 0, b \neq 0$ and $n = 2$, Eq.(1)changes into the functional differential equation

$$x''(z) = \frac{1}{x(az+bx''(z))}, a \neq 0, b \neq 0. \tag{5}$$

We now discuss the existence of analytic solution of Eq. (5) by locally reducing the equation to another functional differential equation with proportional delays. Let

$$y(z) = az + bx''(z), \tag{6}$$

then $x''(z) = \frac{1}{b}[y(z) - az]$, so for any number $z_0 \in C$, we have $x'(z) = x'(z_0) + \frac{1}{b}\int_{z_0}^{z}(y(s) - as)$. Then $x(z) = x(z_0) + x'(z_0)(z - z_0) + \frac{1}{b}\int_{z_0}^{z}\int_{z_0}^{s}(y(t) - at)dtds$, so we have $x(y(z)) = x(z_0) + x'(z_0)(y(z) - z_0) + \frac{1}{b}\int_{z_0}^{y(z)}\int_{z_0}^{s}(y(t) - at)dtds$, therefore, in view of (5) and $x''(z) = \frac{1}{b}[y(z) - az]$, we have

$$\frac{b}{y(z)-az}=x(z_0)+x'(z_0)(y(z)-z_0)+\frac{1}{b}\int_{z_0}^{y(z)}\int_{z_0}^{s}(y(t)-at)dtds. \tag{7}$$

If $z_0$ is a fixed point of $y(z)$, ie., $y(z)-z_0$, we see that

$$x(z_0)=\frac{b}{(1-a)z_0},\; x'(z_0)=-\frac{b[y'(z_0)-a]}{(1-a)^2 z_0{}^2 y'(z_0)}. \tag{8}$$

Furthermore, differentiating both sides of (7) with respect to z, we obtain

$$b^2\{2y'(z)[y'(z)-a]^2-ay''(z)[y(z)-az]\}=y'^3(z)[y(y(z))-ay(z)][y(z)-az]^3. \tag{9}$$

Let

$$y(z)=g(\lambda g^{-1}(z)), \tag{10}$$

We have the auxiliary equation of Eq. (5)

$$\begin{aligned}&b^2\{2g'(\lambda z)[\lambda g'(\lambda z)-ag'(z)]^2-a[\lambda g''(\lambda z)g'(z)-g''(z)g'(\lambda z)][g(\lambda z)-ag(z)]\}\\&=\lambda^2 g'^3(\lambda z)[g(\lambda^2 z)-ag(\lambda z)][g(\lambda z)-ag(z)]^3\end{aligned} \tag{11}$$

To find analytic solution of Eq. (5), we first seek an analytic solution $g(z)$ of auxiliary equation (11) satisfying the initial value conditions

$$g(0)=\mu\neq 0,\; g'(0)=\eta\neq 0, \tag{12}$$

where $\lambda,\mu$ and $\eta$ are complex numbers, and $\lambda$ satisfies one of the following condition: (H1) $0<|\lambda|<1$. We Observe that $\lambda$ is in the unit circle in the case of (H1).

**Theorem 3.1.** Suppose (H1) holds, then, for the initial value conditions (12), Eq. (11) has an analytic solution of the form

$$y(z)=\mu+\eta z+\sum_{n=2}^{+\infty}b_n z^n \tag{13}$$

in a neighborhood of the origin.

*Proof.* Rewrite (11) as

$$\frac{2g'(\lambda z)[\lambda g'(\lambda z)-ag'(z)]^2-a[\lambda g''(\lambda z)g'(z)-g''(z)g'(\lambda z)][g(\lambda z)-ag(z)]}{g'^2(\lambda z)[g(\lambda z)-ag(z)]^3}$$
$$=\frac{\lambda^2}{b^2}g'(\lambda z)\left[g(\lambda^2 z)-ag(\lambda z)\right],$$

or $\left(\frac{ag'(z)-\lambda g'(\lambda z)}{g'(\lambda z)[g(\lambda z)-ag(z)]^2}\right)'=\frac{\lambda^2}{b^2}g'(\lambda z)\left[g(\lambda^2 z)-ag(\lambda z)\right]$. In view of $g(0)=\mu\neq 0$, $g'(0)=\eta\neq 0, a\neq 0$, we have

$$ag'(z)=\lambda g'(\lambda z)+g'(\lambda z)[g(\lambda z)-ag(z)]^2\left\{\frac{a-\lambda}{\mu^2(1-a)^2}+\frac{\lambda^2}{b^2}\int_0^z g'(\lambda s)\left[g(\lambda^2 s)-ag(\lambda s)\right]ds\right\}. \tag{14}$$

This shows that Eq. (11) is equivalent to the integro-differential equation (14). By substituting (13) into (14), we see that

$$b^2\lambda(1-\lambda^{n+1})(n+2)b_{n+2} = \frac{b^2(a-\lambda)}{\mu^2(1-a)^2}\sum_{k=0}^{n}\sum_{i=0}^{n-k}(k+1)(\lambda^i - a)(\lambda^{n-k-i+1}-a)b_{k+1}b_i b_{n-k-i+1}$$
$$+\sum_{k=0}^{n}\sum_{i=0}^{n-k}\sum_{j=0}^{n-k-i}\sum_{l=0}^{n-k-i-j}\frac{(k+1)\lambda^{n-k-i-j+2}(\lambda^i-a)(\lambda^j-a)(\lambda^l-a)(n-k-i-j-l+1)}{n-k-i-j+1}$$
$$\times b_{k+1}b_i b_j b_l b_{n-k-i-j-l+1}, (n=0,1,2,\cdots).(n+2)b_{n+2} \tag{15}$$

So for $b_0=\mu, b_1=\eta$, the sequence $\{b_n\}_{n=2}^{+\infty}$ is successively determined by (15) in a unique manner, so Eq. (5) exists a formal power series solution (13).
Because $0<|\lambda|<1$ so we have

$$\left|\frac{(a-\lambda)(k+1)(\lambda^i-a)(\lambda^{n-k-i+1}-a)}{\lambda(1-\lambda^{n+1})(n+2)\mu^2(1-a)^2}\right| \le \frac{(1+|a|)^3}{|\mu|^2|1-|a||^2|\lambda||1-\lambda^{n+1}|} \le M_1$$

and

$$\left|\frac{(k+1)\lambda^{n-k-i-j+2}(\lambda^i-a)(\lambda^j-a)(\lambda^l-a)(n-k-i+1-a)}{b^2\lambda(1-\lambda^{n+1})(n+2)(n-k-i+1-a)}\right| \le \frac{(1+|a|)^3}{|b|^2|1-\lambda^{n+1}|} \le M_1$$

for some positive number $M_1$. So from (15)

$$|b_{n+2}| \le M_1[\sum_{k=0}^{n}\sum_{i=0}^{n-k}|b_{k+1}||b_i||b_{n-k-i+1}| + \sum_{k=0}^{n}\sum_{i=0}^{n-k}\sum_{j=0}^{n-k-i}\sum_{l=0}^{n-k-i-j}|b_{k+1}||b_i||b_j||b_l||b_{n-k-i-j-l+1}|]. \tag{16}$$

We define a sequence $\{B_n\}_{n=0}^{+\infty}$ by $B_0=|\mu|=|b_0|, B_1=|\eta|=|b_1|$ and

$$B_{n+2} \le M_1[\sum_{k=0}^{n}\sum_{i=0}^{n-k}B_{k+1}B_iB_{n-k-i+1} + \sum_{k=0}^{n}\sum_{i=0}^{n-k}\sum_{j=0}^{n-k-i}\sum_{l=0}^{n-k-i-j}B_{k+1}B_iB_jB_lB_{n-k-i-j-l+1}]. \tag{17}$$

then from (16) and (17) we can show that by induction

$$|b_n| \le B_n, n=0,1,2,\cdots. \tag{18}$$

Now we prove the sequence $\{B_n\}_{n=0}^{+\infty}$ (i.e. governing series) is convergence. Defining

$$G(z) := G(z,\mu,\eta,M) = \sum_{n=0}^{+\infty}B_n z^n, \tag{19}$$

then

$$G^5(z)-2|\mu|G^4(z)+(1+|\mu|^2)G^3(z)-2|\mu|G^2(z)+(|\mu|^2-\frac{1}{M})G(z)+\frac{1}{M}(|\mu|+|\eta|z)=0.$$

Let

$$R(z,G)=G^5-2|\mu|G^4+(1+|\mu|^2)G^3-2|\mu|G^2+(|\mu|^2-\frac{1}{M_1})G+\frac{1}{M_1}(|\mu|+|\eta|z)=0.$$

for $z,G$ in a neighborhood of $(0,|\mu|)$. Since $R(0,|\mu|)=0$ and $R'_G(0,|\mu|)=-\frac{1}{M_1}\neq 0$, then there exists a unique function $G(z)$, analytic in a neighborhood of zero, such that $G(0)=|\mu|, G'(0)=|\eta|$ and $R(z,G(z))=0$, so the power series (19) converges in a neighborhood of the origin, from (18) we obtain the power series (13) converges in a neighborhood of the origin. The proof is completed.

**Theorem 3.2.** Suppose the conditions of Theorem 3.1 are satisfied, Eq. (11) has an analytic solution $y(z)$ of the form (13) in a neighborhood of the number $\mu$, where $y(z)$ is an analytic solution of (11).

*Proof.* In view of Theorem 2.1 or Theorem 2.2, we may find a sequence $\{b_n\}_{n=0}^{\infty}$ such that the function $g(z)$ of the form (9) is an analytic solution of (7) in a neighborhood of the origin. Since $g'(0)=\eta\neq 0$, the function $g^{-1}(z)$ is analytic in a neighborhood of $g(0)=\mu$. Because $y(z)=g(\lambda g^{-1}(z))$, then $y'(z)=\lambda g'(\lambda g^{-1}(z))(g^{-1}(z))' =\dfrac{\lambda g'(\lambda g^{-1}(z))}{g'(g^{-1}(z))}$

and $y''(z)=\dfrac{\lambda^2 g''(\lambda g^{-1}(z))g'(g^{-1}(z))-\lambda g'(\lambda g^{-1}(z))g''(g^{-1}(z))}{(g'(g^{-1}(z)))^3}$, so from (7) we have

$$
\begin{aligned}
&b^2\left\{2y'(z)\left[y'(z)-a\right]^2-ay''(z)\left[y(z)-az\right]\right\}\\
&=b^2\cdot\frac{2\lambda g'(\lambda g^{-1}(z))\left[\lambda g'(\lambda g^{-1}(z))-ag'(g^{-1}(z))\right]^2}{(g'(g^{-1}(z)))^3}\\
&\quad -a\frac{\left[\lambda^2 g''(\lambda g^{-1}(z))g'(g^{-1}(z))-\lambda g'(\lambda g^{-1}(z))g''(g^{-1}(z))\right]\left[g(\lambda g^{-1}(z))-az\right]}{(g'(g^{-1}(z)))^3}\Bigg\}\\
&=\frac{\lambda^3\left[g'(\lambda g^{-1}(z))\right]^3\left[g(\lambda^2 g^{-1}(z))-ag(\lambda g^{-1}(z))\right]\left[g(\lambda g^{-1}(z))-az\right]^3}{(g'(g^{-1}(z)))^3}
\end{aligned}
$$

and

$$
\begin{aligned}
&y'^3(z)[y(y(z))-ay(z)][y(z)-az]^3\\
&=\frac{\lambda^3\left[g'(\lambda g^{-1}(z))\right]^3\left[g(\lambda^2 g^{-1}(z))-ag(\lambda g^{-1}(z))\right]\left[g(\lambda g^{-1}(z))-az\right]^3}{(g'(g^{-1}(z)))^3}
\end{aligned}
$$

as required. The proof is completed.

## References


1. Eder, E.: The functional differential equation x'(t) = x(x(t)). J. Differential Equations 54, 390–400 (1984)
2. Jackiewicz, Z.: Existence and uniqueness of solutions of neutral delay-differential equations with state dependent delays. Funkcial Ekvac. 30, 9–17 (1987)
3. Petahov, V.R.: On a boundary value problem. Trudy Sem Teor Diff Uravnenii Otklon Argument univ. Druzby Narodov Patrisa Lumumby 3, 252–255 (1965)
4. Si, J.G., Liu, T.B.: Local analytic solutions of a functional differential equation with a deviating argument depending on the state derivative near resonance. Comput. Math. Appl. 54, 750–762 (2007)
5. Liu, L.X.: Local analytic solutions of a functional equation. Appl. Math. Compu. 215, 644–652 (2009)
6. Si, J.G., Cheng, S.S.: Note on an iterative functional differential equations. Demonstratio Math. 31(3), 609–614 (1998)

# Analytic Solutions of a Second-Order Functional Differential Equation

LingXia Liu

Department of Mathematics, Weifang University,
Weifang, Shandong 261061, P.R. China
llxmath@126.com

**Abstract.** This paper is concerned with a second-order function differential equation. By constructing a convergent power series solution of an auxiliary equation, analytic solutions for the original differential equation are obtained. We discuss not only those $\alpha$ is a root of unity but also those $\alpha$ near resonance under Brjuno condition.

**Keywords:** iterative functional differential equation; auxiliary equation; resonance; Brjuno condition.

## 1 Introduction

In the last years there has been a growing interest in studying the functional differential equation ([1]-[4]). In this paper, we consider the equation of the form

$$x''(z) = \frac{1}{x(az + bx''(z))}, \; a \neq 0, b \neq 0, \tag{1}$$

where a and b are complex numbers. The equation can be written in the form $x''(z) = f(x(z - \tau(z)))$ with $f(z) = 1/z$ and $\tau(z) = (1-a)z - bx''(z)$. The purpose of this paper is to discuss the existence of analytic solutions to (1) in the complex field. The delay function $\tau(z)$ depends not only on the argument of the unknown function, but also on the state derivative. The functional differential equations with deviating arguments depending on the state derivative have been relatively little researched. Because such equations are quite different from the usual differential equations, the standard existence and uniqueness theorems cannot be applied directly. Therefore, it is of interest to find some or all of their solutions under appropriate conditions. The purpose of this paper is to discuss the existence of analytic solutions to (1) in the complex field.

## 2 Analytic solutions of Eq. (1)

We now discuss the existence of analytic solution of Eq. (1) by locally reducing the equation to another functional differential equation with proportional delays. Let

$$y(z) = az + bx''(z), \tag{2}$$

S. Lin and X. Huang (Eds.): CSEE 2011, Part II, CCIS 215, pp. 236–242, 2011.

then $x''(z)=\frac{1}{b}[y(z)-az]$, so for any number $z_0 \in C$, we have $x'(z)=x'(z_0)+\frac{1}{b}\int_{z_0}^{z}(y(s)-as)$, then $x(z)=x(z_0)+x'(z_0)(z-z_0)+\frac{1}{b}\int_{z_0}^{z}\int_{z_0}^{s}(y(t)-at)dtds$, so we have $x(y(z))=x(z_0)+x'(z_0)(y(z)-z_0)$ $+\frac{1}{b}\int_{z_0}^{y(z)}\int_{z_0}^{s}(y(t)-at)dtds$, therefore, in view of (1) and $x''(z)=\frac{1}{b}[y(z)-az]$, we have

$$\frac{b}{y(z)-az}=x(z_0)+x'(z_0)(y(z)-z_0)+\frac{1}{b}\int_{z_0}^{y(z)}\int_{z_0}^{s}(y(t)-at)dtds. \tag{3}$$

If $z_0$ is a fixed point of $y(z)$, ie., $y(z)-z_0$, we see that

$$x(z_0)=\frac{b}{(1-a)z_0},\ x'(z_0)=-\frac{b[y'(z_0)-a]}{(1-a)^2 z_0^{\ 2} y'(z_0)}. \tag{4}$$

Furthermore, differentiating both sides of (3) with respect to z, we obtain

$$b^2\left\{2y'(z)[y'(z)-a]^2-ay''(z)[y(z)-az]\right\}=y'^3(z)[y(y(z))-ay(z)][y(z)-az]^3. \tag{5}$$

Let

$$y(z)=g(\lambda g^{-1}(z)), \tag{6}$$

We have the auxiliary equation of Eq. (1)

$$\begin{aligned}&b^2\left\{2g'(\lambda z)[\lambda g'(\lambda z)-ag'(z)]^2-a[\lambda g''(\lambda z)g'(z)-g''(z)g'(\lambda z)][g(\lambda z)-ag(z)]\right\}\\&=\lambda^2 g'^3(\lambda z)[g(\lambda^2 z)-ag(\lambda z)][g(\lambda z)-ag(z)]^3\end{aligned} \tag{7}$$

To find analytic solution of Eq. (1), we first seek an analytic solution $g(z)$ of auxiliary equation (7) satisfying the initial value conditions

$$g(0)=\mu\neq 0,\ g'(0)=\eta\neq 0, \tag{8}$$

where $\lambda,\mu$ and $\eta$ are complex numbers, and $\lambda$ satisfies one of the following condition:

(H1) $\lambda=e^{2\pi i\theta}$, where $\theta\in R\setminus Q$ is a Brjuno number ([5] and [6]), i.e., $B(\theta)=\sum_{k=0}^{\infty}\frac{\log q_{k+1}}{q_k}<\infty$, where $\{p_k/q_k\}$ denotes the sequence of partial fraction of the continued fraction expansion of $\theta$, said to satisfy the Brjuno condition.

(H2) $\lambda=e^{2\pi iq/p}$ for some integers $p\in N$ with $p\geq 2$ and $q\in Z\setminus\{0\}$, and $\alpha=e^{2\pi il/k}$ for all $1\leq k\leq p-1$ and $l\in Z\setminus\{0\}$.

We Observe that $\lambda$ is on $s^1$ in the case of (H1) and (H2). $\lambda$ is a p-th unit root (or called p-order resonance) in the case of (H2), while the case (H1) contains a part of $\lambda$ ``near'' resonance.

When $\lambda\in(H1)$ or $\lambda\in(H2)$, we observe that $\lambda$ is on $S^1$, more difficulties are encountered as mentioned in the so-called "small-divisor problem" ([5]}). The Brjuno condition in (H1) is different with the Diophantine condition: $|\lambda|=1,\lambda$ is not a root of unity, and $\log\frac{1}{|\lambda^n-1|}\leq T\log n, n=2,3,\cdots$ for some positive constant $T$.

As stated in [6], for a real number $\theta$ we let $\theta$ denote its integer part and let $\{\theta\}=\theta-[\theta]$. Then every national number $\theta$ has a unique expression of the Gauss' continued fraction $\theta=a_0+\theta_0=a_0+\frac{1}{a_1+\theta_1}=\ldots$, denoted simply by $\theta=[a_0,a_1,\ldots,a_n,\ldots]$, where $a_j$'s and $\theta_j$'s are calculated by the algorithm: **(a)** $a_0=[\theta]$, $\theta_0=\{\theta\}$ and **(b)** $a_n=\left[\frac{1}{\theta_{n-1}}\right], \theta_n=\left\{\frac{1}{\theta_{n-1}}\right\}$ for all $n\geq 1$. Define the sequences $(p_n)_{n\in\mathrm{N}}$ and $(q_n)_{n\in\mathrm{N}}$ as follows $q_{-2}=1, q_{-1}=0, q_n=a_nq_{n-1}+q_{n-2}$; $p_{-2}=0, p_{-1}=1, p_n=a_np_{n-1}+p_{n-2}$. It is easy to show that $p_n/q_n=[a_0,a_1,\ldots,a_n]$. Thus, for every $\theta\in\mathbf{R}\backslash\mathbf{Q}$ we associate, using its convergence, an arithmetical function $B(\theta)=\sum_{n\geq 0}\frac{\log q_{n+1}}{q_n}$. We say that $\theta$ is a Brjuno number or that it satisfies Brjuno condition if $B(\theta)<+\infty$. The Brjuno condition is weaker than the Diophantine condition. For example, if $a_{n+1}\leq ce^{a_n}$ for all $n\geq 0$, where $c>0$ is a constant, then $\theta=[a_0,a_1,\ldots,a_n,\ldots]$ is a Brjuno number but is not a Diophantine number. So, the case (H1) contains both Diophantine condition and a part of $\mu$ "near" resonance.

In order to discuss the existence of the auxiliary equation (3) under (H1), we need to introduce Davie's Lemma. First, we recall some facts in [7] briefly. Let $\theta\in\mathbf{R}\backslash\mathbf{Q}$ and $(q_n)_{n\in N}$ be the sequence of partial denominators of the Gauss's continued fraction for $\theta$. As in [6], let $A_k=\{n\geq 0 \mid \|n\theta\|\leq\frac{1}{8q_k}\}$, $E_k=\max(q_k,\frac{q_{k+1}}{4})$, $\eta_k=\frac{q_k}{E_k}$. Let $A_k^*$ be the set of integers $j\geq 0$ such that either $j\in A_k$ or for some $j_1$ and $j_2$ in $A_k$, with $j_2-j_1<E_k$, one has $j_1<j<j_2$ and $q_k$ divide $j-j_1$. For any integer $n\geq 0$, define $l_k(n)=\max\left((1+\eta_k)\frac{n}{q_k}-2,\ (m_n\eta_k+n)\frac{1}{q_k}-1\right)$, where $m_n=\max\{j \mid 0\leq j\leq n,\ j\in A_k^*\}$. We then define function $h_k:\mathbf{N}\to\mathbf{R}_+$ as follows: $\begin{cases}\frac{m_n+\eta_k n}{q_k}-1, & \text{if } m_n+q_k\in A_k^*,\\ l_k(n), & \text{if } m_n+q_k\notin A_k^*.\end{cases}$ Let $g_k(n):=\max\left(h_k(n),\left[\frac{n}{q_k}\right]\right)$, and define $k(n)$ by the condition $q_{k(n)}\leq n\leq q_{k(n)+1}$. Clearly, $k(n)$ is non-decreasing. Then we are able to state the following result:

**Lemma 1.** (Davie's Lemma [7]) Let $K(n)=n\log 2+\sum_{k=0}^{k(n)}g_k(n)\log(2q_{k+1})$. Then

(a) There is a universal constant $\gamma>0$ (independent of $n$ and $\theta$) such that $K(n)\leq n\left(\sum_{k=0}^{k(n)}\frac{\log q_{k+1}}{q_k}+\gamma\right)$,

(b) $K(n_1)+K(n_2)\leq K(n_1+n_2)$ for all $n_1$ and $n_2$, and (c) $-\log|\alpha^n-1|\leq K(n)-K(n-1)$.

**Theorem 2.1.** Suppose (H1) holds, then, for the initial value conditions (8), Eq. (7) has an analytic solution of the form

$$y(z)=\mu+\eta z+\sum_{n=2}^{+\infty} b_n z^n \tag{9}$$

in a neighborhood of the origin.

*Proof.* Rewrite (7) as

$$\frac{2g'(\lambda z)[\lambda g'(\lambda z)-ag'(z)]^2-a[\lambda g''(\lambda z)g'(z)-g''(z)g'(\lambda z)][g(\lambda z)-ag(z)]}{g'^2(\lambda z)[g(\lambda z)-ag(z)]^3}$$

$$=\frac{\lambda^2}{b^2}g'(\lambda z)\left[g(\lambda^2 z)-ag(\lambda z)\right],$$

or $\left(\frac{ag'(z)-\lambda g'(\lambda z)}{g'(\lambda z)[g(\lambda z)-ag(z)]^2}\right)'=\frac{\lambda^2}{b^2}g'(\lambda z)\left[g(\lambda^2 z)-ag(\lambda z)\right]$. In view of $g(0)=\mu\neq 0$, $g'(0)=\eta\neq 0$, $a\neq 0$, we have

$$ag'(z)=\lambda g'(\lambda z)+g'(\lambda z)[g(\lambda z)-ag(z)]^2\left\{\frac{a-\lambda}{\mu^2(1-a)^2}+\frac{\lambda^2}{b^2}\int_0^z g'(\lambda s)\left[g(\lambda^2 s)-ag(\lambda s)\right]ds\right\}. \tag{10}$$

This shows that Eq. (7) is equivalent to the integro-differential equation (3.10). By substituting (9) into (10), we see that

$$b^2\lambda(1-\lambda^{n+1})(n+2)b_{n+2}=\frac{b^2(a-\lambda)}{\mu^2(1-a)^2}\sum_{k=0}^{n}\sum_{i=0}^{n-k}(k+1)(\lambda^i-a)(\lambda^{n-k-i+1}-a)b_{k+1}b_i b_{n-k-i+1}$$

$$+\sum_{k=0}^{n}\sum_{i=0}^{n-k}\sum_{j=0}^{n-k-i}\sum_{l=0}^{n-k-i-j}\frac{(k+1)\lambda^{n-k-i-j+2}(\lambda^i-a)(\lambda^j-a)(\lambda^l-a)(n-k-i-j-l+1)}{n-k-i-j+1}\times b_{k+1}b_i b_j b_l b_{n-k-i-j-l+1}. \tag{11}$$

So for $b_0=\mu, b_1=\eta$, the sequence $\{b_n\}_{n=2}^{+\infty}$ is successively determined by (11) in a unique manner, so Eq. (1) exists a formal power series solution (9). Because $|\lambda|=1$ so we have

$$\left|\frac{(a-\lambda)(k+1)(\lambda^i-a)(\lambda^{n-k-i+1}-a)}{\lambda(1-\lambda^{n+1})(n+2)\mu^2(1-a)^2}\right|\leq\frac{(1+|a|)^3}{|\mu|^2|1-|a||^2|1-\lambda^{n+1}|}\quad\text{and}$$

$$\left|\frac{(k+1)\lambda^{n-k-i-j+2}(\lambda^i-a)(\lambda^j-a)(\lambda^l-a)(n-k-i-j-l+1)}{b^2\lambda(1-\lambda^{n+1})(n+2)(n-k-i-j+1)}\right|\leq\frac{(1+|a|)^3}{|b|^2|1-\lambda^{n+1}|},$$

then

$$|b_{n+2}|\leq\frac{M}{|1-\lambda^{n+1}|}\Big[\sum_{k=0}^{n}\sum_{i=0}^{n-k}|b_{k+1}||b_i||b_{n-k-i+1}|+\sum_{k=0}^{n}\sum_{i=0}^{n-k}\sum_{j=0}^{n-k-i}\sum_{l=0}^{n-k-i-j}|b_{k+1}||b_i||b_j||b_l||b_{n-k-i-j-l+1}|\Big],\quad n=0,1,2,\cdots, \tag{12}$$

where $M=\max\{(1+|a|)^3\big/|\mu|^2|1-|a||^2,\ (1+|a|)^3\big/|b|^2\}$.

To construct a governing series of (9), we consider the implicit functional equation

$$R(z,G,\mu,\eta,M)=G^5-2|\mu|G^4+(1+|\mu|^2)G^3-2|\mu|G^2+(|\mu|^2-\frac{1}{M})G+\frac{1}{M}(|\mu|+|\eta|z)=0 \tag{13}$$

for $z,G$ in a neighborhood of $(0,|\mu|)$. Since $R(0,|\mu|)=0$ and $R'_G(0,|\mu|)=-1\big/M_1\neq 0$, then there exists a unique function $G(z)$, analytic in a neighborhood of zero, such that $G(0)=|\mu|, G'(0)=|\eta|$ and $R(z,G(z))=0$. So we can prove that (13) has a unique analytic solution $\varphi(z,\mu,\eta,M)$ in a neighborhood of the origin such that $\varphi(0,\mu,\eta,M)=|\mu|, \varphi'_z(0,\mu,\eta,M)=|\eta|$, thus $\varphi(z,\mu,\eta,M)$ in (13) can be expanded into a convergent series

$$\varphi(z,\mu,\eta,M)=\sum_{n=0}^{+\infty}D_n z^n \tag{14}$$

in a neighborhood of the origin. Replacing (14) into (13) and comparing coefficients, we obtain that $D_0=|\mu|, D_1=|\eta|$ and

$$D_{n+2}=M[\sum_{k=0}^{n}\sum_{i=0}^{n-k}D_{k+1}D_iD_{n-k-i+1}+\sum_{k=0}^{n}\sum_{i=0}^{n-k}\sum_{j=0}^{n-k-i}\sum_{l=0}^{n-k-i-j}D_{k+1}D_iD_jD_lD_{n-k-i-j-l+1}]. \tag{15}$$

Note that the series (14) converges in a neighborhood of the origin, hence, there is a constant $T>0$ such that $D_n<T^n, n=0,1,2,\cdots$.

Now we can deduce, by induction, that $|b_{n+1}|\le D_{n+1}e^{K(n)}$ for $n\ge 1$, where $K:N\to R$ is defined in Lemma 2.1. In fact, $|b_1|=|\eta|=D_1$, for inductive proof, we assume that $|b_{j+1}|\le D_{j+1}e^{K(j)}, j\le n.$ From (3.17) we get

$$|b_{n+2}|\le\frac{M}{|1-\lambda^{n+1}|}[\sum_{k=0}^{n}\sum_{i=0}^{n-k}B_{k+1}B_iB_{n-k-i+1}\cdot e^{K(k)+K(i-1)+K(n-k-i)}$$

$$+\sum_{k=0}^{n}\sum_{i=0}^{n-k}\sum_{j=0}^{n-k-i}\sum_{l=0}^{n-k-i-j}B_{k+1}B_iB_jB_lB_{n-k-i-j-l+1}\cdot e^{K(k)+K(i-1)+K(j-1)+K(l-1)+K(n-k-i-j-1)}]$$

Note that $K(n)$ is non-decreasing, so from Lemma 3.1 we have $K(k)+K(i-1)+K(n-k-i)\le K(n-1)\le K(n+1)\le\log|\lambda^n-1|+K(n)$, Then

$$|b_{n+2}|\le\frac{M}{|1-\lambda^{n+1}|}e^{K(n)}[\sum_{k=0}^{n}\sum_{i=0}^{n-k}B_{k+1}B_iB_{n-k-i+1}+\sum_{k=0}^{n}\sum_{i=0}^{n-k}\sum_{j=0}^{n-k-i}\sum_{l=0}^{n-k-i-j}B_{k+1}B_iB_jB_lB_{n-k-i-j-l+1}]=D_{n+1}e^{K(n)}.$$

Moreover, from Lemma (2.1) (a), we know that $K(n)\le n(B(\theta)+\gamma)$ for some universal constant $\gamma>0$. Then from (3.21) we have $\limsup_{n\to+\infty}(|b_n|^{\frac{1}{n}})\le\sup((T^ne^{(n-1)(B(\theta)+\gamma)})^{\frac{1}{n}})=Te^{B(\theta)+\gamma}$. This implies that the convergence radius of (9) is at least $(Te^{B(\theta)+\gamma})^{-1}$. The proof is completed.

When $\lambda\in(H2)$, the constant $\lambda$ is not only on the unit circle in $C$ but also a root of unity. Let $\{C_n\}_{n=0}^{\infty}$ be a sequence defined by
$C_0=|\mu|, C_1=|\eta|$ and

$$C_{n+2}=M[\sum_{k=0}^{n}\sum_{i=0}^{n-k}C_{k+1}C_iC_{n-k-i+1}+\sum_{k=0}^{n}\sum_{i=0}^{n-k}\sum_{j=0}^{n-k-i}\sum_{l=0}^{n-k-i-j}C_{k+1}C_iC_jC_lC_{n-k-i-j-l+1}]. \tag{16}$$

where $M$ is defined in Theorem 2.1 and $\Gamma=\max\{1,|\alpha^i-1|^{-1}:i=1,2,\cdots,p-1\}$.

**Theorem 2.2.** Suppose that (H2) holds, $p$ is given as above. Let $\{b_n\}_{n=0}^{+\infty}$ be determined recursively by $b_0=\mu, b_1=\eta$, and

$$b^2\lambda(1-\lambda^{n+1})(n+2)b_{n+2}=\Theta(n+1,\lambda), n=0,1,2\cdots, \tag{17}$$

where

$$\Theta(n+1,\lambda)=\frac{b^2(a-\lambda)}{\mu^2(1-a)^2}\sum_{k=0}^{n}\sum_{i=0}^{n-k}(k+1)(\lambda^i-a)(\lambda^{n-k-i+1}-a)b_{k+1}b_ib_{n-k-i+1}$$

$$+\sum_{k=0}^{n}\sum_{i=0}^{n-k}\sum_{j=0}^{n-k-i}\sum_{l=0}^{n-k-i-j}\frac{(k+1)\lambda^{n-k-i-j+2}(\lambda^i-a)(\lambda^j-a)(\lambda^l-a)(n-k-i-j-l+1)}{n-k-i-j+1}\cdot b_{k+1}b_ib_jb_lb_{n-k-i-j-l+1},(n=0,1,2,\cdots).$$

If $\Theta(vp,\lambda)=0$ for all $v=1,2,\cdots$, then Eq. (7) has an analytic solution $g(z)$ in a neighborhood of the origin, such that $g(0)=\mu, g'(0)=\eta$ and $g^{(vp+1)}(0)=(vp+1)!T_{vp+1}$, where all $T_{vp+1}'s$ are arbitrary constants satisfying the inequality $|T_{vp+1}|\le C_{vp+1}$ and the sequence $\{C_n\}_{n=0}^{+\infty}$ is defined in (16). Otherwise, if $\Theta(vp,\lambda)\ne 0$ for some $v=1,2,\cdots$, then equation (7) has no analytic solutions in any neighborhood of the origin.

*Proof.* we seek a power series solution of (7) of the form (9) as in the proof of Theorem 2.1, where the equality in (11) or (17) is indispensable. If $\Theta(vp,\lambda)\ne 0$ for some natural number $v$, then the equality in (17) does not hold for $n=vp-1$, since $a^{vp}-1=0$. In such a circumstance equation (17) has no formal solutions.

When $\Theta(vp,\lambda)=0$ for all natural number $v$, for each $v$ the corresponding $b_{vp+1}$ in (17) has infinitely many choices in $C$, that is, the formal series solution (9) defines a family of solutions with infinitely many parameters. Choose $b_{vp+1}=T_{vp+1}$ arbitrarily such that $|T_{vp+1}|\le C_{vp+1}, v=1,2,\cdots$, Where $C_{vp+1}$ is defined by (16). In what follows we prove that the formal series solution (9) converges in a neighborhood of the origin. Observe that $|\lambda^{n+1}-1|^{-1}\le\Gamma$ for $n\ne vp-1$. It follows from (17) and (12) that

$$|b_{n+2}|\le\Gamma M[\sum_{k=0}^{n}\sum_{i=0}^{n-k}|b_{k+1}||b_i||b_{n-k-i+1}|+\sum_{k=0}^{n}\sum_{i=0}^{n-k}\sum_{j=0}^{n-k-i}\sum_{l=0}^{n-k-i-j}|b_{k+1}||b_i||b_j||b_l||b_{n-k-i-j-l+1}|], \tag{18}$$

for all $n\ne vp, v=0,1,2,\cdots$. Let

$$\psi(z,\mu,\eta,\Gamma M_2)=\sum_{n=0}^{+\infty}C_n z^n, C_0=|\mu|, C_1=|\eta|. \tag{19}$$

It is easy to check that (19) satisfies the implicit functional equation $R(z,\psi,\mu,\eta,\Gamma M_2)=0$, where $R$ is defined in (16). Moreover, similarly to the proof of Theorem 2.1, we can prove that the equation $R(z,\psi,\mu,\eta,\Gamma M_2)=0$ has a unique analytic solution $\psi(z,\mu,\eta,\Gamma M_2)$ in a neighborhood of the origin such that $\psi(0,\mu,\eta,\Gamma M_2)=|\mu|$ and $\psi_z'(0,\mu,\eta,\Gamma M_2)=|\eta|$. Thus (19) converges in a neighborhood of the origin. Moreover, It is easy to show that, by induction, $|b_n|\le C_n$, $n=0,1,2,\cdots$. Therefore, the series (9) converges in a neighborhood of the origin. This completes the proof.

Now we give analytic solutions of Eq. (1)

**Theorem 2.3.** Suppose the conditions of Theorem 2.1 or Theorem 2.2 are satisfied, Eq. (1) has an analytic solution $y(z)$ of the form (6) in a neighborhood of the number $\mu$, where $y(z)$ is an analytic solution of (7).

## References


1. Feckan, E.: On certain type of functional differential equations. Math. Slovaca 43, 39–43 (1993)
2. Wang, K.: On the equation x'(t) = f(x(x(t))). Funkcial. Ekvac. 33, 405–425 (1990)

3. Si, J.G., Wang, X.P.: Analytic solutions of a second-order functiontial differential equation with a state derivative dependent delay. Collo. Math. 79(2), 273–281 (1999)
4. Liu, L.X.: Local analytic solutions of a functional equation. Appl. Math. Compu. 215, 644–652 (2009)
5. Bjuno, A.D.: Analytic form of differential equations. Trans. Moscow Math. Soc. 25, 131–288 (1971)
6. Marmi, S., Moussa, P., Yoccoz, J.C.: The Brjuno functions and their regularity properties. Comm. Math. Phys. 186(2), 265–293 (1997)
7. Carletti, T., Marmi, S.: Linearization of Analytic and Non-Analytic Germs of Diffeomorphisms of (C, 0). Bull. Soc. Math., France 128, 69–85 (2000)
8. Davie, A.M.: The critical function for the semistandard map. Nonlinearity 7, 219–229 (1994)
9. Bessis, D., Marmi, S., Turchetti, G.: On the singularities of divergent majorant series arising from normal form theory. Rend. Mat. Appl. 9, 645–659 (1989)

# Synchronization of the Modified Chua's Circuits with a Derivative Filter

Weiming Sun[1,*], Xinyu Wang[2], and Junwei Lei[3]

[1] Ocean School of Yantai University, Yantai, China
[2] Institute of Science and Technology for Opto-electronic Information, Yantai University, China
[3] Dept. of Control Engineering, Naval Aeronautical and Astronautical University, 264001 Yantai, China
{sunweiming1024,wangxinyu1024,leijunwei}@126.com

**Abstract.** A novel derivative filter is firstly proposed in this paper and it is used in the synchronization of Chua's Chaotic system with unknown parameters. It is described with transfer function so it makes use of the advantage of classical control theory. Also, it can absorb the merit of the modern control theory such as Lyapunov stability theorem. But the defect of this paper is that this paper is based on the assumption that three parameters in filters are small enough. So in our future work, a new analysis method will be researched which can relax this assumption.

**Keywords:** Synchronization; Derivative; Filter; Chaos; Control.

## 1 Introduction

Chaos systems have complex dynamical behaviors that possess some special features such as being extremely sensitive to tiny variations of initial conditions[1-4], and having bounded trajectories with a positive leading Lyapunov exponent and so on. In recent years, chaos synchronization has attracted increasingly attentions due to their potential applications of secure communications.

There are many papers investigated synchronization of chaotic systems with adaptive control methods[5-8]. In this paper, a novel kind of derivative filter method is proposed and a transfer function type control law is constructed to synchronize Chua's Chaotic system with unknown parameters.

Obviously, it is also belong to a kind of adaptive method. But it contains the concept of transfer function so it integrates the classic control method with the Lyapunov method perfectly.

## 2 Problem Description

The modified Chua's circuit is often used as the experiment tool of nonlinear research, its model can be written as:

[*] Corresponding author.

S. Lin and X. Huang (Eds.): CSEE 2011, Part II, CCIS 215, pp. 243–247, 2011.

$$\begin{aligned}\dot{x}_1 &= \alpha_1[y_1 - a_1 x_1 - b_1 x_1 |x_1|] \\ \dot{y}_1 &= x_1 - y_1 + z_1 \\ \dot{z}_1 &= -\beta_1 y_1\end{aligned} \quad . \tag{1}$$

Where $x_1, y_1, z_1$ are the states of the system, $a_1 = -\frac{1}{6}, b_1 = \frac{1}{16}$, $\alpha_1, \beta_1$ are unknown constant parameters.

The drive system is chosen as (1), and the response system with the same unknown parameters can be written as:

$$\begin{aligned}\dot{x}_2 &= \alpha_1[y_2 - a_1 x_2 - b_1 x_2 |x_2|] + b_1 u_1 \\ \dot{y}_2 &= x_2 - y_2 + z_2 + b_2 u_2 \\ \dot{z}_2 &= -\beta_1 y_2 + b_3 u_3\end{aligned} \quad . \tag{2}$$

Where the states of drive system and response system are denoted by 1 and 2 respectively.

Our goal is to design a controller $u = [u_1 \quad u_2 \quad u_3]$ such that synchronization between drive system and response system are realized. And the error system can be described as:

$$\begin{aligned}\dot{e}_x &= \alpha_1[e_y - a_1 e_x - b_1 x_2 |x_2| + b_1 x_1 |x_1|] + b_1 u_1 \\ \dot{e}_y &= e_x - e_y + e_z + b_2 u_2 \\ \dot{e}_z &= -\beta_1 e_y + b_3 u_3\end{aligned} \quad . \tag{3}$$

Where $e_x = x_2 - x_1, e_y = y_2 - y_1, e_z = z_2 - z_1$ and $b_i$ are the uncertain coefficients of input.

## 3 Synchronization with Derivative Filter

Considering the above error subsystem, a derivative filter method is proposed as follows.

Consider the first subsystem as

$$\dot{e}_x = \alpha_1[e_y - a_1 e_x - b_1 x_2 |x_2| + b_1 x_1 |x_1|] + b_1 u_1 . \tag{4}$$

Design a control as

$$b_1 u_1 = -\hat{\alpha}_1[e_y - a_1 e_x - b_1 x_2 |x_2| + b_1 x_1 |x_1|] - k_{11} e_x - k_{12} \frac{s}{\tau_1 s + 1} e_x = -\hat{\alpha}_1 f_{11} + f_{12} , \tag{5}$$

Where

$$f_{11} = e_y - a_1 e_x - b_1 x_2 |x_2| + b_1 x_1 |x_1|, \; f_{12} = -k_{11} e_x - k_{12} \frac{s}{\tau_1 s + 1} e_x . \tag{6}$$

It is worthy pointing out that $\frac{s}{\tau_1 s+1}$ is a derivative filter which is described by transfer function. So define a new variable as $\Theta_1$ as

$$\Theta_1 = \frac{s}{\tau_1 s+1} e_x . \tag{7}$$

It satisfies

$$(\tau_1 s+1)\Theta_1 = s e_x . \tag{8}$$

It can be expanded as

$$\tau_1 \dot{\Theta}_1 + \Theta_1 = \dot{e}_x . \tag{9}$$

According to equation (4), it holds

$$e_x \dot{e}_x = \tilde{\alpha}_1 [e_y - a_1 e_x - b_1 x_2 |x_2| + b_1 x_1 |x_1|] e_x - k_{11} e_x e_x - k_{12} \left( \frac{s}{\tau_1 s+1} e_x \right) e_x \tag{10}$$

Where $\tilde{\alpha}_1 = \alpha_1 - \hat{\alpha}_1$ and $\dot{\tilde{\alpha}}_1 = -\dot{\hat{\alpha}}_1$.

Define new variables as

$$\Lambda_1 = \left( \frac{s}{\tau_1 s+1} e_x \right) e_x , \ \Lambda_2 = \left( \frac{s}{\tau_1 s+1} \right) (e_x e_x) \tag{11}$$

It satisfies

$$\Lambda_1 = \left( \frac{s}{\tau_1 s+1} e_x \right) e_x = \Theta_1 e_x . \tag{12}$$

Then it holds

$$\tau_1 \dot{\Lambda}_2 + \Lambda_2 = 2 e_x \dot{e}_x . \tag{13}$$

It is obvious that if $\tau_1$ is small enough, then $\Lambda_1$ is close to $\Lambda_2$, so the above equation can be written as

$$e_x \dot{e}_x = \tilde{\alpha}_1 [e_y - a_1 e_x - b_1 x_2 |x_2| + b_1 x_1 |x_1|] e_x - k_{11} e_x e_x - k_{12} \left( \frac{s}{\tau_1 s+1} \right) (e_x e_x) . \tag{14}$$

Design the adaptive turning law of unknown parameter as

$$\dot{\hat{\alpha}}_1 = [e_y - a_1 e_x - b_1 x_2 |x_2| + b_1 x_1 |x_1|] e_x . \tag{15}$$

Define a Lyapunov function as

$$V_x = \frac{1}{2} e_x^2 + \frac{1}{2} \tilde{\alpha}_1^2 + k_{12} \left( \frac{1}{\tau_1 s + 1} \right) (e_x^2) . \tag{16}$$

Then it is easy to prove that

$$\dot{V}_x \le -k_{11} e_x^2 . \tag{17}$$

With the same method, design the control law and adaptive turning law for the second subsystem and third subsystem respectively as follow.

Design a control law as

$$\begin{aligned} b_2 u_2 &= -[e_x - e_y + e_z] - k_{21} e_y - k_{22} \frac{s}{\tau_2 s + 1} e_y \\ b_3 u_3 &= \hat{\beta}_1 e_y - k_{31} e_z - k_{32} \frac{s}{\tau_3 s + 1} e_z \end{aligned} \tag{18}$$

Design the adaptive turning law as

$$\dot{\hat{\beta}}_1 = e_z e_y . \tag{19}$$

And define Lyapunov functions as

$$\begin{aligned} V_y &= \frac{1}{2} e_y^2 + k_{22} \left( \frac{1}{\tau_2 s + 1} \right) (e_y^2) , \\ V_z &= \frac{1}{2} e_y^2 + k_{32} \left( \frac{1}{\tau_3 s + 1} \right) (e_z^2) + \frac{1}{2} \tilde{\beta}^2 . \end{aligned} \tag{20}$$

Where $\tilde{\beta}$ defined as $\tilde{\beta} = \beta - \hat{\beta}$.

Choose a big Lyapunov function for the whole system as

$$V = V_x + V_y + V_z . \tag{21}$$

It is easy to prove that

$$\dot{V} \le -k_{11} e_x^2 - k_{21} e_y^2 - k_{31} e_z^2 . \tag{22}$$

So the system is stable and $e_x \to 0$, $e_y \to 0$, $e_z \to 0$ and the synchronization can be fulfilled.

## 4 Conclusions

A novel derivative filter is used to synchronize modified Chua's circuits with unknown parameters based on a main assumption that three parameters in filters are small enough. The derivative filter is described by transfer function, so the traditional method is integrated with modern control method perfectly.

## References

1. Elabbasy, E.M., Agiza, H.N., El-Dessoky, M.M.: Chaos, Solitons and Fractals 30, 1133–1142 (2006)
2. Tang, F., Wang, L.: Physics Letters A 346, 342–346 (2005)
3. Hu, M., Xu, Z., Zhang, R., Hu, A.: Physics Letters A 361, 231–237 (2007)
4. Lei, J., Wang, X., Lei, Y.: Communications in Nonlinear Science and Numerical Simulation 14(8), 3439–3448 (2009)
5. Wang, X., Lei, J., Pan, C.: Applied Mathematics and Computation 185, 989–1002 (2007)
6. Ge, Z.-M., Yang, C.-H.: Physica D 231, 87–89 (2007)
7. Kim, S.-H., Kim, Y.-S., Song, C.: Control Engineering Practice 12, 149–154 (2004)
8. Polycarpou, M.M., Ioannou, P.A.: Automatic 32(3), 423–442 (1996)

# Double Integral Adaptive Synchronization of the Lorenz Chaotic Systems with Unknown Parameters

Weiming Sun[1,*], Xinyu Wang[2], and Junwei Lei[3]

[1] Ocean School of Yantai University, Yantai, China
[2] Institute of Science and Technology for Opto-electronic Information, Yantai University, China
[3] Dept. of Control Engineering, Naval Aeronautical and Astronautical University, 264001 Yantai, China
{sunweiming1024,wangxinyu1024,leijunwei}@126.com

**Abstract.** A novel double integral action is firstly proposed in this paper and it is integrated with a novel integral Lyapunov function, which is necessary to analyze the system stability. The synchronization of chaotic system with unknown parameters are fulfilled by the constructing of Lyapunov function and the adopting of double integral action. Also, a filter is applied to make the controller is more flexible. But it is necessary to assume that the filter parameter is small enough when the stability of whole system is analyzed.

**Keywords:** Synchronization; Double Integral; Filter; Chaos; Adaptive Control.

## 1 Introduction

Integral control is an old and useful strategy, which is widely used in many real control systems. It is useful because it can reduce the static error of a system. Double integral strategy is a new technology and it can be powerful if it is integrated with Lyapunov theorem properly. Chaos systems have complex dynamical behaviors that possess some special features such as being extremely sensitive to tiny variations of initial conditions[1-4], and having bounded trajectories with a positive leading Lyapunov exponent and so on. In recent years, chaos synchronization has attracted increasingly attentions due to their potential applications of secure communications. There are many papers investigated synchronization of chaotic systems with adaptive control methods[5-6]. It is difficult to integrate adaptive control method with transfer function method since the latter one is and old and classic method and the former one is a modern method. So in this paper, it is a try good to integrate this two methods to solve the synchronization problem of chaotic system with unknown parameters.

## 2 Problem Description

The famous Lorenz chaos system can be written as:

[*] Corresponding author.

S. Lin and X. Huang (Eds.): CSEE 2011, Part II, CCIS 215, pp. 248–252, 2011.

$$
\begin{aligned}
\dot{x}_1 &= \alpha_1(y_1 - x_1) \\
\dot{y}_1 &= \gamma_1 x_1 - x_1 z_1 - y_1 , \\
\dot{z}_1 &= x_1 y_1 - \beta_1 z_1
\end{aligned}
\tag{1}
$$

where $x_1, y_1, z_1$ are the states of the system, and $\alpha_1, \beta_1, \gamma_1$ are unknown parameters. And model(1) is chosen as the master system and assume the slave system with different unknown parameters can be written as follows:

$$
\begin{aligned}
\dot{x}_2 &= \alpha_2(y_2 - x_2) + b_1 u_1 \\
\dot{y}_2 &= \gamma_2 x_2 - x_2 z_2 - y_2 + b_2 u_2 , \\
\dot{z}_2 &= x_2 y_2 - \beta_2 z_2 + b_3 u_3
\end{aligned}
\tag{2}
$$

where the states of the master system and slave system are denoted by 1 and 2 respectively.

The control objective is to design a controller $u = [u_1 \quad u_2 \quad u_3]$ such that the states of slave system trace to the states of the master system. And the error system can be written as follows:

$$
\begin{aligned}
\dot{e}_1 &= \alpha_2(y_2 - x_2) - \alpha_1(y_1 - x_1) + b_1 u_1 \\
\dot{e}_2 &= (\gamma_2 x_2 - x_2 z_2 - y_2) - (\gamma_1 x_1 - x_1 z_1 - y_1) + b_2 u_2 , \\
\dot{e}_3 &= x_2 y_2 - \beta_2 z_2 - (x_1 y_1 - \beta_1 z_1) + b_3 u_3
\end{aligned}
\tag{3}
$$

Where $e_1 = x_2 - x_1, e_2 = y_2 - y_1, e_3 = z_2 - z_1$ and $b_i$ is known control coefficient.

## 3 Adaptive Synchronization with Hign Order Integral Filter

Considering the above error subsystem, a derivative filter method is proposed as follows.

Consider the first subsystem as

$$\dot{e}_1 = \alpha_2(y_2 - x_2) - \alpha_1(y_1 - x_1) + b_1 u_1 . \tag{4}$$

Design a control as

$$b_1 u_1 = -\hat{\alpha}_2(y_2 - x_2) + \hat{\alpha}_1(y_1 - x_1) - k_{11} e_1 - k_{12} \int e_1 dt + f_{11} . \tag{5}$$

Where

$$f_{11} = -2 \frac{s e_1}{\tau_1 s + 1} k_{13} \iint e_1^2 dt - k_{13} e_1 \int e_1^2 dt . \tag{6}$$

It is worthy pointing out that $\frac{s}{\tau_1 s+1}$ is described by transfer funcion. So $\frac{s}{\tau_1 s+1}e_1$ is approximate to the derivative of error signal $e_1$ if a small enough parameter $\tau_1$ is chosen. According to equation (4) and equation (5), it holds

$$\begin{aligned} e_1\dot{e}_1 &= \tilde{\alpha}_2(y_2-x_2)e_1-\tilde{\alpha}_1(y_1-x_1)e_1-k_{11}e_1^2-k_{12}e_1\int e_1dt \\ &\quad -2e_1\dot{e}_1k_{13}\iint e_1^2dt-k_{13}e_1^2\int e_1^2dt \end{aligned} . \tag{7}$$

Where $\tilde{\alpha}_1=\alpha_1-\hat{\alpha}_1, \tilde{\alpha}_2=\alpha_2-\hat{\alpha}_2$ and $\dot{\tilde{\alpha}}_1=-\dot{\hat{\alpha}}_1, \dot{\tilde{\alpha}}_2=-\dot{\hat{\alpha}}_2$.

Design a adaptive turing law as

$$\dot{\hat{\alpha}}_1=-(y_1-x_1)e_1, \dot{\hat{\alpha}}_2=(y_2-x_2)e_1 . \tag{8}$$

Remember that

$$k_{12}e_1\int e_1dt=\left(\frac{1}{2}k_{12}\int e_1dt\int e_1dt\right)', \tag{9}$$

and

$$2e_1\dot{e}_1k_{13}\iint e_1^2dt+k_{13}e_1^2\int e_1^2dt=\left(k_{13}e_1^2\iint e_1^2dt\right)' . \tag{10}$$

Define a Lyapunov function as

$$V_1=\frac{1}{2}e_1^2+\frac{1}{2}\tilde{\alpha}_1^2+\frac{1}{2}\tilde{\alpha}_2^2+\frac{1}{2}k_{12}\left(\int e_1dt\right)^2+k_{13}e_1^2\iint e_1^2dt . \tag{11}$$

Then it is easy to prove that

$$\dot{V}_x\le -k_{11}e_1^2 . \tag{12}$$

With the same method, design the control law and adaptive turning law for the second subsystem as follow.

$$b_2u_2=-(\hat{\gamma}_2x_2-x_2z_2-y_2)+(\hat{\gamma}_1x_1-x_1z_1-y_1)-k_{21}e_2-k_{22}\int e_2dt+f_{22} . \tag{13}$$

Where

$$f_{22}=-2\frac{se_2}{\tau_2 s+1}k_{23}\iint e_2^2dt-k_{23}e_2\int e_2^2dt . \tag{14}$$

Design a adaptive turing law as

$$\dot{\hat{\gamma}}_1 = -x_1 e_2 , \dot{\hat{\gamma}}_2 = e_2 x_2 . \tag{15}$$

Choose a Lyapunov function as

$$V_2 = \frac{1}{2} e_2^2 + \frac{1}{2} \tilde{\gamma}_1^2 + \frac{1}{2} \tilde{\gamma}_2^2 + \frac{1}{2} k_{22} \left( \int e_2 dt \right)^2 + k_{23} e_2^2 \iint e_2^2 dt , \tag{16}$$

Where $\tilde{\gamma}_1 = \gamma_1 - \hat{\gamma}_1 , \tilde{\gamma}_2 = \gamma_2 - \hat{\gamma}_2$ and $\dot{\tilde{\gamma}}_1 = -\dot{\hat{\gamma}}_1 , \dot{\tilde{\gamma}}_2 = -\dot{\hat{\gamma}}_2$.

Then it is easy to prove that

$$\dot{V}_2 \le -k_{21} e_2^2 . \tag{17}$$

Design the control law and adaptive turning law for the third subsystem as follow.

$$b_3 u_3 = -x_2 y_2 + \beta_2 z_2 + (x_1 y_1 - \beta_1 z_1) - k_{31} e_3 - k_{32} \int e_3 dt + f_{33} . \tag{18}$$

Where

$$f_{33} = -2 \frac{s e_3}{\tau_3 s + 1} k_{33} \iint e_3^2 dt - k_{33} e_3 \int e_3^2 dt . \tag{19}$$

Choose a Lyapunov function as

$$V_3 - \frac{1}{2} e_3^2 + \frac{1}{2} k_{32} \left( \int e_3 dt \right)^2 + k_{33} e_3^2 \iint e_3^2 dt , \tag{20}$$

Choose a big Lyapunov function for the whole system as

$$V = V_1 + V_2 + V_3 , \tag{21}$$

It is easy to prove that

$$\dot{V} \le -k_{11} e_1^2 - k_{21} e_2^2 - k_{31} e_3^2 . \tag{22}$$

So the system is stable and $e_1 \to 0$, $e_2 \to 0$, $e_3 \to 0$ and the synchronization can be fulfilled.

## 4 Conclusions

A novel double integral adaptive strategy is firstly used in the synchronization of Lorenz Chaotic systems with unknown parameters. Unknown parameters of system are identified with adaptive turning law. The whole system is guaranteed to be stable by the constructing of a big Lyapunov function which is composed of three small Lyapunov functions of every subsystem. Also a transfer function filter is used in the double integral action with the assumption that filter parameters are small enough.

## References

1. Elabbasy, E.M., Agiza, H.N., El-Dessoky, M.M.: Chaos, Solitons and Fractals 30, 1133–1142 (2006)
2. Tang, F., Wang, L.: Physics Letters A 346, 342–346 (2005)
3. Hu, M., Xu, Z., Zhang, R., Hu, A.: Physics Letters A 361, 231–237 (2007)
4. Lei, J., Wang, X., Lei, Y.: Communications in Nonlinear Science and Numerical Simulation 14(8), 3439–3448 (2009)
5. Wang, X., Lei, J., Pan, C.: Applied Mathematics and Computation 185, 989–1002 (2007)
6. Polycarpou, M.M., Ioannou, P.A.: Automatic 32(3), 423–442 (1996)

# Actuarial Model for Family Combined Life Insurance

NianNian Jia, PanPan Mao, and ChangQing Jia

College of Science, Harbin Engineering University, Heilongjiang Province, China
jianiannian@hrbeu.edu.cn

**Abstract.** The random interest rate of life insurance is a hot spot in recent literatures on actuarial study and its applications. In order to avoid the risk induced by interest randomness, in this paper, the force of interest accumulation function was modeled with Gamma distribution and negative binominal distribution, we constitute actuarial model for family combined life insurance based on the stochastic interest rate. This model has a wider range of application.

**Keywords:** stochastic interest rate; life insurance; combined life insurance; actuarial model.

## 1 Introduction

As an important part of the financial system the insurance industry is closely related to the public interest. In life insurance, interest-rate risk will bring about immeasurable risk for the insurance company. The theories and methodologies of the life insurance under the random interest-rate have been considered the hot topics of the researchers in recent years. This paper establishes a actuarial model for family combined life insurance under the random interest rate.

ZhiZhong Wang and YiHong Liu (1996) have researched for mortality and insurance premium in combined insurance using the Bayes method [1]. XiaoPing Zhu (1997) have researched for theoretical model of combined life insurance and actuarial principles [2]. YaoHua Wu, XinZhong Cai and ZhiQiang Wu (1998) have researched for couple's combined pension insurance(with death) net premium and reserve[3]. LiYan Wang and EnMin Feng (2003) established a family combined life insurance actuarial model under the interest force with the Wiener process [4]. LiYan Wang, Jing Zhao and DeLi Yang (2007) established continuous couple's combined insurance under the interest force with double-stochastic process, and calculate the life insurance net premium[5]. LiYan Wang, YaLi Hao and HaiJiao Zhang (2010) established endowment assurance model under the interest force with Brownian process and Poisson process, get the calculate formula of all the value of insurance, and furthermore simplify the formula under the hypothesis of de Moivre mortality force[6]. In this paper, the interest force accumulation function was modeled with Gamma distribution and negative binominal distribution, based on this model, we establishes a family combined life insurance actuarial model, and further take into account the death assumed, finally ,a numerical example is presented, explaining the model account with life insurance practice, the model has a wider range of application.

S. Lin and X. Huang (Eds.): CSEE 2011, Part II, CCIS 215, pp. 253–258, 2011.

# 2 Model Basis

## 2.1 Assurer

The members of the family combined life insurance are father, mother and child. We assume that father $(x)$, mother $(y)$ and child $(z)$ are combined insurance, their lives independent and meet the following conditions:

(1) Physically healthy, the legal couples are aged below 35(male above 22 years, while female above 20 years).

(2) Physically healthy, child is younger than 13 years.

## 2.2 Condition Survival Model

(1) The $(\overset{1}{x}y\,|\,z)$ is expressed that the insured aged $x$ die before the insured aged $z$ and $y$. $T(\overset{1}{x}y\,|\,z)$ is the finished time of three combined condition survival state, $f_{T(\overset{1}{x}y|z)}(t)$ is the density function of $T(\overset{1}{x}y\,|\,z)$.

$$ {}_tp_{(\overset{1}{x}y|z)} = P\{T(x)<t, T(y)>t, T(z)>t\} = (1-{}_tp_x)\cdot{}_tp_y\cdot{}_tp_z\,. \tag{1} $$

$$ f_{T(\overset{1}{x}y|z)}(t) = \frac{d}{dt}\left(1-{}_tp_{(\overset{1}{x}y|z)}\right) = {}_tp_{(\overset{1}{x}y|z)}\,\mu_{x+t} = {}_tp_{yz}\cdot\mu_{x+t} - {}_tp_{yz}\cdot f_{T(x)}(t)\,. \tag{2} $$

(2) The $(\overset{1}{y}x\,|\,z)$ is expressed that the insured aged $y$ die before the insured aged $z$ and $x$. $T(\overset{1}{y}x\,|\,z)$ is the finished time of three combined condition survival state, $f_{T(\overset{1}{y}x|z)}(t)$ is the density function of $T(\overset{1}{y}x\,|\,z)$.

$$ {}_tp_{(\overset{1}{y}x|z)} = P\{T(y)<t, T(x)>t, T(z)>t\} = (1-{}_tp_y)\cdot{}_tp_x\cdot{}_tp_z\,. \tag{3} $$

$$ f_{T(\overset{1}{y}x|z)}(t) = \frac{d}{dt}\left(1-{}_tp_{(\overset{1}{y}x|z)}\right) = {}_tp_{zx}\cdot\mu_{y+t} - {}_tp_{zx}\cdot f_{T(y)}(t)\,. \tag{4} $$

(3) The $(\overset{1}{z}(yx))$ is expressed that the insured aged $z$ die before the insured aged $x$ and $y$. $T(\overset{1}{z}(yx))$ is the finished time of three combined condition survival state, $f_{T(\overset{1}{z}(yx))}(t)$ is the density function of $T(\overset{1}{z}(yx))$.

$$ {}_tp_{\overset{1}{z}(xy)} = P\{T(xy) > t, T(z) < t\} = {}_tp_x \cdot {}_tp_y \cdot (1 - {}_tp_z). \tag{5}$$

$$ f_{T(\overset{1}{z}(xy))}(t) = \frac{d}{dt}\left(1 - {}_tp_{\overset{1}{z}(xy)}\right) = {}_tp_{xy} \cdot \mu_{z+t} - {}_tp_{xy} \cdot f_{T(z)}(t). \tag{6}$$

(4) The $((\overset{1}{\overline{xy}})z)$ is expressed that the insured aged $z$ die after the insured aged $x$ and $y$. $T((\overset{1}{\overline{xy}})z)$ is the finished time of three combined condition survival state, $f_{T((\overset{1}{\overline{xy}})z)}(t)$ is the density function of $T((\overset{1}{\overline{xy}})z)$.

$$ {}_tp_{\overset{1}{\overline{(xy)}}z} = P\{T(\overline{xy}) < t, T(z) > t\} = (1 - {}_tp_x - {}_tp_y + {}_tp_{xy}) \cdot {}_tp_z. \tag{7}$$

$$ f_{T((\overset{1}{\overline{xy}})z)}(t) = \frac{d}{dt}\left(1 - {}_tp_{\overset{1}{\overline{(xy)}}z}\right) = (1 - {}_tp_x - {}_tp_y + {}_tp_{xy}) \cdot {}_tp_z \cdot \mu_{\overline{xy}+t}. \tag{8}$$

Because $\mu_x = -\frac{d\,{}_tp_x}{dt} \cdot \frac{1}{{}_tp_x}$ and ${}_tp_{\overline{xy}} = {}_tp_x + {}_tp_y - {}_tp_{xy}$, then (8) formula is

$$ f_{T((\overset{1}{\overline{xy}})z)}(t) = (1 - {}_tp_x \cdot {}_tp_y) \cdot {}_tp_z \frac{{}_tp_x\mu_{x+t} + {}_tp_y\mu_{y+t} - {}_tp_{xy}(\mu_{x+t} + \mu_{y+t})}{{}_tp_x + {}_tp_y - {}_tp_{xy}}. \tag{9}$$

## 3 Model Assumptions

Assume the interest force accumulation function is described as follows:

$$ R(t) = \delta t + \beta N(t) + \gamma G(t) \quad (0 \le t < \infty). \tag{10}$$

Where $N(t)$ follows negative binominal distribution with parameters $(k, p)$, denoted by $Nb(k, p)$ ; $G(t)$ follows Gamma distribution with parameters $(u, v)$ , denoted by $Ga(u, v)$ ; $N(t)$ and $G(t)$ are independent; $\delta$ is the constant interest force, $\beta, \gamma \ge 0$ are independent constant By Eq. (10), the payment at $t$ is

$$ Z(t) = c(t) \cdot e^{-R(t)} = c(t) \cdot e^{-(\delta t + \beta N(t) + \gamma G(t))} \quad (0 \le t < +\infty). \tag{11}$$

**Theorem 3.1.** Let $N(t)$ follows negative binominal distribution with parameters $(k, p)$, $\beta$ is constant, $q = 1 - p$, then

$$E\left(e^{-\beta N(t)}\right) = \left(\frac{p}{1 - qe^{-\beta}}\right)^k .$$

**Theorem 3.2.** Let $G(t)$ follows Gamma distribution with parameters $(u, v)$, $\gamma$ is constant, then

$$E\left(e^{-\gamma G(t)}\right) = \left(\frac{v}{v + \gamma}\right)^u .$$

Denote $B(N, \beta, G, \gamma) = E_N\left(e^{-\beta N(t)}\right) \cdot E_G\left(e^{-\gamma G(t)}\right) = \left(\frac{p}{1 - qe^{-\beta}}\right)^k \cdot \left(\frac{v}{v + \gamma}\right)^u$ (12)

# 4 Actuarial Model

## 4.1 The Continuous Single Net Premium Model under the Condition Survival

(1) The continuous single net premium of the insured aged $x$ die before the insured aged $z$ and $y$. By Eqs. (2) ,(11)and (12),

$$\overline{A}_{\left(\overset{1}{x}y\right)z} = \int_0^{+\infty} c(t) \cdot e^{-\delta t} \cdot B(N, \beta, G, \gamma) \cdot (1 - {}_t p_x) \cdot {}_t p_y \cdot {}_t p_z \cdot \mu_{x+t} dt . \tag{13}$$

(2) The continuous single net premium of the insured aged $y$ die before the insured aged $x$ and $z$. By Eqs. (4), (11)and (12),

$$\overline{A}_{\left(\overset{1}{y}x\right)z} = \int_0^{+\infty} c(t) \cdot e^{-\delta t} \cdot B(N, \beta, G, \gamma) \cdot (1 - {}_t p_y) \cdot {}_t p_x \cdot {}_t p_z \cdot \mu_{y+t} dt . \tag{14}$$

(3) The continuous single net premium of the insured aged $z$ die before the insured aged $x$ and $y$. By Eqs. (6),(11) and (12),

$$\overline{A}_{\overset{1}{z}(xy)} = \int_0^{+\infty} c(t) \cdot e^{-\delta t} \cdot B(N, \beta, G, \gamma) \cdot {}_t p_x \cdot {}_t p_y \cdot (1 - {}_t p_z) \cdot \mu_{z+t} dt . \tag{15}$$

(4) The continuous single net premium of the insured aged $z$ die after the insured aged $x$ and $y$. By Eqs. (9),(11) and (12),

$$\overline{A}_{\overset{1}{(\overline{xy})}z} = \int_0^{+\infty} c(t) \cdot e^{-\delta t} \cdot B(N, \beta, G, \gamma) \cdot (1 - {}_t p_x - {}_t p_y + {}_t p_{xy}) \cdot {}_t p_z$$

$$\cdot \frac{{}_t p_x \mu_{x+t} + {}_t p_y \mu_{y+t} - {}_t p_{xy}\left(\mu_{x+t} + \mu_{y+t}\right)}{{}_t p_x + {}_t p_y - {}_t p_{xy}} dt . \tag{16}$$

### 4.2 The Continuous Life Annuity under the Condition Survival

(1) The continuous life annuity of the insured aged $x$ die before the insured aged $z$ and $y$. By Eq.(1),

$$\bar{a}_{\left(\overset{1}{xy}\right|z} = \int_0^{+\infty} c(t)\cdot e^{-\delta t}\cdot B(N,\beta,G,\gamma)\cdot(1-{}_tp_x)\cdot{}_tp_y\cdot{}_tp_z dt. \tag{17}$$

(2) The continuous life annuity of the insured age $y$ die before the insured aged $z$ and $x$. By Eq.(3),

$$\bar{a}_{\left(\overset{1}{yx}\right|z} = \int_0^{+\infty} c(t)\cdot e^{-\delta t}\cdot B(N,\beta,G,\gamma)\cdot(1-{}_tp_y)\cdot{}_tp_x\cdot{}_tp_z dt. \tag{18}$$

(3) The continuous life annuity of the insured aged $z$ die before the insured aged $x$ and $y$. By Eq. (5),

$$\bar{a}_{\overset{1}{z}(xy)} = \int_0^{+\infty} c(t)\cdot e^{-\delta t}\cdot B(N,\beta,G,\gamma)\cdot{}_tp_x\cdot{}_tp_y\cdot(1-{}_tp_z) dt. \tag{19}$$

(4) The continuous life annuity of the insured aged $z$ die after the insured aged $x$ and $y$. By Eq. (7),

$$\bar{a}_{\overset{1}{(xy)}z} = \int_0^{+\infty} c(t)\cdot e^{-\delta t}\cdot B(N,\beta,G,\gamma)\cdot(1-{}_tp_x-{}_tp_y+{}_tp_{xy})\cdot{}_tp_z dt. \tag{20}$$

### 4.3 The Continuous Level Net Premium under the Condition Survival

(1) The continuous level net premium of the insured aged $x$ die before the insured aged $z$ and $y$. By Eqs. (13) and (17),

$$\bar{P}\left(\bar{A}_{\left(\overset{1}{xy}\right|z}\right) = \frac{\bar{A}_{\left(\overset{1}{xy}\right|z}}{\bar{a}_{\left(\overset{1}{xy}\right|z}} = \frac{\int_0^{+\infty} c(t)\cdot e^{-\delta t}\cdot B(N,\beta,G,\gamma)\cdot(1-{}_tp_x)\cdot{}_tp_y\cdot{}_tp_z\cdot\mu_{x+t}dt}{\int_0^{+\infty} c(t)\cdot e^{-\delta t}\cdot B(N,\beta,G,\gamma)\cdot{}_tp_{\left(\overset{1}{xy}\right|z} dt}.$$

(2) The continuous level net premium of the insured aged $y$ die before the insured aged $z$ and $x$. By Eqs. (14) and (18),

$$\bar{P}\left(\bar{A}_{\left(\overset{1}{yx}\right|z}\right) = \frac{\bar{A}_{\left(\overset{1}{yx}\right|z}}{\bar{a}_{\left(\overset{1}{yx}\right|z}} = \frac{\int_0^{+\infty} c(t)\cdot e^{-\delta t}\cdot B(N,\beta,G,\gamma)\cdot(1-{}_tp_y)\cdot{}_tp_x\cdot{}_tp_z\cdot\mu_{y+t}dt}{\int_0^{+\infty} c(t)\cdot e^{-\delta t}\cdot B(N,\beta,G,\gamma)\cdot{}_tp_{\left(\overset{1}{yx}\right|z} dt}.$$

(3) The continuous level net premium of the insured aged $z$ die before the insured aged $x$ and $y$. By Eqs. (15) and (19),

$$\bar{P}\left(\bar{A}_{\overset{1}{z}(xy)}\right) = \frac{\bar{A}_{\overset{1}{z}(xy)}}{\bar{a}_{\overset{1}{z}(xy)}} = \frac{\int_0^{+\infty} c(t)\cdot e^{-\delta t}\cdot B(N,\beta,G,\gamma)\cdot(1-{}_tp_z)\cdot{}_tp_x\cdot{}_tp_y\cdot\mu_{z+t}dt}{\int_0^{+\infty} c(t)\cdot e^{-\delta t}\cdot B(N,\beta,G,\gamma)\cdot{}_tp_{\overset{1}{z}(xy)} dt}.$$

(4) The continuous level net premium of the insured aged $z$ die after the insured aged $x$ and $y$. By Eqs. (16) and (20),

$$\overline{P}\left(\overline{A}_{\overline{(xy)}z}^{\;1}\right)=\frac{\overline{A}_{\overline{(xy)}z}^{\;1}}{\overline{a}_{\overline{(xy)}z}^{\;1}}=\frac{\int_0^{+\infty}c(t)\cdot e^{-\delta t}\cdot B(N,\beta,G,\gamma)\cdot\left(1-{}_tp_x-{}_tp_y+{}_tp_{xy}\right)}{\int_0^{+\infty}c(t)\cdot e^{-\delta t}\cdot B(N,\beta,G,\gamma)\cdot\left(1-{}_tp_x-{}_tp_y+{}_tp_{xy}\right)\cdot{}_tp_z dt}$$

$$\cdot\frac{{}_tp_x\mu_{x+t}+{}_tp_y\mu_{y+t}-{}_tp_{xy}\left(\mu_{x+t}+\mu_{y+t}\right)}{{}_tp_x+{}_tp_y-{}_tp_{xy}}dt\,.$$

## 5 Conclusion

In this study, the stochastic interest rate used Gamma distribution and negative binominal distribution, established a combined life insurance of continuous life insurance net premium model under the stochastic interest rate. Adjust the parameters in the model, we can get all kinds of traditional life insurance actuarial model. Such as traditional fixed rate model and the single stochastic interest rate model, also we can translate it into discrete life insurance actuarial model under the hypothesis of mortality force. It changed the disadvantages of the anciently life insurance actuarial models can not translate into each other. In addition, the stochastic interest rate with the discrete distribution and continuous distribution has a wider range of practical applications.

**Acknowledgments.** This work was supported by the Fundamental Research Funds for the Central Universities (Grant No. HEUCF20111135).

## References

1. Zhizhong, W., Yihong, L.: Multidimensional Development of Life Insurance. J. Systems Engineering 14(6), 27–31 (1996)
2. Xiaoping, Z.: Actuarial Model for Combined Life Insurance. J. Tong Ji University 25(1), 56–60 (1997)
3. Yaohua, W., Xinzhong, C., Zhiqiang, W.: Calculation of the Insurance for a Couple Against Pension Annuity (Including Death Insurance). Journal of China University of Science and Technology 28(4), 439–445 (1998)
4. Liyan, W., Enmin, F.: Dual Random Model of Family Combine Insurance. Journal of Engineering Mathematics 20(8), 69–72 (2003)
5. Liyan, W., Jing, Z., Deli, Y.: Joint-life insurance under random rates of interest. Journal of Dalian University of Technology 47(6), 920–924 (2007)
6. Liyan, W., Liya, H., Haijiao, Z.: Increasing endowment assurance policy under stochastic rates of interest. Journal of Dalian University of Technology 50(5), 827–830 (2010)

# Fuzzy Optimal Control of Sewage Pumping Station with Flow Concentration

Zhe Xu, Lingyi Wu, Xuetong Zhang, and Anke Xue

Institute of Information and Control, Hangzhou Dianzi University, HangZhou

**Abstract.** This paper aimed to improve the traditional simple water level control for drainage network. Combining with the prediction model of sewage convergence node, the paper proposed a fuzzy optimal control method, tried to solve: (1) Control lag problem caused by a single pump station isolated control; (2) The problem of pump switchd frequently caused by the conservative control; (3) Minimizing sewage overflow problem at the convergence pump station and (4) Energy saving problem of sewage pumping station operation. The actual example showed that this control method had good effect, and it could be applied to the actual drainage operation.

**Keywords:** urban drainage, pipe network convergence, fuzzy control.

## 1 Introduction

The urban sewage system is complicated and nonlinear. The water level of pump station well and the inflow of sewage is influenced by people's behaviors, rainfall and some other unpredicted factors such as channel leakage. With the acceleration of urbanization, the flow of wastewater increased dramatically. There is a serious challenge for the urban drainage system. The existing simple control of water level is easy to cause the sewage overflow and more energy costs, especially for convergence pump station due to the impact of multiple upstream. Therefore, developing the optimal control strategy for convergence station is of great significance.

The scholars in the developed countries have done a lot of work for the existing drainage pumping station control methods and strategies, mainly focused on algorithm, control strategies and implementation method. In China, the scholar Xionghai Wang [6-10] form Zhejiang University devote himself to energy saving and intelligence control of the drainage network. Based on energy-saving mechanism analysis, he adopted the frequency control technology to minimize power consumption and overflow for a single pumping station. But this control exist a lag. Jianzhong Wang [11-12] and others have done some research in energy efficiency and optimal scheduling. Wang Yu[12] established fuzzy control pump station system to make full use of the storage of every pumping station for reducing sewage overflow, but the research does not involve the control of confluence pumping station. Based on predicting model of sewage pumping station[13-14](Xu Zhe etc.), in this paper, the fuzzy control technology is applied to realize the optimal control of confluence node, minimize sewage overflows and reduce energy consumption

S. Lin and X. Huang (Eds.): CSEE 2011, Part II, CCIS 215, pp. 259–266, 2011.

# 2 Fuzzy Controller Design for Confluence Situation

## 2.1 Analysis of Optimal Control of Sewage Pumping Station

The existing simple water level control only according to the current level of pump control regardless of the operational state of its upstream and downstream pumping station.There are several limitations in this control mode, as follows:

(1) Simple water level control system has the drawback of poor coordination performance. It is easy to lead to sewage overflow if the capacity of sewage inflows is large in a short time.

(2)The traditional control method can't adjust pump operations to the future level trend. Moreover, to avoid overflow, water level always be kept within a low level in current control, which result in pump switching frequently and greater energy consumption.

(3) Each pumping station can't take full advantage of its storage function and realize optimal operation in the traditional control method. When the sewage reserves in each upstream pumping station is unbalanced, and the water level of downstream pumping station is high, downstream pumping station likely to overflow if all upstream pumping stations continue drainage.

(4)The full range of control strategies can't be realized. When inflow far exceeds the For the first two cases, according to the level changes trend by the upstream and downstream drainage, we can control early to prevent wasting energy. As for the third case, if the level of downstream pool reach a certain depth ratio and the level of upstream pool is low, lift pump could be turned off to make full use of the storage function of each pumping station. For the last one, outflow should be within downstream discharge capacity, or else the overflow port need to be opened. If the trend of the water level of local pumping station is stable, each station will still runs in traditional mode.

## 2.2 Fuzzy Controller Design

### 2.2.1 Input/Output Variables Selection

Based on the above analysis of optimal control of Sewage pumping station, we know there are three main objectives for the fuzzy controller:1)to realize the optimal control according to the chain relationship of upstream and downstream pumping station, 2)to realize the optimal scheduling of pumping stations, and to make full use of the existing storage equipment, 3)to make Control program From the global perspective to save energy. Considering the above three points, an example of distributed pumping stations was given, as shown in Fig.2. P1 to P5 are the lift pumping station, P4 is local station. In this study, the input variables of fuzzy control are the current water level(L), the trend of water-level prediction ($\Delta L$), the depth ratio difference between a pumping station and its brother pumping stations($D_{4\text{-}3}$), the depth ratio difference between station P3, P4 and downstream station P5 ($D_{3,4\text{-}5}$), and the maximum allowable additional discharge of P5 (safety margin S). The output is the number of opened pump in station P4. If station P4 has no brother node, namely cascade, we only consider the $D_{3,4\text{-}5}$. where $D_{4\text{-}3}$, $D_{5\text{-}(3,4)}$ are formulated as:

$$\Delta L = \frac{L(t+1)+L(t+2)+L(t+3)\cdots L(t+n)}{n} - L(t) \tag{1}$$

$$D_{4\text{-}3} = \frac{L_4}{H_4} - \frac{L_3}{H_3} \tag{2}$$

$$D_{(3\,4)-5} = \max\left(\frac{L_3}{H_3}, \frac{L_4}{H_4}\right) - \frac{L_5}{H_5} \tag{3}$$

$$S_4 = \min\left(\left(N_{total} - N_{before}\right)\times C, S_5\right) \tag{4}$$

Where $t$ denotes predicting time; $n$ is the predicting period; $H$ is maximum forebay water level; $N_{total}$ is the total number of pumps; $N_{before}$ is the opened pump number; $C$ is lift ability of pump.

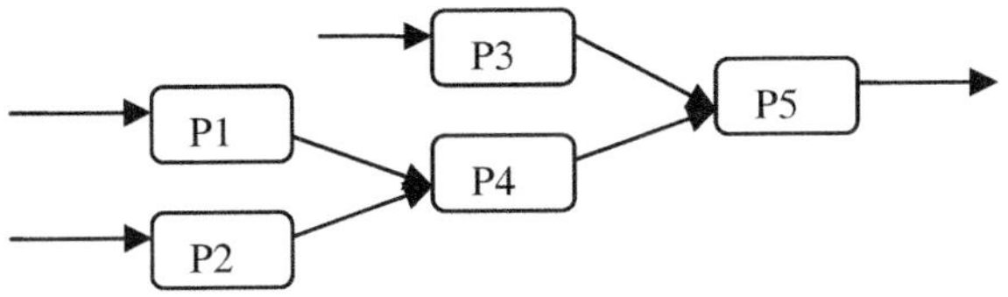


**Fig. 1.** Allocation plan of confluence pumping station

The value of input/output variables are corresponding to the value in the fuzzy linguistic variable sets.

The definition of input fuzzy linguistic variable sets is: L={ZE:zero, LS:less small, S:small, LL:more large, L:large}; △L={NL: negative large, NS:negative small, ZE: zero, PS:positive small, PL:positive large};$D_{4\text{-}3}$={S:small, L:large};$D_{3,4\text{-}5}$={S: small, L:large}; $S_3$={ZE: zero, LS:less small, S:small, LL:more large, L:large}

The definition of output fuzzy linguistic variable sets is:N ={N0:off, N1:open 1 pump, N2:open 2 pumps, N3:open 3 pumps, N4:open 4 pumps};O={U:on, D:off}

It can be seen that this fuzzy controller structure is of 5-input, 2-output. A block diagram of the fuzzy control for confluence pumping station is shown in Fig.3. Where ΔL can be calculated from the predicting model of confluence pumping station, and the output of fuzzy controller N is also treated as a input of predicting model.

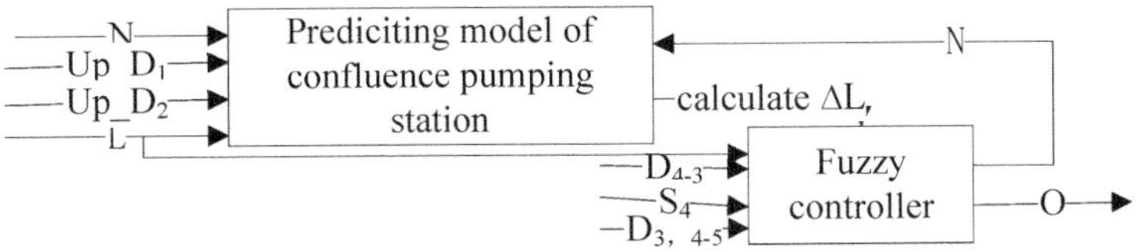


**Fig. 2.** Confluence pumping station fuzzy control beckoned diagram schematic

### 2.2.2 Fuzzification

Fuzzy sets are used to quantify the information in the rule-base, and the inference mechanism operates on fuzzy sets to produce fuzzy sets; hence, we must specify how the fuzzy system will convert its numeric inputs into fuzzy sets so that they can be used by the fuzzy system. The membership function is used to associate a grade to each linguistic term.

Once the fuzzy controller inputs and outputs are chosen, one must think about what are the appropriate membership functions (MFs) are for these input and output variables. In this paper, the membership function selection of input variables is shown in Fig.4.

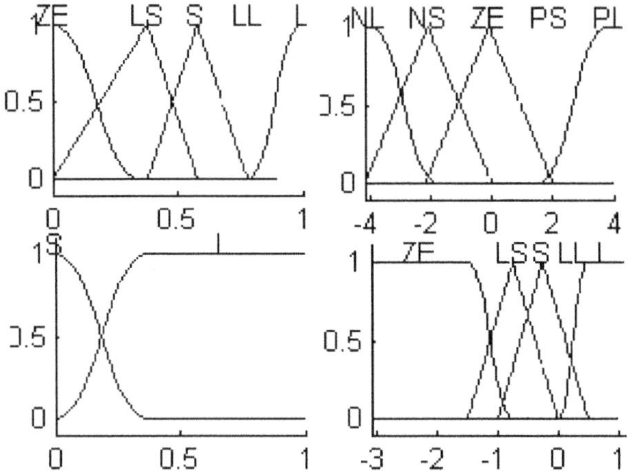


**Fig. 3.** Membership function of L, ΔL, D4-3 ,D3,4-5 and S4

### 2.2.3 EstablishFuzzy Inference

**Table 1.** Control rule table o S4=LS,S,LL,L when If $D_{4\text{-}3}$!=S and $D_{3\ ,\ 4\text{-}5}$! = S

| | S4=LS | | | | | S4=S | | | | | S4=LS | | | | | S4=L | | | | |
|---|---|---|---|---|---|---|---|---|---|---|---|---|---|---|---|---|---|---|---|---|
| | NL | NS | ZE | PS | PL | NL | NS | ZE | PS | PL | NL | NS | ZE | PS | PL | NL | NS | ZE | PS | PL |
| ZE | N0 | N0 | N0 | N1 | N1/U | N0 | N0 | N0 | N1 | N2 | N0 | N0 | N0 | N1 | N2 | N0 | N0 | N0 | N1 | N2 |
| LS | N0 | P0 | N1 | N1/U | N1/U | N0 | P0 | N1 | N2 | N2/U | N0 | P0 | N1 | N2 | N3 | N0 | P0 | N1 | N2 | N3 |
| S | N0 | N1 | N1/U | N1/U | N1/U | N0 | N1 | N2 | N2/U | N2/U | N0 | N1 | N2 | N3 | N3/U | N0 | N1 | N2 | N3 | N4 |
| LL | N1 | N1/U | N1/U | N1/U | N1/U | N1 | N2 | N2/U | N2/U | N2/U | N1 | N2 | N3 | N3/U | N3/U | N1 | N2 | N3 | N4 | N4/U |
| L | N1/U | N1/U | N1/U | N1/U | N1/U | N2 | N2/U | N2/U | N2/U | N2/U | N2 | N3 | N3/U | N3/U | N3/U | N2 | N3 | N4 | N4/U | N4/U |

Fuzzy logical control rules are constructed based on the analysis of optimal control. Within safety margin, if the depth ratio in downstream pumping stations and sibling nodes are high, and the center filled with a low degree, the pump of center station should be shut down to put downward pressure on drainage. If the change trend of the local water level is stable, how to operate for the pump is determined by the real-time value of water level. If $S_4$=ZE, then N=N0,O=U. Under this condition, water level is very high, so the overflow gate shouled be opened. If $D_{4\text{-}3}$=S and $D_{(3\square\ 4)\text{-}5}$=S, the rule is shown in table 1 to 4; If $D_{4\text{-}3}$=S and $D_{3\square\ 4\text{-}5}$ = S, the rule is shown in table1; If $D_{4\text{-}3}$!=S and $D_{(3\square\ 4)\text{-}5}$!=S, the rule is shown in table 2; For example, in table 1, when S4=LSshows that the number of opened pump could add another one in confluence pumping station if the water level of pumping station P4 and P5 are fairly high, and safety margin is relatively small.

**Table 2.** Control rule table of S4=LS,S,LL,L when $D_{4\text{-}3}$=S and $D_{(3，4)\text{-}5}$=S

| Δ L | S4=LS | | | | | S4=S | | | | | S4=LS | | | | | S4=L | | | | |
|---|---|---|---|---|---|---|---|---|---|---|---|---|---|---|---|---|---|---|---|---|
| L LL | NL | NS | ZE | PS | PL | NL | NS | ZE | PS | PL | NL | NS | ZE | PS | PL | NL | NS | ZE | PS | PL |
| ZE | N0 | N0 | N0 | N0 | N0 | N0 | N0 | N0 | N0 | N0 | N0 | N0 | N0 | N0 | N1 | N0 | N0 | N0 | N0 | N1 |
| LS | N0 | N0 | N0 | N0 | N1 | N0 | N0 | N0 | N1 | N2 | N0 | N0 | N0 | N1 | N2 | N0 | N0 | N0 | N1 | N2 |
| S | N0 | N0 | N0 | N1 | N1 | N0 | N0 | N1 | N2 | N2 | N0 | N0 | N1 | N2 | N3 | N0 | N0 | N1 | N2 | N3 |

## 3 An Example of Application

This section presents a sample application of the proposed approach on Wulinmen pumping station. This pumping station is a convergence node of HangZhou drainage pipe network. The elevation of pool bottom is -3.1m, the lowest water level is -0.95m, the highest water level is 1.12m, and the warning water level is 1.6m. To facilitate the simulation, the foresee period is set to 10 minutes. And to ensure the sewage inflows is almost constant in the simulation process, it is assumed that to increase the munber of opened pump have no impact on sewage inflow.

### 3.1 Establish Fuzzy Controller

First, we must identify the input variables and the output variables. The input variables are the depth ratio differences between Wulinmen pumping station and it's brother pumping stations, the depth ratio differences between the bigger of former and Genshanmen pumping stations, the forebay water level, changing trends of water level, and safety margin. The output variables are the number of opened pump and the switch state of overflow port. Then use the fuzzy control toolbox in MATLAB. Firstly, to create a new fuzzy controller through newfis() ; Secondly, to add input/output variables and membership functions through addvar() and addmf(); Thirdly, to add fuzzy control rules through addrule(), to set defuzzification method through setfis(); Finally, to get fuzzy the computational results through evalfis().

### 3.2 Fuzzy Control Results

Fuzzy control output is shown in Fig.6, the corresponding comparison of forebay water level is shown in Fig.7To better illustrate the second point of optimal control in section 3.1, the fuzzy control results of Hangda pumping station is also provided. The result is shown in Fig.8 and Fig.9.

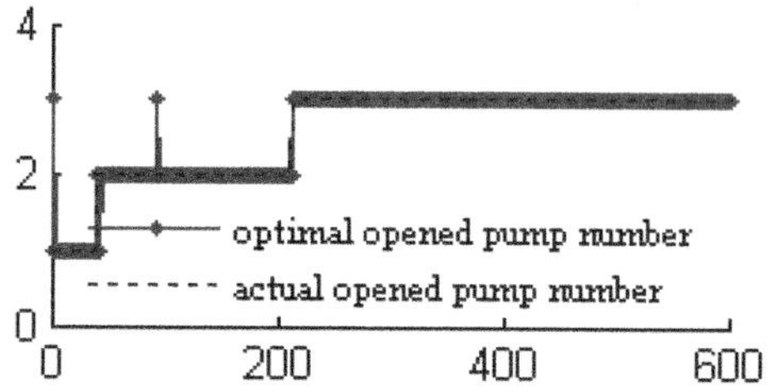


**Fig. 4.** Comparison between actual and optimization control in the number of opened pump

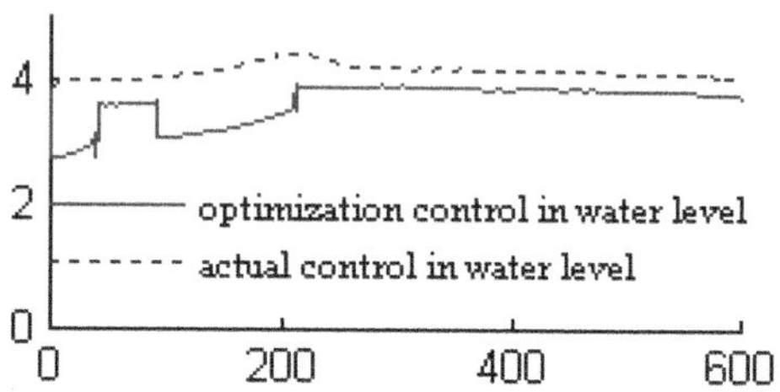


**Fig. 5.** Comparison between actual and optimization control in water level

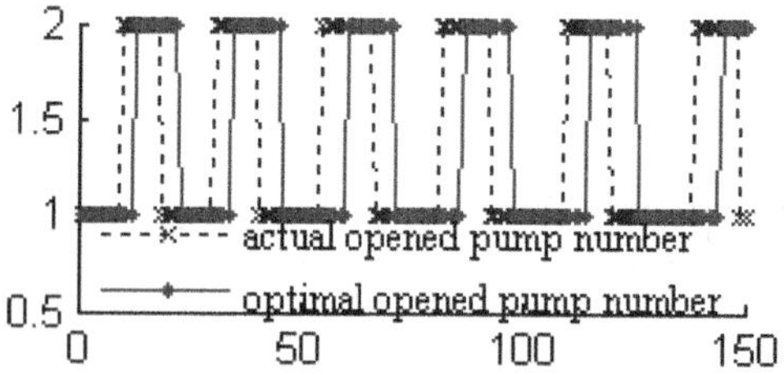


**Fig. 6.** Comparison between actual and optimization control in the number of opened pump

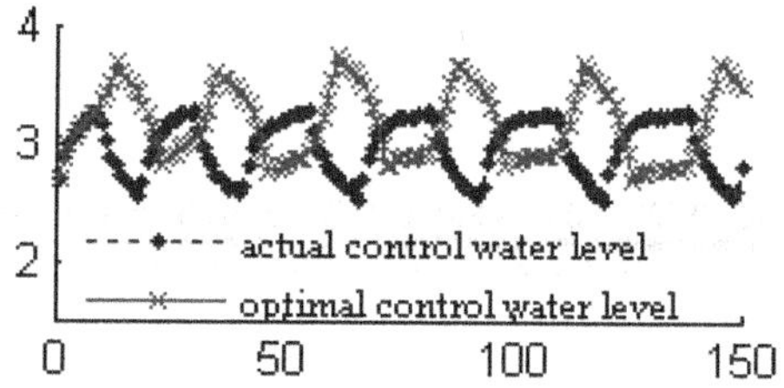


**Fig. 7.** Comparison between actual and optimization control in water level

### 3.3 Results Analysis

According to the Fig.6, it can be seen that one more pump opened at the 100 times under the fuzzy control mode. The Fig.7 shows that the water level of forebay sewage has exceeded the highest level in the traditional control mode, that may cause sewage overflows. But to use of fuzzy optimization control can successfully avoid the warning level and ensure running within the security level. Thus, fuzzy control can reduce the possibility of sewage overflow.

According to the Fig.6 and Fig.7, the added pump work in short duration and the water level decline rapidly at the 100 times. The main reason is assumed that to increase the munber of opened pump have no impact on sewage inflow. Actually, to open more pumps will also increases the sewage inflow. Therefore, in actual operation, the sewage level will more smooth through fuzzy control. Of course, we should establish different membership function to differrent pumping station in the actual application.

Moreover, the water level control can be implement leisurely by the fuzzy controller because of the predict water level can derive from the foecasting model of converge node. Fig.8 and Fig.9 show that through fuzzy optimization control, the pumping station head reduced and the effect of energy-saving is achieved.

## 4 Conclusion and Prospect

In this paper, A fuzzy optimal control method of converge pumping station is proposed, which is validated in Wulinmen pumping station. It can minimize the sewage overflow and effectively reduce the energy consumption. The fuzzy control system applied in drainage system optimization control has a good prospects.

But if it is used in actual operation, further study is needed to discussion how the operation of open pump would affect sewage inflow, improve reliability and precision of predicting model, in order to improve the accuracy of fuzzy control and realize intelligent control of pumping station.

**Acknowledgments.** Supported By the National Natural Science Foundation of China (60974138).

## References

1. Shaheen, H.Q.: Municipal Wastewater Characteristics at Irtah Pumping Station in the Tulkarem City. J. Water Science and Technology 42, 337–340 (2000)
2. Ermolin, Y.A.: Mathematical Modeling for Optimized Control of Moscow's Sewer Network. J. Applied Mathematical Modeling 23, 543–556 (1999)
3. Yagi, S., Shiba, S.: Application of Genetic Algorithms and Fuzzy Control to a Combined Sewer Station. J. Water Science and Technology 39, 217–224 (1999)
4. Tan, P.C., Dabke, K.P., Mein, R.G.: Modelling and Control of Sewer Flow for Reduced Cost Operation of a Sewage Pumping Station. IEEE Transaction on Systems Man and Cybernetics 18(5), 807–813 (1988)
5. Gelormino, M.S., Ricker, M.L.: Model-predictive control of a combined sewer system. Int. J. Control 59(3), 793–816 (1994)
6. He, Z., Wang, X.: Adaptive Parameter Estimation-Based Predictive Multi-Model Switching Control of Drainage Systems. In: Proceedings of the 6th World Congress on Intelligent Control and Automation, vol. (2), pp. 6540–6543 (2006)
7. Wang, X., Zhang, L.: The optimal control policy for Sewage pumping station system. Zhejiang University (Engineering Edition) 36(6), 718–721 (2002)
8. Shi, H., Zhou, X., Wang, X.: The control of the sewage pump stations based on genetic algorithm. China Electric Automation 25(3), 26–27 (2003)
9. Wang, X., Zhou, X.: The energy-saving mechanism and control strategy of Sewage pumping station system. Zhejiang University (Engineering Edition) 3(7), 1068–1071 (2005)
10. He, Z., Wang, K., Wang, X.: Optimal control mechanism and strategies for Sewage system. China Electrotechnical Society 21(5), 79–82, 117–121 (2006)
11. Wang, Y., Zhang, X.: Variable frequency speed PID Control Design for pumping station. China Transport 8, 172–173 (2008)

12. Wang, J., Wang, Y.: A fuzzy control application in reducing the overflow of sewage pipes. Hangzhou Dianzi University Academic 27(6), 79–81 (2007)
13. Zhe, X., Wu, L., et al.: The BP neural network prediction model of urban sewage pumping station. Control Engineering 17(4), 501–503 (2010)
14. Zhe, X., Wu, L.: The Forecasting Model of Urban Pipe Network Convergence Node Pump Station. In: 2010 Second Pacific-Asia Conference on Konwledge Engineering and Software Engineering, ChongQing, pp. 156–159 (2010)

# The Application of Switched Reluctance Motor in Electric Screw Press

Zhang Fangyang and Ruan Feng

School of Mechanical & Automotive Engineering, South China University of Technology
Guangzhou 510640, China

**Abstract.** With computer technology, power electronics, digital control technology, a new type of power system - Switched Reluctance Motor Drives (Switched Reluctance Driver, referred to as SRD) came into being, it is based on modern power electronics and computer CNC technology-based products, mechanical and electrical integration, because of its unique properties make it ideal for press the power system, to solve the problems of existing power systems to improve press performance.

**Keywords:** Switched Reluctance, Electric Screw, Motor.

## 1 Introduction

Widely used at home and abroad are two types of presses, one is screw press, and the other is the mechanical press. Due to technical limitations, most of the existing AC power system presses induction motor, speed requirements for special occasions, models, or use AC variable frequency or DC speed control. Need for high-end home and abroad in recent years, some models use AC servo motor drive.

The press to bear short-term peak load, which presses the load has a strong impact, in a complete working cycle in a short period of time only to work load, and a longer running time is an empty process; also due to the workload of the general press in about 60% of design value, the use of AC induction machine makes presses in the "big horse cart" working condition; and then load the motor in a long time, and the average no-load power factor of motor 0.3, so the energy~0.2 consumption of existing large presses. The use of frequency control, the inverter control difficult, narrow speed range, making its application is much more restricted; AC servo motor can certainly achieve the purpose of digital control, but there is high cost, technology by foreign monopoly.

## 2 Switched Reluctance Motor in the Application of Electric Screw Press

In the press, to make use of SRM drive system to replace the existing system, the need for different types of equipment to improve existing structures and can be used, including screw presses, servo press, speed presses, hot forging press various presses.

CNC screw press with switched reluctance electric screw press structure, can replace the existing friction press, clutch screw press and other models, the whole structure of the switched reluctance motor, belt drive (or gear drive), the flywheel (large drive

S. Lin and X. Huang (Eds.): CSEE 2011, Part II, CCIS 215, pp. 267–271, 2011.

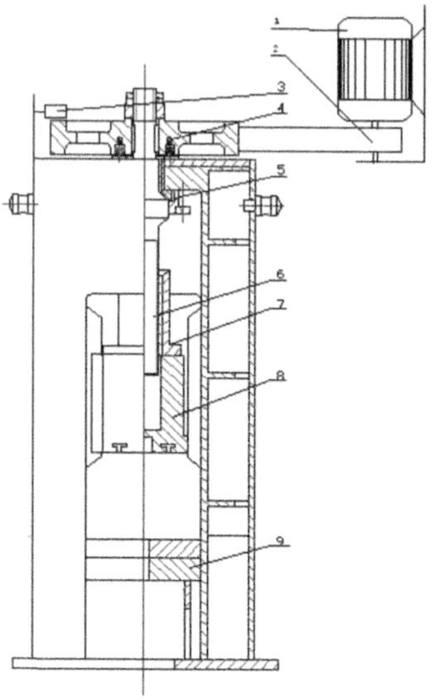


**Fig. 1.** CNC screw press Switched Reluctance

1 - Switched Reluctance Motor 2 - belt drive 3 - Brake 4 - flywheel 5 - thrust bearing 6 - Screw 7 - Nut 8 - slide 9 – Body

wheels), screw, nut, slider, brake and body composition, shown in Figure 1. Switched reluctance motors can be due to frequent starts and stops, and reversing, fast starts and stops, and the starting torque, starting current, friction press to remove the double friction disc, clutch screw press can remove the clutch and enhance the institution.

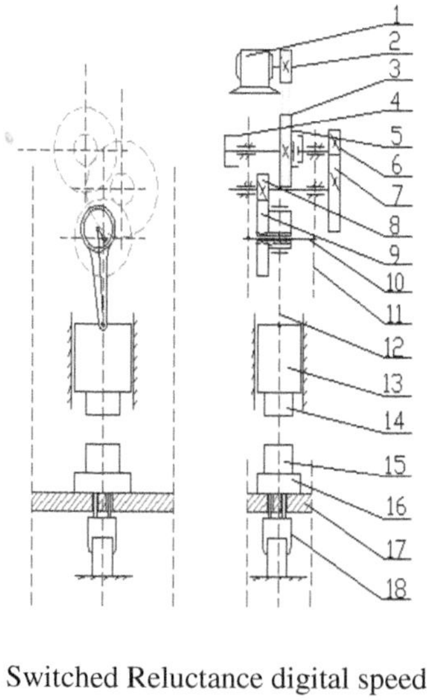


**Fig. 2.** Switched Reluctance digital speed presses

Switched Reluctance digital speed press is a mechanical press based on the existing, arising after replacement of power system. The whole structure of the switched reluctance motor - belt driven - flywheel (big pulley) - Clutch Brake - Gear - executing agencies shown in Figure 2. This model is characterized with a lutch and flywheel.

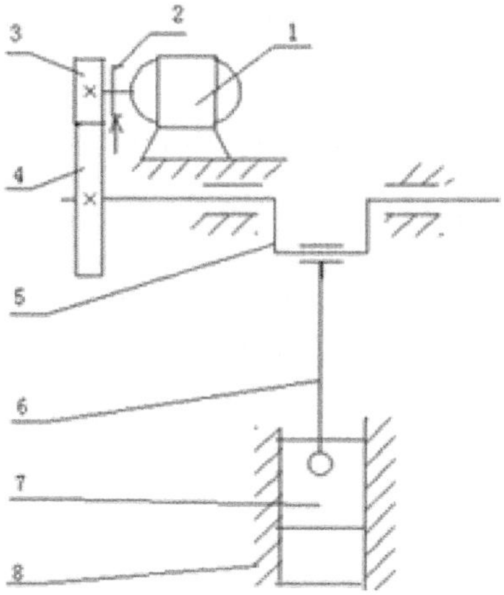


**Fig. 3.** CNC servo press Switched Reluctance

1 - Switched Reluctance Motor 2 - Brake 3 - pinions 4 - gear
5 - Crank 6 - Link 7 - slide 8 – Body

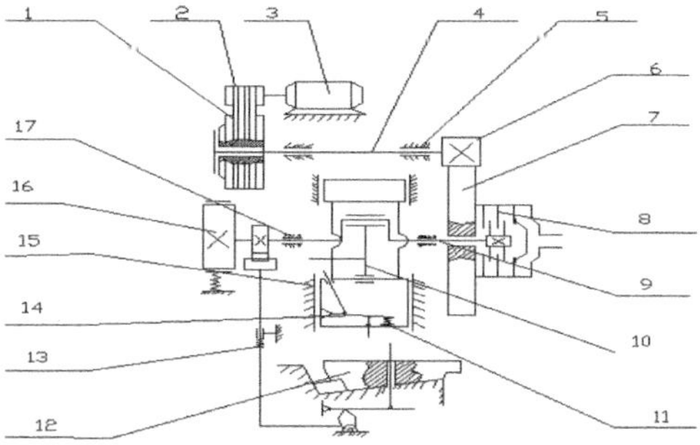


**Fig. 4.** SRM CNC hot forging press

1 - large pulley 2 - small pulley 3 - SRM 4 - shaft 5 - bearing 6 - Small Gear 7 - Big Gear 8 - Clutch 9 - the eccentric shaft 10 - link 11 - slide 12 - wedge work Taiwan 13 - pieces of equipment under the top 14 - on top of pieces of equipment 15 - track 16 - Brakes 17 – Bearings

Switched Reluctance CNC servo press mechanical press, compared with the existing, relatively large structural change, there is no belt drive and clutch, the motor shaft set the brake. The whole structure: switched reluctance motor + + first or second gear actuator, shown in Figure 3. Downhill for switched reluctance motor is much larger than the existing motor, the moment of inertia can be reduced to shorten the time from the stop, this press can be no larger pulley (flywheel), clutch, etc., to streamline its original structure.

CNC hot forging press reluctance is based on the existing hot forging press, the replacement of power generated from the system. The whole structure of the switched reluctance motor + belt + clutch, brake, crank slider executive bodies, as shown in Figure 4.

## 3 Switched Reluctance Features CNC Presses

Switched Reluctance NC press Memec at Shandong Electric Co., Ltd., Zibo Group traction motor, Qingdao YIYOU Metal Forming Machinery Co., Ltd. Rongcheng Dong source Forming Machinery Co., Ltd., Hebei Tianchen Forging Machine Co., Ltd., Hebei Cangzhou Machinery Plant, Tianjin Hao Ding Precision Machinery Technology Co., Ltd., a unit test run, some models have been put into mass production and received good market response, and achieved good social and economic benefits.

Intelligent NC.Different processes, different materials can be stored on different computers is called the working curve, so press the processing performance, processing range greatly. Technology curve, time, yield, power consumption and other data stored and processed by the computer intelligent, easy to implement manufacturing process information management goals.

Normal shutdown, to ensure that the only point on the slide stop.Position sensor feedback position signal, the stop before the motor reducer; When the slider at the top dead point, the motor shaft of the brake start, make sure the slider stopped at the only point. Clutch model can effectively reduce the clutch, brake friction heat.

Safe.In the motor brake (rotor locking feature can be added) and under the brake, the brake block to stop at any position.

## 4 Conclusions

Reluctance by the NC press drive switched reluctance motor drive system, it features high efficiency, energy saving effect, wide speed range, without starting the impact of current, starting torque, a flexible digital control, in addition, also has simple, rugged, reliable, low cost. Tested by the authoritative department, the use of the switched reluctance motor screw press can save energy up to 29.3%.

Application shows that compared with the existing presses, CNC press the advantages of switched reluctance to CNC, speed, speed, transfer energy, energy, intelligence, and high reliability. Therefore, the switched reluctance NC press will be to gradually replace the existing presses, as Press and updating the product of choice.

**Acknowledgment.** The work is supported by Science and Technology Project of Guangdong province of China under Grant Nos. 2008A090300008.

## References


1. Mehta, P.K.: High Performance Concrete for the Future. In: Proceedings International Congresson High Performance Concrete, Brazil University of Santa Catarina, pp. 225–242 (1996)
2. Mehta, P.K.: Durability-critical Issues for the Future. Concrete International, 27–33 (1997)
3. Mindess, S., Young, J.F., Fang, Q., et al.: Translation. Concrete. China Building Materials Industry Press, Beijing (1989)
4. Larrard, F.: A Survey of Recent Researches Performed in the French 'LCPC' (Laboratorial Central desponts of Chaussées) Network on High Performance Concrete. In: Proceedings of High Strength Concrete, Lillehammer, pp. 20–24 (1993)
5. Zhao, T.-j., Li, Q.-y.: High strength and high performance concrete and its application. China Building Materials Industry Press, Beijing (2004)
6. Li, J.-y., Jiang, J.-m., Ge, Z.-s.: New Concrete Technology of Special Features. Chemical Industry Press, Beijing (2007)
7. Wang, D.-m.: High-performance expansive concrete. China Water Power Press, Beijing (2006)

# Software Defects Detecting Method Based on Data Mining

Peng Yang

School of Information Engineering, Guangzhou Panyu Polytechnic, Guangzhou, China
citystars@163.com

**Abstract.** In this paper, we have presented a data mining based defect detection method to detect two kinds of defects from source code: rule-violating defects and copy-paste related defects. Even though the kinds of defects are different, it can identify the defects in one process. Therefore, this method greatly reduces time cost for software testing. In addition, this method is able to mine implicit programming rules and copy-paste segments in software requiring little prior knowledge from programmers. Novel techniques have been also proposed to prune false positives in defect detecting process. Based on the experimental result, the pruning techniques can greatly reduce the false positives and save much effort of manually checking violations.

**Keywords:** terms-software defect detection, data mining, false positive.

## 1 Introduction

Some violations of implicit programming rules are evident. For example, function pair lock and unlock must be used together, that is, a call to lock should be followed by a call to unlock. However, many implicit programming rules are not so evident. For example, in the Linux source code in Fig. 1, a programming rule consists of 4 functions: prepare_to_wait, schedule, signal_pending and finish_wait and a variable wait.

```
Linux2.6.16.2/net/atm/Common.c:
00515    int vcc_sendmsg(...)
00516    {
         ......
00559     prepare_to_wait (..., &wait, ...):
         ......
00566     schedule()
         ......
00567     if (signal_pending(...)) {
         ......
00580     finish_wait (..., & wait);
         ......
00596    }
```

**Fig. 1.** A complex programming rule consisting of four function calls and a variable in the Linux source code

S. Lin and X. Huang (Eds.): CSEE 2011, Part II, CCIS 215, pp. 272–278, 2011.

Such rules are tedious and complicated, especially in large-scale software. Many programmers are reluctant to record such rules in their documents, so that the subsequent programmers are likely to bring defects into the software as they are not aware of the programming rules. Fig. 2(a) shows a programming rule consisting of 3 function calls: IS_ERR, mutex_lock and mutex_unlock, which occurs 43 times in the Linux source code. It is violated only once as shown in Fig. 2(b).

```
Linux2.6.16.2
/drivers/infiniband/core/uverbs_cmd.c:

00325  ssize_t ib_uverbs_post_recv (...)
00326  {
        ......
01341   if (IS_ERR(...))
        ......
01344    mutex_lock (...)
        ......
01365    mutex_unlock(...)
        ......
01374  }
```

**Fig. 2 (a).** A programming rule in function ib_uverbs_post_recv

```
Linux2.6.16.2
/fs/namei, c:

01739  dentry * lookup_create (...)
01740  {
        ......
01743   mutex_lock(...)
        ......
01756   if (IS_ERR (...))
        ......
        // Missing mutex_unlock (...)
01773  }
```

**Fig. 2 (b).** A code segment violating the rule: {IS_ERR, mutex_lock} => {mutex_unlock}

## 2 Related Work

Several methods were proposed to detect rule-violating defects. Engler, Chen and Chou [1] proposed a method to extract programming rules using programmer-specified rule templates, such as function A must be paired with function B. Since such template-based methods only cover the given or explicit rules, it may miss many violations due to the existence of implicit rules. David Evans et al. [2], proposed a tool, LCLint, which is able to report the inconsistency between the source code and the pre-defined specifications.

Locating the copy-paste code segment is the most important task of copy-paste defect detecting method. Some methods were proposed to handle this problem. Dup et al. [3] uses a sequence of lines as a representation of source code and detects line-by-line clones. This method has weakness in the line- structure modification. CCFinder [4] transforms the input source code into tokens and performs a token-by-token comparison. It is not scalable to large-scale software because it consumes large amount of computer memory to store the transformed text. Overall speaking, none of the methods is able to locate the copy-paste defects efficiently.

# 3 Defect Detecting Method

## 3.1 Data Generator

The task of data generator is to transform the source code into the appropriate form for the following processes. Fig. 3 summarizes the flow of the data generation. GCC compiler [5] is used to convert each function in the source code into a set of items. The dump files are obtained by adding -fdump-tree-original option to GCC compiler when compiling the source code. The dump files suffixed with original contain the intermediate representation of the code. The intermediate representation is stored in a tree structure. An abstract tree represents a function, where the leaves and the internal nodes denote different types of elements in the function. Next, we traverse each abstract tree to select the items, such as identifier names, data type specifiers, operators, and control structures. In this way, each function is converted to an itemset.

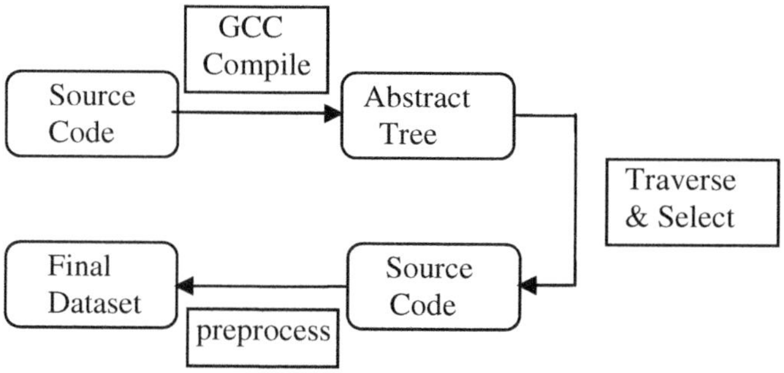


**Fig. 3.** Flowchart of the data generator

## 3.2 Rule-Violating Defect Detecting

1) Frequent Itemsets Mining

Apply Apriori on the preprocessed dataset to mine the frequent itemsets. Any itemset that occurs more frequently than a user-specified threshold will be considered as a frequent itemset. Items in a frequent pattern are highly likely to have some correlation in between. In addition, the location information (where the items locate in the source code) of each frequent itemsets is recorded correspondingly, which is required when evaluating if a violation is a true defect or a false positive.

2) Rule-violating Defect Detecting

In order to measure how seriously a programming rule, $I_{k-1} \Rightarrow I_1$, is violated, we propose a novel term, vio_prob, which is defined as the following:

$$vio\_prob\Big|_{I_{k-1} \Rightarrow I_1} = 1 - \frac{\sup port(I_{k-1} - I_1)}{\sup port(I_{k-1})} \tag{1}$$

where $I_{k-1} \Rightarrow I_1$ is a programming rule generated in Section.2), support $(I_{k-1} - I_1)$ denotes the number of occurrences of the itemset $I_{k-1}$ without $I_1$ in the source code, support ($I_{k-1}$) denotes the number of occurrences of the itemset $I_{k-1}$ in the source code. The higher the vio_prob, the more likely the violation is a defect. Assume t is the minimum vio_prob threshold. If the vio_prob of a violation is higher than but less than 1, it is viewed as a potential defect. Note that if vio_prob=1, the violation is not a defect since it does not happen in the source code at all.

3) False Positive Pruning

No matter how high the vio_prob of $I_{k-1} \Rightarrow I_1$ is, the violations are not necessarily true defects. We will handle three types of false positives (shown in Fig. 4).

Type I: A function F violates a rule by missing calling function A, but it calls function B that calls function A inside. It is not a defect because of the nested calls of functions.

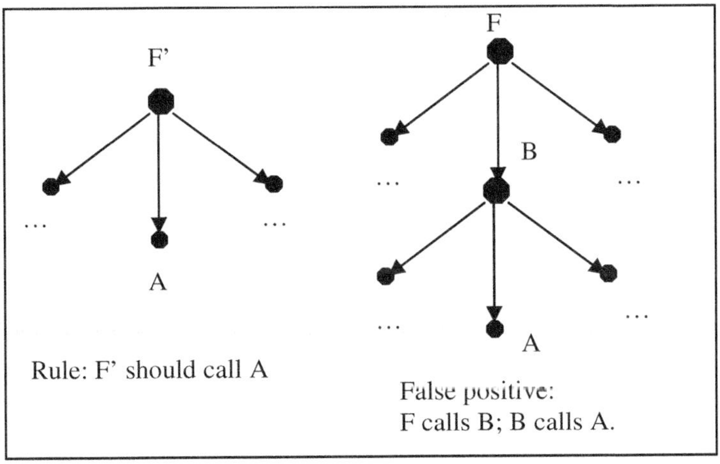


**Fig. 4 (a).** Three types of false positives in rule-violating defect detecting: Type I

Type II: Assuming function pair <A, B> is usually called together, function F calls function A but misses function B. Sometimes it's not a defect because F shares the same functionality with A and has an opposite functionality with B. For example, a function pair <_spin_lock_bh,_spin_unlock_bh > is often called in Linux kernel. However, function ax25_info_start only calls _spin_lock_bh and function ax25_info_stop only calls _spin_unlock_bh. This is because "start" is corresponding to "lock" and "stop" is corresponding to "unlock".

Type III: Like function pairs, struct member pairs can also cause the same false positives. For example, a struct member pair < charset2upper, charset2lower > is often used together in Linux kernel. However, function nls_toupper only uses charset2upper and function nls_tolower only uses charset2lower.

We call the above violations as false positives. In order to prune the false positives from the rule-violating defects detected in the last section, we propose the following methods:

Type I: Prune the Type I false positives by checking the calling paths of the suspicious functions. In the above example, if A is one of the functions called by F and A is called in any function in the path, we prune it as a false positive.

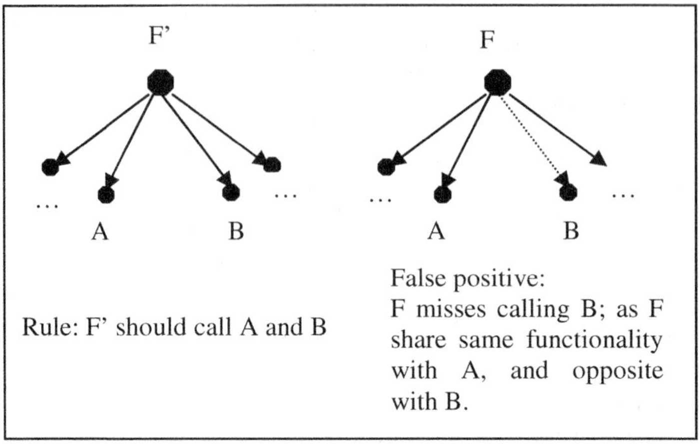


**Fig. 4 (b).** Three types of false positives in rule-violating defect detecting: Type II and III

Type II and III: Since the functions usually emerge in pair, we separate the keywords in each pair into two opposite groups. For example, one group contains keywords like "start", "enter", "acquire", "lock", "up", "read", "begin", etc., and the other group contains "stop", "exit", "release", "unlock", "low", "write", "end", etc. If function F fails to call function A, we will check F and A against the words in the two groups.

## 4 Experimental Results

### 4.1 Experiment Setup

We evaluate our proposed system on the Linux source code. The statistics of the data is shown in Table 1.

**Table 1.** Statistics of Linux Source Code

| Source code | Version | Number of C files | Lines of code | Number of functions |
|---|---|---|---|---|
| Linux | 2.6.16.2 | 6,195 | 4,843,927 | 70,359 |

### 4.2 Result

Throughout our experiments, we vary the following three parameters:

(1) min_support: the number of occurrences of a given itemset in the source code;
(2) vio_prob: the likelihood a violation is a defect;
(3) max_calling_path_depth: the max number of function calls on a calling path.

Observed from our experiments, if min_support is lower than 28, the number of frequent patterns is too many to present; if vio_prob is lower than 95%, the number of false positives is also too many to present. Thus, we only present our results with min_support at 28, vio_prob at 95% and max calling_path_depth at 3, respectively.

Fig. 5 presents the distribution of the frequent item sets discovered from the source code. Totally, 138,273 frequent item sets are obtained. Then, 515 implicit programming rules (with confidence larger than 98%) are generated out of them. These rules can be classified into three categories: function with function (F-F), function with variable (F-V), and variable with variable (V-V). Table 2 shows the distribution of the rules.

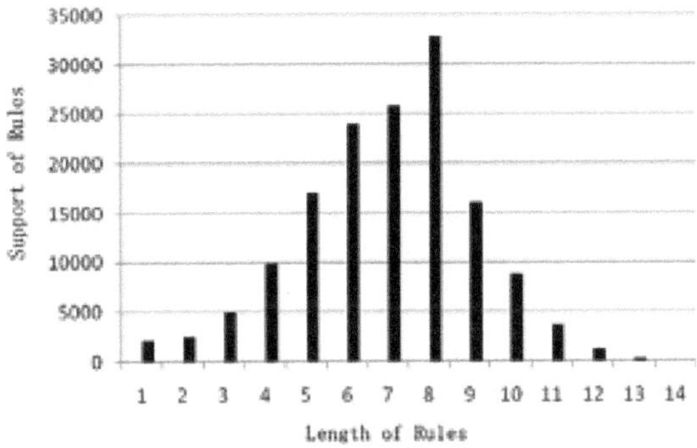


**Fig. 5.** Support of each length frequent pattern

**Table 2.** Distribution of Rules

| Source Code | Total number of Rules | Number of F-F Rules | Number of V-V Rules | Number of F-V Rules |
|---|---|---|---|---|
| Linux | 515 | 276 | 56 | 183 |

Specifically, the rest 23 top violations contain 11 bugs and 12 specifications. The detail of our inspection result is shown in Table 3. As far as our knowledge concerns, these defects are not reported by other existing methods. Actually, we have reported the defects to the Linux community.

**Table 3.** Inspection Result

| Source Code | Inspected (top 50) | | | Uninspected |
|---|---|---|---|---|
| | Bug | Specification | False Positives | |
| Linux | 11 | 12 | 27 | 983 |

## 5 Conclusions

We have evaluated the proposed method with Linux source code which contains more than 4 million lines of code. In the case study, it discovered more than 138 thousand frequent patterns that could be programming rules with high confidence. With vio_prob larger than a specified threshold, we have obtained 515 rules. According to these rules, the method detected 1033 violation code segments. Among the top 50 violations, 11 defects have been confirmed. Most of these defects involve two or more elements and are hard to be detected by other known methods. Due to the efficient data mining techniques, the method is promise to process large-scale source code in a reasonable time. There are many research problems remaining to be explored.

## References

1. Engler, D., Chen, D.Y., Chou, A.: Bugs as inconsistent behavior: A general approach to inferring errors in systems code. In: Proc. of the 18th ACM Symp. on Operating Systems Principles (2001)
2. Evans, D., Guttag, J., Horning, J., Tan, Y.M.: LCLint: A tool for using specifications to check code. In: Proc. of the ACM SIGSOFT 1994 Symp. on the Foundations of Software Engineering (1994)
3. Baker, B.S.: On finding duplication and near-duplication in large software systems. In: Proc. of the Second Working Conference on Reverse Engineering, p. 86. IEEE Computer Society, Los Alamitos (1995)
4. Kamiya, T., Kusumoto, S., Inoue, K.: CCFinder: a multilinguistic token-based code clone detection system for large scale source code. IEEE Transactions on Software Engineering 28(7), 654–670 (2002)
5. Stallman, R.M., The GCC Developer Community: GNU compiler collection internals (GCC)

# A Mechanism of Resource Co-reservation in Large Distributed System Based on Virtual Resource

Jing Li and Qiyun Han

Department of Computer Science, Chongqing Education College,
Chongqing 400067, China

**Abstract.** Resource reservation is an important part of the resource management in Large Distributed System such as grid. An architectural framework for specifying and processing co-reservations in grid environments is presented. Compared to other approaches, the co-reservation framework in this paper is more general. The concept of Virtual Resources is introduced that allows to compose resources by abstracting from their specific features. Virtual Resources enable advanced co-reservation with nested resource levels, resource aggregation, and transparent fault recovery.

**Keywords:** grid, resource reservation, Virtual Resource, QoS.

## 1 Introduction

In order to guarantee the quality of service(QoS) in Large Distributed System such as grid, resource reservation is a very effective approach[1]. The reservation requests of user generally include starting time, insisting time, resource types and reservation types, etc. Because the job on the grid is usually complicated, involving a lot of resource, when the resource is reserved, we can reserve more than one resource, which is called resource co-reservation. For example, interactive data analysis needs to access storage system possessing data replicas, make use of super computer to analyze data, make use of network to transfer data, and meanwhile utilize visualization equipment to communicate between human and machine. Ever resource in the above example requires quality of service. For the sake of that, resource co-reservation is a feasible approach.

Document[2] show that advance reservation may lead to resource fragment and the drop of network performance, putting forward a flexible reservation mechanism, can improve network performance effectively. The document compared four flexible reservation approaches, and the result shows, that some approaches add resource fragment, and some improve network performance. Despite of discussing performance issue, the document provides else the software framework of advance reservation management system.

Globus Architecture Reservation and Allocation(GARA) [3] is the first architecture that want to confirm the Grid QoS by means of resource reservation. It provides uniformed API of preserve process on various resource for users and applications.

S. Lin and X. Huang (Eds.): CSEE 2011, Part II, CCIS 215, pp. 279–283, 2011.

Existing resource reservation approaches focus on the stability of reserved resource, but in the large distributed environment, resource is highly dynamic, which is the inherent character, and so that the above mentioned methods of resource reservation are limited.

This paper combines the resource reservation mechanism with the service-oriented framework, providing a new resource co-reservation mechanism in the large distributed environment, which constructs a visual resource model, shielding the dynamicity of resource, having the property of fault tolerance.

# 2 Visual Resource Model

## 2.1 Hierarchical Model of Visual Resource

Existing resource reservation approaches have definitely defined that what is the resource, and who provides the resource, and who uses the resource, and when sharing the resource. While virtual resource shielding the resource provider and the physical resource itself, is an service-oriented resource reservation architecture.

In grid applications, the construction of VR is transient, having specific demand and target, that is to say, constructing a VR aimed to some user's once resource reservation request. In a lot of cases, VR can be considered as the collection of resouce which is geographic distributed, logically adjacent, and fuction complemental, showing the external characters of standard service interface.

In order to realize resource sharing and coordinate work in distributed heteroganeous network, we must solve the interacitve operation issues. In other grid resource reserve mechanism, the goal is achieved by standarliztion of protocol architecture and realization of API and SDK.

(1) It can realize the abstraction of specific resource, covering the difference of resource's realiation means.

(2) Different services with standard interface can be integrated to form higher services, and higher services themselves also have standard interface and call specification.

(3) It can realze consistent resource reservaion appoach crossover heterogeneous platforms.

(4) It can realize the seperation of logical service instance and phsical resource.

(5) It can map the service to the local running platform, realizing managing the remote services by local running platform.

Besides that, WSDL realize standard denotation of grid service, considering platform's heterogenouce not any more. This method adopts the concept of encapsulation, and it is called Virtualization.

We provide the Visual Resource model is Hierarchical, which is shown an Figure1.

In order to achieve the solution of resource sharing and co-reservation in dynamic and muti-organization system, the resource reservation mechanism must be able to provide services on the base of a variety of local resource, and so that the architecture should firstly cover the difference of various resource, then realize the reservation of any single resource. On that basis, mutiple resources co-reservation is accomplished, and at last many kinds of applications with a virtual resouce are provided. These

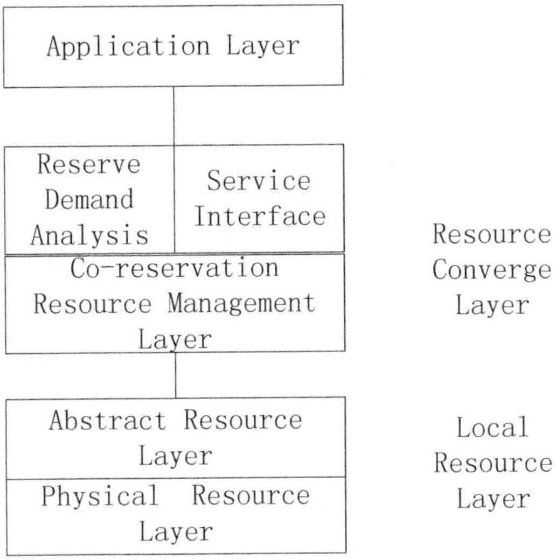


**Fig. 1.** Visual resource hierarchical model

fuctions are brought into effect in sequence by the lays in the architeture respectively. The fuctions of the layers are described as follows.

Local resource layer: including physical resource layer and abstract resource layer. The physical resource layer realizes the access and control of local specific resource, providing resource reservation and sharing-access interface for higher layer. The abstract resource layer realize reservation of single resource, and the protocols are including security negotiation, initialization, detection, control, fee counting, and fee paying.

Resource converge layer: including resource co-reservation management layers, reserved resouce analysis and service interfaces, providing analysis of users' reservation demand, detection of various resource's satate, queue of backing resource, and standard service interfaces of various resource's co-reservation.

Application layer: including application programes of resource reservation in the system, by means of calling underlying API to get coresponding services.

Life cycle of visual resource starts with reserve request provided by specific application, then resource converge layer processes co-reserve of muti-resource, and it does end with the close of specific application. VR shows stable external property, being able to give assurance of quality to specific application.At the same time, it is dynamic inside, not limited by fixed resource provider, being able to combine new resource again continuously according to detection the state of various resource and queue of backing resource, giving up the old depressed resource, getting a dynamic banlance, which is fitting in with the highly dynamics of grid. VR takes on the characteristics of fault tolerance and load balance whose resource state detection mechanism can find the error resource in time taking place with new backing resource, and also can find heavy-loaded resource that will be repalced with light-loaded resource.

### 2.2 Definition of Resource Co-reservation

We define the resource co-reservation R according to following parameters:

--$t_s$ : Reserve start time

--$t_e$ : Reserve end time

--sc : Reserve service type

--$r_i$ : The type of various resource, for example, computing resource, network resource, or memory resource

--c ( r, t ) : Capability fuction, reture the capability provided by resource r at moment t

According to these concepts, we can define the demand of muti-resource co-reservation as follws:

$$R(t_s, t_e, sc, \{( r_1, c(r_1, t)), .., (r_n, c(r_n, t))\})$$

## 3 Simulations and Analysis

We used simulations of a network with a power law link distribution PLOD[4] to validate our analytical results. The prototype system is programmed by java, and tested on Linux.

On the aspect of fault tolerance, we compared the reservation mechanism on the base of VR and fixed resource. Suppose in the system the type of resource is 50, the total number of resource is from 64 to 4096, each resource being invalid with probability from 10% to 20%. Resource reservation experiments are done with group of different resource number. Every group is tested 10 times, and the failure probability is figured out. The result is shown as Figure 1. The abscissa is the number of resource, and the ordinate is the failure probability of resource reservation.

The experiment result shows that the fault tolerance performance of resource preserve mechanism based on VR is better than based on fixed resource. The more amount of resource

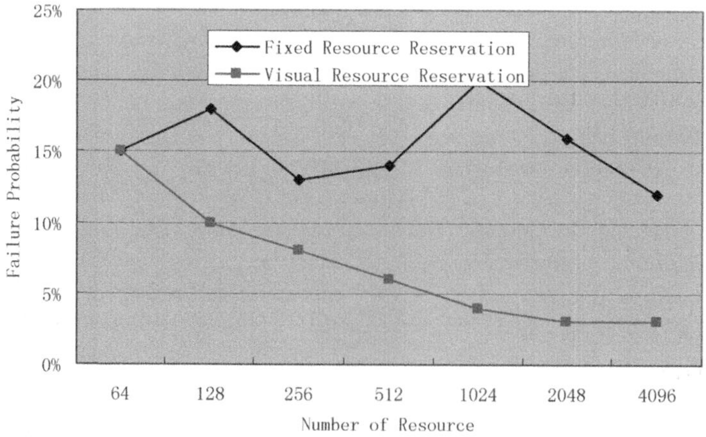


**Fig. 2.** Comparison fault tolerance of VR mechanism and fixed resouce

## 4 Conclusion

In grid, existing resource reservation mechanism is not flexible. We propose a mechanism of resource co-reservation based on visual resource model. In this model, we preserve resource from service-oriented perspective, providing service quality assurance for service consumers. Visual resource having the property of stability, dynamic and user-transparent fault tolerance, is a feasible resource reservation mechanism.

## References

1. Foster, I., Kesselman, C., Lee, C., Lindell, R., Nahrstedt, K., Roy, A.: A Distributed Resource Management Architecture that Supports Advance Reservations and Co-Allocation. In: Proceedings of the International Workshop on Quality of Service (1999)
2. Burchard, L.-O., Heiss, H.-U., De Rose, C.: Performance Issues of Bandwidth Reservations for Grid Computing. In: Proceedings of 15th Symposium on Computer Architecture and High Performance Computing, pp. 82–90 (2003)
3. `http://www.globus.org`
4. Palmer, C., Steffan, J.: Generating network topologies that obey power law. In: Process of GLOBECOM, pp. 434–438. IEEE, San Francisco (2000)

# Characteristics of Risky Merging Driving Behavior on Urban Expressway

Yulong Pei, Weiwei Qi, and Chuanyun Fu

School of Transportation Science and Engineering, Harbin Institute of Technology, Harbin 150090, China

**Abstract.** According to statistical data, risky merging driving behavior has taken adverse impact on traffic safety, and it widely exists in the ramp connecting section of urban expressway. Risky merging driving behavior can be divided into three types considering the difference in merging speed of vehicle, and the statistical characteristics of risky merging are analyzed based on video data. Because different traffic flow parameters have different effect on risky merging, rate of risky merging will rises with the higher traffic volume and lower traffic speed of main road. Results show that the appropriate length of acceleration lane can reduce rate of risky merging, and the security risks on risky merging comes from traffic conflicts initiated by merging gap selecting of ramp vehicle. Furthermore, one times of the most dangerous directness type leads to 3.29 times of traffic conflicts.

**Keywords:** risky driving behavior, risky merging, type classification, statistical characteristics, video data.

## 1 Introduction

Risky driving behavior has become an important cause of vehicle injury on highway safety. The relationship between risky driving habits, prior traffic convictions and motor vehicle injury is examined in New Zealand, and different strategies are required for different high-risk groups (Blows, et al., 2005). Male and teen drivers reported engaging in risky driving behaviors more frequently than female and adult drivers (Rhodes, et al., 2011). Young and male riders were more likely to disobey traffic regulations in Taipei (Chang, et al., 2007). Risky driving is not an isolated behavior, and it always happens with other bad habits for boys or girls (Bina, et al., 2006). The risky merging is an important aspect of risky driving, and the extended acceleration lane could facilitate merging (Waard, et al., 2009). The merging model and travel distance model of ramp vehicles is set up with differential method (Li, et al., 2002). Characteristics and capacities of the on-ramp merging region were analyzed to explore the influence of acceleration lane length (Zhi, et al., 2009). In order to reduce the risky diving behavior, the research can expand from the relative factors: traffic flow parameters and acceleration lane length.

## 2 Categories of Risky Merging Driving Behavior

The risky merging, which is defined as a driving behavior that drivers select the gap less than critical safe gap to merge into main road flow, always occurs on the ramp

S. Lin and X. Huang (Eds.): CSEE 2011, Part II, CCIS 215, pp. 284–290, 2011.

connecting section of urban expressway. The intersection region of ramp connecting section is shown in figure 1. The merging speed of ramp vehicle is related to the traffic security closely, so risky merging driving behavior can be divided into three types based on the merging speed of ramp vehicles.

① Directness type of risky merging driving behavior is that the vehicle from ramp merges into the main road directly without a process for deceleration.

② Buffer type of risky merging driving behavior is that the vehicle from ramp merges into the main road slowly with a process for deceleration.

③ Waiting type of risky merging driving behavior is that the vehicle from ramp merges into the main road very slowly with a process for decelerating to zero km/h.

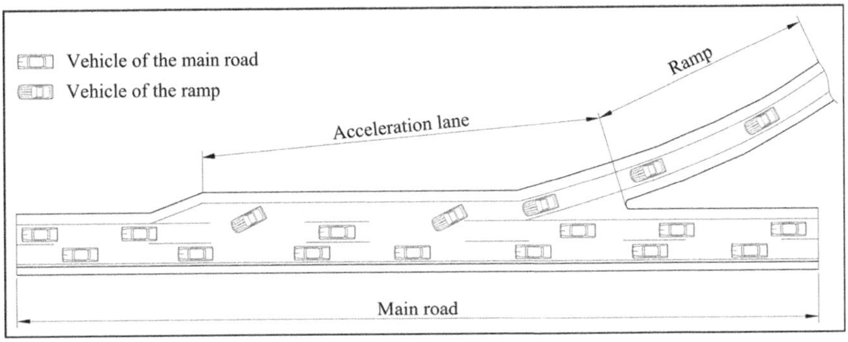


**Fig. 1.** Ramp connecting section on the urban expressway

## 3 Relative Factors of Risky Merging Driving Behavior

The relative factors of risky merging driving behavior can be divided into four aspects in the transportation systems, e.g. driver, vehicle, road and environment. This article focuses on the impact of risky merging driving behavior from the flow parameters and acceleration lane length.

### 3.1 Relationship between Traffic Flow Parameters and Risky Merging Driving Behavior

The experimental data have been got via video capture on traffic flow of the ramp connecting section in urban expressway. Data is divided into different types based on the different traffic volume and traffic speed in main road. Rate of risky merging driving behavior is analyzed by different types of main road traffic flow parameters. Further, the proportion of risky merging driving behavior is counted via directness type, buffer type and waiting type, and specific data is shown in table 1. The analysis result for statistical data about the risky driving behavior in table 1 is shown in figure 2 and figure 3.

**Table 1.** Statistical data of risky merging driving behavior based on traffic flow information

| type of risky merging behavior | | directness type (%) | buffer type (%) | waiting type (%) | rate of risky merging (%) |
|---|---|---|---|---|---|
| traffic volume of main road (pcu/h) | 300 < | 66.67 | 20.00 | 13.33 | 4.5 |
| | 300-600 | 47.62 | 28.57 | 23.81 | 6.3 |
| | 600-900 | 27.78 | 52.78 | 19.44 | 10.8 |
| | 900-1200 | 21.28 | 53.19 | 25.53 | 12.1 |
| traffic speed of main road (km/h) | < 40 | 35.71 | 53.57 | 10.71 | 9.6 |
| | 40-55 | 35.71 | 50.00 | 14.29 | 7.1 |
| | 55-70 | 30.77 | 55.38 | 13.85 | 5.4 |
| | 70-85 | 27.03 | 59.46 | 13.51 | 2.7 |
| | > 85 | 23.53 | 63.53 | 12.94 | 1.5 |

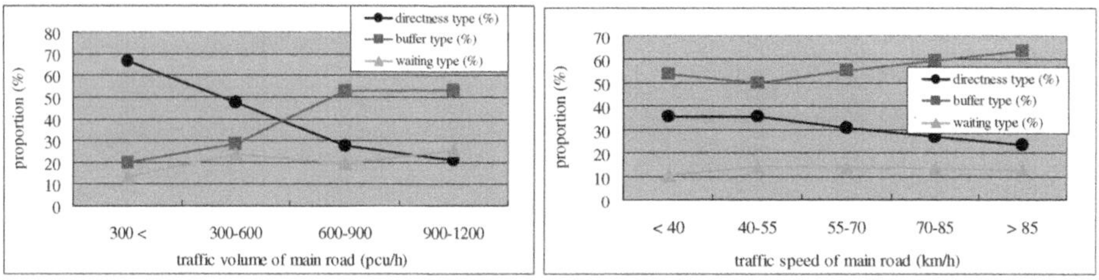


(a) Main road volume (b) Main road speed

**Fig. 2.** Proportion of risky merging under different traffic flow factors of main road

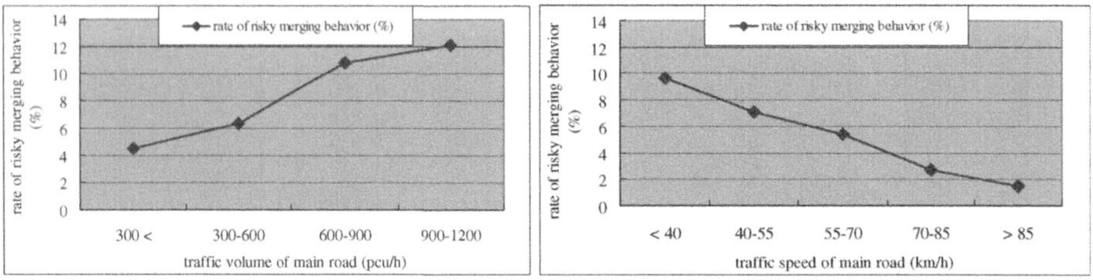


(a) Main road volume (b) Main road speed

**Fig. 3.** Rate of risky merging under different traffic flow factors of main road

The proportion for different types of risky merging driving behavior is shown in figure 2(a). We can get that proportion of directness type cuts down, proportion of buffer type rises, and proportion of waiting type is stable with the augmenting of main road traffic volume. The same conclusion can be got from the figure 2(b), in addition, the different point is that proportion of buffer type > directness type > waiting type, no matter how much the speed of the main road is. Figure 3 indicates that rate of risky merging driving behavior increases gradually with the adding of main road volume, but decreases with the adding of main road speed.

### 3.2 Relationship between Acceleration Lane Length and Risky Merging Driving Behavior

Acceleration lane makes the role of cushion when the vehicles merge into main road from ramp on urban expressway. Length of acceleration lane impacts on traffic flow characteristics directly, and it provides vehicles with space and position to accelerate to an appropriate speed. We divided acceleration lane length into five types based on characteristics of risky merging, and occurrence number of the directness type, buffer type and waiting type were calculated out respectively. Moreover, number of risky merging behavior and traffic conflicts is also shown in table 2.

**Table 2.** Statistical data about risky merging via different length of acceleration lane

| type of risky merging behavior | length of acceleration lane (m) | | | | |
|---|---|---|---|---|---|
| | 50-100 | 100-150 | 150-200 | 200-300 | >300 |
| directness type (times/hour) | 21 | 19 | 7 | 6 | 6 |
| buffer type (times/hour) | 25 | 23 | 19 | 15 | 13 |
| waiting type (times/hour) | 11 | 7 | 5 | 3 | 2 |
| total number of risky merging behavior | 57 | 49 | 31 | 24 | 21 |
| number of traffic conflicts | 91 | 78 | 46 | 36 | 32 |

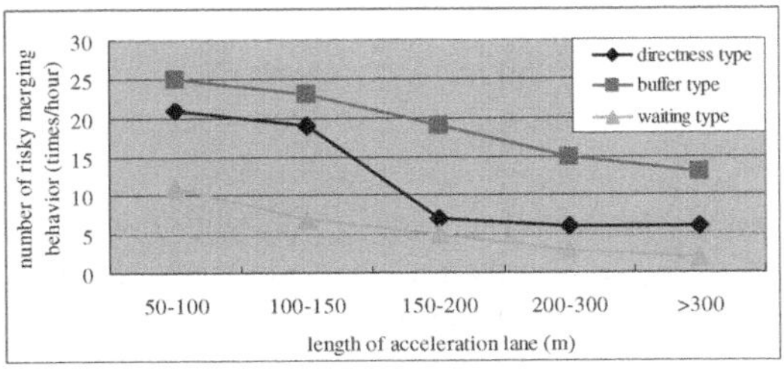


**Fig. 4.** Distribution of risky merging under different acceleration lane length

The figure 4 shows that number for three types of risky merging driving behaviors decreases with length of acceleration lane increasing. The directness type has a catastrophe point; but the buffer type and waiting type changes smoothly. The results show that sufficient length of the acceleration lane can reduce rate of risky merging driving behavior effectively.

The fitting curve between total number of risky merging behavior and length of acceleration lane shows in the figure 5(a), and the fitting curve between number of traffic conflicts and total number of risky merging behavior shows in the figure 5(b). The function model of the two fitting curve is as formula (1) and formula (2).

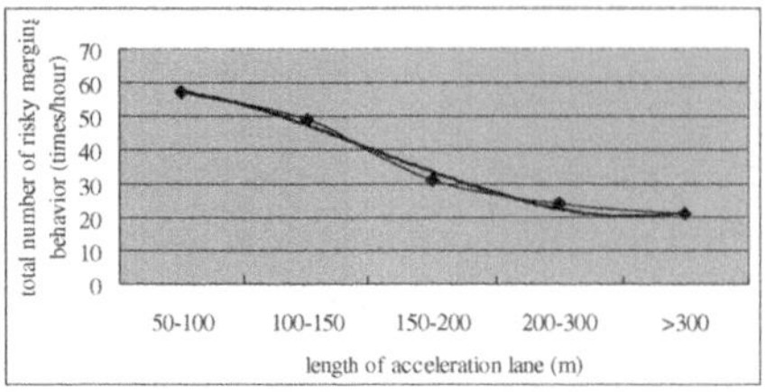


(a) Length of acceleration lane

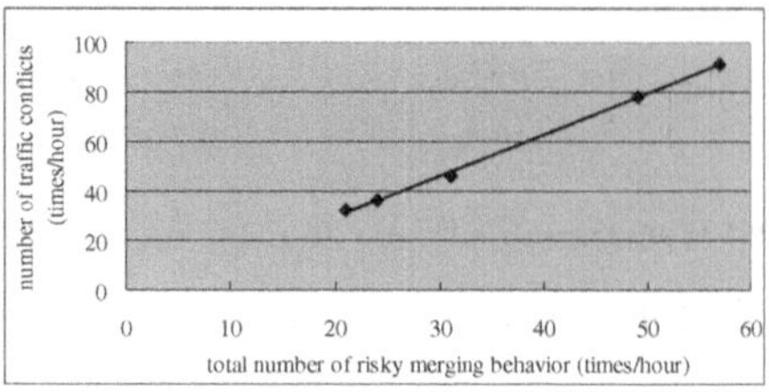


(b) Number of traffic conflicts and risky merging

**Fig. 5.** Fitting curve between total number of risky merging and length of acceleration lane

$$y = 1.1667x^3 - 9x^2 + 8.8333x + 56.4 \quad R^2 = 0.9888 \tag{1}$$

$$y = 1.6645x - 3.9868 \quad R^2 = 0.9986 \tag{2}$$

The research results shows that the connection between total number of risky merging behavior and length of acceleration lane is cubic function, and the connection between number of traffic conflicts and total number of risky merging behavior is linear function. Forever, the design length of acceleration lane is not less than about 300 meters to keep traffic safe.

## 4 Security Risks on Risky Merging Driving Behavior

The impacting degree on traffic safety for different risky merging driving behaviors is different via category index selecting, and the video data of different observation location is shown in table 3, which contains number of risky merging behavior and number of traffic conflicts (times per hours).

**Table 3.** Statistical data for risky merging behavior and traffic conflicts about several samples

| number of sample | number of risky merging behavior (times / hour) | | | number of traffic conflicts (times / hour) |
|---|---|---|---|---|
| | directness type | buffer type | waiting type | |
| 1 | 23 | 18 | 15 | 126 |
| 2 | 12 | 27 | 11 | 103 |
| 3 | 9 | 12 | 26 | 80 |
| 4 | 13 | 10 | 12 | 75 |
| 5 | 19 | 27 | 21 | 137 |

The argument of directness type, buffer type and waiting type is defined as $X_D$, $X_B$ and $X_W$; number of traffic conflicts is defined as $Y$. The relational model about traffic conflicts and risky merging is established from the data in table 3, which is shown as formula (3). Residuals and standard residuals of $Y$ are shown in table 4.

$$Y = 3.29X_D + 1.93X_B + 1.05X_W + 0.06 \tag{3}$$

**Table 4.** Residuals and standard residuals of the regression model

| predictive value ($Y$) | 126 | 103 | 80 | 75 | 137 |
|---|---|---|---|---|---|
| residuals | -0.2427 | -0.1535 | -0.1418 | 0.2536 | 0.2844 |
| standard residuals | -0.9752 | -0.6168 | -0.5697 | 1.0189 | 1.1427 |

The results show that the hazard of directness type is the highest-level via the regression coefficient of formula (3), and one times of directness type leads to 3.29 times of traffic conflicts; the hazard of buffer type is intermediate-level, and one times of buffer type leads to 1.93 times of traffic conflicts; the hazard of waiting type is the lowest-level, and one times of waiting type leads to 1.05 times of traffic conflicts.

## 5 Conclusions

The categories, factors and security of risky merging driving behavior are studied based on measured data, and the conclusions are as follows:

(1) The risky merging driving behavior is divided into directness type, buffer type and waiting type based on the merging speed of ramp vehicle;

(2) The traffic volume and speed of main road impacts risky merging driving behavior significantly, and rate of this driving behavior is in direct proportion with traffic volume, but inversely proportional to traffic speed;

(3) The length of acceleration lane has close relation with risky merging driving behavior, and an adequate length of acceleration lane can reduce rate of risky merging driving behavior efficiently;

(4) The risky merging driving behavior is significantly associated with traffic safety, and one times of risky merging driving behavior leads to about 2 times of traffic conflicts;

(5) The scope of practical application for the research conclusions is limited because of the restricted source for sample data.

**Acknowledgement.** This paper was supported by National Natural Science Foundation of China (51078113).

## References

1. Blows, S., Ameratunga, S., Ivers, R.Q., Lo, S.K., Norton, R.: Risky driving habits and motor vehicle driver injury. J. Accident Analysis and Prevention 37(4), 619–624 (2005)
2. Rhodes, N., Pivik, K.: Age and gender differences in risky driving: The roles of positive affect and risk perception. J. Accident Analysis and Prevention 43(3), 923–931 (2011)
3. Chang, H.-L., Yeh, T.-H.: Motorcyclist accident involvement by age, gender, and risky behaviors in Taipei, Taiwan. J. Transportation Research Part F 10(2), 109–122 (2007)
4. Bina, M., Graziano, F., Bonino, S.: Risky driving and lifestyles in adolescence. J. Accident Analysis and Prevention. 38(3), 472–481 (2006)

5. de Waard, D., Dijksterhuis, C., Brookhuis, K.A.: Merging into heavy motorway traffic by young and elderly drivers. J. Accident Analysis and Prevention 41(3), 588–597 (2009)
6. Li, W.-q., Wang, W.: Merging model of vehicle on freeway acceleration lane. J. China Journal of Highway and Transport 15(2), 95–98 (2002)
7. Zhi, Y.-f., Zhang, J., Shi, Z.-k.: Research on Design of Expressway Acceleration Lane Length and Merging Model of Vehicle. China Journal of Highway and Transport 22(2), 93–97 (2009)

# Research and Design of Direct Type Automobile Tire Burst Early-Warning System

Min Li, JiYin Zhao, XingWen Chen, and YaNing Yang

College of Information and Communication Engineering
Dalian Nationalities University, Dalian, China
limin@dlnu.edu.cn

**Abstract.** To promote the automobile driving safety, A tire burst early-warning system was designed using direct type technology. The paper described the mechanism of automobile tire burst and the principle of automobile tire burst early-warning system. The hardware implementations of the tire sensing module designed with the highly integrated chip MPXY8300 and the central control module based on microprocessor MC68HC908 and UHF transceiver MC33696 were described in detail, and the software flows of two modules and wireless data communications were given as well. The direct type automobile tire burst early-warning system designed in this paper has a high practical value.

**Keywords:** tire burst early-warning system, automobile tire, tire burst mechanism, wireless communication.

## 1 Introduction

Tires are the important part of automobile moving system and play the roles of supporting and running. Tire properties have direct effect on the automobile performance such as safety, smooth, comfort and transport efficiency. According to the joint statistics from the Ministry of Public Security and the Ministry of Transport of the People's Republic of China, 70% traffic accidents occurred in national highway in 2008 were caused by tire burst and the rollover death rate was 100% when any front tire burst at the speed of more than 120km/h [1]. The statistics from the Information Center of the Ministry of Transport of the People's Republic of China implies that tire bursting, fatigue driving and speeding are the three important reasons causing highway accidents, and tire bursting is the primary one because of its complexity and uncontrollability. The survey from the Society of Automotive Engineers showed that the number of traffic accidents caused by tire failure each year in the United States was more than 0.26 millions, of which the rate of the tire failure caused by under-inflation, leaks and too high temperature was 75% [2]. The traffic accidents caused by tire failures bring about huge losses to society, so how to prevent tire bursting has become an important research of automobile electronic technology. The tire burst early-warning system is a key element and component in the in-car electronic system, and one of essential functional elements for future intelligent vehicle. This new technology about the accident prevention out of tire burst prevention reveals a new concept of traffic safety precautions.

S. Lin and X. Huang (Eds.): CSEE 2011, Part II, CCIS 215, pp. 291–297, 2011.

At present, there are two main methods to implement the automobile tire burst early-warning system: direct and indirect. The indirect method, which appeared earlier, monitors the pressure by calculating the tire rolling radius and is used together with automobile Antilock Braking System (ABS) without requiring any additional hardware. Using wheel speed sensors, ABS measures the speed of each wheel. When the pressure of a tire reduces, the rolling radius decreases and the wheel speed increases correspondingly. The purpose of monitoring the tire pressure can be achieved by comparing the speed difference between tires and detecting the relative changes in pressure of the two tires [3-4]. But the indirect method has some limitations. First, the trouble tire can not be determined. Second, the system can not work in some cases, for instance, the pressure of coaxial tires or over two tires are both low, the speed is more than 100km/h and so on [5]. The direct method measures the pressure and temperature of each tire directly through the pressure sensors fitted in each wheel and transmits the detected data to main controller fixed in the cab for processing by means of wire or wireless. Most of the direct type system use wireless means to transmit data of pressure and temperature presently and the system structural framework is shown in Fig. 1. The direct method can measure the pressure of each tire directly and determine the trouble tire, so it becomes the main method of implementing automobile tire burst early-warning system. Starting from researching the mechanism of tire bursting, this paper designed a direct type automobile tire burst early-warning system. The system is scientific in tire burst early-warning threshold, low in power consumption, can directly measure the actual pressure and temperature inside tire real time, is easy to determine the trouble tire and has a high accuracy and reliability.

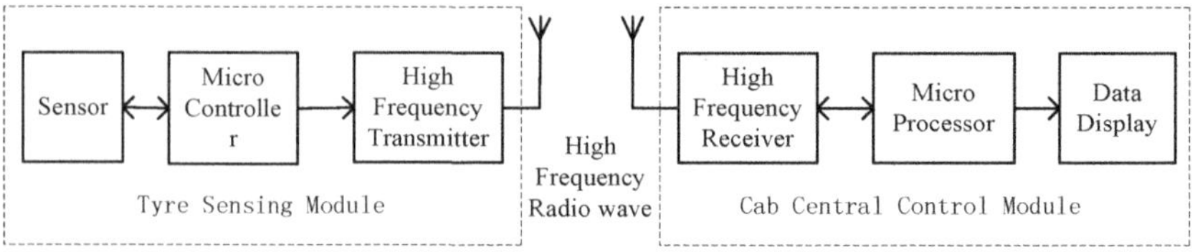


**Fig. 1.** Structural framework of the direct method

## 2 Mechanism of Automobile Tire Bursting

Factors that affect tire burst mainly include tire work pressure, load, speed, temperature, road conditions and chassis technical status and so on. Ultra-load and high speed can cause the tire temperature and pressure rise. Road conditions and chassis technical status directly affect the tire pressure. Tire deformation increases during low tire pressure. The friction coefficient doubles when ground area of tires increases. Tires in high-speed driving accumulate heat rapidly and the tire temperature rises sharply, leading to a burst. Traveling at low speed state, the tire damage due to too large tire deformation becomes larger, which brings hidden risks of tire burst at high speed. Meanwhile, insufficient pressure increases the tire sinkage obviously and driving resistance, reducing the tire life. The study results show that the tire pressure and temperature are the most critical factors of causing tire burst and the constraint relations between them uncertainty. So it is necessary to establish the tire burst modeling strategy which merges the characteristics of pressure gradient, temperature gradient and generalized gradient, using time series analysis to model the tire temperature and pressure. Considering the technical requirements of modeling accuracy and real-time, AR(p) model identification of tire bursting mechanism based on pressures and temperatures should be studied. In order to describe the relations between tire burst and tire temperature and pressure accurately and quantitatively, the automobile tire burst early-warning thresholds should be determined by using immune algorithm and genetic support vector machines, which is tire pressure and temperature thresholds of two-dimensional four values.

## 3 System Working Principle

The direct type automobile tire burst early-warning system was designed according to the structural framework shown in Fig. 1. The system includes four identical tire sensing modules and one cab central control module. In tire sensing modules, the sensor which uses MEMS miniature sensor can measure the pressure, temperature and acceleration inside tires; high-frequency emission part consists of RF mixing transmission circuit, matching network and antenna, realizes the function of data wireless transmission; microcontroller is the core of this module, it controls the sensor to measure the pressure and temperature inside tire, conveys the processed data to high-frequency part and transmits them through antenna. Since the system adopts the active direct method, there is a Lithium battery part in the tire sensing module to power the entire module. In the cab central control module, high-frequency receiving part consists of antenna, matching network and demodulation circuit and receives the pressure and temperature signals from the tire sensing module; data displaying part is composed of display circuits and alarming circuits and shows the data of pressures and temperatures computed and processed by the microprocessor; the microprocessor controls the high-frequency receiving part to receive the tire pressure and temperature signals, computes and processes them, controls data displaying part to display the data of pressure and temperature and realizes sound and light alarm according to the preset tire pressure and temperature early-warning thresholds.

## 4 System Hardware Implementation

### 4.1 Hardware Implementation of Tire Sensing Module

In the tire sensing module, the key component is the sensor responsible for detecting tire pressure and temperature, the core component is single chip microcomputer (MCU). The current sensors used in this area are Infineon SP series, Freescale MPXY series, GE NPX series and so on, as well as their constantly upgrading to improve the various performance. This paper used Freescale MPXY8300 chip shown in Fig. 2. This chip provides a highly integrated, cost-effective solution, with advanced pressure sensor, temperature sensor, acceleration sensor, voltage sensor, signal conditioning circuit, low-frequency receiver/decoder, RF transmitter and 8-bit microcontroller into one, only needs minimal external components, reduces system size, shortens the

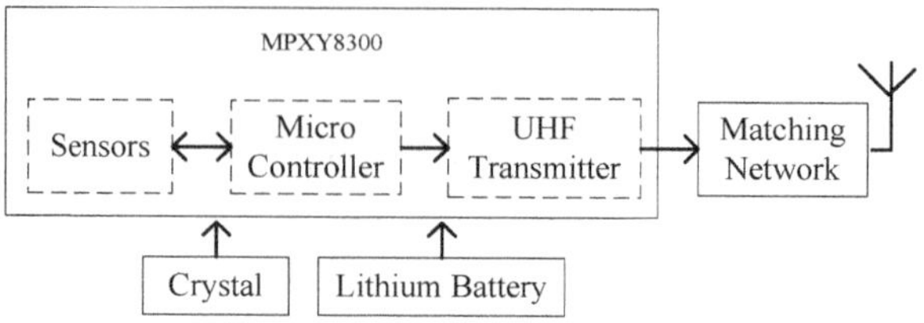


**Fig. 2.** Block diagram of tire sensing module hardware implementation

product design cycle and cuts down the prices greatly [6]. There is only need to design the crystal oscillator, matching network and antenna. The matching network and antenna design as PCB style, with the NX3445GA as the crystal oscillator to provide time base for MPXY8300 and CR2450 as the Lithium cell to power the entire system.

### 4.2 Hardware Implementation of Cab Central Control Module

Fig. 3 shows the block diagram of cab central control module hardware implementation. The system used Motorola chip MC68HC908 as the microprocessor and tuned UHF transceiver MC33696 as the main chip of the high-frequency receiving part. The chip MC33696 corresponds to the UHF transmitter integrated in chip MPXY8300 of the tire sensing module. The matching network and antenna were designed as PCB style. The LCD mode was used to display the data of tire pressures and temperatures as well as the position of the trouble tire. The buzzer and LED were selected for sound and light alarming.

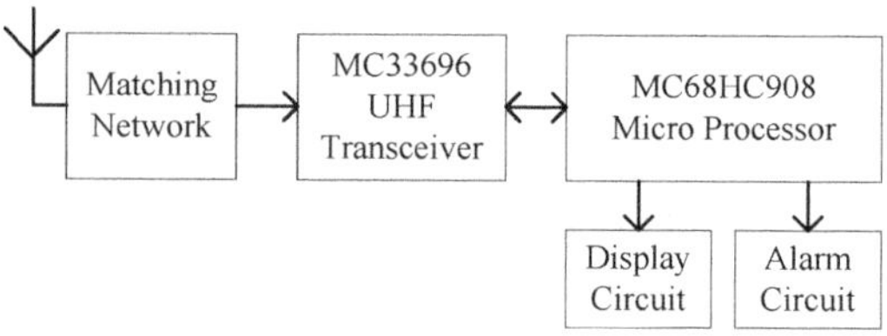


**Fig. 3.** Block diagram of cab central control module hardware implementation

## 5 System Software Design

According to the hardware module constituent, the software design of automobile tire burst early-warning system is also divided into the software design of tire sensing module and cab central control module. Some functions of data wireless transmitting and receiving involved in the two modules can be merged into the software design of wireless communication module.

### 5.1 Software Design of Tire Sensing Module

The main work of tire sensing module program is the system initialization, measure control, the simple processing of measured data and emission control. Fig. 4 shows the main program flow chart and the relevant interrupt service subroutines flow chart.

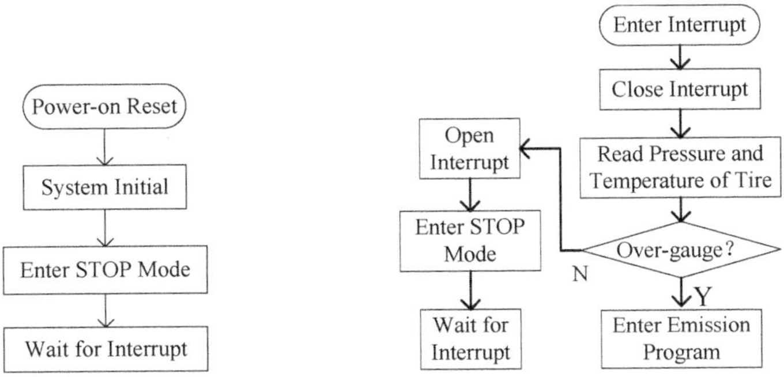


a. Main program flow chart b. Interrupt service subroutine flow chart

**Fig. 4.** Program flow chart of tire sensing module

Analyzing and processing data in tire sensing module mainly judge the data of tire pressure and temperature measured real time. The comprehensive processing of pressure and temperature is implemented by software of central control module. When the program checks an abnormal condition, the status information can be transmitted to central control module through transmitter wirelessly. The microcontroller is in STOP mode most of time and there is the minimum power consumption in this state. Because there are low power wake-up timer and periodic reset driven by LFO in the chip MPXY8300, the microcontroller can be out of STOP mode and starts measuring pressure and temperature immediately after waking up every time.

### 5.2 Software Design of Cab Central Control Module

The main functions of cab central control module program is initializing the central control module, controlling UHF receiver, further processing pressure and temperature information received, displaying data and executing alarm. Fig. 5 shows the main program flow chart and the software programs of displaying circuit and alarming circuit are not discussed here.

The microprocessor can calculate the check sum of the data frame again when receiving the data frame from UHF receiver MC33696 and compare it with the one received. After the frame check sum is verified correct, the tire ID inside the frame will be compared with four tire IDs stored in the memory of microprocessor. If the ID is found matching, the status information of this data frame should be stored in the RAM cell which is reserved for saving the status information of this position tire.

### 5.3 Software Design of Wireless Communication Module

A simple and reasonable communication protocol is needed for the combined program based on UHF transmitter and receiver. In this automobile tire burst early-warning system, data are transmitted at the rate of 9600bps using FSK modulated Manchester encoding. The data transmitted by tire sensing module are packaged as data frame. When the MCU of tire sensing module decides to transmit data frame, it can wake up the receiver, transmit the data and get back to dormancy state again. Table 1 shows the format of data frame [7].

**Table 1.** Format of data frame

| Prefix | Tire ID | Pressure | Temperature | State | Check sum | Stop |
|---|---|---|---|---|---|---|
| 16 bit | 32 bit | 8 bit | 8 bit | 8 bit | 8 bit | 2 bit |

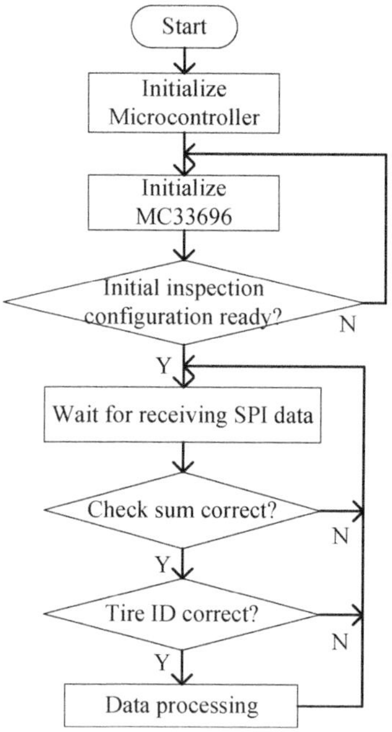


**Fig. 5.** program of central control module

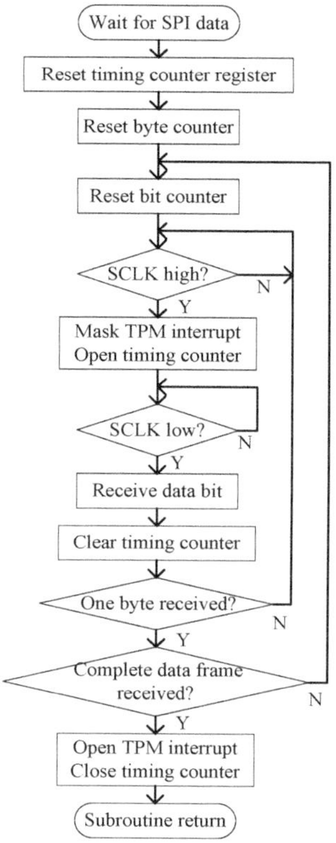


**Fig. 6.** Program of wireless communication module

Fig. 6 shows the program flow chart of data communication between microprocessor MC68HC908 and UHF transceiver MC33696 through the SPI serial protocol interfaces.

## 6 Conclusion

The automobile tire burst early-warning system will become one of the essential functions in automobile safe system. The direct type technique is the main implementing method. Using highly integrated chip MPXY8300, microprocessor chip MC68HC908 and UHF transceiver MC33696, this paper designed a direct type automobile tire burst early-warning system and gave the hardware implementation and software program flow. The system has advantages of high integration and stable performance, can monitor the hidden trouble of automobile tire bursting under driving condition and warns in time. The system has a long service life and adapts to a bad working environment. Application of this system can prevent the occurrence of tire bursting to some extent, so as to improve the safety of high-speed automobile and has a high practical value.

**Acknowledgments.** This paper is supported by "the Fundamental Research Funds for the Central Universities" under Grant No. DC10020117.

## References

1. Wang, G.: Simple Analysis of Tire Burst Early-warning System Technical Application of the Car. Journal of Changchun Institute of Technology(Natural Science Edition) 10(3), 109–112 (2009)
2. Zang, H.Q., Tian, C., Zhao, B.J.: Design of Tire Pressure Monitoring System Based on Embedded Operating System. Transactions of Beijing Institute of Technology 28(10), 870–874 (2008)
3. Zhang, S., Tian, L., Xia, X., et al.: Consideration of the TPMS Design Scheme. Electronic Technology 36(8), 64–65, 61 (2009)
4. Han, J.P., Tang, X.F.: Research on ABS and TPMS Merging System Based on Sliding Mode Variable Structure Control System. International Forum on Computer Science-Technology and Applications, 340–342 (2009)
5. Zhang, Y.H., Zhang, Z.H., Liu, L.T.: Research and Design of TPMS. Chinese Journal of Scientific Instrument 26(8), 441–442 (2005)
6. Lenzen, R.: MPXY8300 Design Reference Manual (Document Number:MPXY8300RM), Toulouse: Freescale Semiconductor (2008)
7. Li, M., Ma, B., Zhang, W.W.: Achieve of Tire Burst Early-warning System Wireless Data Transmission. Instrument Technique and Sensor, 99–101, 104 (2009)

# A Research on Tourism Service Quality: Measurement and Evaluation

Min Wei

The School of Management, Xiamen University
361005 Xiamen, China
xiada2000@126.com

**Abstract.** Tourism service quality in an organization is defined in terms of quality of excellence, quality of value, quality of conformity to some extents, and quality is viewed as a very important for meeting tourism customer expectations. Therefore, excellence in tourism service quality involved using service quality technology is becoming very important. The value of information systems can be realized by improving profit margins for the tourism enterprises, by which the useful applications and maintainable software can be provided. Information systems quality as conformance denotes designing the systems that conform to the end tourism consumers' information requirements. This paper tried to compare with the previous service studies and testify whether the tourism service quality evaluation model could capture the study of tourism service phenomena to find the importance of organizational impact to enhance tourism service quality.

**Keywords:** Information systems, tourism service quality, measurement, Evaluation.

## 1 Introduction

Tourism service quality can be understood defined by Liang (Liang, 2008). Due to the current number of tourism enterprises in the service of each quality system are lack of effective interface between subsystems mechanism, resulting in not to mention tourists for high-quality tourism products. In order to meet the tourism customer expectations, the quality of information systems is accomplished by offering appealing, user-friendly interfaces, entertaining user requests for change.

This study tried to compare with the previous service studies of tourism service phenomena to find the importance of organizational impact in terms of the information systems being applied to tourism service quality. In fact, the information systems applied to tourism service quality represents the quality of information processing itself and the impact of the organization, such as a tourism enterprise. In fact, the software for a tourism consumer is easy to learn, as well as, easy to be maintainable. Information quality, a new concept that is related to the quality of information system for tourism enterprises, can be described in terms of outputs that are useful for business users, relevant for decision making, and easy-to-understand, as well as outputs that meet users' information specifications. Tourism service quality is

S. Lin and X. Huang (Eds.): CSEE 2011, Part II, CCIS 215, pp. 298–304, 2011.

defined as the level of tourism service delivered by information systems, which provide tourism consumers in terms of reliability, responsiveness, assurance, and empathy. These concepts of information systems in terms of tourism service quality are reflected through information systems meeting tourism consumers' expectations.

## 2 Literature Reviews

Tourism service is a very important issue which depends on the aspects with intrinsic attributes of some activities related with this kind of service: activities that happen in tourists' tourism activity and tourism consumption, such as arriving at airport, traveling in scenic zone, experiencing some foods, and so on. In spite the kinds of tourism service, the quality is essential. Therefore, more and more researchers try to study the contents of tourism service quality. See Fig. 1.

**Fig. 1.** Research on tourism service quality

Total Quality Management influences organizational performance as quality of products and services has been found to be the most important factor determining businesses' long-term success (Anderson and Zeithaml, 1984). The productivity, service, knowledge, skills, self-awareness and corporate identity, as well as, the sense of responsibility of the employee in tourism enterprises can be improved(Tu, 2010). Total Quality Management uses a broad definition of quality. It is not only related with the final product and service, but also with organizations which were more important for enhancing the tourism service. So, how to quickly respond to customer complaints, and to provide for the customers to experience better after-sales service for themselves (Zhang, 2008), it should be more paid attention to. Information systems are designed and constructed accuracy method for the system development process (Geoffrey Wall,2005). Nevertheless, application of IS quality management techniques such as the Capability Maturity Model (CMM) has resulted in improved system development productivity. Organizations can achieve improvements in system quality, development cost, and project schedule with IS quality management practices. Some of these practices include institutionalization of quality management practices, senior management leadership, and establishment of performance standards for system development activities and experiences with the employee of tourism enterprises (Choy,D.J.L.,1995). However, an integrated approach to the application techniques of the IS context is lacking. In spite of their importance of the practice, the aspects of information systems quality have not been given adequate emphasis by IS researchers. Petter et al. (2008)believed that there is insufficient empirical evidence to evaluate most of the relationships at the organizational level.

The recent study analyzed the relationship between leadership, information system, service quality, and net benefits of tourism enterprises through a field survey of a municipal model, which showed support for the relationship between overall tourism service quality and overall net benefits of tourism enterprises in this setting. The net benefits construct used is not solely an organizational impact instrument as it has three questions relating to individual satisfaction, individual performance, and organizational performance.

## 3 Organizational Impact

The quality of information systems has been grouped by previous researchers into 6 factors: system quality, software quality, hardware quality, data quality, information quality, as well as, service quality in improving the tourism service quality. See Fig. 2.

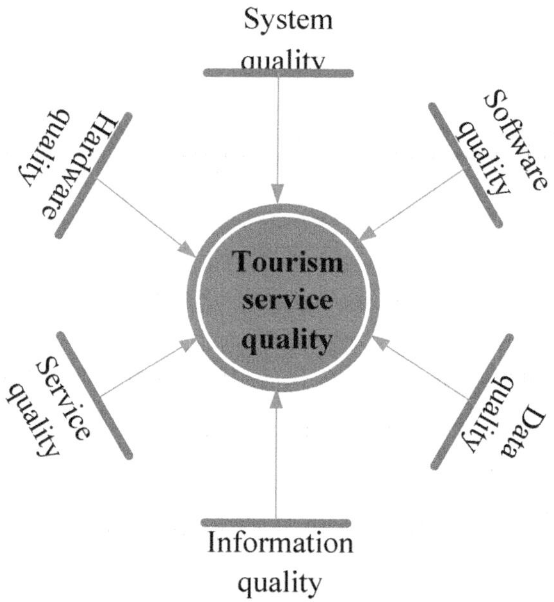


**Fig. 2.** Improving the tourism service quality

As we all know, system quality and software quality for tourism service quality are closely related as both relate to the technical aspects of a tourism management software system. Organizational impact represents the benefits of tourism enterprises received by an organization because of information systems applications. An information system impacts the tourism enterprises by changing the markets, products and services or the economics of tourism service. An instrument for organization has impacted the business value of information system on various business activities within the value chain. The six dimensions used by this paper are process planning and support, supplier relations, service and operations, service enhancement, marketing support, as well as, customer relations. The dimensions mentioned above were intended to span all value chain of activities to meet the premise which the information system impacts are of both primary and secondary activities of a value chain during the tourism activities.

Organizational impacts are always related to organizational performance, which means to improve the tourism service quality in terms of the organization of tourism enterprises. For example, ESPS include the contact number of factors, sense of mission, the evaluation of superiors and colleagues, communication, personal training and development, and five aspects of the workplace to explore how the hotel's internal service quality management(Qi Ling, Shi Yingping,2006) . There are five indicators that compose the construct of organizational impact, two of which are related to internal impacts and three to external impacts. Tourism service cost control is concerned with reduction of costs of new tourism service designs. Internal organizational efficiency is related to efficiency considerations of the decision making process, internal communication, strategic planning, and profit margin. Tourism service supplier search costs reflect the ease with which alternate supply sources and alternate tourism services can be found and the cost effectiveness of the suppliers handling the tourism enterprises' business. Tourism service enhancement was defined in terms of the extent to which information systems improves the quality and availability of tourism services to tourism customers. Market information support was defined as the information provided to the firm with respect to tourism customer needs, market trends, and new markets.

## 4 Tourism Service Quality

### 4.1 System Quality

In tourism enterprises, system quality represents the quality of the information system processing during the business, which includes software and data components. System quality is related to whether there are bugs in the system, the consistency of tourism users' interface, quality of documentation, as well as, sometimes, quality and maintainability of program code. System quality is measured by attributes such as ease of use, functionality, reliability, data quality, flexibility, and integration.

### 4.2 Information Quality

Information quality refers to the quality of outputs the information systems produces, which can be in the form of reports or online screens. It is defined with four dimensions of information quality: accuracy, completeness, consistency, and currency. Accuracy is agreement with an attribute about a real world entity, a value stored in another database, or the result of an arithmetic computation. Completeness is to be defined with respect to some specific application, and it refers to whether all of the data relevant to that application are present. While consistency refers to an

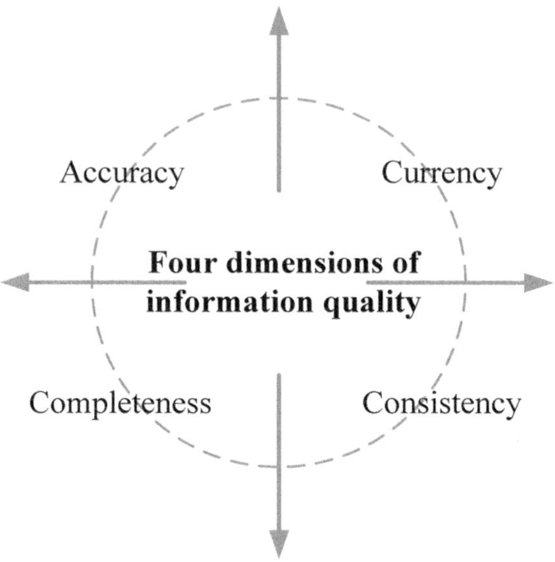


**Fig. 3.** Four dimensions of information quality

absence of conflict between two datasets, currency refers to up-to-date information. Researchers have used a variety of attributes for information quality. See Fig. 3.

In this paper, ease of use and timeliness are included in system quality because they are influenced by the hardware/software system itself. Thus, two broad categories for information quality can be drawn, which are both information content and information format. Information content measures the relevance of the information presented to the user in the report/inquiry screens and the accuracy and completeness of the information. Information format measures the style of presentation of information and whether information is provided in a way of the format easy to be understood.

### 4.3 Service Quality

The construct of tourism service quality has been defined as the degree of discrepancy between tourism customers' normative expectations for tourism service and their perceptions of service performance. A supply chain performance appraisal method that is aimed at the subjective judgment and some direct results when we perform the multi- essential factors appraisal of the supply chain(Zou,2005). According to the fuzzy theory, we establish the fuzzy synthetic performance appraisal model that can handle incorrect information, thus it is helpful to correctly appraise the performance level of the supply chain and diagnose existent questions, which culminated in the development of the SERVQUAL instrument.

## 5 Methodology

The measurements of tourism service quality are applied to this paper are system quality, information quality, service quality and organizational impact. For each construct, the underlying domains of construct are identified to represent each domain. The constructs for their psychometric properties were tested at last.

In addition to background information system, the questionnaire included items that asked respondents about their perceptions of tourism business impact, system quality, information quality, and tourism service quality. Regarding the organizational impact of tourism enterprises, the respondents were asked to give on a 7-point scale (1 means strongly disagree, while 7 means strongly agree) their perceptions regarding a statement. Respondents of tourism enterprises in Xiamen, China were asked to return the questionnaire either in written form. A reminder was sent to non-respondents two weeks just after the initial mailing. Approximately 19 respondents were not usable because the questionnaires were returned through the e-mail as undeliverable, and the completed and returned questionnaires were incomplete too. At last, a total response of 381 usable questionnaires was useful, which is representing approximately a 95% response rate.

For testing the discriminant validity of hypothesized scales, two criteria can be chosen: one is the square root for a construct which should be larger than their corresponding inter-construct correlation coefficients, and the other is the items of construction loadings which might exceed the inter-construct cross loadings at least 0.1. As the tests carried out above, the AVEs range from 0.62 to 0.85 and each AVE were much larger than the corresponding squared inter-construct correlations. See Table 1.

**Table 1.** Testing for discriminant validity

| | System quality | System quality | System quality | System quality |
|---|---|---|---|---|
| System quality | 0.78 | | | |
| Information quality | 0.74 | 0.85 | | |
| Tourism service quality | 0.81 | 0.70 | 0.83 | |
| Organizational impact | 0.72 | 0.74 | 0.62 | 0.65 |

## 6 Conclusion

The analysis model of this paper is then empirically validated using data collected from a field survey of tourism enterprises in Xiamen, China. The study has two key contributions for evaluating and improving tourism service quality. On one hand, it provides a link between information quality and tourism service quality with the organizational impact; On the other hand, the research question for this study was information quality and tourism service quality related to the impact organizational performance measurement and evaluation. In short, the resulting of this paper showed significant direct or indirect links of system quality, information quality, and the service quality.

## References

1. Dick, A.S., Basu, K.: Customer loyalty: Toward an integrated conceptual framework. Journal of the Academy of Marketing Science 22(2), 107–122 (1994)
2. Yu, A.H.-C.: Service Quality in Tourism: A Case Study of the Study Tour of Taiwan. In: Proceedings of the 2005 Northeastern Recreation Research Symposium (2005)
3. Choy, D.J.L.: The Quality of Tourism Employment. Tourism Management (2), 129–137 (1995)
4. Ekinci, Y.: A review of theoretical debates on the measurement of service quality: implications for hospitality research. Journal of Hospitality&Tourism Research 26(3), 199–216 (2002)
5. Galup, S.D., Dattero, R., Quan, J.J., Conger, S.: An overview of IT service management. Communications of the ACM 52, 124–127 (2009)
6. Wall, G.: Human resources development in China. Annals of Tourism Research (3), 689–710 (2005)
7. Gefen, D., Straub, D.: A practical guide to factorial validity using PLS-graph: tutorial and annotated example. Communications of the AIS 16, 91–109 (2005)
8. Getty, J.M., Getty, R.L.: Lodging quality index(LQI):assessing customers' perceptions of quality delivery. International Journal of Contemporary Hospitality Management 15(2), 94–104 (2003)
9. Hendricks, K.B., Singhal, V.R., Stratman, J.K., et al.: The impact of enterprise systems on corporate performance. A study of ERP, SCM, and CRM system implementations. Journal of Operations Management 25, 65–82 (2007)
10. Kahn, B.K., Strong, D.M., Wang, R.Y.: Information quality benchmarks: product and service performance. Communications of the ACM 45, 184–192 (2002)
11. Knutson, B., Stevens, P., Wullaert, C., Patton, M., Yokoyama, F.: LODGSERV: A service quality index for the lodging industry. Hospitality Research Journal 14(2), 277–284 (1990)

# Design and Realization of Real-Time Interactive Remote Video Teaching System Based on FMS

Hua Huang, Yi-Lai Zhang, and Min Zhang

Jingdezhen Ceramic Institute, Jingdezhen, Jiangxi, 333001, China
{jdz_hh,tyzyl,191148074}@qq.com

**Abstract.** The real-time interactive remote video teaching system studied by this paper is designed and developed by FMS technology, which provides a consistent and efficient modern teaching method for remote education. The system is constituted by two modules: courseware teaching module and interactive teaching module. This system breaks through the constraints of time and space of the traditional teaching, which is new and flexible to simulate the classroom teaching environment and to communicate with each other.

**Keywords:** Remote video teaching, Real-time interactive, FMS.

## 1 Introduction

The 21st century is highly information-oriented society, with the rapid development of network information technology and higher quality and quantity of personnel requirements in modern society, information technology enters into a traditional classroom gradually. The traditional teaching model in the current education system still holds an important position, but there are some limitations. The remote education by using network technology [1], makes teaching methods to be of cross-regional nature in the space level, you can use the Internet for real-time interaction, but also upload some excellent educational resources to a network server and share them. The remote education is an important part of online education, the so-called real-time interactive video remote teaching makes use of computer network communication technology and multimedia technology to teach and learn, which breaks through the time and space constraints, so that a user in anywhere is able to learn through the network, and both teachers and students can conduct real-time interactive teaching and learning activities [2].

## 2 Design Idea of This System

### 2.1 The Basic Theory of Design

With the popularity of Internet technology, the remote education is experiencing a changing process from "far" to "close" Although the physical distance still exists, but the actual learning "space" among students is closer. The network teaching can get better interactivity than traditional classroom teaching, not only that, the application

S. Lin and X. Huang (Eds.): CSEE 2011, Part II, CCIS 215, pp. 305–309, 2011.

and promotion of Internet and WWW makes the remote education to change gradually from the behaviorism learning theory to the constructivism learning theory. The J. Piaget's constructivism learning theory: knowledge is not obtained by teachers' teaching, but obtained by way of the meaning construction with the help of others and the use of the necessary learning information when learner is in a certain social and cultural background.

The remote teaching through the Internet, which is not limited in time and place, teachers and students can be on Internet in their own fit time and place, it provides a real-time interaction between teachers and students, also between students and students through the network. Therefore, The internet-based remote education for learners provides a constructivist learning environment and more opportunities to use contexts the knowledge they have learned in different environment, fully embodies students' creative spirit, and the students have a understanding of objective things and can solve practical problem according to their own action's feedback.

### 2.2 Design Principles of the System

The design goal of this system is to meet real-time interactive remote learning activities and do all aspects of teaching and learning activities on the Internet. Basic design principles of this system includes: interactivity, openness, availability, scalability, security, and so on.

Remote teaching method is diverse, the exchange of information is of real-time and asynchronous nature, so streaming media technology can be used to pre-recorded video and audio information on classroom teaching, and a good pre-production of multi-media materials on demand through the network is used to learn at any time for students. This embodies asynchronous nature of remote teaching methods. Remote teaching activities focus on real-time interactivity, and real-time interactive is a advanced methods in the current, such as digital TV or video meeting, which makes use of interactive video and audio, interactive text, graphics, photos and shared whiteboard, that allows teachers and students to participate in remote teaching and learning as like a real-time interaction in the traditional classroom, it is much more convenient than to use BBS, E-mail, message boards and so on.

This system is not only adapted to the different application of domestic bandwidth network environment, but also the overall level of domestic information education. The remote video teaching system on the basis of practical and outstanding features in the classroom teaching should have a large amount of easy to use, useful features, such as synchronized browsing courseware, web functions, and video interaction, etc. This system is necessary to consider the full sharing of information resources, also pay s attention to the protection and isolation of information, such as system security, data access rights control.

## 3 Function Design of This System

By implementing a real-time interactive remote education system based on internet, students and teachers will be able to do remote teaching and learning activities and have a real-time interaction by computer software. Function of this system is divided into two modules: courseware teaching module and interactive teaching module.

### 3.1 Courseware Teaching Function

**Courseware List.** The pictures, a variety of office documents, web pages and other multimedia resources which may be required for teachers are uploaded into courseware list, it is used to achieve a master synchronization browser for them.

**Courseware synchronization.** This system can automatically compress courseware which is uploaded to the server, teachers and students can simultaneously browse the Web pages. In addition, this system can also allow students to automatically download the courseware to local computer; teachers and students can do synchronized browsing and open the local files directly, without accessing the server, which can increase the speed of accessing courseware, save network bandwidth.

**Recording and Playing.** According to their needs, teachers and students can make courses lively; teacher side also supports re-recorded courseware to broadcast. Teachers can allow or prohibit students from recording courseware.

### 3.2 Interactive Teaching Function

In this system, students can study, have examinations, and discuss something. Images, text, graphics, sound and other information can be transferred between students and teachers. The system's main features includes: video and audio interaction, text interaction, whiteboard interaction, file interaction and so on [3].

**Video and audio interactive system.** Eight video sides can be displayed in users' cross-section of this system, According to their needs users can at any time to switch the interface to be displayed; teachers can transmit teaching content and teaching scenarios to the network classroom in real-time, or you can play a student's video. While teachers broadcast their own video, students' video can be watched. Teachers can broadcast their own voice out, and allow a student to speak, or allow two students to broadcast their sound; when the voice broadcast is not opened in this system, teachers and students can whisper between them, it is possible to talk one to one, or one to many. The state of students' raising their hands in the display box of teachers' side, teachers may choose a student who have raised his hands and allow him to speak. When a student is allowed to speak, a dialog box will be popped up to remind students to speak.

**Text interactive system.** In the section of text discussion, teachers can communicate and discuss with all students or a student by typing words. When teachers select a student in the list of personnel, he can send information to the student and have a talk one to one; other members cannot see their talking record. There are some prompts and announcements on system's operational information in this text interactive system, such as how to allow someone to speak, how to record courseware. In addition, the system also has a dictionary filtering function, which is used to filter uncivilized terms.

**Whiteboard interactive system.** On the whiteboard, teachers and students can plot, write, edit or paste existing graphics and pictures. Teachers can control permissions of using the whiteboard, allow or prohibit others from using the whiteboard, select whether to display the object's creator on the board or not, set the created objects' color, font and other attributes. This system currently supports ten whiteboards, which can be expanded as needed.

**File sharing system.** The system is a file sharing system based on FTP server, after the teachers ratified the file-sharing, a variety of resources in FTP folder will appear in the file-sharing area for students who participate in online teaching to upload or download.

## 4 Architecture of This System

**Server side of this system.** The server side of this system includes: Flash Media Server (FMS), IIS and SQL Server, among FMS is as a master server, IIS is a communication bridge between the database server SQL server and FMS server. (Remark: Flash Media Server is the most convenient and efficient solution for develop Web applications of audio and video , e.g., video chat room, video conferencing system, broadcasting system, audio and video message, etc. FMS streaming video technology now achieves carrier-grade level, which is an integrated application system with multi-point voice, video and data communications capability. In the existing network environment, you can easily carry different communication forms of audio, video and other data, it can be used for enterprises' remote office management, online meetings, marketing and sales, business negotiation, online shopping, remote education, telemedicine, online performances and news releases, product demonstrations and promotions, corporate customers' remote support and service and other fields[4].).

**Client side of this system.** The client side of this system includes: Flash and HTML pages, users do not need to download the client, only need to simply open the website's Flash web pages.

**Communication protocols of this system.** The communication protocols of this system includes: HTTP, RTMP and message resolution protocols of this system.

## 5 Interface Design and Realization of the System

The system has been completed by developing and application testing in more than six months, and has won the first prize in the fourth colleges and universities multimedia courseware exhibition game in Jiangxi province, the main function' interfaces are shown in Fig.1 and Fig.2.:

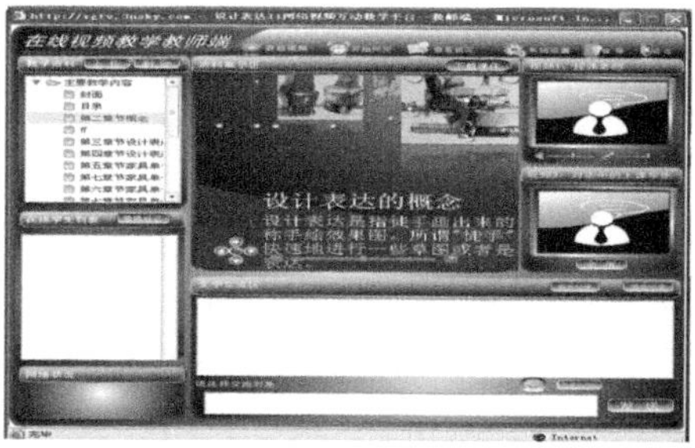

**Fig. 1.** Teachers' function' interface

**Fig. 2.** Students' function interface

The user interface design of this system is of good maneuverability and intuitive, including text, icons, graphics, color and the design of other visual aspects The design of screen interface has a unified interface style by taking into learners' visual psychological characteristics and highlighting the whole nature. Meanwhile focusing on the interface's interaction, the control of content and the super-connected ordering of teaching content. Working region includes five control interface blocks: video display area, whiteboard, class member list, file sharing area and text interaction. According to their need, users can drag, resize, and move the plates to change their relative position in the interface, even to select or hide the plates and so on.

## 6 Conclusion

In this remote real-time interactive multimedia teaching system, some teaching resource flows are created by compressing the video and audio streams of teaching scene with the orders of synchronized browsing courseware, which are real-time transmitted to a remote student's computer by network, students can do remote interactive teaching and learning by asking questions online, sharing teacher side's programs, communicating by text, etc This system can be run in Internet, Intranet, satellite network, campus network, LAN, providing text, audio and video, courseware, electronic whiteboard, interactive broadcast teaching and breaking through the constraints of time and space in virtual reality classroom, as like in the same class for remote students through the real-time interactive audio and video.

## References

1. Wu, J.-Q., Zhao, C.-L., Xu, X.: Comparative analysis between online teaching and classroom teaching. China E-education (June 2000)
2. Chen, X.-n., Zhu, Y.-s., Gong, L.: Research and Design of of the third generation remote education System. Computer Engineering (11) (2003)
3. Zhou, Y.-x.: Web-based collaborative learning and pattern design principles. Research of Modern Remote Education, Chengdu (2004)
4. Lu, Y.: Streaming media technology and modern medicine remote education. Nanjing: Industrial Control Computer 16(8), 6–7 (2007)

# Orthogonal Matching Pursuit Based on Tree-Structure Redundant Dictionary

Song Zhao[1], Qiuyan Zhang[2], and Heng Yang[2]

[1] Zhengzhou Institute of Aeronautical Industry Management, Zhengzhou, 450015
[2] Northwestern Polytechnical University, XiAn, 710068

**Abstract.** Tree based Orthogonal matching pursuit is proposed to overcome the convergence of sparse decomposition. Sparse decomposition can be fast solved by tree based matching pursuit, however, the tree based pursuit is locally best in essence, so it convergences very slowly. We propose the orthogonal matching pursuit algorithm that maintains full backward orthogonality of the residual (error) at every step and thereby leads to improved convergence. Also, it guarantees the sparsity of results and exactly of reconstructed image. Speech signal and earthquake signal are tested via Tree based orthogonal matching pursuit separately, both of which have better convergence performance than tree based matching pursuit.

**Keywords:** Redundant dictionaries, Sparse decomposition, Tree structure, Orthogonal Matching Pursuit.

## 1 Introduction

In signal processing field, sparse decomposition is widely applied in signal, images and video because of its good properties, and this caused wide attention from scholars. It also becomes a research hot spot [1-6]. Different from orthogonal decomposition by wavelet etc, sparse decomposition is done in redundant dictionary, so the decomposed results is not the only solution, however sparse solution which was got though decomposition has great important meaning in compression and denoising. Usually, sparseness of signal decomposition is measured by non-zero number of coefficient vector, so searching best linear expansion of function in redundant dictionary is NP problem.

In redundant dictionary, suboptimal iterative restoration algorithm has been used to develop the best approximation of the recovery function, for example basis pursuit [4] and matching pursuit [7], by relaxing the restrictions to the original optimization problem, both of them found a close answer to the optimal solution. Even so, the computational complexity of basis pursuit and matching pursuit is very high, for some special redundant dictionary, such as Reference [8] using orthogonal basis by means of two-step to approximate the atom, also using fast algorithm of the orthogonal basis to realize the optimization; Reference [9] using separable dictionary, with separable inner product to improve the speed of searching atom at matching. Meanwhile, Reference [10,11] using vector quantization method by eliminating uncommon used atoms to reduce the scale of dictionary and then improving the speed of sparse

S. Lin and X. Huang (Eds.): CSEE 2011, Part II, CCIS 215, pp. 310–315, 2011.

decomposition. Never the less, the above algorithm is proposed for special atomic structure, and in the cost of approximate effect of the image to exchange for computation speed. Reference [2] puts forward an anisotropic atom, due to the good approximation characteristics and visual expression effect, it has a good image compression effect and also gets widespread use in image and video processing. However, such atoms obtained by generation function do not have effective calculation structure, inner product of all the atoms and signal in the redundant dictionary is needed to calculate. In this regard, from the clustering idea of decision tree, Reference [12] proposed a method that the redundant dictionary will be presented in hierarchical tree structure, provides a fast implementation method for image sparse decomposition, but the searching accuracy is not high. Further more, Reference [13] proposed the concept of using molecule that represent each sub-dictionary to improve the accuracy of the algorithm, the tree leaves correspond to the original dictionary of atoms, the other nodes represented by molecules.

Although, dictionary in hierarchical tree structure can achieve fast signal sparse decomposition, but the convergence rate drops, and interferes signal sparse expression. This paper will introduce a new method of combing the orthogonal matching pursuit and the expression of hierarchical tree structure, so that the sparse decomposition of the residual signal can converge to zero after a finite number of iterations.

## 2 Fast Orthogonal Matching Pursuit Method Based on Hierarchical Tree

If the dictionary meets enough irrelevant, so simple greed strategy base pursuit and matching pursuit is able to restore a good approximation. Coherence of dictionary (redundancy) can use the dictionary of coherent coefficient to measure, it is defined as following:

$$\mu = \sup_{\substack{i,j \in D \\ i \neq j}} \left| \langle g_i, g_j \rangle \right| \tag{1}$$

### 2.1 Creating Hierarchical Tree Structure of Atom and Redundant Dictionary

Suppose the element of dictionary $D = \{g_i\}_{i \in \Gamma}$ can be marked by set $\Gamma$ index. If set meets $D_\Lambda = \{g_i\}_{i \in \Lambda}$ $\Lambda \subset \Gamma$ and $\Lambda \neq \phi$ ,we call $D_\Lambda$ as sub-dictionary. Furthermore, if it meets $\bigcup_{i=1}^{n} \Lambda_i = \Gamma$ , $\Lambda_i \bigcap \Lambda_j = \phi$ , $\forall i \neq j$ , and then sub-dictionary $\{D_{\Lambda_i}\}$ is called as a partition of dictionary D. Given $\lambda_\Lambda$ as sub-dictionary's minimum coherent ratio.

$$\lambda_\Lambda = \min_{i,j \in \Lambda} \left| \langle g_i, g_j \rangle \right| \tag{2}$$

Minimum coherent ratio $\lambda_\Lambda$ must be positive, so the molecules can express as sub-dictionary. We can see by the condition, optimization standard expressed by sub-dictionary depends on the average distance measurement to define a convex set, and it also means being able to use standard optimization tools to find optimal molecular. Sub-dictionaries expression that can be defined as the molecular and its equivalent definition is :

$$m_\Lambda^{opt} = \arg \min_{\substack{m \\ \|m\|=1}} \sum_{i\in\Lambda} d(m, g_i) \tag{3}$$

By the condition (2) we can know, simple molecular definition implies that molecules and appropriate sub-dictionary meet $\sigma_\Lambda \geq \lambda_\Lambda$. In other words, to increase molecules into sub-dictionary $D_\Lambda$ does not change the minimum coherence of sub-dictionary. Condition (3) also allows the molecules included in the subspace formed by dictionary $\{D\}$. Most clustering algorithms use centroid to represent a cluster, and centroid is the smallest average distance in the whole cluster. Using atomic distance to measure $d(g_i, g_j) = 1 - \left|\langle g_i, g_j \rangle\right|^2$ ,with regard to dictionary $D_\Lambda$ , optimized centroid also is standardized molecules $m_\Lambda^{opt}$

$$m_\Lambda^{opt} = \arg \min_{\substack{m \\ \|m\|=1}} \sum_{i\in\Lambda} d(m, g_i) \tag{4}$$

Block incoherent dictionary is a special dictionary which can be reduction. In reduction dictionary, and we can find a partition, and for this partition, it has a small block coherent ratio $\mu_B$ , defined as:

$$\mu_B = \max_{i \neq j} \max_{\substack{k\in\Lambda_i \\ l\in\Lambda_j}} \left|\langle g_k, g_l \rangle\right| \tag{5}$$

If D can be reduced representation, D's coherent ratio is large. The contrary is not established.

Dictionary D coherent ratio $\mu$ can be large. For the block non-coherent dictionary, the structure redundancy is low, and the coherence of block structure provides redundant measurement of sub-dictionary. Redundant dictionary of hierarchical tree structure links high degree of coherence, blocks of non-coherent with incoherent dictionary.

## 2.2 Orthogonal Matching Pursuit Algorithm Based on Tree Structure

Pursuit algorithm make atoms produced at every step orthogonal to atoms produced before, so the signal projection is in the orthogonal space, and improve the sparse decomposition the convergence speed of the sparse decomposition.

In practical applications, Dictionary D may traversal from different location of signal space though the translation operator. Suppose that $T_p$ shifts generating function to the location of p, and it is an operator to maintain the support set and

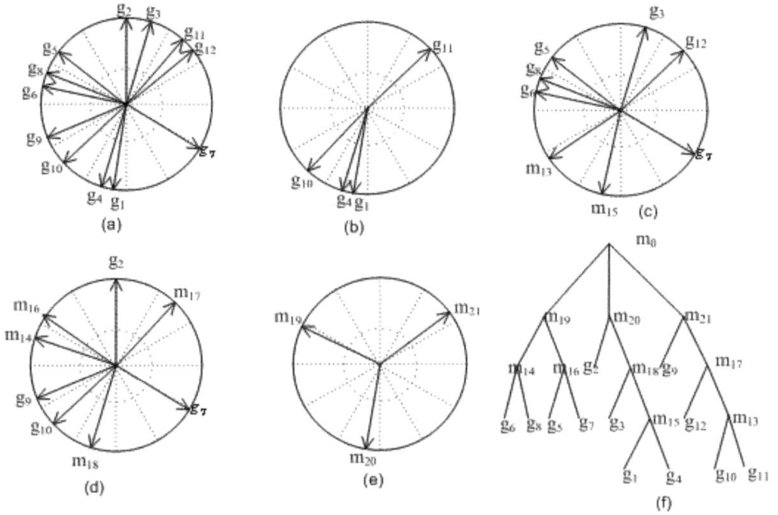

**Fig. 2.** 2-D hierarchical tree structure (a) atoms in dictionary; (b)-(e) forming of molecular; (f) hierarchical tree structure of redundant dictionary

energy constant of atoms, which can make generating function approximating or translating to any location of signal supported. In the root node, needing to calculate residual signal $R_n$ and the product molecules of the first layer node of the tree, this step is equivalent to matching pursuit using tree's first layer of the molecular as a dictionary.

To sum up, in this paper, a dictionary which bases on hierarchical tree structure of orthogonal matching pursuit algorithm is as follows:

Input: $T = \{c_i, m_i\}$ , tree structure dictionary
W, the size of the local search window
f, signals needing to be addressed

Output: $\{g_n\}$ ,subset of selected atom
$\{a_n\}$ ,atom coefficient of the corresponding projection

Initialization: $R_0 = f$ , $n = 0$

$$[p_{opt}, c_{opt}] = \arg\max_{p,c\in c_0} \left|\langle R_n, T_p m_c \rangle\right|$$

while $|c_{opt}| > 1$ do

$$[p_{opt}, c_{opt}] = \arg\max_{p,c\in c_0} \left|\langle R_n, T_p m_c \rangle\right|$$

$$|p - p_{opt}| < w$$

end while

$$g_{n+1} = T_{opt} g_{copt}$$
$$R_{n+1} = f - P_{span\{g_{\lambda 1}, g_{\lambda 2}, \ldots g_{\lambda n+1}\}} f$$
$$n = n + 1$$

Until algorithm satisfies the termination condition.

## 3 Experiment Results

Redundant dictionary's generating function commonly uses Gabor functions. For the signal of 512 length, scale s often takes nine different values, and frequency $\xi$ takes 50 discrete points, phase $\phi = 0$, $u = 256$; the number of atoms contained in the dictionary is 450(do not consider translation). It can be seen from figure 3, the convergence rate of orthogonal matching pursuit is often faster than the speed of matching pursuit, and figure 3 (b) shows that in the later iterations, convergence of matching pursuit is almost at a standstill, however, the orthogonal matching pursuit is still able to converge to zero in accordance with L2-norm.

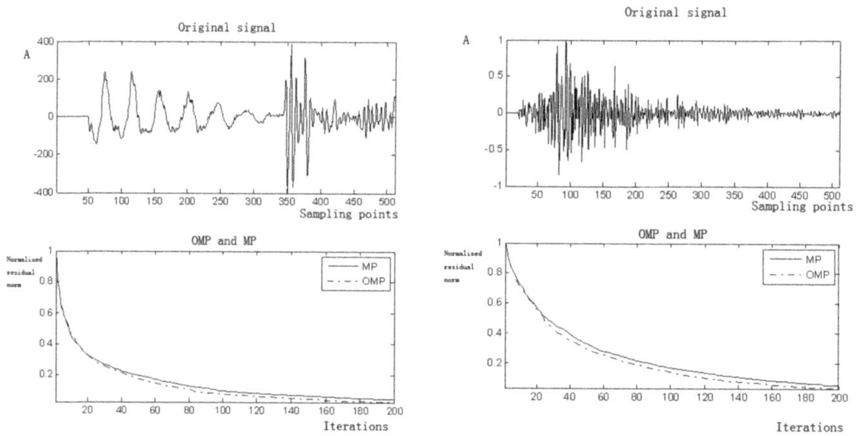


(a) Part of the Voice signal "Greasy" (b) Seismic signal "earthquake"

**Fig. 3.** The original signal and the orthogonal matching pursuit approach based on hierarchical tree structure

## 4 Conclusion

In this paper, we propose the orthogonal matching pursuit algorithm that maintains full backward orthogonality of the residual (error) at every step and thereby leads to improved convergence. It ensures that the results of decomposition are sparse. However, how to optimize the structure of dictionary and enhancing the ability of atom to express the signal is the direction to strive in the next step. This work is supported by the Natural science research project foundation of Education Department of Henan Province 2011A510024.

## References

1. Iventura, R.F., Vandergheynst, P., Frossard, P.: Lowrate and flexible image coding with redundant representations. IEEE Trans. Image Process 15(3), 726–739 (2006)
2. Neff, R., Zakhor, A.: Matching pursuit video coding, I. Dictionary approximation. IEEE Trans. Circuits Syst. Video Technol. 12(1), 13–26 (2002)
3. Chen, S.S., Donoho, D.L., Saunders, M.A.: Atomic decomposition by basis pursuit. SIAM J. Scientific Comput. 20(1), 33–61 (1999)
4. Elad, M., Aharon, M.: Image denoising via sparse and redundant representations over learned dictionaries. IEEE Trans. on Image Processing 15(12), 3736–3745 (2006)
5. Zhang, C.: Redundant dictionary based overcomplete signal representation and sparse decomposition. Chinese Science Bulletin 51(6), 628–633 (2006)
6. Olshausen, B.A., Field, D.J.: Sparse coding with an overcomplete basis set: A strategy employed by V1. Vision Research 37(23), 3311–3325 (1997)
7. Mallat, S., Zhang, Z.: Matching pursuit with time-frequency dictionaries. IEEE Trans. Signal Process. 41(12), 3397–3415 (1993)
8. Cheung, K.-P., Chan, Y.-H.: A fast two-stage algorithm for realizing matching pursuit. In: Proc. IEEE Int. Confe. Image Processing, vol. 2, pp. 431–434 (October 2001)
9. Bin, L.: Low Bit-Rate Video Coding Based on Undecimated Wavelet Dictionary. Journal of Software 15(2), 221–228 (2004)
10. Chou, Y.-T., Huang, W.-L., Huang, C.-L.: Gain-shape optimized dictionary for matching pursuit video coding. Signal Processing 83(9), 1937–1943 (2003)
11. Schmid-Saugeon, P., Zakhor, A.: Dictionary Design for Matching Pursuit and Application to Motion-Compensated Video Coding. IEEE Trans. Circuits Syst. for Video Technology 14(6), 880–886 (2004)
12. Monaci, G., Jost, P., van der Gheynst, P.: Image Compression with learnt Tree-structured dictionaries. In: Proc of IEEE MMSP, Siena, Italy, pp. 35–38 (2004)
13. Jost, P., van der Gheynst, P., Frossard, P.: Tree-based Pursuit: Algorithm and Properties. IEEE Trans. on Signal Processing 12(54), 4685–4697 (2006)

# Formal Methods for Aspect-Oriented Specification of Cyber Physical Systems

Lichen Zhang

Faculty of Computer Science and Technology
Guangdong University of Technology
Guangzhou 510090, China
Zhanglichen1962@163.com

**Abstract.** This paper proposes an approach for specifying cyber physical systems based on aspect-oriented formal method, which exploits the diversity and power of existing formal specification languages. There is no requirement that different aspects of a system should be expressed in the same language. So the different aspects can be specified by one formal specification technique or different formal specification techniques. Cyber physical systems can be divided into different aspects such as functional aspects, timing aspect, sequential and concurrent behavior and synchronous communication. In this paper, we present a combination of the formal methods Timed-CSP and Object-Z. Each method can describe certain aspects of a cyber physical system: CSP can describe behavioral aspects and real-time requirements; Object-Z expresses complex data operations. This aspect oriented formal specification method simplifies the requirement analysis process of cyber physical systems. A case study of train control system illustrates the specification process of aspect-oriented formal specification for cyber physical systems.

**Keywords:** Aspect-oriented, Real Time, Cyber Physical Systems, Formal Method, Object-Z, Timed-CSP.

## 1 Introduction

Cyber-physical systems (CPS)[1] are systems in which computational (cyber) processes interact closely with physical dynamical processes. Applications of Cyber-Physical Systems include, among others, critical infrastructure control (electric power, water resources, gas and fuel distribution, transportation, etc.), process control and manufacturing, highly dependable medical devices and systems, traffic control and safety, advanced automotive systems, energy conservation and environmental control. The design and verification of cyber physical systems requires a good understanding of formal mathematical methods that are found in both computer science and the traditional engineering disciplines. These formal methods are used to model, verify, and design complex embedded systems in which the interaction of computational and physical processes must be approached in a holistic manner. Formal Description Techniques FDTs like Object-Z[2]and Timed-CSP [3] have been successfully applied to the specification of ''traditional'' communication protocols, services and network applications.

S. Lin and X. Huang (Eds.): CSEE 2011, Part II, CCIS 215, pp. 316–322, 2011.

Aspect-oriented programming (AOP) [4] is a new software development technique, which is based on the separation of concerns. Systems could be separated into different crosscutting concerns and designed independently by using AOP techniques. The design and analysis of various types of systems, like real time cyber physical systems or communication protocols, require insight in not only the functional, but also in the real-time and performance aspects of applications involved. Research in formal methods has recognized the need for the additional support of quantitative aspects, and various initiatives have been taken to accomplish such support.

In this paper, we provide some ideas for the aspect –oriented formal specification of cyber physical systems and one well known case study to validate aspect-oriented formal specification.

## 2 Aspect-Oriented Formal Specification

Aspect-oriented approaches use a separation of concern strategy, in which a set of simpler models, each built for a specific aspect of the system, are defined and analyzed. Each aspect model can be constructed and evolved relatively independently from other aspect models. This has three implications:

- An aspect model can focus on only one type of property, without burden of complexity from other aspects. Hence an aspect model is potentially much simpler and smaller than a traditional mixed system model. This is expected to dramatically reduce the complexity of understanding, change, and analysis.
- Different levels of detail or abstraction can be used in the construction of different aspect models. This allows us to leverage of existing understanding of certain aspect of the system to reduce the complexity of modeling and analysis. For example, if the timing property of a component/subsystem is well understood, we can build an extremely simple timing model for the component.
- Existing formal notations normally are suitable for describing one or a few types of system properties. By adopting the aspect concept, we can select the most suitable notation to describe a given aspect. Likewise, we can select the most suitable analysis techniques to analyze a given property.

In many ways, Object-Z and Timed CSP complement each other in their capabilities. Object-Z has strong data and algorithm modeling capabilities. The Z mathematical toolkit is extended with object oriented structuring techniques. Timed CSP has strong process control modeling capabilities. The multi-threading and synchronization primitives of CSP are extended with timing primitives. Moreover, both formalisms are already strongly influenced by the other in their areas of weakness. Object-Z supports a number of primitives which have been inspired by CSP notions such as external choice and synchronization. CSP practitioners tend to make use of notation inspired by the Z mathematical toolkit in the specification of processes with internal state. The approach taken in the TCOZ notation[5] is to identify operation schemas (both syntactically and semantically) with (terminating) CSP processes that perform only state update events; to identify (active) classes with non-terminating CSP processes; and to allow arbitrary (channel based)communications interfaces between objects. An example of timing expression[5] is shown as Fig.1.

**TimedCollection**[$X$]

*left*, *right* : **chan**
*mems*: $F(T \times X)$
$\Delta$
$t$ : T
*oldest* : $X$

$mems \neq \emptyset \Rightarrow (t, oldest) \in mems \wedge t = min \text{ dom } mems$

*INIT*

$mems = \emptyset$

$Add_0$

$\Delta(mems)$
$e? : X$
$t_i? : T$

$\delta = t_a \wedge mems' = ps(t_i? + t_a, mems) \cup \{(t_o, e?)\}$

*Delete0*

$\Delta(mems)$
$t_i : T$

$mems \neq \emptyset \wedge \delta = t_d$
$mems' = ps(t_i + t_d, mems \setminus \{(t, oldest)\})$

$Add \mathrel{\hat{=}} [e : X; t_i : T] \bullet left?e@t_i \rightarrow Add_0$
$Delete \mathrel{\hat{=}} [t_i : T | mems \neq \emptyset] \bullet right!oldest@t_i \rightarrow Delete_0$

*Purge*

$\Delta(mems)$

$\delta = t_p \wedge mems' = ps(t_i + t_p, mems)$

$Main \mathrel{\hat{=}} \mu TC \bullet [mems = \emptyset] \bullet Add; TC \Box [mems \neq \emptyset] \bullet ((Add \Box Delete) \rhd \{t\} \ Purge); TC$

**Fig. 1.** Timed Object-Z Model of the TimeCollection

Aspect-oriented specification is made by extending TCOZ notation with aspect notations. The schema for aspect specification in has the general form as shown in Fig.2, Fig.3 and Fig.4.

**AspectName** [*generic parameters*]

*Type and Constant Definitions*
*Inherited Aspects*
*PointCut State*
*PointCut Operations*
*InterType State*
*InterType Operations*
*Composite Schema*

**Fig. 2.** Aspects of Model Structure

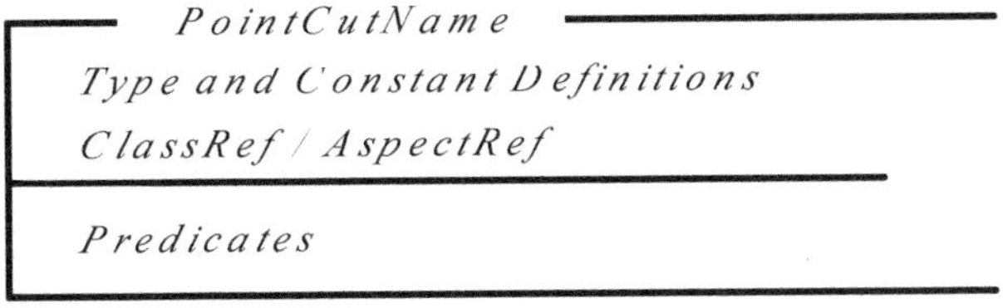


**Fig. 3.** PointCut Operation Schema of Structure

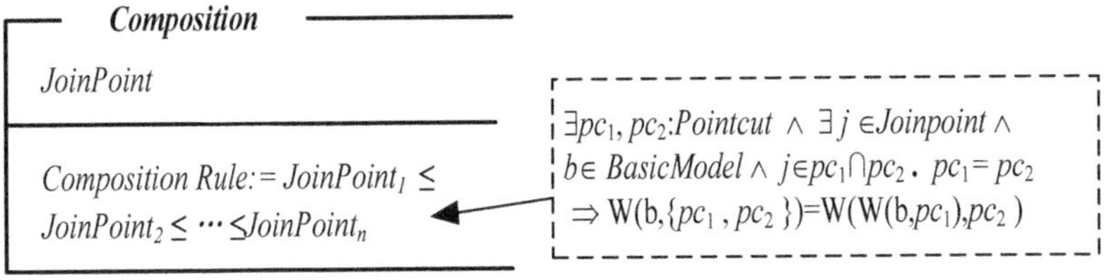


**Fig. 4.** Composition Schema of Structure

## 3 Case Study: Aspect-Oriented Formal Specification of a Train Control System

The Problem that must be addressed in operating a railway are numerous in quantity, complex in nature, and highly inter-related. For example, collision and derailment, rear-end, head-on and side-on collisions are very dangers and may occur between trains. Trains collide at level crossing. Derailment is caused by excess speed, wrong switch position and so on. The purpose of train control is to carry the passengers and goods to their destination, while preventing them from encountering these dangers. Because of the timeliness constraints, safety and availability of train systems, the design principles and implementation techniques adopted must ensure to a reasonable extent avoidance of design errors both in hardware and software. Thus a formal technique relevant to design should be applied for train systems development. The purpose of our exercise is to apply aspect -oriented formal method s to develop a controller for train systems that tasks as input: a description of track configuration and a sequence of description of the moves of each of these trains.

The controller should take care of trains running over the track. It should control the safety of the configuration, ie. No two trains may enter the critical section. When one critical section is occupied, some others, which share some part of section with this one, should be locked. The controlled can control the status, speed, position of trains.

In order to keep the description focused, we concentrate on some particular points in train control systems rather than the detailed descriptions of all development process. The specification is made by integrating Object-Z and Timed CSP[5][6][7].

A train controller [8][9][10]limits the speed of the train, decides when it is time to switch points and secure crossings, and makes sure that the train does not enter them too early as shown in Fig.5.

**TrainController**

*getPosition* : [*p* ? : *Position*]; *getSpeed* : [*s* ? : *Speed*] **chan**
*setMaxSpeed* : [*s* ! : *Speed*] **chan**
*setTrackElement* : [*id* !: *Identifier*; *dir* !: *Direction*] **chan**
*securedTrackElement* : [*id* ?: *Identifier*] **chan**
*insecureTrackElement* : [*id* : *Identifier*] **Local_chan**
*setDangerPosition* : [*id* : *Identifier*] **Local_chan**
*calcMaxSpeed* : [*maxs* : *Speed*] **Local_chan**

*trackdata* : ***TrackDate***
*dangerPositions* : **P *Position***
*dynProf* : *Position* → *Speed*
*notifiedElems* : ***P Identifier***
*position* : ***Position***
*speed* : ***Speed***
*t* : T

*dynProf* = *calcProfile* (*trackatlas.staticprof*, *dangerPosition*)

**Init**

*dangerPositions* = {*elem* : *ran trackdata.elems* • *elem.pos*}

**Fun_getPosition**

Δ(*Position*)
*pos* **? : *Position***

*Position'* = *pos* ?

**Fun_getSpeed**

Δ(*speed*)
*speed* **? : *Speed***

*speed'* = *speed*?

**Fun_setDangerPosition**

Δ(*dangerPositions*, *dynProf*)
*id* **? : *Identifier***

∃ *elem* : **ran** *trackdata.elems* | *elem.id* = *id* ? •
*dangerPosition'* = *dangerPostions* ∪ { *elem.pos*}

**Fun_clearDangerPosition**

Δ(*dangerPositions*, *dynProf*)
*id* **? : *Identifier***

∃ *elem* : **ran** *trackdata.elems* | *elem.id* = *id* ? •
*dangerPosition'* = *dangerPostions* \ { *elem.pos*}

**Fun_insecureTrackElement**

Δ(*notifiedElements*)
*id* ? : ***Identifier***

∃ *toNotify* : {*elem* : **ran** *trackdata.elems* | *elem.id* ∉ *notifiedElements* ∧ *elem.setpos* ≤ *position*} • *id* ! = *toNotify.id*
*notifiedElements'* = *notifiedElements* ∪ *id*!

**Fun_setTrackElement**

*id* ? : ***Identifier***
*dir* ! : ***Direction***

*dir* ! = *μ elem* : *trackdata.elems* | *elem.id* = *id* ! • *elem.dir*

**Fun_calcMaxSpeed**

*mars* ! : ***Speed***

**let** *endp* == *position* + *reactDistance*(*speed*) • *maxs* ! = *min dynProf*([*position*, *endp*])

[*count* (*setMaxSpeed*) = *n*] $\xrightarrow{t}$ [*count* (*setMaxSpeed*) > *n*]
[*enabled* (*insecureTrackElement*) ∧ *count* (*insercureTrackElement*) = *n*] $\xrightarrow{t}$ [*count* (*insecureTrackElement*) > *n*]

*SuperviseSpeed* $\hat{=}$ *getSpeed* ? *Speed* → *getPos* ? *Pos* → *calcMaxSpeed* ? *Maxs* → *setMacSpeed* ! *Maxs* → *SuperviseSpeed*
*SecuredNotifier* $\hat{=}$ *securedTrackElement* ? *id* → *clearDangerPosition* ! *id* → *SecuredNotifier*
*SuperviseTrack* $\hat{=}$ *insecureTrackElement* ? *id* → *setTrackElement* ! *id* ? *Dir* → *SuperviseTrack*
*Main* $\hat{=}$ *SuperviseSpeed* || *SecuredNotifier* || *SuperviseTrack*

**Fig. 5.** Control Model of Train Dispatching System

In automatic railway crossing system, The sensors detect the presence of the train near rail crossing and barrier shuts down when train is approaching to the railway crossing. Once train crosses the rail crossing barriers opens by itself as shown in Fig.6 and Fig.7.

*TrainCrossController[X]*

$leave : [t\_p\,?\,Train\_Position;\ t\,? : Train] : \mathbf{method}$

$enter : [t\_p\,?\,Train\_Position;\ t\,? : Train] : \mathbf{method}$

$up : [t\_p\,?\,Train\_Position;\ t\,? : Train] : \mathbf{method}$

$down : [t\_p\,?\,Train\_Position;\ t\,? : Train] : \mathbf{method}$

$alloc, req, updPos, updSpd : \mathbf{local\_chan}$

$\forall t : Train \bullet tid(t) > 0$

$\forall t_1, t_2 : Train \mid t_1 \neq t_2 \bullet tid(t_1) \neq tid(t_2)$

$\forall t\_p : Train\_Position \bullet prevs(nexts(t\_p)) = t\_p$

$\forall t\_p : Train\_Position \bullet nexts(prevs(t\_p)) = t\_p$

$\forall t\_p : Train\_Position \bullet sid(nexts(t\_p)) \geq sid(t\_p)$

$\forall t\_p_1, t\_p_2 : Train\_Position \mid t\_p_1 \neq t\_p_2 \bullet tid(t\_p_1) \neq tid(t\_p_2)$

$\forall t\_p : Train\_Postion \mid t\_p \neq nil \bullet length(t\_p) \geq d\_safety + max\_spd \bullet \Delta t$

*Init*

$\forall t : Train \bullet trian(Train\_Position(t)) = t \wedge next(prevs(t)) = t \wedge prevs(next(t)) = t \wedge 0 \leq pos(t) \leq length(Train\_Position(t)) \wedge 0 \leq spd(t) \leq max(Train\_Position(t)) \wedge alloc(Train\_Position(t)) = tid(t)$

$\forall t : Train \bullet alloc(nexts(Train\_Position(t))) = tid(t) \vee length(Train\_Position(t)) - bd(spd(t)) > pos(t)$

$\forall t\_p : Train\_Position \bullet Train\_Position(trian(t\_p)) = t\_p$

*effect_update*

$\Delta (spd)$

$\Delta t : Trian \mid pos(t) < length(Train\_Position(t)) - d \wedge spd(t) - max\_speed \bullet \Delta t > 0 \wedge max\{0, spd(t) - max\_speed \bullet \Delta t\} \leq spd'(t) \leq d\_max(Train\_Position(t))$

$\Delta t : Trian \mid pos(t) \geq length(Train\_Position(t)) - d \wedge alloc(nexts(Trian\_Position(t))) = tid(t) \wedge max\{0, spd(t) - max\_speed \bullet \Delta t\} \leq spd'(t) \leq d\_min(d\_max(Train\_Position(t)), d\_maxnexts(Trian\_Position(t)))$

$Main = ((updSpd \rightarrow state)) \Box (leave \rightarrow main) \Box (enter \rightarrow up \mid down) \Box (up \mid down \rightarrow main)$

**Fig. 6.** Trains through the intersection mode

```
TrainCrossAspect
@{TrainController}
  PointCut PC1:{Fun_setDangerPosition, Fun_clearDangerPosition,Fun_insercureTrackElement, ⋯}
  PC1
  Mutex = 1: TrianCrossController[mutex]
```

**Fig. 7.** Cross roads aspects model

# 4 Conclusion

In this paper we proposed to use aspect-oriented formal specification for cyber physical systems, we present a combination of the formal methods Timed-CSP and Object-Z. A case study of train control system illustrates the specification process of aspect-oriented formal specification for cyber physical systems.

The further work is devoted to integrated aspect-oriented formal development tool.

**Acknowledgments.** This work is supported by the Major Program of National Natural Science Foundation of China under Grant No.90818008.

# References

1. Lee, E.A.: Cyber Physical Systems Design Challenges. In: 11th IEEE International Symposium on Object Oriented Real-Time Distributed Computing (ISORC), January 23 (2008)
2. Smith, G.: The Object-Z Specification Language. Software Verification Research Centre University of Queensland (2000)
3. Reed, G.M., Roseoe, A.W.: A timed model for communicating sequential processes. In: Kott, L. (ed.) ICALP 1986. LNCS, vol. 226. Springer, Heidelberg (1986)
4. Aspect-Oriented Software Development, `http://aosd.net/`
5. Mahony, B.P., Dong, J.S.: Timed Communicating Object Z. IEEE TSE 26(2), 150–177 (2000)
6. Dong, J.S., Zucconi, L., Duke, R.: Specifying parallel and distributed systems in Object-Z. In: Agha, G., Russo, S. (eds.) The 2nd International Workshop on Software Engineering for Parallel and Distributed Systems, pp. 140–149. IEEE Computer Society Press, Boston (1997)
7. Mahony, B.P., Dong, J.S.: Blending Object-Z and Timed CSP: An introduction to TCOZ. In: The 20th International Conference on Software Engineering (ICSE 1998). IEEE Computer Society Press, Los Alamitos (1998)
8. Hoenicke, J.: Specification of Radio Based Railway Crossings with the Combination of CSP, OZ, and DC, `http://citeseerx.ist.psu.edu/viewdoc/summary?doi=10.1.1.21.4394`
9. Faber, J., Jacobs, S., Sofronie-Stokkermans, V.: Verifying CSP-OZ-DC Specifications with Complex Data Typesand Timing Parameters. In: Davies, J., Gibbons, J. (eds.) IFM 2007. LNCS, vol. 4591, pp. 233–252. Springer, Heidelberg (2007)
10. Janota, A.: Using Z Specification for Rail Way Interlocking Safety. Periodica Polytechnica Ser. Transp. Eng. 28(1-2), 39–53 (2000)

# MDA Approach for Non-functional Characteristics of Cyber Physical Systems Based on Aspect-Oriented Method

Lichen Zhang

Faculty of Computer Science and Technology
Guangdong University of Technology
Guangzhou 510090, China
Zhanglichen1962@163.com

**Abstract.** In this paper, we propose an aspect-oriented MDA approach for non-functional properties to develop cyber physical systems. An aspect-oriented UML profile is built to develop cyber physical systems. Aspect-oriented UML models are designed as Platform Independent Models (PIM) for target-platform implementation, which deal with non-functional properties. OCL formal language is used to restrict the model in every stages of MDA, and the real-time extension of OCL formal language is made to describe the timing constraints of cyber physical systems. Finally, the model- based development and aspect-oriented approach, the formal methods and the cyber physical system are integrated effectively. A case study illustrates the aspect oriented MDA development of cyber physical systems.

**Keywords:** Non-Functional Properties, Aspect-Oriented, MDA.

## 1 Introduction

Cyber-physical systems (CPSs) are physical and engineered systems whose operations are monitored, coordinated, controlled and integrated by a computing and communication core. Recent years have witnessed the growing applications of CPSs in numerous critical domains including healthcare, transportation, process control, factory automation, smart building and spaces etc. By seamlessly integrating sensing, networking, and computation components with the control of physical processes, CPSs are expected to transform how we interact with and manipulate the physical world.

Model Driven Architecture (MDA)[1] is based on a series of industry-standard software development frameworks, model drives the software development process, and using support tool model can to achieve automatic conversion among the models, between the model and the code. Its core idea is to establish a Platform Independent Model (PIM) with complete description of system requirements and specific platform implementation technology, through a series of model transformation rule set, the platform independent models to be able to transfer to complete presentation system requirements, and specific implementation techniques related to platform specific model (PSM), finally, using MDA tools will be making platform specific model automatically transferred to code.

S. Lin and X. Huang (Eds.): CSEE 2011, Part II, CCIS 215, pp. 323–328, 2011.

Aspect-oriented software development methods [2] make up object-oriented software development methods in system development needs of non-functional characteristics of the existing limitations question problem. Use separate technology of concerns separates all the crosscutting concerns of the system, and then analyzed, designed, modeled for each cross-cutting concerns, to address crosscutting concerns in object-oriented software development, the code tangling and scattering problems, enhancing the system's modular degree, lowering coupling between modules [3].

In this paper, we propose an aspect-oriented MDA for non-functional properties to develop dependable and distributed real-time systems.

## 2 Applying AOP and MDA to Non-functional Requirements

In the design phase, cyber physical system will be designed considering both the requirements and the constraints posed by the system. Using the MDA approach[4] to produce the platform specific models includes five steps (see Fig. 1):

Step one: Create the PIM for the cyber physical system.

Step two: Select the target system and create the generic system aspects.

Step three: Transform PIM to enhanced PIM using the application converter.

Step four: Transform the generic aspects to enhanced aspects using the aspect converter.

Step five: Weave the enhanced aspects into the enhanced PIM to produce the PSM.

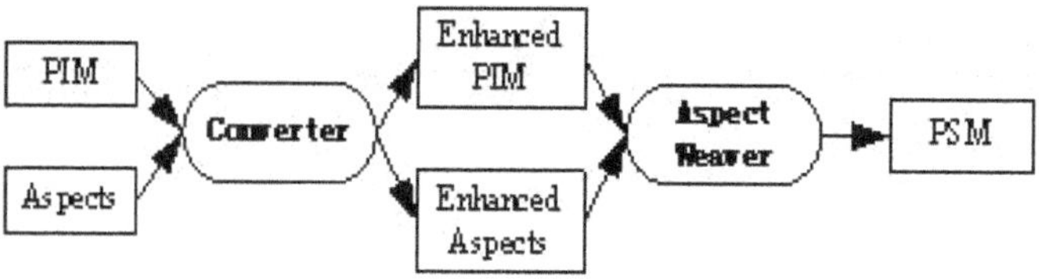


**Fig. 1.** Process view of the PSM generation

To address the system development, principled methods are needed to specify, develop, compose, integrate, and validate the application and system software used by DRE systems. These methods must enforce the physical constraints of cyber physical system systems, as well as satisfy the system's stringent functional and non-functional requirements. Achieving these goals requires a set of integrated Model Driven System (MDM) tools [5]that allow developers to specify application and system requirements at higher levels of abstraction than that provided by low-level mechanisms, such as conventional general-purpose programming languages, operating systems, and system platforms. Different functional and systemic properties of cyber physical system systems via separate system and platform-independent models are applied[5].

## 3 Case Study: Intelligent Transportation Systems

Intelligent Transportation systems(ITS)[6] – automotive, aviation, and rail – involve interactions between software controllers, communication networks, and physical devices. These systems are among the most complex cyber physical systems being designed by humans, but added time and cost constraints make their development a significant technical challenge. MDA approach can be used to improve support for design, testing, and code generation[7]. MDA approach is increasingly being recognized as being essential in saving time and money. Transportation systems consist of embedded control systems inside the vehicle and the surrounding infrastructure, as well as, the interaction between vehicles and between vehicle and the infrastructure. The dynamics of these interactions increases complexity and poses new challenges for analysis and design of such systems[8][9].

The modeling process of non functional requirements of ITS by aspect –oriented MDA is shown as Fig.2, Fig.3, Fig.4.

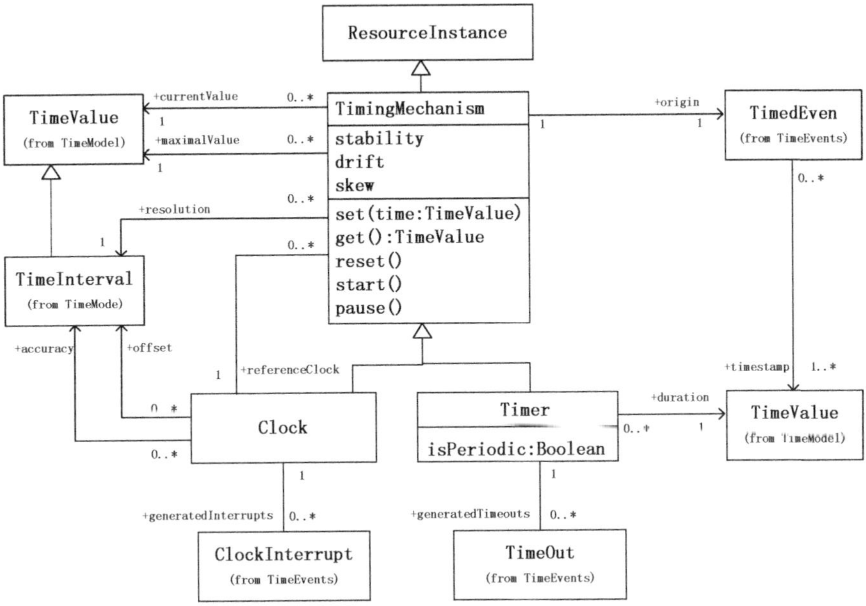


**Fig. 2.** Time Mechanism mode[10]

OCL [11]supplements UML by providing expressions that have neither the ambiguities of natural language nor the inherent difficulty of using complex mathematics. Time aspect is specified by OCL as folows.

Context TimeAspect:
inv: *--the isSingleton is not equal isPrivileged*
if self.isSingleton then self.isPrivileged=false
else self.isPrivileged=true endif
self.isSingleton=not self.isPrivileged
*--maxTime always large than mixTime*
self.maxTime>=self.mixTime
*--the Clock has only one instance*
self.setClockAdvice.Clock.allInstances()->size()=1

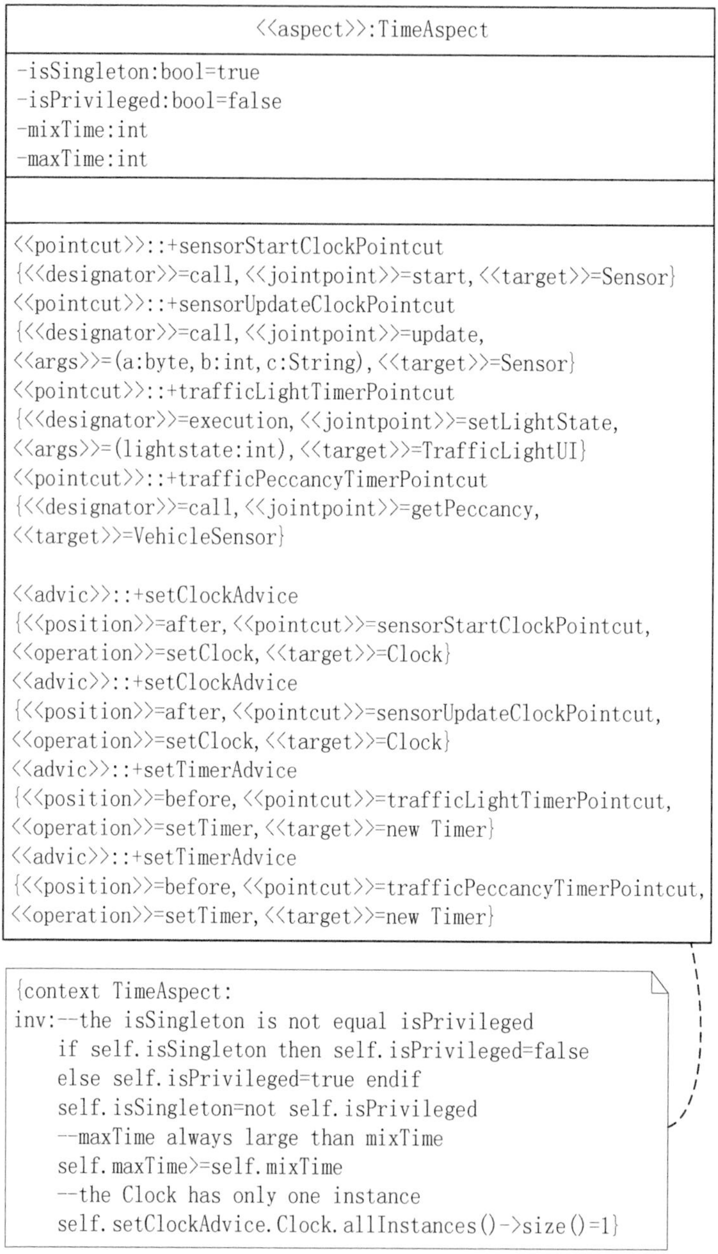


**Fig. 3.** Aspect oriented model of Non-Functional Property Time

```
package com.aspect.its;

public aspect TimeAspect {
    private boolean isSingleton = true;
    private boolean isPrivileged = true;
    private int mixTime;
    private int maxTime;

    public boolean isSingleton() {
        return isSingleton;
    }

    public void setSingleton(boolean isSingleton) {
        this.isSingleton = isSingleton;
    }

    public boolean isPrivileged() {
        return isPrivileged;
    }

    public void setPrivileged(boolean isPrivileged) {
        this.isPrivileged = isPrivileged;
    }

    public int getMixTime() {
        return mixTime;
    }

    public void setMixTime(int mixTime) {
        this.mixTime = mixTime;
    }

    public void setMixTime(int mixTime) {
        this.mixTime = mixTime;
    }

    public int getMaxTime() {
        return maxTime;
    }

    public void setMaxTime(int maxTime) {
        this.maxTime = maxTime;
    }
    pointcut sensorStartClockPointcut():
        call(* Sensor.start(..));

    pointcut sensorUpdateClockPointcut(byte a, int b, String c):
        call(* Sensor.update(byte ,int,String))&&args(a,b,c);

    pointcut trafficLightTimerPointcut(int lightstate):
        execution(* TrafficLightUI.setLightState(int))&&args(lightstate);

    pointcut trafficPeccancyTimerPointcut():
        call(* VehicleSensor.getPeccancy(..));

    after():sensorStartClockPointcut(){
        Clock.setClock();
    }

    after(byte a, int b, String c):sensorUpdateClockPointcut(a,b,c){
        Clock.setClock();
    }

    before(int lightstate):trafficLightTimerPointcut(lightstate){
        new Timer().setTimer();
    }

    before():trafficPeccancyTimerPointcut(){
        new Timer().setTimer();
    }
}
```

**Fig. 4.** Aspect Code of Non-Functional Property Time

Now we return to the model transformation, whose essential point is mapping to the special programming language code as shown in Fig. 4.

## 4 Conclusion

In this paper, we proposed an aspect-oriented MDA for non-functional properties to develop real-time cyber physical systems. We illustrated the proposed method by the development of ITS and demonstrated aspect-oriented MDA approach that can be used for modeling non-functional characteristics of complex system, effectively reduce the complexity of software development and coupling between modules to enhance the system's modular.

The further work is devoted to developing tools to support the automatic generation of model and code.

**Acknowledgments.** This work is supported by the Major Program of National Natural Science Foundation of China under Grant No.90818008.

## References

1. Object Management Group.OMG MDA guide v1.0.1 (EB/OL), `http://www.omg.org/docs/omg/03-06-01.pdf`
2. Wehrmeister, M.A., Freitas, E.P., Pereira, C.E., et al.: An Aspect-Oriented Approach for Dealing with Non-Functional Requirements in a Model-Driven Development of Distributed Embedded Real-Time Systems. In: 10th IEEE International Symposium on Object and Component-Oriented Real-Time Distributed Computing, Santorini Island, Greece, May 7-9, pp. 428–432. IEEE Computer Society, Los Alamitos (2007)
3. Liu, J., Zhong, Y., Zhang, L., Chen, Y.: Applying AOP and MDA to middleware-based distributed real-time embedded systems software process. In: 2009 Asia-Pacific Conference on Information Processing, APCIP 2009, pp. 270–273 (2009)
4. Liming, Z., Gorton, I.: UML Profiles for Design Decisions and Non-Functional Requirements. In: Second Workshop on Sharing and Reusing Architectural Knowledge - Architecture, Rationale, and Design Intent, May 20-26, pp. 8–9 (2007)
5. Frankel, D.S.: Model Driven Architecture: Applying MDA to Enterprise Computing. OMG Press
6. Ranjini, K., Kanthimath, A., Yasmine, Y.: Design of Adaptive Road Traffic Control System through Unified Modeling Language. International Journal of Computer Applications 14(7) (February 2001)
7. Clemente, P.J., Sánchez, F., Perez, M.A.: Modelling with UML Component-based and Aspect Oriented Programming Systems. In: Seventh International Workshop on Component-Oriented Programming at European Conference on Object Oriented Programming (ECOOP), Málaga, Spain, pp. 1–7 (2002)
8. Madl, G., Abdelwahed, S.: Model-based Analysis of Distributed Real time Embedded System Composition. In: Proceedings of the 5th ACM International Conference on Embedded Software, New Jersey, USA (2005)
9. Weis, T., Becker, C., Geihs, K., Plouzeau, N.: A UML Meta-model for Contract Aware Components. In: Gogolla, M., Kobryn, C. (eds.) UML 2001. LNCS, vol. 2185, p. 442. Springer, Heidelberg (2001)
10. A UML Profile for MARTE: Modeling and Analysis of Real-Time Embedded systems, Version 1.0.OMG Document Number:formal (November 02, 2009), Standard document, `http://www.omg.org/spec/MARTE/1.0`
11. Lavazza, G.Q., Venturelli, M.: Combining UML and formal notations for modeling real-time systems. ACM SIG.

# QoS Specification for Cyber-Physical Systems

Lichen Zhang

Faculty of Computer Science and Technology
Guangdong University of Technology
Guangzhou 510090, China
Zhanglichen1962@163.com

**Abstract.** Cyber-physical systems (CPSs) are physical and engineered systems whose operations are monitored. Cyber-physical systems having quality-of-service (QoS) requirements driven by the dynamics of the physical environment in which they operate, the description, control, management, consultation and guarantee of QoS are very complex and challenging work, Quality of Service(QoS) is directly related to system's performance. This paper proposes an aspect-oriented QoS modeling method based on UML and formal methods. We use an aspect-oriented profile by the UML meta-model extension, and model the crosscutting concerns by this profile. Finally, we illustrate QoS aspect-oriented specification via an example of real-time fire alarm system.

**Keywords:** QoS, Real-Time systems, Aspect-Oriented.

## 1 Introduction

A cyber-physical system (CPS)[1] is a system featuring a tight combination of, and coordination between, the system's computational and physical elements. Today, a pre-cursor generation of cyber-physical systems can be found in areas as diverse as aerospace, automotive, chemical processes, civil infrastructure, energy, healthcare, manufacturing, transportation, entertainment, and consumer appliances. The dependability of the software [1] has become an international issue of universal concern, the impact of the recent software fault and failure is growing, such as the paralysis of the Beijing Olympics ticketing system and the recent plane crash of the President of Poland. Therefore, the importance and urgency of the digital computing system's dependability began arousing more and more attention. A digital computing system's dependability refers to the integrative competence of the system that can provide the comprehensive capacity services, mainly related to the reliability, availability, testability, maintainability and safety. With the increasing of the importance and urgency of the software in any domain, the dependability of the distributed real-time system should arouse more attention.[2]

Fundamental limitations for Cyber-Physical Systems (CPS) include:

• Lack of good formal representations and tools capable of expressing and integrating multiple viewpoints and multiple aspects. This includes lack of robust formal models of multiple abstraction layers from physical processes through various layers of the information processing hierarchy; and their cross-layer analyses.

S. Lin and X. Huang (Eds.): CSEE 2011, Part II, CCIS 215, pp. 329–334, 2011.

• Lack of strategies to cleanly separate safety-critical and non-safety-critical functionality, as well as for safe composition of their functionality during humanin-the-loop operation.

• Ability to reason about, and tradeoff between physical constraints and QoS of the CPS.

Aspect-oriented programming (AOP) [3] is a new software development technique, which is based on the separation of concerns. Systems could be separated into different crosscutting concerns and designed independently by using AOP techniques. The QoS of dependable cyber physical system [4]is very complex, currently the QoS research still hasn't a completely and technical system, and there isn't any solution meeting all the QoS requirements. We design the QoS of dependable real-time system as a separate Aspect using AOP, and proposed the classification of complex QoS, divided into the timing, reliability and safety and other sub-aspects. These sub-aspects inherit t members and operations from the abstract QoSaspect. We design each sub-aspects through aspect-Oriented modeling, to ensure the Quality of dependable real-time system meeting the requirements of the dependability.

This paper proposes an aspect-oriented QoS specification method based on UML, mainly form an aspect-oriented profile by the UML meta-model extension[5], and model the crosscutting concerns by this profile.

## 2 Aspect-Oriented Specification of QoS

AOP provides mechanisms for decomposing a problem into functional components and aspectual components called aspects[4]. An aspect is a modular unit of crosscutting the functional components, which is designed to encapsulate state and behavior that affect multiple classes into reusable modules. Distribution, logging, fault tolerance, real-time and synchronization are examples of aspects. The AOP approach proposes a solution to the crosscutting concerns problem by encapsulating these into an aspect, and uses the weaving mechanism to combine them with the main components of the software system and produces the final system.

UML is acquainted to be the industry-standard modeling language for the software engineering community, and it is a general purpose modeling language to be usable in a wide range of application domains. So it is very significant to research aspect-oriented real-time system modeling method based on UML[6]. However they didn't make out how to model real-time systems, and express real-time feature as an aspect. In this section, we extend the UML, and present an aspect-oriented method that model the real-time system based on UML and Real-Time Logic (RTL). Real Time Logic is a first order predicate logic invented primarily for reasoning about timing properties of real-time systems. It provides a uniform way for the specification of both relative and absolute timing of events. The specification of the Object Constraint Language (OCL) [7] is a part of the UML specification, and it is not intended to replace existing formal languages, but to supplement the need to describe the additional constraints about the objects that cannot be easily represented in graphical diagrams, like the interactions between the components and the constraints between the components' communication. Since OCL is an expression language, it can be checked without an executable system. All these features turn out to be useful in representing QoS properties, which can be represented by the combination of precondition, post-condition and invariant in OCL. The QoS attributes are represented by the member variables of the class, and the QoS actions are represented by the methods[8]. They are checked at run time, before and after the calls so that the change of the QoS parameters of the system is monitored in a timely basis.[9]As the QoS concern needs to be considered inmost parts of the system, it is a cross-cutting concern.[10] Cross-cutting concerns are concerns that span multiple objects or components. Cross-cutting concerns need to be separated and modularized to enable the components to work in different configurations without having to rewrite the code.

## 3 Aspect Oriented Model of QoS of Fire Alarm System

An automatic fire alarm system is designed to detect the unwanted presence of fire by monitoring environmental changes associated with combustion. In general, a fire alarm system is either classified as automatically actuated, manually actuated, or both. Automatic fire alarm systems can be used to notify people to evacuate in the event of a fire or other emergency, to summon emergency services, and to prepare the structure and associated systems to control the spread of fire and smoke.

QoSConstraint Q1,Q2, Q5 of the fire alarm system is expressed as follows with formal technique RTL[11]:

[Q1]: $\forall i \exists j @(\uparrow data.collect,j) @(\downarrow stop,i) \geq COLLECT\_MIN\_TIME \wedge @(\downarrow data.open,j) - @(\downarrow stop,i) \leq COLLECT\_MAX\_TIME$

[Q2]: $\forall i \exists j @(\uparrow data.process,j) - @(\downarrow stop,i) \geq DATA\_PROCESS\_MIN\_TIME \wedge @(\downarrow data.process, j) - @(\downarrow stop,i) \leq COLLECT\_MAX\_TIME$

[Q5]: $\forall i \exists j @(\uparrow alarm.process.,j) - @(\downarrow command.send,i) \geq ALARM\_PROCESS\_MINTI ME \wedge @(\uparrow alarm.process,j) - @(\downarrow command.send,i) \leq ALARM\_PROCESS\_MAXTIME$

QoSConstraint Q3 and Q4 of the fire alarm system is expressed as follows with XML[12]:

[Q3]:
```
<QoS type="Level">
  <Firelevel   val = "FIRE_MAX_LEVEL"/>
</QoS>
```
[Q4]:
```
<QoS type="Contraint">
  <frame_rate val = "FRAME_RATE_CONSTRAINT"/>
  <audio_sample_rate val = "FRAME_RATE_CONSTRAINT"/>
</QoS>
```

We separate QoS from real-time fire system as an aspect, the aspect-oriented model of QoS of real-time fire system is shown as Fig.1.

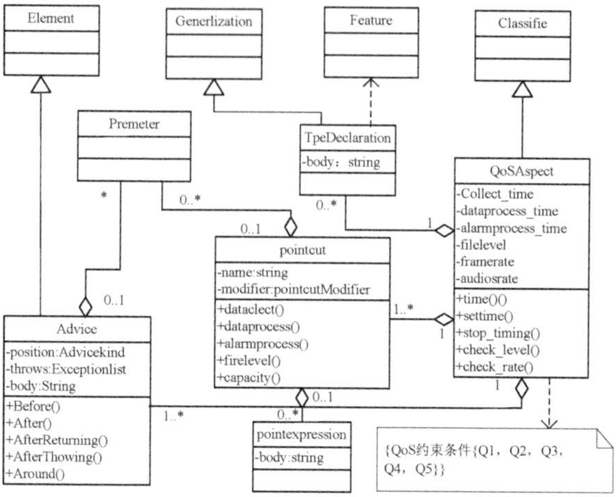


**Fig. 1.** QoS aspect-oriented model of Fire real-time system

We use QoSAspect to express QoS of real-time fire alarm system. The class diagram of Fire Real-time System with the aspect-oriented extension is shown as Fig.2.

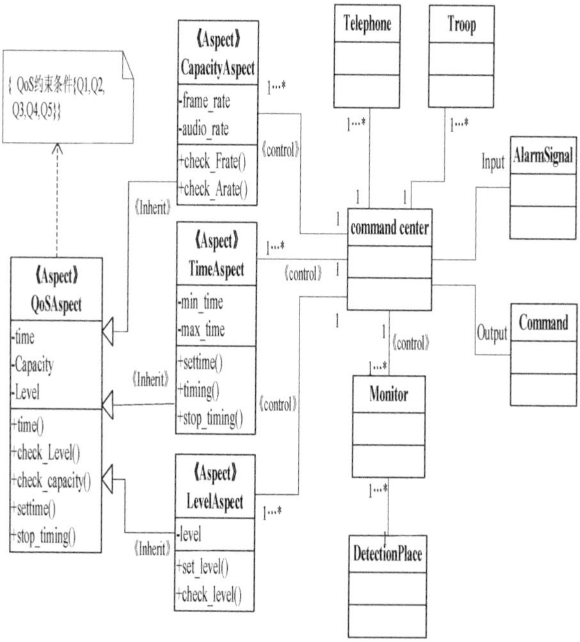


**Fig. 2.** Class Diagram of Fire Real-time System

The QoS Aspect Weaving Diagram of real-time fire alarm system is shown as Fig.3.

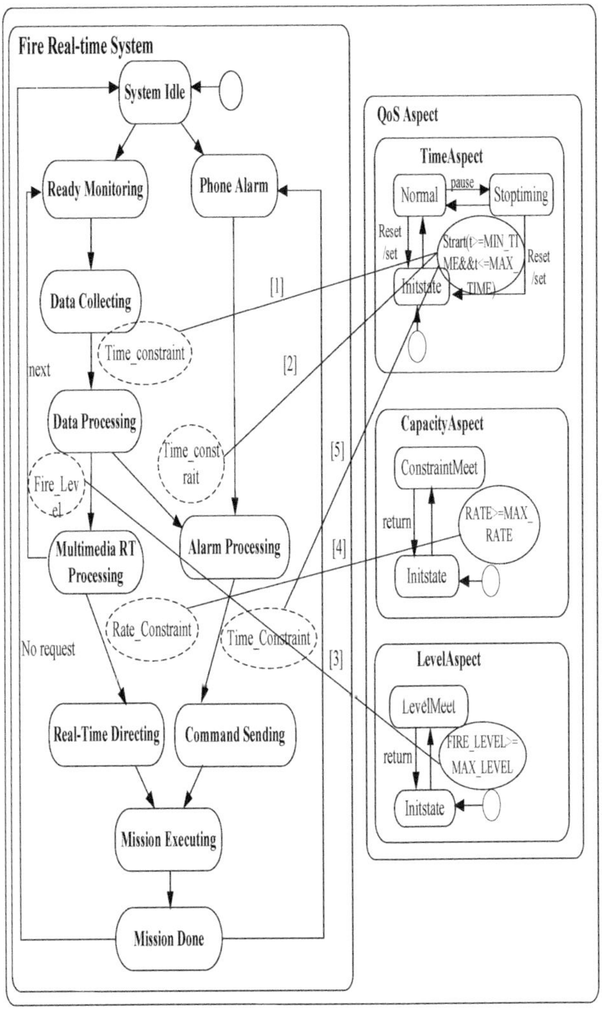


**Fig. 3.** QoS Aspect Weaving Diagram of Fire Real-time System

# 4 Conclusion

With the deepening of the dependable computing research, the system's dependability has becoming a important direction of the distributed real-time system, the modeling and design of dependable and distributed real-time system has become a new field. Dependable and distributed real-time systems need to be able to specify and control the QoS provided them, monitor it, and adapt as QoS changes. In this paper proposed an aspect-oriented QoS modeling method based on UML and formal methods, we illustrated QoS aspect-oriented modeling via an example of real-time fire alarm system.

Future works will focus an automatic weaver for aspect oriented model of QoS.

**Acknowledgments.** This work is supported by the Major Program of National Natural Science Foundation of China under Grant No.90818008.

## References

1. Lee, E.A.: Cyber physical systems: Design challenges. In: Proc. of the 11th IEEE Int Symp on Object Oriented Real-Time Distributed Computing, pp. 363–369. IEEE, Piscataway (2008)
2. Svizienis, A., Laprie, J.C., Randell, B.: Dependability of computer systems: Fundamental concepts, terminology, and examples. Technical report, LAAS-CNRS (October 2000)
3. Kiczales, G., et al.: Aspect-Oriented Programming. In: Proceedings of the 11th European Conference on Object-Oriented Programming (June 1997)
4. Frolund, S., Koistinen, J.: Quality of Service Specification in Distributed Object Systems. IEE/BCS Distributed Systems Engineering Journal 5, 179–202 (1998)
5. Aldawud, T.E., Bader, A.: A UML Profile for Aspect Oriented Modeling. In: Workshop on AOP (2001)
6. Clemente, P.J., Sánchez, F., Perez, M.A.: Modelling with UML Component-based and Aspect Oriented Programming Systems. In: Seventh International Workshop on Component-Oriented Programming at European Conference on Object Oriented Programming (ECOOP), Málaga, Spain, pp. 1–7 (2002)
7. Lavazza, G.Q., Venturelli, M.: Combining UML and formal notations for modeling real-time systems. ACM SIGSOFT Software Engineering Notes 26, 196–206 (2001)
8. Object Management Group. UML Profile for Modeling Quality of Service and Fault Tolerance Characteristics and Mechanisms Joint Revised Submission, OMG Document realtime/03-05-02 edition (May 2003)
9. OMG. UMLTM Profile for Schedulability, Performance, and Time Specification formal (January 02, 2005), `http://www.omg.org/cgi-bin/doc?`
10. Wehrmeister, M.A., Freitas, E.P., Pereira, C.E., et al.: An Aspect-Oriented Approach for Dealing with Non-Functional Requirements in a Model-Driven Development of Distributed Embedded Real-Time Systems. In: 10th IEEE International Symposium on Object and Component-Oriented Real-Time Distributed Computing, Santorini Island, Greece, May 7-9, pp. 428–432. IEEE Computer Society, Los Alamitos (2007)
11. Jahanian, F., Mok, A.K.: Safety Analysis of Timing Properties in Real-Time Systems. IEEE Trans. Software Eng. 12(9), 890–904 (1986)
12. Frolund, S., Koistinen, J.: QML: A Language for Quality of Service Specification. Technical Report HPL-98-10 (February 1998)

# The Exploration of Cloud Computing*

Yan-Xiao Li , Yue-Ling Zhang, Jun-Lan Liu, Shui-Ping Zhang, and Feng-Qin Zhang

Telecommunication Engineering Institute
Air Force Engineering University
Xi'an, Shaanxi, China
{YanXiaoLi,YueLingZhang,JunLanLiu,
Shui-PingZhang,Feng-QinZhang}@hotmail.com

**Abstract.** Cloud computing comes into focus as a way to increase capacity or add capabilities without investing in new infrastructure, training new personnel, or licensing new software. This paper gave an encompassing look at the history of cloud computing and the growing cloud economy. Shedding light on its infrastructure, service and technology, it introduced and displayed the basics of cloud computing fully and clearly as well as provided a through analysis of the virtualization, the key feature of the cloud computing. Cloud computing is at an early stage, still a lot need to be explored to expand it application spectrum and to bring advantages to the corporations.

**Keywords:** cloud computing, private cloud, public cloud, virtualization.

## 1 Cloud Computing Basics

A Cloud computing is Internet-based computing, whereby shared resources, software and information are provided to computers and other devices on-demand [1]. The name cloud computing was inspired by the cloud symbol that's often used to represent the Internet in flowchart and diagrams. It is a general term for anything involves delivering hosted service over internet.

The term of Cloud computing is not made overnight but evolves from some following concepts.

- Parallel computing

  Parallel computing, or high performance computing, super computing, as an important component of cloud computing [2], separates a problem into multiple computing tasks and executes them simultaneously in a parallel computer to achieving efficiency.
- Grid computing

  As a distributed computing pattern, gird computing connects the idle server, storage system and network together, forming a integrated system to provide end users with excellent computing power and storage capacities for the special tasks [3]. In essence it concentrates on managing the heterogeneous resources and guarantee sufficient service for the computing tasks. The difference between the

* Supported by Natural Science Foundation of Shaanxi Province under Grant 2009Jm8035.

S. Lin and X. Huang (Eds.): CSEE 2011, Part II, CCIS 215, pp. 335–340, 2011.

grid computing and the cloud computing is the resource belonging. In gird computing multiple disparate resources provide running environment for a single task while in cloud computing one integrated resource serves for multiple users.

- Utility computing

It emphasizes IT resources can be provided according to the needs of the users and be paid by the actual usage. It aims at reducing the cost of using and managing these resources, without spending a lot of money in investing.

The concept of cloud computing displays the features information service vividly in this ever changing information age. As a business computing model, Cloud computing allows consumers and businesses to use applications without installation and access their personal files at any computer with internet access. Cloud computing technology allows for much more efficient usage of information technology services over the internet at a dramatically lower cost and with dynamic scalability by centralizing storage, memory, processing and bandwidth.

## 2 Cloud Computing Infrastructure

Anyone who uses technology today expects things to be immediate, interactive, connected and fluid. Delivering technology-enabled services is more important than ever. Cloud computing is a technology that uses the internet and central remote servers to maintain data and applications. The services cloud computing provide are divided into following three categories, as shown in Fig.1.

- Infrastructure as a service (IaaS)
- Platform as a service (PaaS)
- Software as a service (SaaS)

Cloud computing offers an on-demand access to an elastic pool of assets—services, applications, servers, storage and networks. It's the elasticity that makes a cloud a cloud. You scale up or down as needed. And you pay only for what you use.

The advantages could be concluded as: scalability, availability, reliability, security, flexibility and agility, serviceability, efficiency.

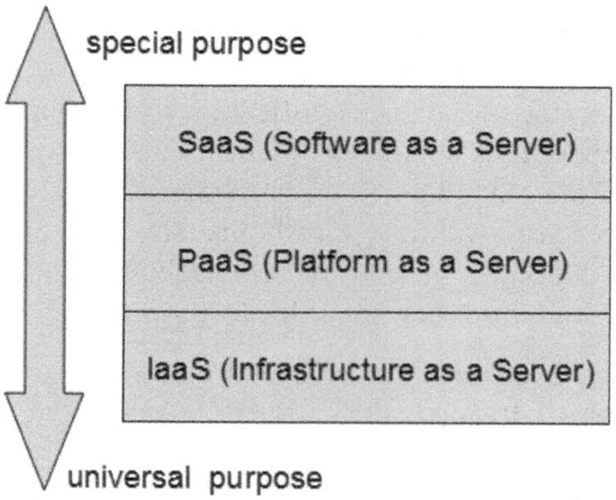


**Fig. 1.** Service category of cloud computing

The technology infrastructure of the cloud computing is divided into four layers: physical layer, resource pool layer, management midware layer and SOA layer, as shown in Fig.2.

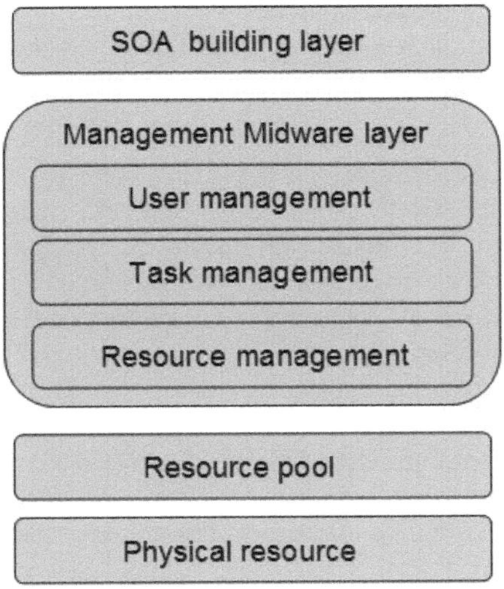


**Fig. 2.** Technology infrastructure of the cloud computing

The cloud model has four deployment models: private cloud, community cloud, public cloud, and hybrid cloud. Here is a definition for each deployment model.

- Private cloud

  The cloud infrastructure is operated solely for an organization. It may be managed by the organization or a third party and may exist on premise or off premise.
- Community cloud

  The cloud infrastructure is shared by several organizations and supports a specific community that has shared concerns (e.g., mission, security requirements, policy, and compliance considerations). It may be managed by the organizations or a third party and may exist on premise or off premise.
- Public cloud

  The cloud infrastructure is made available to the general public or a large industry group and is owned by an organization selling cloud services.
- Hybrid cloud

The cloud infrastructure is a composition of two or more clouds (private, community, or public) that remain unique entities but are bound together by standardized or proprietary technology that enables data and application portability (e.g., cloud bursting for load-balancing between clouds.

# 3 Cloud Computing and Virtualization

Virtualization is the key feature of the on-demand, self-service and pay-by-use model of cloud computing. As the focus of the data center, Virtualization is the logical representation of the resources. It is not constrained by the physical limitation, targeting the simplification of the representation, access management of the infrastructure, system and software, presenting standard interfaces for the system virtualization.

- Infrastructure virtualization
  Infrastructure virtualization contains network virtualization and storage virtualization. Network virtualization means integrating the hardware and the software of the network to providing visualizing technologies for the virtual network. Storage virtualization provides an abstract logical view of the physical storage devices to the users as that they can access the integrated storage resources by the view's standard logical interfaces.
- System virtualization
  The key of the system virtualization is to virtualize one or more virtual machine by using virtualization software. Virtual machine is a logical computer system which has complete hardware functions while running in a isolated environment by using system visualizing technologies, including guest operating system and its applications. The value of the system virtualization lies in the server virtualization which contains the following three key points: CPU, storage, device and I/Os. Live migration is required to integrate the dynamical system in a better way.
- Software virtualization
  The industry takes application virtualization and high level language virtualization. The former decoupling the application and the operating system, provides a virtual running environment to the applications. The latter aims solving the migration of the executable program among different computer architectures. The programs written by the high level language is complied as standard midware instructions which can be executed in an interpreted way.

# 4 Cloud Computing Benefits

The Cloud Computing concept is one based on the outsourcing of computing resources. Rather than purchasing certain hardware or software at relatively high capital expenditure, companies simply rent applications from service providers and access them over the Internet. As an alternative to managing traditional IT resources, a host company takes care of the background technicalities and you simply connect to your services through a secure Web-browser - using them whenever needed. These services are hosted online on secure servers - a network of computers collectively referred to as 'the cloud'.

Cloud computing is the business realization of the concepts of parallel computing, distributed computing and grid computing.

The benefits of deploying applications using cloud computing include reducing run time and response time, minimizing the risk of deploying physical infrastructure, lowering the cost of entry, and increasing the pace of innovation.

Cloud computing is a better way to run for the small and medium business. One of the advantages of cloud computing are those both small and medium sized businesses can instantly obtain the benefits of the enormous infrastructure without having to implement and administer it directly.

Cloud computing promises the agility CIOs need. It helps you deliver the precise portfolio of services your people need, each from the best source—public cloud, private cloud and traditional network.

Enterprises not only use public cloud computing services, they build private clouds to offer services to their business users. Successful Hybrid Delivery Cloud computing solutions are:

- Secure—they protect corporate and customer data.
- Open—they enable choice.
- Automated—they enable end-to-end service quality.
- Resilient—they meet defined service levels for availability, quality and performance.
- Seamless—they combine public and private clouds with traditional delivery to create a unified IT service portfolio.

Cloud Computing offers efficiency and utilization of resources based on a 'pay-as-you-go' model for improved scalability. The Cloud model can help companies avoid return on investment risk and uncertainty. Businesses can allocate more or less funds to services depending on how much they actually need, growing or reducing over time. This can be a massive benefit to the smaller company. With no initial cost barrier holding them back they can drastically decrease time to market.

The advantage of location independence is another key benefit of Cloud services. With the advancing of the Wireless Broadband technology users can connect to their software applications and store data and information instantly over the net, having the freedom from in-office infrastructure. For companies with multiple users in the field, applications no longer need to be installed directly onto the user's machine making them run more efficiently and securely.

Cloud computing promises to increase the velocity with which applications are deployed, increase innovation, and lower costs, all while increasing business agility.

## 5 Cloud Computing Dynamics

The research of cloud computing is initiated by the industry and later on getting increasing attention of the academia. Its far reaching effect will shape the way we use information. A paradigm shift to cloud computing will affect many different sub-categories in computer industry such as software companies, internet service providers (ISPs) and hardware manufacturers.

The cloud computing strategy of IBM fully integrates the On Demand (OD) concept, Service Oriented Architecture (SOA) design and Dynamic Infrastructure idea. IBM is customizing the cloud computing solution for different users and helping them to build their own computing environment. IBM presents the concept of Ensembles to remove the boundary of the physical resource pools.

Amazon Web Service (AWS) is the service which Amazon provided to users which can help them be free from the complex data center management and maintain and enjoying reliable, scalable and low cost information. Amazon realized the pay-by-use model.

Google cloud computing technology includes: Google File System (GFS) which provide the ability of data access and storage, MapReduce which makes the data parallel processing easy and available, Chubby which guarantees the simultaneous processing in the distributed environment and Bigtable which makes data management convenient.

The platform Microsoft provided is a PaaS cloud computing model which contains: Windows Azure, Microsoft.Net, Microsoft SQL, Live. Although Microsoft has released several versions, they are still under testing and at the stage of community technology previewing.

## 6 Conclusions

Cloud Computing is constantly evolving. It is a model for enabling convenient, on-demand network access to a shared pool of configurable computing resources (e.g., networks, servers, storage, applications, and services) that can be rapidly provisioned and released with minimal management effort or service provider interaction. Cloud computing is not only an advanced concept and technology but also a successful business computing model.

Begun life as "Grid Computing" which later matured to offer computing resources as a metered service, known as "utility computing", cloud computing is deeply and broadly effecting the IT infrastructure. In the future cloud computing will incorporate gird computing and show up as a new trend, which is Gloud, to overcome some of its defects. And unquestionably the era of Gloud will be more beautiful.

## References

1. Armbrust, M., Fox, A., Griffith, R., et al.: Abovethe Clouds: A Berkeley View of Cloud Compuitng, mimeo, UC berkeley, RAD Laboratory (2009)
2. Darlington, J., Guo, Y.-k., To, H.W.: Structured parallel programming: theoru meetspractice. Computing Tomorrow: Future Research Directions in Comuter Science Book Contents, 49–65
3. Foster, I., Kesselman, C., Tuecke, S.: The Anatomy of the Grid: Enabling Scalable Virtual Organizations. International Journal of High Performance Computing Applications 15(3) (2001)

# Application of SPAD and Vis/NIR Spectroscopy to Estimation Nitrogen Status of Oilseed Rape Plant

Dengsheng Zhu[1], Fei Liu[2], Yongni Shao[2], and Yong He[2,*]

[1] School of Information Engineering, Jinhua College of Profession & Technology, Jinhua 321007

[2] College of Biosystems Engineering and Food Science, Zhejiang University, Hangzhou 310029, China

yhe@zju.edu.cn

**Abstract.** The estimation of nitrogen status non-destructively in oilseed rape was performed using spectral reflectance with visible and near infrared reflectance spectroscopy, and SPAD values of the oilseed rape leaves of 30 plots were measured by a SPAD 502 chlorophyll meter, and the research was carried out at experiment field in Zhejiang University during growing season from 2007 to 2008. The SPAD 502 chlorophyll meter was applied to investigate the distribution rule of chlorophyll concentration in the oilseed rape. Regression model between the spectral reflectance and SPAD value was built by partial least squares (PLS). The correlation coefficient (r), root mean square error of prediction (RMSEP) and bias in prediction set were 0.9368, 3.4992 and 1.834e-07. The correlation between the first derivative of spectral reflectance of oilseed rape leaves and SPAD value were analyzed, and the results showed that good correlation coefficient was obtained in the range from 510 to 640 nm and 685 to 720 nm, and the maximum value for correlation coefficient was at the wavelength 707 nm. The linearity equation between the red edge index and chlorophyll concentration was also analyzed, with the correlation coefficient of 0.986. It is concluded that Vis/NIRS combined SPAD 502 chlorophyll meter was a promising technique to monitor nitrogen status in oilseed rape.

**Keywords:** Visible/near infrared spectroscopy, oilseed rape, Nitrogen, SPAD.

## 1 Introduction

Since the middle of the last century, oilseed rape growing has been increased from 2.7 million $hm^2$ to over 30 million $hm^2$. The rapeseed oil becomes an important source of the edible oil [1]. Nitrogen is one of the most critical nutrients for oilseed rape growth. An appropriate nutrient management can improve the oilseed rape yield and quality. Several methods are available for assessing the nitrogen status of the crop, such as Kjeldahl method and chlorophyll meter (soil plant analysis development, SPAD). The SPAD meter is a fast and nondestructive diagnostic tool, which is based on the regression between the light transmittance and foliar N of crop leaf [2,3]. Hence, most of the studies developed regression models based on vegetative indices

* Corresponding author.

S. Lin and X. Huang (Eds.): CSEE 2011, Part II, CCIS 215, pp. 341–347, 2011.

derived from several spectral bands [4,5]. Some researchers reported that the light absorption by chlorophyll at the wavelengths of 430, 450, 650 and 660 nm and light reflectance at 550 nm made leaves look green [6]. However, it is believed that the unique information of the unused bands in the case of hyper-spectral or spectroscopic techniques is largely lost in a specific narrow band. The acquired data for one object by spectral scanning always contain several hundreds of variables. Hence, it is very crucial to use multivariate calibration methods to extract the relevant part of the information for the very large data and produce the most reliable models compared to only regression model with several bands. Several common methods including principle component regression (PCR), SMLR, and partial least squares (PLS) regression have been developed for data mining in agricultural applications [7,8].

In this paper, the he correlation between chlorophyll concentration (SPAD value) and the spectral reflectance of oilseed rape leaves were analyzed using regression model built by PLS. The visible and near infrared reflectance spectroscopy (Vis/NIRS) method for nitrogen estimation was developed.

## 2 Materials and Methods

### 2.1 Checking the PDF File

The field experiment was conducted at the experiment farm of Zhejiang University in the 2007/2008 oilseed rape-growing season. The experimental design was the quadratic regression orthogonal design, with one variety of oilseed rape (*Brassica Napus*), three factors including nitrogen (N), phosphorus (P) and potassium (K), five rates of N, P and K, and two replicates. In this paper, only N was investigated. The five N concentration rates were 0, 90, 135, 180 and 270 kg N/ha, and the standard level was 180 kg N/ha. The number of plots was 30 (including replicates) and the plot size was 1.6 5 m.

### 2.2 Field Data Acquisition

Five plant leaf samples of each plot were selected and a total of 150 samples were obtained for spectral measurement. All leaf samples were used for Vis/NIRS analysis at 325-1075 nm. For each sample, three reflecting spectra were taken with a field spectroradiometer [FieldSpec® HandHeld (HH) (325-1075 nm), Analytical Spectral Devices (ASD), Inc., Boulder, CO], using RS2 V4.02 software for Windows designed with a Graphical User Interface (GUI) from ASD. The scan number for each spectrum was set to 10 at the same position, thus a total of 30 individuals were properly stored for later analysis. Considering its 25° field-of-view (FOV), the spectroradiometer was placed at a height of approximately 200 mm and 45° angle away from the surface of the certain oilseed rape leaf. A Minolta SPAD-502 chlorophyll meter was used to measure the chlorophyll concentration of the oilseed rape leaf, and SPAD readings for each sample were used as the referenced concentration of nitrogen status.

### 2.3 Data Pretreatment

Due to the potential system imperfection, obvious scattering noises could be observed at the beginning and end of the spectral data. Thus, the first and last 75 wavelength data were eliminated to improve the measurement accuracy. In addition, the influence of chlorophyll concentration to spectra was primary at the visible region, so the wavelengths from 400 to 800 nm were analyzed in the following. After that, the spectral data was preprocessed by the Savitzky-Golay smoothing with a window width of 7 (3-1-3) points, and then was the use of the multiplicative scatter correction (MSC). The pretreatments were implemented by "The Unscrambler V9.6" (CAMO PROCESS AS, OSLO, Norway).

### 2.4 Partial Least Squares (PLS)

PLS is a bilinear modeling method where the original independent information (X-data) is projected onto a small number of LVs to simplify the relationship between X and Y for predicting with the smallest number of LVs. In the development of PLS model, full cross-validation was used to evaluate the quality and to prevent overfitting of calibration models. The optimal number of latent variables (LVs) was determined by the lowest value of predicted residual error sum of squares (PRESS). The prediction performance was evaluated by the correlation coefficient ($r$), the root mean square error of calibration (RMSEC) or prediction (RMSEP) and bias. The ideal model should have higher $r$ value, lower RMSEC and RMSEP values, and lower bias. The models were carried out by "The Unscrambler V 9.6".

## 3 Results and Discussion

### 3.1 Chlorophyll Concentration in Oilseed Rape Plant and Leaves

The SPAD values were varied by the different growing periods of oilseed rape and the position of the growing leaf. In order to be able to choose a position as a representative, the distribution of chlorophyll concentration in the entire oilseed rape plant was interesting to be known. Fig. 1 shows the distribution of SPAD values in two randomly chosen oilseed rape plants measured on March 10th, 2008. SPAD values varied from the different ramus in the same oilseed rape plant, and were different from the top leaf to the middle one in the same ramus, which follow the rule of the top SPAD values was larger than the middle one. In the different ramus, the SPAD values of the bottom ramus were larger than the top one, and it follows the transportation rule of nitrogen in the oilseed rape plant. The nitrogen in the oilseed rape always transform from the bottom leaf to the top one.

The distribution of SPAD values in the same leaf was different (Fig. 2). The top SPAD values were larger than the bottom ones. The distribution of SPAD values at both sides was asymmetric, with one side approximately larger than the other side. In the same side, the SPAD values became smaller from the top to bottom. During measurement, the SPAD value for the entire leaf was obtained as an average value. The trend of chlorophyll concentration was diminished as time went on. In individual plot the SPAD values measured on March 31st were larger than those measured on March 21st, such as plot 3, 5, 12, 14. After the grouting period of oilseed rape, the chlorophyll concentration of leaf was tend to diminish in all plots from April 6th to the end of the measurement. Generally, the growing periods for oilseed rape can be divided into six phases, including planting, seeding, developing, budding, flowering and grouting. On the first and middle ten days of March, oilseed rape was approximately in the flowering phase. The oilseed rape plant developed persistently, and the ability for photosynthesis was increased ceaselessly. In the grouting phase, the nutrition of leaves started transporting to fruit corner, and the chlorophyll concentration diminished ceaselessly. So during the whole growing period of oilseed rape, the chlorophyll concentration follows the rule of increasing and then diminishing. In our study, the beginning of the measurement was from flowering phase (March 21st), so the chlorophyll concentrations of most plots were already reached maximum or exceeded maximum, the chlorophyll concentrations was at diminishing phase. The SPAD values for March 31st were larger than March 21st in some plots may affect by the factors, like the flowering phase was relatively long, the leaves were still at the developing phase and so on.

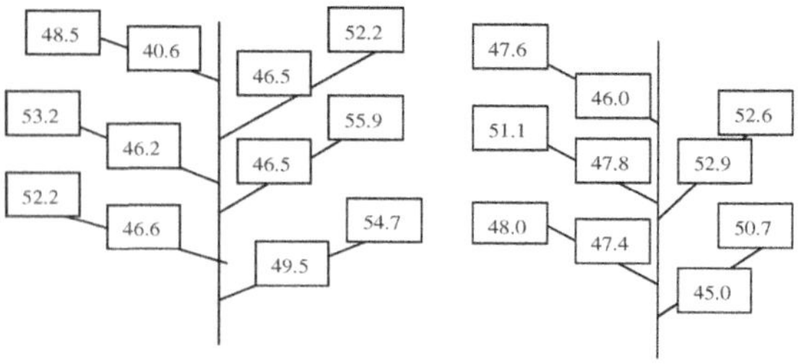


**Fig. 1.** The distribution rule of SPAD values in two randomly choosing oilseed rape

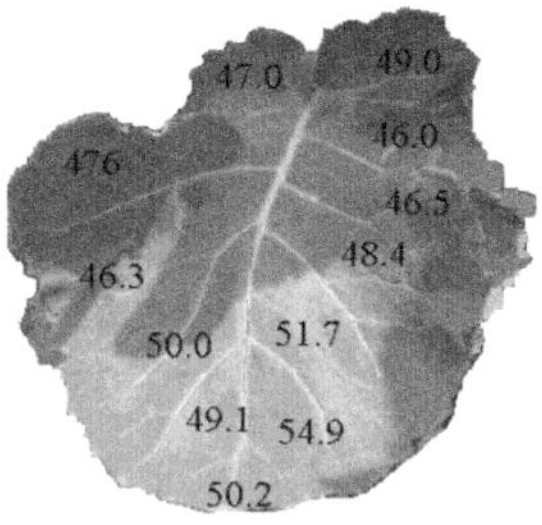


**Fig. 2.** The distribution rule of SPAD values in oilseed rape leaf

## 3.2 Vis/NIR Spectral Analysis and Method Development

### 3.2.1 Spectral Reflectivity Property of Oilseed Rape Leaves

Fig. 3 shows the spectral reflectivity of leaves in different growing periods, measured on March 10th, March 24th, April 13th, and April 28th from one plot. In the visible region, the ability for photosynthesis increased from planting phase to flowering phase, because of the ceaselessly growing of oilseed rape plants. After flowering, the nutrition of leaves started transporting to fruit corner, and the chlorophyll concentration diminished gradually. The reflectivity at red and blue ranges started rising. After the fruit corner matured, the lower leaves started shedding continuously, and the nutrition of leaves transported to fruit corner. At the same time, the chlorophyll started decomposing, and the photosynthesis ability for leaves was weakened. The chlorophyll reduced rapidly after supplying nutrition to fruit corner, and the reflectivity at red and blue ranges rose rapidly. At this time, the reflectivity at green waveband was still larger than the reflectivity at red and blue ranges, and there existed a small reflective apex in the visible region. The reflectivity at red and blue wavebands increased gradually as the oilseed rape growing boosted. From the red range to near infrared one, the reflectivity increased gradually as the leaf area index augmented. When the leaf area index increased to a certain value, the reflectivity went to stable. The internal structure of leaves started changing when fruit corner developed. At the same time, the near infrared reflectivity started declining gradually, until the oilseed rape plant became maturity.

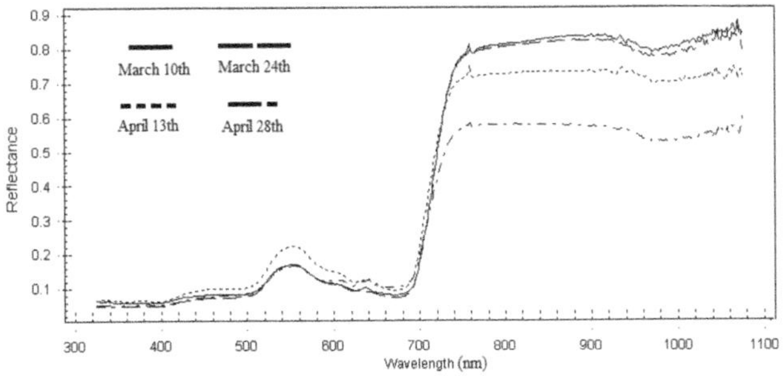


**Fig. 3.** The Spectral reflectivity of oilseed rape leaves in one plot from different growing periods

### 3.2.2 Data Pretreatment

Generally, the SPAD 502 chlorophyll meter measures the chlorophyll concentration around the verge of oilseed rape leaf and it is not corresponding to the spectral curves collected. Thus, ruinous experiment was done to find the relationship between the chlorophyll concentration for verge and middle of the leaves. SPAD 502 chlorophyll meter was used to measure the verge of eight random choosing leaves from different plots. In each leaf, it measured 30 points and used the averaging values as the final concentration. After that, the linear regression between the fringe and middle chlorophyll concentration values was analyzed. The result shows the correlation was high. The correlation coefficient was 0.981. The middle SPAD values could be obtained by measuring the SPAD values of verge. So, the SPAD values used later were preprocessed by this formula. The final data used to regression analysis were the average values of the fringe and middle SPAD values.

### 3.2.3 PLS Regression Model

PLS model between spectral reflectivity and chlorophyll concentration was developed after the spectral data preprocessed by S.Golay smoothing and MSC. Different LVs were applied to build the calibration models, and no outliers were detected in the calibration set during the development of PLS models. Among 150 samples, 125 samples were used as calibration sets and the left 25 samples as the prediction sets. Among calibration models, the models with 5 LVs turned out to be the best for prediction SPAD value by comparison using aforementioned evaluation standards in section 2.4. In the prediction models, the correlation coefficient ($r_p$), RMSEP and bias by optimal PLS models were 0.9368, 3.4992 and 1.834e-07. Hence, an acceptable prediction performance was achieved by this PLS calibration model.

### 3.2.4 Correlation Analysis between Spectra and Chlorophyll Concentration

The reflectance data of oilseed rape leaf may be affected by the height of spectroscopy, the variety of illumination intensity, and background factors of different regions. In order to eliminate the influence of background, and clearly reflect the spectral variation properties, the original spectral reflectivity was disposed with first derivative. The reflectivity values after first derivative were averaged by ten from wavelengths 400 to 800 nm, and 40 data were obtained in all. The correlation between chlorophyll concentrations (SPAD values) was built by SPSS12.0 software, and the result was shown in Fig. 4. The SPAD value changed dramatically over wavelengths from visible spectral region to near infrared spectral region. Generally, the SPAD value showed negative correlation with reflectance mainly at the wavelength region from 430-600 nm and 670-700 nm, and positive correlation mainly from 580-775 nm. Wavelength near 460 nm, 500 nm, 550 nm, 690 nm has higher negative correlations, and near 670 nm, 720-760 nm have higher positive correlations. From Fig. 4, it included two wavelengths just matched with green peak and red edge, which took an important part in the assessing of nitrogen status. Wavelength regions showing high correlation indicated that reflectance at these wavelengths might be important for the SPAD value.

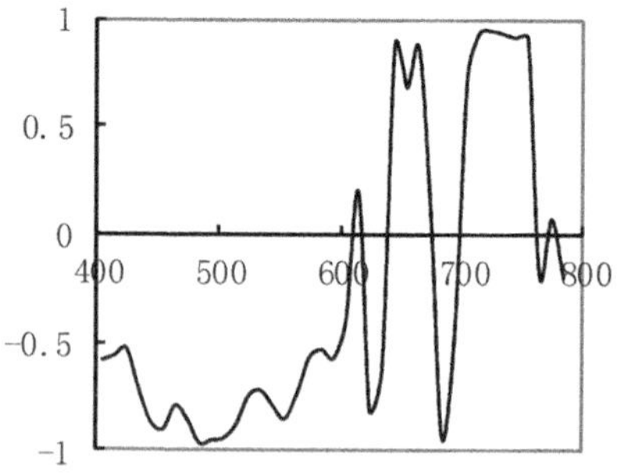


**Fig. 4.** Relationship between the first derivative of spectral reflectivity and chlorophyll concentration

### 3.3 The Relationship between the Red Edge and Chlorophyll Concentration

The red edge was the wavelength corresponding to the maximum slope of reflected spectral curve, which was caused by intensive absorption of red waveband and intensive reflection of near infrared wavebands. Set the chlorophyll concentration values as independent variable X, and red edge inflexion as dependent variable Y, the linear regression equation by linear regression analysis was obtained. The relativity between red edge inflexion and chlorophyll concentration was high, as the correlation coefficient was obtained at 0.986. So it was feasible to analyze chlorophyll concentration of oilseed rape leaves quantitatively using the spectroscopy.

## 4 Conclusions

The distribution rule of chlorophyll concentration in the oilseed rape was measured by the SPAD 502 chlorophyll meter, and the third ramous of the oilseed rape from top was measured, and the variation rule for the chlorophyll concentration in different growing periods was analyzed. The spectral reflectivity property of oilseed rape leaves in different growing periods was analyzed, and it indicated that the spectral reflectance for the leaves were smaller in the visible region and bigger in the near infrared region before flowering, but with reverse after flowering. Regression model between the spectral reflectance and chlorophyll concentration was built by PLS, and the $r_p$, RMSEP and bias in prediction set were 0.9368, 3.4992 and 1.834e-07. The correlation between the first derivative of spectral reflectance of oilseed rape leaves and chlorophyll concentration were analyzed, and the results showed that good correlation coefficient was obtained in the range from 510 to 640 nm and 685 to 720 nm, and the maximum value for correlation coefficient was at the wavelength 707 nm. The linearity equation between the red edge index and chlorophyll concentration was also analyzed, with the correlation coefficient of 0.986. It is concluded that Vis/NIRS combined SPAD 502 chlorophyll meter was a promising technique to monitor nitrogen status in oilseed rape.

**Acknowledgments.** This study was supported by 863 National High-Tech Research and Development Plan (2011AA100705), National Special Public Sector Research of Agriculture (200803037) and Agricultural Science and Technology Achievements Transformation Fund Programs (2009GB23600517).

## References

1. Yang, Z.C., Sun, L., Zhang, Y.: The Present Status of World Oilseed Rape Industry and Development Policy in China. World Agriculture 3, 10–12 (2003)
2. Chapman, S.C., Barreto, H.J.: Using A Chlorophyll Meter to Estimate Specific Leaf Nitrogen of Tropical Maize During Vegetative Growth. Agronomy Journal 89, 557–562 (1997)
3. Wu, F.B., Wu, L.H., Xu, F.H.: Chlorophyll Meter to Predict Nitrogen Sidedress Requirements for Short-Season Cotton (Gossypium hirsutum L). Field Crops Research 56, 309–314 (1998)
4. Best, R.G., Harlan, J.C.: Spectral Estimation of Green Leaf Area Index of Oats. Remote Sensing of Environment 17, 27–36 (1985)
5. Aparicio, N., Villegas, D., Casadesus, J., Araus, J.L., Royo, C.: Spectral Vegetation Indices as Non-Destructive Tools for Determining Durum Wheat Yield. Agronomy Journal 92, 83–91 (2000)
6. Min, M., Lee, W.S., Kim, Y.H., Bucklin, R.A.: Nondestructive Detection of Nitrogen in Chinese Cabbage Leaves Using Vis-NIR Spectroscopy. Hortscience 41, 162–166 (2006)
7. Martens, H., Naes, T.: Multivariate Calibration. Wiley, Chichester (1989)
8. Cen, H.Y., Yong, H., Huang, M.: Measurement of Soluble Contents and pH in Orange Juice Using Chemometrics and Vis-NIRS. Journal of Agricultural and Food Chemistry 54, 7437–7443 (2006)

# DOA Estimation of UCA Based on Reconstructing Matrix

ChangPeng Ji and MeiYu Wu

School of Electronic and Information Engineering,
Liaoning Technical University, Huludao, China
ccp@lntu.edu.cn, xiaoyuer860214@163.com

**Abstract.** A decorrelation algorithm UCA-TOEP for DOA estimation of uniform circular array (UCA) is proposed. After mode excitation method being used to change UCA into virtual uniform linear array, a special Toeplitz matrix is reconstructed in UCA-TOEP algorithm, and then DOA is estimated with MUSIC algorithm. Thus DOA of coherent signals could be effectively distinguished. Compared to virtual spatial smoothing algorithm, computation of UCA-TOEP is reduced and UCA-TOEP is with better performance in the condition of low SNR and few snapshots. Simulation results show that UCA-TOEP algorithm is feasible and it has better performance than spatial smoothing algorithm.

**Keywords:** DOA estimation; uniform circular arrays; coherent signals; UCA-TOEP; spatial smoothing.

## 1 Introduction

With people's pressing need for personal communication and high-speed data services, the achievement of wireless location for mobile terminal is still one of important research topics. Among DOA estimation algorithms, multiple signal classification (MUSIC) algorithms are classical algorithms of spatial spectrum estimation theoretical system [1]. In ideal conditions or of unrelated sources, MUSIC algorithm can work effectively and estimate azimuth of signal source. But when the sources are coherent, performance of MUSIC algorithm will deteriorate [2].

UCA is with planar array structure, allowing estimating two-dimensional direction-of-arrival (DOA); the pattern of UCA is relatively regular, and the main lobes of static pattern within 360° omni-directional are same. While the main lobes and side lobes of ULA formed by different incident angle are different, and the main lobes are growing with the increase of incident angle. Therefore, literatures [3-4] propose that UCA is transformed from array space to mode space by array preprocessing technology, to obtain virtual uniform linear array with translation invariance as same as ordinary ULA, thus spatial smoothing algorithm can be used to solute coherent signals. However, calculation of spatial smoothing method is relatively large due to smoothing, and in smart antenna systems, the relevant circumstance of signal sources is not learned in advance, so it is difficult to determine times of smooth and uneasy to achieve [5].

S. Lin and X. Huang (Eds.): CSEE 2011, Part II, CCIS 215, pp. 348–353, 2011.

This algorithm (here referred to as UCA-TOEP) estimates DOA (with MUSIC) by reconstructing a special Toeplitz matrix. Compared to DOA estimation of mode spatial smoothing, UCA-TOEP estimation algorithm has following advantages: because smoothing calculation isn't required, so the amount of calculation reduces; estimation is relatively effective with a small number of snapshots; be able to work better at low SNR.

## 2 Model and Virtual Preprocessing

In the assumption of narrowband signals, we suppose that there are $D$ signal sources impinging on M-element uniform circular arrays of antenna array, $\phi_i$ is the direction of incident signal. For simplicity, we assume that all sources are coplanar with arrays, and received signals are expressed as

$$X(t) = A(\phi)S(t) + V(t) \quad . \tag{1}$$

where $S(t)= [s_1(t), s_2(t),\ldots, s_D(t)]^T$ is vector of incident signal, $s_i(t)$ is complex envelope of the $i$th signal; $V(t)= [v_1(t), v_2(t),\ldots, v_{M-1}(t)]^T$ is noise vector in communication. Suppose that each array noise is spatial white noise that is independent, power equivalent to each other, and is not related to signal sources. $A(\phi)= [a(\phi_1), a(\phi_2), \ldots, a(\phi_D)]$ is steering vector of antenna arrays, $a(\phi_i)= a(\eta,\phi_i)= [\, e^{-j\eta\cos(\varphi_i - r_1)}, e^{-j\eta\cos(\varphi_i - r_2)}, \ldots, e^{-j\eta\cos(\varphi_i - r_{M-1})}\,]^T$ is uniform circular array direction vector corresponding to the $i$th signal. And $r_m=2\pi m/M$ is azimuth angle of the $m$th array, $\eta=2\pi r\sin\theta$, as we assume that all signal sources are coplanar with arrays, so $\theta=\pi/2$.

In order to make many DOA estimation methods of uniform linear array be able to be applied in circular array, mode excitation method is used to transform the output of arrays, getting

$$\tilde{X}(t) = QX(t) = \tilde{A}(\phi)S(t) + \tilde{V}(t) \quad . \tag{2}$$

where $Q = J^{-1}FIM$ is transfer matrix, $J = \mathrm{diag}\{1/j^n J_n(\beta)\}$, $F=[z_{-n}, z_{-n+1},\ldots, z_n]^H$, $n=-N,\ldots,0,\ldots, N$, $z_n=[1, e^{j2\pi n/M}, \ldots, e^{j2\pi n(M-1)/M}]^H$. $N=2\pi r/\lambda$ is the largest mode number of mode excitation; $J_n(\beta)$ is the first kind Bessel function of N-order.

When the number of array elements meets $M>2N+1$, the transformed array manifold of virtual uniform linear array has following approximate form

$$\tilde{A}(\phi) = \begin{pmatrix} e^{-jn\varphi_1} & \cdots & e^{-jn\varphi_D} \\ \vdots & \ddots & \vdots \\ e^{jn\varphi_1} & \cdots & e^{jn\varphi_D} \end{pmatrix} . \tag{3}$$

From Eq. 3, we can see that mode space of circular arrays is converted to virtual uniform linear arrays with $2N+1$ mode through preprocessing. Virtual ULA also has translation invariance as same as ordinary ULA. Namely, virtual ULA possesses Vandermonde form, which makes them being able to use techniques such as spatial smoothing to solute coherent signals.

## 3 UCA-TOEP Algorithm

Above analysis shows that central array element is the symmetry point of virtual array. Suppose there are $K$ coherent signals and $L$ irrelevant signals impinging on array. Data received from the $n$th virtual uniform linear array can be expressed as follows.

$$y_n(t) = s_1(t)\sum_{i=1}^{K} \rho_i e^{jn\phi_i} + \sum_{i=K+1}^{K+L} s_i(t)e^{jn\varphi_i} + v_n(t), n = -N,\cdots,0,\cdots,N \ . \tag{4}$$

where $\rho_i$ is the relative ratio of complex envelope of each signal and the first signal ($s_1(t)$), then $\rho_1$=1. From Eq. 4, we know that the data received from the central array of virtual uniform linear array is

$$y_0(t) = s_1(t)\sum_{i=1}^{K} \rho_i + \sum_{i=K+1}^{K+L} s_i(t) + v_0(t) \ . \tag{5}$$

Correlation operation of the above data received from array is defined as follows

$$r(n) = E[y_0(t)y_n^*(t)] = \sum_{i=1}^{K+L} g_i e^{-jn\varphi_i} + \tilde{\sigma}^2\delta_{0n} \ . \tag{6}$$

where

$$g_i = \begin{cases} (\sum_{q=1}^{K} \rho_q)P_{1,1}\rho_i^* + \sum_{q=K+1}^{K+L} P_{q,1}\rho_i^*, & i = 1,\cdots,K \\ (\sum_{q=1}^{K} \rho_q)P_{1,i} + \sum_{q=K+1}^{K+L} P_{q,i}, & i = K+1,\cdots,K+L \end{cases} . \tag{7}$$

And $\tilde{\sigma}^2 = E[v_0(t)v_0{}^*(t)]$ is noise covariance of the central array, $P_{m,d}$ $=E[s_m(t)s_d^*(t)]$, $m$, $d$=1,2,…,$K$,…,$K$+$L$.

Toeplitz matrix containing direction of arrival information of every signal can be constituted by $r(n)$

$$R = \begin{bmatrix} r(0) & r(1) & \cdots & r(N) \\ r(-1) & r(0) & \cdots & r(N-1) \\ \vdots & \vdots & \vdots & \vdots \\ r(-N) & r(-N+1) & \cdots & r(0) \end{bmatrix} = \tilde{A}_1 G \tilde{A}_1^H + \tilde{V}_1 \ . \tag{8}$$

where

$$\tilde{A}_1 = \begin{pmatrix} 1 & \cdots & 1 \\ \vdots & \ddots & \vdots \\ e^{-jn\varphi_1} & \cdots & e^{-jn\varphi_{K+L}} \end{pmatrix}. \tag{9}$$

$G$=diag($g_1,\ldots, g_{K+L}$), $\tilde{V}_1 = \tilde{\sigma}^2 I_{N+1}$.

$\tilde{A}_1$ will be full rank in row as long as DOA of signals is different from each other. Similarly, $G$ will be full rank as long as $g_i$ of $G$ isn't equal to zero. Then, rank($R$) =rank( $\tilde{A}_1 G \tilde{A}_1^H$ )=$K$+$L$, reaching the purpose to solute coherence. However, mode spatial smoothing algorithms can not solute coherence entirely. So, decoherence properties of TOEP algorithm is better than that of mode spatial smoothing algorithm.

Assumed that $K+L<N+1$, eigen-decomposition of $R$ is

$$R = \sum_{i=1}^{N+1} \lambda_i e_i e_i^H = U_S \Sigma_S U_S^H + U_N \Sigma_N U_N^H . \tag{10}$$

where $\lambda_1 \geq \lambda_2 \geq \ldots \geq \lambda_{K+L+1} = \ldots = \lambda_{N+1} = \tilde{\sigma}^2$ are $N$+1 eigenvalues corresponded. $e_i$ ($i$=1,2,…,$N$+1) is corresponding eigenvector. $U_S=[e_1,e_2, \ldots, e_{K+L}]$ is signal subspace corresponding to big eigenvalues, $\Sigma_S$ =diag[$\lambda_1, \lambda_2,\ldots, \lambda_{K+L}$]; $U_N=[e_{K+L+1},\ldots, e_{N+1}]$ is noise subspace corresponding to small eigenvalues, $\Sigma_N$ =diag[$\lambda_{K+L+1},\ldots, \lambda_{N+1}$]. To get estimation of direction of arrival, spatial spectrum is given by [6]:

$$P_{MUSIC}(\phi) = \frac{1}{a^H(\phi) U_N U_N^H a(\phi)} . \tag{11}$$

Do spectral peak searching for Eq. 11 to find the angles corresponding to the maximum points, and that is the incident direction of signals.

## 4 Simulation Experiment

Computer simulation is carried out with following conditions: Assume 8-element uniform circular array is used, radius is 0.7λ. There are three incident sources, angles of arrival being respectively 40°, 50° and 70°. 40° and 50° are coherent signal sources, while 70° is unrelated with the others. Fig. 1 shows that DOA estimation for 3 signals of conventional MUSIC algorithm, spatial smoothing algorithm of virtual linear array and the algorithm in this paper, and the input signal to noise ratio (SNR) is 10dB. Fig. 2 shows the relationship of output SNR changing with the number of snapshots, similarly, the input signal to noise ratio is 10dB.

Seeing from Fig. 1, conventional MUSIC algorithm can only reach one peak, but is unable to resolve coherent sources. Virtual linear array spatial smoothing method can clearly distinguish three sources. Now look at UCA-TOEP algorithm in this article also can well resolve them, and the peaks are slightly sharper.

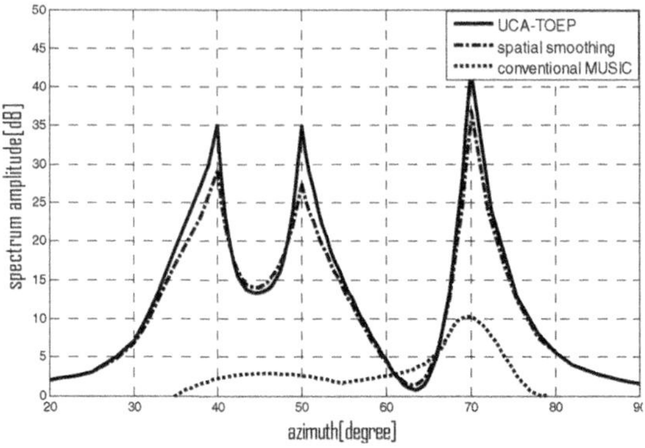


**Fig. 1.** Spatial spectrum of the three algorithms

Fig. 2 shows that under the condition of relatively small number of snapshots, the output SNR of conventional MUSIC algorithm is much lower than that of virtual spatial smoothing and UCA-TOEP algorithm. Virtual spatial smoothing algorithm has a short fluctuation when the number of snapshots is small, while UCA-TOEP algorithm is more robust than virtual spatial smoothing.

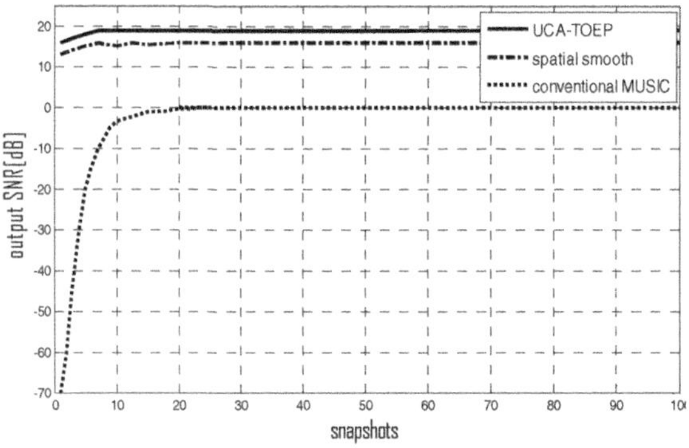


**Fig. 2.** The relationship of snapshots and output SNR

From Fig. 1, Fig. 2 and analysis above, we can see that most DOA estimation errors of the algorithm in this paper occur within small angle. Therefore, large deviation appearing twice continuously is with small probability. Hence, DOA estimation of this paper can be improved as following in order to obtain more accurate estimation.

(1). Calculate angle of arrival $\phi_a$ of user signal;
(2). estimate angle of arrival $\phi_b$ of user signal again;
(3). If $|\phi_a - \phi_b| < \xi$, then go to the last step, otherwise estimate $\phi_c$ again;
(4). two smallest difference values are given to $\phi_a$ and $\phi_b$, and then go to (3);
(5). take $\phi = (\phi_a + \phi_b)/2$ as the final results of DOA estimation for this time.

After taking above improvements, DOA estimation accuracy will increase, while calculation will increase 3 to 5 times. Fig. 3 shows that when SNR is less than -2dB, DOA estimation standard deviation of UCA-TOEP before and after improvement declines nearly to be linear with the increase of SNR.

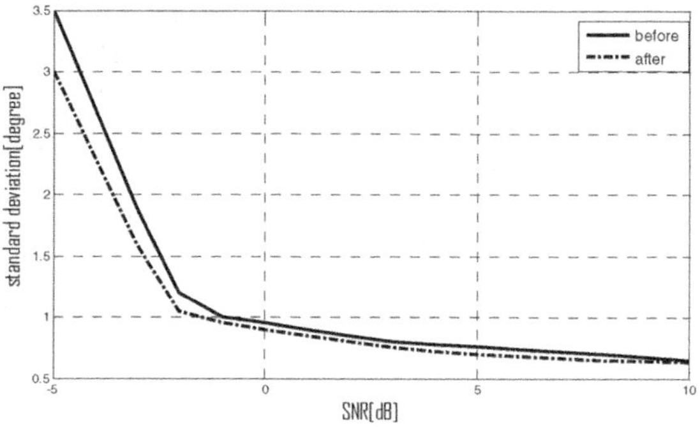


**Fig. 3.** The relationship of SNR and DOA estimation standard deviation

# 5 Conclusion

In this paper, Toeplitz matrix reconstructed instead of original covariance matrix of virtual linear arrays is used for DOA estimation. UCA-TOEP algorithm can solute coherent signals quite well, and can effectively resolve azimuth of coherent sources. Compared to spatial smoothing algorithm of virtual linear arrays, UCA-TOEP algorithm reduces computation as it doesn't need to do smoothing, and is slightly better in DOA estimation. Probability theory is also put forward to improve the algorithm, which makes accuracy and stability of DOA estimation improved further.

# References

1. Joseph, C.L., Theodore, S.R.: Smart Antennas for Wireless Communications IS-95 and Third Generation CDMA Applications, pp. 360–400. Prentice Hall PTR, Upper Saddle River (1999)
2. Lavate, T.B., Kokatek, V.K., Sapkal, A.M.: Performance Analysis of MUSIC and ESPRIT DOA Estimation Algorithms for Adaptive Array Smart Antenna in Mobile Communication. In: ICCNT, pp. 308–311 (2010)
3. Ren, W.D.: Research on DOA Estimation Algorithm in Smart Antenna. Shandong University, Jinan (2008)
4. Maheswara, R.K., Reddy, V.U.: Analysis of Spatial Smoothing with Uniform Circular Arrays. IEEE Trans. on S P, 1726–1730 (1999)
5. Gao, S.Y., Wang, Y.L., Chen, H.: Research on the MODE-TOEP Beamforming Algorithm. Journal of Electronics & Information Technology, 1096–1099 (2008)
6. Jin, R.H., Geng, J.P., Fan, Y.: Smart Antennas for Wireless Communication. Beijing University of Posts and Telecommunications Press, Beijing (2006)

# Determination of the Feasibility Regions of Control Variables in Coordinated Control System of Coal-Fired Unit

JunJie Gu, YanLing Ren, and Zhi Yang

School of Energy Power and Mechanical Engineering
North China Electric Power University, 071003, Baoding, China
renyanling0525@163.com

**Abstract.** In order to improve the quality of coordinated control system, set point value with multi-objective optimization has been widely cause the attention of researchers. However, the accuracy of the feasibility regions for the decision variables in fitness function is the foundation of the multi-objective optimization, and have an important influence to multi-objective optimization results. Take a 600 MW coal-fired unit as an example, according to the system mechanism builds a boiler-turbine dynamic model. And present a method, in this paper, which is using the iteration way and observing its physical constrain to determine the feasibility regions of control variables. The simplified model uses fuel quantity, turbine valve value and feedwater valve value as the control variables, to get the region of the pressure and control variables during 50%-100% load.

**Keywords:** multi-objective optimization; coordinated control; unit model; control variable; feasibility regions.

## 1 Introduction

The current operating context of a coal-fired power unit is characterized by many needs and requirements. So this situation may require a coal-fired power unit to achieve optimal operation under multiple operation objectives, such as minimization of load tracking error, minimization of throttling loss in the main steam, etc. In [1],[2].[3],many optimization techniques is presented for solving a multi-objective optimization problem. All study shows the importance of accuracy of the feasibility regions for control variables in power unit model. The first step toward multi-objective optimization is to identify the power unit's feasibility regions of pressure and control variables, defined by the set of all permissible operating points. The feasibility regions of pressure and control variables is the foundation of the multi-objective optimization, and have an important influence to multi-objective optimization results. In[1],[2].[3], the nonlinear mathematical model of a 160MW oil-fired drum type boiler-turbine-generator unit is used to obtain the feasibility regions of pressure and control variables. However, there are all large volume coal-fired units in China. Then, we must find a appropriate coal-fired unit model, and calculate the feasibility regions of control variables. Then, the coordinate control system of coal-fired unit could be optimized usefully. In this paper, based on [4,5] the model theory and design, and

S. Lin and X. Huang (Eds.): CSEE 2011, Part II, CCIS 215, pp. 354–360, 2011.

operation data of a 600MW unit to establish its dynamic model. And in the condition of satisfying theoretical significance, in 50%-100% load, the power unit's feasibility regions of pressure and control variables could be obtain from using a method of iteration. It would provide an accurate feasibility regions for multi-objective optimization of coordinated control system.

## 2 Unit Model

The unit model is mainly divided into two aspects, boiler model and turbine model. Due to the big volume and high complicacy of the boiler model, compared to turbine model, and it becomes hard to establish the accurate dynamic model. In [4,5,6],many scholars of the world get study into it. In [7,8] some scholars research unit plant nonlinearity dynamic model. This article based on the literature [4,5,6,7,8] in the theory side, and use the structure of model in literature [4,7], to confirm a 600MW unit dynamic model.

In the MCR condition, the power is 655.811MW, the stream flow is 2028 t/h, the stream pressure is 18.84MPa, the feedwater temperature 305 °C .

### 2.1 Dynamic Model of the Unit

$V_{wt}$, $V_{st}$, the volume of water, steam in the system( $m^3$ )

$\rho_w$, $\rho_s$, the density of saturated water, saturated( $m^3/kg$ )

$h_w$, $h_s$, the enthalpy of saturated water and saturated steam( $kJ/kg$ )

$q_f$, $q_s$, the mass flow rate of feedwater and stream flow( $kg/s$ )

$p_b$, the pressure of drum( $Pa$ )

$V_t$, the total volume of the evaporation system $m^3$, ($V_t = V_{st} + V_{wt}$)

$h_f$, the enthalpy of feedwater ( $kJ/kg$ )

$Q$, heat absorbed by water wall in the furnace( $kJ/s$ )

$m_m$, mass of evaporation system ( $kg$ )

$C_p$, specific hear capacity of metal ( $kJ/kg\cdot{}^oC$ )

$t_m$, metal temperature ( ${}^oC$ )

$E$, electric power ( MW)

$u_1$, the revolving speed of coal feeder that control the mass flow rate of fuel(%)

$u_2$, the position of steam valve actuator that control the mass flow rate of steam to turbine(%)

$u_3$, the revolving speed of pump that control the mass flow rate of feedwater to the drum(%)

To setup the boiler model, we assumed, see [6]:

(1) The work medium, in the whole evaporation system, is at saturation state, the pressure of anywhere in the system is the same. The density, enthalpy and other parameters of saturated water and saturation steam is only the function of drum pressure.

(2) The temperature of metal is the same as saturation temperature.

(3) The velocities of water and steam in the riser are equal.

(4) ignore delay and inertia of the unit.

The state equations are as follow:

$$e_{11}\frac{dV_{wt}}{dt}+e_{12}\frac{dp_b}{dt}=q_f-q_s \tag{1}$$

$$e_{21}\frac{dV_{wt}}{dt}+e_{22}\frac{dp_b}{dt}=Q+q_f h_f-q_s h_s \tag{2}$$

$$12\frac{dE}{dt}=0.537\,p_t\mu_2-E \tag{3}$$

$$p_t=p_b-0.00122\,(19.1\mu_1)^{1.5} \tag{4}$$

Let:

$$e_{11}=\rho_w-\rho_s \quad ; \quad e_{12}=V_{st}\frac{\partial\rho_s}{\partial p_b}+V_{wt}\frac{\partial\rho_w}{\partial p_b} \quad ; \quad e_{21}=\rho_w h_w-\rho_s h_s \quad ;$$

$$e_{22}=V_{st}(h_s\frac{\partial\rho_s}{\partial p_b}+\rho_s\frac{\partial h_s}{\partial p_b})+V_{wt}(h_w\frac{\partial\rho_w}{\partial p_b}+\rho_w\frac{\partial h_w}{\partial p_b})-V_t+m_m C_p\frac{\partial t_m}{\partial p_b}$$

## 2.2 Static Model of the Unit

Another purpose of this paper is to confirm the variables variation scope of the inputs. So the critical thing is to establish the static model, which bases on system mechanism. To setup the static model, we assumed:

(1) The steam flow equal to water flow.

(2) Heat absorbed by water wall equal to the decrease of stream heat and feedwater heat.

The static model are below:

$$q_f=q_s \tag{5}$$

$$Q=\left(h_s-h_f\right)q_f \tag{6}$$

$$0.537\left(p_d-0.00122(19.1u_1)^{1.5}\right)u_2-E=0 \tag{7}$$

Let $q_f = c_3 u_3$ , $q_s = c_2 u_2\left(-c_2 u_2 + \sqrt{(c_2 u_2)^2 + 4c_1^2 p_d}\right)/(2c_1)$ , $Q = c_4 u_1$ denotes specific feedwater flow, stream flow, heat absorbed by water wall. $c_1$ , $c_2$ , $c_3$ , $c_4$ can be calculated from data of a known running condition.

The above static equation can still be simplified, and get the relationship between the stream pressure and power. But for it lacks of physical condition limit, so it can not get a reasonable feasibility regions of the control variables. The additional constrain need to be added, that is the deviation of the drum water level[6]. In the operation condition of the unit, the deviation of the drum water level is needed to be control during ±50mm, so:

$$dh = \frac{1}{V_w I_L}(B_1 W - B_2 D + B_3 Q) \tag{8}$$

Let:

$r = h_s - h_w$ ; $I_L = I_p(\rho_w - \rho_s)$ ; $I_p = K_1 V_w + K_2 V_s + K_3 m_m$ ;

$K_1 = \rho_w \frac{dh_w}{dp_d} + \frac{r\rho_s}{\rho_w - \rho_s}\frac{d\rho_w}{dp_d}$ ; $K_2 = \rho_s \frac{dh_s}{dp_d} + \frac{r\rho_w}{\rho_w - \rho_s}\frac{d\rho_s}{dp_d}$ ;

$K_3 = C_p \frac{dt_m}{dp_d}$ ; $B_1 = V_w(\Delta i \frac{d\rho_w}{dp_b} + \rho_w \frac{dh_w}{dp_b}) + V_s\left[(\Delta i + r)\frac{d\rho_s}{dp_b} + \rho_s \frac{dh_s}{dp_b}\right] + m_m C_p \frac{dt_m}{dp_b}$ ;

$B_2 = V_w(\rho_w \frac{dh_w}{dp_b} \; r\frac{d\rho_w}{dp_d}) + V_s\rho_s \frac{dh_s}{dp_d} + m_m C_p \frac{dt_m}{dp_b}$ ; $B_3 = -V_w \frac{d\rho_w}{dp_d} - V_s \frac{d\rho_s}{dp_d}$ ;

Let $V_w$ denote evaporation system water volume, $V_s$ denote evaporation system steam volume, $I_L$ is the inertial coefficient of drum water level, , $B_1$ , $B_2$ , $B_3$ can be calculated from data of a known operating condition.

But this constrain is overly complex. It is better to have a simple constrain which is the upper pressure limit. It could be obtained from the actual operation situation.

## 3 Calculation of the Feasibility Regions of Control Variables

$u_1$ , $u_2$ , $u_3$ are the control variables and the inputs in the model. Calculation of the feasibility regions is divided into two parts, them are the upper limits of the pressure and inputs and the lower limits of the pressure and inputs. The limits were determined through an iterative process using the static model of power unit along the whole power range, one power value at a time. At any given power, the lower pressure limit is calculate as follow. First, giving a power value and a pressure value. Then increase the pressure value, and calculate all other model variables using the static model, and verify them to get physically meaningful values. Keep increasing the pressure value till physically reasonable equilibrium points could not be obtained[7]. Similarly, the lower limit is found, but the pressure is decreased at each iteration. The process is reduplicated over the whole power range. Meanwhile, the above method also can get the feasibility regions of control variables. The flow chart of Matter calculation is below:

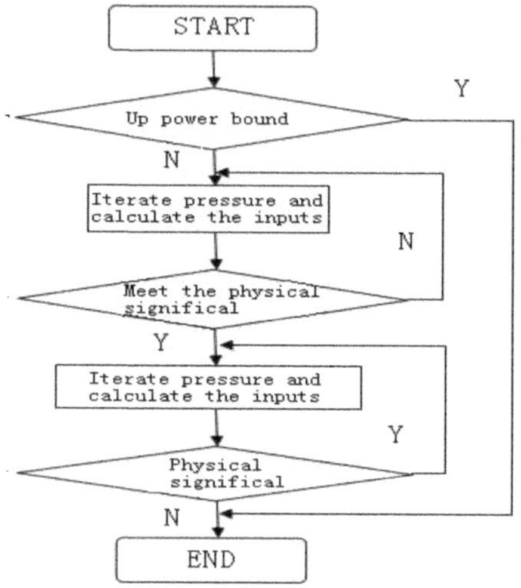


**Fig. 1.** Flow chart of Matter solve process

It contains two parts of logic iteration loop. One solves the lower limit of pressure, which iterates from a low pressure and gets the lower limit of pressure when control variables do not satisfy the physical constrain. It is called anti-logic. Another solves the upper limit of pressure, which iterates from the lower limit of the pressure and gets the upper limit of pressure, until control variables do not satisfy the physical constrain. At the same time the feasibility regions of control variables are obtained.

Figure two to five show the feasibility regions of pressure and control variables:

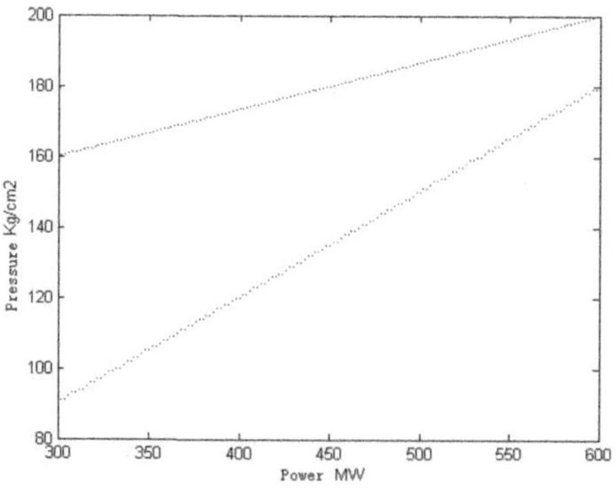


**Fig. 2.** Power-pressure operating window

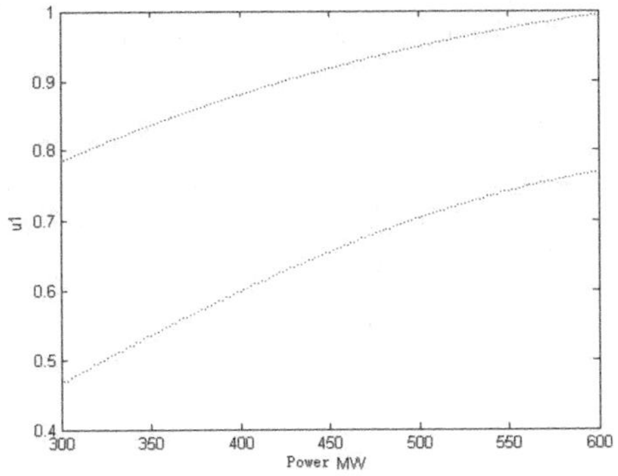


**Fig. 3.** The feasibility region of revolving speed of coal feeder

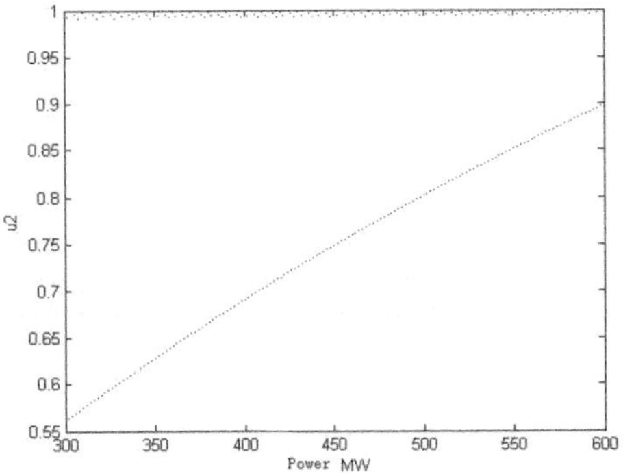


**Fig. 4.** The feasibility region of position of steam valve actuator

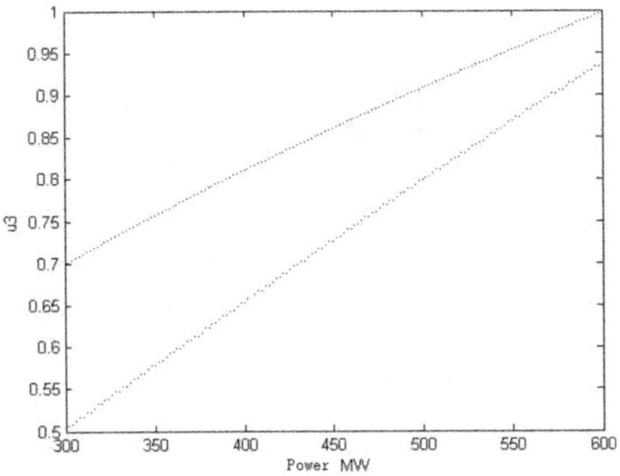


**Fig. 5.** The feasibility region of revolving speed of pump

It can be seen from the figures clearly that the feasibility regions of pressure and control variables will be generated, when a unit load demand is given. Then mufti-objective optimization can be made. From the actual operation point of view, it meets the operating condition, and also offers an accurate feasibility regions for the mufti-objective optimization of the unit.

## 4 Summary

Based on the model of 600MW coal-fired units as an example, this paper shows a method of iteration to Calculate the feasibility regions of the control variables. This method is simple and practical, and can get accurate feasibility regions. Furthermore, it also establishes a better basis for multi-objective optimization of the coal-fired unit.

## References

1. Gardens-Admirer, R., Lee, K.Y.: Multiobjective optimal power plant operation through coordinate control with pressure set point scheduling. IEEE Transactions on Energy Conversion 16, 115–122 (2001)
2. Heo, J.S., Lee, K., Garduno-Ramirez, R.: Multiobjective control of power plants using particle swarm optimization techniques. IEEE Transactions on Energy Conversion 21, 552–561 (2006)
3. Ying, G., Jianhong, L., Tiejun, Z.: Study on multi-objective optimization for control system of the thermal process and its application in boiler-turbine coordinated control. Thermal Power Generation 37, 35–42 (2008)
4. Astrom, K.J., Bell, R.D.: Drum-boiler dynamics. Automatica 36, 363–378 (2000)
5. Astrom, K.J., Ecklund, K.: A simplified nonlinear model of a drum boiler-turbine unit. Int J. Control 69, 145–169 (1972)
6. Liu, C., Niu, Y., Liu, J., Jin, X.: A boiler mode of 300MW power unit for control system performance studies. Journal of System Simulation 13, 59–61 (2001)
7. Liang, T.: Research of unit plant nonlinearity dynamic model. North China Electric power University (2005)
8. Liu, J., Tian, L., Zeng, D., Liu, X.: Analysis on the nonlinearity of load-pressure characteristics of a 660MW unit. Journal of Power Engineering 25, 533–540 (2005)
9. Garduno-Ramirez, R.: Overall intelligent hybrid control system for a fossil—fuel power unit. The Pennsylvania State University, Philadelphia (2000)

# Research and Practice on CAD / CAM Highly Skilled Personnel Training Based on Web

Qibing Wang

Engineering training center, Huaihai Institute of Technology,
Lianyungang, P.R. China, 222005
wqb123926@163.com

**Abstract.** In order to train applied and innovative compound talents with high skill and quality and to enrich the content of teaching practice, the practice base in the school should be established and improved. And the researches on the teaching model of CAD 3-D innovative design and CAM advanced manufacturing in engineering training center involved with undergraduate students, graduate students, teachers with double certificates (technical certificates and teacher certificate) or the related professional people in the surrounding areas should be executed. Building the open teaching practice laboratory for CAD / CAM CNC machining highly-skilled talents based on Web platform can better meet the demand of the society for high-quality and highly-skilled talents. After nearly two years of teaching and research, the author has successfully conducted many CAD / CAM high quality and highly skilled open trials based on network platform, resulting in realizing the desired purpose.

**Keywords:** Network, CAD / CAM, Highly-skilled talents, training model, Research and practice.

## 1 Introduction

At present, with the rapid development of the manufacturing sector, active involvement in high-tech research and developing distinctive applied talents have become an important trend in the development of science and engineering colleges [1,2]. In order to respond the "Education and Training Program on Excellence Engineer" project of the Ministry of Education and to meet the demand of times, our institution has established and improved the practice base in school , and focused on training students' practical ability of using new technologies and new techniques to solve practical engineering problems so as to train the applicative and innovative compound talents with strong learning ability and stored stamina as well as enrich the content of teaching practice[3]. There are many earth-shaking changes in practice environment, teaching equipment, teachers and practical training through the investment of thirty million RMB coming from joint projects of the central and local governments and the funs from the institute. The past single training level with basic training (metalworking) has developed into an initially established four-level training with engineering cognition, basic engineering training, general engineering training

S. Lin and X. Huang (Eds.): CSEE 2011, Part II, CCIS 215, pp. 361–366, 2011.

and mechanical and electrical integrated innovation. We have launched the research and practice on CAD / CAM training highly-skilled talents based on web platform in order to strengthen the content of the construction, expansion of training content, training highly qualified and highly skilled talents.

## 2 Exploration on CAD / CAM Highly-Skilled Talents Training Model

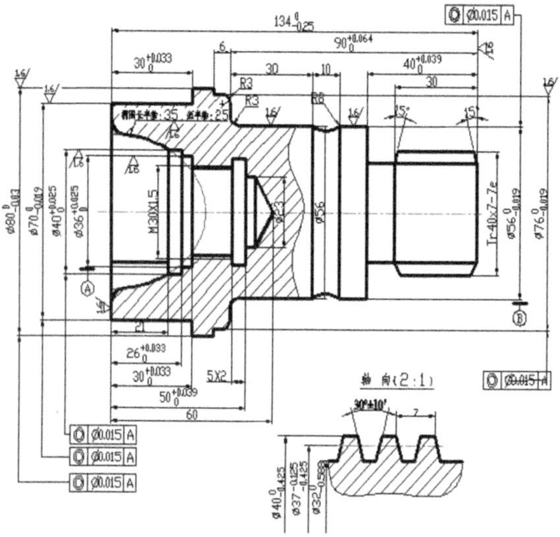


**Fig. 1.** The plan of part 1

How to satisfy the demands for further improving skills and techniques of the students who have already obtained the advanced certificate in CNC lathe (the current level of CNC lathe technician)? How to meet the demand for improving the techniques of advanced manufacturing industry? How to meet the demand for improving capability and quality of the teachers with double certificates in vocational college? All these change us in the aspect of advanced manufacturing practice teaching!

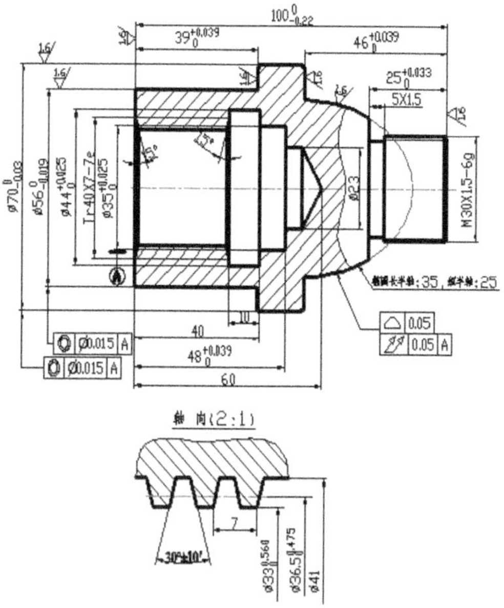


**Fig. 2.** The plan of part 2

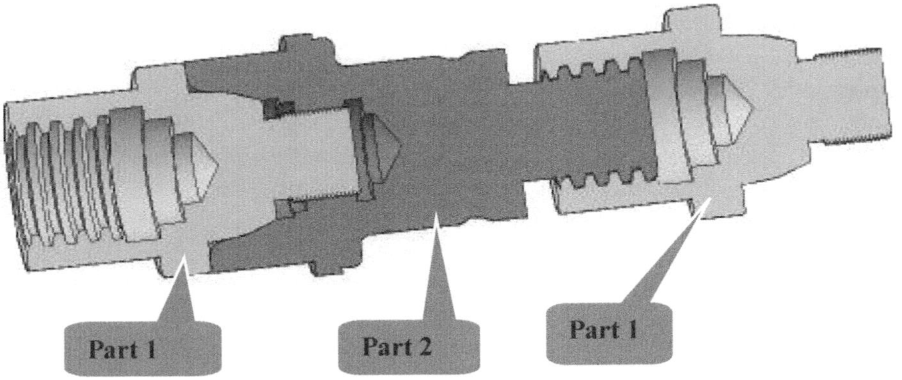


**Fig. 3.** Assembly section drawing in three dimensional concept

## 2.1 CAD Model of Capacity and Improve Training and Practice

Firstly, we must determine the training content and training drawings, the right side of part 1 in Figure 1 and the left cavity of part 2 in Figure 2 are trapezoidal threads cooperating with each other. The left side of part 1 in Figure 1 and the right side of part 2 in Figure 2 commonly have a spiral surface coordination as well as elliptical surface coordination. The assembly section in three-dimensional concept is shown in Figure 3. The technical requirements are very high in the assembly, dimensional tolerance of parts, geometric tolerance. Since most students have a basic AutoCAD knowledge that is widely used. The AutoCAD is installed in each computer in the CNC equipment room, and the students are required to familiarize the software applications. The students are demanded to not only draw outlines of the plane in Figure 1 and Figure 2, but also draw three-dimensional map. What's more, they should know the assembly and checking whether there is any interference between the parts just as what is shown in Figure 3 so that hidden quality problem in the upstream can be controlled. Then, the students are required to design an assembly, practice design and drawing and test assembly so that the students can discover engineering problems in and solve them timely in the product design phase.

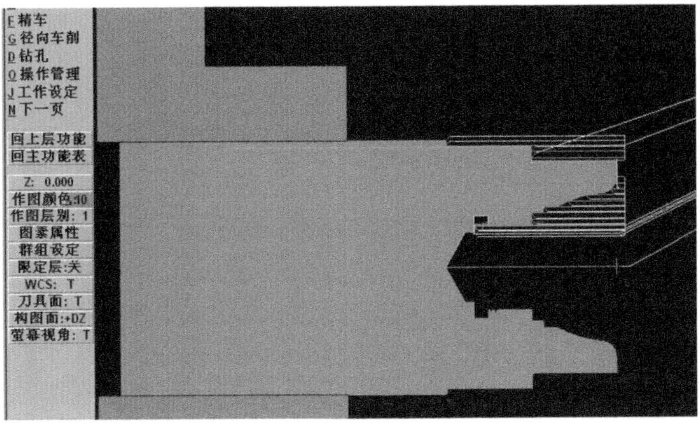


**Fig. 4.** The path simulation drawing of The left part of the cutting tool in part 1

## 2.2 Research and Practice on Improving CAM Ability Training Mode

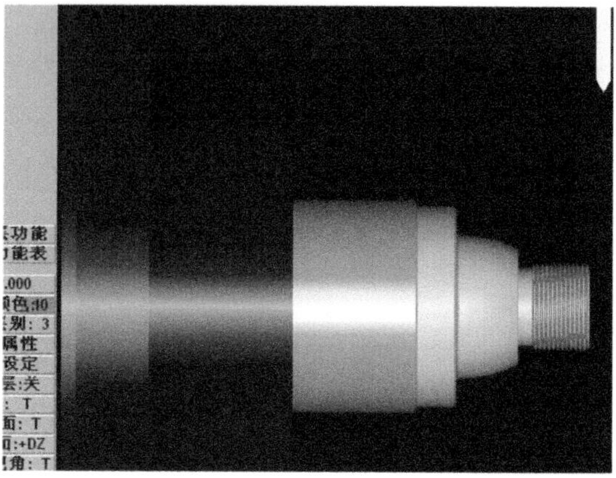

**Fig.5.** The verification drawing of the Entity cutting of the right side of part 2

```
%
O0001
G21
(PROGRAM NAME - O1030   DATE=DD-MM-YY - 19-02-11  TI
(TOOL - 1 OFFSET - 1)
(CENTER DRILL - 6. DIA.)
G0 T0101
G97 S300 M03
G0 G54 X0. Z5. M8
Z2.
G98 G1 Z-60. F100.
G0 Z5.
M9
G28 U0. W0. M05
T0100
M01
(TOOL - 2 OFFSET - 2)
(OD ROUGH RIGHT - 80 DEG.  INSERT - CNMG 12 04 08)
G0 T0202
G97 S500 M03
G0 G54 X80.4 Z3.2 M8
G1 Z2.7 F100.
Z-59.8
X84.
X86.828 Z-58.386
```

**Fig. 6.** Automatically generated NC machining program

Most students have used Master CAM software in early open trial and early integrated training. The Master CAM combines functions two-dimension such as 2-D drawing, three-dimensional solid modeling, surface design, voxel merging, NC programming, tool path simulation and simulation processing, demanding little in the operating environment, producing the best results in design and CNC machining operations for the users[3-7]. Master CAM is very extensively applied, and it is compatible with other CAD / CAM software. Therefore, we ask participants to be proficient in Master CAM software, and execute skillfully graphic conversion between soft wares in CAD / CAM, rough work piece, tool, process parameters design such as rough and finishing machining on the parts in Figure 1 and Figure 2. In addition, they should master the tool path simulation shown in Figure 4 and the verification of entities cuts shown in Figure 5, and they also should know how skilfully modify process parameters, automatically generate the NC machining program shown in Figure 6 (Note: Due to space limitations, only a small portion is shown). The participants indicate that the automatic programming is more efficient and more accurate than the manual programming. Practice shows that the more complex the part surface is , the more superior the automatic programming is.

# 3 Research and Practice on CAD / CAM Highly Skilled Talents Training Based on Web

How to input the 1,000 segments of processing procedures generated by the parts in Figure 1 and Figure 2 into the CNC machine? Manual input? Clearly the efficiency is too low, so this scenario is not desirable. Then copy them through the memory card? or through the network transmission? In recent years, we have converted the numerical control room computers, servers, switches, CNC machine tools into the network DNC system through technology research and CNC equipment modification. Thus, the students can open folder with the program, select the machine code, and then send the program through the network, which is shown in Figure 7.

So how to process the parts above successfully by CNC machine tools? While most students have pre-intermediate and advanced NC level training exercises, the technical requirements of these parts are quite high. Therefore, the precision accuracy of machine tools such as the precision of guide, spindle and tool holder should be verified and the measuring tools, cutting tools, rough, and chip solution should be also verified before the process. Then you can process the parts step by step according to the operation rules and flow of work. In the processing, the process testing should be paid attention to prevent the parts being scrapped. As for the processed parts, the participants can test the qualities of the parts in the drawing, and then, analyze the causes, propose measures for improvement and the process optimization solutions according the defects to further improve the quality of the parts. Through the above process of practice, students' ability to process difficult high-precision parts can be improved, and so be their abilities to analyze and address quality defects. The students with interest can also participate in the level appraisal of vocational skills organized by labor personnel department, and the students with good results can obtain the CNC lathe technician certificate.

**Fig. 7.** The path drawing of program transmission

# 4 Conclusion

In short, in order to respond the "Education and Training Program on Excellence Engineer" project of the Ministry of Education and to train the "Double-identity" teachers and to cultivate the applicative and innovative compound talents with strong learning ability and stored stamina in our college, we have launched several researches and practices in the above aspects. Through nearly two years of teaching,

CAD / CAM highly skilled talents training and open trial are successfully held based on web, developing the students' ability of comprehensive utilization of previously learned professional theory and practical experience. It has also improved the ability of processing the complex parts, increased the capacity of analysis and processing quality defects, cultivated the practice innovation capacity of students and enhanced the overall quality, which enriches the content of teaching practice and improves the practice base in college. The lesson plans, courseware, training atlas and examination question bank related to CAD / CAM highly skilled talents training based on web platform have been initially established.

**Acknowledgements.** Teaching Reform Fund for Huaihai Institute of Technology (5509002); Ministry of Finance Central and local governments jointly-built fund projects.

## References

1. Smith, T.F., Waterman, M.S.: Identification of Common Molecular Subsequences. Mol. Biol. 147, 195–197 (1981)
2. Yuan, X.F., Liu, J.F.: Journal of Jiangsu Teachers University of Technology (2), 94 (2004) (in Chinese)
3. Kuang, W.H., Xia, J.Y.: Journal of Chengdu College of Education (8), 76 (2006) (in Chinese)
4. Wang, Q.B., Wen, X.Q., Wang, Z.M., et al.: Research and Exploration in Laboratory (4), 116 (2009) (in Chinese)
5. Wang, Q.B., Wen, X.Q., Wang, J.S., et al.: Manufacturing Automation (1), 21 (2008) (in Chinese)
6. Bie, H.L., Zhu, C.M.: Machinery Design and Manufacture (12), 83 (2006) (in Chinese)
7. Ding, G., Zhang, J.Q.: Machinery (12), 52 (2004) (in Chinese)
8. Wang, Z.C., He, W.P., Li, T.C.: Journal of Zhengzhou Railway Vocational &Technology (6), 45 (2007) (in Chinese)

# On the Edge-Balance Index Sets of the Network Graph

Yuge Zheng and Hongjuan Tian

School of Mathematics and Information Science,
Henan Polytechnic University, Jiaozuo 454003, China
zhengyuge@hpu.edu.cn, tianhj1985@126.com

**Abstract.** Let $G$ be a graph with vertex set $V(G)$ and edge set $E(G)$, and let $Z=\{0,1\}$ .Any edge labeling $f:E(G)\to Z_2$ of a graph induces a vertex labeling $f^+:V(G)\to Z_2$. Defined by $f^+(v)=i$ if $v$ is incident to more $i$-edges than $(1-i)$-edges, and $f^+(v)$ is unlabeled if $v$ is incident to an equal number of 0-edges than 1-edges. Denote by $e_f(i)$ and $v_f(i)$ the number of edges and vertices, labeled $i$ .We call edge-friendly if $|e_f(0)-e_f(1)|\le 1$. Define the edge-balance index set of $G$ as: $\{|v_f(0)-v_f(1)|: f$ is an edge friendly labeling of $G\}$. In this paper, we will study the edge-balance index sets of the network graph $C_{2^m}\times P_{m_2}$ $(m\ge 2)$, and solve formula proof and graphic tectonic methods.

**Keywords:** edge-friendly labeling; edge-balance index set, $C_{2^m}\times P_{m_2}$.

## 1 Introduction

The means of using the vertex and edge of graph labeling function theory to study the graphs is firstly introduced by B.M. Stewart in 1966. Over the years, many domestic and foreign researchers work on some studies in this field, and have gained a series of achievements, such as graphical construction method of edge-magic graphs and edge-graceful graphs, their theoretical study and so on. Boolean index sets of graphs are that make the vertex sets and the edge sets of graphs through the mapping function with $Z_2$ corresponding, to study the characteristics of various types of graphs and the inherent characteristics of graphs, Boolean index set theory can be applied to information engineering, communication networks, computer science, economic management, medicine, etc. The edge-balance index set is an important issue in Boolean index set.[1]

Let $G(V,E)$ be a graph with vertex set $V$ and edge set $E$. Integers set $Z=\{0,1\}$. Given an edge labeling functions $f:E(G)\to Z_2$ .That is to say, $\forall e\in E(G)$, $f\{e\}=0 or 1$. According to the edge labeling $f$, we define an associated partial vertex labeling $f^+:V(G)\to Z_2$ as follow. Define $f^+(V)$ to be 0 if it is incident to more 0-edges than 1-edges and 1 if it is incident to more 1-edges than 0-edges, if the vertex $v$ is incident to an equal number of 0-and 1-edges, leave it unlabeled. Hence $f^+$ is a partial function. For each it $\{0,1\}$, let $e_f(i)=|\{uv\in E:\ f(uv)=i\}|$, and

S. Lin and X. Huang (Eds.): CSEE 2011, Part II, CCIS 215, pp. 367–372, 2011.

$v_f(i)=\left|\{v\in V: f^+(v)=i\}\right|$. If no ambiguity occurs, we could omit the subscript and simply write $e(i)$ and $v(i)$ respectively.[1].

**Definition 1[1].** An edge labeling $f: E(G)\rightarrow\{0,1\}$ of the graph $G$ is said to be edge-friendly if $\left|e_f(0)-e_f(1)\right|\leq 1$ where $e_f(i)$ is the cardinality of the set $\{e\in E(G): f(e)=i\}$ for $i\in\{0,1\}$.

**Definition 2[1].** The value $\left|v_f(0)-v_f(1)\right|$ is called on edge-balance index of $G$ under an edge-friendly labeling $f$. The edge-balance index set of the graph $G$, denoted by $EBI(G)$, is defined as the set of all possible edge-balance indices of $G$, that is the set $\{\left|v_f(0)-v_f(1)\right| : f$ is an edge friendly labeling of $G\}$.

Specifically, the maximum of edge-balance index denoted by $\max\{EBI(G)\}$

**Definition 3.** The sign $P_{m_2}$ denotes a path graph containing paths with $m$ vertex and every vertex has 2 bifurcates except the last vertex of path.

**Definition 4.** The sigh $C_{2^m}$ denotes a circle graph, which is the inner circle containing 2 vertex and the vertex number of $m$ circle from inside to outside successively increase by power exponent.

**Definition 5.** The sigh $C_{2^m}\times P_{m_2}$ denotes a power circle nested graph by the circle graph $C_{2^m}$ and the path graph $P_{m_2}$.

**Example:** The figure 1 gives several $C_{2^m}\times P_{m_2}$ graphs

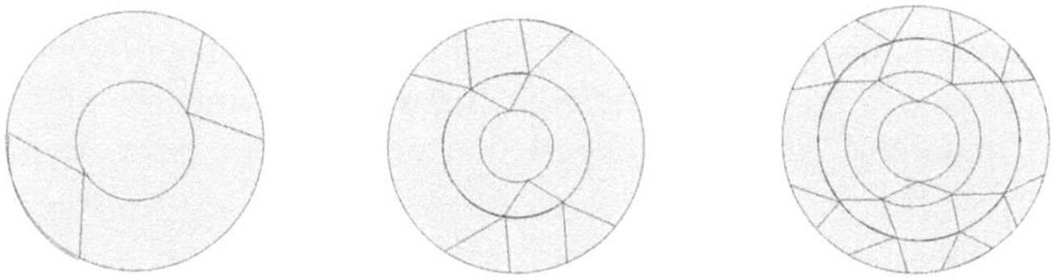

$C_{2^2}\times P_{2_2}$ $C_{2^3}\times P_{2_3}$ $C_{2^4}\times P_{2_4}$

**Fig. 1.**

**Definition 6.** There are m circles in the graph $C_{2^m}\times P_{m_2}$ $(m\geq 2)$, respectively using $\{(1)_1,(1)_2\},\{(2)_1,(2)_2,(2)_3,(2)_4\},\cdots,\{(m-1)_1,(m-1)_2,\cdots,(m-1)_{2^{m-1}-1},$

$(m-1)_{2^{m-1}}\}, \{(m)_1, (m)_2, \cdots, (m)_{2^m-1}, (m)_{2^m}\}$ as the vertices label from inside to outside $m$ circles clockwise; Symbols $(j)_i$ says points in the ring $j$, I says the point $i$ of the circle.

Among them:

$$(1)_1 \to (2)_1 \to (3)_1 \to \cdots \to (m-1)_1 \to (m)_1;$$
$$(1)_1 \to (2)_1 \to (3)_1 \to \cdots \to (m-1)_1 \to (m)_2;$$
$$(1)_1 \to (2)_1 \to (3)_1 \to \cdots \to (m-1)_2 \to (m)_3;$$
$$(1)_1 \to (2)_1 \to (3)_1 \to \cdots \to (m-1)_2 \to (m)_4;$$
$$\vdots \qquad \vdots \qquad \vdots$$
$$(1)_1 \to (2)_{2^1} \to (3)_{2^2} \to \cdots \to (m-1)_{2^{m-1}} \to (m)_{2^{m-1}-1};$$
$$(1)_1 \to (2)_{2^1} \to (3)_{2^2} \to \cdots \to (m-1)_{2^{m-1}} \to (m)_{2^{m-1}};$$
$$(1)_2 \to (2)_{2^1+1} \to (3)_{2^2+1} \to \cdots \to (m-1)_{2^{m-2}+1} \to (m)_{2^{m-1}+1};$$
$$\vdots \qquad \vdots \qquad \vdots$$
$$(1)_2 \to (2)_{2^2} \to (3)_{2^3} \to \cdots \to (m-1)_{2^{m-1}} \to (m)_{2^m};$$

The above says that the $C_{2^m} \times P_{m_2}$ $(m \geq 2)$ have $2^m$ paths.

We call an edge labeling of $f(xy)$ is 1 to be 1-edge and an edge labeling of $f(xy)$ is 0 to be 0-edge, if an vertex labeling is 1 to be 1-vertex and an vertex labeling is 0 to be 0-vertex when the context is clear.

## 2 Proof

For the edge-friendly labeling of the graph $C_{2^m} \times P_{m_2}$ $(m \geq 2)$, $\gamma$ stands for the total edges of the graph .we discuss the edge-balance index set of this graph from the following tow parts: $m \equiv 0(\text{mod } 2); m \equiv 1(\text{mod } 2)$.

But, in this paper, we only researcher it when $m \equiv 1(\text{mod } 2)$.

**Lemma 1.** For the graph $C_{2^m} \times P_{m_2}$ $(m \geq 2)$, when $m \equiv 1(\text{mod} 2)$, $\max\left\{EBI\left(C_{2^m} \times P_{m_2}\right)\right\} = \dfrac{2^{m+2} - 2^3}{3}$.

*Proof.* First, structure the graphics with the maximum edge - balance index:

We mark all edges for 0-edge which are linked to $(j)_i$ $(j = 2k, 1 \leq j \leq m; i = 2k+1, 1 \leq i \leq 2^j, k \in \mathrm{N})$. And let the label of edges $(j)_{i-1}(j)_i$, $(j = 2k+1, 1 \leq j \leq m; i = 4k, 1 \leq i \leq 2^j, k \in \mathrm{N})$ is signed 0-edge, the other edges is labeled as 1 except . The $(1)_1(1)_2$ is 0-edge.

In the graph, there are $\gamma=2+(2^2+2^3+\cdots+2^{m-1}+2^m)\times 2=2^{m+2}-6$ edges, of which $2^{m+1}-3$ are the 0-edge. According to the definition of edge-friendly labeling, $|e(0)-e(1)|\le 1$, we can get the edge-friendly labeling of the labeled graph by calculation. In this labeled graph, all the vertices $(j)_i$ $(j=2k, 1\le j\le m; i=2k+1, 1\le i\le 2^j, k\in \mathrm{N})$ are 0-vertex, others are 1-vertex except the vertices $(1)_1(1)_2$. Thus, we can gain $V(0)=\{(2)_1,(2)_3,(4)_1,(4)_3,(4)_5,(4)_7,\cdots,(m-1)_1,(m-1)_3,$ $\cdots,(m-1)_{2^{m-1}-1}\}=2^1+2^3+\cdots 2^{m-4}+2^{m-2}=\dfrac{2^m-2}{3}$.

Due to $2^1+2^2+\cdots 2^{m-1}+2^m=2^{m+1}-2$ vertices in this graph which is defined by the $f^+$, two of the vertices is not defined, then $V(1)=2^{m+1}-2-\dfrac{2^m-2}{3}-2$. After this $|v(0)-v(1)|=\left|\dfrac{2^m-2}{3}-\left(2^{m+1}-2-\dfrac{2^m-2}{3}-2\right)\right|=\dfrac{2^{m+2}-2^3}{3}$.

We can prove $EBI(C_{2^m}\times P_{m_2})=\dfrac{2^{m+2}-2^3}{3}$ is the maximum of edge-balance index in the graph $C_{2^m}\times P_{m_2}$ $(m\ge 2)$.

In this labeled graph, the vertices of the outside circle have 3 adjacent edge, and the edges adjacent to the vertices of the inner circle are 4, the edges next to the vertices of others circles are 5. The degree of 0-virtex is five except four degree point in the inner circle. what's more, the adjacent edge with 0-vertex is 0-edge, the adjacent edge with 1-vertex whose degree is three is only one, the adjacent edge with 1-vertex whose degree is five is two, another four degree is not defined. when we change the 0-vertex into 1-vertex or undefined vertex, we must take out 2 0-edges.thus, wherever we put the 2 0-edge, the number of the 0-vertex will increase, meanwhile the number of the 0-vertex will not increase. So the difference value is maximum, that's to say: $\max\{EBI(C_{2^m}\times P_{m_2})\}=\dfrac{2^{m+2}-2^3}{3}$.

**Lemma 2.** In the graph $C_{2^m}\times P_{m_2}$ $(m\ge 2)$, when $m\equiv 1(\mathrm{mod}\,2)$, $\left\{\dfrac{2^{m+2}-2^3}{3}-1,\right.$ $\left.\dfrac{2^{m+2}-2^3}{3}-2,\cdots,2,1,0\right\}\subset EBI(C_{2^m}\times P_{m_2})$

*Proof.* In the lemma 3, when $m\equiv 1(\mathrm{mod}\,2)$, we get $\max\{EBI(C_{2^m}\times P_{m_2})\}=$ $\dfrac{2^{m+2}-2^3}{3}$.

Below transform is based on the biggest index labeled graph constructed form lemma 1, and every step all proceeds is based on the previous step to transform, which in turn constructs labeling graphs corresponding with the index.

Among the $(2k-1)_r(2k)_s \leftrightarrow (2k)_s(2k)_{s+1}$ means that we exchange $(2k-1)_r(2k)_s$ and $(2k)_s(2k)_{s+1}$ which are 0-edge and 1-edge respectively.

First, build an odd index set:

Step 1: $(1)_1(2)_1 \leftrightarrow (1)_1(1)_2$, we can obtain the edge-balance indexes which are $|v(0)-v(1)| = \frac{2^{m+2}-2^3}{3}-1$.

Step 2: $(2)_1(3)_1 \leftrightarrow (3)_1(3)_2$, we can obtain the edge-balance indexes which are $|v(0)-v(1)| = \frac{2^{m+2}-2^3}{3}-3$.

Step 3: $(2)_3(3)_5 \leftrightarrow (3)_5(3)_6$, we can obtain the edge-balance indexes which are $|v(0)-v(1)| = \frac{2^{m+2}-2^3}{3}-5$.

Step 4: $(2)_3(3)_6 \leftrightarrow (3)_6(3)_7$, we can obtain the edge-balance indexes which are $|v(0)-v(1)| = \frac{2^{m+2}-2^3}{3}-7$.

Step 5: Exchange the 0 – edge connecting the t forth and fifth circles with the right 1-edge of the same vertices in the fifth circle in turn, we can obtain 16 labeled graphs of odd edge-balance indexes which are $\left\{\frac{2^{m+2}-2^3}{3}-9, \frac{2^{m+2}-2^3}{3}-11, \frac{2^{m+2}-2^3}{3}-13, \frac{2^{m+2}-2^3}{3}-15, \cdots; \frac{2^{m+2}-2^3}{3}-35, \frac{2^{m+2}-2^3}{3}-37, \frac{2^{m+2}-2^3}{3}-39\right\}$.

Step 6: Similar to step 5, from inside to outside exchange the 0-edge connecting the circles $2k$ and $2k+1$ $\left(1<k\le\frac{m-1}{2}\right)$ with the right 1-edge of the same vertices in the circle $2k$ in turn, we can obtain these labeled graphs of even edge-balance indexes which are $\left\{\frac{2^{m+2}-2^3}{3}-41, \frac{2^{m+2}-2^3}{3}-43, \frac{2^{m+2}-2^3}{3}-45, \cdots, 7, 5, 3, 1\right\}$.

Even index collection structure as follows:

Step 1: Exchange the 0–edge connecting the second and third circles with the right 1-edge of the same vertices in the third circle in turn, we can obtain 4 labeled graphs of even edge-balance indexes which are $\left\{\frac{2^{m+2}-2^3}{3}-2, \frac{2^{m+2}-2^3}{3}-4, \frac{2^{m+2}-2^3}{3}-6, \frac{2^{m+2}-2^3}{3}-8\right\}$.

Step 2: Similar to step 1, from inside to outside exchange the 0-edge connecting the circles $2k$ and $2k+1$ $\left(1<k\le\frac{m-1}{2}\right)$ with the right 1-edge of the same vertices in the circle $2k$ in turn, we can obtain these labeled graphs of even edge-balance indexes which are $\left\{\frac{2^{m+2}-2^3}{3}-10, \frac{2^{m+2}-2^3}{3}-12, \frac{2^{m+2}-2^3}{3}-14, \cdots, 8, 6, 4, 2, 0\right\}$.

In conclusion, we can prove $\left\{\frac{2^{m+2}-2^3}{3}-1,\frac{2^{m+2}-2^3}{3}-2,\cdots,2,1,0\right\} \subset EBI(C_{2^m}\times P_{m_2})$.

**Theorem 1.** In the graph $C_{2^m}\times P_{m_2}$ $(m\geq 2)$, when $m\equiv 1(\bmod 2)$, $EBI\left(C_{2^m}\times P_{m_2}\right)=$ $\left\{\frac{2^{m+2}-2^3}{3},\frac{2^{m+2}-2^3}{3}-1,\frac{2^{m+2}-2^3}{3}-2,\cdots,2,1,0\right\}$.

**Acknowledgments.** Thanks for project supported by Applied Mathematics Provincial-level Key Discipline of Henan Province and Operational Research and Cybernetics Key Discipline of Henan Polytechnic University. Thanks for project supported by the Ph.D. Foundation of Henan Polytechnic University Grant No.B2011-085.

## References

1. Zheng, Y., Lu, J., Lee, S.-M., Wang, Y.: On the perfect index sets of the chain-sum graphs of the first kind of K4-e. In: 2009 2nd International Conference on Intelligent Computing Technology and Autination, ICICTA 2009, vol. 4, pp. 586–589 (2009)
2. Yu, G., Zeng, Q., Yang, S., Hu, L., Li, X., Che, Y., Zheng, Y.: On the intensity and type transition of land use at the basin scale using RS/GIS: A case study of the Hanjiang River Basin. Environmental Monitoring and Assessment 160(1-4), 169–179 (2010)
3. Kwong, H., Ng, H.K.: On Friendly Index Sets of 2-regular graphs. Iscrete Mathematics 308, 5522–5532 (2008)
4. Salehi, E., Lee, S.-M.: Friendly index sets of trees. Congressus Numerantium 178, 173–183 (2006)
5. Kim, S.-R., Lee, S.-M., Ng, H.K.: On Balancedness of Some Graph Constructions. Journal of Combinatorial Mathematics and Combinatorial Computing 66, 3–16 (2008)
6. Chen, B.L., Huang, K.C., Liu, S.-S.: On edge-balanced multigraphs. Journal of Combinatorial Mathematics and Combinatorial Computing 42, 177–185 (2002)
7. Lee, A.N.-T., Lee, S.-M., Ng, H.K.: On The Balance Index Set of Graphs. Journal of Combinatorial Mathematics and Combinatorial Computing 66, 133–150 (2008)

# The Simulation of Applying Wavelet Transform to Identify Transformer Inrush Current

Hang Xu[1], Xu-hong Yang[1], and Yu-jun Wu[2]

[1] School of Electric and Automatic Engineering, Shanghai University of Electric Power, Shanghai, 200090, China
[2] Wuxi New-net Information System Co., Ltd. Wuxi, Jiangsu, 214028, China
tianyaxuhang@126.com

**Abstract.** In this paper, an efficient method based on wavelet transform is presented for detection of magnetizing inrush current in transformer. The author uses power system toolbox(PSB) in simulink, establishes the simulation model of magnetizing inrush current and fault current, and completes the feature extraction of two kinds of waveform by using wavelet transform. By lots of simulation results, it shows that this method can identify magnetizing inrush current well, with high accuracy.

**Keywords:** Wavelet Transform, Transformer, Magnetizing inrush current, Matlab simulation.

## 1 Introduction

As one of the most important and expensive electrical equipments in the power system, power transformer beares an important role of electricity transmission and voltage transform . Its reliable operation is directly related to that whether the power system can work safely and steadily.Therefore,the protection research of power transformer has always been an important topic in power system protection[1].

The recognition of inrush current is always the key problem in transformer differential protection, and it is of important significance in improving the accuracy of transformer protection movement and the quality of power supply of power system. Therefore domestic and overseas scholars have done a lot of study work, and many magnetizing inrush current identification methods are proposed, mainly include second harmonic, discontinuous angle principle, waveform symmetry principle, the magnetic flux characteristics identification method etc. Among them second harmonic method is the most widely used. However, various identification methods have their respective advantages and disadvantages, and they still cannot meet the requirements of modern microcomputer relay protection. So using the new principle and method to identify transformer magnetizing inrush current has reality urgency.

Wavelet transform is a new kind of time-frequency transform theory truly rising from the late 1980s, which is developed from Fourier transform, with multiple-scales analysis and good time-frequency localization characteristics, and can accurately

S. Lin and X. Huang (Eds.): CSEE 2011, Part II, CCIS 215, pp. 373–379, 2011.

extract signal characteristics. In recent years, many scholars apply wavelet algorithm to the transformer protection combined with traditional methods. The author of paper [3] decomposes the current signal on both sides of transformer by using wavelet transform, and gets the sign of fault jump's product to accurately discriminate inrush current and fault current. For transformer inrush current and internal fault current recognition problem, a wavelet neural network solution is put forward in paper [4], which combines the advantages of wavelet transformation and neural network. Paper [5] uses wavelet analysis to extract the discontinuous angle features to identify inrush current and short-circuit current, but its reliability is low in serious interference site.

## 2 Wavelet Transform Theory and Multi-resolution Analysis

Wavelet analysis, which has been developed in recent years, can achieve high resolution in time domain and frequency domain, so it is called "Math Microscope", has been more and more widely used in digital signal processing area. Wavelet analysis has excellent characteristics in signal mutating (edge) and singular detection, which can "focus" to any detail of the signal and can achieve the perfect description of signal singularity.

Assuming the function $\Psi(t) \in L^2(R)$, and it meets the following condition:

$$\int_R \Psi(t)dt = 0 \tag{1}$$

$\Psi(t)$ will be considered as base wavelet or mother wavelet. Introduce scale factor a and shift factor b, and they meet: a, b ∈ R and a≠0. Flex and translate base wavelet, the following family of functions will obtained:

$$\Psi_{a,b}(t) = |a|^{-1/2}\Psi\left(\frac{t-b}{a}\right) \tag{2}$$

Say $\Psi_{a,b}(t)$ as wavelet analysis.

In practical applications, computer handling signal f (t) is usually discrete sequence, so a and b shall be discretized, and wavelet analysis becomes discrete wavelet transform.

Usually, scale factor a and shift factor b of the continuous wavelet transform only need to be respectively discrete sampled as following: a= $a_0^j$, $b = k a_0^j b_0$, $j, k \in Z$ and $a_0 > 1$, then the corresponding wavelet function will be :

$$\Psi_{j,k}(t) = a_0^{-\frac{j}{2}}(a^{-j}t - kb_0) \tag{3}$$

Discrete wavelet transform can be defined as:

$$W_f(a_0^j, ka_0^j b_0) = \int_R f(t)\overline{\Psi_{j,k}(t)}dt \tag{4}$$

In 1988, S.Mallat proposed the concept of Multi-Resolution Analysis which shows the multi-resolution frequency characteristics of wavelet vividly from the concept of space when he built orthogonal wavelet base, and gave the construction method and fast algorithm (Mallat algorithm) of orthogonal wavelet. In Mallat algorithm, each signal and system has a resolution and the resolution of the signal and system is separately associated with frequency content in signal and bandwidth of the system. The wider bandwidth is, the higher resolution is and vice versa.

Figure 1 gives the structure of three-layer multi-resolution analysis.

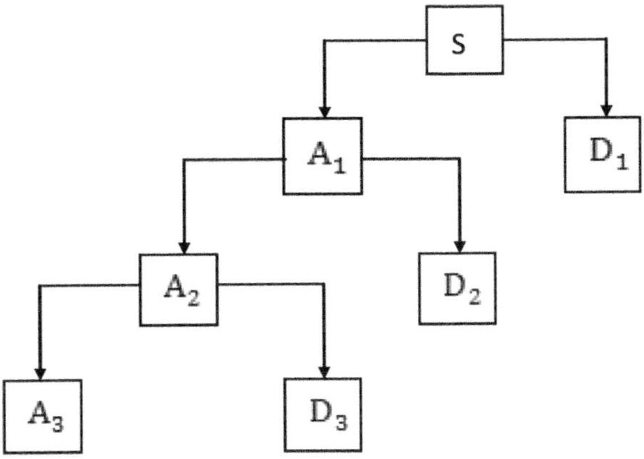


**Fig. 1.** The structure of three-layer multi-resolution analysis

As can be seen obviously from the figure, low frequency part is further decomposed in multi-resolution analysis, and high frequency part then will not be considered. The relationship of decomposition is $S = A_3 + D_3 + D_2 + D_1$. This paper uses wavelet and corresponding filters group provided by S. Mallat etc. The wavelet can accurately locate mutations position of the signal.

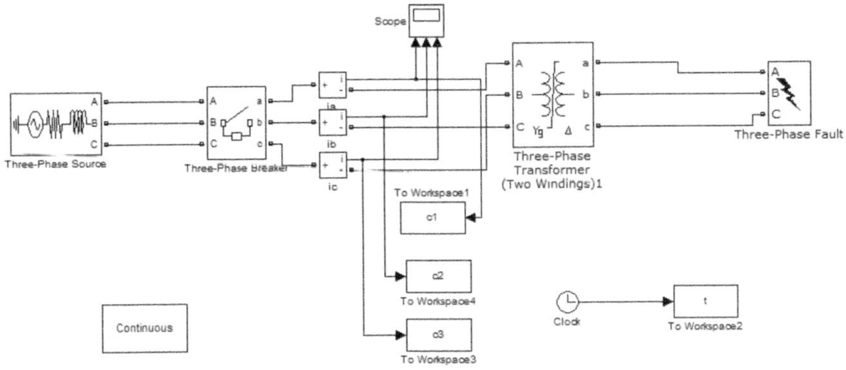


**Fig. 2.** The figure of transformer simulation model

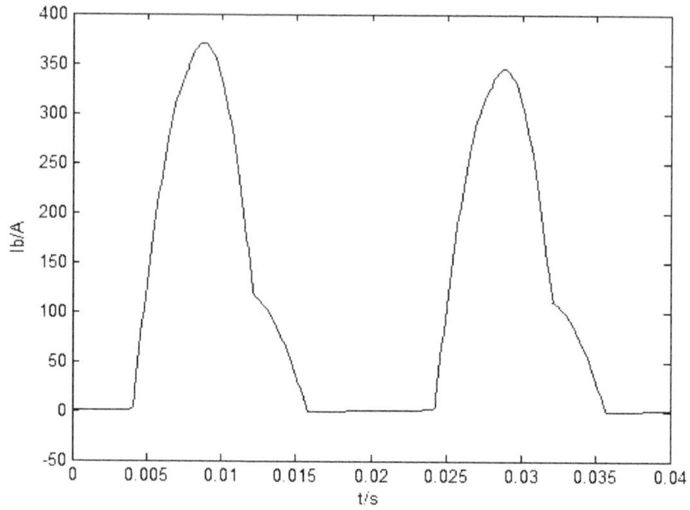


**Fig. 3.** The simulation waveform of inrush current

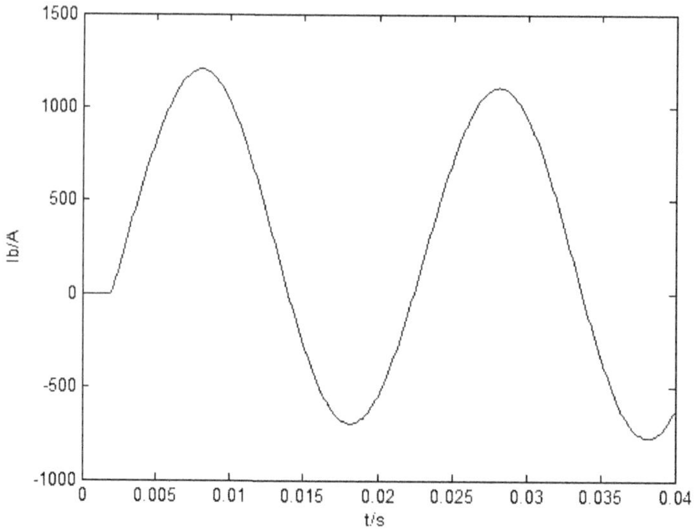


**Fig. 4.** The simulation waveform of three-phase short-circuit current

## 3 Power Simulation Modeling

Simulink, which is a visualization simulation tool in MATLAB and a MATLAB-based diagram design environment, can provide the integrated environment for dynamic system modeling, simulation and comprehensive analysis. This paper uses power system toolbox (PSB) in simulink to implement the digital simulation for transformer inrush current process. Specific circuit model is shown in figure 2.

Modules in the simulation model are directly adopted from SimPowerSystems module base in simulink of Matlab. Equivalent source: $V_N$=121kV. The main parameters of transformer: $S_N$=40MVA, Frequency is 50Hz, three-phase duplex winding, the voltage class is 110kV／11kV, no tapping, discontinuous angle varies between 0 and 360°.

This study achieves inrush currents under different circumstances by changing initial switching-in angle of the voltage and residual magnetism of the transformer, and short-circuit currents by setting up different faults. Figure 3 shows A phase inrush current when initial angle $\alpha = 0$ and residual magnetism is $(0, 0, 0)$. Figure 4 shows three-phase short-circuit current. The horizontal axis shows sales time, the vertical axis represents the value.

## 4 Wavelet Analysis Method to Identify Inrush Current

Because there are many kinds of wavelet function and it can be built according to actual needs, how to choose the wavelet function is a very important issue. According to characteristics of the signal, small amplitude, high frequency oscillation, decay faster, short-term mutation, this paper selects DB4 of Daubechies series constructed by famous wavelet analysis scholars Daubechies Ingrid to transform the original waveform to scale 1, 2, and 3.

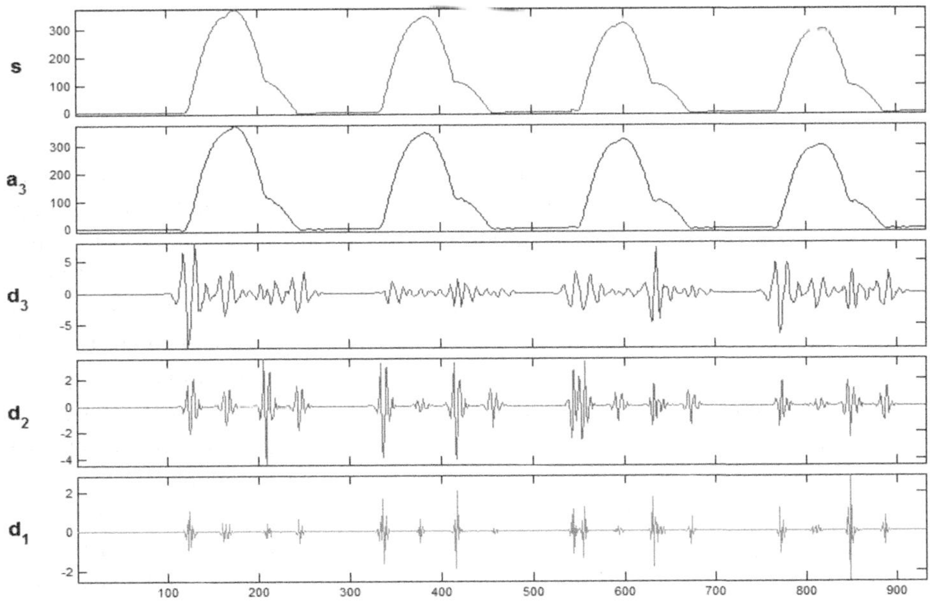


**Fig. 5.** Wavelet transform waveform of inrush current

Figure 5 and 6 respectively shows wavelet transform waveform of inrush current and three-phase short-circuit current. Among them, d1, d2, d3 represents the high coefficients of wavelet transform waveform in the first, second, third scale, and A3 shows wavelet discrete signal.

According to the simulation waveform, it can be seen obviously that there is much difference between wavelet transform waveform of inrush current and of three-phase short-circuit current.

First, as figure 5 shows, there are four mutations each industrial frequency cycle in detail 1 and detail 2, which indicates that mutations happen in electromagnetic quantity of transformer; In figure 6, there is a larger pulse component in fault beginning, but this component quickly decay, with no other obvious change. Thus, it is possible to identify inrush current and the fault current well.

Then, the coefficient of magnetizing inrush current mutational site is larger, and the waveform of short circuit current is quite gentle, without apparent mutations.

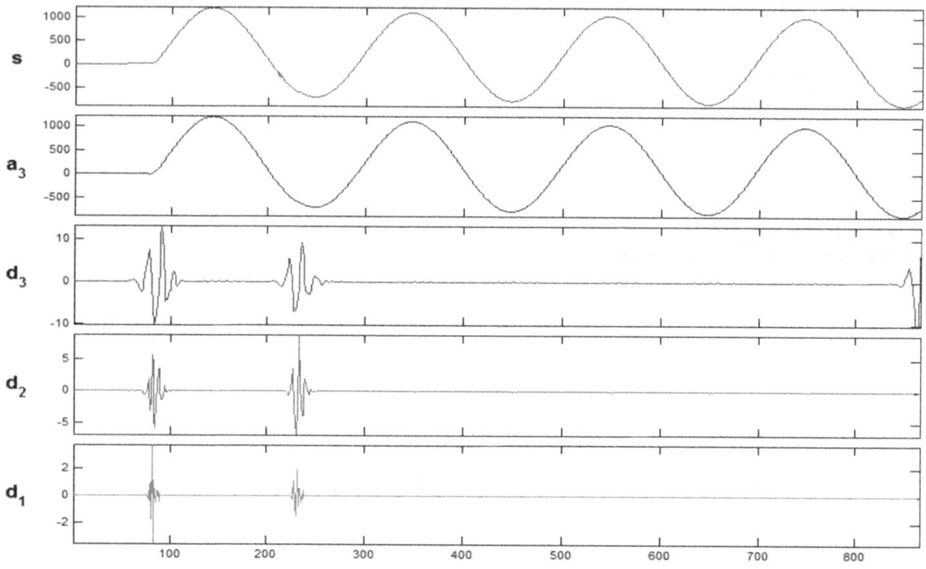


**Fig. 6.** Wavelet transform waveform of three-phase short-circuits current

## 5 Conclusions

According to the mechanism of transformer magnetizing inrush current, this paper establishes relevant model in MATLAB software, carries out the digital simulation and puts forward a method of using wavelet transform to identify inrush current and the fault current. This method can effectively identify magnetizing inrush current, without quantitative calculation on second harmonics content and the size of the continuous angle. It is feasible, with less computation.

**Acknowledgement.** This paper is supported by Project 61040013 supported by National Natural Science Foundation of China, Leading Academic Discipline Project of Shanghai Municipal Education Commission (Project Number: J51301), Innovation Program of Shanghai Municipal Education Commission (Project Number: 09YZ347), and the foundation of shanghai university of electric power ( project number: K-2007-06).

## References

1. Yabe, K.: Power Differential Method for Discrimination between Fault and Magnetizing Inrush Current in Transformers. IEEE Transaction on Power Delivery 12(3) (July 1997)
2. He, J.-l., Song, C.-j.: Power System Protection Principle. China Power Press, Beijing (1999)
3. Kulidjian, A., Kasztenny, C.B.: New Magnetizing Inrush Restraining Algorithm for Power Transformer Protection. In: Development in Power System Protection, Conference Publication, vol. (479), pp. 181–184. IEEE, Los Alamitos (2001)
4. Robertson, D.C., Camps, O.I., Mayer, J.S., et al.: Wavelets and Electromagnetic Power System Transients. IEEE Trans. On Power Delivery 11(2), 1050–1058 (1996)
5. Wang, L.-l., Rong, Y.-j.: The Identification of Transformer Magnetizing Inrush Current and Internal Fault Current Based on Wavelet Neural Network. Relay 31(7), 20–22 (2003)
6. Jiao, S.-h., Liu, W.-s., Liu, J.-f.: The New Principle Using Wavelet Theory in Identifying Transformer Magnetizing Inrush Current and Short-circuit Current. Journal of Chinese Electrical Engineering 19(7), 1–6 (1999)
7. Feng, X.-l., Guo, N.: The New Method of Wavelet transform in Identifying transformer Magnetizing inrush current and fault current. Relay 33(11), 31–36 (2005)

# Responses of the Non-structural Carbohydrates in *Fargesia nitida* to Elevation in Wolong Nature Reserve

Hongli Pan[1,2,3], Yu Tian[1,2], Sangen Wang[4], Zhong Du[1], Xingliang Liu[3], and Maihe Li[1,5,*]

[1] Institute of Mountain Hazards and Environment, Chinese Academy of Sciences and Ministry of Water Conservancy, Chengdu 610041, China
[2] Graduate School of Chinese Academy of Sciences, Beijing 100049, China
[3] Sichuan Academy of Forestry, Chengdu 610081, China
[4] College of Agronomy and Biotechnology, Southwest University, Chongqing 400716, China
[5] Swiss Federal Research Institute WSL, Zuercherstrasse 111, CH-8903 Birmensdorf, Switzerland
maihe.li@wsl.ch

**Abstract.** Changes in environmental factors (air temperature, soil conditions, etc.) with elevation can greatly impact plant growth, leading to morphological and physiological differences in plants. Non-structural carbohydrates (NSC) is an important physiological indicator for plant growth. In this paper, the biomass-weighted tissue NSC concentration of Fargesia nitida along an elevational gradient (2480 m, 2700 m, 2940 m and 3170 m a.s.l.) in the Wolong Nature Reserve was studied. Results indicated that the biomass-weighted NSC concentrations of F. nitida showed a wavy trend of "low - high - low - high" with increasing elevation in April, and a humped curve in October. The biomass-weighted NSC concentrations in plants at each elevation increased from the beginning (April) to the end (October) of a growing season, with higher increase magnitude at the middle elevations. The increased NSC accumulated mostly in roots, followed by stem and shoots. The non-linear responses of NSC to elevation reflect the survival strategies of F. nitida at different elevations associated with different environmental conditions.

**Keywords:** bamboo, elevation gradient, NSC, survival strategy, the Wolong Nature Reserve.

## 1 Introduction

Many environmental factors, such as air temperature, precipitation, soil conditions, and sunshine, vary with elevation, leading to the ecological and physiological differences of plants [1].In recent year, more and more attention has been paid to research on the relationship between plant growth and environmental gradient.

*Fargesia nitida* (Mitford) Keng f. ex Yi, a staple bamboo species for the food of the Giant Panda (*Ailuropoda Melanoleuca*), is widely distributed in the Wolong Nature Reserve. Metabolism of non-structural carbohydrates (NSC) has great

* Corresponding author.

S. Lin and X. Huang (Eds.): CSEE 2011, Part II, CCIS 215, pp. 380–385, 2011.

influence on the growth and development of plants [2, 3]. The NSC concentration in plants reflects a level of materials available for plant growth and assimilation [2, 3]. To study the concentration of carbohydrates in plants and its change with growing seasons can better know its status in plants [4].

In this paper, we investigated the biomass-weighted NSC and tissue NSC concentration in *F. nitida* along an altitudinal gradient from 2480 m a.s.l. (the lowest distribution of *F. nitida*) to 3170 m (the uppermost limit of its distribution) in the Wolong Nature Reserve, compared their changes with growing seasons and explained these results based on environmental conditions and plant survival strategy. Our works will contribute to understanding the physiological responses of *F. nitida* to elevation, obtaining the knowledge of physiological changes of *F. nitida* with growing seasons, and acquiring the adaptive strategy of bamboos to environmental stress.

## 2 Materials and Methods

### 2.1 Study Sites

The Wolong Nature Reserve (200 000 ha; 102°52′~103°24′ E; 30°45~31°25′ N) is located in a transitional area from the Sichuan Basin to the Tibetan Plateau. The precipitation-rich reserve has hot summers and cold winters, and distinct arid and humid periods, affected by the Qinghai-Tibet plateau climate system. The soil of the study transect has developed from granite, classified as mountain yellow soil [5].

*F. nitida* is widely and densely distributed under the canopy of evergreen broadleaved forest at 2480 m a.s.l. to evergreen and deciduous broadleaved mixed forest at 3170 m in the reserve [6]. According to nearly-equal interval of elevation, a 400-m$^2$ plot (20 m × 20 m) was established at 2480 m (the lowest plot), 2700 m, 2940 m, and 3170 m (the uppermost plot), respectively. The detailed characteristics of plots are listed in table 1.

**Table 1.** Characteristics of plots [6]

| Elevation (m a.s.l.) | 2480 | 2700 | 2940 | 3170 |
|---|---|---|---|---|
| Aspect | southeast | southeast | southeast | southeast |
| Slope degree (°) | 15 | 10 | 10 | 15 |
| Air temperature (°C) | 7.4 | 6.2 | 4.9 | 3.6 |
| Precipitation (mm/a) | 700 | 862 | 1040 | 1210 |
| Soil temperature (°C) | 6.9 | 5.6 | 4.3 | 3.0 |
| soil organic carbon (g/kg) | 92.7 | 104.7 | 110.8 | 134.6 |
| soil available nitrogen (mg/kg) | 584.7 | 630.4 | 701.2 | 775.0 |
| Overstory canopy coverage (%) | 62 | 60 | 50 | 38 |

### 2.2 Field Sampling

The field sampling was carried on in April, 2008. Three bamboo clumps in each plot were randomly selected, and all tissues (leaf, shoot, stem and root) samples were collected respectively. In October, tissue samples at every elevation were taken again from another three bamboo clumps. These samples were deenzymed and dehydrated at 70°C to a constant weight.

### 2.3 Chemical Analysis

Dried leaf, shoot, stem and root samples were ground to pass a 1-mm sieve. Total soluble sugars concentration was determined by the anthrone method [3], and starch concentration was determined by the anthronecolorimetric method [3].

### 2.4 Data Analysis

Non-structural carbohydrate is defined as the sum of the starch plus the total soluble sugars for each tissue. The biomass-weighted NSC concentration is given by

$$C = \frac{\sum M_i \bullet C_i}{\sum M_i} \times 100\% \cdot \qquad (1)$$

Where, $C$ stands for the biomass-weighted NSC concentration (%), $M_i$ stands for the biomass of a tissue of *F. nitida* (g), $C_i$ stands for the NSC concentration in the tissue(%),$\sum M_i \cdot C_i$ stands for the total NSC mass of all the tissues(g) and $\sum M_i$ stands for the total biomass of all the tissues of *F. nitida* (g).

One-way analysis of variances (ANOVAs) was used for all variables to test the differences among plots at different elevations, and followed, if significant, by Tukey's HSD test at α=0.05 to evaluate differences between pairs of means. Statistical analysis was done with SPSS 12.0 statistical software packages (SPSS 12.0 for Windows, Chicago, USA).

## 3 Results

The biomass-weighted NSC concentration of *F. nitida* growing at 4 elevation were significantly different ($p$<0.001) both in April and in October (Table 2). It revealed a wavy trend of "low - high - low - high" with increasing elevation in April, and a one-hump trend in October. During the growing season of *F. nitida* (from April to October of 2008), the biomass-weighted NSC concentration increased but it increased slightly at the boundary elevations and dramatically at the middle elevations (Table 2).

**Table 2.** Biomass-weighted NSC concentration (soluble sugars + starch) of *F. nitida* at different elevation and its change with sampling time in the Wolong Nature Reserve, SW China. Different letters indicate statistically significant differences ($p$ < 0.05, n = 3) between elevations within each category and sampling time.

| elevation | 2480 m | 2700 m | 2940 m | 3170 m | $F$ |
|---|---|---|---|---|---|
| $C_0$ | 7.14±0.02 a | 7.90±0.06 b | 7.24±0.06 a | 8.03±0.08 b | 29.39*** |
| $C_1$ | 7.83±0.17 a | 10.37±0.22 b | 9.63±0.06 b | 8.13±0.02 a | 43.08*** |
| $\Delta C$ | 0.69 | 2.47 | 2.39 | 0.10 | |

Note: $C_0$ is the NSC concentration in April (%), $C_1$ is the NSC concentration in October (%), $\Delta C = C_1 - C_0$. the same below.

**Table 3.** Tissues NSC concentration of *F. nitida* at different elevation and its change with sampling time in the Wolong Nature Reserve, SW China. Different letters indicate statistically significant differences ($p < 0.05$, n = 3) within each tissue category and sampling time.

| elevation | | 2480 m | 2700 m | 2940 m | 3170 m | F |
|---|---|---|---|---|---|---|
| | C0 | 12.09±0.06a | 12.69±0.09b | 11.55±0.11c | 11.54±0.17cd | 22.71*** |
| leaf | C1 | 12.01±0.30a | 13.43±0.17b | 11.82±0.14ac | 12.66±0.30ab | 9.19** |
| | ΔC | -0.08 | 0.72 | 0.27 | 1.12 | |
| | C0 | 8.48±0.02 a | 9.81±0.06 b | 8.71±0.06ac | 10.40±0.08d | 293.39*** |
| shoot | C1 | 8.53±0.17 a | 10.84±0.22b | 9.95±0.06 c | 8.29±0.02 a | 73.08*** |
| | ΔC | 0.05 | 1.03 | 1.24 | -2.11 | |
| | C0 | 6.07±0.10 a | 6.34±0.08ab | 5.67±0.06 c | 6.73±0.11 b | 24.92*** |
| stem | C1 | 6.86±0.13 a | 9.43±0.38 b | 8.66±0.21 b | 6.80±0.10 a | 32.26*** |
| | ΔC | 0.79 | 3.09 | 1.99 | 0.07 | |
| | C0 | 9.68±0.10 a | 10.39±0.01b | 8.08±0.06 c | 9.66±0.19ad | 79.09*** |
| root | C1 | 11.37±0.14 a | 13.25±0.40b | 12.64±0.27b | 10.45±0.16a | 22.88*** |
| | ΔC | 1.69 | 2.86 | 4.56 | 0.79 | |

For each tissue of *F. nitida* growing at all elevations, the NSC concentration had significantly differences ($p<0.05$) both in April (before growing season of *F. nitida*) and in October (after growing season of *F. nitida*). NSC in all tissues revealed the same trend of "low - high - low - high" with increasing elevation in April. Similarly, leaf NSC concentration revealed the same trend in October. Whereas, in other tissues (shoot, stem and root), it revealed a one-hump trend in October. During the growing season, tissue NSC of the bamboo increased except in the leaf at 2480 m and shoot at 3170 m (Table 3).

## 4 Discussion

Photosynthesis products can be classified into structural carbohydrates and non-structural carbohydrates (NSC). Non-structural carbohydrates includes soluble sugars (sucrose, glucose, fructose, fructan, etc.), starch and fat, which are important materials for plants metabolism and main carbon providers for plant growth, especially when plant photosynthesis keeps at a low rate [7].

Li et al. [2, 3] compared concentration of soluble carbohydrates of the same tree species and found concentration of soluble carbohydrates at treeline was obviously higher than at low elevation. However, our works didn't reveal the same trend. NSC concentration of *F. nitida* revealed a wavy trend of "low - high - low - high" in April and a one-hump trend in October. However, the natural environmental factors, such as air temperature, precipitation, soil temperature, soil pH, SOC and soil available nitrogen didn't reveal the same trend (table 1). We think the growth of the bamboo is determined by not a single environmental factor, but the comprehensive environmental conditions.

The biomass-weighted NSC reveals the comprehensive NSC level in *F. nitida*. The biomass-weighted NSC increased at all elevations, but the increase at the middle elevations exceeds the increase at the boundaries (table 2). The result shows the middle elevations has better conditions for the growth of *F. nitida*. In April, the bamboo growing at the uppermost elevation showed extremely high biomass-weighted NSC concentration (table 2). It is possibly caused by low temperature-inducing growth limitation [8]. Studies showed low temperature stress was advantageous for NSC accumulation in plant tissues [2, 3, 8]. At low temperature, plant cannot make full use of carbon assimilation substance for cellular reproduction and growth, leading to the accumulation in plant tissues [9]. In addition, plant photosynthesis at high elevation is no less than at low elevation [4]; on the contrary, low temperature in nights and in cold winter at the uppermost elevation reduces plant respiration [10], which can also lead to NSC accumulation.

Soluble sugars and starch are two main components of NSC. Soluble sugars was helpful for plant to live through low-temperature season by decreasing the freezing-point of cytoplasm to prevent plant cells from freezing [11]. Sakai & Yoshida [12] observed soluble sugars concentration in plants body was in proportion with frost-resistance capability of plants. Starch, main storage materials of plant, can supply for plant growth through converting into sugars. Generally, plants growing harsh condition at high altitude or in cold winter need sufficient carbohydrates to maintain their growth, as well as enough soluble sugars to promote their survival capability. So, higher NSC concentration at the uppermost elevation played a positive role on resisting low temperature threat. This may be a survival strategy for *F. nitida* growing at 3170 m.

During the growing season of the bamboo (from April to October in 2008), the biomass-weighted NSC increased at each elevation, but the increase was different for each tissue. The increased NSC accumulated first in the root, then in the stem and shoot, and then in the leaf (table 3). The NSC distribution is beneficial to translocation of water and other nutrients. The abundant NSC in the root can also provide adequate materials for the growth of whole bamboo clump and the new bamboo shoots in the coming year. So, the tissue NSC distribution of the bamboo may be also an important survival strategy of *F. nitida*.

## 5 Conclusion

Our results show the growth of *F. nitida* is determined by the comprehensive environmental conditions, such as air temperature, precipitation, soil temperature, soil pH, SOC and soil available nitrogen. The response of NSC in *F. nitida* to elevation is caused by the natural conditions and it is also the survival strategies of the bamboo.

## References

1. Pan, H.L., Li, M.H., Tian, Y., et al.: Responses of the Morphological Peculiarities and Above-ground Biomass of Fargesia angustissima to the Altitudinal Gradients in Wolong Nature Reserve. J. Sichuan Forestry Sci. Technology 31(3), 30–36 (2010)
2. Li, M.H., Xiao, W.F., Shi, P.L., et al.: Nitrogen and Carbon Source-sink Relationships in Trees at the Himalayan Treelines Compared with Lower Elevations. Plant Cell Environ 31(10), 1377–1387 (2008)
3. Li, M.H., Xiao, W.F., Wang, S.G., et al.: Mobile Carbohydrates in Himalayan Treeline trees I. Evidence for Carbon Gain Limitation but not for Growth Limitation. Tree Physiol. 28(8), 1287–1296 (2008)

4. Körner, C.H.: Alpine Plant Life-Functional Plant Ecology of High Mountain Ecosystems, 2nd edn. Spirnger, Berlin (2003)
5. Pan, H.L., Liu, X.L., Cai, X.H., et al.: Growth and Morphological Responses of Fargesia angustissima to Altitude in the Wolong Nature Reserve, Southwestern China. Acta Ecol. Sinica. 29, 144–149 (2009)
6. Pan, H.L.: Ecophysiological Responses of Dwarf Bamboos to Altitude in the Wolong Nature Reserve, Southwestern China. Institute of Mountain Hazards and Environment, CAS, Chengdu (2009) (in Chinese)
7. Holl, W.: Seasonal Fluctuation of Reserve Materials in the Trunkwood of Spruce (Picea abies(L.) Karst). J. Plant Physiol. 117, 355–362 (1985)
8. Körner, C.: Carbon Limitation in Trees. J. Ecol. 91, 4–7 (2003)
9. Shi, P.L., Körner, C.H., Hoch, G.: A Test of the Growth-limitation Theory for Alpine Tree Line Formation in Evergreen and Deciduous Taxa of the Eastern Himalayas. Funct. Ecol. 22, 213–220 (2008)
10. Wieser, G.: Carbon Dioxide Gas Exchange of Cembran Pine (Pinus cembra) at the Alpine Timberline duirng Winter. Tree Physiol. 17, 473–477 (1997)
11. Vagujfalv, A., Kerepesi, I., Galiba, G., et al.: Frost Hardiness Depending on Carbohydrate Changes during Cold Acclimation in Wheat. Plant Sci. 144, 85–92 (1999)
12. Sakai, A., Yoshida, S.: The Role of Sugar and Related Compounds in Variations of Freezing Resistance. Cryobiology 5, 160–174 (1968)

# Eliminates the Electromagnetic Interference Based on the PLC Configuration Software

Zu Guojian

Loudi Vocational and Technical College,
Loudi Hunan 417000

**Abstract.** In view of various kinds of PLC, mostly are in the bad electromagnetic environment in which electric circuit with the strong electricity and the strong electricity equipment form, thus its reliability is not high, affecting the system stable movement. From the three basic principles including suppression noise source, cut-off or weaken electromagnetic interference dissemination way and enhancement installment and system antijamming ability suppression electromagnetic interference obtaining, this article discusses the system in the electromagnetic interference production, the dissemination and the suppression measure, proposes to use the software resources to eliminate the failure detection, analyses the measures of reduced numeral quantity input perturbation through counting law and detention input method and the mold by the quantity input perturbation procedure through the restriction law, the delay filtering restriction law and the delay filtering comparison test. These methods have merits such as less investment, unceasing promotion and enhancing the reliability of system greatly, making up the insufficiency of the hardware designs to a certain extent, which are promoted to apply to PLC system.

**Keywords:** three principles; digital input disturbance; modelquantity disturbance; measures.

## 1 Introduction

PLC control system is featured as powerful function, simple program design, good scalability, easy maintenance, high reliability, adaptable to harsh industrial environments and so on, it is widely used in industry production control system, some PLCs are installed in control room(for example PLC in coal washing workshop of coal separating plant), some are installed at on electrical devices at production sites, most of them are in the harsh electromagnetic environment with strong electrical circuit and equipments which will easily be affected by disturbance or malfunction. To improve PLC reliability, apart from enhance controller performance, also need to solve disturbance problems in project design, installation and maintenance, take use of PLC software to reduce disturbance is a critical aspect of keeping a PLC control system regularity and stability. The following analyzes methods using PLC configuration software to reduce interference and enhance system interference resistance capacity.

S. Lin and X. Huang (Eds.): CSEE 2011, Part II, CCIS 215, pp. 386–391, 2011.

## 2 Three Principles to Suppress Electromagnetic Interference

In order to keep system from electromagnetic environment or reduce internal and external electromagnetic interference, three inhibit measures should be taken from design stage: ① interference suppression; ② cut or attenuate transmission of electromagnetic interference; ③ improve devices and systems anti- interference performance. These three points are basic principles of electromagnetic interference suppression.

The anti-jamming function of PLC control system is a systematic engineering, it requires not only products with strong anti-interference ability but also all-round consideration for user departments from engineering design, installation, operation to maintenance. Methods should be taken to control external interferences during the design stage, such as shield PLC system and external lead to prevent spatial interference of electromagnetic radiation; external leads isolation or filtering, especially power cable should be arranged in layers to prevent accidental electromagnetic interference from an external lead wire transmission, proper landing point and grounding design to improve the ground system hardware, also use software and the PLC configuration software to further improve system security and reliability. In complex industrial electromagnetic environment, sometimes hardware measures only cannot completely eliminate interferences, software should also be utilized. Some software measures can be used as follows:

(1), delay recognition: software can postpone input for 20ms, read the same signal for two or more times, only valid when results are the same.

(2), block interference: some interferences are predictable, for example when programmable controller output action commands to the execution agency (such as high-power motor, solenoid), it often accompanied by interferences signals as sparks, arc and so on, their signals may interfere with PLC and cause programmable controller to receive wrong information. During these disturbance period, take use of software to block some input signals of programmable logic controller, and unblock when the interference-prone period is over.

(3), software filter: take use of software filter for analog signals, nowadays, most large-scale PLC programming supports for SFC and Structured Text programming, which can be easily compiled for complex procedures and to complete appropriate functions.

(4), fault detection and diagnosis: highly reliable programmable logic controller can perform perfect self-diagnosis function, if programmable logic controller has a failure, it can easily find fault parts and the specific location with self-diagnostic program, then replace it and resume normal work.

Large number of engineering practices have shown that the external programmable logic controller input and output devices, such as limit switches, solenoid valves and contactors, their failure rate is much higher than the programmable logic controller itself, if these components appear failure, the programmable logic controller generally cannot perceive, and cannot cause automatically shut down, then failure may extend until strong electrical protective devices shutdown, sometimes it even cause equipment and personal accidents. After device shutdown, there needs a lot of time to find out fault. In order to detect faults, make programmable logic controller shutdown and alarm before any accident happens, and to facilitate troubleshooting, improve maintenance efficiency, take use of ladder program to achieve self-diagnosis and self-failure treatment. Modern programmable controller has a large number of software resources, such as in FX2N Series PLC, there are thousands of supplementary relays, hundreds of timers and counters with a considerable margin. Utilize these resources for fault detection.

# 3 Electromagnetic Interference Elimination Method with PLC Configuration Software

## 3.1 Reduce Digital Input Perturbation

Among which CON is counter, NOT non-gate, RS prior flip-flop, IN is the input, OUT is the output, N is the number of pulse samples. When there is an external signal input, N consecutive pulses which control system acquires makes RS flip-flop output as "1", only when the external input signal change from "1" to "0", RS flip-flop reset terminal is "1", reset output of RS flip-flop to "0." There are transient interference pulses, CON counter will collect less than N consecutive pulses, and CON counter cannot output, which reduces interferences.

The advantage of this method is fast response, which inhibits periodical transient disturbance, and the disadvantage is that it cannot eliminate the CON interference of counter sampling time.

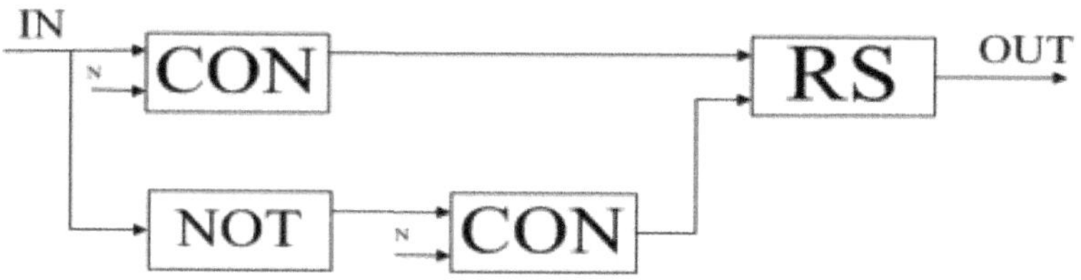


**Fig. 1.** Counter-law

**Fig. 2.** Delayed input method

IN is the input; OUT is the output, TIME (ET) for the delay time, TON for the delay output. When input IN = 1, the start counter until time (PT) = delay time, OUT = 1. When counter time <delay time, OUT = 0. The delay time is best within 1S.

The advantage of this method is eliminating periodic disturbance in short time period, and the disadvantage is slow response which is against speed signal transmission.

### 3.2 Reduce Analog Input Perturbation

MOVE command is to maintain instruction (Enable terminal EN = 1, OUT = IN. EN = 0, OUT keeps previous value), GE is greater than or equal to instruction (OUT = 1, IF IN1 ≥ IN2), LE is less than or equal to instruction (OUT = 1, IF IN1 ≤ IN2), HL value for the cap, LL for lower limit. When analog input signal is between HL and LL, OUT = IN. When IN-AI signal exceed or equal to HL or LL, GE, or LE to determine IN-AI signal, which makes OUT1 or OUT2 to output "1" and seal the MOVE, thereby maintaining the MOVE output as HL or LL set value which limits the range.

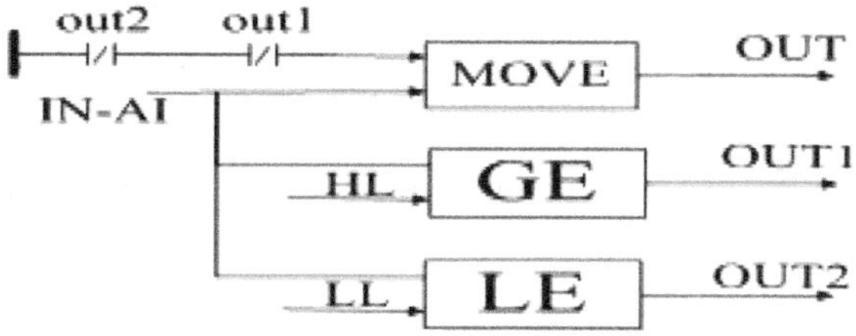


**Fig. 3.** Limiting law

The advantage of this method is effectively overcome pulse interference caused by accidental factors, and disadvantage is poor smoothness.

MOVE is to maintain instruction (Enable terminal EN = 1, OUT = IN. EN = 0, OUT keeps previous value), GE is greater than or equal to instruction (OUT = 1, IF IN1 ≥ IN2), LE is less than or equal to instruction (OUT = 1, IF IN1 ≤ IN2), HL value for the cap, LL for lower limit, LG as delay filter instruction, TIME for the delay filter time.

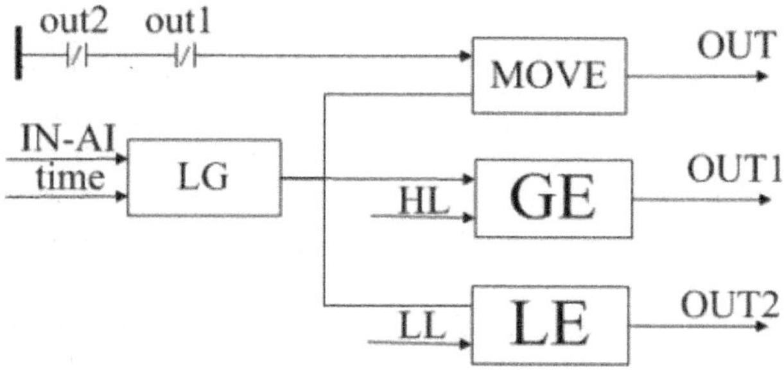


**Fig. 4.** Filter Delayed limiting law

The advantage of this method is effectively inhibiting periodic pulse interference, smoothness improves compared with limiting law, and the disadvantage is slow signal response speed.

LG is delay filter, SUB is subtract instruction, ABS is instruction absolute value, GE is greater than or equal to instruction, HL is the maximum deviation, TIME is delay filter time. Normally, the input signal IN-AI will output directly through first order delay filtering, OUT = IN-AI values; when there is a jump signal, take the absolute value subtraction of input signal IN-AI which goes through the first-order delay filter and the jump signal of input signal which contains IN- AI (no matter it is negative deviation or positive deviation), then compares with HL value, if it is greater than or equal HL default, OUT1 = 1, then switch LG-delay filter to tracking status, and OUT maintains the input signal IN-AI value before jump . Continue to reduced until jump signal weakened, OUT1 = 0, OUT = IN-AI.

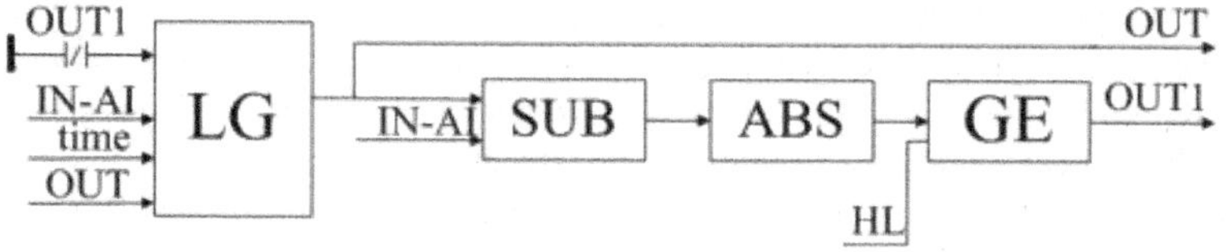


**Fig. 5.** Filtering delay Comparative Law

The advantage of this method is a good inhibition of periodic disturbance with high smoothness, and the disadvantage is that sensitivity mainly depends on TIME-delay filter time.

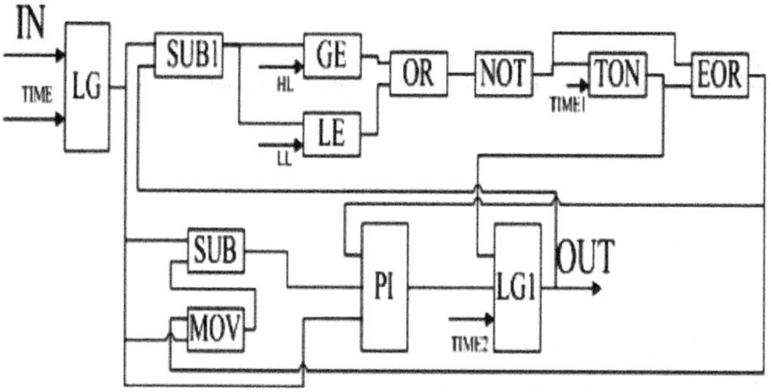


**Fig. 6.** Points shake consumer filtering

LG is for delay filter, SUB for subtract instruction, GE is greater than or equal to instruction, LE is less than or equal to instruction, OR is OR gate (DFB function of its own), NOT is a non-door, TON is the delay output, EOR is XOR gate, MOV is to keep commands, PI is the proportional integral regulator, HL is the maximum positive deviation, LL is the maximum negative deviation, TIME is time delay filter time, TIME1 for delay output time, TIME2 for delay filter time. Parameter settings: LG (TIME = 1S), TON (TIME1 = 10S), LG1 (TIME = 30S), HL = 0.2, LL =- 0.2, PI (TI = 10S, release P and block it as a pure integrator regulator), use other PLC software design and configuration as interference approaches and: add time delay shielding function, digital filtering and sample shaping frequency (effectively eliminate periodic interferences), positioning correction (0) (eliminate interference and error accumulation). For demanding control systems, there is also two-out-of-three method to improve safety, and for extensive control system, take use of indirect jump, software trap, fault-tolerant software and other technologies to improve reliability.

## 4 Conclusions

It is impossible to all eliminate disturbances because of electromagnetic complexity, above methods are all come from practicality and they have made some achievements. With the growth of actual production needs and experiences, it will constantly improve interference software processing performance.

## References

1. Chen, P., Zhao, X.-B.: Programmable controller technology and application system design. China Machine Press, Beijing (2002)
2. Pi, Z.x., Gong, Z.M., et al.: Programmable logic controller system design and its application. China Machine Press, Beijing (2000)
3. Song, B.: PLC programming theory · Algorithms and techniques. China Machine Press, Beijing (2005)
4. Hu, M.: Analysis on filtering algorithm for decaying DC component in microcomputer-based relay. Proceedings of the CSU-EPSA 17(4), 36–40 (2005)
5. Zhang, Y.: Study of adaptive window length algorithm based on linear differential equation. Proceedings of the CSEE 20(7), 24–27 (2000)
6. Zhang: Inverter use of the source of interference and anti-jamming measures. Metallurgical Series (1), 13–14 (June 1, 2007)
7. Liu, Q., Yu, S., Ma, W., et al.: An investigation for increasing the power system stabilityby using powersystemstabilizer. Journal of Tsinghua University 11(2), 75–85 (1979)
8. Qing, D., Yushan, H., Yuqian, H.: The dynamic simulating system for testing governor power system stabilizer. Journal of North China Electric Power University 26(2), 49–54 (2009)
9. Jin, X., He, S., Qi, X.: The design of wireless network in substation based on blue-tooth technology. Electric Power 35(8), 61–64 (2010)
10. Fang, G., Hu, B., Song, G.: Detection of three typical disturbances in power system. Electric Power Automation Equipment 28(6), 74–77 (2010)

# Research on Fault Detection System of Marine Reefer Containers Based on OC-SVM

Jun Ji and Houde Han

Merchant Marine College, Shanghai Maritime University, Shanghai, China
jennyji35@163.com

**Abstract.** A fault detection model based on One-Class Support Vector Machine (OC-SVM), a new SVM method, is established to solve the large difference in sample size between the normal data and fault data of reefer containers. During the model training process, only the normal samples are needed to be learned, and an accurate identification of abnormal can be achieved, which may solve the problem of lack of fault samples in practice. By comparison experiments between different kernel functions and kernel parameter optimization, a fault detection model of reefer containers based on OC-SVM is established, and the test results show that the model has a high recognition rate against abnormal and low false alarm rate.

**Keywords:** Reefer Container, Fault Detection, OC-SVM, Parameter Optimization.

## 1 Introduction

Typically, equipment fault detection is dealt with as a pattern recognition issue of two type classification, however, there is little or no abnormal samples but a lot of normal samples in practice. Due to the lack of abnormal training samples, commonly used classification methods can not achieve good results, so setting alarm thresholds of a certain parameter that can reflect the equipment operation status is used in fault detection. In fact, only use one parameter is not accurate, because it is likely to result in miss alarm and false alarm. For example, though supply air temperature of reefer containers may reflect the operation conditions of the refrigeration unit most of time, when some faults happen, the supply air temperature is still within the normal range, while other parameters such as the exhaust air temperature of compressor change a lot, if threshold alarm method is yet used, miss alarm will take place.

The lack of abnormal samples of reefer containers determines its fault detection can not follow the conventional way and new solving ways and methods are needed. New methods can only use a large number of existing normal samples to achieve the early identification of abnormal conditions and failures. OC-SVM (One-Class SVM) is a new kind of SVM, which only need one class of samples as training samples, through adaptive learning of their distribution, effective recognition of different modes and states may be realized[1]. OC-SVM is introduced to condition assessment of reefer containers, and fault detection system of reefer containers based on OC-SVM is established in this paper.

S. Lin and X. Huang (Eds.): CSEE 2011, Part II, CCIS 215, pp. 392–397, 2011.

## 2 OC-SVM Anomaly Detection Algorithm

OC-SVM is a new kind of SVM method, which is originally used as high-dimensional distribution estimation [2], and one class data classification problem is proposed on the basis by Tax and others. First, the original samples were projected to high dimensional feature space through kernel mapping. Then, the distribution range model of study samples was established in the feature space, and the distribution area was asked to cover the training sample as compact as possible to construct the classification decision function. When unknown samples fall into the decision area, they are judged as normal. Otherwise they are judged as abnormal [3].

The idea that using hyper spheres instead of hyper planes to divide the data, which was proposed by Tax has changed the data set description. Suppose the sample set X ( $x_i \in R$ , $i$ =1, ..., $l$ ) corresponds to the normal state, the so-called anomaly detection is to find a data set covering X and construct a decision function. For any one sample, there is the relationship:

$$\begin{cases} f(x) \geq 0, & x \in C(x) \\ f(x) < 0, & x \notin C(x) \end{cases} \tag{1}$$

Then, the samples falling within C(x) will be determined as normal, otherwise determined as abnormal.

First, project the sample set into high dimensional feature space through kernel function. In order to reduce false rate, we should find a compressed sphere in the feature space that contains as much training samples as possible, which is called hyper sphere. Then introduce the relaxation variable $\xi_i$ to make the training samples included as much as possible in the hyper sphere during the guarantee that the hyper sphere is most compressed. This problem can be expressed as an optimization problem:

$$\begin{cases} \min\limits_{R,\xi_l,c} & R^2 + \dfrac{1}{vl}\sum\limits_{i=1}^{l}\xi_i \\ s.t. & \|\varphi(x_i) - c\|^2 \leq R^2 + \xi_i \\ & \xi_i \geq 0, \quad i = 1, \ldots\ldots l \end{cases} \tag{2}$$

R for hyper sphere radius, c for circle center, $v \in (0,1]$

In order to compromise between the hyper sphere radius and the number of the training samples it contains, when $v$ is small, restrict the samples in the sphere to the greatest extent and when $v$ is large, compress the size of the sphere as much as possible. Lagrange function is utilized to solve this optimization problem.

$$L(R, c, \xi_i, \alpha_i, \beta_i) = R^2 + \frac{1}{vl}\sum_{i=1}^{l}\xi_i - \sum_{i=1}^{l}\alpha_i(R^2 + \xi_i - \|\varphi(x_i) - c\|^2) - \sum_{i=1}^{l}\beta_i\xi \tag{3}$$

Seek the partial differential equations for R, c and $\xi$ respectively, and make them equal to zero.

Dual form of this optimization problem is as follows:

$$\begin{cases} \sum_{i=1}^{l} \alpha_i = 1 \\ c = \sum_{i=1}^{l} \alpha_i x_i \\ 0 \le \alpha_i \le \frac{1}{vl} - \beta_i \le \frac{1}{vl} \end{cases} \tag{4}$$

$$\begin{cases} \min \quad \sum_{i=1}^{l} \alpha_i \alpha_j k(x_i, x_j) - \sum_{i=1}^{l} \alpha_i k(x_i, x_i) \\ s.t. \quad 0 \le \alpha_i \le \frac{1}{vl}, \quad \sum_{i=1}^{l} \alpha_i = 1 \end{cases} \tag{5}$$

We can see that the constraint condition of this dual problem is linear. The samples corresponding to $\alpha_i \neq 0$ are called support vectors. Use KKT condition to find the sample points $x_i$ landed on the optimal hyper sphere, These $x_i$ are satisfied with the constraint condition: $0 < \alpha_i < \frac{1}{vl}$, $\sum_{i=1}^{l} \alpha_i = 1$. On the condition $\alpha_i = \frac{1}{vl}$, quite a few of the sample points are outside of the hyper sphere, which are considered as abnormal sample points, while most of the sample points are within the sphere.

The decision function is as follows:

$$f(x) = R^2 - \|\varphi(x) - c\|^2 \tag{6}$$

When $f(x) \ge 0$, $\varphi(x)$ is within the hyper sphere, and is considered normal. Otherwise, $\varphi(x)$ is outside the sphere, and is considered abnormal.

# 3 Fault Detection of Marine Reefer Containers Based on OC-SVM

## 3.1 Data Sources

The data come from Fault Analysis Experiment of the reefer containers. Among them, there are 1505 sets of normal condition and 426 sets of abnormal. Each set contains 14 variables and the data format is as follows:

$$X_i = \begin{bmatrix} T_{suc}^1 & T_{dis}^1 & T_{kin}^1 & T_{kout}^1 & T_{be}^1 & T_{af}^1 & T_{0in}^1 & T_{0out}^1 & P_{suc}^1 & P_{dis}^1 & P_{be}^1 & P_{af}^1 & T_{ain}^1 & T_{aout}^1 \\ T_{suc}^2 & T_{dis}^2 & T_{kin}^2 & T_{kout}^2 & T_{be}^2 & T_{af}^2 & T_{0in}^2 & T_{0out}^2 & P_{suc}^2 & P_{dis}^2 & P_{be}^2 & P_{af}^2 & T_{ain}^2 & T_{aout}^2 \\ \cdot & \cdot & \cdot & \cdot & \cdot & \cdot & \cdot & \cdot & \cdot & \cdot & \cdot & \cdot & \cdot & \cdot \\ T_{suc}^n & T_{dis}^n & T_{kin}^n & T_{kout}^n & T_{be}^n & T_{af}^n & T_{0in}^n & T_{0out}^n & P_{suc}^n & P_{dis}^n & P_{be}^n & P_{af}^n & T_{ain}^n & T_{aout}^n \end{bmatrix}_{14\times n}$$

1000 sets data randomly selected from the 1505 normal samples are used to train the OC-SVM model and the rest 505 sets are used to check the model. 426 sets of abnormal samples are used to check the abnormal recognition effect of the OC-SVM model.

The samples are normalized before training and testing to eliminate the influence of different physical dimensions and the normalization method is as follows:

$$X^* = [X - \min(X)]/[\max(X) - \min(X)] \tag{7}$$

$X^*$ and $X$ are the value after and before normalization respectively. $\min(X)$ and $\max(X)$ are the minimum and maximum value of the samples.

### 3.2 Determination of OC-SVM Kernel Parameters

Xu et al [4] found that RBF was the most suitable function for OC-SVM, so in the experiment, RBF is introduced as the nonlinear mapping function to project the original data space to the feature space. Vapnik et al [5,6] considered that when RBF is used as SVM kernel function, its parameters can automatically determined and SVM has rapid training speed. But in our experiment, OC-SVM fault detection rate varied widely when the parameters $\gamma$ and $v$ of RBF have different value. Therefore, we change the value of the parameters $\gamma$ and $v$ until the optimal parameters pack is found.

The parameters optimization algorithm of OC-SVM's RBF kernel function is designed in this paper. The training effect of OC-SVM is optimized through adjusting the kernel parameter $\gamma$ and control parameter $v$. The algorithm flow chart is shown in Fig.1.

Set 1 and Set 2 are the normal samples set and abnormal samples set.

Using this algorithm, optimal kernel parameters can be automatically determined within the setting range. The results of OC-SVM experiment show that when $\gamma$ ranges from 0.01 to 0.2 and $v$ ranges from 0.01 to 0.1, OC-SVM fault detection model has an accuracy rate of above 90% for training samples, a total detection rate of above 95% for testing samples, an acceptance rate of above 90% for normal samples in testing sets and a recognition rate of above 95% for abnormal samples in testing sets.

Therefore, after $\gamma$ and $v$ are primarily selected, we use the above algorithm to further optimize them with $\gamma$ in the range of 0.01 to 0.2 and $v$ in the range of 0.01 to 0.1. The experiment results are shown in Fig.2.

Fig.2 shows that when $\gamma \in [0.02, 0.03]$ and $v \in [0.01, 0.02]$ ,the accuracy rate for training samples and total detection rate for testing samples reach above 99% and 98% respectively, which are the highest and remain unchanged, the acceptance rate for normal samples can reach 100% and the recognition rate for abnormal samples reduces with the increase of $\gamma$ and $v$ has a highest value of 97.4% when $\gamma$ takes 0.02 and $v$ takes 0.01.

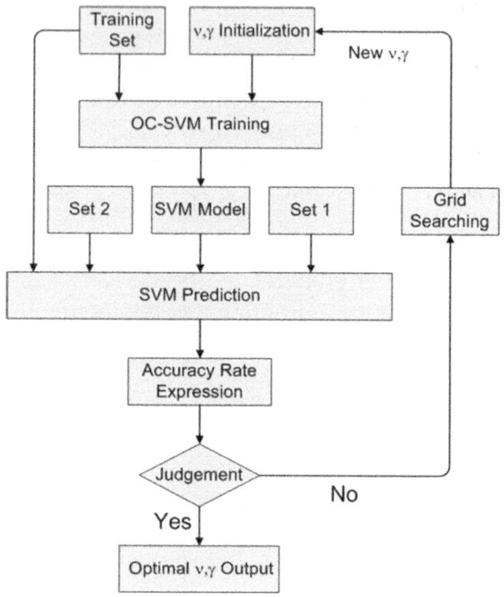


**Fig. 1.** OC- SVM Parameter Optimization Algorithm Flow

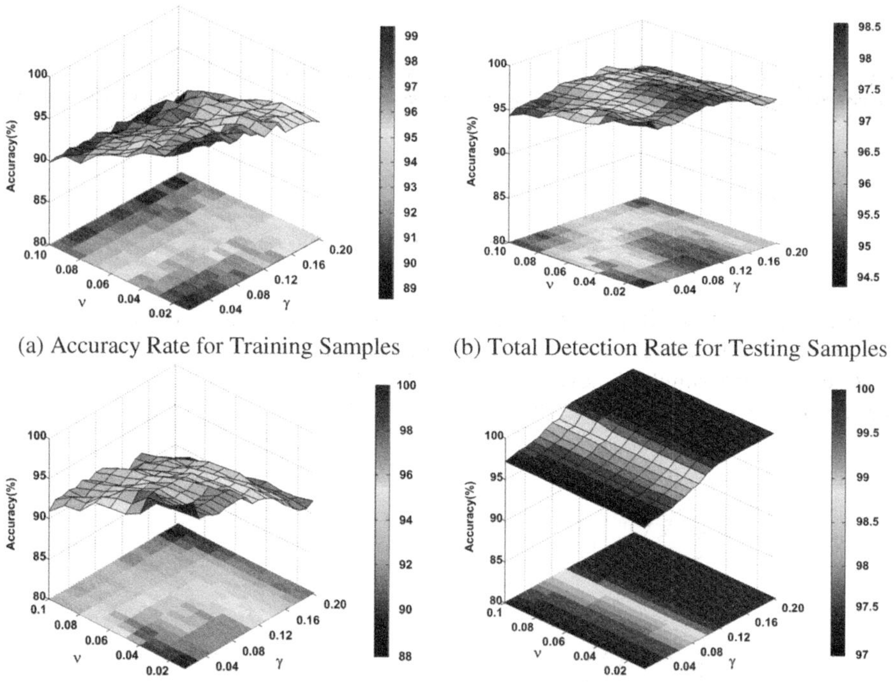


(a) Accuracy Rate for Training Samples (b) Total Detection Rate for Testing Samples

(c) Acceptance Rate for Normal Samples (d) Recognition Rate for Abnormal Samples

**Fig. 2.** OC-SVM Fault Detection Rate

The research results also show that with the increase of $\gamma$, the training model has an increasing abnormal recognition rate, but a decreasing acceptance rate for normal samples. That is to say the detection rate for abnormal samples is inversely proportional to the acceptance rate for normal samples. To get a high abnormal detection rate must be at the expense of the normal recognition rate, i.e., increasing the possibility of misjudging normal condition for abnormal.

According to the experiment results and the feature of the above parameters, the selected RBF kernel function parameter $\gamma$ is 0.02 and model control parameter $\nu$ is 0.01. Meanwhile, the OC-SVM model has a recognition rate for abnormal samples of 97.4% and an acceptance rate for normal samples of 100%, that is to say, the abnormal detection rate of the reefer container fault detection model is 97.4% with the missed alarm rate of 2.6%, and the normal recognition rate is 100% with false alarm rate of 0.

## 4 Conclusions

The latest research results of statistical learning SVM is applied to the fault detection of reefer containers in the paper, and the fault detection model based on OC-SVM is put forward. OC-SVM is a new SVM method developed on the basis of basic binary SVM classification algorithm and its training process needs only one class learning samples which is adapted to solving the problem of reefer containers that in actual operation there are only a large amount of normal samples, but few abnormal samples. The reefer container fault detection model based on OC-SVM is established in the paper and parameter optimization algorithm of RBF kernel function is designed. With the optimal model parameters, good experiment results are achieved that the recognition rate of abnormal samples reaches 97.4% and the acceptance rate of normal samples comes up to 100%, i.e., zero false alarm rate.

## References

1. Zhong, Q.L., Cai, Z.X.: Sensor Fault Diagnosis based on One-Class SVM. Computer Engineering and Application 19, 1–3 (2006)
2. Scholkopf, B., Platt, J.C., Taylor, J.S., et al.: Estimating the support of a high-dimensional distribution. Neural Computation 13, 1443–1471 (2001)
3. Tax, D.M.J., Duin, R.P.W.: Support vector domain description. Pattern Recognition Letters 20, 1191–1199 (1999)
4. Xu, T., Luo, Y., He, D.K.: SMO Training Algorithm for Hyper-sphere One-class SVM. Computer Science 5(6), 178–180 (2008)
5. Vapnik, V.: SVM Method of Estimating Density, Conditional Probability, and Conditional Density. In: IEEE International Symposium on Circuits and Systems, May 28-31(2000)
6. Gary, G.L., Nelson, N.H., Norden, E.H.: Application of the Hilbert-Huang transform to machine tool condition/health monitoring. In: AIP Conference Proceedings, vol. 615(1), pp. 1711–1718 (2002)

# H.264 Interframe Coding Mode Decision Based on Classification and Spatial Correlation

Qin Huanchang

Department of Physics and Electronics Information Engineering,
Baise University, Baise, China

**Abstract.** To select the best model for the current macroblock, the encoder need to traverse all the possible models, each model calculated the cost of coding needed, select from the best model for minimum cost, this method is called full analysis. Full analysis of course be able to choose the best coding mode, it is clear that computation is enormous, according to experimental statistics, H.264/AVC motion estimation and compensation is calculated in the standard module of the largest, accounting for the entire Encoder 1/3, and mode decision algorithm as a core, a direct impact on the efficiency of motion estimation and compensation. It is noteworthy that the decision based on rate-distortion model and the model based on absolute error and the decision there are some differences, because the decision of this thesis is concerned with modeling algorithms, and therefore to the discussion of simplicity, this all ideas will be discussed and algorithms are based on the absolute error and the standard model of decision.

**Keywords:** H.264; Interframe Coding Mode; Spatial Correlation.

## 1 Introduction

Used in H.264/AVC variable block size motion compensation, decision making model greatly enhance the computational complexity, as a matter of concern. Coupled with Skip Mode and two intra modes, P frames may have a macro block motion compensation of 10 candidates split mode. Segmentation means that every small piece of a small residual amount, but the encoder needs to spend more bits to describe the segmentation method; large partition will have a greater residual volume, while the split method takes a description of a smaller[1] .

## 2 Strategy of Hierarchical Models and Simplify the Decision Comparison of Classification Policy Is Now One of a Wide Range of Mode Decision

The main idea of this method is divided into four sets of all the modes, as shown in Table 1: Class0 only the Skip mode, Class16x16 segmentation model consists of three larger, the smaller included in the Class8x8 in split mode, ClassIntra including Intra4x4 and Intra16x16 both intra mode. Hierarchical mode decision algorithm can be encoded as the first test for Skip mode Skip code if the conditions are the best model for the Skip

S. Lin and X. Huang (Eds.): CSEE 2011, Part II, CCIS 215, pp. 398–404, 2011.

mode, and end the current macro block mode decision analysis to the next macroblock; if not prepared for the Skip mode The current model is to determine which inter mode is set (Class16x16 or Class8x8), and the frame in the collection set to make another mode of analysis, the best model for encoding. Should be noted that, because the Skip mode means the macroblock motion vector with a 0, no residual data is encoded, does not require motion search, so this is one of the most simple model. General algorithm is first check the Skip mode[2]. Class16x16 and Class8x8 choice is a simple integer pixel motion compensation analysis, costs were calculated and compared between the two: cost16x16, cost8x8 and implementation. cost16x16, cost8x8 were used Inter16x16, Inter8x8 pattern required when the cost of encoding. If cost16x16 than cost8x8 we believe that the current image segmentation for motion compensation more small, so choose the best model Class8x8 collection; whereas in Class16x16 to choose. From the theoretical description of the algorithm, we can see through the classification method of handling the number of candidate modes is reduced, the computational complexity and therefore reduced to a certain extent.

However, the pattern determines the complexity of large, hierarchical thinking that the decrease of the initial scope of the candidate model, but Skip, Intra model still must be analyzed simultaneously in a pattern set within several different macroblock is also divided To choose the best one by one mode of detection. Also note that the grading policy of the current macro block using only its own characteristics, ignoring the space between the macro block correlation. Although the main frame encoding video sequences in order to eliminate redundant information in time, but as an independent image, each image within a specific region have a strong spatial correlation. That is, two adjacent macroblocks will be greater the probability of the segmentation method using the same inter prediction coding. To solve this problem, we propose a classification strategy and method of combining space-related. The method combines the advantages of both models policy decisions, and more fully utilize the image itself to the user, to a certain extent, reduce the complexity, and to ensure better image quality[3].

**Table 1.** Model divided into 3 sub-set of

| Mode set Category | Specific prediction model | |
|---|---|---|
| Class0 | skip model | |
| Inter prediction mode | Class16x16 set | Inter16x16、Inter16x8、Inter8x16 |
| | Class4x4 set | Inter8x8、Inter8x4、Inter4x8、Inter4x4 |
| ClassIntra | Intra16x16、Intra4x4 | |

In the spatial correlation and classification algorithm combines, first on the classification algorithm is simplified. H.264/AVC standard allows the P frames appear macroblock intra mode coding, P frames so each macro block, must be tested for the intra mode. Detection range includes two models Intra4x4 and Intra16x16 collection, which can be seen that the frame is detected large amount of computation. Through the study found, occurs in the P frame intra-coded macroblocks low probability, has little effect on image quality, this is the theoretical basis of a simplified grading[4].

Grading policy is applied to the current thinking among many of the fast algorithm, played an important role[5]. Because of space constraints, these algorithms can not be at each of the specific practices. [6] proposed a use of the current macroblock has been encoded and reconstructed around the macroblock mode of current macroblock mode decision narrowed the scope of the strategy. The idea of this fast algorithm good inspiration for us, that is inter coded macroblocks can also use the spatial correlation of the current frame. Policy decisions considering two models, although they considered the different entry points, each with different strengths of the Department, but does not affect the integration of these two ideas together to form a more perfect fast algorithm.

**Table 2.** P frame intra-coded without the impact on image quality

| Test sequence | The percentage of intra coding | Original PSNR Value (dB) | PSNR Value (dB) | ΔPSNR(dB) |
|---|---|---|---|---|
| Akiyo | 0.1% | 43.56 | 43.55 | 0.01 |
| Bus | 3.7% | 39.77 | 39.7 | 0.07 |
| Foreman | 5.0% | 40.61 | 40.47 | 0.14 |
| Mobile | 0.5% | 36.09 | 36.08 | 0.01 |
| Stefan | 8.0% | 38.18 | 38.00 | 0.18 |

simplify the classification algorithm and the mode of combining space-related decision.

After the decision of the pattern we found, P frames in a small number of intra-coded macro blocks, accounting for 5%. The intra-mode analysis is more complex, while its little effect on image quality. Based on this theory, a classification algorithm first simplification is simplified classification algorithm; and will simplify the classification and spatial correlation algorithm to combine the P frame mode decision.

Policy decisions based on hierarchical model, you can narrow the range of ways to reduce the candidate pattern complexity of the algorithm, but it ignored the video sequence itself provides a large number of redundant information, these redundant information is spatial and temporal correlation . Spatial correlation here means that the current macro block and around the adjacent macroblocks in the motion compensation mode of macroblock segmentation similarity. Based on the findings of the background or in a smoother image for the macroblock area, Inter16x16 mode is often the best mode; the other hand, those in the fast-moving target boundary or a macroblock region, more inclined to use Inter8x8 or Inter4x4 model as the best model for encoding. Figure 1 shows a residual frame (without motion compensation). H.264/AVC encoder every part of the frame to select the best partition size, which can produce the minimum amount of coded residual and motion vectors of the partition. Can be seen almost no change in areas such as the division of background to be used 16x16; in some changes, including some small details, such as clothes and more selected 16x8, 8x16, 8x8, etc. to predict the separation; most abundant in the details the human face, the book uses some of the smaller number of partitions, such as 8x4, 4x8 and 4x4 and so on. Therefore, it is obvious to see that the macro blocks in the same region is likely to have the same best model. Can think of a macroblock partition mode to a great probability is equal to its adjacent macroblocks of a segmentation model. On the point of view, this paper carried out in the third part of the experiment [6].

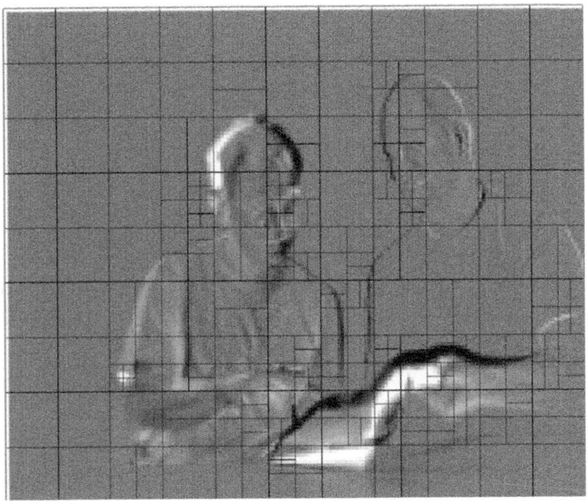

**Fig. 1.** The best inter prediction mode selection

Video sequence has a strong spatial correlation, the proposed method is to make full use of images of this nature, reduce the scope of mode selection to reduce the complexity of mode decision algorithm. That is the encoding of the current macroblock prior to its four neighboring macroblock coding modes used can get. According to the theory of spatial correlation, the current macro block coding mode of C around a large extent with the same coded macroblocks, if the macro has been around the block encoding mode case, is still on the current macro block using the Full Search coding strategy is clearly unreasonable. Therefore, we narrow the current range of macroblock mode selection, ie that the current macro block mode and its surrounding four macro blocks of a macro block in the same pattern, which makes the range of candidate models is limited to around 4 macroblocks Among the model types covered. Of course, this approach must be based on image quality and the rate of stream at an acceptable range.

At this point, we thought the decision on classification policy and spatial patterns of thinking have been associated with a clear understanding of, and propose two strategies will be used in conjunction with ideas through a variety of combinations in the experiments, the following specific algorithm has been is to better integrate the two methods[7].

Note that, if the current macroblock after detection by Skip Skip mode that can not be encoded, and the surrounding four references for the Skip mode macroblock they are nothing, because no suitable model for analysis, the program may enter the dead circulation, so the algorithm added in step 3 of the judge. That is, when that happens, the direct model to simplify the classification decision[8].

Whether it is relative to the full analysis or the decision is a simple method of classification model, you need to reduce the scope of the analysis of the candidate models. At the same time in the same region as macroblock partition mode to ensure a high degree of spatial correlation of image quality, so the method is of practical significance.

## 3 Based on Classification and Spatial Correlation Fast Algorithm for H.264 Inter Mode Decision and Performance Analysis Test

### 3.1 Test Environment and the Evaluation Criteria

This article uses the x264 software reference model, 5 groups in Table 3 video sequences in the experiment. Experiments done under the PC, using Intel Petium4 processor, 3.0GHZ, memory 512MB. CIF format for the five video sequences to test the first 300 frames, QP is fixed at 30, using CAVLC encoding. With TS that the space-based classification and correlation with the pattern determines the method compared to the original time savings, it's the formula defined as follows:

$$TS = (TO-TR) / TO * 100\% \text{ formula} \tag{1}$$

It is noteworthy that where TO grading policy that uses only the time required for full search algorithm rather than the time required, TR, said the decision to use this algorithm to model the time required.

### 3.2 Test Results and Analysis

To verify the macroblock inter mode decision has a high degree of spatial correlation, we focus on Table 3 in the 5 groups were tested video sequences, Table 3 for the experimental results. The results show that the segmentation of the current macro block mode and one of the surrounding four blocks of the same situation macro average of 91% of the entire test group, the best case, this figure reached 96%, while the worst-case scenario has 88%. It should be noted that we have chosen both slow motion test sequences, the background single sequence, including the movement intense, the background is complicated, so it has universal significance.

**Table 3.** Test results of spatial correlation

| Test sequence. | Ratio of the same model （%） |
|---|---|
| Akiyo | 96% |
| Bus | 90% |
| Foreman | 91% |
| Mobile | 90% |
| Stefan | 88% |

To illustrate the inter coded macroblocks for intra coding mode without detection, or image quality will not have a significant impact on stream rate, there has also joined the Intra encoding does not use inter mode decision, namely, simplified Experimental classification model policy decisions. 5 for the experimental group of the same sequence in the use of decomposition strategies without the intra coding mode detection, and classification of the data obtained with the original strategy of data compared. Table Δ PSNR_Y, Δ PSNR_U, Δ PSNR_V YUV component signal to noise ratio representing the decline. The results of Table 4, PSNR, the average decline in value of 0.096dB, the rate of increase of stream .986 percent, which is totally acceptable in the range of. One can imagine, do not use Intra mode coding, the interframe prediction coding complexity will decrease, mode decision time will be shortened. But because of our hierarchical model investigated here is the combination of space-related decisions and efficiency after the results of this study, no classification of the simplified mode decision strategy in the time savings[9].

**Table 4.** Intra mode is not used for inter prediction of coding

| Test sequence... | ΔPSNR_Y (dB) | ΔPSNR_U | ΔPSNR_V | ΔBitrate (%) |
|---|---|---|---|---|
| Akiyo | 0.01 | 0.01 | 0.02 | 3.30 |
| Bus | 0.08 | 0.18 | 0.14 | 0.68 |
| Forman | 0.11 | 0.16 | 0.16 | 0.76 |
| Mobile | 0.01 | 0.01 | 0.01 | 0.01 |
| Stefan | 0.19 | 0.18 | 0.17 | 0.18 |

Table 5 shows the classification of space-related policy decisions combining the model results. Should be noted that, x264 reference model has been determined using a classification model strategy, so here is saving time compared to simply using a hierarchical model of policy decisions in the case. ΔPSNR and ΔBitrate respectively on the image quality of the algorithm and rate the impact of these effects is due to a narrower range of candidate patterns caused. Δ FPS frame rate of increase expressed, TS is the time savings, this is because the decision to shorten the pattern by the benefits of time. The results show that: the classification decisions and the spatial correlation model used in combination, the average savings of 41.7% of the computing time, while PSNR lower 0.12dB, bitrate increased by about 3.7%, these effects are completely in practical applications acceptable.

**Table 5.** Comparison of coding performance

| Test sequence... | ΔPSNR(dB) | ΔBitrate(%) | ΔFPS | TS(%) |
|---|---|---|---|---|
| akiyo | 0.026 | 4.46% | 7.87 | 26.0% |
| bus | 0.177 | 4.96% | 5.77 | 51.7% |
| foreman | 0.196 | 6.43% | 8.56 | 54.3% |
| mobile | 0.026 | 1.30% | 4.27 | 39.4% |
| stefan | 0.193 | 1.33% | 4.27 | 35.4% |

## 4 Conclusions

In this paper, a hierarchical model of policy decisions, while thinking of a combination of spatial correlation, and further narrow the scope of the candidate model was proposed based on classification and spatial pattern of correlation with the decision method. The method takes advantage of redundant information provided by video sequences, without affecting the image quality and bit rate under the premise of effectively improve the efficiency of the encoder. Experimental results show that the algorithm can save 41.7% average of the time, conducive to practical applications, while PSNR drop of about 0.12dB, will not affect the image of the visual effects.

P frame is currently only used in the encoding of the fast mode decision strategy, if the B frame coding is also used in this way, I believe will further enhance efficiency. Also noted that the video sequence provided by the correlation between the time information is not used effectively if the time into the use of image correlation method by combining space-time, should be able to achieve better results.

**Acknowledgment.** The work is supported by the Natural Science project of Guangxi province Education Department of China under Grant Nos. 201012MS191 and the Joint Science project of Guangzhou University & Baise University under Grant Nos. GBK2010002.

## References

1. Joint Video Team (JVT) of ISO/IEC MPEG & ITU-T VCEG: Draft ITU-T Recommendation and Final Draft International Standard of Joint Video Specification (ITU-T Rec. H,264|ISO/IEC 14496-10 AVC), Doc. JVT-G0505r1 (2003)
2. Wiegand, T., Sullivan, G., Bjbtegaard, G., et al.: Overview of the H.264/AVC video coding standard. IEEE Transactions on Circuits and Systems for Video Technology 13(7), 560–576 (2003)
3. Zhou, M.: Evaluation and simplification of H.26L baseline coding tools. Joint Video Team (JVT) of ISO / IEC MPEG & ITU-T VCEG . Geneva, CH (2002)
4. Wang, Y., Ostermann, J., Zhang, Y.-Q.: Video processing and communication. Electronics Industry Press (2003)
5. Richardson, I.E.G.: H.264 and MPEG-4 Video Compression: video coding for Next-genration Multimedia (2003)
6. Tekalp, A.M.: Digital Video Processing. Printice Hall, Englewood Cliffs (1995)
7. Castleman, K.R.: Digital image Processing. Printice Hall, Englewood Cliffs (1996)
8. Shen, L.-S., Zhuo: Ability. Low-rate video coding and transmission. Electronics Industry Press, Bingjing (2001)
9. Bing, Y., Zhang, T.-Y.: Based on the classification and spatial correlation of the H.264-P frame mode decision fast algorithm. Microelectronics and Computers 24(5), 45–48 (2007)

# Algorithm of Multiple Sensitive Attributes Privacy Based on Personal Anonymous

Shen Guozhen

Education Technology Center, Zhejiang Shuren University, Hangzhou, China

**Abstract.** This paper analyzes the (k, l) - Anonymous model need to address two key questions: re-encoding methods and information loss metrics, and in-depth analysis of the (k, l) - Anonymous model and similar models that exist in many sensitive attributes privacy disclosure and attack the problem while giving a certain amount of improvement. In order to avoid multiple records corresponding to a single individual in the situation over in anonymous generalization, this paper presents a new connection based on lossy re-encoding attribute more sensitive method, experimental results show that the coding method can keep the same person more sensitive to possible link between property information. Meanwhile, in order to solve the existing multi-anonymous model sensitive information, especially due to its relevance leak, the paper related to the privacy of individuals Anonymous technological trends, a relational database to solve more sensitive attribute data released disclosure of private information model, the paper gives the formal description of this model and the corresponding algorithms.

**Keywords:** Personal Anonymous; Privacy; multiple sensitive attributes.

## 1 Introduction

Personal information is information that can be used directly or indirectly identify the data on natural conditions, including name, gender, age, date of birth, ID number, health and so on. The European Union in 1992 "The Council Data Protection Act," the draft proposed amendments states: "Personal data is information about a natural person can be identified with any information, not limited to handle information in the form, which includes any kind and any form of information, so long as such information is related to the individual, whether living or dead people who, as long as this or these people can identify the "Personal data includes personal natural conditions, social and political background, life experience and habits, basic family situation [1-2].

## 2 The Information Loss Measure

Makes the quasi-identifier data generalization accuracy on the property value has decreased, will bring some loss of information. This paper shall examine the data attribute values before and after changes in the level of uncertainty, to quantify the loss of information to define the size. Specifically, we order the property divided into two

S. Lin and X. Huang (Eds.): CSEE 2011, Part II, CCIS 215, pp. 405–410, 2011.

types of property and disorderly property. Ordered attribute values relations between the natural order, such as age, height, education level and other properties; unordered attribute values do not exist between the natural order of relations, such as nationality, marital status. For ordered and unordered attributes attributes, the data caused by the loss of information generalization are defined as follows [3]:

**Definition 1 (generalized ordered attribute information loss).** Tuple t in the ordered set X, the value of property x = [a ~ b], generalization and the corresponding property value$x^*$=[$a^*$ ~ $b^*$], then the tuple t, its properties in order on X Information loss

$$L(x, x^*)=\begin{cases} \frac{b^*-a^*+1}{b-a+1}, x \neq x^* \\ 0 \quad , x = x^* \end{cases}$$

**Definition 2 (generalized disorder attributes the loss of information).** Let tuple t in the disordered set of attributes X, the value of y, generalization and the corresponding property value is a collection of y *, then the tuple t, its properties in the disordered information on the loss of Y

$$L(y,y^*)=\begin{cases} \frac{|y^*|}{|y|}, y \neq y^* \\ 0 \ , y = y^* \end{cases}$$

Where, I y I and I I denote a collection of generalized attribute values before and after the y and the number of elements.

On this basis, we can further define the data generalization brought about in a particular tuple of information loss, and bring in the whole data set of information loss.

**Definition 3 (generalized loss of information per group).** Based quasi-identifier QI, if the tuple t after treatment by the generalization, then t takes into losses arising from the information

$$\sum_{i=1}^{u} L(t[X_i], t^*[X_i]) + \sum_{j=1}^{v} L(t[Y_j], t^*[Y_j])$$

Which $X_i(i = 1,2, \dots, u), Y_j(j = 1,2, \dots, v)$were ordered for the QI attributes and disorder in the property, $t[X_i]t[Y_j]t^*[X_i]t^*[Y_j]tt^*X_iY_j$

**Definition 4 (generalized loss of information data set).** Let the original data set is D, Overview of the results of the data sets, then D takes into losses arising from the information $L(D, D^*) = \sum_{i=1}^{n} L(t_i, t_i^*)$,

Among them, $t_i(i = 1,2, \dots, n)Dt_i^*t_iD^*$

**Definition 5 (almost complete data set).** Let the original data set is D, quasi-identifier QI, if the D takes into on the QI, so all the tuples in QI values are the same, is referred to as D, fully generalized data set.

Taking into account the protection needs to meet the almost anonymous data sets can have more than one, which correspond to the information loss may vary. This section considers the perspective from the loss of information, to define the distance between data objects [4].

**Definition 6.** (distance between the tuple) has a tuple and, if = the generalized equivalence group formed by representatives of element, the distance between the

$DS(t_1, t_2)=L(t_1, t^*)+L(t_2, t^*)$,

Where, L () and L (), respectively, and almost into the resulting loss of information.

**Definition 7.** (class representative per group) with class G, the G of all tuples most probable treatment, so as to form equivalent group EQ. If the representative element for the EQ, then G is also known as class representatives per group [5].

**Definition 8.** (per group to the class distance) is set to class representatives of G tuple, if the tuple t G, t is the distance to the class G $DS(t, G)= DS(t, t^*)+|G|* DS(t_g, t^*)$,

Where, | G | for the class G contains the number of tuples, = t and equivalent to the representative group element.

**Definition 9.** (between-class distance) has two classes represented by G and E respectively, and tuples, classes G and H is the distance between

$DS(G, H)=|G|* DS(t_g, t^*)+|H|* DS(t_h, t^*)$,

Where, | G | and | H | classes G and H were included in the number of tuples, = is equivalent to and the group formed by representatives of dollars.

## 3 Overview of Multi-dimensional Approach

Anonymous existing models, such as k-anonymity and l-diversity, often remove individual identity attributes, and then generalized quasi-identifier property to satisfy the anonymity requirement. These models typically assume a single anonymous individual in the data table at most once, and only a sense of property Weimin vector. But in the actual environment, this premise may be difficult to satisfy. For example, in a hospital's patient table, because while some patients suffering from various diseases, has led to a number, and different drugs for different diseases have been well treated. In Table 1, Mike also suffering from Bronchitis (Bronchitis), Coryza (rhinitis) and Flu (influenza), resulting in 3 times. According to the traditional, anonymous methods (to meet the 3 - diversity), if an attacker finds Mike in a QI group, then it will be Mike reasoning with Bronchitis (bronchitis) or Coryza (rhinitis) or Flu (influenza). In fact, the two speculate at this time are correct. So when the probability of leakage of privacy is 100%. This is a single instance in case of multiple records corresponding to the lower level of privacy protection phenomenon. Moreover, removal of identity information is lost between the same person may be more sensitive to attributes associated information, are more likely to lose the association between the different attributes of dimensional information. And this information is very important to many studies, such as complications of medical research and drug effects between the study match.

**Table 1.** The table according to the distance

| NO | Name | Age | Postcode | Disease | Medicine |
|---|---|---|---|---|---|
| 1 | Mike | 55 | 10085 | Bronchitis | An |
| 2 | Mike | 55 | 10085 | Coryza | Bn |
| 3 | Mike | 55 | 10085 | Flu | Gn |
| 4 | Emily | 43 | 10075 | Pneumonia | Cn |
| 5 | Emily | 43 | 10075 | Coryza | En |
| 6 | Tim | 45 | 10075 | Gastritis | Dn |
| 7 | Jane | 49 | 10076 | Dyspepsia | En |
| 8 | Ella | 46 | 10079 | Gastric ulcer | Fn |
| 9 | Johny | 47 | 10077 | Dyspepsia | Dn |
| 10 | Tom | 48 | 10078 | Flu | Fn |
| 11 | Fermous | 48 | 10089 | Bronchitis | Gn |
| 12 | Linda | 48 | 10071 | Flu | Fn |
| 13 | Boly | 49 | 10084 | Bronchitis | An |
| 14 | Rose | 5 | 20084 | Bronchitis | An |

To avoid this privacy leak, In this section we analyze a single individual, and multi-dimensional multiple records corresponding to the case sensitive property, proposed to maintain the anonymity of identity. Retention of information can improve the effectiveness of identity information. For example, in Table 2, to retain the identity information can help researchers find the situation complicated by different diseases.

## 4 Personalized Anonymous

Currently, the data published in addressing privacy generally used when K-anonymous or l-diversity method for the two models, but if you take into account the individual requirements of anonymous or more data sources, the previously published data, there may be private information discovered by an attacker. K-anonymity can only be prevented because of the individual and the connection between tuples, and can not stop the individual and the connection between the sensitive values. In [2] proposed the concept of personalized anonymity, not only to publish the view to meet the k - anonymity, but also considering the attributes of each individual's privacy confidentiality requirements.

Personal privacy (personalized privacy preservation) the principle of anonymity [3], in order to meet the requirements of different individuals and the level of privacy protection, and to overcome the anonymity of the uniform data caused by "excessive" protection and protection of the "inadequate."

### 4.1 The One-Dimensional Tree

Single Weimin sense of property classification tree can be a number of simple one-dimensional tree composition. One-dimensional classification tree for each leaf node is sensitive to this dimension attribute values, and these leaf nodes is sensitive to maintain a certain relationship between the property value, summed up their relations for the root of the tree-dimensional classification.

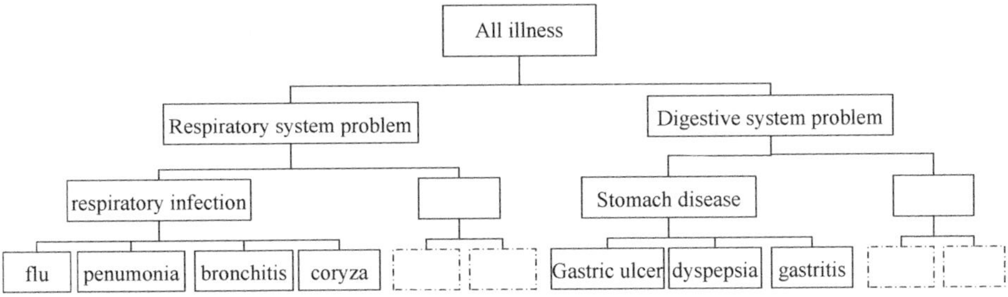


**Fig. 1.** Disease classification tree single Weimin sense of property

For example, mike sensitive private property Bronchitis bronchitis, so the tree can be sensitive to the mike record Bronchitis replace attribute values in the general value of the classification tree that respiratory infection. To ensure that the requirements of their individual anonymity, but you can see through the release of the data, even if disease-dimensional properties are summarized in sensitive attributes, but attributes the record-dimensional medicine is not sensitive to changes in property values, according to Chapter III of the general method can be sensitive to the multidimensional relationship between property values, so mike if the attacker knows the value of the individual privacy bronchitis Bronchitis medicine propertics and the corresponding dimension of the sensitive attribute value is An there is a certain relationship, then the dimension sensitive property by medicine However, any value can be inferred mike in respiratory infection cases most likely to suffer from Bronchitis, so mike's personal privacy will be leaked. Based on correspondence between the multidimensional case of sensitive attributes in this section, a new general method is proposed to generalize.

## 4.2 Semantic Classification Trees and More Sensitive Property

Node attributes is based on the concept of a virtual structure, the tree is reflected through the property, does not maintain the appropriate connections. The system can be carried out gradually by the nodes until the leaf nodes of the cluster members to its neighbor can easily be familiar with so far, through the collection of cluster nodes dynamically updated to ensure that each child leaves Results point correlation in real time. Although in principle the individual characteristics of each node's unique, and eventually the system can be refined to a node of a cluster level.

Tree structure of each sensitive attribute more than a semantic classification trees, each semantic root of the tree was the largest broad-sensitive property value, the root node the left subtree undcr the original one-dimensional tree. Second from the left the left subtree of the original one-dimensional classification tree sensitive attribute value attribute dimension to this sensitive relationship between the value of other sensitive, sub-tree root for the collection of sensitive property value, sub-tree leaf node point is the relationship between different clusters of sensitive attribute values. The third child from a tree to the left of the right subtree, each sub-tree is the original one-dimensional tree in the sensitive attribute value attribute with other Weimin sense of the relationship between the different sub-trees corresponding to different dimensions of the sensitive attribute values in the These sub-tree, the first sub-root of the tree i is expressed as the original one-dimensional classification tree-sensitive property value with the first i Weimin sense of the relationship between the set of attributes, the leaf node is the root of the relationship between the corresponding sub-tree clusters.

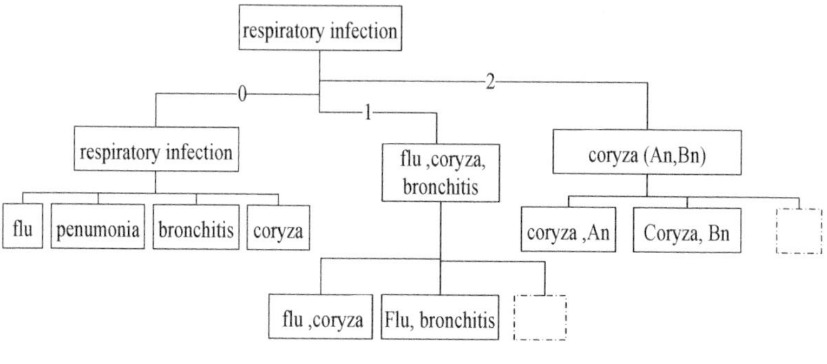


**Fig. 2.** More sensitive to semantic classification tree properties

## 5 Conclusions

This paper analyzes the requirements for personal privacy more sensitive dynamic data dissemination, discussion of the original single anonymous individual properties of the defect sensitivity, a sensitive property of a general semantic tree and the new algorithm, to solve the more sensitive the model Property under the personal anonymous attacks on the privacy of data release and disclosure issues.

## References

1. Qin, X., Men, A., Zou, Y.: Based on K-anonymity privacy protection algorithms. Chifeng University, Natural Science (May 2010)
2. Liu, T., Ni, W., Chong, Z., Zhang, Y.: Multi-dimensional numerical sensitive attributes privacy preserving data dissemination methods. Southeast University, Natural Science (April 2010)
3. Hu, X., Zhang, Z., Zhang, B., Dong, J.: Tuple based on the privacy of sensitive data protection method. Southeast University, Natural Science (May 2010)
4. Song, J., Huang, L., Zhou, G., Li, Y.: View of the safety release (review). Hebei Normal University (February 2008)
5. Song, J., Liu, G., Huang, L., Zhu, C.: K-anonymous method and the quasi-identifier set of views related to the solution algorithm. Computer Research and Development (January 2009)

# Research of Embedded Real-Time System Security Level Assessment Techniques

Wu Yufeng

College of Computer and Information Engineering,
Lishui University, Lishui, China

**Abstract.** This paper briefly describes the main ideas and implementation of technology strategy of Embedded real-time system security level assessment which is important part of mechanism to system flexibility based on embedded real-time system, designs the embedded real-time system security level assessment system based on pattern matching for the characteristics of embedded real-time systems. By building security level assessment system and simulation testing environment based on RT-Linux real-time operating system, the impact of Embedded real-time system security level assessment strategy on performance of RT-Linux was tested.

**Keywords:** security, Pattern matching, Embedded Operating System (EOS), Real-time System, RT-Linux.

## 1 Introduction

With the extensive application of real-time computer technology, embedded real-time systems are widely used in the field of industrial control and defense systems. How to ensure the reliability of real-time system in order to avoid disastrous consequence, currently is a very important area of embedded systems research, mechanism to system flexibility based on embedded real-time system came into being.

Embedded real-time systems usually arise in the implementation process some of the problems, such as some exception caused by malicious attacks of common suspicious or malicious code on the system. Moreover, the changes of system hardware state may also lead to system exception, such as equipment failure may make the system don't know what to do. Therefore, we can enhance the flexibility of real-time system itself and then improve the security and adaptability of the system. Embedded real-time system security level assessment techniques as a core part of Mechanism to System Flexibility is essential. Based on this, the design and implementation of Embedded real-time system security level assessment system were gived by this paper, which Improved the core functions of Mechanism to System Flexibility Based on Embedded Real-time System, so that the Mechanism to System Flexibility can effectively prevent exceptions caused by attacks of malicious code or hardware problems.

By using Embedded real-time system security level assessment techniques, real-time system can control computer program dialogue or Objects to work together and the system hardware more effectively, also can identify and prevent suspicious or malicious exploratory behavior, so as to prevent the malicious code to explore the

S. Lin and X. Huang (Eds.): CSEE 2011, Part II, CCIS 215, pp. 411–417, 2011.

system defect. Thus, that can make the system flexible. By analyzing the security of embedded real-time operating system process, system status which is unsafe, defective, or hidden in the process can be find. Therefore, according to their characteristics of activities and the severity of impact on the system, The Embedded real-time system security level can be evaluated. After that, protection measures of different levels will be taken so as to ensure the safe operation of embedded real-time system. All of above is valuable and has prospects of a wide range of applications.

## 2 Overview of Embedded Real-Time System Security Level Assessment Strategy

Embedded real-time system security level assessment module is embedded real-timesystem core of mechanism to system flexibility, it can work out comprehensive assessment of collected information based on pattern matching method, then give the system's security level and action proposal which will be send to the module of explaination and implementation. The module will decompose the information after received them, then pass the decomposition information to the different module. After that, every module will start appropriate services according to the information received. Architecture of mechanism to system flexibility is shown in Figure 1:

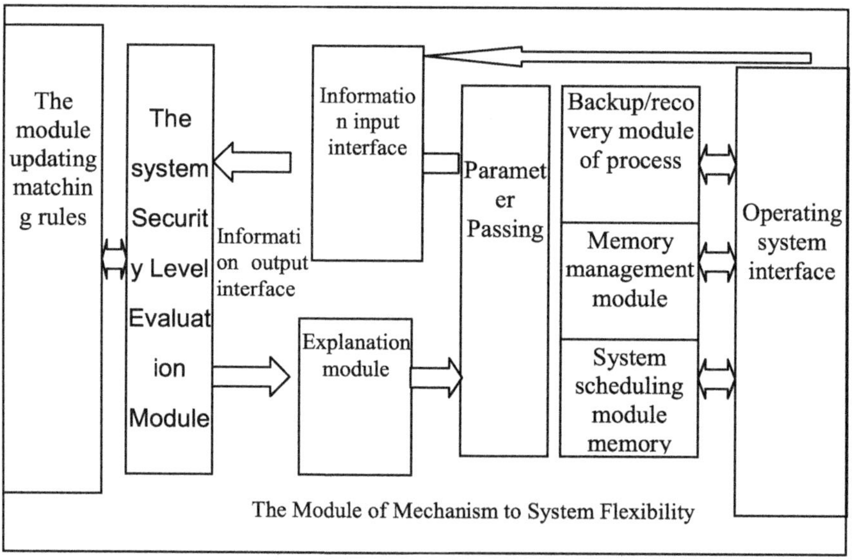


**Fig. 1.** Architecture of the Mechanism to System Flexibility

The Module of Mechanism to System Flexibility use pattern-matching principle, it collects the behavioral characteristics of various types of information, and then establishes Features Library related to them. If the input information that indicating the system behavior matches the records in the Features Library, then the system will give the corresponding system security rating. So the system security level evaluation can be realized. The input information is a formatted string that make up of process information, system information (including hardware information), it will be analyzed by pattern matching. After the information is inputted into system, the matching result will be getting by pattern matching, which will be output to the module of explanation and implementation. Parameters will be sending to distributed module of process security, memory management, and system scheduling though Parameter passing module, the distributed modules decide which coping strategies of mechanism to system flexibility will be use according to the data received.

## 3 Embedded Real-Time System Security Level Assessment Based on Pattern Matching

Pattern matching is to compare the collectted information with the known pattern in the database of the system security level so as to give the current Security Level Status of the system. Security level assessment system receives a string consists of operating system processes information, operating system information and embedded system hardware information through the information input interface, then compares it with the pattern in the database of the system security level. If the match is successful then the system security level can be get, else the current string pattern will be recordded for pattern update. The task of system mainly covers the complete pattern matching and pattern updates.

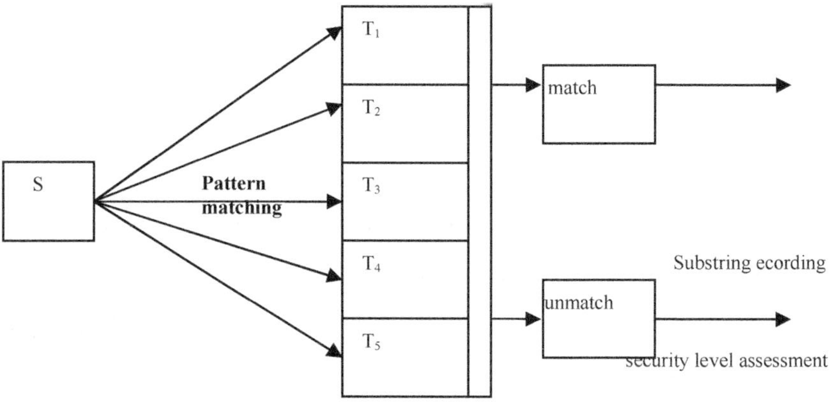


**Fig. 2.** Process of pattern matching

Pattern matching is the location of substring, which is an important string operations. For two strings T (length of n) and S (length m), the process of finding substring S (also called pattern) in the string T is called pattern matching. When S substring is found in the string T, that called a success matching, otherwise a failure matching. The security level assessment system based on pattern matching has several strings, which can be classified into five security level sets, denoted by T1, T2, T3, T4 and T5 respectively, a level of security for each. T1 represents The highest level of security, T5 represents the lowest level of security. Each collection contains the strings of its security and there are more than one string in each security level. When pattern matching is successful, system security level can be got according to the matched string. Otherwise, the pattern of substring will be recorded in order to form a new string pattern of security level. Figure 2 shows this process:

According to the above description of matching rules Algorithm, two-dimensional array can be selected to store the strings (security level pattern). Each line of the two-dimensional array stores a set of security level pattern separately. Because the number of each security level pattern strings is not fixed, the column number of two-dimensional array is uncertain, and the row number is 5(5 security levels).

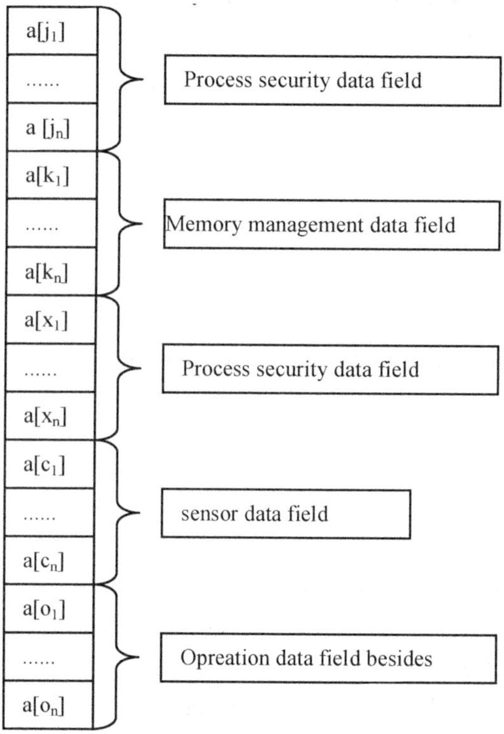


**Fig. 3.** Data structure definition of substring S(system state pattern)

Pattern matching, just the information sub-string and each element of two-dimensional array can be compared and matched. If there is a pattern matched, the corresponding values of the array element will return to determine the current security level. If no matched pattern, the current information substring will be recorded. When a certain amount of information sub-string is accumulated, through manual processing, the unmatched sub-string pattern can be processed to a a new security level pattern string by using rtificial intelligence algorithms.

The Substring includes not only the data information from process security, System scheduling and Memory management, but also the status infirmation of system hardware coming from the sensor and operating system status information which the distribution modulecan not provide. Specific structure is shown in Figure 3.

Security level assessment pattern library stored information pattern of a variety of security level by using two-dimensional array data structure. Because the system has fixed five security level, the row number of the two-dimensional array is 5. The pattern number of each security level is not fixed, so the column number of two-dimensional array is not fixed. After each pattern update, the column number of two-dimensional array are likely to increase, depending on the column number of the longest row updated. The whole security level assessment pattern library is a two-dimensional array with unfixed number of column. To facilitate future pattern updates, the maximum column number of the two-dimensional array will not be limitted. Specific structure is shown in Figure 4.

| | | | | | |
|---|---|---|---|---|---|
| $T_1$ security level | $a[b_1]$ | ...... | $a[b_n]$ | ...... | $a[b_m]$ |
| $T_2$ security level | $a[c_1]$ | ...... | $a[c_n]$ | ...... | $a[c_p]$ |
| $T_3$ security level | $a[d_1]$ | ...... | $a[d_n]$ | ...... | $a[d_q]$ |
| $T_4$ security level | $a[e_1]$ | ...... | $a[e_n]$ | ...... | $a[e_r]$ |
| $T_5$ security level | $a[f_1]$ | ...... | $a[f_n]$ | ...... | $a[f_s]$ |

**Fig. 4.** Data structure definition of security level assessment pattern library

## 4 Experimental Simulation and Performance Evaluation

Experiment emulation is based on RT-Linux real-time operation system. RT-Linux is open source and multi-tasking operating system, which has the characteristics of hard real-time. RT-Linux places a real-time kernel between the Linux kernel and hardware interrupts, regards standard Linux kernel as a process of the real-time kernel to manage together with user's process. The standard Linux kernel has the lowest priority; it can be interrupt by the real-time process, the normal Linux process still can run on Linux kernel. In this way, RT-Linux not only can use all services of Linux, but also can offer the real-time environment.

First, cut Linux kernel code and compile it according to the system's need, which will miniaturize Linux. And then transform them in real-time. Combination of the standard Linux kernel with real-time RT-Linux kernel provide real-time guarantee and extended functionality. Adding the Module of System Flexibility (Msf_1.5) software package with security level assessment module to the combination, the Embedded real-time system security level assessment system will be realized.

Here we choose Linux-2.4.4 and RTLinux-3.1, patch the RTLinux-3.1 source code and The Module of System Flexibility (with security level assessment module, Msf_1.5) source code to the miniaturization of the Linux-2.4.4 kernel, recompile the kernel, then the kernel with real-time performance and mechanism to system flexibility is formed. Concrete steps are as follows:

1. Copy Linux-2.4.4.tar.gz、rtlinux_3.1.tar.gz and Msf_1.5to /usr/src and unpack.
2. Build links.
   # cd /usr/src
   # ln –s rtlinux-3.1 rtlinux
   #cd rtlinux
   # ln –sf /usr/src/linux ./linux
3. Patch Real Time Kernel and Module of System Flexibility(with security level assessment module, Msf_1.5)
   #cd /usr/src/linux
   #patch –p1 < /usr/src/rtlinux/kernel_patch-2.4.4
   #patch –p1 < /usr/src/frm_1.2/kernel_patch-2.4.4
4. Configure and compile the kernel.
5. Boot real-time kernel.

Then reboot：#reboot

6. Install Real-Time Module to RT-Linux.

Install rtl_sched.o，rtl_fifo.o and rtl_time.o to /lib/modules/2.4.4-rtl.

After the completion of the above steps, the operating system with real-time performance and mechanism to system flexibility is completed. The architecture of RT-Linux with the Mechanism to System Flexibility(with security level assessment module, Msf_1.5) operating system is shown in Figure 5.

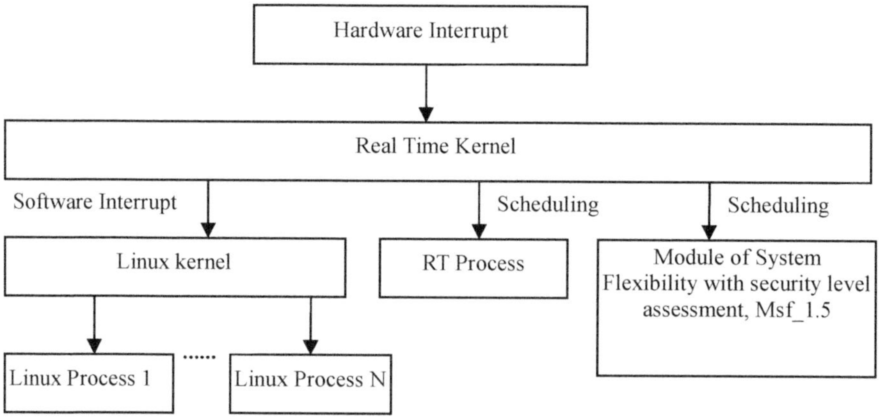


**Fig. 5.** RT-Linux with the Mechanism to System Flexibility (with security level assessment module, Msf_1.5)operating system

## 5 Conclusions

The embedded real-time system security issue has been hot in academic study circles. In order to improve the security and reliability of embedded real-time system, we can enhance the real-time system flexibility itself. The core problem of System Flexibility is system security level assessment, this paper deeply studies and discusses Embedded real-time system security level assessment techniques according to current development status, designs and implements security level assessment System based on pattern matching. Simulation results show that the security level assessment system can effectively prevent the system from exception without affecting system performance and improve the capacity of flexible response.

## References

1. Kang, K., Son, S.: Systematic security and timeliness tradeoffs in real-time embedded systems. In: Proceedings-12th IEEE International Conference on Embedded and Real-Time Computing Systems and Applications, RTCSA 2006, pp. 183–189 (2006)
2. Hwang, D.D., Schaumont, P., Tiri, K., Verbauwhede, I.: Securing embedded systems. IEEE Security & Privacy 4(2) (2006)
3. Zhao, G., Zhang, G., Yao, A.: Research of Mechanism to System Flexibility Based on Embedded Real-time System. In: 2010 International Conference on Advanced Measurement and Test, AMT 2010 (2010)
4. Ravindran, K., Kwiat, K.A., Sabbir, A.: Adapting distributed voting algorithms for secure real-time embedded systems. In: Proceedings of the 24th International Conference on Distributed Computing Systems Workshops (2004)
5. William Beck, R., Mark Vanfleet, W.: High Assurance Security/Safety For Deeply Embedded, Real-time Systems. In: 2006 Embedded Systems Conference - Silicon Valley (ESC 2006) (2006)
6. Kerstan, T., Oberthür, S.: Configurable Hybridkernel for Embedded Real-Time Systems. Embedded System Design (2007)
7. Kitani, T., Takamoto, Y., Yasumoto, K., Nakata, A., Higashino, T.: A flexible and high-reliable HW/SW codesign method for real-time embedded systems. In: 2004 IEEE First Symposium on Multi-Agent Security and Survivability (2004)

# A Kind of Data Stream Clustering Algorithm Based on Grid-Density

Zhong Zhishui

Department of Mathematics and Computer Science
Tongling University
Tongling, China

**Abstract.** One target on this thesis is to study and realize a kind of data stream clustering algorithm with quick running rate and high clustering accuracy. In order to reach this, we have done some work as follows. Background and relevant work on data stream mining is discussed. Popular traditional clustering algorithms are summarized and the data stream clustering algorithms are researched. On the basis of these, we propose GD-Stream (Grid-Density based Evolving Stream) algorithm, which is a framework based on grid-density. By modifying the synopsis data structure, This algorithm has the following characteristics. Borrowing the framework from CluStream algorithm, GD-Stream is divided into online layer and offline layer, using density-decaying skill Online layer reads data stream rapidly, and stores relative information by synopsis data structure. With this, offline layer provide accurate clustering. The two layers work together to achieve the balance of accuracy and speed..

**Keywords:** Data Stream, Clustering, Grid-Density.

## 1 Introduction

In traditional cluster analysis, k-means algorithm is the most classic. Similarly, in the data stream clustering analysis, k-means also play a huge role. There are many current clustering algorithms are based on k-means clustering based on the idea, and on this basis to improve them to suit the characteristics of streaming data, and have achieved good results. But the decision was based on the method by which have a common drawback, that is, the determination of the number of clusters k. Different data for the data distribution is likely to be a natural class apart. In addition, because k-means is used in the measurement method based on distance, so the clustering preferences in the sphere [1].

In the flow of data clustering, many current clustering algorithms are based on CluStream online / offline double frame. This framework can solve the two-tier data stream clustering problem in real-time requirements and the conflict between clustering quality. Online store periodically flow through the layer data stream summary statistics, and these data micro-clustering. Line layer in accordance with the time frame pyramid time series generated by storing a snapshot in time to meet the different time periods off layer summary information about the recovery. The offline layer is the input of the different parameters on the micro-cluster to gather once again class to achieve the final

S. Lin and X. Huang (Eds.): CSEE 2011, Part II, CCIS 215, pp. 418–423, 2011.

clustering results. In other words, the final clustering result is based on online clustering on the micro level, on-line layer of the early clustering quality and clustering effect will eventually have a huge impact [2-5].

There are many current clustering algorithms are based on k-means clustering based on the idea, and on this basis to improve them to suit the characteristics of streaming data, and have achieved good results. But the decision was based on the method by which have a common drawback, that is, the determination of the number of clusters k. Different data for the data distribution is likely to be a natural class apart. In addition, because k-means is used in the measurement method based on distance, so the clustering preferences in the sphere. In this paper, an improved density grid-based data stream clustering algorithm GD-Stream, any shape can be a very good data mining. Algorithm to higher dimensional space is divided into multiple cells, and to introduce the concept of density to describe the weight data point x at some point the impact on the degree of clustering. In order to improve the clustering quality and speed, GD-Stream method proposed strategy for dealing with isolated points, according to the dynamic changes in the density to distinguish the real outlier and removed it, and this knockout on the back does not affect the clustering results. The algorithm does not compute and compare the distance, greatly improving the efficiency of the implementation. Experiment proved, compared with CluStream algorithm GD-Stream Algorithm clusters of arbitrary shape can be tapped with higher accuracy and speed.

## 2 Basic Concepts and Related Description

In this article, assume that the input data is n-dimensional, each input data are defined in the space R:. Which is the definition of i-dimensional space. Now n-dimensional space R space divided into a number grid, assuming For each dimension, the will of its corresponding i-dimensional space (i = 1,2,3, ..., n) is divided into ri section, then. so the grid formed by the data space was divided into grids . and for which any of a, (ji = 1, ..., ri) the density of the grid composed of g, can be expressed as: g = (j1, j2 ..., jn). If a data record x = (x1, x2, ..., xn), mapped to the grid by the density g of good will have g (x) = (j1, j2 ..., jn), where (xi).

**Definition 1 (density weighted).** Let x be a data stream of data points, then x is a density of weight change with time t function:

$$D(x,t) = \lambda^{t-t_c} = \lambda^{t-T(x)} \tag{1}$$

Where is the attenuation factor, for the moment of arrival of data points x for x corresponding timestamp.

Macro perspective, the weight density of data points x describes the time in t affect the degree of clustering. The new data points the initial density of weight 1, as time increases exponentially. In fact, any set of data points here the timestamp of a data item that can be used.

**Corollary 1.** Let X (t) is [0, t] between the set of all data items, then X (t) the density of all the weights of the data and satisfy:

$$\lim_{t\to\infty}\sum_{x\in x(t)} D(x,t)=\frac{1}{\lambda} \tag{2}$$

*Proof.* Let X (t) in all [0, t] time to reach the data item as:, where the data points in the first i time to arrive. In t times their density weights were:, their, and as, when t take the limit goes to infinity.

**Definition 2 (mesh density).** G in the cell, the data point set of any data item at time t corresponds to a density of weight, before the time t (including t time) all the weights and the known density of mesh density :

$$D(g,t)=\sum_{i=1}^{n} D(x_i,t) \tag{3}$$

**Corollary 2.** Let g at the time the cell receives a new data item is the time of the last receive data (>), the cell density of the grid update g

$$D(g,t_{\mathrm{n}})=\begin{cases} 0, \\ \lambda^{t_n-t_l}D(g,t_l)+1, \end{cases} \tag{4}$$

*Proof.* g in that moment all the data in the grid point set, which arrives at the i time for the time received a new data item, then there is:

$$D(g,t_l)=\sum_{i=1}^{n} D(x_i,t_l) \tag{5}$$

According to (1), available:

$$\begin{aligned} D(x_i,t_n) &= \lambda^{t_n-T(x_i)} = \lambda^{t_n-t_l}\lambda^{t_l-T(x_i)} \\ &= \lambda^{t_n-t_l}D(x_i,t_l),(i=1,\cdots,n) \end{aligned} \tag{6}$$

By the formula (3) we can see the infinite as time goes on, if a cell has not been a new data point arrives, then the mesh density will gradually decay, dense grid, the grid for the transition may be attenuated or sparse grids; the previous cell does not affect the subsequent data clustering, can be 0 depending on the mesh density. On the contrary, when a large data set arrives, the sparse grid, the grid can be upgraded to a transitional or dense grid. by the This may reflect the density of different regions of data space changes over time.

**Definition 3.** (dense grid and sparse grid) is a dense grid set the threshold, the threshold is sparse grid is the most recent moment of receiving data.

(1). If $D(x,t_n)\geq C_m$, then we say the cell is dense grid.

(2). If $D(x,t_n)\leq C_l$, and $t_n >> t_l$, called the cell for the sparse grid.

(3) If $C_l \leq D(x,t_n)\leq C_m$, then we say the cell is sparse grid.

Some cells, if it is a long time no data arrives, causing the cell is very small amount of data, raw data on the impact of clustering is negligible, such as evacuation mesh grid; Accordingly, if From a certain moment, the data in the cells gradually increased up to a certain number, called the cell for the dense grid.

By definition 2 we can see, the grid will sparseness and change over time. Thus, over time, should be timely adjusted the viewing grid density clustering.

**Definition 4 (feature vector).** A subnet grid g in the feature vector at time t is a four-tuple $V < G, D_g, t_1, t_m >$ where G is the center coordinates of the cell, the cell g is the most recent update time for the g from the hash as a sparse cell Delete the last time the table is the last time the grid density.

# 3 GD-Stream Algorithms

GD-Stream is the basic idea: First, read each line of a continuous layer of newly arrived data items, and map it to the cell has divided, and update the feature vector of cells. Second algorithm to define a time interval gap, after a gap of each layer in the off-line time to do a cluster. Algorithm after the first gap. Initialize cluster, computing the feature vector of each cell to identify the dense cells, transitional cells, and they gathered for a cell cluster or class, the corresponding class number is given as the initial cluster. Then, the algorithm after a gap of time in each cluster must be adjusted. Adjustment, first check the arrival of the new data caused by changes in cell density, and then check under the isolated point strategy to remove those cells isolated points only, in order to reduce the workload of the algorithm. Then adjust the clustering, based on changes in the density of all in this gap period adjusted for the feature vector associated with the cell type of adjustment. Will have become sparse in the cells removed from the class, but also will be transformed into dense or transitional cells into the best neighborhood class.

GD-Stream uses three parameters: gap, Dl, Dm. gap for the clustering process of reviewing and updating the cell density and other attributes, adjust the cluster time interval. Dl was to evaluate whether the sparse cell density of the cell cap, and Dm for the evaluation of whether each cell density, cell dense limit, which by definition 3 calculations.

# 4 Experimental Analysis

## 4.1 Comparison of Speed of the Algorithm

Speed of the algorithm to reflect the actual implementation of the algorithm efficiency. In this chapter experiments to compare the data sets used KDDCUP99 GD-Stream Algorithm and CluStream algorithm speed. The results shown in Figure 1.

It can be seen from Figure 1, with the increasing length of the data stream, two algorithms were tested linear increasing trend. Specifically, GD-Stream Algorithm for more rapid growth rate, almost 4 times CluStream algorithm, which GD-Stream.

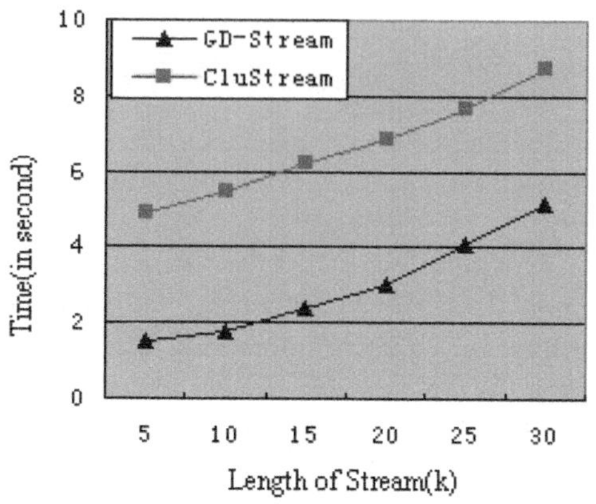


**Fig. 1.** Comparison of algorithms speed

### 4.2 Clustering Algorithm That Is More Efficient

GD-Stream algorithms achieve faster clustering speed, because GD-Stream algorithm simply add new data points corresponding to spatial cell, need not calculate the distance, easy to operate. Moreover, GD-Stream Algorithm for the time intervals in each gap, according to adaptively adjust the density of the cluster, and by the decay factor to distinguish between "new" "old" data, taken a solitary point deletion technology, greatly reducing the work, improving the clustering speed. However CluStream algorithm for each data point to calculate the distance to all the micro-cluster and compare operations, also need to deal with micro-clusters, which have greatly increased the amount of CluStream computation.

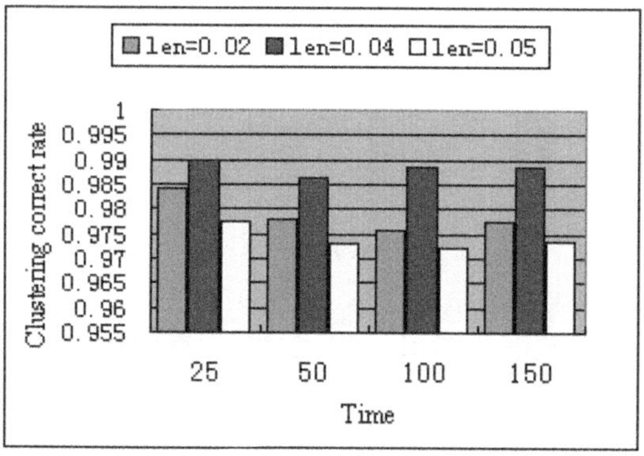


**Fig. 2.** Comparison of the accuracy of two algorithms

### 4.3 Comparison of the Accuracy of Algorithms

Algorithm GD-Stream is still used to represent the data set KDDCUP99 accuracy. The data set is divided into three categories accuracy, with a precision of each value len said, here were taken len = 0.02, len = 0.04, len = 0.05. Specifically that the results shown in Figure 2.

It can be seen from Figure 2, not the more detailed the better mesh, fine mesh, the more easier it will be a cluster of data points in the same division to the other clusters, resulting in lower accuracy. But also can not divide being too rough, it will make the original does not belong to the same cluster of data points is divided into the same bunch of errors in the. len = 0.04 by experiment when the clustering accuracy is the highest. while we can see that the correct clustering results rate above 97%, the highest close to 99%, also shows that the clustering algorithm GD-Stream high quality

## 5 Conclusions

This chapter presents a grid-based density data stream clustering algorithm GD-Stream, the density decay algorithm in the grid data on the partitioning of different density, can be clusters of arbitrary shape and make up the traditional grid-based ignored by the data stream clustering isolated point on the grid boundary defects. At the same time interval as the clustering gap reference time period, reducing the amount of computation. Compared with CluStream algorithm, and proved to GD-Stream algorithm has a faster speed, while a higher clustering accuracy.

## References

1. Zhang, X., Zeng, W.: Research and advances of real-time data stream clustering. Computer Engineering and Design 30(9), 2177–2186 (2009)
2. Tu, L., Chen, L.: Stream data clustering based on grid density and attraction. ACM Transactions on Knowledge Discovery from Data (TKDD), 12–39 (2009)
3. Zheng, Y., Ni, Z., Wu, S., et al.: Data stream cluster algorithm based on mobile grid and density. Computer Engineering and Applications 45(8), 129–131 (2009)
4. Mahdiraji, A.R.: Clustering data stream: A survey of algorithms. International Journal of Knowledge-based and Intelligent Engineering Systems, 39–44 (2009)
5. Guha, S., Rastogi, R., Shim, K.: CURE:An efficient clustering algorithm for large database. Information Systems 26(1), 35P–58P (2001)

# The Research on Algorithm of Multi-sensitive Properties Privacy Based on Personal Anonymity

He Yali and Peng Guoxing

School of Computer and Communication
Hunan University of Technology
Zhuzhou, China

**Abstract.** Paper first analyzes the major technical database of existing privacy and privacy requirements in more sensitive attribute data released under the scenarios studied in the publishing process more sensitive data privacy issues, pointing out that a single individual more than the corresponding records and the corresponding personal data published anonymously in the privacy protection model number of key issues need to be addressed, including information disclosure risk measure, information loss measure and control the dynamic semantic tree, and induction, summed up the release of existing multi-attribute data privacy-sensitive technology characteristics.

**Keywords:** personal anonymity; status maintained; semantic classification trees; multi-sensitive properties.

## 1 Introduction

Personal information privacy is about your personal identifiable information to control and use the removal of unlawful interference. With the rapid development of information technology, people enjoy the convenience and the resulting fast, but the information technology of the "double-edged sword" makes the collection of personal information is becoming increasingly easy to improper use of such information or be personal property will result in an open and spiritual loss. Therefore, the protection of personal privacy can not just stop at the so-called protection of the right alone, but should be moving in the direction of the protection of personal information. Privacy from the traditional "individual to live in peace without interference," the negative rights of the evolution of positive significance for the modern "information privacy." Performance of personal privacy on private affairs and control over the private information.

Demand from the perspective of privacy protection, privacy protection can be divided into for the current user's privacy and data-oriented privacy protection. Privacy protection for users from the user point of view, the main information on the protection of personal privacy, which is involved in protecting the privacy of the database user-related sensitive data and some of the sensitive behavior (such as query or delete some data.) Privacy protection for users with different users, the specific application

S. Lin and X. Huang (Eds.): CSEE 2011, Part II, CCIS 215, pp. 424–428, 2011.

environment, a variety of laws and regulations have a close relationship. Comprehensive privacy protection for users needs include four aspects: the user's anonymity, pseudonym of the user, observation of user behavior is not

Data-oriented database, to consider how to protect the privacy of data mainly expressed in the privacy of information about some of the privacy protection of sensitive data used to eliminate data access caused by the leakage of privacy issues, generally through the adding of these data some of the labels or by some special treatment to achieve the above objectives. Privacy protection for data privacy protection is the focus of the study, this discussion is for data privacy protection, to consider how to protect stored in a database of personal privacy information, that can directly or indirectly reflect the private information of the data to establish some privacy information on the relevant mechanisms.

## 2 Related Technologies

During the general algorithm on minimum loss, because of the record alone during the merge, the resulting loss of information beyond the scope of minimum information loss, the loneliness it was recorded as an independent element group.
General algorithm:

1. In the table to meet the rest of the records to find properties in this category, generalized disorder of records, from the disorder of the cost of the property value within the group, consider ordering property records, find records that meet the minimum distance join group
2. If there is no record of the disorder to meet the group of generalized attribute generalization, to continue the generalization properties of the disorder in this group, repeat one, until the meet the l-diversity
3. Alone tuple tuples ordered directly through the minimum distance to find the corresponding broad group, but if the combination of individual and group alone element caused by the loss of information when a great degree, you do not have to merge the records, saying the record is Independent alone tuple, its operations remain in the multi-sensitive property sheet to reconsider the merger.

## 3 Almost One-Dimensional Method

Alignment of generalized method of generalized identity property, have a relational table, which identifies the value of the property the same quasi-tuples form a QI group. In order to reduce the level of generalization anonymous table can be used to allow the right amount of tuples inhibition (suppression) of the method, but also reduced the tuple release table suppression of information validity.

Almost re-encoding method into global and local recoding categories. Requirements of the global re-encoded on the same quasi-identifier property values are generalized to almost the same level of the hierarchy tree. [3] proposed a global re-encoding algorithm, it can also be applied to maintain the identity of the generalization method. The algorithm is simple, it is time to select a quasi-generalized identity property, until the meet the needs of anonymity. Global re-encoding is usually over-generalization.

Partial re-encoding does not require the same values of the property to the same level of generalization. Wong et al proposed a top-down partial re-encoding algorithm [5]. This method will first of all tuples are generalized to a wide Overview of the group, and then in maintaining the principle of anonymity, based on the specialization of tuples (specialization). This cycle continues until the special needs of anonymity until the break. This paper proposes a bottom-up partial re-encoding algorithm.

In order to merge into a single multi-dimensional table table produce the least sensitive property of the duplicate records, this article limits the sense of property of the traditional single Weimin anonymous conditions. For each single Weimin sense that the property sheet, in addition to a single record corresponding to the record number of sensitive attributes and the last shall be independent of tuple alone, the rest of the record must meet the traditional anonymous methods. Here we first summarize the new approach through the sense of a single Weimin property sheet to meet the constraints of the anonymous condition.

**Table 1.** Overview of the Disease attribute table after an anonymous (to meet the limit of 3 - Anonymous and 3 - diversification)

| Goup ID | NO | Age | Postcode | Disease |
|---|---|---|---|---|
| 1 | 1 | 55 | 10085 | Bronchitis |
| 1 | 2 | 55 | 10085 | Coryza |
| 1 | 3 | 55 | 10085 | Flu |
| 2 | 4 | 43 | 10075 | Pneumonia |
| 2 | 5 | 48 | 10075 | Coryza |
| 3 | 6 | [45~47] | 1007* | Gastritis |
| 3 | 8 | [45~47] | 1007* | Gastric ulcer |
| 3 | 9 | [45~47] | 1007* | Dyspepsia |
| 4 | 7 | [48~49] | 100** | Dyspepsia |
| 4 | 10 | [48~49] | 100** | Flu |
| 4 | 11 | [48~49] | 100** | Bronchitis |
| 4 | 12 | [48~49] | 100** | Flu |
| 4 | 13 | [48~49] | 100** | Bronchitis |
| 5 | 14 | 5 | 20084 | Bronchitis |

Described in Table 1, the first corresponding to a single record for multiple sensitive attributes impoverished group, and then a single case from the first record of a non-start, first of all minimal generalized disorder properties, then the last chapter summarizes grouping algorithm 1, the remaining records 12 and 13 can not meet the 3 - tuple diversity of anonymity for the lonely, according to algorithm 1.1, the first record minimum of 12 properties disorder generalized as 1007 *, corresponding to the third group, then consider ordering attribute values, but the record of 12 48 Ordered attribute value, not the third group, ordered range and are therefore incorporated into the record 12, the fourth group. Similarly, Algorithm 1 was recorded there should also be incorporated into the fourth group of 13. Hop into any of the records of 14 information packet loss caused by too much, so lonely as an independent element group 5.

**Table 2.** Overview of Medicine attributes of the table after an anonymous

| Goup ID | NO | Age | Postcode | Medicine |
|---|---|---|---|---|
| 1 | 1 | 55 | 10085 | An |
| 1 | 2 | 55 | 10085 | Bn |
| 1 | 3 | 55 | 10085 | Gn |
| 2 | 4 | 43 | 10075 | Cn |
| 2 | 5 | 48 | 10075 | En |
| 3 | 6 | [45~48] | 1007* | Dn |
| 3 | 8 | [45~48] | 1007* | Fn |
| 3 | 9 | [45~48] | 1007* | Dn |
| 3 | 12 | [45~48] | 1007* | Fn |
| 4 | 7 | [48~49] | 100** | En |
| 4 | 10 | [48~49] | 100** | Fn |
| 4 | 11 | [48~49] | 100** | Gn |
| 4 | 13 | [48~49] | 100** | An |
| 5 | 14 | 5 | 20084 | An |

3 describes the general process that, according to algorithm 1.1, the first three records will be assigned to a 6,8,9 record group, this time recording the minimum order is [45 47], but at this point does not meet the three principles of diversity, Therefore, the minimum distance algorithm to find records based on 12 into the group to meet after the anonymous request, so this range of four after an orderly record of generalization into one group. Similarly, alone recorded 13 into group 4. Hop into any of the records of 14 information packet loss caused by too much, so lonely as an independent element group 5.

## 4 The General Algorithm for Anonymous Individuals

Mike more sensitive properties through a personal anonymous algorithm summary, due to their personal privacy and property values Bronchitis Bronchitis medicine An attribute has an associated dimension, so to change the value of the corresponding private property the same dimension also changes the sensitive records of the corresponding attributes of other dimensions value, in this case is the generalization for the respiratory tablet. In addition, private property in the same dimension, the cluster identified by the leaf node property values and personal privacy of the property value associated with flu, flu the same reason the properties of this algorithm is summarized respiratory infection.

Generalization based on anonymous mike after the individual data sheets for the SST `, note 1 GroupID and Name the first and third records.

Data Sheet for the SST `

## 5 Conclusions

Mike more sensitive properties through a personal anonymous algorithm summary, due to their personal privacy and property values Bronchitis Bronchitis medicine An attribute has an associated dimension, so to change the value of the corresponding private property the same dimension also changes the sensitive records of the corresponding attributes of other dimensions value, in this case is the generalization for the respiratory tablet. In addition, private property in the same dimension, the cluster identified by the leaf node property values and personal privacy of the property value associated with flu, flu the same reason the properties of this algorithm is summarized respiratory infection.

## References

1. Litai, Y., Tang, C., Wu, J., Zhou, M.: Two clustering based on k-anonymity privacy protection. Jilin University (Information Science) (February 2009)
2. Han, J., Cen, T., Yu, H.: Data table k-anonymous study of the micro-clustering algorithm. Electronics (October 2008)
3. Li, J.I., Lv, Y., Zhao, W., Liu, G., Li, H.J.: Barrels of K-based anonymity multidimensional table incremental update algorithm. Yanshan University (May 2009)
4. Qi, R., Wang, K., Guo, X., Li, J., Tang, J., Liu, G.: Priority strategy based on the maximum Ye Zizi tree protection method of multi-sensitive properties. Yanshan University (May 2009)
5. Teng, J., Zhong, C.: Anonymous method based on data published privacy leaks Control Technologies. Guangxi Academy of Sciences (April 2009)

# Analysis of Embedded Real-Time System Security

Ma Jingjing

College of Computer and Information Engineering, Lishui University, China

**Abstract.** Embedded real-time systems are facing more and more security problems. Malicious attacks on the system from suspicious or malicious code and the change of system hardware state could lead to system exception, resulting in system reliability and security deteriorated. This paper summarizes the characteristics of embedded real-time system and its special requirements for security, analyses security problems faced by embedded real-time system and defects of programming languages. And then put forward methods to improve the safety of embedded real-time system, providing a new security idea for solving the embedded Real-time system security.

**Keywords:** Security, embedded system, real-time system, RTOS.

## 1 Introduction

Embedded systems are becoming increasingly large and complex as software complexity increases, more and more software defects, resulting in system reliability and security is getting worse. Embedded real-time systems often arise in the implementation process issues, such as some common suspicious or malicious code on the system caused by malicious attacks and other anomalies. Hardware state changes may also lead to system abnormalities. For example, equipment failure and high temperature, vibration and other factors may cause the system to know what to do. In an increasingly widely used in real-time computing aerospace, defense, transportation, nuclear energy and health, and many other safety critical systems (SCS) in the context, to reduce or prevent the occurrence of catastrophic accidents, the need to improve real-time operating system, SCS reliability and security.

Currently, the researchers embedded real-time systems for the analysis of the security system is often not enough, the safety of embedded real-time system analysis and solution remains at the local, the lack of total system solutions. Based on this, this study will focus on the special requirements of embedded real-time systems, a detailed analysis of the system and summarizes the impact of insecurity embedded systems, embedded real-time systems to solve security issues basis.

## 2 Special Requirements of Embedded Real-Time Systems

Real-time systems generally refers to a class of highly time-sensitive application requirements. Such as industrial process control (such as steel making, continuous chemical process, etc.), spacecraft control systems, weapons guidance systems, strategic defense systems. These systems often require multiple computers to the

S. Lin and X. Huang (Eds.): CSEE 2011, Part II, CCIS 215, pp. 429–433, 2011.

information collected in order of priority in a number of seconds, respectively, a number of milliseconds, or even within several microseconds to process and respond appropriately.

Real-time system is a timely response to external random events occur, and fast enough to complete the event handling computer applications. In real-time systems, system correctness depends not only on the correctness of the results system, but also on the correct results to time. Therefore, system designers need to care about embedded real-time behavior of the system uncertainty.

## 2.1 Characteristics of Embedded Real-Time System

Although the argument for embedded real-time systems vary, the main meaning is the same. Embedded real-time system has the following characteristics:

(1) embedded real-time systems are often a computer or microprocessor core, with other machinery, electronic equipment, one complete certain specific features, is a computer application system.

(2) embedded real-time systems is usually a reactive system. Links with other parts of the computer through the appropriate combination of sensors and actuators to complete, the system inside and outside the state an important way to report the event was modeled to the computer;

(3) embedded real-time systems often have real-time requirements. Real-time refers to the function of the system depends not only on the correctness of the logic of generating the correct results, but also on the results to time. That is, the system must be scheduled within the time limit to respond to some input. "In the book (or required) time to complete", which is the key to understanding real-time requirements. Real-time requirements can be divided into: strong (hard) real-time and weak (soft) real-time two. Strong real-time is the real time requirement of the system can not be satisfied if, it may lead to catastrophic consequences (such as nuclear reactor control, aircraft control, etc.); weak real-time is the real time requirement of the system can not be satisfied if only will make the system performance degradation, and will not lead to serious consequences (such as laser printer control, etc.);

(4) embedded real-time systems must meet stringent reliability and safety requirements;

(5) embedded real-time systems are often subject to application environments in the size, shape, weight, heat dissipation and power consumption constraints, etc.;

(6) embedded real-time systems typically have a distributed structure, and may be heterogeneous;

(7) embedded real-time system must have sufficient flexibility to meet the rapid expansion of the upgrading needs of even the running function.

## 2.2 The Special Requirements of Embedded Real-Time Systems

Embedded real-time system is an important class of computer applications. In contrast, such a system, for the time characteristics, reliability, and security has a high, and even the requirements of load moment:

(1) systems often contain a variety of sensors and multi-channel digital to analog conversion equipment, real-time system that is sometimes the output data from the analog sensor output, it is necessary to convert it to digital only after processing by a computer; the other On the one hand, the results of computer processing is sometimes necessary to convert the analog output to control the implementation of the action agencies.

(2) respond to external events must be completed within a certain time. Similarly, the requirements of the various output must also be completed within a certain time, in fact obtain, process data and processed data output, need to be completed within a specified time. This time the sum is called the system response time. Guarantee within the prescribed time limits to respond to the critical real-time system design.

(3) the requirements of environmental adaptation. Systems often used in industrial field environments, even in the worse environment. The former requires calculations in a wide temperature or humidity range, and can adapt to the industrial site electromagnetic interference environment; the latter often refers to the field or vehicle, shipboard, airborne, satellite and other applications, it is often also called ultra- Wide operating temperature range and moisture, shock, anti-shock, anti-salt spray and fungus. These require the use of special computer components and special structure, this requirement is called "hardening" requirements.

(4) high reliability of software requirements. Such as in military and aerospace applications, often using a specific programming language (such as the Ada language) and high-quality compiler, designed to ensure the correctness of a good program, and may take advantage of hardware features, the speed of the target program has a very high quality. In addition, many real-time applications involving a large number of communication links, especially wireless communications, in addition to the high sensitivity of the reception facilities, the use of redundant coding to ensure that adverse weather conditions in the high reliability of data communication are often used.

(5) must meet certain peak load requirements. A real-time system load may be uneven, and sometimes heavy load, sometimes the load light. To most of the time may not be fully utilized. But the whole system must satisfy certain requirements of the peak load, that heavy load, even under overload conditions, some of the critical tasks can also be successfully implemented.

(6) to guarantee the necessary computational speed (sometimes at very high speeds) to achieve high reliability under the premise, which is crucial. Real-time systems often operate in harsh conditions not suitable for people involved in the environment. Face temperature changes, strong shock and strong electromagnetic interference problems and real-time systems often need to use industrial grade components, sometimes the design at the system level using fault-tolerant or hot backup mechanisms.

## 3 Programming Languages and Software Defects Defects

Currently, the most widely used high-level language is C language, which is an open language, grammar, free-form, different people can write code to achieve the same functionality of a variety of procedures. Using the C language to write high performance programs, including system program and application, but C language and its support libraries are essentially unsafe, unintended mistakes can lead to very serious security vulnerabilities.

This is the C language in the two most common and most difficult to check the run-time error. Memory access error mainly refer to: the malloc and free functions memory leaks caused by improper use and have access to free memory once again; and mainly refers to a pointer reference error: Use of undefined memory and it has been defined uninitialized memory references both. Can be seen that the main reason causing the pointer reference error is due to the corresponding memory area is not able to correctly use the result. So, in a static test can take a similar approach in these two errors were detected. Specific methods are:

1) Use a pointer check mark pointer reference error

Annotation defines a pointer to a pointer may be null if the constraint, each pointer state is running the program will continue with the change pointer status by checking the consistency between the label and a pointer, can be found on the program possible null pointer reference.

2) Check the memory using the memory allocation errors marked

C language is not garbage collection mechanism, therefore, dynamically allocated memory when in use do not pay attention if the maintenance is easy to forget a memory leak caused by the release. Memory leak can not be found not only at compile time, and, in the run-time errors often occur without obvious, but steady decline in available memory until the memory occurred, causing the entire system "down." At the same time, such errors is difficult to reproduce, is found in the debugging process very difficult. Label-based detection model, the memory allocated by the state to clearly mark the release of the responsibility for the dynamic memory, the model of consistency through the detection of the release of the responsibility to detect possible memory leaks.

This is a regular in the embedded software development errors, as in the C language source code, there is no direct statement to operate on the stack, so such errors at compile time can not be detected out. In order to detect whether there is a stack overflow error occurs, in fact, is to monitor the stack for each function is called when the stack frame on the existing number of frames in each stack memory requirements. However, due to the increase or decrease the stack frame is shown dynamically at run time, so the static stack overflow checking is very difficult. In the following two aspects:

1) How to get every function of the size of the stack frame, that is, the memory requirements of the stack frame.

Each function can be defined any number and type of temporary variables and input parameters, how can accurately calculate all these variables and parameters used by the memory size.

2) how the statistics of the whole process function call situation.

Stack frame on the stack, with the number and order of sequence of function calls and dynamic changes, how can accurately track all program functions from the main start of the call sequence for the overflow detection has a decisive influence. Written by C language source code distributed in different files, each file contains several functions, how to analyze the code sequence for each function call stack overflow error detection need to adopt to complete the stack overflow checker.

3) embedded real-time system security improvements

High security design of embedded real-time systems are complex, by strengthening the security of embedded real-time operating system protection can be designed to reduce the burden of application developers. In fact, the processor speed under the conditions of steady growth, for cost considerations, design engineers often expect the same critical level of the processor can run multiple different applications, but will also increase the risk accordingly. To the extent the same system the important threads of different (non-critical tasks and critical tasks) "peaceful coexistence", the management processor and other resources of the operating system must be able to appropriate partitioning software to ensure the effective allocation of resources.

## References


1. Chen, L.: Component-based embedded real-time software reliability assessment model and application. Electronic Science and Technology University (2008)
2. Fei, C.H.: Real-time operating system, a number of key issues. China's outstanding full-text database, Dissertation (Ph.D.) (April 2004)
3. William Beck, R., Mark Vanfleet, W.: High Assurance Security / Safety For Deeply Embedded, Real-time Systems. In: Embedded Systems Conference - Silicon Valley, ESC2006 (2006)
4. Mao, J.: Embedded real-time systems on key technologies. China's outstanding full-text database, Dissertation (Ph.D.) (February 2005)
5. Wang, B.-J.: Embedded real-time systems and resource sharing scheduling model and algorithm. China's outstanding full-text database, Dissertation (PhD) (April 2006)
6. Liu, H., Qing, S.-H., Liu, W.: Secure Operating System Design and Implementation of Audit. Computer Research and Development (October 2001)

# Proper Contactless Switch Selection in Control System

Zu Guojian

Loudi Vocational and Technical College, Loudi, Hunan, 417000

**Abstract.** The article introduces some kinds of contact less switch. It includes voltage regulator, silicon controlled rectifier, photocoupler, MOSFET and IGBT. Different structures have different switch characters The theory and character of each kind are analysed, parameters of representativechip are compared According to this, engineer can choose the right switch

**Keywords:** contactless swich, SCR, photocoupler, OCM OS FET, IGBT.

## 1 Introduction

Microcontroller and embedded systems are becoming more and more widespread, mechanical relays are still the most commonly used execution control switch. As the control system puts up with higher quality requirements, which mechanical relays often cannot meet in control speed, electromagnetic compatibility, isolation and other aspects. On the other hand, increased integration with integrated circuits, it is common to see devices with power supply voltage less than 3 V, especially a lot of battery-powered portable devices which require low voltage, low-power devices. Under this circumstance, mechanical relays are hard to meet new requirements.

Traditional integrated analog switch circuit such as ADG211 series has resistance from tens to hundreds Ω range, which cannot meet requirements form power switch control devices. With the development of semiconductor technologies, there are already many contactless switches for designers. Contactless switch is superior to mechanical relays in electromagnetic compatibility, reliability, security and other aspects. This article describes several new contactless switches. Different applications have different requirements in applications. Such as control subject is divided in to AC and DC; magnitude differences of control objects in load current and contact resistance; operation voltage ct may vary from a few V to 380 V or even higher. New devices provide various applications to meet performance requirements, readers can choose appropriate devices, or use simple circuit to make contactless switch.

## 2 Common Contactless Switch

### 2.1 Contactless Switch with Three Terminal Regulator

Designers are very familiar with the common seen low- price three-terminal regulator. Figure 1 is a designed circuit with three-terminal switching regulator circuit. The signal

S. Lin and X. Huang (Eds.): CSEE 2011, Part II, CCIS 215, pp. 434–439, 2011.

from control side decides whether to ground the three terminal regulators. If it is grounding, then output terminal is power-up, or it is disconnected. This circuit is very simple and easy to debug, and have a variety of voltage regulators suitable for controlling DC load. The disadvantage is that the decrease of pipe pressure weakens regulator output voltage, which is not suitable for battery-powered devices. So take use of three-terminal low dropout voltage regulator will improve the capacity.

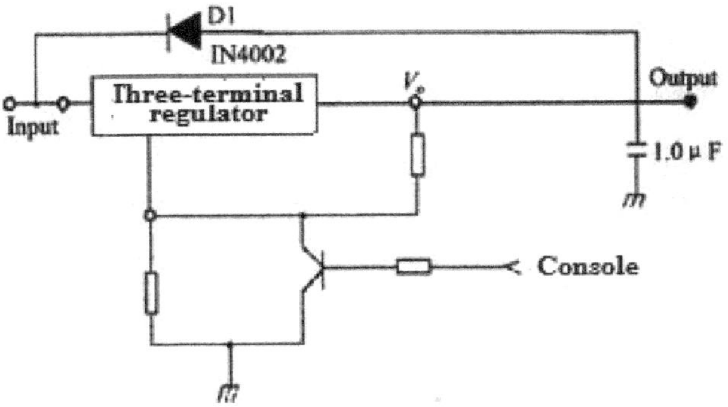


**Fig. 1.** The control switch of three-terminal regulator

## 2 Contactless Switch with SCR Devices

There are many such devices are available, like STMicroelectronics (ST) ACS series. It can be directly used to control fans, washing machines, motor pumps and other equipments, and the isolation voltage is up to 500 V or more. Figure 2 is a typical application circuit. Such devices is low price, but can only be used for the exchange of load switch control.

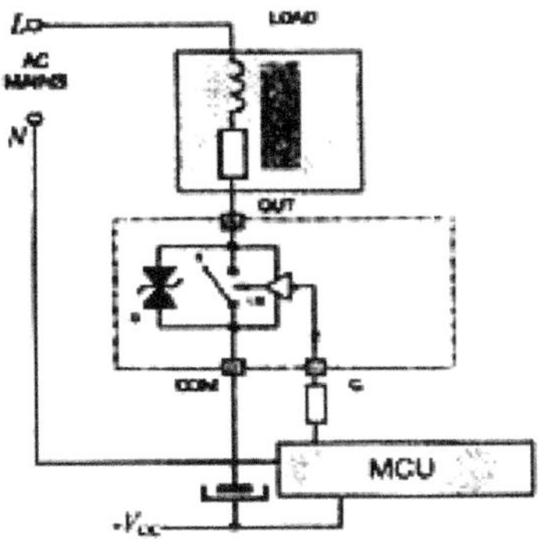


**Fig. 2.** Non-contact switch Typical Application Circuit based on SCR devices

## 3 Contactless Switch with Optocoupler Transistor and Darlington Transistor

Phototransistor-based contactless switch is known as optical couplers (photocoupler) [1], such as Sharp PC817 Series, NEC PS2500 Series, Agilent HCPL260L/060L and so on. Its working principle is shown in Figure 3. When forward voltage in the input terminal increases, the light-emitting diode (LED) is lit, phototransistor produces photocurrent to supply load from collector. When there is reverse voltage at input terminal, LED does not light, so that phototransistor is off, the equivalent load is an open circuit. From works principle, these devices are used in DC load and also in transmit the same pulse signal from current direction. The device's operating speed is relatively high, generally in the microsecond or faster.

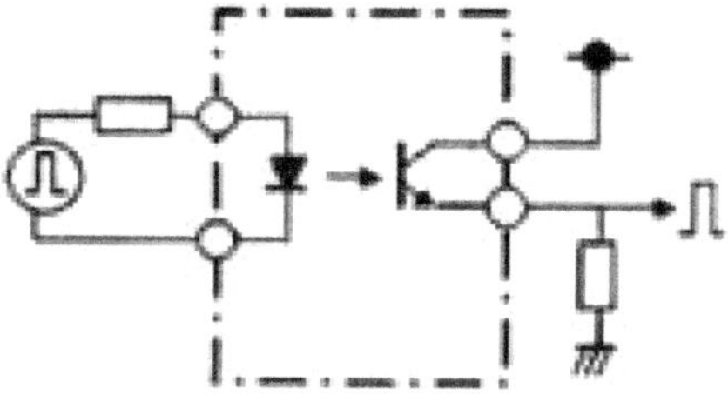

**Fig. 3.** Transmission map of Square wave

Darlington is a composite of two bipolar transistors. The greatest advantage is current multi-stage amplification, as shown in Figure 4.

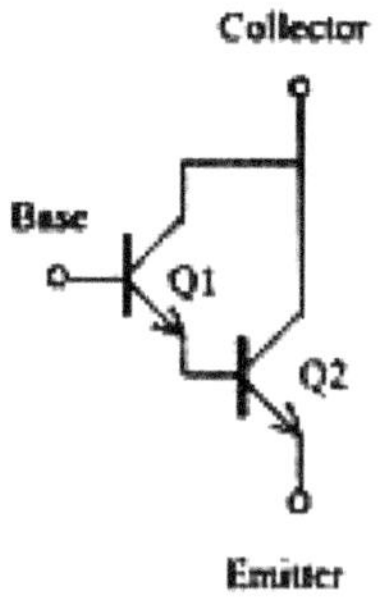


**Fig. 4.** The equivalent circuit of Darlington

The disadvantage is large capacity pipe pressure. Since the two transistors have a same collector, the saturation voltage of Darlington transistor is equal to the addition of

Q2 forward voltage and saturation voltage of transistor Q1, while the forward voltage is much higher than saturation voltage, so that the whole Darlington transistor saturation voltage is particularly high, and power consumption is higher during conductance.

The Darlington optocoupler of Fairchild Semiconductor (fairchild) isolates Darlington output configuration, which separate input photodiode with the primary gain and output transistors in order to get lower saturation voltage (0.1 V) and higher operating speed compared with traditional Darlington phototransistor. The company's latest 5 products have single or double channel configurations to provide low power consumption of 3.3 V or 5 V operating voltage. Double channel HCPL0730 and HCPL0731 optocouplers provide 5 V operation voltages and SOIC8 package achieves optimum mounting density. The operation voltage of single-channel FOD070L, FOD270L and double channel FOD 073 L devices is 3.3 V, which saves 33% consumption than traditional 5 V parts. NEC's chip PS2802.1 / 4, PS27021, PS25021/2/4, PS25621 / 2 are also belongs to Darlington optocouplers.

## 4 Contactless Switch with MOS or IGBT

Contactless switches based on MOS FET are coupled in different ways because of coupling styles, such optical coupling MOS FET (OCMOS FET) is the optical coupling style, the principle is shown in Figure 5, and within dotted frame is OCMOS FET internal schematic diagram.

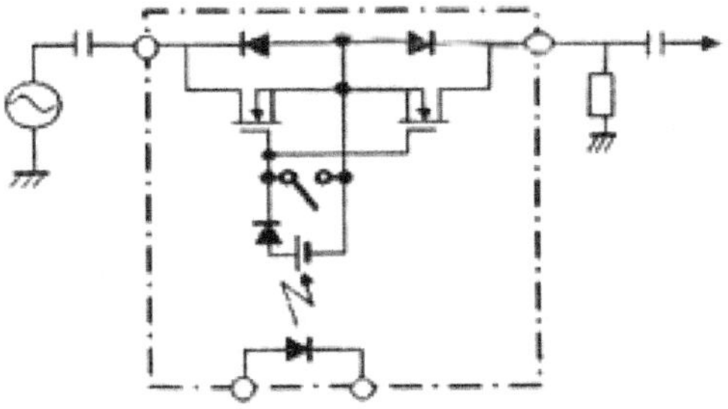

**Fig. 5.** Typical Application Circuit of OCMOS FET

There is photovoltaic cell within circuit, when light-emitting diode is lit, the cell charges FET gate capacitance and increases voltage between gate and source, which conduit MOS FET, and the switch is closed. When the LED goes out, photovoltaic cell is no longer charging gate capacitance, and internal discharge switch is automatically closed, which forces gate to discharge, the gate-source voltage decreased rapidly, and off with and switch. OCMOS FET has two types: one is maketype, normally it is disconnected; the other is breaktype; and usually it is conducted. This dissertation focuses on the maketype. MOS FET optical coupling is universal AC and DC, which working speed is milliseconds and not as fast as optocouplers, its maketype characteristics is irrelevant with input current parameters. OCMOS FET can control strong current by weak one which means to drive a-level current by aA input current. Because FET is a bilateral maketype with low-resistance, so it is mainly used to interrupt AC signals, as shown in Figure 5, and OCMOS FET is also known as solid-state relay (SSR) [2].

There are a lot of MOS FET based countless switching devices, such as Nippon Electric Company (NEC) of the PS7200 Series, Toshiba TLP351 series, Panasonic Nais AQV series. Usually low on-resistance is suitable for larger load current practice, such as N EC PS710B1A: on-resistance Ron = 0.1Ω (max), load current IL = 2.5 A (maximum), on-time Ton = 5 ms. Low CR product type MOS FET is for optical high-speed signals switch applications, such as test terminal of measuring instruments. CR product refers to the product of output capacitance and on-resistance at output stage MOS FET, which is a parameter index to evaluate a MOS FET characteristic. Like NECPS7200H1A: resistance Ron = 2.2 Ω, CR product is 9.2 pF • Ω, on-time Ton = 0.5 ms, the load current IL = 160 mA. Insulated gate bipolar transistors IGBT [3], and the structure is shown in Figure 6.

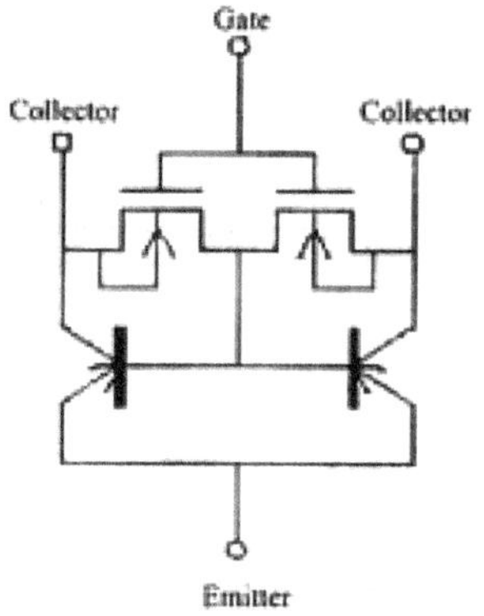


**Fig. 6.** The equivalent circuit of IGBT

This structure makes IGBT acquired not only the advantage of MOS FET IGBT of larger DC current, but also bipolar transistor current handling capacity of large, high blocking voltage. It can be connected to switching circuit, as NPN type bipolar transistor, the significant difference between them is that IGBT does not need gate to maintain conduction. GBT-based contactless switch is as Agillent HCPL3140/HCPL0314 series.

In order to facilitate designers with appropriate circuit choice, Table 1 shows different types of reference chips for contactless switch, and compares various chips' switch characteristics.

## References

1. Differences between Photocoupler and Opticalcoupled MOS FET NEC (2004)
2. Van Zeghbroeck, B.: Principles of Semiconductor Devices (2004)
3. Uyemura, J.P.: First Course in Digital Systems Design: An Integrated Approach. Brooks/cole, Monterey (2000)

# The Design of Distributed Simulation Platform for Magnetic Suspension Rotor System Based on HLA

HaiNing Fang, JinGuang Zhang, and JiaYing Niu

School of Mechanical and Electronic Engineering,
Wuhan University of Technology
430070, Wuhan, China
fanghaining2006@126.com

**Abstract.** On the basis of analysis on current situation of magnetic suspension rotor system, the paper have put forward a research approach that the distributed simulation platform for magnetic suspension rotor system based on HLA(High Level Architecture),conceived simulation scenarios on account of the basic function of magnetic suspension rotor system, established federal members of HLA simulation system, designed FOM and SOM, finished the main thread for simulation system, and provided a feasible method to the virtual prototype of magnetic suspension rotor system for the research.

**Keywords:** magnetic suspension rotor; HLA; simulation; the virtual pototype; modularized.

## 1 Introduction

Magnetic suspension rotor system is the rotor system based on the magnetic bearing, which involves many science fields. It has no mechanical contact, no friction, no lubrication, etc, and can be widely used in aerospace, machine processing in which industry have special requirements for speed and rotating precision. [1]

At present, the main problems of the magnetic suspension rotor system at home and abroad in the simulation design process are as follows:

1) System, the study of involves multidisciplinary domain knowledge, such as structural dynamics, control system, magnetic field analysis which are respectively in different fields. Since the software matching problems, the research achievement is hard to interactive.

2) The existing system design adopts the traditional serial design style, making the simulation between control system and rotor structure's dynamic characteristics too hard to coordinate and interact.

3) Because of fixed structure, the existing virtual prototypes are difficult to achieve modular distributed simulation, replace, or change simulation members flexibly, which limit the function of virtual prototype. [2-4]

HLA (High Level Architecture) is the general technical framework, which is used to produce computer simulation system. Through the interface standard and bottom

S. Lin and X. Huang (Eds.): CSEE 2011, Part II, CCIS 215, pp. 440–445, 2011.

transmission service provided by RTI (Run-Time Infrastructure), the simulation system can flexibly replace, increase or decrease simulation members, realizing the modularized and distributed simulation of magnetic suspension rotor system, completing the optimal integrated design and coordination match between structural characteristics and control system.

## 2 The Structure Priciple and the Research Status

The study object in this paper is the magnetic suspension rotor system supported by active magnetic bearings. Figure 1 shows the schematic diagram for active magnetic bearings.

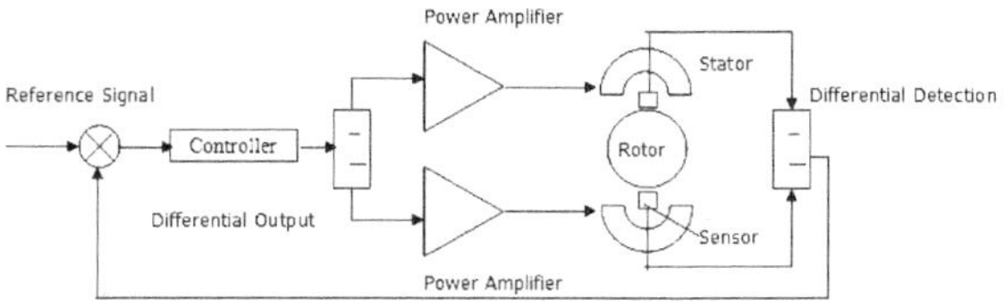


**Fig. 1.** The schematic diagram for active magnetic bearings

Active magnetic bearing is composed of five main parts which are rotor, electromagnet stator, sensors, controller and power amplifier. Its working principle is as follows: the electromagnet stator produce suction on the rotor because of current, making the rotor in a particular balance position. During the rotor operation process, when it receives the interference and deviates the balance position, displacement sensor will send this migration signals to controller, then the controller adjust the size of control current to change the size of stator suction through power amplifier, finally, the rotor will return to the original balance position.

## 3 The Design of Simulation Platform for Magnetic Suspension Rotor System

In order to use a few parameters to describe the simulation platform's design process clearly, magnetic suspension rotor system is simplified in this paper.

### 3.1 Simulation Scenario

Assuming a smooth operation of magnetic suspension rotor system, the stator currents receive external interference and fluctuate (Control system provides the interference). We observe the dynamics analysis for the mechanical system and the changes of magnetic field, to make the control system work out corresponding control based on the information collected by sensors, finally enabling the rotor return to the initial balance position.

### 3.2 The Development Environment of Simulation Platform

The simulation system adopts Windows XP as its operation system and DMSO RTI 1.3-NG as its operation support framework. LibRTI of RTI is encapsulated in every federate. Federate can interconnect with each other according to LAN and exchange information based on the service set by RTI 1.3. The software build environment is Visual Studio 2008, while the programming language is C++ and database is SQL 2005.

### 3.3 The Structure of Simulation Platform Structure and Design of Federal Member

All the federates is designed partly according to the system structure and property. In the process of writing programs, we design each instruction subroutines to code modules, which can not only realize coordinated and matching simulation effect, but also make functional components achieve a nice reusability and expansibility, to facilitate future system development with different control systems and different support methods etc.

In order to simplify the platform, it is divided into the following 5 federates:

1) Controller Federate (Controller): Controller Federates is the control part of the entire system, containing the functions of controller, power amplifier and sensor. Its function specifically expressed as: subscribing the positional relationship between the rotor and electromagnet stator issued by Mechanical Structure Federate; publishing the current signal processed by the power amplifier. The function is achieved by MATLAB/Simulink.

2) Mechanical Structure Federate (Mechanical Structure): the mechanical structure federate is the structure basis of simulation system. The specific expression of its function as: subscribing control signals published by Controller Federate; publishing the positional relationship between rotor and stator, the axial position of rotor, the motor's speed and so on. The function is achieved by Solid Works.

3) Magnetic Field Analysis Federate (Magnetic Analysis): Magnetic Field Analysis Federate is the technological key of the magnetic suspension rotor system, with the function of analyzing the magnetic field produced by the system. The specific expression of its function as: subscribing the positional relationship between the rotor and stator, the axial position of rotor published by Mechanical Structure Federate, and the current signal published by Controller Federate; publishing magnetic information and the magnetic field distribution. The function is achieved by ANSYS.

4) Dynamics Analysis Federate (Dynamics Analysis): Dynamics Analysis Federate analyzes the system's dynamic. Its function specifically expressed as: subscribing mechanical structure, the motor's speed and so on; publishing dynamics analysis numeric. The function is achieved by ADMS.

5) Data Base Federate (Data Base): Data Base Federate provides storage space for the simulation operation parameters of the magnetic suspension rotor system and supports for the bottom. Its function specifically expressed as: subscribing initial parameters and the information in operation. The function is achieved by SQL Server 2005.

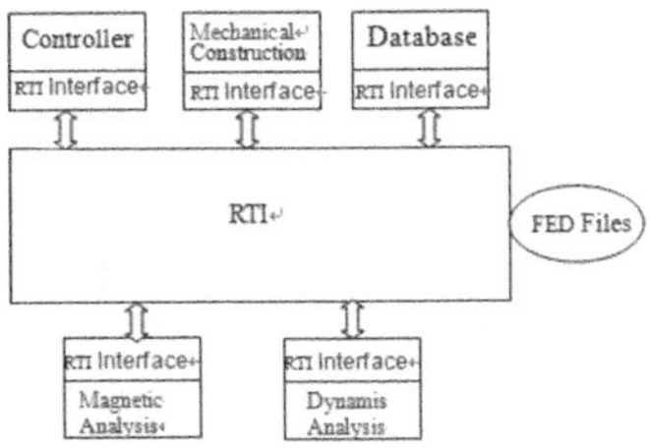


**Fig. 2.** The schematic diagram of simulation system structure

Figure 2 shows the schematic diagram of simulation system structure. FED documents in the figure is the executive data file of the federation.

### 3.4 The Design for SOM/FOM in Simulation Platform

HLA rules request federation and federates to build their own object models, namely SOM and FOM. SOM (Simulation Object Model) as a standardized object model reflects that the federates have the disclosure of publishing information and subscribing information outside. FOM (Federation Object Model) represents the agreement between federates' data and the collection of object class as well as interaction class

**Table 1.** The attribute of object class

| Object Class | Attribute | Subscribe or not | | | | |
|---|---|---|---|---|---|---|
| | | 1 | 2 | 3 | 4 | 5 |
| Controller（PS） | Current | | √ | √ | | √ |
| Mechanical Construction（PS） | AirGap | √ | | √ | √ | √ |
| | Speed | √ | | √ | √ | √ |
| Magnetical Analysis（PS） | MagneticInfo | | | | | √ |
| | MagneticForce | √ | √ | | √ | √ |
| Dynamics Analysis（PS） | AnalysisResult | | | | | √ |
| Data Base（S） | | | | | | |

**Table 2.** The parameters of interaction class

| Interaction Class | Parameter | Description |
|---|---|---|
| CurrentChange | ToChangeCurrent | Change the size of control current |
| SpeedChange | ToChangeSpeed | Change rotor's speed |

published by all federates. Table 1 and table 2 are given respectively the attribute of object class, the parameters of interaction class and the published/subscribed relationships between every federate. (The serial numbers in the table are one-one correspondence to the federates' code listed in section 3.3).

## 4 Operate Simulation Platform

Based on the above demand and capacity analysis of each function module in the magnetic suspension rotor simulation system, we develop the complete federate code of the local RTI component based on LibRTI in C++ library for the goal of the reuse of components and scalability system. The main behavior code of all federates can achieve real-time data updates in local computer. When simulation requirements change, changing federates' main behavior code can reconstruct and expand federates without changing other parts.

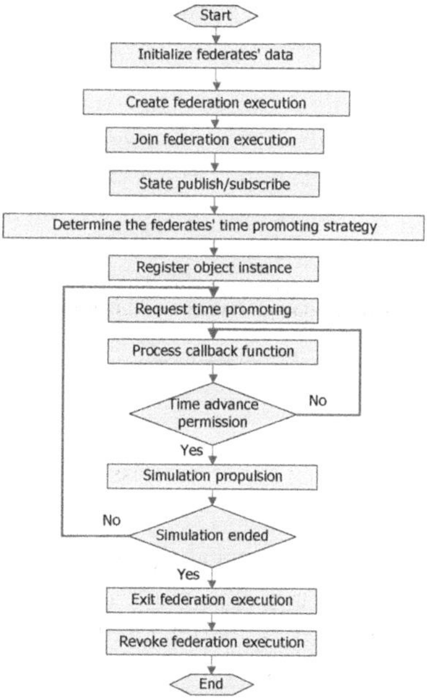


**Fig. 3.** The main thread of simulation system based on HLA

Figure 3 gives the main thread of magnetic suspension rotor simulation system based on HLA.

## 5 Conclusion

HLA has the features of new concept, rich connotation and wide field. In recent years, the development is gradually mature, but still has many technical difficulties in time management strategy, data distribution rules and engineering practice. According to our previous research achievements, and research on the analysis of magnetic rotor system's current situation, we brought forward the research idea that constructing distributed simulation platform for magnetic suspension rotor system. On the basis of simulation scenario, we tried to build the simulation platform based on HLA. Simultaneously, these were also a beneficial expanded attempt to the engineering field of HLA. It can be believed that simulation platform for magnetic suspension rotor system based on HLA will be a efficient way to the simulation research on magnetic suspension rotor system, by means of refining object class and interaction class to improve time management strategies and data process.

## References

[1] Hu, Y.-f., Zhou, Z.-d., Jiang, Z.-f.: The basic theory and application of magnetic bearings, pp. 8–12. Chia Machine Press, Beijing (2006)
[2] Zhang, J.-g., Hu, Y.-f., Qian, J., Ding, G.-p.: Flexible Rotor Modeling and Co-simulation in Magnetic Suspended System. Journal of Wut( Information & Management Eng Ineering) 32(1), 8–12 (2010)
[3] Sun, S.-q., Yu, L.: Dynamic Modeling and Optimum Simulation of a Flexible Rotor Magnetic Bearing System. Journal of System Simulation 15(3), 365–368 (2003)
[4] Li, K.-l., Xie, Z.-y.: Model analysis of the magnetic suspension rotor based on Ansys. Mechanical & Electrical Engineering Magazine 25(1), 1–3 (2008)
[5] Zhou, Y., Xie, J.-w.: HLA simulation program design, pp. 55–157. Publishing House of Electronics Industry, Beijing (2002)
[6] Satoshi, K., Taku, M., Tatsumi, T.: A multi-disciplinary distributed simulation environment for mechatronic system design enabling hardware-in-the-loop simulation based on HLA. International Journal on Interactive Design and Manufacturing 1(3), 175–179 (2007)
[7] Fan, J., Dong, T.-y., Sun, S.-a.: Research and Design of a Dynamic Forest Growing Simulation System Based on HLA. In: Corchado, J.M., et al. (eds.) DCAI 2008. ASC, vol. 50, pp. 11–18 (2009)
[8] Li, X.-y., Zhang, Z.-l., Long, Y.: Design of Surface-to- Surface Missile Launching Control System Based on HLA. Computer Simulation 25(10), 54–58 (2008)
[9] Liu, B., Zhang, H.-j., Yang, X.-r.: Design and Realization of Unified RTI Interfaces Based on Distributed Objects. Computer Engineering 34(23), 71–73 (2008)

# A Method of Distributed Data Storage Based on Integrated Threshold

Cao Hong

School of Computer and Information Engineering, Lishui University,
Lishui 323000, China

**Abstract.** Wireless sensor networks existing in the data storage methods exist to some extent, fixed and stored in hot storage node problem, the performance there are still large room for improvement. This paper proposes a threshold based on an integrated data storage method, the threshold to the consolidated dispatching node, shut down redundant nodes, the nodes are scheduled to enter the work state and the other nodes into the suspended state. Scheduling mechanism through which not only balance the network load, to eliminate storage hot issues, but also work to reduce the number of nodes, to save energy and prolong network lifetime.

**Keywords:** wireless sensor networks, distributed data storage, integrated threshold.

## 1 Introduction

In the past, using data-centric storage methods, the most classic is Sylvia Ratnasamy et al in 2002, the geographical hash function table proposed method GHT: Geographic Hash Table. It is based on attribute name, the perception of the data attribute set name for the event. The specific process is as follows: the perception of certain attributes of data collection as the key k, through a hash function h (x) the key k is mapped to a network location h (k), the data for <k, h (k)> is stored in the adjacent h (k) of the node. User issues a query request, the direct route to the neighboring h (k) of the node to access data. GHT exist in the following two questions: First, all queries are concentrated in a recent column address discrete nodes, the query often occurs when hot issues; second storage nodes fixed and can not be adjusted dynamically. If an event type occurs more frequently, due to storage nodes fixed and the nodes of storage space is limited, it can not effectively collect the data storage node, resulting in data loss.

## 2 Distributed Data Storage Method on Integrated Threshold

### 2.1 Methodological Thought

Node scheduling storage threshold achieved by the integrated, comprehensive, including storage space threshold value S and energy E in two parts. Maintain the

S. Lin and X. Huang (Eds.): CSEE 2011, Part II, CCIS 215, pp. 446–451, 2011.

same grid in the active state of a node to act as a storage node, the data storage work, the other nodes into the suspended state. Node can sense pending state environment and the implementation of packet delivery, but can not be stored sensory data. Nodes to store data in the process, because the consumption of energy and storage space, energy and storage space, there is a threshold level to achieve integrated in the corresponding threshold, the node is suspended, no longer work for data storage. Then grid scheduling factor of each node p, then select the other nodes take turns to act as storage nodes. All storage nodes to achieve the comprehensive threshold value, adjust the threshold to the next level threshold, the next round of data storage.

## 2.2 Scheduling Factor

Let node i, Ei is the current residual energy, E energy threshold for the current level of energy threshold, Si is the current node i remaining storage space, S is the current threshold level stored in memory threshold, then the scheduling factor p is defined as :

$$p = (E\text{-}Ei) * (S\text{-}Si) \tag{1}$$

## 2.3 General Threshold

General threshold value is stored by the energy threshold and the threshold composition. In theory, the initial energy value of each node and the storage space should be the same. But in fact a large number of nodes in wireless sensor networks of many, and throw then in the form of self-organization form through the network; the initial neighbor discovery phase of the grid, the grid nodes in the exchange of information between nodes with the neighbors in the process, will consume out a small part of the energy and storage space. Therefore, the actual initial energy of each node and the storage space is not the same. Let node i, Ei is the current residual energy, E energy threshold for the current level of energy threshold, Si is the current node i remaining storage space, S is the current threshold level stored in memory threshold, threshold with the remaining energy and storage space decreases. We assume that these variables are the percentage of the nodes in the storage. Node's remaining energy through the AD converter (ADC) to the power supply voltage acquisition MCU (microprocessor) and get to the node can monitor their own energy situation.

Node thresholds using an integrated scheduling approach, the beginning of each round is the first work node node Ei * Si value of the decision, that is, from the current residual energy of each node and the storage space of the product decisions. The initial formation of self-organizing network, in the neighbor discovery phase, each node in the grid by the finite radio, broadcasting a message to the exchange of information. Specific process is as follows:

Step1: In the beginning of each round, each node i its initial calculation of the value of Ei * Si, in order to determine their own activities or in the current round is to be suspended. Ei * Si value of the largest node as the one in the first active node, other nodes into the suspended state.

Step2: Node i in the active work as a data store, and calculate the p value, when that p <0, when sending a packet to the host Storage_Full other nodes within the grid, into a suspended state.

Step3: the other node then calculated the value of their Ei * Si, Ei * Si value of the largest node as the active node in this round, the other nodes into the suspended state.

Step4: until all nodes in the grid to meet the p <0, the grid reaches a threshold value of the last node sends a data packet to the same Change_Threshold grid all the other nodes. Then all nodes in the grid will be their energy threshold and store threshold E S also change the threshold to the next level, the threshold level T = T +1.

Scheduling factor p is calculated taking into account the node to the energy and the storage of two factors, it is very necessary. For example, in Figure 1, the node only consider the storage space without considering the residual energy. Storage space is divided into two thresholds, respectively, 50 and 100. It is prone to this situation: When a node B stores 48 events, node failures due to battery energy exhaustion, with the node failure, previously stored event data will be lost.

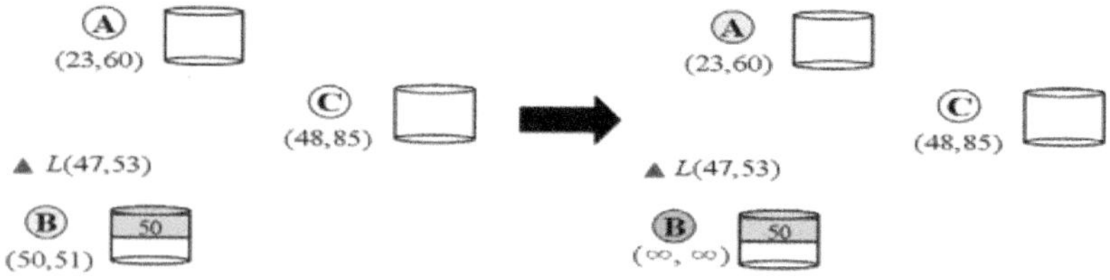


**Fig. 1.** The single-threshold case

## 3 Extended Grid

In all nodes within a grid of energy and storage have reached the highest level under the threshold, then the resulting data will be stored because they did not lose. Can use the previously mentioned concept of the virtual grid, the grid expansion adjacent to form a virtual grid.

Grid to reach the highest threshold of the last node broadcasts a packet to a neighbor Grid_Extend grid. Neighbor grid nodes receive data packets Grid_Extend continue to broadcast the packet to its neighbors Grid_Extend node. Each packet received Grid_Extend node, through the following four equations to determine if they are not in the expansion of the grid range.

Abscissa of the grid extension $(X_{extended\_grid}, Y_1)$ (2)

Vertical extension of the grid coordinates $(X_1, Y_{extended\_grid})$ (3)

Diagonal grid extension $(X_{extended\_grid}, Y_{extended\_grid})$ (4)

$$Y_{extended\_grid} = Y \pm G_L \tag{5}$$

Where x and y are the x-axis of the grid extension and the y-axis. Nodes must satisfy the above formula in the extended grid; does not meet the formula node packets dropped Grid_Extend to stop transmission. Nodes within the grid is extended, modified form of the mesh nodes Gvid and Gl. Generated by the above formula, a merger a number of new virtual grid mesh. This virtual grid each node through limited radio to exchange information, update the node's own node form the grid. When a node detects an event, it sends a packet to put a virtual grid, made under the previous threshold value based on node scheduling integrated approach, looking for a storage node as the active node to store the event data.

The following example to illustrate the expansion of the grid through the implementation steps: Suppose, node 4 and node 5 to the highest level after the threshold has been suspended. Now node 6 after receiving a data packet will be put into suspended state. Because grid (1,2) within the node into the suspended state, it can not be stored sensory data. In this case node 6 to the adjacent grid (0,1), (0,2), (0,3), (1,1), (1,3), (2,1), (2 , 2) and (2,3) sends a data packet Grid_Extend will form the grid nodes in the Gl = Gl +1, as shown in Figure 2.

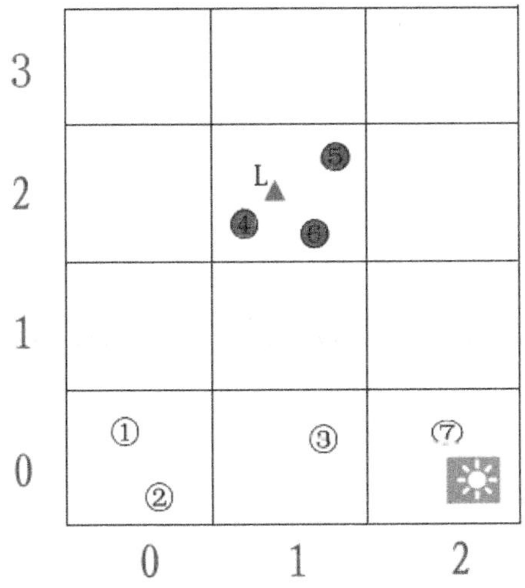


**Fig. 2.** Grid expansion

Adjacent to all nodes within the grid using Grid_Extend packet Gvid, by (2) to formula (5) to determine whether the grid in the extended range. In the extended grid of nodes to update their grid nodes table Gvid to Grid_Extend packet Gvid, Gl = Gl +1, and Grid_Extend forward the packet to the adjacent grid. Not in the extended grid nodes within the transmission and discard Grid_Extend stop packets. Figure 3, the grid (1,2) with the grid (0,1), (0,2), (0,3), (1,1), (1,3), (2,1) , (2,2) and (2,3) to form a new extended grid. New extension nodes within the grid, sending messages to each other to exchange information, update their grid node form.

Thus, the grid will be expanded into one of several mesh grid to store the perception data, the first to solve the problem of data storage nodes fixed to realize the dynamic data storage; second, to solve the storage node the problem of limited space, can maximize the effective data storage, avoiding the loss of valid data.

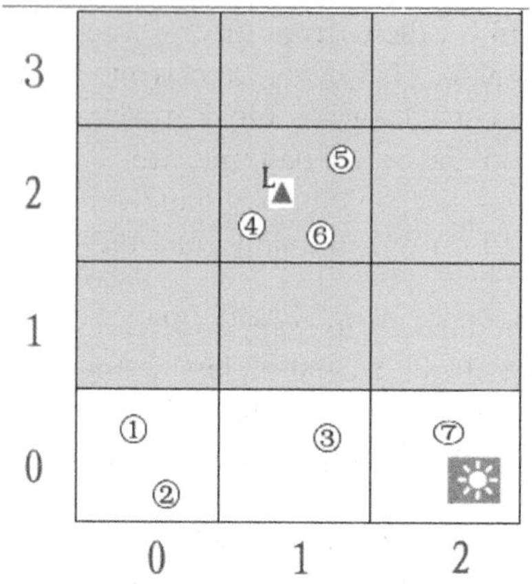


**Fig. 3.** After extended grid (1,2)

## 4 Conclusions

This paper analyzes the existing data storage methods shortcomings, this paper proposes to solve the problem. Second, the threshold proposed the concept of integrated and applied to the node scheduling approach, and then the whole region is divided into two-dimensional perception of the grid, the grid nodes within the implementation schedule based on the data storage method: the threshold level under the Comprehensive Threshold calculation scheduling factor, scheduling factor to determine the status of the node. First, avoid the multiple nodes to store the same data, information redundancy situation occurs, the second, to solve all the queries are concentrated in the individual nodes, the query occurs frequently and hot issues. Extended by a grid into the grid formed by the number of virtual grid, the first to solve the problem of data storage nodes fixed to achieve the dynamic storage, and second, to solve the limited storage node problem, the maximum Effective data storage limits to avoid the loss of valid data.

## References

1. Yick, J., Mukherjee, B., Ghosal, D.: Wireless sensor network survey. Computer Networks (52), 2292–2330 (2008)
2. Akyildiz, I.F., Su, W., Sankarasubramaniam, Y., Cayirci, E.: Wireless sensor networks: a survey. Computer Networks (38), 393–422 (2002)
3. Bonnet, P., Gehrke, J., Seshadri, P.: Towards Sensor Database Systems. Mobile data managemet., 3–14 (2001)
4. Abbas, A.M., Kure, O.: Quality of Service in mobile ad hoc networks: a survey. International Journal of Ad Hoc and Ubiquitous Computing 6(2), 75–98 (2010)
5. Padmanabhan, P., Gruenwald, L., Vallur, A., Atiquzzaman, M.: A survey of data replication techniques for mobile ad hoc network databases. The International Journal on Very Large Data Bases 17(5), 1143–1164 (2008)

6. Dietrich, I., Dressler, F.: On the Lifetime of Wireless Sensor Networks. ACM Transactions on Sensor Networks 5(1), 1–39 (2009)
7. Wu, M., Xu, J., Tang, X.: Processing Precision-constrained Approximate Queries in Wireless Sensor Networks. In: IEEE International Conference on Mobile Data Management, MDM 2006 (2006)
8. Tat, W.C.: Along & Across algorithm for Routing Events and Queries in Wireless Sensor Networks. In: Proceedings of Intelligent Signal Processing and Communication Systems, pp. 725–728 (2005)
9. Anastasi, G., Conti, M., Francesco, M.D.: Extending the lifetime of wireless sensor networks through adaptive sleep. IEEE Transactions on Industrial Informatics 5(3), 351–365 (2009)
10. Ha, R.W., Ho, P.-H., Shen, X.S., Zhang, J.: Sleep scheduling for wireless sensor networks via network flow model. Computer Communications 29, 2469–2481 (2006)
11. Heinzelman, W.R., Chandrakasan, A., Balakrishnan, H.: Energy-efficient communication protocol for wireless microsensor networks. In: Proceedings of the 33rd Hawaii International Conference on System Sciences (HICSS) (January 2000)

# Assess Impact of Expo 2010 on House Price Based on Mathematical Model

Yue Jia, Long Zuo, Zheng-gang Zhao, and Wen-shan Ren

Dalian Neusoft institute of Information, Department of Computer Science and Technology, Dalian Liaoning 116023
jiayue@neusoft.edu.cn

**Abstract.** The urban house price is one of the biggest concerns of the present-day China. Shanghai World Expo 2010 has created a huge economic and social benefit. Meanwhile the Expo keeps house price unstable. Based on the research on house price in past 18 months and the tourist flow during the Expo, the article builds three models. The first model describes the change of house price in the whole country, the second describes the impact of Expo's tourist flow on house price, and the third describes the trend of Shanghai's house price. Thus the impact of Expo on house price is assessed quantitatively. At last, the article further expounds the rationality and expansibility of these models.

**Keywords:** Expo 2010, house price, quantitative assessment, regression equation, linear regression.

## 1 Introduction

Short-term fluctuations of house prices are affected by multiple factors in China, and the mathematical model of its basic trend can be established. The house prices of Shanghai in the short term trend, based on it in country by hosting the World Expo impact of this factor can increase the corresponding parameter variable, the corresponding model, and then extended to the major events on short-term price movements of the mathematical model of the impact. Model and solution in the process, make the following notations in Table 1.

**Table 1.** Symbol definition explains

| Symbol | Definition | Symbol | Definition |
|---|---|---|---|
| t | Used to describe the time( February 2009 means 0 point) | $N_0$ | Initial house price |
| N | The national average house price index (February 2009 means base 1) | H | The average number of daily visits |
| W | Shanghai local house price index (2009 February means base 1) | E | The enthusiasm degrees of populace to the world expo |
| A | House demand index | G | People's consumption ability |
| B | House supply index | K | The amount of the world expo impact on house price |
| D | The timing of expo impact on house price | C/k | Constant/coefficient constants |

S. Lin and X. Huang (Eds.): CSEE 2011, Part II, CCIS 215, pp. 452–457, 2011.

# 2 Analysis

## 2.1 Trend of National House Prices in Short-Term

In the past 18 months, national house price level is in the rapid rise early, and the national regulation stabilized it in April 2010. Some of the results of prediction which made by Gray Theory show, the house price will maintain a sustained upward trend.

However, the data used in this prediction is less and there's a larger time span, so the results do not fit the characteristics of short-term price fluctuations, and do not coincide with the data that the National Bureau of Statistics released.

It can be seen from the analysis in the change of house price that the price increased exponentially before April 2010, and the growth rate has dropped since April 2010 and has been closed to a certain value. According to economic principles: during k time period, commodity price y (k) is determined by the number of products x (k) during the same period, reflecting the demand of consumer for this product, the greater the number is the lower the price is. During K +1 period, the number of products x (k +1) is determined by commodity price in the previous period, reflecting the relationship between the supply and the demand, because the higher the price will be more production. Through this theory, we established the Model I by the analysis of detail data in the past 18 months, this model shows the trend of National house prices in short-term.

## 2.2 Factors That Will Impact House Price in Shanghai

Through researching the house price of 70 cities in the past 18 months, we can see that the house price of Shanghai has a large fluctuation before and after the Expo. But in this time, in Dalian which is the coastal city and economic developed city and tourist city the same as Shanghai, and Tianjin which is he coastal city and municipalities the same as Shanghai, the house price was consistent with the national price movements without short-term fluctuations. We can see that the price fluctuation is a result of the Shanghai World Expo.

## 2.3 Trend of House Price in Shanghai in Short-Term

So we can consider that he trend of house price in Shanghai is determined by two factors, the trend of national house price and the impact of the World Expo. Then we can establish the model of short-term change of house price in Shanghai.

# 3 Establishment and Solution of Model

## 3.1 Model of the Trend of National House Prices in Short-Term

The trend curve of national house price is coincide with the Logistic curve, that is: when t is very small it grows exponentially, when t increases its growth rate declines, and more and more closes to a certain value. This type of problem can be solved by the Logistic equation. Logistic growth model, also known as self-restraint of the equation, is commonly used in population growth and commodity sales forecasting problem.

Supposed in t time the house price index is $N = f_N(t)$, demand index is A, the supply index is B, after Δt time, the demand increases $AN\Delta t$, supply increases $BN\Delta t$.

Price changes: $\Delta N = (A-B)N\Delta t$, if $\Delta t \to 0$ then $dN = (A-B)N_{dt}$ (1)

Supposed both A and B are linear, then: $A = m - n \cdot f_N(t), B = p + q \cdot f_N(t)$, m, n, p, q are constants

$$\frac{dx}{dt} = (m-q)N - (n+q)N^2 \to \frac{dN}{dt} = (n+q)N(\frac{m-q}{n+q} - N)$$

Supposed $\alpha = n+q, \beta = \frac{m-q}{m+q}$

$$\frac{dN}{dt} = \alpha N(\beta - N) \tag{2}$$

$\alpha dt = \frac{1}{N(\beta - N)} \cdot dN$, Simplification $\alpha\beta dt = \frac{1}{N}dN + \frac{1}{\beta - N}dN$

Both sides have points,

$$\alpha\beta t = \ln N - \ln(\beta - N) + \ln C, \quad \alpha\beta t = \ln\frac{CN}{\beta - N}, \quad e^{\alpha\beta t} = C\frac{CN}{\beta - N} \tag{3}$$

$t = 0, N = N_0$ are brought into (3),

then $e^{\alpha\beta t} = \frac{N(\beta - N_0)}{N_0(\beta - N)} \to N = \frac{\beta}{1 + \frac{\beta - N_0}{N_0}e - \partial\beta t}$ (4)

This is the national average house price in the Logistic growth model.

In order to calculate the national average house price of natural growth, we need determine the value of $N_0$, β and αβ in (4). Analysis data in November 2009 and August 2010, then: $t_1 = 9, N_1 = 1.065814767$, $t_2 = 18, N_2 = 1.134672965$,

Bring them into (3), then

$e^{9\alpha\beta} = \frac{1.065814767\ (\beta - 1)}{(\beta - 1.065814767\ )}$ (5) $e^{18\alpha\beta} = \frac{1.134672965\ (\beta - 1)}{(\beta - 1.134672965\ )}$ (6)

Combine (5) with (6), $\beta = 4.81548810792$, $\beta$ is brought into (5), then: $\alpha\beta = 0.00901549110728$

Therefore, final house price of the logistic model is:

$$N = \frac{4.81548810792}{1 + \frac{4.81548810792 - N_0}{N_0}e - 0.2705t}$$ (Model I)

### 3.2 Model of the Impact of Expo 2010 on House Price in Shanghai

Supposed the average daily number of visitors is H, then $H = f_H(t)$. We define the enthusiasm of the people on the Expo is E, the spending power is G, the Expo will affect the amount of house price is K.

We can simplify the problem through the relation between supply and demand, the house price N is proportional to demand and inversely proportional to supply, if the supply keeps not been changed in short time, the house price is only related with the demand. We only consider the most important factors affecting demand (demand number), the demand is proportional to the number of demand H. Therefore, house price N is proportional to number of demand H. We assume that the number of changes in demand is related with the degree of enthusiasm E, spending power G, and the number of current demand H.

Then $E = k_0 \cdot \frac{G}{H}$, elapsed time: $\Delta t$: $f_H(t+\Delta t) = \frac{E}{Ht}$。

Above all, house price affect volume $f_K(t) = k_2 \cdot \frac{G}{(H_{(t-\Delta t)})^2}$. Based on the report of traffic in Shanghai world expo, through EUREQA official second function fitting, we get more optimal function:

$f_K(t) = k_2 \cdot (2.61001 + 27.9483t - 4.63839t^2)$ (Model II)

## 3.3 Model of the Trend of Short-Term House Price in Shanghai

The model of trend of Shanghai house $f_W$ (Model Ⅲ) should be the result of Model I and Model II, namely: $f_W(t) = f_N(t) + f_K(t)$

$f_K(t)$ affects house price in Shanghai only during the expo, namely:

$$f_W(t) = \begin{cases} f_N(t) & t \le 15, t \ge 19 \\ f_N(t) + f_K(t) & 15 \le t \le 19 \end{cases}$$

Because the Shanghai expo was released in advance, and is not the sudden incident, the expo was held in Shanghai will be advanced affect house price, set this value of advancement as D. Namely: $f_W(t) = f_N(t) + f_K(t-D)$

Then $W = \dfrac{4.81548810792}{1 + \dfrac{4.81548810792 - N_0}{N_0} e - -0.2705t} + k_2(2/61001 + 27.9483t - 4.63839t^2)$ (Model III)

Meanwhile we can get the result: $k_2 \approx 0.001$

Analysis the trend of national house price, the trend of house price in Shanghai, and the traffic in Expo, the result is shown in Fig. 1:

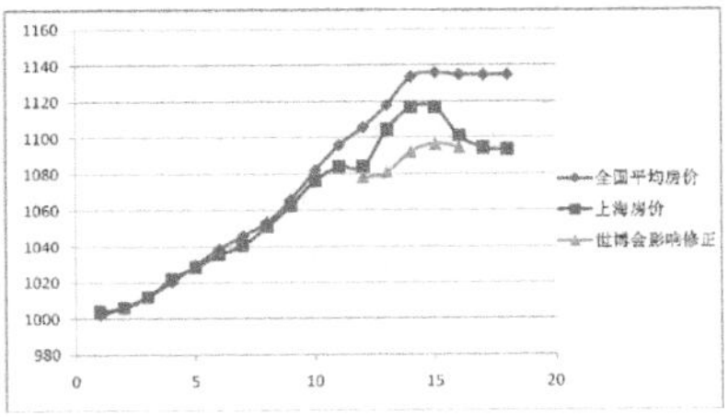


**Fig. 1.** The analysis of house price and passenger flow

# 4 Application and Analysis of Model

## 4.1 Application and Analysis of Model I

Fig. 2 is the comparison between the actual trend and the predicted trend of the national house price, generated by MATLAB:

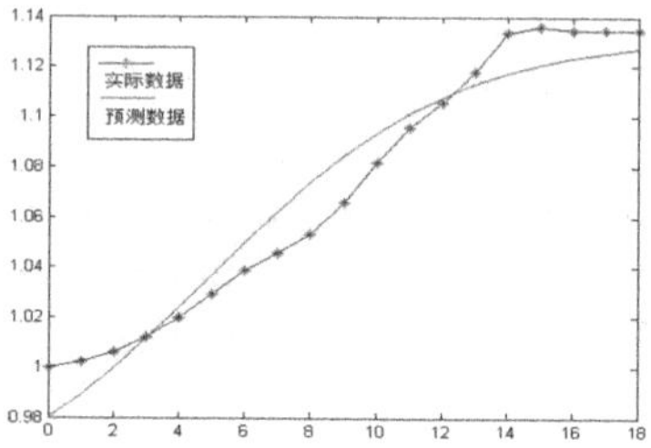


**Fig. 2.** The comparison between actual trend and predicted trend of the national house price

## 4.2 Application and Analysis of Model Ⅱ and Model Ⅲ

The error of the result calculated by Model Ⅲ is about 4%, the maximum error is 5.7%. The actual value and forecast value are shown in Fig. 3.

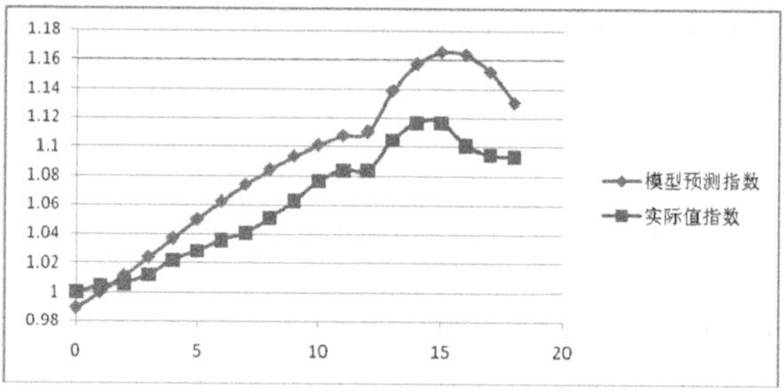


**Fig. 3.** The comparison between actual index and predicted index of house price in Shanghai

Through the above analysis, the model accurately predicts the recent trend of house price in Shanghai.

As the model predicts, house price will reach a plateau after the world expo.

# 5 Improvement of Model

The government's policy also has influence on house pricing, if there's time, we can do more survey and collect more data around government policies to build more accurate model.

This paper discussed the impact on house pricing from World Expo 2010. If we gather more data, we can cover more important events which have impacted on the house price.

## References

1. Deng, O., Li, Y.: Research on House Price in China by Gray Prediction. Economic Research Guide. J. (July 2008)
2. Wang, G., Wang, M.: Modern Method on Mathematical Model. Science Press, Beijing (2008)
3. MBA lib, Logistic's equation (October 2010), http://wiki.mbalib.com/wiki/%E9%80%BB%E8%BE%91%E6%96%AF%E8%92%82%E6%96%B9%E7%A8%8B
4. Expo 2010 Shanghai China, Statistics for Passenger Flow in Expo 2010 Shanghai (October 2010), http://www.expo2010.cn/yqkl/indexn.htm

# Study of Anti-lock Brake System Control Strategy in Automobile

Jifei Chen and Gangping Tan

School of machinery and transportation, Southwest Forestry University, Kunming 650224, China
cjf100cjf@yahoo.com.cn, tgpfreemail@126.com

**Abstract.** Anti-lock Brake System (ABS), which is the advanced system of brake, was developed to base on the adhesion characteristic between the tire and the road in automobile. ABS can automatic amend wheel press to prevent wheel from locking at braking and can improve braking efficiency. In this paper, the control strategy was present by analyzing principle of ABS and establishing the kinetics model of tire and model of the brake.

**Keyword:** Anti-lock Brake System, control strategy, kinetic model, automobile.

## 1 Introduction

Anti-lock Brake System (ABS) is the advanced system of brake in automobile and it was developed to base on the adhesion characteristic between the tire and the road. ABS can automatic amend wheel press to prevent wheel from locking at braking. ABS can improve braking efficiency, shortened braking distance, prevent wheel sideslip, decrease to braise tire.

## 2 Principle of ABS

The principle of ABS is base on kinetics among tire, braking disc, ground. We will analysis single wheel kinetics (figure 1). The kinetic equation is

$$I_{\omega}\dot{\omega} = F_x r - T_b \tag{1}$$

Where $F_x$ is brake power of ground, $T_b$ is moment of a force from the braking disc, $I_{\omega}$ is moment of inertia of the wheel, $\omega$ is angular velocity of the wheel, $r$ is the radium of the wheel.

The running state of the wheel was decided by the moment of force and brake power of ground in figure 1. Where $V_{\omega}$ and $V_x$ are rotating speed of the wheel and rapidity of vehicle. $F_x$ was controlled by adhesion characteristic. The wheel was locked and slipped when moment of a force from the braking disc exceeded limiting

S. Lin and X. Huang (Eds.): CSEE 2011, Part II, CCIS 215, pp. 458–465, 2011.

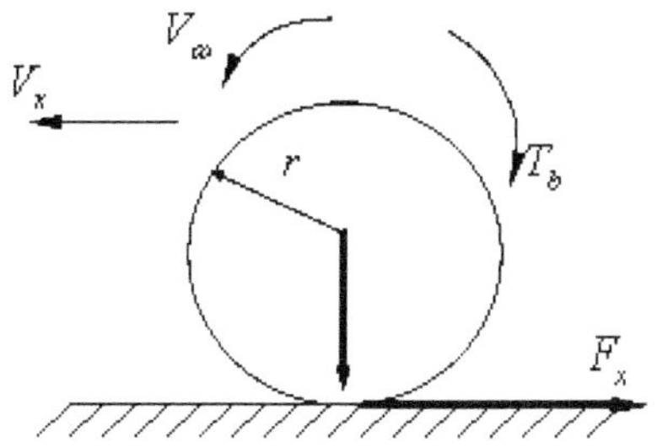


**Fig. 1.** The kinetic chart of the wheel

value. We should control $T_b$ right range to prevent the wheel from slipping. The wheel slip rate was defined between $V_\omega$ and $V_x$.

$$S_b = \frac{(V_x - V_\omega)}{V_x} \tag{2}$$

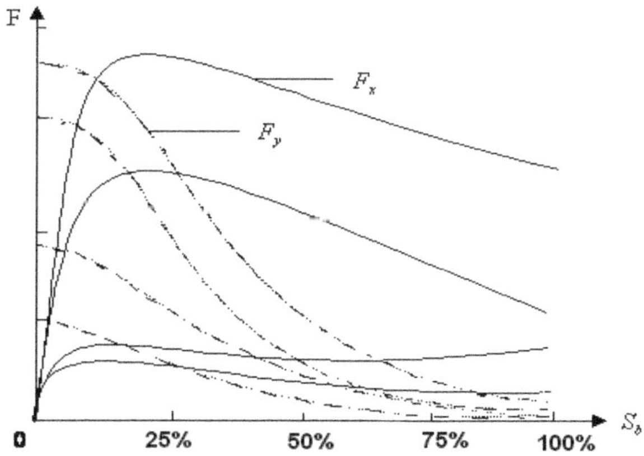


**Fig. 2.** Curve of slip rate and brake force

Where $S_b$ is longitudinal slip rate. Numerical value of the slip rate affected the wheel running state whether it is scrolling rate, slipping or scrolling and slipping [1]. $S_b$ is bound up with $F_x$ and $F_y$ while the wheel was braked (figure 2).

## 3 The Kinetics Model of Tire

In the all kinds of kinetics l model, the semi-empirical model of E Index put forward by professor of GUO KONGHUI have got good effect in the actual use [2,3].

$$\begin{cases} S_x = \dfrac{V_{sx}}{\omega R} = \dfrac{V_x - \omega R}{\omega R} \\ S_y = \dfrac{V_{sy}}{\omega R} = \dfrac{V_y}{\omega R} = (1 + S_x) tg\beta \end{cases} \tag{3}$$

Where $V_x$ and $V_y$ are Longitudinal and lateral speed of the tire, $S_x$ and $S_y$ are longitudinal slip rate and lateral slip rate, $V_{sx}$ and $V_{sy}$ are Longitudinal and lateral tire slip velocity, $\omega$ is tire rotation angular velocity (rad / s), R is tire rolling radius (m), $\beta$ is tire side slip angle;

$$tg\beta = \frac{V_x}{V_y} \tag{4}$$

$$\begin{cases} \Phi_x = K_x S_x / (\mu_x F_z) \\ \Phi_y = K_y S_y / (\mu_y F_z) \\ \Phi = \sqrt{\Phi_x^2 + \Phi_y^2} \end{cases} \tag{5}$$

Where $\Phi_x$ and $\Phi_y$ are relative longitudinal slip rate and relative lateral Slip rate, $\Phi$ is relative to the total Slip ratio, where $F_z$ is vertical load of the tire; $K_x$ and $K_y$ are the longitudinal slip degree and lateral stiffness of the tire; $\mu_x$ and $\mu_y$ are the longitudinal and lateral friction coefficient between the tread and the road. With the increase of slip velocity, $\mu_x$ and $\mu_y$ will reduce, so the formula can correct as fellow:

$$\mu_x = \mu_{x0} \exp(-\frac{V_{sx}}{V_{cx}}) \tag{6}$$

$$\mu_y = \mu_{y0} \exp(-\frac{V_{sy}}{V_{cy}}) \tag{7}$$

Where $\mu_{x0}$ and $\mu_{y0}$ are the static longitudinal and lateral friction coefficient between the tread and the road; $V_{cx}$ and $V_{cy}$ are characteristic velocity. According to the semi-empirical tire kinetic model, longitudinal force ($F_x$) and Lateral force ($F_y$), aligning torque ($M_z$) can be expressed by E index as follows:

$$\begin{cases} F_x = \mu_x F_z \cdot (1-\exp(-\Phi - E_1\Phi^2 - (E_1{}^2 + \frac{1}{12})\Phi^3)) \cdot \Phi_x / \Phi \\ F_y = \mu_y F_z \cdot (1-\exp(-\Phi - E_1\Phi^2 - (E_1{}^2 + \frac{1}{12})\Phi^3)) \cdot \Phi_y / \Phi \\ M_z = F_y(D_x + X_b) - F_x Y_b \end{cases} \tag{8}$$

Where $D_x$ is the distance, which the tire was dragged as follows;

$$D_x = (D_{x0} + D_e)\exp(-D_1\Phi - D_2\Phi^2) - D_e \tag{9}$$

$X_b$ and $Y_b$ are horizontally deformation cased by longitudinal force( $F_x$ ) and lateral force( $F_y$ ).

$$\begin{cases} X_b = F_x / C_{bx} \\ Y_b = F_y / C_{by} \end{cases} \tag{10}$$

Where $C_{bx}$ and $C_{by}$ are the longitudinal and lateral stiffness of the tire. The formula from 3 to 10 is the E index of semi-empirical model in longitudinal slip and side. The vertical load ( $F_z$ ) is a relative with parameters of $E_1$, $K_x$, $K_y$, $\mu_x$, $\mu_y$, $D_{x0}$, $D_e$, $D_1$, $D_2$. The parameters can be determined according to the test.

## 4 The Kinetic Model of the Brake

The kinetic model of the brake describes the kinetic properties from the pressure input of the brake wheel hydraulic pressure jar to the brake torque output [4,5]. Braking torque of automotive wheel is produced by friction between the rotating element and a fixed element. According to friction principle, Wheel braking torque can be described as follow:

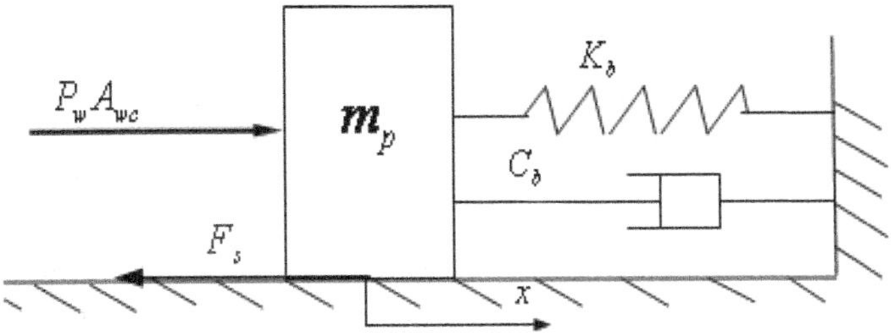


**Fig. 3.** Schematic chart of kinetic model of the wheel cylinder piston

$$T_b = GN_b \tag{11}$$

Where $G$ is the constant factor, which relative with brake Friction Materials, Structure and temperature, can be measured by the test; $N_b$ is the pressure of wheel cylinder piston pushed out. In generally, $N_b$ and the pressure $P_w$ of Wheel cylinder meet the following static relations (figure 3):

$$N_b = P_w A_{wc} \tag{12}$$

Where $A_{wc}$ is square measure of Wheel cylinder piston.

The static kinetic of the brake during braking applied the general requirements of the simulation. However, the variety of brake pressure can make the force and the state of movement of the piston cylinder change among the process of anti-lock braking. So, when we compute the pressure that the piston was pushed out we should consider the influence cased by the kinetic characteristics to establish the kinetic model of the wheel cylinder piston.

The dynamic equations of the piston can be described as follows:

$$m_p \ddot{x} + C_b \dot{x} + K_b x = P_w A_w - F_s \tag{13}$$

Where $x$ is the displacement of cylinder piston; $m_p$ is quality of the piston movement; $K_b$ and $C_b$ are the stiffness and damping of brake; $F_s$ is the friction force which the piston suffered when it is moving at out of lubricating. Then the friction force can be described as follows:

$$\dot{F}_s = \frac{|\dot{x}|}{X_s}[\mathrm{sgn}(\dot{x}) F_{sf} - F_s] \tag{14}$$

$$F_{sf} = C_0 + C_1 |x| + C_2 x^2 \tag{15}$$

Where $X_s$ is relaxation length of the friction; $C_0$, $C_1$, $C_2$ are the coefficient of friction which change with piston displacement.

The dynamic pressure of the piston is change force as follow:

$$N_b = C_b \dot{x} + K_b x \tag{16}$$

## 5 The Description of Control Strategy

The torque of braking was modulated by signal of the wheel running state [5]. The control process on High-road adhesion coefficient was described as figure 4. At beginning phase of braking, with the increase of brake pressure, The angle acceleration of the wheel increases, until the wheel angular acceleration attain to the control threshold that was set ($-a_1$) (Phase 1) . However, in order to avoid wheels to enter into the phase of brake pressure increasing, it is still in the stable region of the slip rate range. At same time, we should compare threshold $\lambda_1$ between reference of wheel slip rate and controller setting slip rate. If the reference of wheel slip ratio is less than the threshold $\lambda_1$. We hold pressure to make the wheels into sufficient brake until the reference wheel slip rate greater than the threshold $\lambda_1$, when wheel entered the unstable region, the control system will increase pressure (phase 3) .

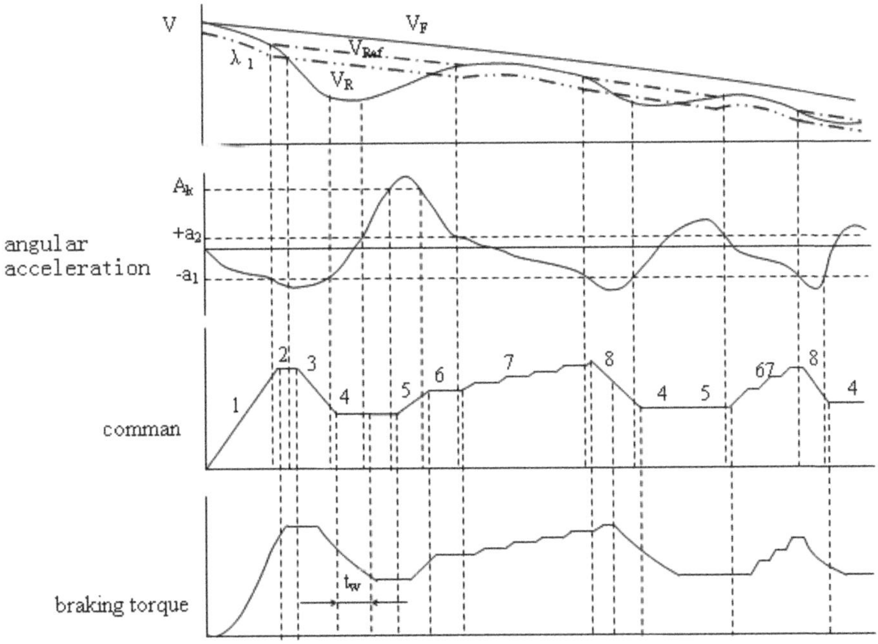


**Fig. 4.** Control process on High-road adhesion coefficient

But because of dynamic characteristics of Hydraulic system, Wheel brake pressure will decrease after a period of time. When the wheel angular acceleration is more than the threshold ( $-a_1$ ), the control system put out the command of maintain phase 4 . In the high-road adhesion coefficient the, wheel cylinder pressure of work is high in the anti-lock braking process, brake pressure drops quickly. Wheel angular acceleration in a very short time will exceed the angular acceleration threshold ( $+a_2$ ). The low friction coefficient road condition can be determined, after this the control process will maintain in the conditions of low road adhesion coefficient. When the wheel angular acceleration exceeds the threshold ( $+a_2$ ), the control system will continue to maintain brake pressure. In order to adapt decompression over in the high road anti-lock braking adhesion, we should set the second acceleration threshold as (+ A) to control increasing the braking force (phase 5) until angular acceleration of the wheel is lower than the threshold (+A). Then the system returns to the stage of brake pressure holding (phase 6) . When the angular acceleration of the wheel is lower than the control threshold ( $+a_2$ ) once again, the wheel returns to the stable region again. In order to make the wheel in a stable region for a longer time, the step pressurization will be taken (phase 7) . The wheel angle acceleration is lower than the control threshold ( $-a_1$ ) again, the control system began to enter reduction phase of the brake pressure (phase 8) , and it enter the next cycle anti-lock brake process of pressure regulating.

The control process on low-road adhesion coefficient was described as figure 5. The phase 2 and 3 of the anti-lock brake pressure modulation is the same with High-road adhesion coefficient

When the system output the command to hold pressure (phase 5) , wheel brake pressure will continue to decrease. However, wheel cylinder pressure is low work and the brake pressure decline slowly, Wheel angular acceleration can not attain the threshold ( $+a_2$ )in waiting time ($t_w$), the control logic determine the wheel is in a low friction coefficient road. The system will control to reduce the brake pressure until wheel angular acceleration exceeds the threshold ( $+a_2$ ). The system entered the phase of brake pressure again (phase 6). When the wheel angular acceleration below the threshold ( $+a_2$ ), wheel returned to the stable region and the brake pressure rise via the relatively low rate of pressure (phase 7) until the wheel angle acceleration is lower than the control threshold ( $-a_1$ ) and entered the next Anti-lock control loop phase of pressure reduction (phase 8) . So the wheels can quickly return to the stable area.

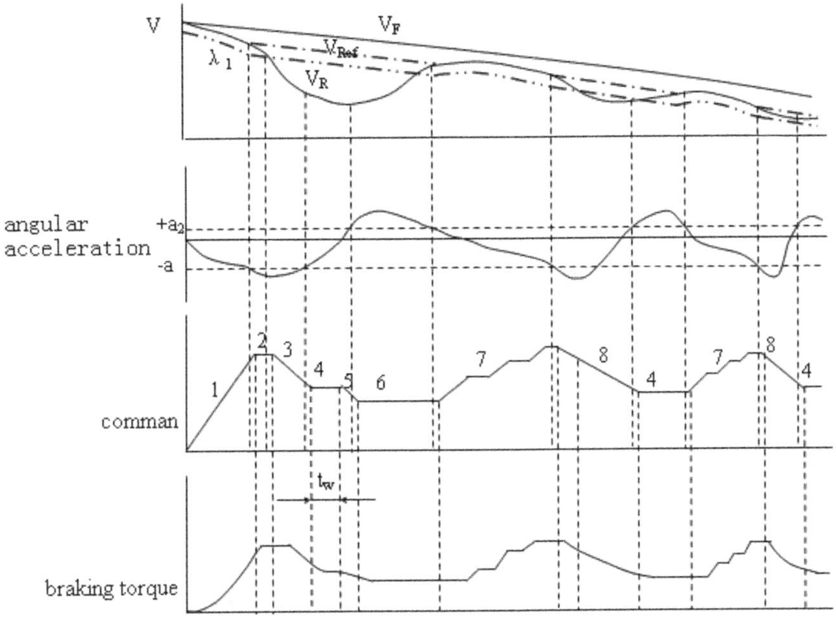


**Fig. 5.** Control process on low-road adhesion coefficient

## 6 Conclusion

Anti-lock Brake System (ABS) was developed to base on the adhesion characteristic between the tire and the road in automobile. ABS can automatic amend wheel press to prevent wheel from locking at braking and can improve braking efficiency. The torque of braking modulated the wheel running state. This paper present control strategy by variety of wheel rotating speed, angular acceleration on the torque of braking at high road and low road adhesion coefficient.

## References

1. Guo, K.: Kinetic of automobile (1991) (in Chinese)
2. Guo, K., Ren, L.: A Non-steady and Non-linear Tire Model Under Large Laterl Slip Condition. SAE paper, 2000-01-0358
3. Loh, W.-Y., Basch, R.H., Harlka, T.D., Hartsock, D.: Development of a Brake Dynamometer-Vehicle Model Hardware-in-the-Loop System. SAE paper, 2003-01-3337
4. Brach, P.M.: Vehicle dynamics model for simulation on a microcomputer. Int. J. of Vehicle Design 12 (1991)
5. Rogers, K.J.: Passenger car testing: Acceleration measurement on dynamometer and road. Int. J. of Vehicle Design 5 (1984)

# Subjective Trust Evaluation Model Based on Preferences

Ruizhong Du[1,2] and Pengyuan Zhao[2]

[1] Computer School of Wuhan University, 430079, Wuhan, Hubei
[2] Institute of Network Technology, Hebei University, 071002, Baoding, Hebei
durz@hbu.edu.cn

**Abstract:** In the open network environment, in the face of vast amounts of resources and services, due to the fraud and the existence of unreliable service, the user increased choice, at the same time how to identify and select efficient and secure resources or services. So the trust evaluation has become one of the focuses of current research. Because the trust relationship between entities is highly subjective, uncertainty, ambiguity, drawing on the trust model of human sociology, according to preference entities interested in, considering the various attributes and characteristics, adopts the fuzzy comprehensive evaluation method to trust the entity evaluation. In order to increase the recommended credibility and reliability of recommended entities, the introduction of services similarity algorithm in the Web semantic network, the weighted harmonic and recommendation trust value, in order to improve the reliability of recommended entities, and finally considering the direct trust and recommendation trust of entities, accordingly conclude both meet the current needs of the requesting entity safe and reliable service entities. Finally, simulation experiments performed to verify the validity of model credibility assessment.

**Keywords:** trust evaluation model, subjective trust, fuzzy comprehensive evaluation, similarity.

## 0 Introductions

With the computer technology and communication technology continues to evolve, the network environment has been transformed from relatively static, specific organizations and user groups for the closed network early, into publicly accessible, a large number of dynamic user-oriented open network. In the face of vast amounts of resources and services, due to the fraud and the existence of unreliable service, the user increased choice, at the same time how to identify and select efficient and secure resources or services. The current solution is an effective trust evaluation system, and become one of the focuses of current research [1,2].

Trust evaluation system to collect, analyze substantive historical information to predict their behavior in future transactions that may act, thus to select a higher trust entity to transact for user, to reduce the risk of transaction failure, and to avoid losses. Thus, the nature of the trust evaluation system is to assess the behavior of the entity, according to certain algorithms, to get the trust value of the behavior of an entity, for the parties to the transaction reference.

S. Lin and X. Huang (Eds.): CSEE 2011, Part II, CCIS 215, pp. 466–472, 2011.

# 1 Related Work

In the online world, the establishment of the relationship between entities is based on the judgment of mutual trust. In 1996, Blaze et al address Web services security, first coined the "Trust Management" term [3]. If using the classical precise reasoning method, can not fully reflect the reality of things, because, trust or not is evaluated by the people, is a subjective judgments, and highly subjective, uncertainty, ambiguity [4,5].

Tang Wen et al [6,7] introduce linguistic variables and fuzzy logic to trust management study, adopt a fuzzy measure of trust mechanisms, provide a valuable new ideas to the trust management research. Wang Shouxin et al [8] propose a subjective trust evaluation methods based on cloud model, which evaluates the credibility of entity by the expectation of subjective trust cloud and hyper entropy, then design a trust change cloud model to describe the trust change of entity, and thus provide the basis for the trust decision. Manchala [9] proposed a trust model based on fuzzy logic, proven by introducing the concept of fuzzy authentication to describe the fuzziness of the trust relationship, which depicts the complex and variable relationship between the entities in the model, and solve the problem of fuzzy modeling of trust. The disadvantage is the lack of a direct measure of trust; it would be an exploit of no great difficulty to threatening event and increase the possibility of the system to be attacked. Chen Jiangang et al[10] propose the trust mechanism to access grid resources based on fuzzy set.

Consider the trust relationship between entities is highly subjective, uncertainty, ambiguity in the distributed network environment, according to entities preference, considering the various attributes and characteristics, adopt the fuzzy comprehensive evaluation method to evaluate the trust of entity, introduce services similarity algorithm in the Web semantic network, the weighted harmonic and recommendation trust value, in order to improve the reliability of recommended entities, and consider the direct trust and recommendation trust of entities, accordingly conclude both meet the current needs of the requesting entity safe and reliable service entities.

# 2 Trust Management Model

## 2.1 Definition

**Definition A.** Let X be non-empty domain, x is the element of X, for any x ∈ X, given the following mapping: $X \to [0,1]$ , $x \mapsto \mu_A(x) \in [0,1]$ , the following set composed by the ordered pair $A = \{x | (\mu_A(x))\}$, is called fuzzy subset on the X, $\mu_A(x)$, is called membership function of x on A(Also can be expressed as A (x)) $\mu_A(x)$, is called membership degree of x on A, x in terms of a specific.

**Definition B.** Direct Trust: means the trust relationship that established under the direct interaction occurred between of two entities of the system in the past.

**Definition C.** neighbor entity: entity that has had directly historical transactions with the request entity, is called neighbor entity.

**Definition D.** service attribute set: the vector composed by the element which can affect the result of inter-entity transaction, is called service attribute set, indicated by $ATTR = \{attr_1, attr_2, ......, attr_n\}$ ,and $attr_i$ is the No.*i* attribute of the service. To facilitate the calculation, the method using normalized values of all properties into a dimensionless quantity and is in [0, 1].

**Definition E.** trust level evaluation sets: the set composed by various evaluation results concluded by subjective entity according to the trust level of the objective entity, indicated by $V = \{v_1, v_2, \ldots, v_m\}$ .

**Definition F.** Preference: the subjective desire of service requester entity to the attribute of the service.

**Definition G.** weight set of service attribute: Different attributes corresponding to different weight coefficients. The set composed by different weight coefficients, indicated by $W = \{\omega_1, \omega_2, \ldots, \omega_n\}$ , and $0 \le \omega_i \le 1$, $\sum_{i=1}^{n} \omega_i = 1$ .

## 2.2 Satisfactions

After each transaction, integrating the service properties, using the Comprehensive Evaluation Method to calculate the value of trust evaluation when the service provider entity $x_j (x_j \in X)$ provided the transaction service to the service requester entity $x_i (x_i \in X)$ indicated by $S_k$ , and *k* is the No. *k-time* transaction. As the following:

(1) Determine the evaluation set, indicated by $V = \{v_1, v_2, \ldots, v_m\}$ , any property of service that can be used several fuzzy subsets. The definition as follows:

$v_1 (0 \le R < T_1)$ : "no confidence" subset; $v_2 (T_1 \le R < T_2)$ "critical trust " subset; $v_3 (T_2 \le R < T_3)$ "trust" subset; $v_4 (T_3 \le R \le 1)$ "full confidence " subset;

(2) To evaluate single factor of each service attribute, then get the membership vector $r_i = (r_{i1}, r_{i2}, \ldots, r_{im})$.

(3) To calculate the integrated membership, $B = W \circ R$ , of which $\circ$ is the fuzzy composition operator, then get satisfaction vector;

## 2.3 Value of Direct Trust

If there were n times transactions between entity $x_i$ and entity $x_j$, when the No.k transaction is completed, taking into account transaction time $t_k$ of the No. $k$ transaction, the transaction amount $m_k$ and the total number of transactions n between entity $x_i$ and entity $x_j$, the direct trust degree is calculated as:

$$T_d(x_i,x_j)=\begin{cases}0.5 & n=0\\ \alpha\times\dfrac{\sum_{k=1}^{n} f(t_k)\times q(m_k)\times S_k}{n} & n>0\end{cases} \quad (1)$$

There into, $\alpha=\sqrt{n/(n+1)}$ is the function about transaction time. $f(t_k)=e^{-\lceil (t_0-t_k)/T\rceil}$ is attenuation coefficient of trust about time, $t_0$ is the time of this transaction. The function $q(m_k)=e^{-1/m_k}$ $(m_k>0)$ is used to adjust the impact of the transaction amount to the evaluation value. This makes it difficult to obtain high confidence with small transactions, to avoid malicious entities to obtain the trust by small transactions then fraud in the large transaction.

### 2.4 Similarity of Recommendation

To ensure that the recommendation content be credible, we introduce similarity in the Web semantic network, which was assessed by calculating the similarity between the service provider entity provided and the service the requester entity requested, only services provided to meet user requirements be considered as a reliable service. We adopt the model provided by reference[11] to calculate the similarity, denoted by WS.

To adjust the trust value with similarity weighted, so that the higher the similarity, the higher trust value after adjusted. Calculated as follows:

$$ST(x_i,x_k)=\frac{2\times WS(x_i,x_k)\times T(x_i,x_k)}{\delta\times T(x_i,x_k)+(1-\delta)\times WS(x_i,x_k)} \quad (2)$$

$\delta$ is adjusted factor, $WS(x_i,x_k)$ is the similarity between the entity i and entity j, $T(x_i,x_k)$ is the level of the trust, $ST(x_i,x_k)$ is the final trust value which is adjusted with similarity weighted.

### 2.5 Trust Level

The trust value of entity $x_j$, which is calculated through collecting and integrating the recommendation of the neighbor entities by entity $x_i$, is called recommendation trust, denoted by $T_r(x_i,x_j)$. In calculating the recommendation values, this paper considered the similarity of recommendation, trust value of the recommender entity, the transaction amount of the recommender entity, the number of transactions.

When computing the recommendation value in the domain, according the direct transaction along with the entity $x_i$, searching the set of recommendation entity L( $x_1,x_2,\cdots,x_{Num}$ ). The value of recommendation in domain, $T_r(x_i,x_j)$, is calculated as:

$$T_r(x_i, x_j) = \sum_{k=1}^{Num} \frac{ST(x_i, x_k)}{\sum_{k=1}^{Num} ST(x_i, x_k)} T_d(x_k, x_j) \tag{3}$$

There into, *Num* is the number of the entity in the set of recommendation entity L.

Integrated direct trust and recommendation trust, come to the entity $x_j$'s general trust of the entity $x_i$ .

$$T(x_i, x_j) = \beta T_d(x_i, x_j) + (1-\beta) T_r(x_i, x_j) \tag{4}$$

The confidence factor $\beta$ is the direct trust and recommendation trust comprehensive weight, to enhance system reliability. Confidence factor adjustment formula is as follows:

$$\beta = 1 - \rho^k, \rho \in (0, 1) \tag{5}$$

## 3 Simulations

In order to verify the validity and accuracy of FTSTEM model, we design the simulation experiment of the model based on the PeerSim[13] simulation platform provided by the project team BISON[12]. According to the characteristics of the network entity, the entity node is divided into four types: absolute trust, trust, critical trust and distrust that the partners after a long-term exchange.

### 3.1 The Variance of Trust Value in Four Types of Entity with the Increase in the Number of Transactions

Suppose: $T_1 = 0.3$ , $T_2 = 0.6$ , $T_3 = 0.9$ . Figure 1 shows the variance of the average value of trust with the increase in the number of transactions.

**Fig. 1.** The variance curve of trust value in four types of entity with the increase in the number of transactions

With the increase in the number of transactions; the new reputation value of each node is updated. The global reputation value of the absolute trust node and the trust node will increase as the number of transactions increase, the trend of the trust value will be increased gradually, and the trust value of the critical trust node showed wavy, the value of the distrust node trust is a decreasing trend. Figure 1 shows that, FTSTEM model reflects well with the trust value of node entities while change the number of transactions, meet the expectations of the analysis.

### 3.2 Success Rate of Transactions in Model FTSTEM VS. EigenRep

Model simulation transaction is successful or not, is judged according to the user's satisfaction feedback. When the satisfaction is greater than 0.6, we will consider the transaction is successful, otherwise fail. Transaction success rate is expressed as the number of successful transactions in all proportion.

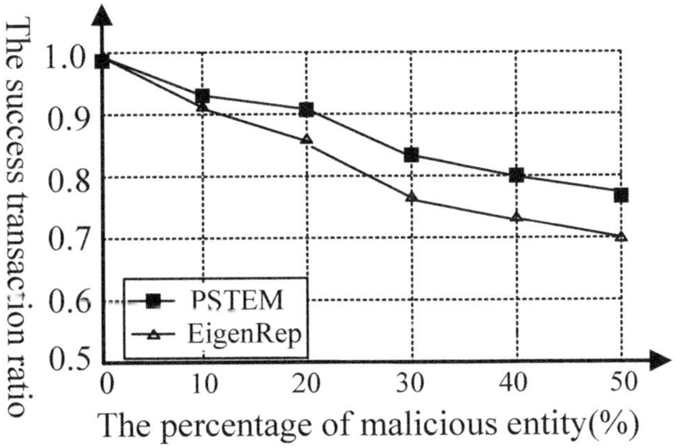


**Fig. 2.** Transaction success rate comparisons

It can be seen from Figure 2, when there is no malicious node in the system, the success rate of transaction is 95%. With the increase of malicious nodes, the success rate of the transaction dropped significantly in EigenRep model, and when the malicious nodes in the system reaches 50%, the success rate of the transaction is only about 60% in EigenRep model. In simulation experiment of our model, while there are 50% malicious nodes in the system, this model still has a 80% success rate of transaction, experimental results confirm the accuracy and effectiveness of the model.

## 4 Conclusions

Because the trust relationship between entities is highly subjective, uncertainty, ambiguity, drawing on the trust model of human sociology, according to preference entities interested in, considering the various attributes and characteristics, adopt the fuzzy comprehensive evaluation method to trust the entity evaluation, the introduction of services similarity algorithm in the Web semantic network, the weighted harmonic and recommendation trust value, in order to improve the reliability of recommended entities, and finally considering the direct trust and recommendation trust of entities, accordingly conclude both meet the current needs of the requesting entity safe and reliable service entities. Finally, simulation experiments performed to verify the validity of model credibility assessment.

This research was supported by the National Natural Science Foundation of China (GrantNo.60873203), Natural Science Foundation of Hebei Province of China(GrantNo. F2010000331).

## References

1. Youwen, Z.H., Lousheng, H., Guoliang, C.H., Wei, Y.: Dynamic Trust Evaluation Model under Distributed Computing Environment. Chinest Journal of Computers 34(1), 55–63 (2011)
2. Changxiang, S.H., Huanguo, Z.H., Huaimin, W.: Research and Development of the Trusted Computing. Science China 40(2), 139–166 (2010)
3. Blaze, M., Feigenbaum, J., Lacy, J.: Decentralized trust management. In: Proceedings of the 17th Symposium on Security and Privacy, pp. 164–173 (1996)
4. Beth, T., Borcherding, M., Klein, B.: Valuation of Trust in Open Networks. In: Proceedings of the European Symposium on Research in Secutity, pp. 3–18. Springer, Brighton (1994)
5. Jiangang, C.H., Ruchuan, W., Lin, Z.H., Haiyan, W.: The Resource Access Mechanism in Grid Based on Fuzzy-Trust. Chinese Journal of Computers 32(8), 1676–1682 (2009)
6. Wen, T., Zhong, C.H.: Research of Subjective Trust Management Model Based on the Fuzzy Set Theory. Journal of Software 14(9), 1401–1408 (2003)
7. Wen, T., Jianbin, H., Zhong, C.H.: Research on a Fuzzy Logic-Based Subjective Trust Management Model. Journal of Computer Research and Development 42(10), 1654–1659 (2005)
8. Shouxin, W., Li, Z.H., Hesong, L.: Evaluation Approach of Subjective Trust Based on Cloud Model. Journal of Software 21(6), 1341–1352 (2010)
9. Manchala, D.W.: Trust Metrics, Models and Protocols for Electronic Commerce Transaction. In: 18th International Conference on Distributed Computing Systems, pp. 3–12 (1998)
10. Chao, C.H., Ruchuan, W., Lin, Z.H.: The Research of Subjective Trust Model Based on Fuzzy Theory in Open Networks. Chinese Journal of Electronics 38(11), 2505–2509 (2011)
11. Qing, X.: Algorithms of Service Discoving Based on Semantic Distance. In: Shan Dong: A Dissertation for the Master's Degree of Shang Dong University, pp. 17–26 (2006)
12. BISON (2011.5.3), http://www.cs.unibo.it/bison
13. PeerSim (2011.5.3), http://peersim.sourceforge.net

# Research on Client Detection of BitTorrent Based on Content

Ying-xu Qiao and Hong-guo Yang

College of Computer Science & Technology
Henan Polytechnic University
JiaoZuo, China
qiaoyingxu@hpu.edu.cn
Department of Computer Science & Technology
JiaoZuo Teachers College
JiaoZuo, China
yanghg@hpu.edu.cn

**Abstract.** Among all available peer-to-peer applications for content distribution, BitTorrent has become the most popular by dominating approximately half of the all Internet traffic. It uses dynamic ports and masquerade its traffics as HTTP traffics .Traditional client detections of BitTorrent are unable to indentify the BitTorrent download traffic for their incapability of applition layer data analysis.In this paper , a new method of BitTorrent traffic identification is presented. Finally, this paper presents an experiment to testify the availability.

**Keywords:** application layer, client, content.

## 1 Introduction

BitTorrent is a protocol for distributing files. It identifies content by URL and is designed to integrate seamlessly with the web. Its advantage over plain HTTP is that when multiple downloads of the same file happen concurrently, the downloaders upload to each other, making it possible for the file source to support very large numbers of downloaders with only a modest increase in its load.however, BitTorrent application uses dyanmic ports to camouflags its traffics, So it becomes more difficult to measure BitTorrent traffic. In this paper, a measurement approach based on content and application level signatures, which reliably detects and measures BitTorrent traffic is described in details.

The rest of the paper is organized as follows. In Section 2, we analysises the coretechnology of the BitTorrent in details, In Section 3, we describe the proposed measurement approach on content of Bittorent .Section 4 gives detailed analysis of files characterstics and user behavior in BitTorrent ,and we conclude in Section 5.

## 2 Analysis of BitTorret Protocols Mechanism

A BitTorrent file distribution consists of these entities: an ordinary web server a static 'metainfo' file, a BitTorrent tracker, an 'original'downloader, the end user web

S. Lin and X. Huang (Eds.): CSEE 2011, Part II, CCIS 215, pp. 473–476, 2011.

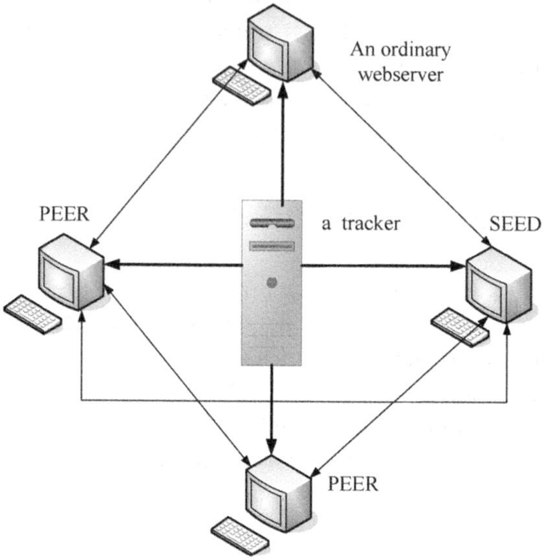


**Fig 1.** A BitTorrent file distribution

browsers and the end user downloaders. There are ideally many end users for a single file. The structure of A BitTorrent file distribution is shown in Fig 1.

To start serving, a host goes through the following steps: (1) Start running a tracker. (2) Start running an ordinary web server, such as apache. (3) Associate the extension .torrent with mimetype application/x-BitTorrent on their web server. (3) Generate a metainfo (.torrent) file using the complete file to be served and the URL of the tracker. (4) Put the metainfo file on the web server. (5) Link to the metainfo (.torrent) file from some other web page. (6) Start a downloader which already has the complete file (the 'origin').

A more detailed description of the BT protocol and algorithms is found in [2].

## 3 Measurement Strategy of BitTorrent Client

Tracker responses are bencoded dictionaries. If a tracker response has a key failure reason, then that maps to a human readable string which explains why the query failed, and no other keys are required. Otherwise, it must have two keys: interval, which maps to the number of seconds the downloader should wait between regular rerequests, and peers. peers maps to a list of dictionaries corresponding to peers, each of which contains the keys peer id, ip, and port, which map to the peer's self-selected ID, IP address or dns name as a string, and port number, respectively. Note that downloaders may rerequest on nonscheduled times if an event happens or they need more peers. The measure model of Bt client is shown in Fig 2.

The peer protocol refers to pieces of the file by index as described in the metainfo file, starting at zero. When a peer finishes downloading a piece and checks that the hash matches, it announces that it has that piece to all of its peers. Connections

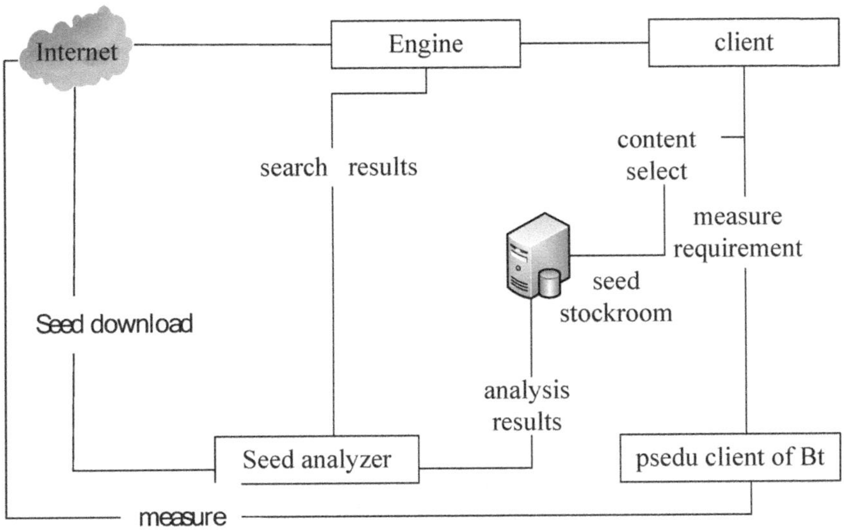


**Fig. 2.** The measue model of Bt client

contain two bits of state on either end: choked or not, and interested or not. Choking is a notification that no data will be sent until unchoking happens. The reasoning and common techniques behind choking are explained later in this document. Data transfer takes place when everone side is interested and the other side is not choking. Interest state must be kept up to date at all times - whenever a downloader doesn't have something they currently would ask a peer for in unchoked, they must express lack of interest, despite being choked. Implementing this properly is tricky, but makes it possible for downloaders to know which peers will start downloading immediately if unchoked.

## 4 Measurement Results

We analyze 400 million share files and 100 daylog of user behavior which we collected at the measurement system of BitTorrent, there are several lessons to be learned from our measurement results (1) few video files occupy a disproportionately high fraction of space and consume a disproportionately high fraction of bandwidth. (2) a large percentage of shared files have never been downloaded by any user, (3) a file's extension contains little information about whether it is useless.Fig 3 illustrates the breakdown of files with respect to their categories, for example, movie, tv, education and science.

As the graph shows,a high percentage of files ,abouts 39% of total files,belong to other users,there are two causes for this phenomenon.First, a part of users shares many system files. Second,we use extension as the criterion for classification ,and neglect some special file format.

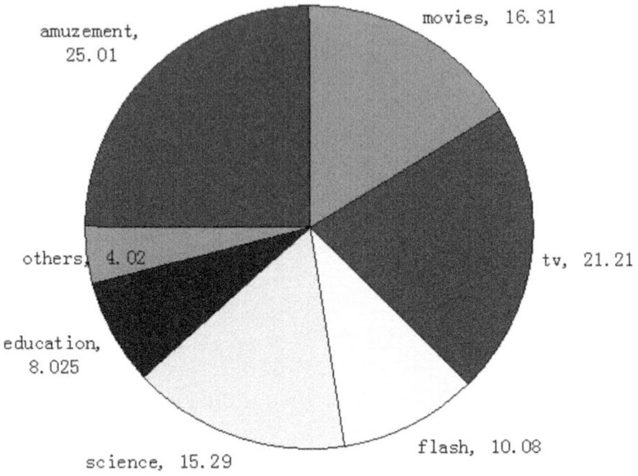


**Fig. 3.** File categories of Bt

# 5 Conclusions

In this paper, a measurement approach based on application level signatures, which reliably detects and measures BitTorrent traffic is described in details. A presentation of the results from a successful field test in a large university network indicated that this approach is not only accurate, but also provided the performance and scalability required for practical applications.

# References

1. Yu, J.-d.: Improvement of Incentives Mechanism on BitTorrent-like Peer-to-Peer FSS. Journal of DongHua University 24, 523–5277 (2007)
2. Qian, M.: BitTorrent traffic detection based on rule. Computer Engineering and Design 29, 357–359 (2008)
3. BitTorrent Specification v1.0 http://wiki.theory.org/BitTorrent Specificion

# The Research of Dynamic Mining Technology in the Application of E-Commerce

Rencai Gao

Baicheng Normal College 137000

**Abstract.** As large amount of Web data in the network and up to date, how to dig out the useful parts from the massive data, as the traditional mining methods can only dig out the data from the out dated and old parts, in this way, it can not make a rational predict of the future result. On the basis of this, the paper proposes a concept of a dynamic data mining and through a sliding window for data collection, the excavation of dynamic processing algorithms, thus arrives at a dynamic mining algorithm DDMA.

**Keywords:** Data Mining,Sliding Window, Dynamic Data Mining, Dynamic Data Processing.

With the development of network technology, especially the popularity of the Internet, people transit from the traditional business model to the current electronic commerce model..In this way the distributors and the customers make transactions via the Internet, saving a lot of the cost and time. How to take a better and faster advantage of this modern means of exchange to to shorten the distance between dealers and customers, which is currently e-commerce urgent problems. We know that if we can act in the business of customer historical data generated for the data warehouse structure, constructed a reasonable view of business behavior and ontology databases for model analysis, can be a good solution to this problem. This is definitely one area of the data mining. We can extract customer-related knowledge from the log files of every customer left on its server. As the customer's access behavior, frequency, content, time, etc., taking advantage of the dynamic mining technology to carry out targeted e-commerce and reasonable action.

## 1 Common Method of Web Data Mining

Traditional mining techniques are mainly used to analysis historical data , the results are out dated, but information is constantly changing, dynamic updates, so how to maintain a smooth transition with the historical data, more accurate data set for follow-up becomes the main problem of dynamic mining technology to solve . In e-commerce activities, all the acts of customers will be marked on the server, which would normally be stored as the format of standard common log file or standard combined log file in the cookie. If we called the data in the log files as dynamic data source (Dynamc Data Soumes), Marking the data in DDS as dm (m for the data symbol number, $i \in z$). Before the data analysis, setting a current time point as T, and

S. Lin and X. Huang (Eds.): CSEE 2011, Part II, CCIS 215, pp. 477–481, 2011.

making collection of the data before T, then organized them into an effective transaction database which will provide a reliable database to ensure dynamic data mining. In general, we can use the following method for static data collection, while taking advantage of the dynamic sliding window processing.

### 1.1 The Analysis of Association Rules

If we do not consider the support and confidence of association rule, there will be endless association rules.But in fact, the users only have the interest on certain support and confidence of association rules. In order to find meaningful association rules, we need to set two threshold: minimum support and minimum confidence. The analysis of association rule is the method which is used to identify the database in the transaction from the given minimum support and minimum confidence of the users.

For example:When a customer visits a web page, usually through the interest term to find relevant links to access, if we make a mining analysis on the basis of some strategy of the association data and get the rates of the statistics of the pages that customers access and the pages that the customers have more interest, we can well organized a site, and implement effective marketing strategies. To this end, we should identify the relationship of the nodes on the page, the link between nodes, interest and interest nodes described and find the relationship between them. Define page nodes as two-tuples (t, s), a link of the page can be represented as triples (nod e, L, t _ no e), the node is the source page node, L is the chain node of source node and the target page page node , t _ node for the target page node. Nodes of interest and the interest in the relationship between nodes are defined as follows:

Set interest set as

$$I=\{i_1,i_2, . . . i_n\} \text{ ,} \tag{1}$$

the $i_1,i_2, . . . I_{n,}$ are interest entries,then:

**Definition 1:** Defining interest nodes as two-tuples ( i ,r i g h t ), N( i ) for short, the i $\in I$, right for the weight of interest nodes.

**Definition 2:** Defined the relationship of interest nodes as interest association rules, showing as triples ( N $(i_s)$ , r i g h t , N ( $i_l$ ) ), Rule (N( $i_s$) , N( $i_l$ ) ) for short, right is for the weight of interest node N( $i_s$) goes to N ( $i_l$ ), $0 < r i g h t < 1$ .

Obviously when the weight, right> 0.5, the chance of client chooses to view the node is greater.As there are a lot of interest in association rules, so all the interest can be expressed as a set of association rules R. Finding out the valuable rules as knowledge in set R, when customers visit again, we can find the related interest entries from the interest association rules. In this way, we may find link that the customers will visit.

### 1.2 Clustering Acquisition Method

Clustering Clustering is divided into groups of customers and Web page clustering. To understand the groups of customer clustering, we first must give a correct description of browsing behavior of customers. Any server log can be expressed as the following:

$$L = (IP\ , ID, URL\ , browsetime\ ) \tag{2}$$

The IP, ID, URL, browsetime are IP address, the customer identity, customer requested URL and the corresponding browsing time.Therefore, the customer's IP address, the customer ID and the time of the pages that customers browsed made up of the browsing behavior of customers. Therefore, the information above we can construct a incidence matrix of the customer who visit a page within a period of time. We can set up a incidence matrix Mmxn of n customers visit m URL and its related URL—ID.

$$M_{mxn=} \begin{Bmatrix} m11 & m12 & m13 & m14 \\ m21 & m22 & m23 & m24 \\ m31 & m32 & m33 & m34 \\ m41 & m42 & m43 & m44 \end{Bmatrix} \tag{3}$$

Here, m is the number of visiting the page of a customer within a period of time.By the matrix Mmxn, each column shows the vector representation of a client's URL in the site visits. As it is the individual subgraph of the customers access to the site,similar customer base has a similar subgraph . We can check whether the customer trades or not by the registration.If the customer made a visit and did not deal with the distributor,column vector is the number of not traded.then the similar customer base for potential customers; otherwise, for the trading customer base.

you can Hamming distance. That is, for M [i, j]> 0, so

$$M\ [i, j] = 1. \tag{4}$$

Then calculate the Hamming distance between vectors, Hamming distance smaller, the higher their similarity. Hamming distance based on the actual situation of setting threshold value, then the customer cluster. Cluster analysis is a very important e-commerce, through clustering a group of customers with similar browsing behavior and to analyze the common characteristics of customers, to better help users understand their own e-commerce customers to provide more appropriate services.

## 2 Dynamic Data Acquisition Technology (SW)

Sliding Window (SW) has application in many areas such as time series data mining, computer network communications and mobile data flow data mining. This article also draws on the technology to enable data obtained dynamically. Of all things is in time and space, production, development and destruction of the thing are time-linked. Therefore, for dynamic data obtained dynamically to a data source, measure of the sliding window can be determined using the time. While the division of data window is based on time paragraph, in order to get data from the DDS quickly and timely, if the data's temporal correlation is not very strong or the data is a discrete data, it often stored through a database, This requires the database to store data stored in DDS generation time, so that we can use database query language fast retrieved to meet requirements of Web data. For sliding window defined following: defined 1: in DDS, in accordance with the data di (i for data identifies, i $\in$ z+) of generated time into window size for t (t for time paragraph, and t= $\Upsilon$ $\delta$ (n $\in$ z+, and $\Upsilon \geq 1$)) of data paragraph (k $\in$ n), each data paragraph for a data window, $\Upsilon$ for data threshold value. Defined 2: for positive $\omega$,

$$\omega = n\ \Upsilon \tag{5}$$

(n $\in$ z+, and n $\geq$ 1),

a moments t has data paragraph set

$$D=\{D_1,D_2, \text{sth}, D\ n\} \tag{6}$$

fell to window size for $\omega$ of window in SW , the window forward mobile s (s $\in$ z+, 1 $\leq$ s $\leq$ n) a data window size of location each isolation time $\Upsilon$, said window SW for sliding window

## 3 Dynamic Data Processing

Dynamic Data Processing (Dynamic Data Processing, DDP) is compared with the traditional data processing data mining purposes. Traditional data mining data is fixed for a particular data set; the use of dynamic data mining, in order to find the most recent, interesting data, such as the URL and so many visits in the data processing is also required to dynamically The real-time data processing. Dynamic data processing including the elimination of noise, missing data handling, conversion, feature extraction and data reduction processing. Traditional data processing can be pre-data transformation, protocol and other methods, mainly due to the dynamic process of dealing with dynamic data acquisition of dynamic real-time data coming. As data from the dynamic data acquisition time period are based on real-time data, taking into account data processing, the boundary data may be ignored, combined with overlapping windows technique, dynamic data window to select a dynamic real-time data processing.

## 4 The Dynamic Implementation of Data Mining

In order to achieve dynamic target data set in the dynamic mining, you need to find the required knowledge. But what kind of reasonable mining algorithm, when started digging. At the same time the results of different types of data and different mining results of this set is only the R space analysis, the test is given a dynamic data mining algorithms (Dynamic Data MningTesting Algo rithm, DDM A). For the mining result set is R, then actual results and mining using the average error between the results and set the maximum error and restart the mining to compare the time to determine. Algorithm is as follows:

Input: mining result set

$$R = (r_1, r_2, \ldots, r_i), \tag{7}$$

the actual results of

$$Z = (z_1, z_2, ..., z_i), \tag{8}$$

maximum error δmax, restart the mining time t. Output: Ri +1,
1) Calculate the mining error:

$$\delta = 1/t\sum_{i=1}^{t} | z_i - r_i | \tag{9}$$

2) if (δ ≤ δmax) then start the next mining, the output Ri +1, time t;
else adjust the mining process, start digging, back to 1);
3) if

$$t = \theta t \tag{10}$$

then start the next excavation, back to 1);

Algorithm for the resumption of mining time θ threshold set by the user of its value equal to the current data should generally be greater than the threshold 6. Time too small will cause mining idling, that is able to detect a large number of mining process more innovative knowledge and rules; time had caused the General Assembly could not have timely access to new knowledge.

## 5 Conclusion

In this article, traditional data mining can not meet the dynamic requirements of the data source data analysis presented in the form of dynamic data mining, given the method of dynamic data mining; in order to reflect and adapt to the dynamic data mining, dynamic data acquisition for the process dynamic data processing, data mining have been analyzed to achieve a dynamic and gives the realization of ideas; in the dynamic data acquisition in the smooth sliding window data collection, and through the dynamic data window with data collection process dynamic sent dynamic real-time data; In the data mining process, through a combination of data sets to the dynamic evaluation of the follow-up mining results, given the dynamic spatial data mining test algorithm.

## References


[1] Lin, J., Liu, M.: Mining and OLAP Theory and Practice THU (2003)
[2] Wang, D.: Time series data mining research and application ZJU (2004)
[3] Teng, M., Xiong, Z.: Zhang Yufang Dynamic data mining research. Computer Application (2008)
[4] Si, K., Mao, Y.: A new kind of based on data stream data model Computer technology and development (2007)
[5] Tang J.: Electronic commerce of data mining Web. Nanchang University Journal (2003)
[6] Wu, H., Zhao, J., Peng, L., et al.: Based on the fluctuation characteristics of sequential data mining. Control and decision (2007)

# Translating Artifact-Based Business Process Model to BPEL

Dong Li[1] and QinMing Wu[2]

[1] School of Software Engineering, South China University of Technology,
Guangzhou 510006, China
cslidong@scut.edu.cn
[2] School of Computer Science and Engineering, South China University of Technology,
Guangzhou 510006, China
wqm869@qq.com

**Abstract.** The artifact-based modeling can provide flexible and intuitive descriptions to the conceptual workflow. In the physical execution level, BPEL has been widely supported by the industrial sector as a language. Building the bridge between them will help business people to define and improve process model efficiently. In this paper, we present a method to translate artifact-based business process model to BPEL. The Read/Write services of artifact are proposed to solve data exchange. And some mapping rules are proposed to convert the flow type to the control structure. After that, the algorithm of translating artifact-based business process model to BPEL is proposed and an experimental example is provided.

**Keywords:** Artifact, Business Process, BPEL.

## 1 Introduction

How to quickly design and modify business process to make business process management flexible, which can rapidly response to the change of business environment, has become more and more important. Artifact-based business process modeling is gaining more and more attention from the academy and the industry, whose advantage is that it can make the business data an important role in the process of design.

BPEL has become the standard of the execution process definition [1]. But BPEL code is too complicated and cumbersome. So BPEL is not suitable for business people to design business process directly. In contrast, BPEL is more suitable as a language in the physical execution level.

From the perspective of business people, they are more willing to use the graphical modeling tool [2]. They need the modeling method which can describe business needs clearly and completely, improve the flexibility of business processes. So, we try to find a way translate a graphical conceptual business process model to an executable process model. To this end, we developed a SVG based business process modeling tool for

S. Lin and X. Huang (Eds.): CSEE 2011, Part II, CCIS 215, pp. 482–489, 2011.

artifact-based process design. The advantage of SVG is that it can be easily used for graphical design and display of the process model. Additionally, SVG is of a XML-based format, which can be conveniently converted into BPEL which is based on XML. The core of our approach is to adopt SVG to describe the artifact-based business process model, then translates the SVG file to a BPEL program, so we can realize translating a conceptual model to a executable physical model. Also this approach converts a data-centric business process model to task-centric business process model.

The remainder of this paper is organized as follows. Section 2 surveys the related work in the area. Section 3 presents the elements of artifact-based business process model. Section 4 describes the conversion rules and algorithm of translating artifact based business process model to BPEL, Section 5 provides the experimental example and Section 6 concludes the paper.

## 2 Related Work

[3] introduced the concept of business artifact and demonstrated that the artifact is a familiar concept for business people. [4] summarized 9 kinds of operation in the operating model, and converted them into a model based on petri-net for analyzing and verifying. [5] established BALSA basic model and used the ER diagram describing the structure of artifact. The paper described the life cycle of artifact by a finite state machine, and used the ECA rules to describe the relationship between artifact and service. Some syntax rules were proposed in [6] for verifying the syntactical correctness of the conceptual workflow. And the paper also proposed several optimization rules for optimizing the conceptual workflow, which can simplify the conceptual workflow thus will improve the performance of the execution of the workflow future.

Model transformation is the key for business process modeling, so some researches focused on converting the high-level abstract models to BPEL such as BPMN to BPEL. [7] pointed out that BPMN creates a standardized bridge for the gap between the business process design and process implementation. The paper presented a simple, yet instructive example of how a BPMN diagram can be used to generate a BPEL process. [8] demonstrated that a method for end-to-end development of process-oriented systems and translating BPMN models to BPEL definitions for subsequent refinement.

## 3 Artifact-Based Business Process Model

An artifact is a concrete, identifiable, self-describing chunk of information that can be used by a business person to actually run a business [3]. Artifact-based business process model creates a flexible presentation method for business people to manage and analyze the business process.

Our laboratory develops a tool supporting artifact-based process modeling, which borrowed many ideas of modeling from [9]. The modeling units are as the following:

1) Artifact: It is the set of business records which represents the information content of the business.

2) Service: It constructs localized functions, which operates on artifacts.
3) Repository: It is used to save artifacts. Repository provides the method of archiving artifacts, including artifacts which are processing and completed.
4) Connector: It represents a service reads artifacts from the repository or saves artifacts into the repository.
5) Event: Events are created by outside requests or services, and are received by services.
6) Message: It is sent or received along with the Event.

# 4 Conversion of the Artifact Based Model to BPEL

## 4.1 Read/Write Services of Artifact

In the artifact-based business process, services transmit data mainly through the form of artifacts. As shown in Figure1, we design two internal common Web services for artifacts, one service is named Read, which reads artifacts from the database according to conditions; the other service is named Write, which saves artifacts to the repository according to the rules.

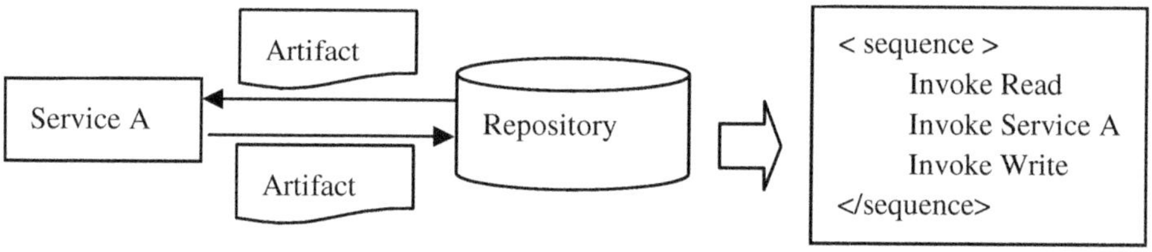


**Fig. 1.** A service reads and writes artifacts through the services of artifact transmission

BPEL itself doesn't provide logging mechanism, but through the two common services we can record log information, including web services, information, exchanging of data.

## 4.2 Conversion Rules between Control Structures

There are mainly sequence、flow、switch、pick and while in BPEL control structures, and the conversion rules between our model and BPEL are as follows:

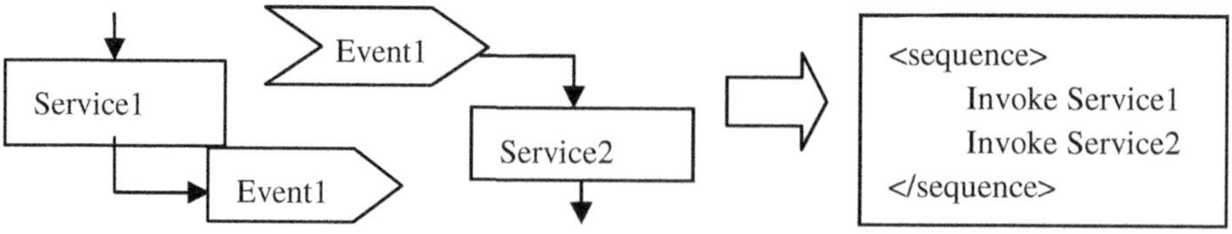


**Fig. 2.** Sequence activity

1) In Figure 2, the sequence activity is used to define a collection of activities which are executed sequentially in lexical order.

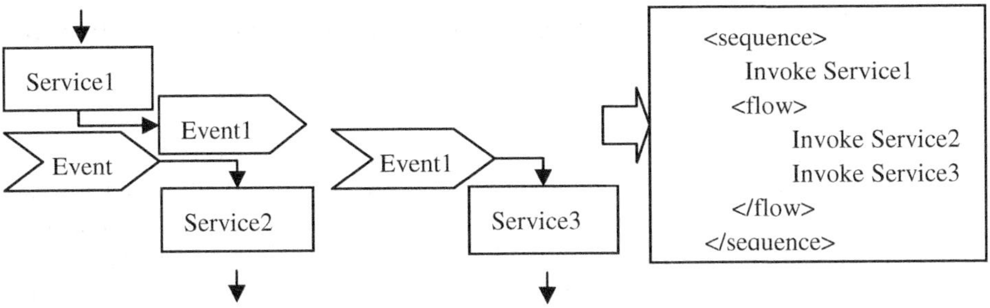


**Fig. 3.** Flow activity

2) In Figure 3, the flow activity is used to define a collection of activities which are executed parallelly.

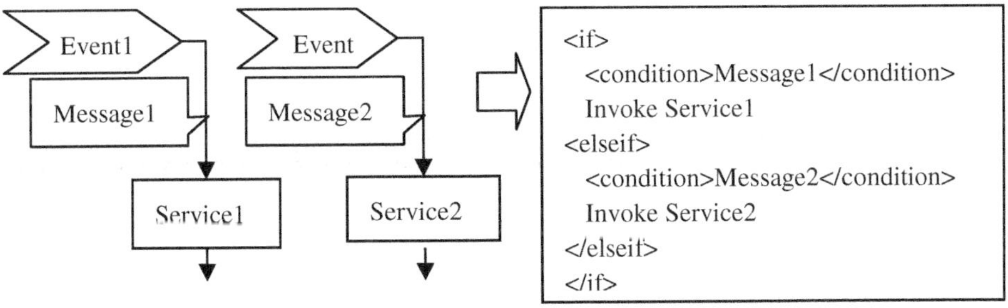


**Fig. 4.** If-else activity

3) In Figure 4, the if-else activity allows you to select exactly one branch of the activity from a given set of choices.

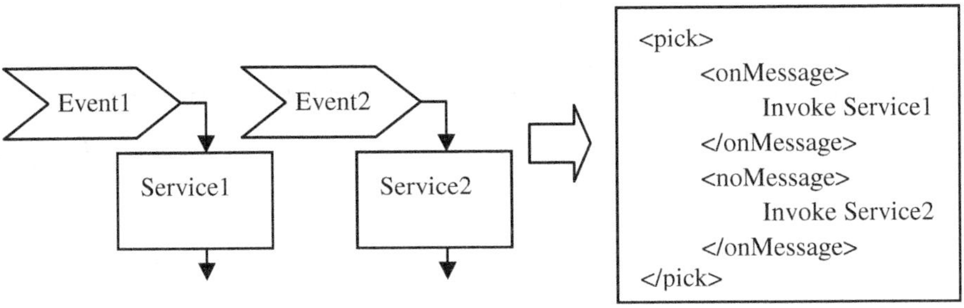


**Fig. 5.** Pick activity

4) In Figure 5, the pick activity is used to define a collection of activities which are triggered by some of multiple suitable messages.

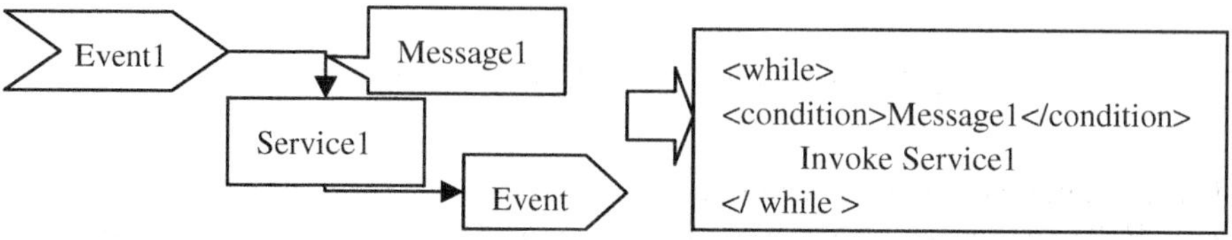


**Fig. 6.** While activity

5) In Figure 6, the while activity allows you to repeatedly execute the child activity as long as a given condition evaluates to true.

### 4.3 Process of Conversion

Generally, the conversion to BPEL needs scanning process, parsing process and generating BPEL codes, the specific process is described as follows:

1) The program scans the model file first time, searches service components, and reads the related attributes. It retrieves the corresponding WSDL file of the service, reads the properties and saves them to the service hash table.
2) It scans the model file second time, searches connector components. If a connector's direction is from a service component to a repository, then the write service is added after the service; if a connector's direction is from a repository to a service, then a read service is added. And other components' attributes are read and saved to their corresponding component hash table.
3) The program analyzes the event components to determine the order of services.
4) According to the conversion rules in Section 4.2, the program chooses BPEL structural components to parse the process.
5) And finally, the program generates BPEL code.

## 5 An Example

We provide a practical scenario of a flight booking process. John wants to book two one-way tickets from San Francisco to New York City. He submits his traveling plan to a travel agency. On receiving his request, the travel agency searches the available flight seats and sends the optional choices to John; John selects the best ones and sends a confirmation with other further information to continue the flight booking. The travel agency records the further information and then contacts the credit card payment center to check balance and to pay for the tickets.

The detailed design procedure is omitted and the diagram of the business operational model of the flight booking process is shown as Figure 7.

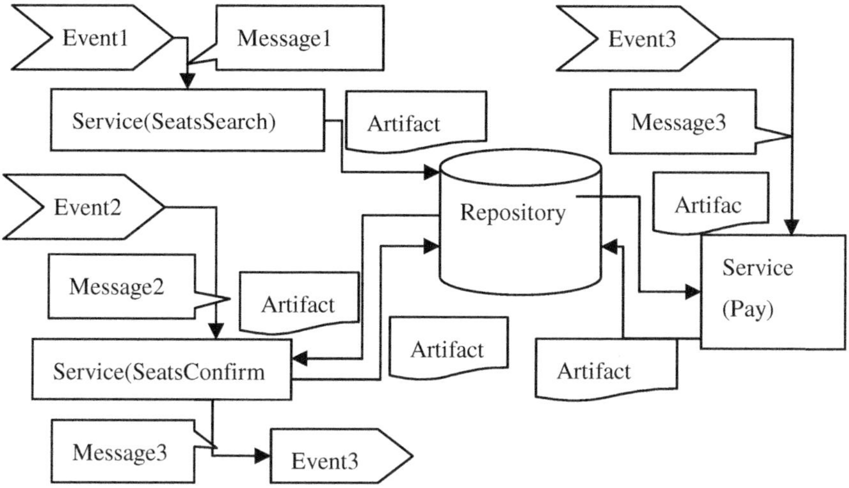


**Fig. 7.** Graphical workflow diagram of flight booking process

Due to space limitations, here only shows the main BPEL code after the translation, as follows:

```
<process>
  <sequence>
    <pick>
          <onMessage   operation="SeatsSearchOperation" >
              <sequence>
                  <invoke operation="SeatsSearch"/>
                  <invoke operation="Write"/>
              </sequence>
          </onMessage>
          <onMessage   operation="SeatsConfirmOperation" >
              <sequence>
                  <invoke operation="Read"/>
                  <invoke operation="SeatsConfirm"/>
                  <invoke operation="Write"/>
                  <invoke operation="Read"/>
                  <invoke operation="Pay"/>
                  <invoke operation="Write"/>
              </sequence>
          </onMessage>
      </pick>
  </sequence>
</process>
```

We run the instance of the simulation in the BPEL engine of *NetBeans*, and the log fragment generated is as follows:


```
<?xml version="1.0" encoding="UTF-8" standalone="no"?>
<WorkflowLog xmlns:xsi="http://www.w3.org/2001/XMLSchema-instance"
description="Booking log"
xsi:noNamespaceSchemaLocation="http://is.tm.tue.nl/research/processmining/WorkflowLog.xsd">
  <Source program="Booking"/>
    <Process discription="Booking process" id="DEFAULT">
      <ProcessInstance discription="Booking process instance" id="1">
        <AuditTrailEntry>
          <WorkflowModelElement>SeatsSearch</WorkflowModelElement>
          <EventType>complete</EventType>
          <Timestamp>2011-05-04T19:31:09</Timestamp>
        </AuditTrailEntry>
          ...
```


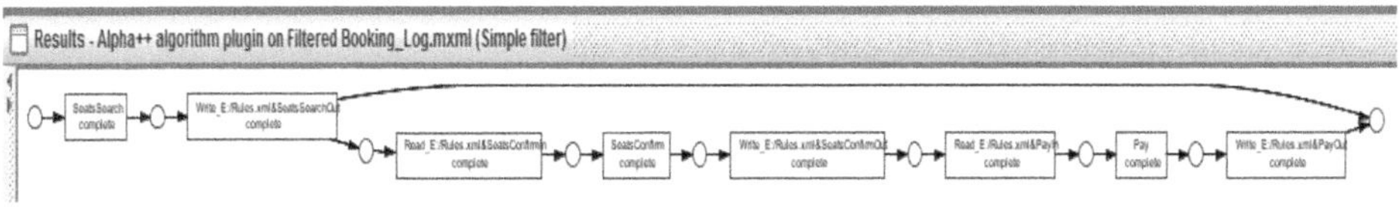


**Fig. 8.** The Petri-net graph

We analyze the log through the ProM, which is a generic open-source tool for implementing process mining tools, and we get the Petri-net graph showed in Figure 8. The results show our translation is correct and reasonable.

## 6 Conclusion

Artifact-based business processes, which focus on changes of critical data in the business process, respond to real world behavior more and are more familiar to business people. The algorithm, which translates artifact-based business processes model to executable BPEL code, can improve the efficiency of runtime simulation and rapid deployment for business process. How to automatically find the most suitable services, match and optimize is the next step in the research.

**Acknowledgments.** This research was supported partially by the State Natural Sciences Foundation Projects under grant number 71090403 and Ministry of Education of returned overseas students to start research and fund projects under grant number B7110020.

## References


[1] Wohed, P., Aalst, W., Dumas, M.: Analysis of Web Services Composition Languages: The Case of BPEL4WS. In: Song, I.-Y., Liddle, S.W., Ling, T.-W., Scheuermann, P. (eds.) ER 2003. LNCS, vol. 2813, pp. 200–215. Springer, Heidelberg (2003)

[2] Georgakopoulos, D., Hornick, M., Sheth, A.: An Overview of Workflow Management: From Process Modeling to Workflow Automation Infrastructure. Distributed and Parallel Databases 28(3), 119–153 (1995)

[3] Nigam, A., Caswell, N.S.: Business Artifacts: An Approach to Operational Specification. IBM Systems Journal 42(3), 428–445 (2003)

[4] Liu, R., Bhattacharya, K., Wu, F.Y.: Modeling Business Contexture and Behavior Using Business Artifacts. In: Proc. of International Conference on Advanced Information Systems Engineering, Trondheim, Norway (2007)

[5] Bhattacharya, K., Gerede, C.E., Hull, R., et al.: Towards Formal Analysis of Artifact-centric Business Process Models. In: Proc. of International Conference on Business Process Management, Brisbane, Australia (2007)

[6] Li, D.: Conceptual Workflow Verification and Optimization for Artifact-centric Business Process. International Forum on Information Technology and Application (2009)

[7] White, S.: Using BPMN to model a BPEL process. BP Trends 3(3), 1218 (2005)

[8] Ouyang, C., Marlon, D., Wm, P.A., et al.: From business process models to process-oriented software systems: the BPMN to BPEL way (EB /OL) (2006210) (2007211), http://eprints.qut.edu.au/archive/00005266/

[9] Bhattacharya, K., Hull, R., Su, J.: A data-centric design methodology for business processes (the draft version) (2008)

# A Personalized E-Commerce Recommendation Method Based on Case-Based Reasoning

Huailiang Shen

Zhejiang Business Technology Institute, Ningbo 315012, China
huailiang_shen@163.com

**Abstract.** Web recommender prediction is a popular personalized service on the Internet and has attracted much research attention. In recent years, personalized e-commerce recommendation has received increasing attention. The foundation problem of personalized recommendation system is to track users' interests and their changes. There are mainly two solutions to solve the problem. One is content-based filtering and the other is collaborative filtering. Collaborative filtering technique identifies other users that have shown similar preferences to the items and recommends what they have liked to the target user. Collaborative filtering has been the prevalent recommendation approach. However, the applicability of collaborative filtering is limited due to the sparsity problem, which refers to a situation that user-item rating data are lacking or are inadequate. In order to alleviate the sparsity, a personalized e-commerce recommendation method based on case-based reasoning is presented. This approach produces prediction using case-based reasoning and produces recommendation based on collaborative filtering.

**Keywords:** personalized, recommendation method, e-commerce, collaborative filtering, case-based reasoning.

## 1 Introduction

With the development of Internet and network technology, people need to take a lot of time and effort to get information they want from the Internet. Traditionally, a variety method of information retrieval can be used to access through the home address. People can use multiple search engine queries. Companies can take advantage of specialized databases to query the information [1,2,3].

Currently, the resources on the Internet increasing at an exponential rate, people from this vast ocean of information to find information of interest is undoubtedly and very difficult. Traditional search engines such as information retrieval technologies can meet certain requirements of the people needs. But because of its keyword matching by the information retrieval, there is still a lot of hard manual processing of information that the target user does not care. This one appears every time a large amount of online information resources, on the one hand and the problems caused by information overload, and the accumulation rate constant to reach beyond the user's manual processing.

S. Lin and X. Huang (Eds.): CSEE 2011, Part II, CCIS 215, pp. 490–495, 2011.

Information filtering technique is proposed for this problem. Collaborative filtering has been the prevalent recommendation method [4,5,6]. However, the applicability of collaborative filtering is limited due to the sparsity problem, which refers to a situation that user-item rating data are lacking. In this paper, a personalized e-commerce recommendation method based on case-based reasoning is presented. This approach produces prediction using case-based reasoning and produces recommendation based on collaborative filtering.

# 2 Prediction Using Case-Based Reasoning

## 2.1 Case-Based Reasoning

Firstly, there is a large number of pre-existing and mature cases in the case-base reasoning. Then, based on the customer needs and characteristics, retrieve a set of related cases in accordance with the similarity from the case database. Then, make the necessary modifications, combinations and treatment. Finally, the formation of different customers can be recommended. The basic work flow of CBR [7,8,9] in the recommended system is shown as Figure 1.

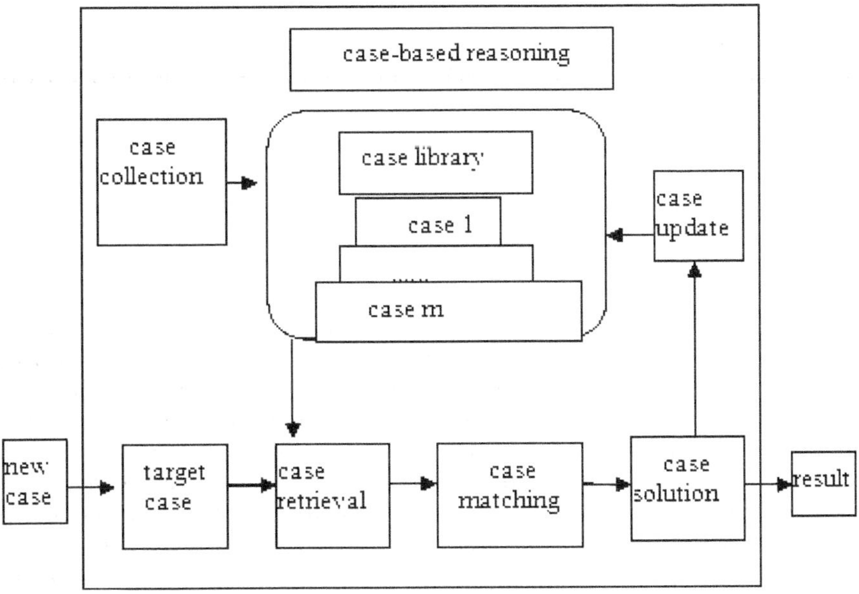


**Fig. 1.** Case-based reasoning

CBR systems work in the specific recommendations includes the following steps:

(1) Case collection. Collect the typical case from the customers' personal characteristics, needs and consumer behavior information.

(2) Case retrieval. According to customer demand information, case retrieval dynamically adjusts the weights of relevant features and characteristics of properties to provide recommendations related to the results.

(3) Case match. Case similarity computation, and sorting through the results of similarity measure, retrieved from the case-based matching cases.

(4) Case solution. On the best program for solving similar cases, make changes and adjustments to meet the target case solution.

(5) Case update. Case of the adjusted target case the recommended solution as the goal added to the case base, in order to accumulate experience to solve new problems in the future.

## 2.2 Prediction

Case-based reasoning is an important part of establishing indexing mechanism similar cases. Case retrieval is the essence of the problem cases and the similarity between target case match. Similar to the match can be quantified similarity, the current similarity measures are significantly dependent on the applications, but they are for the same purposes, means that a case of a characteristic value and with it the demand characteristics of the cases related to the value of the similarity cased.

The algorithm is as follows:

(1) Case database: Case database = (case 1, case 2 ,..., case n) = (C1, C2, C3, ···, Cn)

(2) The distance of two cases

Ca = {Pa1, Pa2, ···,Pan}

Cb = {Pb1, Pb2, ···,Pbn}

The distance is:

$$dis(C_a, C_b) = \sum_{i=1}^{n} w_i \times |P_{ai} - P_{bi}|$$

(3) The similarity of two cased.

The similarity is:

$$sim(C_a, C_b) = \frac{1}{1 + dis(C_a, C_b)}$$

(4)The prediction

Select the max similarity of the target case as the prediction case.

# 3 Producing Recommendation Based on Collaborative Filtering

Collaborative filtering is based on an assumption: if the user ratings on some items were similar, then they score on other items is relatively similar [10,11]. Implementation of collaborative filtering usually has two steps: firstly, access to user information, that access to certain information items the user evaluation; and then analyze and predict the similarity between users of the specific user preference information.

### 3.1 Obtaining Data

We obtain the user-item data mainly through user evaluation of the given information. Evaluation is divided into two kinds of explicit and implicit. Explicit evaluation requires the user to consciously express their level of acceptance of a message, usually expressed with an integer value like different levels. The implicit evaluation is to obtain information from the user behavior.

### 3.2 Calculating Similarity

There are some similarity algorithms that have been used in collaborative filtering recommendation algorithm [12,13].

Pearson's correlation measures the linear correlation between two vectors of ratings. The formula is as following:

$$sim(i,j)=\frac{\sum_{t\in I_{ij}}(R_{it}-A_i)(R_{jt}-A_j)}{\sqrt{\sum_{t\in I_{ij}}(R_{ic}-A_i)^2\sum_{t\in I_{ij}}(R_{jt}-A_j)^2}}$$

The cosine measure looks at the angle between two vectors of ratings. The formula is as following:

$$sim(i,j)=\frac{\sum_{c=1}^{n}R_{ic}R_{jc}}{\sqrt{\sum_{c=1}^{n}R_{ic}^2\sum_{c=1}^{n}R_{jc}^2}}$$

The adjusted cosine is used in some collaborative filtering methods for similarity among users where the difference in each user's use of the rating scale is taken into account. The formula is as following:

$$sim(i,j)=\frac{\sum_{t\in I_{ij}}(R_{iu}-A_u)(R_{ju}-A_u)}{\sqrt{\sum_{t\in I_{ij}}(R_{iu}-A_u)^2\times\sum_{t\in I_{ij}}(R_{ju}-A_u)^2}}$$

### 3.3 Producing Recommendation

User interest degree of prediction can be calculated as follows:

$$P_{u,i}=\overline{R_u}+\frac{\sum_{m=1}^{n}\left(R_{m,t}-\overline{R_m}\right)\times sim(u,m)}{\sum_{m=1}^{n}sim(u,m)}$$

## 4 Conclusions

Web recommender prediction is a popular personalized service on the Internet and has attracted much research attention. In recent years, personalized e-commerce recommendation has received increasing attention. The foundation problem of personalized recommendation system is to track users' interests and their changes. There are mainly two solutions to solve the problem. One is content-based filtering and the other is collaborative filtering. Collaborative filtering technique identifies other users that have shown similar preferences to the items and recommends what they have liked to the target user.

Collaborative filtering has been the prevalent recommendation approach. However, the applicability of collaborative filtering is limited due to the sparsity problem, which refers to a situation that user-item rating data are lacking or are insufficient. In order to alleviate the sparsity, in this paper, a personalized e-commerce recommendation method based on case-based reasoning is presented. This approach produces prediction using case-based reasoning and produces recommendation based on collaborative filtering.

## References

1. Lee, J.-S., Jun, C.-H., Lee, J., Kim, S.: Classification-based collaborative filtering using market basket data. Expert Systems with Applications 29, 700–704 (2005)
2. Ahn, H.J.: A new similarity measure for collaborative filtering to alleviate the new user cold-starting problem. Information Sciences 178, 37–51 (2008)
3. Lekakos, G., Giaglis, G.M.: Improving the prediction accuracy of recommendation algorithms: Approaches anchored on human factors. Interacting with Computers 18, 410–431 (2006)
4. Papagelis, M., Plexousakis, D.: Qualitative analysis of user-based and item-based prediction algorithms for recommendation agents. Engineering Application of Artificial Intelligence 18, 781–789 (2005)
5. Gong, S.: A Personalized Recommendation Algorithm on Integration of Item Semantic Similarity and Item Rating Similarity. Journal of Computers 6(5), 1047–1054 (2011)
6. Gong, S.: Privacy-preserving Collaborative Filtering based on Randomized Perturbation Techniques and Secure Multiparty Computation. International Journal of Advancements in Computing Technology 3(4), 89–99 (2011)
7. Jiang, L., Xu, B., Xi, J.: CBR filtering algorithm and intelligent information recommending system. J. Tsinghua Univ. (Sci. & Tech.) 46(S1), 1074–1077 (2006)
8. Chedrawy, Z., Abidi, S.S.R.: An Intelligent Knowledge Sharing Strategy Featuring Item-Based Collaborative Filtering and Case Based Reasoning. In: 5th International Conference on Intelligent Systems Design and Applications (ISDA 2005), pp. 67–72 (2005)
9. Roh, T.H., Oh, K.J., Han, I.: The collaborative filtering recommendation based on SOM cluster-indexing CBR. Expert System with Application 25, 413–423 (2003)
10. Gong, S.: Employing User Attribute and Item Attribute to Enhance the Collaborative Filtering Recommendation. Journal of Software 4(8), 883–890 (2009)

11. Gong, S.: A Collaborative Filtering Recommendation Algorithm Based on User Clustering and Item Clustering. Journal of Software 5(7), 745–752 (2010)
12. Gong, S.: An Efficient Collaborative Recommendation Algorithm Based on Item Clustering. In: Luo, Q. (ed.) Advances in Wireless Networks and Information Systems. LNEE, vol. 72, pp. 381–387. Springer, Heidelberg (2010)
13. Huang, q.-h., Ouyang, w.-m.: Fuzzy collaborative filtering with multiple agents. Journal of Shanghai University (English Edition) 11(3), 290–295 (2007)

# The Exact Solutions of Variable Coefficient Auxiliary High Order Generalized KdV Equation

Bo Lu, Yuzhen Chen, and Qingshan Zhang

Department of mathematics, Henan Institute of Science and Technology
Xinxiang 453003, P.R. China
cheersnow@163.com
cyz_dlut@126.com
qingshan11@yeah.net

**Abstract.** Selecting appropriate trial functions, the nonlinear variable coefficient PDE can be convert transformed to algebraic equations. The exact solutions to a class of nonlinear variable coefficient partial differential equations are successfully derived. In this paper the second type KdV equation with variable coefficients is studied, and their exact solutions are obtain. The method is obviously suitable to solving other nonlinear partial differential equations with variable coefficients of the solution.

**Keywords:** the trial function, the Variable Coefficient KdV equation, the Hopf - Cole transformation, exact solutions.

## 1 Introduction

As we known, the KdV equation as the basic model of nonlinear wave theory, is to become one of the basic equation in the mathematical physics, the research on it is also very active. Many methods to solve it is given, such as the Anti scattering method[1], the Hirota's, the Back-lund's and the Darboux's transformation[2], The homogeneous balance method[3], the Hyperbolic function method[4], the directly reduction method[5], and the Jacobi elliptic function method[6]. The methods are used to solve the equations with constant coefficients, but variable coefficient differential equations can describe the physical phenomenon actually. So the variable coefficient PDEs are more important. In this paper, we use the Hopf - Cole transformation and trial function method to study the variable coefficient auxiliary high order generalized KdV Equation. And the exact solution of it is given.

## 2 Exact Solution of Variable Coefficient KdV Equation

We consider the following the variable coefficient KdV Equation:

$$u_t + [\sigma(t) + \mu(t)x]u_x + \alpha(t)\, uu_x + \beta(t)u_{xxx} + \gamma(t)u = 0. \qquad (1)$$

S. Lin and X. Huang (Eds.): CSEE 2011, Part II, CCIS 215, pp. 496–499, 2011.

The Eq.(1) includes the Generalized KdV Equation

$$u_t + [\alpha(t) + \beta(t)x]u_x - 3c\gamma(t)uu_x + \lambda(t)u_{xxx} + 2\beta(t)u = 0. \quad (2)$$

Let, $y = \frac{1}{1+e^{\xi}}$, then, $y' = y^2 - y$, and we have:

$$u = \frac{\partial v}{\partial x} = A_0(t) - A(t)y^2 f(t), \quad (3)$$

Take the partial derivative with respect to x and t in Eq.(3),and take them in Eq.(1), we arrive at:

$$\begin{aligned}&(2\alpha A^2 f^3 + 24\beta Af^4)y^5 - (5\alpha A^2 f^3 + 60\beta Af^4)y^4 + (2Aff_t x +\\ &2Afg_t + 2Af^2\sigma + 2\alpha AA_0 f^2 + 2Af^2\mu x + 4\alpha A^2 f^3 + 50\beta Af^4)y^3 +\\ &(A_t f + Af - 3Aff_t x - 3Afg_t - 3Af^2\sigma - 3Af^2\mu x - 3\alpha Af^2 A_0 - \alpha A^2 f^3\\ &-15\beta Af^4 + \gamma Af)y^2 + (-A_t f - Af + Aff_t x + Afg_t + Af^2\sigma + Af^2\mu x\\ &+\alpha AA_0 f^2 + \beta Af^4 - \gamma Af)y + A_0 t + \gamma A_0 = 0.\end{aligned} \quad (4)$$

The Eq.(4) is identically equal for any $y$, we can get the following system of algebraic equations

$$\begin{cases} A_0 t + \gamma A_0 = 0, \\ -A_t f - Af + Aff_t x + Afg_t + af^2\sigma + af^2\mu x \\ \qquad + \alpha AA_0 f^2 + \beta Af^4 - \gamma Af = 0, \\ A_t f + Af - 3Aff_t x - 3Afg_t + -3f^2\sigma - 3af^2\mu x \\ \qquad -3\alpha AA_0 f^2 - 15\beta Af^4 + \gamma Af = 0, \\ 2Aff_t x + 2Afg_t + 2Af^2\sigma + 2\alpha AA_0 f^2 + 2Af^2\mu x \\ \qquad + 4\alpha A^2 f^3 + 50\beta Af^4 = 0, \\ 50\alpha A^2 f^3 + 60\beta Af^4 = 0, \\ 5\alpha A^2 f^3 + 6\beta Af^4 = 0, \end{cases} \quad (5)$$

Solving the system, we conclude:

$$A_0(t) = c_0 e^{-\int \gamma(t)dt}, \quad (6)$$

$$\begin{cases} A(t) = -12\dfrac{\beta(t)}{\alpha(t)} f(t), \\ f(t) = c_1 e^{-\int \mu(t)dt}, \\ g(t) = \int [-\sigma(t) f(t) - \alpha(t) A_0(t) f(t) - \beta(t) f^3(t)] dt, \end{cases}$$

where $c_0, c_1$ is the integral constants.

Thus, we get the solution of Eq.(1):

$$u = A_0(t) + A(t)(y^2 - y) f(t)$$

$$= c_0 e^{-\int \gamma(t)dt} - 12c_1^2 \frac{\beta(t)}{\alpha(t)} e^{-2\int \mu(t)dt} [\frac{1}{(1+e^{\xi})^2} - \frac{1}{1+e^{\xi}}]$$

$$= c_0 e^{-\int \gamma(t)dt} + 3c_1^2 \frac{\beta(t)}{\alpha(t)} e^{-2\int \mu(t)dt} \frac{e^{\xi}}{(1+e^{\xi})^2}$$

$$= c_0 e^{-\int \gamma(t)dt} + 3c_1^2 \frac{\beta(t)}{\alpha(t)} e^{-2\int \mu(t)dt} \sec h^2 (\frac{1}{2}\xi) \quad . \tag{7}$$

## 3 Inspection of the Exact Solution

Let

$$y = \frac{1}{\xi},$$

Thus, we get

$$y' = -y^2,$$

and

$$u = \frac{\partial v}{\partial x} = A_0(t) - A(t) y^2 f(t).$$

Similarly to section 2, put the above-mentioned equations into (2), we find that:

$$A_0(t) = c_0 e^{-2\int \beta(t)dt},$$

$$A(t) = -\frac{4}{c} f(t) = -\frac{4c_1}{c} e^{-\int \beta(t)dt},$$

$$f(t) = c_1 e^{-\int \beta(t)dt},$$

$$g(t) = \int [3c\gamma(t)A_0(t)f(t) - \alpha(t)f(t)]dt.$$

So we can get the exact solution of the Eq.(2).

$$u = e^{-2\int \beta(t)dt}[c_0 - \frac{4c_1}{c}\frac{1}{\xi^2}] \quad (8)$$

Clearly, Eq. (8) is the exact solution of the Eq. (2), and it's identical with the conclusion of the section 2.

## 4 Conclusion

In this paper, we combine the Hopf - Cole is a transform method and trial function method. We solve the variable coefficients nonlinear PDE concisely, and prove the existence indirectly. This method can be used to solve other variable coefficients nonlinear PDE, and the exact solution applies on physics and engineering.

## References

1. Ablowitz, M.J., Clarkson, P.A.: Solitons. In: Nonlinear Evolution and Inverse Scattering. Cambridge University Press, Cambridge (1991)
2. Gu, C.: The Soliton Theory and Application. Zhejiang technology Press, Hang zhou (1990)
3. Wang, M.L.: Solitary Wave Solution for Varian Boussinesq Equations. Phys. Lett. 199(A), 169–172 (1995)
4. Guo, G.: Note of Hyperbolic Function Method of Finding Solitary Wave Solutions. Physics Journals 51(6), 1159–1162 (2002)
5. Yan, Z., Zhang, H.: Nonlinear Wave Agenda of Similar Reduction With Damping Term. Physics Journals 49(11), 2113–2117 (2000)
6. Han, J.: Exact Analytical Solutions of KdV Equation. Journal of Anhui University 6(3), 44–50 (2002)
7. Shi, L., Han, J., Wu, G.: Trial Function method and Exact Solution of Generalized Variable Coefficient KdV Equation. Journal of Anhui Normal university 28(4), 418–421 (2005)

# The Design of Police-Using Information System Based on GIS

Xiaotian Wang and Hui Zhou

Dalian Neusoft Institute of Information,
Computer Science and Technology Dept.
Dalian, China
wangxiaotian@neusoft.edu.cn

**Abstract.** After deeply analyzing the applications, current development and existing problems of the police-using GIS, according to the requirement of Dalian City Public Security Bureau , this paper presents a Police Geographic Information System which was developed by the MapGIS 7.3 platform. The system was designed by three-layer architecture and adopted the Dijkstra based on genetic algorithm to solve the shortest path problem. Through using this system can provide the statistics and analysis for the police-using data to improve the use efficiency.

**Keywords:** Police-using GIS, Three-layer Architecture, Dijkstra, Genetic Algorithm.

## 1 Introduction

With the rapid development of GIS technology, it was widely used in more and more areas. The Police Geographic Information System was a kind of management system with GIS technology. It integrates the software like Geographic information management system and Database management system and the data which is about the Public Security Administration for capturing, managing, analyzing, and displaying all forms of geographically referenced information. Through this can help the public Security organs to improve the speed of reaction and the cooperative engagement capability for the crime.

After analyzing the applications, development and existing problems of the police-using GIS, according to the requirement of Dalian City Public Security Bureau , this paper proposed a Police Geographic Information System which was developed by the MapGIS 7.3 platform. It can improve the use efficiency, save the time and decrease the strength of work and has the characteristics of friendly interface, reasonable layout , easy maintenance, high reliability and convenient operation. Through using this system can ensure the accuracy of the data which the policemen inputted or checked. It can also provide the statistics and analysis for the data.

S. Lin and X. Huang (Eds.): CSEE 2011, Part II, CCIS 215, pp. 500–506, 2011.

## 2 System Architecture Design

### 2.1 Conventional System Architecture

Recently more and more software were designed by multi-layer architecture. The two-layer architecture was a very popular one which divided the whole service application of the system into two layers, one is the presentation layer and another is the data access layer. It is shown as Fig 1.

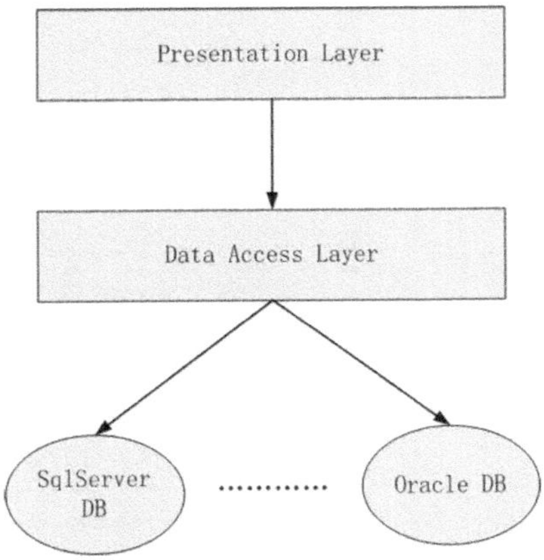


**Fig. 1.** As the figure shown, there are presentation layer and data access layer in the two-layer architecture

The presentation layer is in charge of the user interaction. Generally speaking it is the system interface that can display data, input the data , check the validation verification of the system and assure the robustness of the running program. The data access layer is to access the database to add , delete, edit or display the data .It can also verify the correctness of the data format and type and user authority. Through the above verifications can decide whether to continue the operation or not and assure the normal operation.

The two-layer architecture has some advantages like high development efficiency, simple program and high updating data speed. It is very good for the small project. However, there are also some disadvantages like it is not easy to maintain and edit the software which was designed by the two-layer architecture. It is not a good choice for the police-using GIS.

### 2.2 Three-Layer Architecture

As the Dalian Police Geographic Information System is a large project which is for the Dalian City Public Security Bureau, so the three-layer architecture is more suitable for the software than the two-layer one. In the three-layer architecture there are presentation layer, business logic layer and data access layer. With the independent presentation layer it will do not need to rewrite the whole codes for the system when there are some frequent changes for the interface. This implements the low coupling of the software design idea. The centralized management of data operation can decrease the repeated codes for the App and improve the rate of the codes reuse. The unified planning for the business logic can bring convenience for the development and maintenance, it can also make the system layer more clearly. This system adopted the architecture which extended the three-layer architecture and based on .net development platform. It is shown as Fig 2.

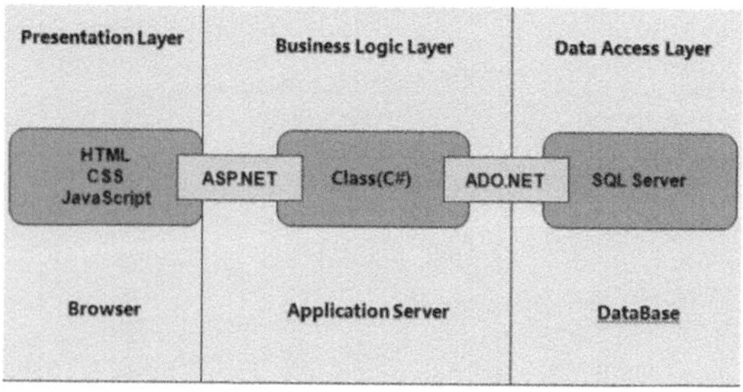


**Fig. 2.** As the figure shown, there are presentation layer, business logic layer and data access layer in the three-layer architecture

The presentation layer is the interface of the system which is in charge of the interaction between the user and system. Different users have different authorities, the normal policemen users who do not have the fundamental knowledge about GIS can only do some simple operations to assist the daily case in handling, and the users which are from the information department or the leaders have higher authorities and can do more operations about the system. The main technology for this layer is Asp.net which can separate the pages and the codes to make the presented part be independent from the control logic part.

The data access layer implements the connection and some other operations of the database.

The business logic layer is the core of the system which contains the implementation of every business logic .It can access the presentation layer and the data access layer and realize the whole system function. Such as confirming the spot where the accident happened, taking the query analysis and assistant decision, forming the thematic maps and report forms and so on. The classes that encapsulate the business logic and data access logic were written by C# language. The operation process of the Three-Layer Architecture is shown as Fig 3.

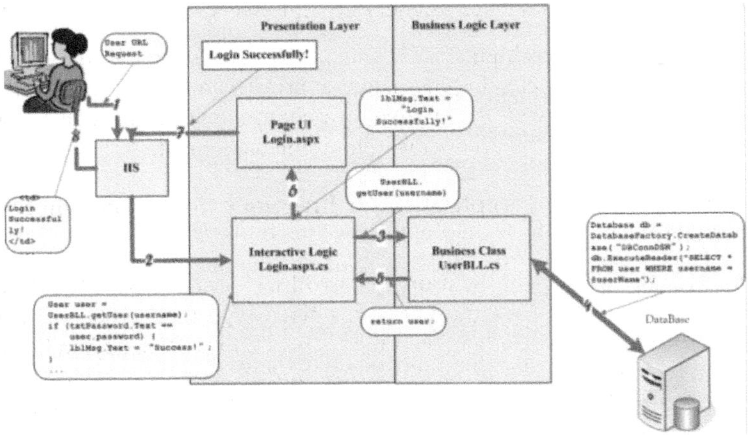


**Fig. 3.** As the figure shown, it is the operation process of the three-layer architecture

# 3 Business Process Analysis and Function Design

## 3.1 System General Structure

According to the above architecture design, the system's general functions included the file operation, data operation, map layer management, map operation, graphic editing, topology editing, GDB management, space analysis, attribute operation, three-d navigating and output. It also has some basic functions of GIS system and some special functions of Police Geographic Information System like query, statics, receive alarm, case analysis and GPS positioning. The system structure is shown as Fig 4.

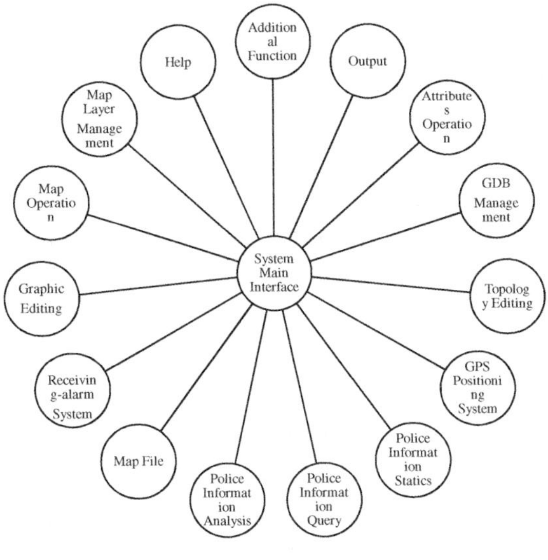


**Fig. 4.** As the figure shown, this is the system general structure

The system adopted the component GIS technology to divide the modules. The basic thought of the Component GIS is to divide the GIS into several controls and every control has different functions. It is very convenient to do final GIS application through the visual software development tool to integrate the GIS Controls or the GIS Controls and non-GIS Controls. The controls like the building blocks which can implement different functions. The Application System is composed of which controls is based on the requirement.

With the components GIS technology these functions can be divided into different modules. The operations of file and data belong to the map file module; The analysis of space and case belongs to the police information analysis module; The GDB management belongs to the map layer management module. In order to accommodate to the users the system provides the functions like the tooltip massages and shortening the distance of the mouse movement to improve the user friendliness of the interface.

## 3.2 Main Function Module Design

With the Structure design and functional partitioning, there are three special modules compared with the other police-using GIS. They are the police information analysis module, and receiving-alarm module. The following contents mainly analyze the two modules and will focus on the police information analysis module which can get the best path through adopting the Dijkstra based on genetic algorithm.

### 3.2.1 Police Information Analysis Design

This module is mainly in charge of the analysis of the police information included the analysis of space and data. It is shown as Fig 5.

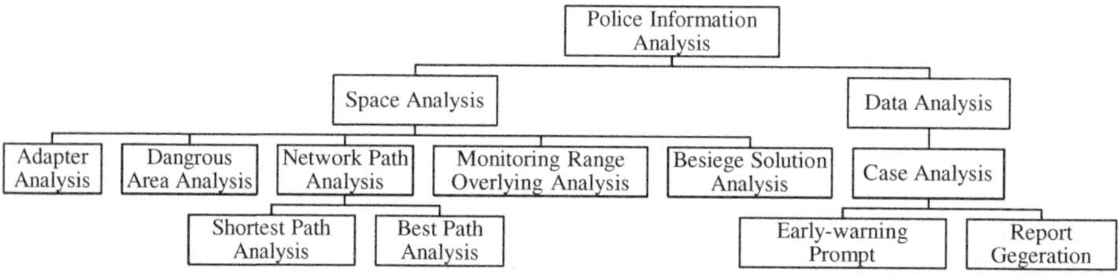


**Fig. 5.** As the figure shown, this is the Police Information Analysis module

The space analysis contains the adapter analysis, besiege solution design, dangerous area analysis, network path analysis and monitoring range overlying analysis; The data analysis contains the case analysis; The case analysis contains two sub-function which are early-warning prompt and report generation; The network path analysis contains the shortest path analysis and the best path analysis.

Among the above functions route selection is one of the most important problems for the system. Given a pair of origin and destination, there are many possible routes.

Selecting the shortest path is a hard problem, we have found that the classical Dijkstra algorithm can be used to find subjectively the shortest path, but when there are too many nodes will make the data increase sharply to lead the combination explosion. So this paper adopted the Dijkstra based on genetic algorithm to solve the problem.

Assume that all of the positions in the map form a list as P and assign every position a number from 1~n, make the array as P too. It is :

$$\begin{array}{c} P=(v1,v2,v3,\ldots,v_n) \\ | \quad | \quad | \qquad | \\ P=(1,\ 2,\ 3,\ldots,\quad n) \end{array} \tag{1}$$

Use the coded string : 12345678…,n to present the route from position $v_2,v_3,\ldots,v_n$ to the beginning position $v_1$ . 2,5,4,…,n is the number of every pass through position and the sequence of the number like 2-5-4-…n can be presented as the route. The number of the position can be stored in an array, pass one position then put the number of the position into the array. At last through traversing the array to get every number and the sequence number is the route.

Through adopting the fitness function to evaluate every route. The fitness function has two situation which are the ideal state and the usual state. The function in the ideal state is $F=F_i/f$ , F is the fitness for a random path, $F_i$ is the length of the whole path and f is the average length of all of the paths. In the usual state the function can be formulated as:

$$f = 1/(1 + k * g) \tag{2}$$

k means the number of times of the break circuit, g means the punishment coefficient and f is the fitness coefficient.

At last confirm the path collection by the selection algorithm. The first one is to eliminate the worst and keep the best solution. The other is sample selection, accumulate the expected survival number of every path in the next generation :

$$\frac{Ni}{Ni} = M * F/\sum_{i=1}^{M} Fi\ (i = 1,2, \ldots M) \tag{3}$$

Use the integral part of Ni to confirm the survival number of every path. Through this can know the$\sum_{i}^{M}$ Ni paths in the next generation. And descending order the decimal part then take the first $M\sim\sum_{i}^{M}$ Ni paths into the collection. Until now the M paths have been confirmed in the next generation.

### 3.2.2 Receiving-Alarm Design

Through the receiving-alarm module the policemen can get the case position through pointing the coordinate of the case position in the map directly. It will be very helpful to manage the police strength to let the policemen to handle the case who are nearest to the case position. And it can also show the latest case information.

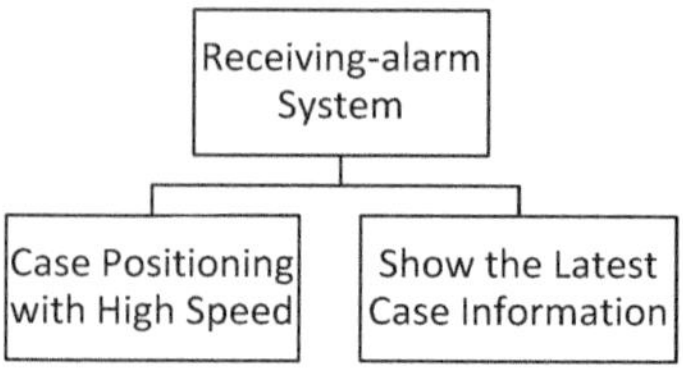


**Fig. 6.** The receiving-alarm module contains two main functions, one is showing the latest case information and the other is confirm the case position and the closest police station to the case with high speed

## 4 Conclusion

This paper designed and implemented the Police-using GIS for Dalian City Public Security Bureau, it adopted the three-layer architecture and designed the functions to meet the requirement of Dalian City Public Security Bureau. Especially adopting the Dijkstra based on genetic algorithm to get the shortest path. The design has been applied in the actual development and achieved the scalable goal. And the development practice improved that through using this design, the shortest path could be selected exactly, the reuse of codes is in a high degree with great effectiveness.

## References

1. Kang, X.-j., Wang, M.-c.: Shortest path algorithm implements based on genetic algorithm. Computer Engineering and Applications 44, 22–23 (2008)
2. Xu, S.-l., Wang, S.-q., Bi, Z.-k.: Design of logistics information system architecture based on GIS. Computer Engineering and Design 31(6), 1259–1263 (2010)
3. Wang, X.-p., Li, S.-q., Liu, Z.-y.: SOA application of research in the logistics system. Computer Engineering and Design 29(2), 303–305 (2008)
4. Wu, X.-C.: Data center integrated development platform. Electronic Industry Press, Beijing (2010)
5. Zhou, J.-j., Liu, G.: Implementation of Auxiliary GIS System for Police. Computer Engineering 34(13), 280–282 (2008)

# Knowledge Management Capability of Library: Scale Development and Validation

YangCheng Hu

School of Business and Management, Nanchang Institute of Technology, Nanchang, China
hyczju@126.com

**Abstract.** Despite the importance of knowledge management to library, there has been very little research in this area. The current paper intends to develop a scale of knowledge management capability for library. Initial scale items are generated from interviews and extant research. Exploratory factor analysis and confirmatory factor analysis are conducted on a survey finished by staff from library. The final library knowledge management capability (LKMC) scale includes 4 dimensions, 14 measurement items with good reliability and validity. Meanwhile, the results of Pearson's correlation analysis indicate that knowledge management is a critical success factor for innovation performance in China's library settings. The proposed LKMC scale can be used as a diagnostic tool for libraries to identity areas where specific improvements are needed. Finally, this paper discusses the implications of the findings and highlights the future research directions.

**Keywords:** Knowledge management, Innovation, Scale, Library.

## 1 Introduction

In this environment of rapid changes and uncertainty, where the demands of the customers keep changing, the only way for any organization to obtain superior performance is via its knowledge. Thus, the capture, sharing, retention and reuse of knowledge has become of vital importance. Providing correct knowledge to the right places at the right time is an essential requirement for the introduction of knowledge management [1].

The library's primary functions are to act as a knowledge repository and an agent for the dissemination of knowledge [2]. The rapid changes of information technology have placed ready information at everyone's fingertips, and this information is readily available outside of a library [3]. As libraries struggle with the fallout of the digital age, they must find a creative way to remain relevant to the twenty first century user who has the ability and means of finding vast amounts of information without setting foot in a brick and mortar library [4]. For libraries, as for other types of organizations, one of the most useful solutions that can be adopted in order to survive and to be successful in a society dominated by knowledge, is to implement a knowledge management strategy [5].

In spite of the fact that knowledge management has great potential value to library, to date the evaluation on library knowledge management capability (LKMC) has seldom been addressed. This paper tries to remedy the gap mentioned between theory

S. Lin and X. Huang (Eds.): CSEE 2011, Part II, CCIS 215, pp. 507–512, 2011.

and practice by developing and empirical testing a measurement scale of LKMC. The reminder of this paper is organized as follows. We first examine some relevant literatures. This is followed by discussions of the research methodology and results of the empirical study. We conclude the paper with the implications of the findings and give some suggestions for future research.

# 2 Literature Review

## 2.1 Knowledge and Its Manament

Although the concept of knowledge has been addressed by scholars for a long time, the concerns about managing organizational knowledge has been introduced and gained spectacular acceleration during the last few decades [6]. Broadly, knowledge management is a purposeful and systematic management of knowledge and the associated processes and tools with the aim of realizing fully the potential of knowledge in making effective decisions, solving problems, facilitating innovations and creativity and achieving competitive advantage at all levels [7].

Darroch (2003) develops three scales to measure behaviors and practices for each component of knowledge management: knowledge acquisition, dissemination and responsiveness [8]. According to Cavaleri (2004), knowledge processes are composed of a series of activities that facilitate the evolution of knowledge toward greater reliability in producing desired results [9]. Knowledge management involves the formalization of experience into knowledge and expertise that create new capabilities [10]. Similarly, Turban et al. (2003) highlight that knowledge management is the process of accumulating and creating knowledge, and facilitating the sharing of knowledge, so that it can be applied effectively and efficiently throughout the organization [11].

## 2.2 Innovation Outcomes of Knowledge Management Capability

Knowledge is a fundamental factor, whose successful application helps organizations deliver creative products and services [12]. Effective knowledge management facilitates knowledge communication and exchange required in the innovation process, and further enhances innovation through the development of new insights and capabilities [13][14]. Innovation process highly depends on knowledge, especially on tacit knowledge. A firm with a capability in knowledge management is also likely to be more innovative [15]. Knowledge management competencies are fundamental to innovation, enabling it to survive competitively and to grow [16]. Therefore, we expect the following hypothesis:

H1. Knowledge management capability is related positively to library innovation.

# 3 Research Methodology

## 3.1 Sample and Data Collection

Samples for this study were library staff from Zhejiang Province of China. Following previous studies, the single-informant method was used. Some precautions were taken to minimize the problems associated with such a method. First, care was taken to select measurement items that have proven to be valid and reliable in previous studies. Second, a pilot test was conducted and the questionnaire was further revised using the feedback from the pilot study. Adopting convenience sampling method, a total of 450 questionnaires were sent to potential respondents. A number of 286 questionnaires were returned. Therein, 29 are invalid due to too many omissions and unqualified answers, and 257 questionnaires are valid. The samples include 10 academic libraries and 5 public libraries with a staff above 50 on average.

### 3.2 Statistic Methods

The SPSS13.0 and LISREL8.7 statistics software was used for data processing. According to statistical requirements for factor analysis, different samples shall be used in exploratory factor analysis (EFA) and confirmatory factor analysis (CFA). 129 of the 257 valid questionnaires were randomly selected for EFA, while the remaining 128 were used for CFA and hypothesis testing.

## 4 Data Analysis and Results

### 4.1 Content Validity and Exploratory Factor Analysis

We first conducted six in-depth interviews with top managers from library to generate a list of key attributes associated with knowledge management capability. We compared these qualitative data to the items proposed by previous research. This process resulted in 25 measurement items which were then submitted for a pilot study and seven items were removed. In this way, we retained 18 items. We then conducted EFA with varimax rotation. Items with loadings (>0.50) on their hypothesized factor and low cross-loadings (<0.40) were retained. Three items were dropped and a four-factor solution of LKMC was proved (see Fig. 1). These four factors include 15 items with 77.015% accumulative variance explained, and each item has a comparatively high loading on the corresponding factor, ranging from 0.560 to 0.902 (see Table 1). In accordance with previous analysis, these factors were labeled as knowledge capture (KCA), knowledge storage (KST), knowledge sharing (KSH), and knowledge application (KAP).

### 4.2 Confirmatory Factor Analysis

The CFA was conducted and one item was eliminated for non-significant factor loadings (<1.96). Thus the four-factor structure of LKMC with 14 items was finally developed. The standardized factor loadings for the indicator variables were all above the recommended 0.60 (Bagozzi and Yi, 1988), and their t-scores indicated that all item loadings were significant ($p<0.01$). Fit indices shows that the model fit the data well ($GF=0.89$, $NFI=0.94$, $CFI=0.96$, $IFI=0.96$, $RMR=0.054$, $X^2/df=1.65$). The results mean that the four-factor structure model of LKMC is supported.

**Table 1.** Rotated Component Matrix

| | KCA | KST | KSH | KAP | Communalities |
|---|---|---|---|---|---|
| KCA1 | 0.859 | | | | 0.841 |
| KCA2 | 0.853 | | | | 0.818 |
| KCA3 | 0.842 | | | | 0.828 |
| KCA4 | 0.742 | | | | 0.723 |
| KCA5 | 0.730 | | | | 0.620 |
| KST1 | | 0.902 | | | 0.891 |
| KST2 | | 0.850 | | | 0.849 |
| KST3 | | 0.805 | | | 0.815 |
| KST4 | | 0.649 | | | 0.656 |
| KSH1 | | | 0.883 | | 0.818 |
| KSH2 | | | 0.828 | | 0.799 |
| KSH3 | | | 0.560 | | 0.577 |
| KAP1 | | | | 0.842 | 0.864 |
| KAP2 | | | | 0.738 | 0.685 |
| KAP3 | | | | 0.706 | 0.769 |

Rotation Method: Varimax with Kaiser Normalization.

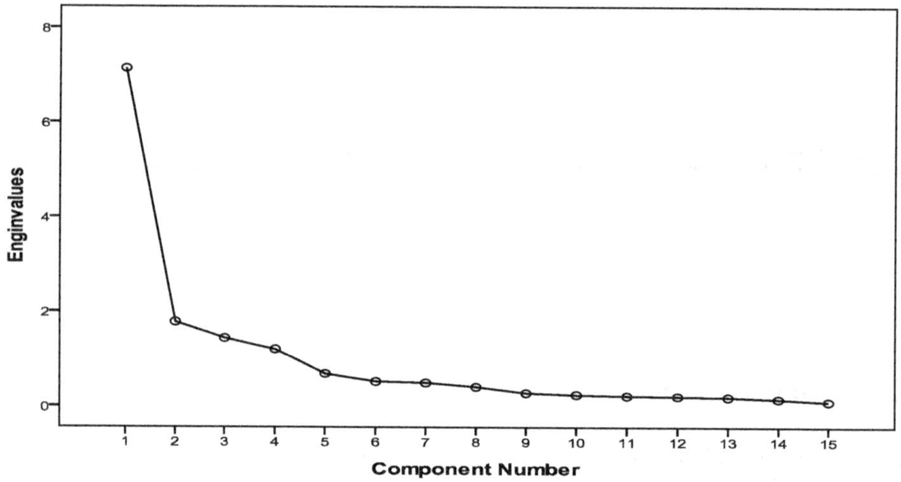


**Fig. 1.** Scree Plot of the Exploratory Factor Analysis

## 4.3 Construct Validity Assessment

**Discriminant Validity.** First the CFA is performed on selected pairs of constructs, allowing for correlation between the two constructs and then the test is rerun again by constraining the correlation between the pairs by fixing it to 1. A statistically significant value of $\triangle X^2$ demonstrates that the two constructs are distinct. In this way, six constrained models were developed to compare with freely estimated model. The results show the lowest $\triangle X^2$ is much higher than 10.328 ($p$=0.001). Therefore, the correlation coefficients between each pairs are not equal to 1 (Hinkle et al., 1994), implying the LKMC scale has good discriminant validity.

**Predictive Validity.** Predictive validity shows the ability of a scale to behave as expected with respect to some other constructs to which it is related (Churchill, 1995). To analyze this type of validity, Pearson correlation coefficients were considered among the analyzed variables and other representative variables of a concept that might be influenced by this first variable (Flavian and Lozano, 2003). In this case, the correlations were calculated between LKMC and library innovation performance (LIP). As may be seen in Table 2, the correlations between the four dimensions of LKMC and the LIP indicators are relatively high and highly significant, and this shows the possible predictive capacity of this measurement instrument in the library settings.

**Table 2.** Pearson Correlation Coefficients

| | KCA | KST | KSH | KAP | LIP |
|---|---|---|---|---|---|
| KCA | 1.000 | | | | |
| KST | 0.582** | 1.000 | | | |
| KSH | 0.529** | 0.690** | 1.000 | | |
| KAP | 0.688** | 0.655** | 0.562** | 1.000 | |
| LIP | 0.263** | 0.460** | 0.294** | 0.355** | 1.000 |

Notes: **means p<0.01, CUS=customer satisfaction.

## 5 Conclusion and Discussion

The findings in this study support our assertion that LKMC is a multidimensional construct including knowledge capture, knowledge storage, knowledge sharing, and knowledge application. Our data indicate that these components are convergent on a common construct and the relationship of the scale developed in this study has acceptable discriminant validity. The predictive validity of the scale indicates that LKMC is positively associated with library innovation performance. The findings support prior theoretical research which suggests that knowledge management is a factor in determining organizational innovation. The practical implication is that libraries in China can improve their innovation performance through the adoption of knowledge management strategy.

**Acknowledgments.** It is a project supported by the National Natural Science Foundation of China (71062004).

## References

1. O'Dell, C., Grayson, C.J.: If only we knew what we know: identification and transfer of internal best practices. California Management Review 40(3), 154–173 (1998)
2. Kim, Y.-M., June, A.: Adoption of Library 2.0 functionalities by academic libraries and users: A knowledge management perspective. Journal of Academic Librarianship 36(3), 211–218 (2010)

3. Yi, Z.: Knowledge management for library strategic planning: Perceptions of applications and benefits. Library Management 29(3), 229–240 (2008)
4. Parker, K.R., Nitse, P.S., Flowers, K.A.: Libraries as knowledge management centers. Library Management 26(4/5), 176–189 (2005)
5. Porumbeanu, O.-L.: Implementing Knowledge Management in Romanian Academic Libraries: Identifying the Elements that Characterize their Organizational Culture. Journal of Academic Librarianship 36(6), 549–552 (2010)
6. Vandaie, R.: The role of organizational knowledge management in successful ERP implementation projects. Knowledge-Based Systems 21(8), 920–926 (2008)
7. Kebede, G.: Knowledge management: An information science perspective. International Journal of Information Management 30(5), 416–424 (2010)
8. Darroch, J.: Developing a measure of knowledge management behaviors and practices. Journal of Knowledge Management 7(5), 41–54 (2003)
9. Cavaleri, S.A.: Leveraging organizational learning for knowledge and performance. The Learning Organization 11(2/3), 159–176 (2004)
10. Beckman, T.J.: The current state of knowledge management. In: Liebowitz, J. (ed.) Knowledge Management Handbook. CRC Press, New York (1999)
11. Turban, E., Rainer, R., Potter, R.: Introduction to Information Technology. John Wiley & Sons, Inc., Hoboken (2003)
12. Gupta, B., Iyer, L.S., Aronson, J.E.: Knowledge management: practices and challenges. Industrial Management & Data Systems 100(1), 17–21 (2000)
13. Madhavan, R., Grover, R.: From embedded knowledge to embodied knowledge: New product development as knowledge management. Journal of Marketing 62(4), 1–12 (1998)
14. Chen, C.-J., Huang, J.-W.: Strategic human resource practices and innovation performance-The mediating role of knowledge management capacity. Journal of Business Research 62(1), 104–114 (2009)
15. Darroch, J.: Knowledge management, innovation and firm performance. Journal of Knowledge Management 9(3), 101–115 (2005)
16. Pitt, M., MacVaugh, J.: Knowledge management for new product development. Journal of Knowledge Management 12(4), 101–116 (2008)

# A Fast Algorithm for Outlier Detection in Microarray

You Zhou[1], Chong Xing[1], Wei Shen[1,2], Ying Sun[1], Jianan Wu[1], and Xu Zhou[1,3,*]

[1] College of Computer Science and Technology, Jilin University, Key Laboratory of Symbol Computation and Knowledge Engineering of Ministry of Education, 130012 Changchun, China
[2] College of Computer Science and Technology, Beihua University, 132021 Jilin, China
[3] Center for Computer Fundamental Education, Jilin University, 130012 Changchun, China
zhouxu@jlu.edu.cn

**Abstract.** A Fast Outlier Sample Detection(FOSD) algorithm is proposed in this paper which can be used to recognize mislabeled samples or abnormal samples in microarray datasets. The proposed algorithm uses CL-stability alorithm as a basic operator. The Machine Learning method is used as classifier in the FOSD. The outlier samples are detected depending on the gobal stability of samples. Experimental results show that the FOSD algorithm is not only better than other existing algorithms, but also robust for detecting outlier samples in microarray dataset.

**Keywords:** Outlier sample detection, Microarray, Pattern recognition.

## 1 Introduction

Microarray is a high-throughput technique for measuring the expression level of thousands of genes simultaneously that is widely adopted in biological and medical research. In general, there are some challenges for analyzing microarray data like outlying samples. The procedure of generating microarray data is complex; outliers can be experimental artifacts or genuine consequences of biological diversity. However, an outlier sample caused by biological diversity may be different from the other outliers, possibly harder to detected and questionable if it an outlier at all. For example, West et al. analyzed 49 breast tumor samples and identified 9 samples as possibly having the wrong labels [1].

It is important to clarify the definition of outlier samples. The definitions of outlier proposed over the years are different in some aspects, but it can be described as 'an observation that deviates so much from other observations as to arouse suspicions that it was generated by a different mechanism.' [2]. A vast literature exists on outlier detection and robust analysis methods, namely methods insensitive to the presence of outliers [3-6]. Sometimes, a outlier sample is also a sample which is erroneously classified in the original dataset. Outlier samples can influence successive analysis and some approaches have been proposed to solve this problem, but they used different names. For example, references [7] and [8] called it as 'outlier' sample, references [9] and [10] called it as 'mislabeled' sample. In fact, mislabeled samples may be sometimes detected as outliers, but a mislabeled sample does not need to be an outlier, and an outlier is not necessarily mislabeled. In this paper, we focus on the bigger region which

* Corresponding author.

S. Lin and X. Huang (Eds.): CSEE 2011, Part II, CCIS 215, pp. 513–519, 2011.

is called outlier detection problem. A mislabeled sample is a sample which is classified into the wrong class. An outlier sample is a sample which is suspiciously different from the other samples. For example, a cancer patient tissue sample is erroneously classified as a normal case, and some technical experimental mistake alters some features of the data of the corresponding microarray producing an outlier.

Outlier detection for high-dimensional data has received a lot of attention in recent years. For instance, *Knorr and Ng* considered distance-based outlier detection, where the notion of outlier is strongly dependent on the underlying distance metric [6]. *Aggarwal and Yu* studied the problem of outlier detection for high-dimensional data using projections into subspaces [11]. *Yan et al.* used genetic algorithm combined with a distance-based outlier detection method to find outliers in microarrays [12]. However, these approaches are clearly not scalable for large p. Furthermore, to use a distance-based approach, one has to define a distance function between samples. In a high-dimensional space, it is often difficult to do so.

There are few papers that specifically focus on mislabeled sample detection. Furey et al. used a SVM on the full datasets and on a specified top-ranked features [9]. Samples which have been consistently misclassified in all tests are identified as suspects. Li et al. used a genetic algorithm in order to select subsets of genes that can potentially discriminate between tumor and normal tissue samples and then each sample is classified according to the class membership of its k nearest neighbors [13]. Kadota et al. tackled the problem of detecting outliers using a technique based on Akaike's Information Criterion [14]. Lu et al. used a SVM as base classifier on a leave-one-out iterative strategy to find outliers [7]. Malossini et al. proposed two algorithms CL-stability and LOOE-sensitivity based on the generate of the leave-one-out perturbed classification matrix. The output is a list of suspects for further analysis [10].

In this paper, our goal is to detect outlier samples and automatically correct them, avoiding the pitfall of pointing to outliers as suspects of being misclassified. We propose an fast algorithms based on CL-stability which is called FOSD(Fast Outlier Samples Detection). We provide an empirical evaluation of the two algorithms using real and synthetic datasets. We evaluate precision, recall, and percent of outliers found to show the performance of our new algorithm to find outlier samples. We also evaluate the relative number of outliers that our new algorithms detect as abnormal that were detected as suspects by CL-stability. From the empirical evaluation of FOSD, we find that the FOSD not only can correct the results of CL-stability but also can detect outlier samples and better than CL-stability.

# 2 Methodology

## 2.1 CL-Stability Algorithm

CL-stability is an algorithm for detecting mislabeled samples. The basic and important part of CL-stability is the construction of the Leave-One-Out Perturbed Classification matrix L (LOOPC matrix). The elements of the matrix whose dimensions are $n \times n$ where n is the number of the samples, are computed as follows. First, the j-th sample is left-out from the dataset, then the label of the i-th sample is flip and the perturbed dataset is used to construct a classifier, finally the resulting classifier is used to classify the j-th left-out sample. The result of the classifier is the Lij element of the L matrix. The principle of CL-stability is to assess the stability of a sample with respect to a small perturbation (just one flip) of the other samples of the dataset. A good or stable sample should be consistently classified, with respect to its original label, and not be easily affected by the correctness of 1 flipped sample elsewhere in the dataset. The CL-stability algorithm identifies a list of samples failing this requirement. A sample is predicted as suspect if it is not stable when other samples are perturbed. The algorithm to construct LOOPC matrix and CL-stability could be found in the paper [10].

## 2.2 Fast Outlier Sample Detection

Fast Outlier Sample Detection(FOSD) uses further the idea of stability used by CL-stability. In CL-stability the stability of a sample under perturbation of the data set is very important for determining whether the sample is a suspect or a normal sample. The main strategy of FOSD is based on two assumptions: (i) CL-stability can correctly detect some outlier sample and (ii) if correctly detected and flipped, a sample should be stable under successive runs of CL-stability itself. As a whole, it is an iterative strategy and some outlier samples which cannot be detected by CL-stability. may be detected. When we flip all the labels of suspect samples and run CL-stability again, if a flipped suspect sample is stably classified into the other class then FOSD will output it as outlier sample. If a flipped suspect sample is detected again as suspect in the following run of CL-stability then FOSD will output it as an abnormal sample. Finally, if one iterative step gives the same results of the previous step then the algorithm will finish. The algorithm of FOSD is shown in Fig 1.

**Fast Outlier Sample Detection**

```
i := 0
D_0 := CLS (S_0)
if (D_0 ≠ φ)
      do
         A := A ∪ D_i
         i := i + 1
         if (a_n ≥ n/2)
               a_n := a_n − b_n
         endif
         S_i := flipset (S_{i−1}, D_{i−1})
         D_i := CLS (S_i)
   endif   for ((x_k ∈ D_i) ∧ (x_k ∈ D_{i−1}))
               B := B ∪ {x_k}
         endfor
   while (D_i = D_{i−1})
```

**Fig. 1.** Fast Outlier Sample Detection. Where $S_i$ is the dataset and the meaning of i is ith iterative step. Where A is the mislabeled list of FOSD. Where B is the abnormal list of FOSD. Where $D_i$ is a temporary suspect list of ith iterative step of FOSD. The CLS($S_i$) function is used to run CL-stability with dataset $S_i$. The flipset($S_i$, $D_i$) function is used to flip the label of sample which is included in $D_i$. The formula of $a_n$ is used in [10].

# 3 Experimental Results

## 3.1 Synthetic Dataset

In order to test the performance of the algorithms we generated synthetic datasets with known about which samples are respectively correct, outlier.The synthetic datasets are generated following the principle of simulating microarray data with some differences with respect to the synthetic datasets described in [10]. We followed some simplicist hypotheses: the gene expression is distributed in different levels for normal and tumor samples, most of the genes have similar expression values in the two conditions, and only a few genes will distinguish normal samples from tumor samples.

The dataset simulates microarray data represented as a matrix, each row represents a microarray sample and each column represents a gene in microarray. The dataset is generated feature by feature, namely gene by gene. For a two-classes classification problem, most of the features which are called normal feature are generated by sampling from a Gaussian distribution with variance $\sigma = 0$ and mean $\mu = 1$ sampled from a uniform distribution in the interval [-20,20]. In this way the dataset simulates gene levels on a logarithmic scale and it is more similar to real data than the data in Malossini et al. who used mean $\mu = 0$ . The two classes differ only for a limited number of features which are called special feature, whose value are sampled from Gaussian distributionswith parameters $\mu_+ = 3$, $\sigma_+ = 1$ and $\mu_- = -3$, $\sigma_- = 3$ respectively. Outliers are randomly selected from the dataset and a substantial number of feature values are changed, this is done independently from the class of the sample. Finally, outlier samples are randomly selected from the dataset and their class is flipped controlling the proportion of normal and outliers that are outlier.

**Table 1.** The results on synthetic dataset of CL-stability and ROSD

| | CL-stability | | | | ROSD | | | | |
|---|---|---|---|---|---|---|---|---|---|
| Dataset | P | R | $f_{s/o}$ | $f_{s/mo}$ | P | R | $f_{m/o}$ | $f_{m/mo}$ | $f_{oa/o}$ |
| S-0-0-0-000 | 100 | 100 | ----- | ----- | 100 | 100 | ----- | ----- | ----- |
| S-3-0-0-000 | 99.6 | 99.3 | ----- | ----- | 100 | 100 | ----- | ----- | ----- |
| S-0-1-0-800 | 95.0 | ----- | 5.00 | ----- | 95.0 | ----- | 5.00 | ----- | 0.00 |
| S-0-2-0-800 | 92.0 | ----- | 4.00 | ----- | 92.0 | ----- | 4.00 | ----- | 0.00 |
| S-3-1-0-800 | 97.7 | 99.6 | 8.00 | ----- | 97.0 | 100 | 5.00 | ----- | 75.0 |
| S-0-1-1-800 | 99.0 | 63.0 | 1.00 | 63.0 | 96.0 | 64.0 | 8.00 | 64.0 | 0.00 |
| S-3-1-1-800 | 98.0 | 82.2 | 8.00 | 30.0 | 96.7 | 88.2 | 15.0 | 53.0 | 25.0 |
| S-3-0-2-800 | 100 | 95.5 | ----- | 89.0 | 100 | 99.2 | ----- | 98.0 | ----- |

Table 1 shows the average performance measures of CL-stability and FOSD on 100 runs on different instances of the synthetic datasets. In Table 1, the first row shows that CL-stability and FOSD produce no false negatives when there is nothing to detect. The second row of Table 1 show us that CL-stability has a very high precision and recall in detection of outlier samples with a few false negatives, and FOSD is even better than CL-stability with no false negative points. Rows 3, 4 and 5 show that in presence of outliers and in absence of outlier points both CL-stability and FOSD have some false positives. FOSD does not correct the false positives of CL-stability with no outlier samples datasets and in presence of outliers. Rows 6-7 show that CL-stability predicts outliers as outlier samples sometimes and increases the percent while the number of outliers increases. FOSD can correct some parts of these false positives and with a slightly higher recall. In rows 8,9 and 10, FOSD has an higher recall and percent of detecting outliers than CL-stability. The results confirm that CL-stability is good at detecting outlier samples and shows that it is somehow sensitive to outliers and the presence of outliers decreases the recall. The conclusions of the comparison of the two algorithms present three aspects. First, FOSD is better than CL-stability to find outliers. Second, if the outliers can be detected by CL-stability as suspect samples, FOSD can sometimes correct it as abnormal samples mostly. Thirdly, FOSD is better than CL-stability to find outliers, even though the result is not exact but it can give more information to the user. Both the algorithms are sensitive to the presence of outliers when there are non outlier samples.

### 3.2 Real Datasets

We use three well-known real datasets to test our algorithms: (1) A breast cancer dataset was first presented in *West et al.*, which consists of 49 tumor samples classified as positive to estrogen receptor ER+ or negative ER-.The expression levels of 7129 genes are given for each sample [1]. (2) A colon tissue dataset from *Alon et al.* which consists of 40 tumor samples and 22 normal samples. A selection of 2000 genes with highest minimal intensity across the samples has been made by *Alon et al.* [15]. (3) A leukemia dataset from *Golub et al.*, which consists of 25 acute myeloid leukemia and 47 acute lymphoblastic leukemia. The expression levels of 7129 genes are given for each sample [16].

For establishing the outlier samples of the three datasets we followed what done in [10]. In particular, the samples identified by *West et al.* and *Alon et al.* are used, whereas for the leukemia dataset no ground truth is available but there is consent in the literature about one sample be outlier.

**Table 2.** The results on three real datasets

| Datasets | Algorithm | Positive | | | | | | | | | Negtive |
|---|---|---|---|---|---|---|---|---|---|---|---|
| Colon | Alon et al | T2 | T30 | T33 | T36 | T37 | N8 | N12 | N34 | N36 | |
| | FOSD | √ | √ | √ | √ | √ | √ | | √ | √ | N2,N28 |
| | CLstability | √ | √ | √ | √ | | | | √ | √ | N2,N28 |
| Breast | West et al | 11 | 14 | 16 | 31 | 33 | 40 | 43 | 45 | 46 | |
| | FOSD | | √ | √ | √ | | √ | √ | √ | | |
| | CLstability | | √ | √ | √ | | √ | √ | | | |
| Leukemia | Golub et al | 66 | | | | | | | | | |
| | FOSD | √ | | | | | | | | | |
| | CLstability | √ | | | | | | | | | |

Table 2 presents the results of our new algorithm when run on the real datasets. One of our goals is how to correct the results of CL-stability. It should be better that the abnormal samples should be got rid of and the outlier samples should be flipped. But the ground truth is unknown. From this table about colon dataset, N2 and N28 are false positive points by CL-stability, but they are predicted as outlier samples. The outlier samples have a higher confidence which can be predicted as outlier samples of FOSD. This point of view is also illustrated by synthetic datasets. It has proved the correction effect of FOSD.

## 4 Conclusion

In this paper, we propose a fast outlier sample detection(FOSD) algorithms base on CL-stability for detecting possibly outlier samples. The principle of this algorithm is based on the global stability of dataset. We test this algorithm with synthetic dataset and real datasets, the results show that the new algorithms are useful tool for detecting outlier sample and better than CL-stability.

**Acknowledgments.** This work was supported by the National High-Tech R&D Program of China 863 (2009AA02Z307); the NSFC (60803052, 60903097); the Science-Technology Development Youth Research Project from Jilin Province of China (20090116, 20101589); the project of (20090194) support by Jilin University, "211" and "985" project of Jilin University.

## References

1. Smith, T.F., Waterman, M.S.: Identification of Common Molecular Subsequences. J. Mol. Biol. 147, 195–197 (1981)
2. West, M., et al.: Predicting the clinical status of human breast cancer by using gene expression profiles. Proceedings of the National Academy of Sciences of the United States of America 98(30), 11462–11467 (2001)
3. Hawkin, D.: Identification of outlier. Chapman and Hall, London (1980)
4. Barnett, V., Lewis, T.: Outliers in statistical data. John Wiley & Sons, Chichester (1994)
5. Tucakov, V., Ng, R.: Identifying unusual people behavior: A case study of mining outliers in spatio_temporal trajectory databases. In: Proc. SIGMOD Workshop on Research Issues on Knowledge Discovery and Data Mining (1998)
6. Johnson, T., et al.: Fast Computation of 2-Dimensional Depth Contours. In: Proc. KDD, pp. 224–228 (1998)
7. Knorr, E.M., Ng, R.T.: Algorithms for mining distance-based outliers in large datasets. In: Proceedings 24th International Conference Very Large Data Bases, VLDB, NY, USA, pp. 392–403 (1998)
8. Lu, X., et al.: A simple strategy for detecting outliers in microarray data. In: 8th Conference on Control, Automation, Robotics and Vision, Kunming, China, pp. 1331–1335 (2004)
9. Kadota, K., et al.: Detecting outlying samples in microarray data: a critical assessment of the effect of outliers on sample classification. Chem.-Bio. Inform. J. 3, 30–45 (2003)
10. Furey, T.S., et al.: Support vector machine classification and validation of cancer tissue samples using microarray expression data. Bioinformatics 16, 906–914 (2000)
11. Malossini, A., Blanzieri, E., Ng, R.: Detecting potential labeling errors in microarrays by data perturbation. Bioinformatics 17, 2114–2121 (2006)
12. Aggarwal, C.C., Yu, P.S.: Outlier detection for high dimensional data. In: Proceedings of ACM SIGMOD 2001, Santa Barbara, CA, pp. 37–46 (2001)

13. Yan, C., et al.: Outlier analysis for gene expression data. J. Computer Sci. & Technol. 19, 13–21 (2004)
14. Li, L., et al.: Gene assessment and sample classification for gene expression data using a genetic algorithm/k-nearest neighbor method. Comb. Chem. High Through. Scr. 4, 727–739 (2001)
15. Kadota, K., et al.: Detecting outlying samples in microarray data: a critical assessment of the effect of outliers on sample classification. Chem.-Bio. Inform. J. 3, 30–45 (2003)
16. Alon, U., et al.: Broad patterns of gene expression revealed by clustering analysis of tumor and normal colon tissues probed by oligonucleotides array. Proc. Natl. Acad. Sci. USA 96, 6745–6750 (1999)
17. Golub, T.R., et al.: Molecular classification of cancer: class discovery and class prediction bye gene expression monitoring. Science 286, 531–537 (1999)

# Modeling and Solution for Multiple Chinese Postman Problems

Jing Zhang

Basic Courses Department
Beijing Union University
Beijing, China
zhang1jing4@sina.com

**Abstract.** The current study for the Chinese postman problem has reached a mature stage, but multiple Chinese postman problems for the study also relatively small. Multiple Chinese postman problem is starting from the post office messenger, requires area at least through every street once a return to the post office. Multiple Chinese postman problems in the study the characteristics of solutions based on the mathematical model, multiple Chinese postman problem into a single Chinese postman problem, simulated annealing method described in detail the design process, but also gives the ant colony algorithm and hybrid genetic algorithm for multiple Chinese postman problem solving ideas. The results show that these algorithms on multiple Chinese postman problem solving are effective, can be a larger probability of obtaining global optimal solution.

**Keywords:** Multiple Chinese Postman Problems, Simulated Annealing, Algorithm, Ant colony Algorithm.

## 1 Introduction

Chinese postman problem is a typical combinatorial optimization problem, which in many areas have a wide range of applications. Chinese postman problem is the multi-m a postman from the same street (or a different street) starting to take a walking route, respectively, so that each street has one and only one after the postman (except for starting the street), and the shortest total distance. Chinese postman problems are: an n-street, and asked the postman to reach each street once, and only once, and return to the starting point, requiring the shortest walking route. In order to effectively address the minimum completion time, distance, symmetric or asymmetric multiple Chinese postman problems, this paper, genetic algorithm optimization and matrix decoding method is proposed in order to determine by which each street and each postman postman's walk through the line, that is, the postman to find an optimal allocation and walking routes, walking in after the postman, the postman who spent most of time is minimized. Simulation results show that the proposed algorithm is effective.

In practical problems, because the speed of walking each postman may be different, the time limit to complete the postman is a certain practical sense, therefore,

S. Lin and X. Huang (Eds.): CSEE 2011, Part II, CCIS 215, pp. 520–525, 2011.

it is necessary to complete all of the time the postman to minimize the maximum value of multiple Chinese postman problems studied. Chinese postman problems related to the research problems in the real world there are great practical value, such as transportation, pipeline, route selection, and design of computer network topology, the postman messengers and so on, can be abstracted into Chinese postman problems or multiple Chinese postman problems.

# 2 Simulated Annealing

## 2.1 The Basic Principle of Simulated Annealing

Simulated annealing algorithm is based on the Monte Carlo method for solving a heuristic iterative random search algorithm, which simulates the thermal annealing process of solid material balance problems with random search optimization to find the similarity to the global optimum optimal or near global optimal objective. Simulated annealing algorithm is a problem for the minimum can be applied to the previous update or basic learning process (random or directed). Then, the principle is similar with the metal annealing, in the beginning to minimize or to learn more quickly, the temperature was risen high, then (slowly) to stabilize the temperature.

In this process, the length of each step of the update process with the corresponding parameters is proportional to the temperature of these parameters play a role. Optimal solution in the search process, in addition to receiving optimal solution, there is a random acceptance criteria (metropolis criterion) to accept worse solutions limited, and the probability to accept worse solutions gradually tends to 0, which makes the algorithm may out from the local minimum area, that is possible to find the optimum solution and to ensure the convergence of the algorithm.

## 2.2 The Basic Steps of Simulated Annealing

Let $(T, g)$ is an instance of combinatorial optimization problems, find $I \in T$, so $g(I) = \min g(i)$, $i \in T$ of the simulated annealing calculation steps can be described as follows: (from textbooks)

1° Set the given initial temperature $t_0 > 0$ and the final temperature $t_g$, temperature coefficient of $\alpha$ $(0 < \alpha < 1)$;

2° The number of iterations $n = 1$;

3° The temperature $t_n$, from the neighborhood of solution $i$ randomly generate a new feasible solution $j$; Calculate the evaluation function $g(i)$ and $g(j)$ The difference $\Delta g = g(j) - g(i)$;

4° If $\min \{(1, \exp(-\Delta g / t_n)\} > p$, which $p$ is [0, 1] random number between, then accept new solution;

5° In accordance with the temperature coefficient of $t_{n+1} = \alpha_{tn}$ annealing;

6° Increasing the number of iterations $n$, if $n$ reaches the maximum number of iterations, or to terminate the temperature, stop the iteration, otherwise return 3°.

### 2.3 Solving Multiple Chinese Postman Problems of the Simulated Annealing Algorithm Model

The objective function is access to all of the objective function of the total path length of the street Minimum and m a postman as a balanced task. Not necessarily the shortest total distance is the task of balancing the various postmen; it is required to "balance" to make certain demands, so we define

$$\frac{\min_{i,j}(g_i - g_j)^2}{\max_{i,j} g_i^2}$$

To characterize the equilibrium level, it is clear that the smaller the value the better.

Simulated annealing, when the temperature is high enough, the system can freely change the configuration, you can move or in the surface free energy as the rules do not walk, that is, the freedom to choose a feasible solution. When the neighborhood solutions of the objective function value than the current solution when the objective function value better to replace the current solution to neighborhood solutions. When the temperature decreases, the system configuration in the energy surface movement will be restricted, and gradually concentrated to the low energy region, in every iteration process, are based on current solution as the center of the neighborhood and then randomly generate a new solution. It also makes simulated annealing method has the ability to escape local optimal solution, through the cooling action to control the speed of convergence. If the resulting solution than the current, the simulated annealing method will make use of probability functions and control the temperature parameters to determine whether to accept the new solution. As the temperature dropped, accept a lower probability of getting smaller and smaller solutions.

## 3 Genetic Algorithm

Genetic algorithm is developed in recent years a new global optimization algorithm. This reflects the nature of "natural selection, survival of the fittest" evolutionary process. It borrows the biological point of view of genetics, through natural selection, genetic variation and other mechanisms to achieve the improvement of the adaptability of each individual. When the genetic algorithm to solve the problem, we must first deal with problem-solving model structure and parameters of encoding, usually with a string that will issue symbolic of this process. Professor Holland in 1962, the GA algorithm was first proposed the idea to attract a large number of researchers, quickly extended to the optimization, search, machine learning, etc., and laid a solid theoretical foundation.

Chinese postman problems because the local structure of the problem is extremely complex algorithm to slow convergence near local optimal solution, if the increase in mutation rate, they do not get excellent offspring. Genetic algorithms applied to the Chinese postman problems to be overcome on the issue of convergence speed and the main difficulty is the contradiction between the maturities.

### 3.1 Encoding

For these reasons, this paper uses a dynamic mapping approach to coding, to ensure that the various genetic manipulation of genes obtained are valid. There is two ways postman route, record or records in order to access the street line of the order. If access to the order of 123, the number is 111, easy to show that this encoding and access route is one mapping. Access sequence set $r_1$, $r_2$ , ... ,$r_n$, then the requirement that a street map directly to the street number, street maps for a later visit to the street without the serial number list. We adopt the latter in the specific code, if the list directly to the street map with the street number of the method will create difficulties for crossover and mutation operation such as access to the order of the street 1, street 2, street 3, if its code is 123, then the access road intersection with the 321 genes that are ineffective, and mutation of its genes are not valid any gene.

### 3.2 Evaluation Function

Evaluation function can be directly by a transform path length, the longer the path; the evaluation function is given a lower score, the more likely that the genotype be eliminated. Meanwhile, the evaluation function should be some discrimination; this paper is given to develop appropriate indicators of the strength of the comparison matrix of all individuals, and normalization methods.

### 3.3 Initial Population

Considering the representation of the initial population, the specific emulated bit up in the first $n+m$ to $n$-$m$+1 is a random number, the complexity of computing experience and genetic algorithms to select the number of initial population of 10 individuals.

### 3.4 Genetic Operators

1° Mutation operator

The evaluation function design based on the mutation probability

$$\frac{\max f - f}{\max f - \min f}$$

after appropriate scaling and the constant increase in the minimum point has a certain mutation probability.

2° The selection operator

In order to solve the problem of single source, the use of demo program, the following is a demonstration of several different situations screenshots, for simplicity, we directly demonstrate a single source point of the Chinese postman problem, respectively, in 10 streets, 100 streets, 500 streets were calculus. In order to retain under the optimum of the parent gene, after the parent and offspring to join scores roulette, make a selection.

3° Crossover operators

The right side of the street the shortest distance corresponds to the distance a small street to the offspring in the right and began to consider the parent in the street right in the street, and so on. Direct access point crossover, you can choose the greedy-based crossover operator, that is randomly selected in the parent of a street as the offspring of the original street from two parent to find the right side of the street streets, and street to calculate.

## 4 Ant Colony Algorithm

Ant colony algorithm by the Italian scholar, who first proposed a novel simulated evolutionary algorithm. More specifically, the ants in the course of the campaign can leave it on the path through the pheromone, but also in the process of movement that pheromone perception of the existence and strength, and to guide the direction of their movement.

After much careful observation of bionics home and study discovered that ant process, through a substance called pheromone mutual transmission of information. It is the study of ant colony behavior generated. In order to avoid the ants took the same path twice, for each ant to set up a taboo table to record the path it traveled. Toward the pheromones of ants tend to move towards high intensity, so the composition of the large number of ant colony behavior will show a positive feedback effect of information: a path through the more ants, then later chose the greater the probability of the path.

Ant group is through exchange of information between individuals that collaborate with each other mechanism to achieve the purpose of the search of food. Ant colony algorithm has the advantage: it is an adaptive, self-organization, in essence, parallel methods, and positive feedback is a way to promote optimal solution to the evolution of the system has strong robustness, slightly modified model of the ant colony algorithm to be applied to other problems, and it can be combined with a variety of heuristic algorithms to improve the performance of the algorithm.

Chinese postman problems in the field of combinatorial optimization problem is a typical problem to solve this problem have a greater practical significance, and this problem can be testing new algorithms as a criterion, so this issue has been a research focus. But the algorithm also has a slow convergence and easy to fall into local optimum and other shortcomings. There is no theoretical basis, adjusted and determined by experiment.

## 5 Conclusion

The gene encoding algorithm is simple and can clearly reflect the relationship between the postman and the streets and visit the order, and of the decoding method, optimizing distance for symmetric and asymmetric multi-Chinese postman problem. Minimizing this distance all the postman conducted a study and proposed hierarchical genetic algorithm to optimize the number of Chinese postman problem.

This is because large-scale multiple Chinese postman problems is difficult to find the optimal solution, but also requires a lot of time, so research in this area should also continue to deal with multiple Chinese postman problems algorithm further improved, try using a new algorithm for find a better solution as soon as possible. Using this algorithm, the simulation test, effectively solve the problem of multiple

Chinese postman problems, and achieved a quite satisfactory solution, verify the feasibility and effectiveness. However, the results on the schedule for further analysis results are still found scattered. Hybrid genetic algorithm of this paper is better than the genetic algorithm alone has improved a lot.

**Acknowledgment.** The work is supported by Funding Project for Academic Human Resources Development in Institutions of Higher Learning under the Jurisdiction of Beijing Municipality (PHR (IHLB))(THR201108407). Thanks for the help.

## References

1. Eiselt, H.A., Gendreau, M., Laporte, G.: Arc Routing Problems, Part I: The Chinese Postman Problem. Operations Research 43(2), 231–242 (1995)
2. Thimbleby, H.: The directed Chinese Postman Problem. Software: Practice and Experience 33(11), 1081–1096 (2003)
3. Assad, A.A., Pearn, W.-L., Golden, B.L.: The Capacitated Chinese Postman Problem: Lower bounds and solvable cases. American Journal of Mathematical and Management Sciences 7(1-2), 63–88 (1987)
4. Minieka, E.: The Chinese Postman Problem for Mixed Networks. Management Science 25(7), 643–648 (1979)
5. Nobert, Y., Picard, J.-C.: An optimal algorithm for the mixed Chinese postman problem. Networks 27(2), 95–108 (1996)
6. Corberán, A., Martí, R., Sanchis, J.M.: A GRASP heuristic for the mixed Chinese postman problem. European Journal of Operational Research 142(1), 70–80 (2002)
7. Pearn, W.L., Liu, C.M.: Algorithms for the Chinese postman problem on mixed networks. Computers & Operations Research 22(5), 479–489 (1995)
8. Ghiani, G., Improta, G.: An algorithm for the hierarchical Chinese postman problem. Operations Research Letters 26(1), 27–32 (2000)
9. Pearn, W.L., Chou, J.B.: Improved solutions for the Chinese postman problem on mixed networks. Computers & Operations Research 26(8), 819–827 (1999)
10. Pearn, W.L.: Solvable cases of the k-person Chinese postman problem. Operations Research Letters 16(4), 241–244 (1994)
11. Ralphs, T.K.: On the mixed Chinese postman problem. Operations Research Letters 14(3), 123–127 (1993)

# Damage of Earthquake and Tsunami to Coastal Underground Works and Engineering Fortifying Methods

YanRu Li, ZhongQing Cheng, and HaiBo Jiang

Department of Logistics Command and Engineering,
Naval University of Engineering, Tianjin 300450, China
yanru_li@126.com

**Abstract.** Though it was generally agreed that the seismic damage of underground works was less than that of ground buildings, some underground structures were found to be damaged seriously in several past strong earthquakes. The damage mechanism of earthquake and seismic tsunami to coastal underground works are analyzed, and the characteristic and position of damage are studied. Lastly, methods to fortify coastal underground works are proposed. It is concluded that the damage can be reduced to the least by strengthening the structure of underground works in proper ways and improving the prediction of earthquake and tsunami.

**Keywords:** Earthquake, Tsunami, coastal underground works, engineering Fortifying methods.

## 1 Introduction

The 2011 Tōhoku earthquake, also known as the Great East Japan Earthquake, was a magnitude 9.0 (Mw) undersea megathrust earthquake off the coast of Japan that occurred at 14:46 JST (05:46 UTC) on Friday, 11 March 2011. The earthquake triggered extremely destructive tsunami waves of up to 38.9 meters (128 ft) that struck Japan, in some cases traveling up to 10 km (6 mi) inland, causing heavy casualties and property losses. The 2004 Indian Ocean earthquake and tsunami were occurred at December 26, 2004, killing over 230,000 people in fourteen countries, and inundating coastal communities with waves up to 30 meters (100 feet) high. Such tragic disasters not only bring great damages and disasters to people, but also prompt us to ponder the threats of earthquake and tsunami to coastal underground works, and the engineering methods to fortify them.

## 2 Threats of Earthquake to Coastal Underground Works

Earthquake is common natural disaster resulted by a sudden release of energy in the Earth's crust that creates seismic waves. Mainland China, located between two of the world's major seismic belts -- the Circum-Pacific Seismic Belt and the Eurasia

S. Lin and X. Huang (Eds.): CSEE 2011, Part II, CCIS 215, pp. 526–531, 2011.

Seismic Belt, suffers from frequent continental earthquakes. Movements of the Circum-Pacific Plate, the India Plate and the Philippine Plate have created large seismic fault zones in China. All of China's 23 main seismic belts have a history of violent earthquakes.

When an earthquake occurs, the structural vibration and distortion of underground works are bounded by rock and soil around, while buildings on the ground shake or tremble freely. So it is generally agreed that the seismic damage of underground works is lower than ground buildings. However, it has been demonstrated that some underground structures damaged seriously in past several strong earthquakes. In general, the degree of seismic damage of underground works can be analyzed as follows.

### 2.1 Mechanism of Seismic Damage

In general, rock failure and action of seismic inertial force are two reasons lead to the seismic damage of underground works. As the main failure reason accepted currently, the former is because the constrained force or carrying capacity of Rock and soil on the underground structure reduce or fail under cyclic seismic loading, which will result in the structural failure of underground works. These kinds of failure mostly occurred in poor geological conditions (such as liquefaction, slope failure, fault slip, etc) or shallow-buried part. The latter is because the seismic inertial force acts on structure caused by strong stratum movement, which mostly occurred in shallow-buried part or open-cut structures. As a result, the underground structure with lower stiffness is better for seismic design [1].

### 2.2 Factors Effecting Seismic Damage

**The seismic factors.** These factors relate to seismic waves such as swing, frequency, duration, wavelength, and incidence direction.

**The geologic factors.** These factors relate to the geologic environment in which the underground works constructed, including the type, characteristic of stratum, depth of embedment, ground stress and poor geological conditions. For example, it is approved that the seismic damage is less with increase of buried depth by lots of cases. As shown in diagnosis of underground works damaged Tangshan Earthquake in 1976, the seismic damage in highly seismic region was reduced rapidly with increase of buried depth increasing in the law of an exponential curve, but kept a constant under depth of 500 meters [2]. By studying 192 cases of seismic damaged underground works, Sharma and Juddlel came to the conclusion that the rate of serious damage of works within a hundred meters depth and five hundred meters depth were 17% and 18% separately, but rare damage were found over three hundred meters depth [3]. However, when works was in soft rock, the law of embedment depth and seismic damage was not evident under the influence of fault, ground stress and rock characteristic [4].

**The structural factors.** These factors relate to the characteristic of underground works used in analysis of seismic response, such as the size, supporting structure, structural stiffness and dynamic characteristics of underground works.

### 2.3 Common Seismic Characteristic and Position of Underground Works

The common seismic damage in underground works are landslides, collapse of entrances and water gushing, soil liquefaction and so on are caused by structure twisted, lining off, rock loose in earthquakes.

Generally entrance of underground works is in shallow depth. When earthquake occurred, the access of underground works was the easiest to block or Collapse by landslide and collapse of access, the liquefaction of basement and surface subsidence. In addition, it have been found that the concentrated stress of varied structural forms in access and the deformation or off track of doors were easily occurred in strong earthquake.

The corner and around the corner are the weak links of underground works which are easy to damage by stress concentration, such as the position of intersection, bend, arch, structure and shape transformation.

It had been founded that the ancillary facilities, equipments and separate lining often damaged severely when the earthquake occurs. Furthermore, secondary disasters associated with earthquake such as fire and electrical short circuit always occurred.

## 3 Threats of Tsunami to Coastal Underground Works

At the Earth's surface, earthquakes manifest themselves by shaking and sometimes displacement of the ground. When the epicenter of a large earthquake is located offshore, the seabed may be displaced sufficiently to cause a tsunami. Earthquakes can also trigger landslides, and occasionally volcanic activity.

Compared with buildings on the ground, the harm of tsunami to underground works is smaller. When the tsunami struck, the sea water carries collapsed buildings and other floating objects would damage to the entrance of underground projects. In general, if the protective doors have enough tightness and resistance, the damage will not be severe. However, if the entrance in the mountain collapses by the tsunami, the internal part of project will damage seriously. Furthermore, damages of roads and bridges outside, water supply, drainage, telecommunications, electricity and other infrastructure caused by the tsunami will affect the engineering support.

## 4 The Fortifying Methods against the Earthquake and Tsunami

### 4.1 Site Selection of Project

Most of the seismic damage of the underground engineering occurred in bad geological area such as lithology changed greatly, faultage zone, shallow underground project site, which should be avoided by geological survey before construction. If the underground engineering must be built in such bad areas, it is necessary that the underground faultage recent activity, activity modes, activity levels and other characteristics would be surveyed. A special monitoring system arranged for the active faultage, and the corresponding measures should be adopted in the design.

The import and export section of underground engineering are destroyed easily, so a better the geological location should to be choose to avoid the fractured zone, weathering and unloading zone, landslide and the saturated sand areas. The underground works near to mountain slope should be set inside the mountain, maintaining a thick outer cover. The centerline of the underground engineering should be designed above the ground water level, in particular, should avoid large areas of saturated sand zone and the water-blocking fault. The centerline should be avoid the maximum principal stress direction, to prevent ground water pressure sharply increase by the earthquake and seismic effect amplification by loose soil, and to avoid rock burst and inducing earthquake.

In response to the tsunami hazards of underground works, the regional background of strong earthquakes (the frequency of the tsunami disaster, the tsunami intensity), the distance from the coast, the terrain height and other factors should be considered in planning. Planting High-density cocos, Palmetto, betel trees and mangrove is effective in reducing and blocking the influx of tsunami waves in lowlands along the coast. In addition, constructing a serial of channels in land to reduce waves could be an effective method.

## 4.2 Increasing Structural Strength of Underground Works

**Enhancing the earthquake-resistance performance of surrounding rock.** One of the main paths to reduce the seismic damage is injecting slurry or setting anchor rod to surrounding rock, which would reduce the response of lining by way of increasing the rock stiffness. At the same time, the control organization and the structure can bear the earthquake function together by means of the earthquake-resistance system to moderate the earthquake reaction in the structure.

Isolation shake techniques developed in recent years as a damping shake technology, which uses a special measures to isolate the impact of the earthquake on the structure, and its purpose is to separate two lining and rock media by use damping layer, so that reducing and changing the intensity and methods of earthquakes on structures in order to achieve the purpose of reducing vibration. Currently rubber seismic isolation layer and the foam concrete layer were studying [5]. Among them, the rubber layer at the top and the bottom bear greater tensile stress. The rubber is easy to aging at prolonged higher stress state, and will gradually lose their seismic performance. Because rubber is also temperature sensitive material, whether to maintain the inherent performance is not known at harsh climate. But tensile stress has not occurred in the foam concrete layer, which stress state is in good condition. It is a good light absorbing materials, and its cost is lower.

**Improving earthquake-resistance of structure.** Earthquake effect on the underground structure could be reduced by changing the stiffness, mass, strength and damping characteristics of the structure. There are several measures can be used. The first is to reduce the overall quality of underground structures by using lightweight aggregate concrete, such as haydite concrete; the second is to increase intensity of the lining. The earthquake reinforcement measures in the stress concentration sites should be taken; the third is to increase the ductility and damping of underground structures by using flexible joints, polymer reinforced concrete or other measures; the fourth is to improve the structure shape. Smoother structure and shape to avoid sharp corners is as far as possible used; the last is to adjust the stiffness of the underground structure. There are two directions in underground structure stiffness adjustment: increasing the stiffness of the structure, the structure relative to the surrounding rock is a rigid structure. After the surrounding rock deformation, the structure can completely resist deformation of the surrounding rock; reducing the structural stiffness, to make the structure flexible or ductility increases to be deformed as the surrounding rock deformation. The approach of adjusting the stiffness should be estimated in advance whether it is economic and reasonable.

Numerical analysis shows [6]: seismic response of underground structure is mainly affected by the soil cover, and the relationship with its own inertia force and stiffness is not large. It is unrealistic that underground structures would be designed to resist the movement and the deformation of the surrounding soil medium. During the seismic design of underground structures, it should be appropriately increased flexibility and ductility, to increase the seismic performance. The traditional view is unreasonable that to improve the seismic performance of underground structure only rely on enhancing the structural strength. In addition, the import and export should be reinforced.

### 4.3 Strengthening Earthquake Prediction

In east Japan earthquake, people gain valuable escape time because the use of earthquake prediction and tsunami warning system. The meaning of the earthquake prediction is the advance forecast about the time, place and magnitude of earthquake. Domestic and foreign scholars believe that the occurrence of the earthquake has a certain gestation process, in which the stress state of the media will be changed significantly in the source area and the surrounding area. People may look into these changes by a variety of observations near the ground.

## 5 Conclusions

In conclusion, rock failure and action of seismic inertial force are main reasons lead to the seismic damage of underground works. When earthquake and tsunami occur, the access of underground works and structural stiffness varied will be easier to damage. The loss induced by earthquake and tsunami is huge. Compared with earthquake and tsunami, the power of human is limited. However, by strengthening the structure of underground works and improving prediction of earthquake and tsunami, we can reduce the damage to the least.

## References

1. Shao, G., Luo, W., Li, F., et al.: An Investigation on Aseismic Behaviours of Railway Tunnel Lining During Earthquake. China Railway Science 12, 92–108 (1992) (in Chinese)
2. Wang, J.: Distribution of Underground Seismic Intensity in The Epicentral Area of The 1976 Tnagshan Earthquake. Acta Seismologica Sinica 2, 314–320 (1980) (in Chinese)
3. Sharma, S., Judd, W.R.: Underground opening damage from earthquakes. Eng. Geol. 30, 263–276 (1991)

4. Li, T.: Faliure Characteristics and Influence Factors Analysis of Mountain Tunnels at Epicenter Zones of Great Wenchuan Earthquake. Journal of Engineering Geology 16, 742–754 (2009) (in Chinese)
5. Huang, S., Chen, W., Yang, J., et al.: Research on Earthquake-induced Dynamic Responses and Aseismic Measures for Underground Engineering. Chinese Journal of Rock Mechanics and Engineering 28, 483–490 (2009) (in Chinese)
6. Chang, H.: The Anti-Seismic Research of Underground Constructures. Jilin University, Changchun (2005) (in Chinese)

# Intelligent System of Circumstance Monitoring and Safety Evaluation for Underground Works

Zhongqing Cheng, Yunpeng Zhao, and Yanru Li

Department of Logistic Command and Engineering,
Naval University of Engineering, Tianjin, 300450, China
bace@tom.com

**Abstract.** An intelligent system of circumstance monitoring and security evaluation for underground works is established in this paper integrating sensing technology, computer network technology and intelligent data processing technology. Under the system, underground works security is automatically evaluated based on circumstance and video parameters collected, processed and transferred. Underground works security evaluation model is formulated in this paper, with each evaluation object categorized into different levels with respective standards for evaluation. And steps for intelligent system development are expounded and structural model of the system is also proved in this paper.

**Keywords:** Intelligent system; safety evaluation; circumstance monitoring, underground works.

## 1 Introduction

Accurate monitoring of underground works plays an important role in ensuring their regular performance and in making sound and economical plans for their preservation and reinforcement. After years of service, Underground works, due to factors like material aging, erosion and ground base wall-rock condition alternation, inevitably suffer damages in structure, the worst of which may even endanger the whole project. Effective monitoring of underground works is also a must in guaranteeing the health of the people working there and in preserving the quality of the materials and the equipment stored there. Underground works, for natural limitations, are usually in a state of high temperature, intense humidity and bad internal ventilation. However, material and equipment storage demands a high requirement on humidity and temperature. And set standards of temperature and humidity as well as those of the percentage of oxygen and density of harmful gas in the air are also required to ensure a friendly working environment for the staff there. If people stay for a long time in a low-oxygen circumstance, they will suffer chest stress and headache or even get intoxicated or suffocated [1]. Based on sensing technology and computer network technology, the intelligent system proposed in this paper effectively monitors the harmful gas in the underground works. With information structure of underground works evaluation system and unified model description of data, the whole process of security evaluation is automatically carried out through computers.

S. Lin and X. Huang (Eds.): CSEE 2011, Part II, CCIS 215, pp. 532–538, 2011.

## 2 System Structure

The system is mainly composed of three parts, namely field data collecting equipment, network technology and center for monitoring and evaluation.

The system is integrated in nature, involving sensor measuring, autocontrol and air purging technology, and electric facilities for civil decoration, damp proofing, dehumidification, ventilation and air conditioning. Analog images undergo A/D conversion and compression through video code devices and are further transferred through network to the monitoring center in the form of a data package blending images, voice code stream and data circulation. The monitoring center then, though the corollary video decoder, decodes the image code stream, generates analog signals and transfers to display devices. The data are exported to the superposed facilities either to realize online image transference through hard disc recorder or to realize coordinated warning by installing more long-period delay recorders. Additionally, video code devices are ready to receive control signals from image control board and relay them to Platform-lens camera lens controller. In this way, Platform-lens and camera lens can always be controlled, directing the cameras in focusing, positioning and the beginning and ending of the data collection process.

### 2.1 Field Data Collection Facilities

Field data collection involves collection and processing of working condition parameters and circumstance parameters and video surveillance over important sites. Working conditions parameters includes the structure of the underground works, protection devices, ventilation and air-conditioning facilities, water supply and drainage, heating and oil feeding system and electric facilities, while circumstance parameters includes wind speed, humidity and the density of carbon monoxide, carbon dioxide, hydrogen sulfide, sulfur dioxide, marsh gas. Given the special conditions in the underground works, a remote image collecting and transferring system that can be installed in the job sites is designed in this paper integrating the functions of video signals collection and transference. The data collected undergo real-time analysis through advanced DSP devices before being transferred to the central facilities. Fig. 1 is a schematic diagram of the field data collection.

The analog video input signals from cameras are transferred to the A/D converter via amplification circuit. Then they are stored in frame storage units after A/D conversion. The Digital Signal Processor (DSP) helps achieve soft synchronism by sorting out both line and field sync signals, and the video image data of the sorted sync signals are rewritten back to frame storage units after accurate sampling and image cutting by CPLD. DSP computes the video image data, sending video pictures when conditions are met [2].

The executive sector can automatically switch the three ventilation modes(namely natural ventilation, purifying and filtering ventilation and closed ventilation), making the system more economical and reliable.

Sensor layout also accounts for another major field of research in data collection. This includes: ①.the number of the sensors, i.e. in what way the number of sensors can be reduced without affecting the access to the dynamic information of the project;

②.precision of data collection, for mismatched sensor configuration will affect the precision of parameter identification. In order to digitalize the parameters and transfer them through network to monitoring center, some data call for manual input [3].

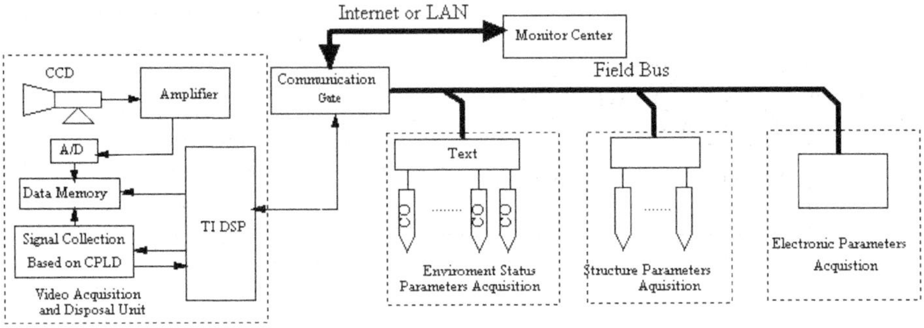


**Fig. 1.** Schematic diagram of the field data collection

## 2.2 Network Transference Facilities

The monitoring system adopts the combination of field bus, Internet, TCP/IP and OPC as its transferring method, as is shown in Fig. 2.

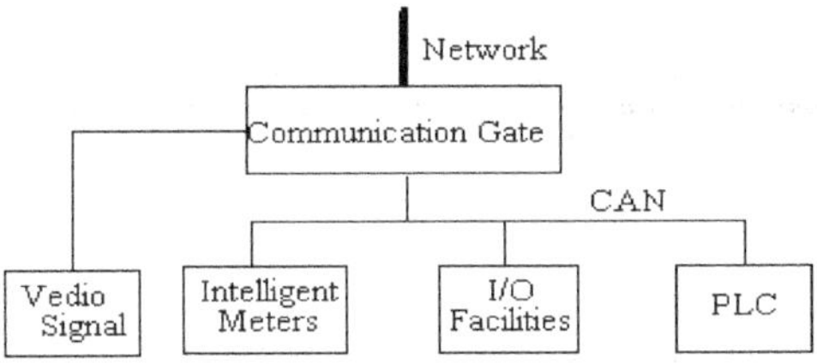


**Fig. 2.** Network between field bus and Internet

The major feature of the monitoring network is that system communication is based on combined communication protocols between the Internet and field bus and interlinkage and mutual operation between computer systems and field meters and facilities are realized by linking the high-speed computer network to the relatively low-speed field bus through gateway. With corresponding TCP/IP protocol and network parameters (Parameters like IP address, mask code and gateway are assigned beforehand by the center terminal) in the network configuration, the far-end central facilities will automatically establish linkage when they are turned on.

## 2.3 Monitoring and Evaluation Center

Monitoring and evaluation center is composed of hard disk recorder and servers like database, image control, operation, monitoring and data collections, as is shown in Fig. 3. The center decodes the video and acoustic information from several underground works, sending video information to the monitors and acoustic information to audio facilities like loudspeakers. People in the center are free to watch any images of all underground works. They can also control the moving and turning of cameras and their focusing, dimming, zooming in and zooming out. Besides, the system enjoys several other functions. First, the system enjoys the capacity for mutual dialogue. People in the center can choose to monitor and conduct dialogue with those in any of the underground works. The sound from the center can also be broadcast to all underground works. Video images are converted into digital data through digital compression technology and uploaded in the intranet so that people in charge and relevant departments can have rapid and real-time access to what is going on in the underground works. Data from other monitoring systems can also be shared through the transparent package, forming an integrated monitoring system of images and data. Secondly, the system has the function of coordinated warning. When the system gives an alarm, the lighting system will be automatically initiated so as to get clearer recorded images of the alarm-initiating spots. At the same time, the images are switched and long-period delay recorders are automatically set to work.

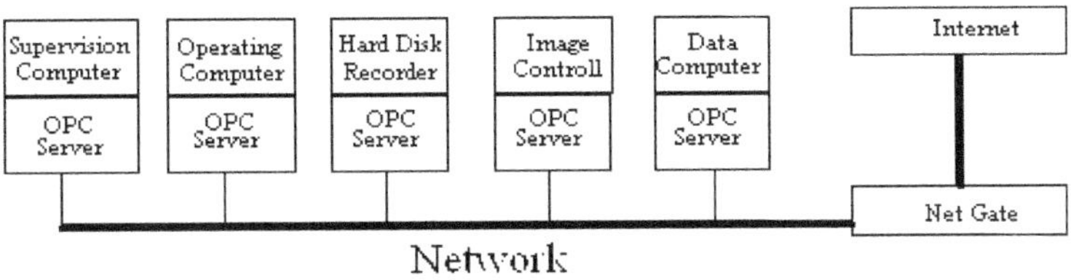


**Fig. 3.** Structure of monitoring and evaluation center

The system can monitor all images from the underground works and their function mode and performance parameters. Fourthly, parameters of the system can be configured and modified. Operators have to log in before they can enter the control panel. And all operations are recorded and stored for future check. Fifthly, the system has the function of warning filtering. When an alarm-initiating accident causes a series of warnings, the system filters all other related warnings presenting only the one directly by the accident. There are several branch command boards to the system, which makes management more convenient.

Field monitoring, data computation, data analysis and evaluation can be simultaneously carried out in the monitoring center. And the system is composed of knowledge base, structure contra-analysis base, intelligent reasoning subsystem and multimedia interactive subsystem. Various table and literal reports on underground works security are brought out through the effective performance of intelligent evaluation system.

## 3 Setup of Security Evaluation Models

The security evacuation mode is hiberarchy, as is shown in Fig. 4. In order to set up a security evacuation model for underground constructions, several layers are defined to evaluate the target, and evaluation levels standard are also defined.

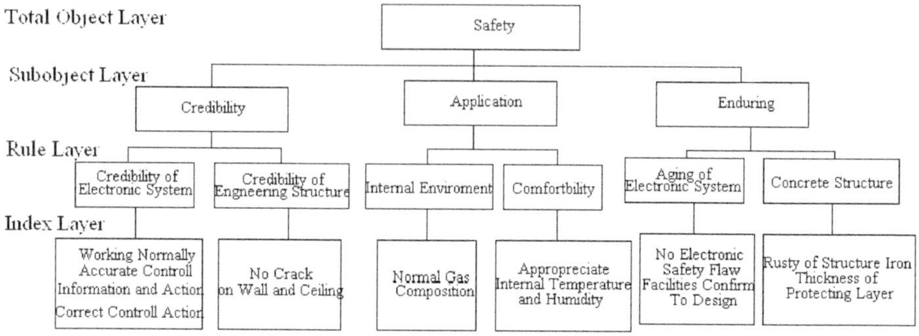


**Fig. 4.** Hiberarchy of evaluation model

### 3.1 Definitions of Evaluation Layers

Object layer: the security of underground constructions is the evaluation objective. This sets up a overall requirement for the underground constructions and provides basis for the final evaluation.

Object sublayer: it is a evaluation of the security, adaptability, and endurance of the underground constructions.

Rule layer: it's a classification of influence index for evaluating sub-targets.

Index layer: it refers to the evaluation index used for direct examination, quantification, and qualitative descriptions.

### 3.2 Definitions of Examination Index

Reliability index: in the regular circumstances, to the level of underground constructions facilities should not be under the project requirement as well as the construction criteria.

Adaptability index: under the regular use, the underground constructions facilities should have good working performances.

Endurance index: under the regular use and maintenance, the underground constructions facilities should possess sufficient endurance within the designed lifespan.

### 3.3 Evaluation Level Standards

Terms of Level are defined to evaluate the security of underground constructions facilities. Four levels are defined for object layer, object sublayer, rule layer and index layer. The level will be determined with the evaluation index and the evaluation value obtained in the fieldwork as per the following standards:

Level one: compliance with existing criteria and project requirement, secure and enduring, no other measures needed.

Level two: slightly under existing criteria and project requirement, basically adaptable, with reasonable endurance, no other measures needed at the moment.

Level three: not compliant with existing criteria and project requirement, security and regular use are damaged, extra measures is supposed to be applied.

Level four: not compliant with existing criteria and project requirement, regular use and endurance are rarely possible, extra measures must be applied.

## 4 Development of Security Auto-evaluation System

This system is developed on the OS platforms of Windows, and a WYSISWYG programming language, Visual C++, is used for it. The powerful object-oriented features of Visual C++ were fully utilized in the development process and helped carry out a user-friendly interface, which is compatible with those of windows and combines graphics and texts. And, in order to make the evaluation system easy to deploy, install and uninstall programs are created and packaged within the system.

Security of underground constructions can be evaluated, analyzed and compared according to the construction security index. In order to get a more reliable security evaluation index, a relevant model system must be set up to support the obtaining of security index. On the basis of the study on security evacuation index system, structure of the quantitative model of security evaluation index and it supportive relationship with the security evaluation index can be determined.

Features of auto-evaluation system include:

Imports raw data and process them, automatically select and compute evacuation parameters;

Automatically assign evaluation index values;

Determines the evaluation values and grades for the underground construction facilities that are being evaluated;

Gives out the evaluation conclusion and suggests measures that should be adopted;

The system is mainly composed of the following:

Evaluation parameters processing module. Human-machine communication methods are used to determine all the parameters that the evaluation needs. Dialogs prompt users to input original data, which is processed right after.

Evaluation and grading module. This module will automatically perform evaluation per evaluation model, raw data, and evaluation index, and then, on the basis of the evaluation results, determining the grade of the underground constructions facilities.

Solution determining module. This module use AI to select and determine the solution, per evaluation results and the grade of the underground construction facilities.

System control module. This module includes sub-modules such as menu control, human-machine communication control, inputs and outputs control, etc.

Multimedia module. In development of the system, the multimedia functions of Visual C++ have been fully used. Voice prompts are enabled, which inquest the user-friendliness.

Online help module. This system has a powerful online help feature, so as to help users use the system correctly.

## 5 Conclusions

The intelligent system is applicable in various underground projects. It can provide scientific basis for monitoring and improvement on factors like structure state parameters, temperature, humidity, percentage of oxygen and harmful gas in the underground works as well as for analysis on the preservation of the works and on the existence conditions of staff working there. Round-day monitoring of the evaluation system makes it possible to get timing knowledge of the security state in the underground works. Thus, the system will help a lot in underground works management improvement and in scientific decision making.

**Acknowledgement.** The authors gratefully acknowledge to the PLA naval research plan for support.

## References

[1] Ren, X.-h., Wang, G.-h., Zhang, S.-j.: Study on harmful gases detecting and warning technique. Journal of Transducer Technology 20(3), 6–8 (2001) (in Chinese)
[2] Sun, J.-p., Guan, Y., Zhang, J.: Mine long-range video supervision system based on DSP. Journal of Xiangtang Mining Institutes 18(1), 7–9 (2003) (in Chinese)
[3] Jia, Y., Cao, D., Yang, J.: Development and application of an intelligent monitoring system for underground works. Railway Standard Design (8), 52–54 (2004) (in Chinese)

# Research on Application of Collaborative Filtering in Electronic Commerce Recommender Systems

Wangjun Zhang

Zhejiang Business Technology Institute, Ningbo 315012, China
wangjun_zhang1@163.com

**Abstract.** In electronic commerce era, personalized recommender systems are popularly being employed to help users in selecting suitable items to meet their personal requirements. These systems learn about user interests over time and automatically suggest items that fit the learned model of user interests. It is important for companies to develop web-based marketing strategy such as product bundling to increase revenue. Recommendation system is a platform that can be used to reduce the searching cost of users, increase the effectiveness of promotion strategies and enhance loyalty. The core technology implemented behind this type of recommender systems includes content analysis, collaborative filtering and some hybrid variants. Collaborative filtering is a data analysis task appearing in many challenging applications and it can often be formulated as identifying patterns in a large and mostly empty rating matrix. In this paper, firstly, the principle of collaborative filtering recommendation is introduced. Then, describes the workflow of the collaborative filtering algorithm. Unresolved issues of collaborative filtering technology and research directions are pointed out finally.

**Keywords:** electronic commerce; recommender system; collaborative filtering; personalized.

## 1 Introduction

E-commerce from the physical environment is to achieve the transfer of the virtual environment. An important advantage of its network is to provide a more convenient messaging and information services channel. But e-commerce has also brought an important issue. Online merchants providing the type and quantity of goods are very large, but the user via a small computer screen can not easily find products of interest. Users do not want to spend too much time online to find the product in the endless, as in the physical environment as checking the quality of goods. Therefore, how to provide customers with product information and advice to help customers successfully complete the purchase process, that becomes an urgent need for e-commerce site [1,2,3,4,5].

Personalized recommender systems are popularly being employed to help users in selecting suitable items to meet their personal requirements. These systems learn about user interests over time and automatically suggest items that fit the learned model of user interests. It is important for companies to develop web-based marketing

S. Lin and X. Huang (Eds.): CSEE 2011, Part II, CCIS 215, pp. 539–544, 2011.

strategy such as product bundling to increase revenue. Recommendation system is a platform that can be used to reduce the searching cost of users, increase the effectiveness of promotion strategies and enhance loyalty. The core technology implemented behind this type of recommender systems includes content analysis, collaborative filtering and some hybrid variants. Collaborative filtering [6,7,8] is a data analysis task appearing in many challenging applications and it can often be formulated as identifying patterns in a large and mostly empty rating matrix.

In this paper, firstly, the principle of collaborative filtering recommendation is introduced. Then, describes the workflow of the collaborative filtering algorithm. Unresolved issues of collaborative filtering technology and research directions are pointed out finally.

## 2 Collaborative Filtering

Collaborative filtering systems use statistical techniques to search for the target customers of certain nearest neighbor, and neighbors of the project according to a recent target of the project score in predicting the score, resulting in the corresponding list of recommendations[9,10,11]. Collaborative filtering process is shown in Figure 1.

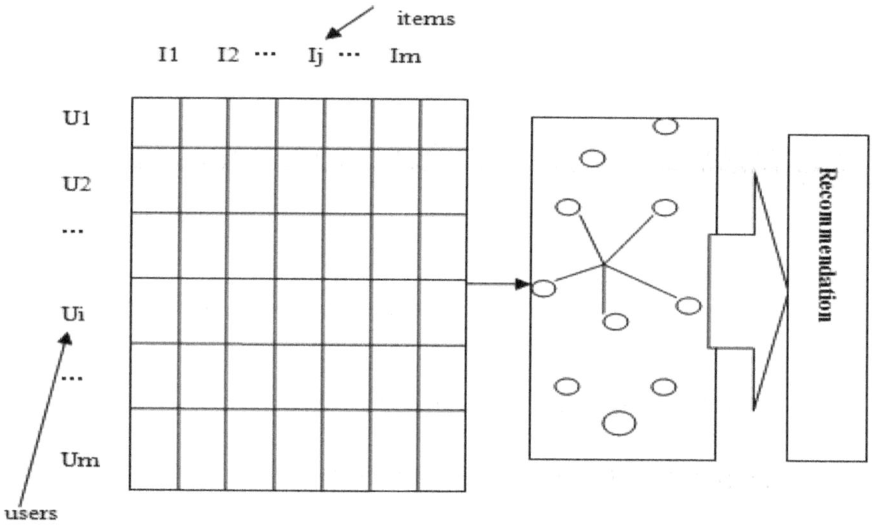


**Fig. 1.** The workflow of collaborative filtering

### 2.1 The Idea of Collaborative Filtering Algorithms

Collaborative filtering is based on the user's interests neighbor prediction preference for the target user's interest. Firstly, using statistical techniques to find the target user preferences with the same neighbors. Then depending on the target user's preferences neighbors have recommended to the target users. If the user ratings on some of the projects were similar, their scores on other projects are also similar. If most of the users rating of some of the projects were similar, the current user rating of these projects is also more similar. Collaborative filtering system uses statistical techniques to search for the target number of nearest neighbor users, then the nearest neighbor of the project score in predicting the target user's rating on the project, resulting in the corresponding list of recommendations.

## 2.2 Data Representation

The user already had a project evaluation model can effectively measure the similarity between users. User score data can be a m × n-order user - item rating matrix, m-line users on behalf of a m, n column represents the n items, the first i elements in row j column i represents the user j's rating on the project value, as shown in Table 1.

**Table 1.** Data Representation

| | Item1 | ... | Itemj | ... | Itemn |
|---|---|---|---|---|---|
| User1 | R11 | ... | R1j | ... | R1n |
| ... | ... | ... | ... | ... | ... |
| Useri | Ri1 | ... | Rij | ... | Rin |
| ... | ... | ... | ... | ... | ... |
| Userm | Rm1 | ... | Rmj | ... | Rmn |

## 2.3 Similarity Measure

There are many ways of similarity measure between users [12,13,14,15]: Cosine-based Similarity, Adjusted Cosine Similarity and Correlation-based Similarity. First, the user i and user j score over all items, and then measure the similarity of different user i and user j calculate the similarity between, denoted by sim (i, j).

(1) Cosine-based Similarity
User Rating seen as the n-dimensional vector space entry, if the user did not rate their items, then the score of the users is set to 0, the similarity between users through the cosine of the angle between the measurement vector. Let user i and user j in the n-dimensional space on the item scores were expressed as a vector of user i and user j the similarity between the sim (i, j) is:

$$sim(i,j) = \cos(\bar{i},\bar{j}) = \frac{\bar{i},\bar{j}}{\|\bar{i}\|\|\bar{j}\|}$$

(2) Adjusted Cosine Similarity
Fixed full account of the cosine correlation score scale of the problem of different users, by subtracting the user ratings on the project to realize its advantages. Let user i and user j score set off phase, the user i and user j is the similarity between the sim (i, j) is:

$$sim(i,j) = \frac{\sum_{c\in I_{ij}} (R_{ic} - \overline{R}_i)(R_{jc} - \overline{R}_j)}{\sqrt{\sum_{c\in I_{ij}} (R_{ic} - \overline{R}_i)^2}\sqrt{\sum_{c\in I_{jc}} (R_{jc} - \overline{R}_j)^2}}$$

(3) Correlation-based Similarity

Pearson correlation coefficient, also known as the relevant measure of similarity, let user i and user j have a common collection of items with scores Ii, j = I1 ∩ I2 said that the user i and user j, the similarity sim (i, j) is:

$$sim(i, j) = \frac{\sum_{c \in I_{i,j}} (R_{i,c} - \overline{R}_i)(R_{j,c} - \overline{R}_j)}{\sqrt{\sum_{c \in I_{i,j}} (R_{i,c} - \overline{R}_i)^2} \sqrt{\sum_{c \in I_{j,c}} (R_{j,c} - \overline{R}_j)^2}}$$

### 2.4 The Formation of a Neighbor Set

The formation of a collection of commonly used is neighbor Top-N and K nearest neighbor method. Core of the algorithm is a need to recommend the service to find the most similar to the target user nearest neighbor set. According to a predetermined number of neighbors N, the algorithm using the above similarity descending order according to selected users as neighbors before the N sets of users. Or according to a predetermined similarity threshold, select all the similarity greater than the threshold value as the neighbor sets of users.

### 2.5 Recommendation

Nearest neighbors based on the current user information on the item score predicts the current user does not score the item score, producing Top-N item recommendations. Through the above similarity measure proposed by the target user's nearest neighbors, the next step required to generate the appropriate recommendation. The formula is as following:

$$P_{ui} = \overline{R}_u + \frac{\sum_{j \in NN_u} sim(u, j)(R_{j,i} - \overline{R}_j)}{\sum_{j \in NN_u} |(sim(u, j))|}$$

## 3 Further Work

### 3.1 Sparsity

E-commerce sites often have a large number of commodities, and buy or make an assessment of each user only a very small part of it. Usually less than 1% of the evaluation data, the data will lead to sparse algorithm accuracy rate. The correlation coefficient of at least two users need to evaluate the same product to more than two terms, both in fact similar users are likely to result in a lack of the same products are not similar. A solution is currently in the use of specific collaborative filtering algorithms before using other technologies to fill the original data. On the other hand through the data to fully exploit the potential implicit in the model, breaking the collaborative filtering needs related to the user marked the same project limits.

### 3.2 Cold Start

For a new user, the system does not have any purchase information of the user record and can not find its nearest neighbor, which can not be recommended. Similarly, when the system by adding a new project, the project did not score records, you can not find out the nearest neighbor and make recommendations or ratings predictions. Collaborative filtering systems that exist in this kind of problem is called cold start problem [16,17,18]. In order to solve the cold start problem, widely used is content-based nearest neighbor search technique.

### 3.3 Scalability

The use of e-commerce is a huge amount of data involved, and such as nearest neighbor algorithm efficiency with the increase in the number of users and products decreased [19,20]. How to improve the algorithm to make it more responsive to large-scale computing is one focus of the study. Often used to solve this problem is the user screening, on the one hand can be different for different given the weight recommended by the other hand, the user can select the nearest neighbors to make certain improvements.

### 3.4 Real-Time

Problem of real-time online service with the need for substantial increase in the customer becomes more and more prominent, effective way to solve this problem is to use distributed computing technology. Another project-based clustering collaborative filtering algorithm can significantly reduce the nearest neighbor of the query space, in order to effectively solve large-scale data processing recommendation system problems faced by real-time.

## 4 Conclusions

In this paper, we introduce the principle of collaborative filtering recommendation. Then, we describe the workflow of the collaborative filtering algorithm. At last, we point out unresolved issues of collaborative filtering technology and research directions.

## References

1. Goldberg, D., Nichols, D., Oki, B.M., Terry, D.: Using collaborative filtering to weave an information tapestry. Communications of the ACM 35(12), 61–70 (1992)
2. Sarwar, B.M., Karypis, G., Konstan, J.A., Riedl, J.: Analysis of Recommendation Algorithms for E-Commerce. In: Proceedings of the ACM EC 2000 Conference, Minneapolis, MN, pp. 158–167 (2000)
3. Gong, S.: Employing User Attribute and Item Attribute to Enhance the Collaborative Filtering Recommendation. Journal of Software 4(8), 883–890 (2009)
4. Resnick, P., Iacovou, N., Suchak, M., Bergstrom, P., Riedl, J.: Grouplens: An open architecture for collaborative filtering of netnews. In: Proceedings of the ACM CSCW 1994 Conference on Computer-Supported Cooperative Work, pp. 175–186 (1994)
5. Herlocker, J.: Understanding and Improving Automated Collaborative Filtering Systems. Ph.D. Thesis, Computer Science Dept., University of Minnesota (2000)

6. Breese, J., Hecherman, D., Kadie, C.: Empirical analysis of predictive algorithms for collaborative filtering. In: Proceedings of the 14th Conference on Uncertainty in Artificial Intelligence (UAI 1998), pp. 43–52 (1998)
7. Balabanovic, M., Shoham, Y.: FAB: Content-based collaborative recommendation. Commun. ACM 40, 3 (1997)
8. Basu, C., Hirsh, H., Cohen, W.: Recommendation as Classification: Using Social and Content-based Information in Recommendation. In: Recommender System Workshop 1998, pp. 11–15 (1998)
9. Gong, S.: A Collaborative Filtering Recommendation Algorithm Based on User Clustering and Item Clustering. Journal of Software 5(7), 745–752 (2010)
10. Sarwar, B., Karypis, G., Konstan, J., Riedl, J.: Item-Based collaborative filtering recommendation algorithms. In: Proceedings of the 10th International World Wide Web Conference, pp. 285–295 (2001)
11. Billsus, D., Pazzani, M.J.: Learning Collaborative Information Filters. In: Proceedings of ICML 1998, pp. 46–53 (1998)
12. Jin, R., Si, L.: A bayesian approach toward active learning for collaborativefiltering. In: Proceedings of the 20th Conference on Uncertainty in Artificial Intelligence, pp. 278–285. AUAI Press, Banff (2004)
13. Gong, S.: Privacy-preserving Collaborative Filtering based on Randomized Perturbation Techniques and Secure Multiparty Computation. International Journal of Advancements in Computing Technology 3(4), 89–99 (2011)
14. Gong, S.: An efficient collaborative recommendation algorithm based on item clustering. In: Luo, Q. (ed.) Advances in Wireless Networks and Information Systems. LNEE, vol. 72, pp. 381–387. Springer, Heidelberg (2010)
15. Resnick, P., Varian, H.R.: Recommender Systems. Special Issue of Communications of the ACM 40(3) (1997)
16. Schafer, J.B., Konstan, J., Riedl, J.: Recommender Systems in E-Commerce. In: Proceedings of ACM E-Commerce 1999 Conference (1999)
17. Schafer, J.B., Konstan, J., Riedl, J.: Electronic Commerce Recommender Applications. Journal of Data Mining and Knowledge Discovery 5(1/2), 115–152 (2001)
18. Karypis, G.: Evaluation of Item-Based Top-N Recommendation Algorithms. Technical Report CS-TR-00-46, Computer Science Dept., University of Minnesota (2000)
19. Gong, S.: A Personalized Recommendation Algorithm on Integration of Item Semantic Similarity and Item Rating Similarity. Journal of Computers 6(5), 1047–1054 (2011)
20. Xie, B., Han, P., Yang, F., Shen, R.-M., Zeng, H.-J., Chen, Z.: DCFLA: A distributed collaborative-filtering neighbor-locating algorithm. Information Sciences 177, 1349–1363 (2007)

# A Model of Real-Time Supply Chain Collaboration under RFID Circumstances

Yi Tao[1] and Youbo Wu[2,*]

[1] Institute of Modern Service, Zhejiang Shuren University, Hangzhou, 310015, China
[2] Institute of Manufacturing Engineering, Zhejiang University, Hangzhou, 310027, China
willwub@163.com

**Abstract.** In the increasing requirements of quick responses to the market, for the market competition became more fiercely and the customers' diversified needs of product increased. Enterprises realized the importance of collaboration between suppliers and partners, especially for saving response time to order. More and more enterprises feel the time pressure during the process of customer's need when they should satisfy customer's individual demands. Lowering down the logistics time means lowering logistics cost, especially the stock cost. Thus the market competition ability of business enterprises could be raised. While RFID(Radio Frequency Identification) techniques could reduce the dealing time on supply chain. How to use RFID techniques to decline inventory and waste is a tough issue. This article targeted on collaboration of supply chain in clothing industry for research object, a valid method for modeling of real-time supply chain collaboration was proposed. RFID technology was applied in stock management, as optimization method was adopted for inventory control based collaboration. A collaboration model was build to decline time, cost and waste. Finally, the application method was verified according emulation.

**Keywords:** RFID, Real-Time Supply Chain, Collaboration, Optimization.

## 1 Introduction of RFID Technology and Its Application

RFID (Radio Frequency Identification) technique is a kind of non-touch automatically identification technique. The basic theory is making use of the radio frequency signal and space coupling to deliver and realize the automatically identification of products [1]. Because of the characteristics of RFID with non-touch and high speed automatically identification, we can establish logic connection from the RFID label information to the DBMS systems for tracking real-time status of products in Supply Chain.

In the logistics and the supply chain management systems, RFID is extensively been applied in collecting data, especially in situation that needs high accuracy of data. RFID can carry out fast logistics information obtaining, which could support management and controlling decision. Yi Wei-Ming[2] added the price factor for

*Corresponding author.

S. Lin and X. Huang (Eds.): CSEE 2011, Part II, CCIS 215, pp. 545–551, 2011.

reference and made use of a RFID technique to carry on the stock control and optimization research. Ye Jianfang[3] used RFID in clothing warehouse management. Reference [4] to reference [9] showed models in telecommunication systems that the cooperation how to be carried out. From which we saw that RFID technique has been successful widely applied in supply chain, then it inspired us to use RFID technology in real-time supply chain collaboration.

## 2 The Model of Real-Time Supply Chain Collaboration in a Clothing Manufacturing Enterprises

### 2.1 Problem Description

The product warehouse is the bottleneck of supply chain in our case. There usually have dozens of thousands suits, while the highest number could reach 100,000 sets around, the lowest number in off season also have 70,000 sets or so in the product warehouse. Currently the existent key problems in the product warehouse management are as follows.

(1) The great capacity goods for check-in and check-out. According to produced product reached to store in warehouse, there was generally more than 2000 clothing for store everyday. When check-in, the information of every suit should be correct as well as the stored location. While check-out, the operator should find out each suit to pack from each position of the product warehouse according to customer's order. While picking to order, one should check each suit packed the right price according to customer's needs. If not, the operator must change the price correct. Because the information of each price was not in the information system, it needs manpower to ensure the right price.

(2)The workload for stocktaking during delivery is huge. While delivering goods in addition to check price, the goods number, outer materials, inner material also need to check to ensure whether it accords with the customer's order or not. The stocktaking needs 5 or 6 people spend 10 days to complete at once.

(3)The returned goods from monopoly stores puzzled. Once there are returns of goods, the price should be checked as check-in or delivery.

Therefore, the RFID technique needs to apply for decrease the workload of warehouse management in the clothing manufacturing warehouse. The mainly work is to cut down the workload of the product warehouse management and rise up the efficiency and accuracy of the product warehouse management.

### 2.2 The Proposal of RFID Application for Inventory Control in Supply Chain Collaboration

**The proposal of RFID application in product warehouse stocktaking.** On consideration of the wide warehouse area, the fixed-type long-distance RFID reader was abandoned, although that plan could raise stocktaking efficiency. Whatsoever, on consideration of the efficiency of facilities, the workload amount, cost, etc. The plan of mobile-type RFID stocktaking was adopted. In which the staff used the long-distance RFID reader & notebook to walk along aisle in warehouse as Fig. 1 showed.

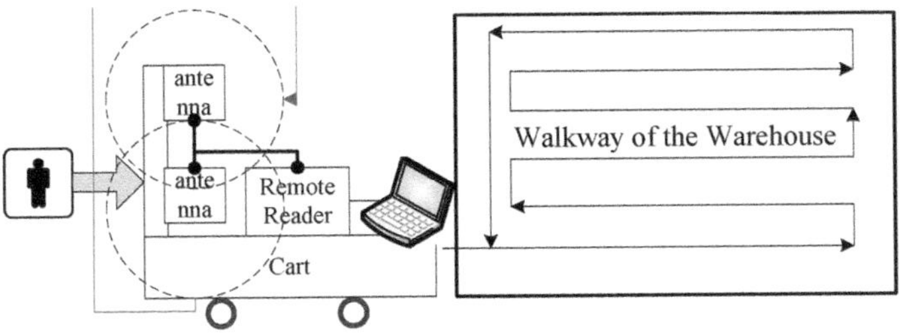


**Fig. 1.** The plan of RFID application in stocktaking in warehouse

This plan doesn't need to carry on a reformation to the production warehouse management. Meanwhile many existed small trolleys could be used. It only need to put the antennas, long-distance RFID card reader machines and notebook on the small trolley, as long as the staff pushes the car to walk over each position of the production warehouse, the half-automation of stock stocktaking could been carried out.

**The facility plan of RFID in the product warehouse.** The RFID facilities in the product warehouse included RFID equipments, computers, WAN, etc. as Fig. 2 show.

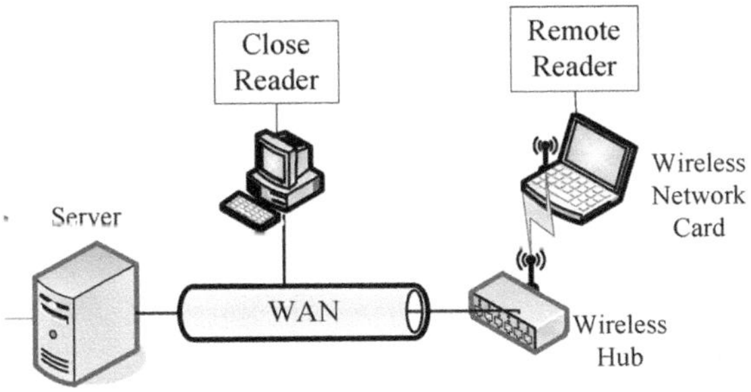


**Fig. 2.** RFID infrastructure in product warehouse

In this project, through connecting the data from close RFID reader and the remote reader to DBMS system within server, the logic contact could be built. Therefore, it can immediately carry out real-time data collection and management in product warehouse, the management function such as check-in, re-packing, check-out, delivery, could be carried out, etc.

## 2.3 The Model of Real-Time Supply Chain Collaboration in the Distribution Warehouse

The distribution warehouse connected the factories and retailers, while the inventory level lied on the producing capacity of the factories and demand of retailers. The stock capacity and transportation capacity are limited, thus the supply time was limited. One could not get what desired in time and so on. According to these, we assumed the model of real-time supply chain collaboration among factories, distribution warehouse and retailers as follow.

a) During the set time period $T$, the requirement of retailers was normal distribution.
b) One retailer could accept service from only one distribution warehouse in time period $T$. However, during each $T$, the retailer could choose other distribution warehouse for services.
c) The distribution warehouse kept safety storages and (Q, R) check strategy to serve retailers.
d) The storage level of retailers was ignored.

$T$ was a collection of time set, one time set was described as $t$. $I$ was a collection of retailers, one retailer was described as $i$. $J$ was a collection of distribution warehouses, one distribution warehouse was $j$. $d_{it}$ means during period $t$, the average requirement of retailer $i$. $v_{it}$ means the standard deviation during period $t$. $D_{jt}$ means the average demand of distribution warehouse to factory. $V_{jt}$ means the standard deviation of distribution warehouse to factory. $h_{jt}$ means the unit inventory cost in period $t$. $F_{jt}$ means the one purchase cost of distribution warehouse $j$ in $t$. $\alpha$ means the ratio of availability in distribution warehouse. $Q_{jt}$ means the purchasing volume of distribution warehouse $j$ in $t$. $R_{jt}$ means the purchasing point of distribution warehouse $j$ in $i$. $S_{jt}$ means the safety storage volume of distribution warehouse $j$ in period $t$. $L_{jt}$ means the lead time of purchasing in distribution warehouse. $CI_{jt}$ means the biggest storage volume of distribution warehouse $j$. $CV_{jt}$ means the biggest transportation volume from the factory to distribution warehouse at once. $a_{jt}$ means the unit cost of transportation from the factory to distribution warehouse $j$. $b_{jt}$ means the unit cost of transportation from distribution warehouse $j$ to retailers. $FA_{jt}$ means the opening cost of distribution warehouse $j$ in period $t$. $FB_{jt}$ means the re-opening cost of distribution warehouse $j$ in period $t$. $FC_{jt}$ means the closing cost of distribution $j$ in period $t$. $FO_{jt}$ means the operation cost of distribution $j$ in period $t$. $\lambda$ means the efficient days for requirement. $X_{jt}$ means whether the distribution warehouse $j$ was closed or not in period $t$, if open, the value is 1, otherwise the value is 0. $Y_{ijt}$ means the retailers were served by distribution warehouse $j$, if yes, the value is 1 or the value is 0. So we can get the following equations.

$$Q_{jt} + R_{jt} \le CI_{jt} \tag{i}$$

$$0 \le Q_{jt} \le CV_{jt} \tag{ii}$$

$$\begin{aligned} &\left( \sum_{t=i}^{i} \sum_{j} FA_{jt} X_{jt} + \sum_{t=2}^{T} \sum_{j} FA_{jt} X_{jt} \left(1 - X'_{jt}\right) \right) + \sum_{t}^{T} \sum_{j} FO_{jt} X_{jt} + \sum_{t=1}^{T-1} \sum_{j} FC_{jt} X_{jt} \left(1 - X_{j,t+1}\right) + \\ &\min \sum_{t=3}^{T} \sum_{j} FB_{jt} X_{jt} \left(1 - X_{j,t-1}\right) X''_{jt} + \sum_{t} \left( \lambda \sum_{i} \sum_{j} \left(a_{jt} + bi_{jt}\right) d_{it} Yi_{jt} \right) + \\ &\sum_{t} \left( F_{jt} \frac{\lambda \sum_{i} d_{it} Yi_{jt}}{Q^{*}_{jt}} \right) + \frac{h_{jt} Q^{*}_{jt}}{2} + h_{jt} z_{\alpha} \sqrt{L_{jt} \sum_{i} v_{it}^{2} Yi_{jt}} \end{aligned} \tag{iii}$$

$$s.t. \sum_{j \in J} Yi_{jt} = 1 \qquad i \in I, t \in T \tag{iv}$$

$$Yi_{jt} \le X_{jt} \quad i \in I, j \in J, t \in T \tag{v}$$

$$X'_{jt} = \max\{X_{j1}, X_{j2}, \ldots, X_{j,t-1}\} \tag{vi}$$

$$X''_{jt} = \max\{X_{j1}, X_{j2}, \ldots, X_{j,t-2}\} \tag{vii}$$

$$Xjt \in \{0,1\} \quad j \in J, t \in T \tag{viii}$$

$$Yi_{jt} \in \{0,1\} \quad i \in I, j \in J, t \in T \tag{ix}$$

While among the equations, equation (**iii**) was the objective function. In equation (**iii**), the first item means the first opening cost, and the second item means the operation cost of the distribution warehouse. The item three in equation (**iii**) means the closing cost, as the forth item was reopening cost, the fifth item was the transportation cost from factory to distribution warehouse as well as transportation from distribution warehouse to retailers. The sixth item means the keeping cost of the distribution warehouse.

After the model establishment, we use Particle Swarm Optimization (PSO) algorithm to calculate out the best result. The equation was as follows.

$$v_i^{t+1} = \varpi v_i^t + c_1 r_1 (p_{best} - x_i^t) + c_2 r_2 (g_{best} - x_i^t) \tag{x}$$

$$x_i^{t+1} = x_i^t + v_i^{t+1} \tag{xi}$$

In equation (**x**) and (**xi**), $\varpi$ means inertia weight, $c_1$ and $c_2$ are constant; $r_1$ and $r_2$ are random number between 0 and 1. Thus we could used this algorithm to calculate the result out in MATLAB.

## 3 The Control Method of Real-Time Supply Chain under RFID Circumstances in the Specified Clothing Industries

For the supply chain collaboration, the stock management of product warehouse was mainly connected to market sale variety and the production plan of ready-to-wear clothes. In spite of the cow whip effect, the RFID is widely used in each sub- system to decrease the influence. Particularly during the retail business, the precision information collection & prediction with market is the foundation and premise of stock optimization of production warehouse.

### 3.1 The Application Proposal of RFID in Retailing

The main work of applying RFID technique in the clothing retailing is to add logistics information during packaging, transportation, delivery, sale or storing, etc., among them the packaging mainly occurs at the door of production warehouse. However, the transportation and distribution could be carried out by itself, as well as by the 3rd Part Logistics (like railroad, postal service, etc.) to complete. Then the goods are sent to stores for sales, the figure 3 showed the activity process.

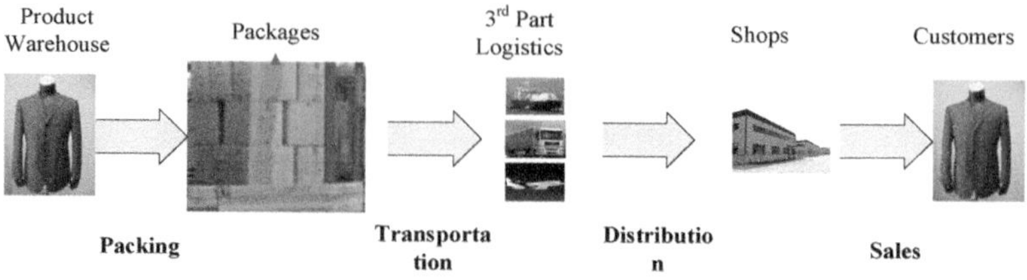


**Fig. 3.** Activities of clothes distribution after producing

## 3.2 Information Model in Retailing Logistics Based on RFID

From the point of supply chain collaboration, process of clothing retailing, we can see that there are information like delivery order, packaging order, transportation order, etc. While applying RFID technique in the process of clothing retail and sale, it needs to build up the related information model which describes the above-mentioned voucher. In which except for transporting list, the other all relates to clothing product.

The model mainly accords to the sale request, mainly was used for sort in the product warehouse, and part of information (order number, date of goods, goods number, dimensions, amount, etc.) is automatically born. When the 3rd logistics was confirmed, the trunk RFID tags needs to write in information like carrier, consigner, reception, transportation date, etc.

The clothing RFID label was mainly used for inner management, which includes monopoly stores' sale. The related information about delivery order and delivery goods within RFID label was also used for the inner management. However the information about carrier, consigner and reception was mainly served for the 3rd logistics. So except these delivery data, all others could be secret for business protection.

## 3.3 RFID-Based Control Model of Stock for Supply Chain Collaboration

Because the supply chain was mainly affected by dynamic market needs, the control model of the product stock mainly depends on sale of season variety, sale strategy of monopoly store, etc. So we can use sales forecasting model to decide the control model of product stock. From the sale profits control model we can get:

$$P(s_t,(a,y_t)) = \sum_{i=0}^{D}\left[pr(u_{ta}=i)\{G(a,y_t)-(h[k_t-i])-(b[x_t-i])\}\right]-G_a \qquad \text{(xii)}$$

*P* stands for sales profit in the formula (**xii**), $P(s_t,(a,y_t))$ means the sales amount in period $S_t$., $G_a$ is the sale cost of the each one period. *h* is stock cost, *G* is a selling price, and *b* is depreciation cost. For certain products, the stock cost and depreciation cost are constant. We can get $\text{Max}\{P(s_t,(a,y_t))\}$, at the same time the inventory control model of production warehouse is:

$$C_{all} = \frac{\sum_{i=1}^{N} Ci}{N} = \frac{1}{N}\sum_{i=1}^{N}\left(C_i^{steady}+C_i^{per}+C_i^{order}+C_i^{trasfer}\right) \qquad \text{(xiii)}$$

In equation (**xiii**), $C_{all}$ is all cost of product stock, which includes fixed stock cost $C_{steady}$, the cost of unit stock $C_{per}$, order cost $C_{order}$, stock conversion cost $C_{transfer}$. The targeted function was $T(P,C)$. Then we could get the control model of the product stock on purpose of supply chain collaboration in equation (**xiv**).

$$T(P,C) = \min\{P(s_t,(a, y_t)) - C_{all}\} \tag{xiv}$$

## 4 Conclusion

This article researched the real-time supply chain collaboration under RFID circumstances. The model of factory-distribution warehouse-retailer supply chain was established. Optimization method of inventory control strategy in Many-Varieties-Small- Batches clothing manufacturing production warehouse management was analyzed. RFID technology was adopted to collect real-time data, decrease cow whip effect, optimization control of production warehouse was realized, and the sketch map was successfully applied in a specified clothing enterprise.

**Acknowledgments.** This paper is supported by Zhejiang Shuren University Research Fund No. 2011A22006 and Zhejiang Education Department Fund for Excellence in Universities, No. 200804ZJU01.

## References

1. You, Z., Li, S.: Radio Frequency Identification (RFID) Theory and Application. Publishing House of Electronics Industry, Beijing (2004)
2. Yi, W.-M., Zuo, A.-p.: Dynamic Pricing For The Perishable Inventory Management Using RFID Technology. In: 2009 International Conference on Management of e-Commerce and e-Government, ICMeCG 2009, pp. 351–354 (2009)
3. Ye, J., Wang, Z., Pan, X.: Application of RFID technology to the raw material warehouse of apparel enterprise. Journal of Textile Research 29(3), 132–136 (2008)
4. Dadjo, L., Girard, A., Gregoire, J.-C.: A model of user behavior in multi-service priced networks. Telecommunication Systems 35(1-2), 43–53 (2007)
5. McCalla, C., Whitt, W.: A time-dependent queueing-network model to describe the life-cycle dynamics of private-line telecommunication services. Telecommunication Systems 19(1), 9–38 (2002)
6. Seah, W.K.-G., Takahashi, Y., Hasegawa, T.: Analysis of demand-priority access using a priority queue with server vacations and message dependent switchover times. Telecommunication Systems 10(3-4), 243–268 (1998)
7. Khan, M.A., Toker, A.C., Sivrikay, F., et al.: Cooperation-based resource allocation and call admission for wireless network operators. Telecommunication Systems, Online First, January 26 (2011)
8. Cox Jr., L.A.: Data mining and causal modeling of customer behaviors. Telecommunication Systems 21(2-4), 349–381 (2002)
9. Massey, W.A.: The analysis of queues with time-varying rates for telecommunication models. Telecommunication Systems 21(2-4), 173–204 (2002)

# Cooperative Packet Transferring in MANET

Yujie Yan and Bitao Ma

Wuhan Second Ship Design and Research Institute
450 Zhongshan Road, Wuhan, China
yanyujiecn@163.com

**Abstract.** This paper focuses on the packet transferring in MANET and provides a solution with game theory. It designs a model for packet transferring and derives the node's pareto optimal utility. A NEPT algorithm is proposed for propelling the nodes to operate at the pareto optimal utility and constitutes a Nash Equilibrium. The simulation results verify that NEPT assures the nodes to achieve the pareto optimal utility by stimulating cooperation; deviation at any one node will decrease utilities of all related nodes.

**Keywords:** MANET, Packet transferring, Cooperative, Nash Equilibrium.

## 1 Introduction

MANET is a distributed system in which these nodes have to forward packets for each other in order to enable multi-hop communication [1]. There is the optimal tradeoff between the energy expenditure and packet transferring. Researchers have identified the problem of stimulating cooperation in ad hoc networks and proposed solutions to give nodes incentive to contribute to common network service [2]. However, some researchers propose a framework for cooperation without any incentive mechanism and prove that the network topology and communication patterns have a significant impact on the spontaneous cooperation [3].

In MANET, since nodes are self-interested and rational, it is reasonable to assume that each node wants to maximize its own benefit by enjoying relay service from other nodes and at the same time minimizing its contribution on account of saving energy. This selfish behavior can significantly damage network performance. It is necessary to build a cooperative packet transferring mechanism. Game theory has been used as an inspiration for solving cooperation problems in many applications [4]. It would be a novel approach to introduce ideas of Game theory into MANET for cooperative packet transferring.

This paper focuses on the packet transferring in MANET and makes four contributions. Firstly, it designs a system model for packet transferring with the idea of microeconomics. Secondly, it uses rational arguments to derive the node's pareto optimal utility. Thirdly, it proposes a packet transferring algorithm named NEPT which propels the nodes to operate at the pareto optimal utility and constitutes a *Nash Equilibrium*. Finally, the simulation results confirm that NEPT assures the nodes to achieve the pareto optimal utility by stimulating cooperation; deviation at any one node will decrease utilities of all related nodes.

S. Lin and X. Huang (Eds.): CSEE 2011, Part II, CCIS 215, pp. 552–559, 2011.

## 2 Packet Transferring Model

We consider a MANET system with N nodes distributed among $K$ energy classes. Let $n_i$ be the number of nodes in class i (i=1, 2 ...$K$). Each node in class i has a power constraint $\rho_i$ which is the ratio of the average energy constraint to expected lifetime in class i. We assume $\rho_1 > \rho_2 ... > \rho_{K-1} > \rho_K$. The system operates in discrete time and in each slot, a source and some relay nodes are randomly chosen out of the $N$ nodes to carry out a session.

The type of the session is determined by the nodes in the highest energy class. We use $M$ to denote the possible maximum number of relays that the source can use to reach its destination, and q(m) the probability that the source requires m relays. So there is $m \leq M$ and we assume q(0)=0, that is, there is at least one relay in each session.

Each node decides its action according to the history information of the sessions it has joined. We use $b_h^j(k)$ to denote the number of relay requests initiated by node h for a session of type j till time k and $a_h^j(k)$ the number of successful relay requests initiated by node h for a session of type j till time k. Similarly, we use $d_h^j(k)$ to denote the number of relay requests that node h has received from a session of type j till time k and $c_h^j(k)$ the number of relay requests that node h has successfully relayed for a session of type j till time k. we define $\alpha_h^j(k) = a_h^j(k) / b_h^j(k)$ which is an indication of the throughout experienced by h till time k, with respect to the session of type j. Also $\beta_h^j(k) = c_h^j(k) / d_h^j(k)$ is an indication of the relay rate experienced by h till time k, with respect to type j session.

For the relay help provided by other nodes, we define the average amount of throughout per unit of energy expenditure as the node's utility described as:

$$U_h^j(k) = \alpha_h^j(k) / \rho_j \tag{1}$$

For the cost spent on transferring packet for other nodes, we define the average amount of relay rate per unit of energy expenditure as the node's cost described as:

$$C_h^j(k) = \beta_h^j(k) / \rho_j \tag{2}$$

## 3 Pareto Optimal Utility of Node

We assume that nodes are rational, that is, each node wants to maximize its own utilities at the least cost. After many sessions, the node's utilities will converge to the pareto optimal point such that a node can not improve its utilities without decreasing some other node's utilities.

In multiple relay, we assume that the node in class i in a session of type j accepts a relay request with probability $\sigma_{ij}$. By rationality, a node in a session has no incentive to behave any differently from the other nodes in this session. For example, there is a system with $n_1$ node in class 1 and $n_2$ in class 2. The node in class 1 will not expend more energy than the node in class 2. So the node in class 1 will behave as the nodes in class 2 and all these nodes in the session will have the same probability to accept a relay request, that is, $\sigma_{ij} = \sigma_{jj}, 1 \le i \le j \le K$, and we denote $\sigma_{jj}$ by $\sigma_j$.

Consider a node h participating in a type j session, the average energy efficiency per slot spent by the node as a source can be denote by $\Gamma_{hj}^{(s)}$ described as follows:

$$\Gamma_{hj}^{(s)} = \frac{1}{N} \cdot \lim_{k\to\infty} U_h^j(k) = \frac{1}{N} \cdot \frac{1}{\rho_j} \sum_{m=1}^{M} q(m) \sum_{p_1,\dots,p_j} \theta(m; p_1, \dots p_j) \sigma_j^{(p_1+\dots p_j)} \tag{3}$$

Where $1/N$ is the probability that node h is the source. $\theta(m; p_1, \dots p_j)$ is an abstractive multivariable probability function conditioned on the fact that the type j session has m relay nodes and the number of the selected nodes in class 1, 2,…, j is as follows: $p_1, p_2, \dots, p_j$. $\sigma_j^{(p_1+\dots p_j)}$ represents the probability that all relay nodes accept the request.

Similarly, the average energy efficiency per slot spent by node h as a relay can be denoted by $\Gamma_{hj}^{(r)}$ described as follows:

$$\Gamma_{hj}^{(r)} = \frac{1}{N} \cdot \frac{1}{\rho_j} \sum_{m=1}^{M} m q(m) \sum_{p_1, \dots p_j} \theta(m-1; p_1, \dots p_j) \sigma_j^{(p_1+\dots p_j)} \tag{4}$$

Where $m/N$ is the probability that node h is chosen as one of the m relays.

As the average energy per slot expended by node h on participating in all types of session should not exceed its power constraint. For the simple system with two nodes both distributed in energy class 1 mentioned at the beginning, we get an inequality for each node as follows:

$$\frac{1}{2} \lim_{k\to\infty} U_1^1(k) + \frac{1}{2} \lim_{k\to\infty} U_2^1(k) \le 1 \tag{5}$$

When (5) comes to an equation, the utility improvement of one node will decrease the other's utility. So the pareto optimal utility can be derived by imposing the equality relation in (5). Consider $K$ classes and N nodes with $p_i$ nodes in class i, and $q(1) = 1$, $M = 1$. For a node in class i, the average energy per slot expended by the node on participating in all types of sessions as a source can be denoted by $\Gamma_i^{(s)}$:

$$\Gamma_i^{(s)} = \sum_{j=i}^{K} \rho_j \Gamma_{hj}^{(s)} = \frac{1}{N(N-1)} \cdot [\sum_{k=1}^{i-1} p_k \sigma_i + (p_i - 1)\sigma_i + \sum_{l=i+1}^{K} p_l \sigma_l] \tag{6}$$

When the relay belongs to a class lower than (or equal to) i, the session is of type i and if the relay belongs to a class higher than i, the session type is the same as the class of the relay. The same expression holds for the average energy per slot $\Gamma_i^{(r)}$, expended by the node on participating in all types of session as a relay. The pareto optimal probability $\sigma_i^*$ can be derived from the set of equations:

$$\Gamma_i^{(s)} + \Gamma_i^{(r)} = \rho_i \quad 1 \le i \le K \tag{7}$$

$$\sigma_i \in [0,1] \quad 1 \le i \le K \tag{8}$$

Substituting for $\sigma_i^*$ into the following Equation (9), we can get the node's pareto optimal utility $U_i^*$ in type i session.

$$U_i^* = \lim_{k\to\infty} U_h^i(k) = \frac{1}{\rho_i}\sum_{m=1}^{M} q(m) \sum_{p_1,\ldots,p_i} \theta(m; p_1,\ldots.p_i)\sigma_i^{*(p_1+\ldots p_i)} \tag{9}$$

In particular, for $K$=1, we can solve $\sigma_i^* = N\rho_1/2$ and $U_i^* = N/2$. We can observe that the pareto optimal utility is also only related with the type of sessions as the $\sigma_{ij}$ behaves. Furthermore, for the case of multiple relays, we can solve the node's pareto optimal utilities in each type of sessions.

## 4 NEPT Algorithm

For propelling the nodes to operate at the pareto optimal utility and constitute a *Nash Equilibrium* [5][6], we propose a Nash Equilibrium based packet transferring algorithm named NEPT which can effectively stimulate cooperation among nodes. In our algorithm, each node maintains a record of its past experience by using the two variables $U_h^j(k)$ and $C_h^j(k)$.

The decisions are taken by the relay nodes based only on their $U_h^j(k)$ and $C_h^j(k)$ values. Firstly, consider the single relay case with $N$ nodes, $K$ classes, $q(1)=1$, $M=1$. Assume that a node h receives a relay request for a type j session, and it will take decision according to NEPT algorithm as follows:

If $U_h^j(k) \ge C_h^j(k) - \varepsilon$ and $C_h^j(k) \le \sigma_j^*/\rho_j$ Accept ; Else Reject

Where $\varepsilon$ is a small positive number. Thus, a request for a type j session is accepted if the node's utility has almost exceeded its cost which has not reached the corresponding pareto optimal cost. Since $\varepsilon$ is a small positive number, nodes are a little generous by agreeing to forward packets for others even if they have not received a reciprocal amount of help. For the above single relay case, NEPT aims to equalize the amount of cooperation a node provides with the amount of cooperation it receives. When multiple relays are used, the amount of help rendered is always more than the amount of help received. This is because a node is a relay more often than it is a source. Assume that a relay request for a type j session arrives at node h, the NEPT algorithm becomes as follows:

If $C_h^j(k) \le \sigma_j^*/\rho_j$ and $U_h^j(k) \ge \rho_j U_j^* C_h^j(k)/\sigma_j^* - \varepsilon$ Accept; Else Reject

We now prove that the NEPT algorithm constitutes a *Nash Equilibrium*. Considering a system of $N$ nodes, with all nodes belongs to the same class and having energy constraint $\rho$, $q(1)=1$, $M=1$ and $K$=*1*. If *N-1* nodes excluding p are employing NEPT, we know that a node h employing NEPT rejects a relay request whenever

$C_h(k) > \sigma^* / \rho$. Thus we have: $\limsup_{k\to\infty} C_h(k) \le \sigma^* / \rho = N/2, h \ne p$. Since the acceptance mechanism in NEPT is independent of the source identity, each user receives the same amount of help. Hence $\limsup_{k\to\infty} U_h(k) / \rho \le N/2 = U^*, h = 1,...,N$. This shows that if node p tries to deviate from NEPT, then it can not achieve utility greater than the pareto optimal utility, which forms a *Nash Equilibrium* among nodes.

Next, we prove that $U_h(k)$ and $C_h(k)$ converge to the pareto optimal utility $U^*$. For the generic node h, we define $\gamma_h(k)$ as $(\text{No. of successful sessions generated by h till k})/k$ and $\eta_h(k)$ as $(\text{No. of sessions relayed by h till k})/k$.

Recall that the source and the relay are chosen randomly from the $N$ nodes, we can derive:

$$\gamma_h = \lim_{k\to\infty} \gamma_h(k) = \lim_{k\to\infty} \frac{\text{No. of successful sessions generated by h}}{k} \cdot \frac{\text{No. of sessions relayed by h}}{\text{No. of sessions relayed by h}} = \frac{\lim_{k\to\infty} \rho U_h(k)}{N(N-1)} \tag{10}$$

Similarly, we have:

$$\eta_h = \lim_{k\to\infty} \eta_h(k) = \frac{\lim_{k\to\infty} \rho C_h(k)}{N(N-1)} \tag{11}$$

Since the total number of successful initiated requests must be equal to the total number of relayed requests, we have:

$$\sum_{i=1}^{N} (\gamma_i(k) - \eta_i(k)) = 0 \tag{12}$$

According to the corollary from [7], $\gamma_h(k) - \eta_h(k)$ converges to zero. So we have: $U_h(k) - C_h(k) = 0$. Node h will always accept a relay request whenever $C_h(k) \le \sigma^* / \rho$, thereby increasing $C_h(k)$, thus we have $\liminf_{k\to\infty} C_h(k) \ge \sigma^* / \rho$. Recall the former conclusion that $\limsup_{k\to\infty} C_h(k) \le \sigma^* / \rho$, we can conclude that: $\lim_{k\to\infty} C_h(k) = \sigma^* / \rho$. Since $U_h(k) - C_h(k)$ goes to zero, $\lim_{k\to\infty} U_h(k) = \sigma^* / \rho$.

As the node's behavior is only related with the type of sessions but not with its energy class. So the conclusion above can extended to the case of $K \ne 1$. Furthermore, by appropriately scaling equation (13) and adding these variables with appropriate weights $q(m), m = 1,...,M$, the conclusion above can be also extended to the multiple relay case.

## 5 Performance Evaluation

We focus on the single relay case. Consider a system with 30 nodes distributed among 5 energy classes. Each class has 6 nodes. Also we assume $q(1) = 1$, $M = 1$. The energy constraints are given by ρ1=0.026, ρ2=0.022, ρ3=0.018, ρ4=0.014, ρ5=0.01. The pareto optimal utilities associated with different session types can be calculated as follows: $U_5$*=15, $U_4$*=16, $U_3$*=18, $U_2$*=22, $U_1$*=32.

The node's average utilities associated with different session types as the time elapses are recoeded. The results in Fig. 1 shows that the utility for each session type converges to the desired pareto optimal utility. This verifies that NEPT assure the nodes to achieve the pareto optimal utility by stimulating cooperation.

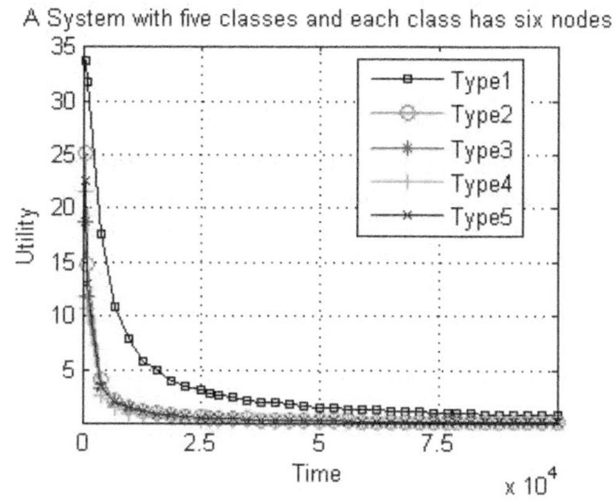


**Fig. 1.** Utilities versus time when all nodes employ NEPT

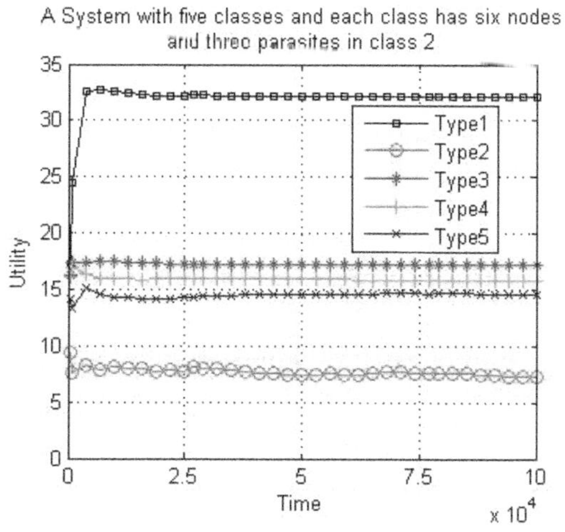


**Fig. 2.** Utilities versus time when nodes are not slightly generous

To investigate the importance of node's generosity ( $\varepsilon > 0$ ) in NEPT, we set $\varepsilon = -0.1$ and observe the utilities vs. time as described in Fig. 2. It shows that these utilities converge to zero, or equivalently the network throughout goes to zero. This is because the miserliness ( $\varepsilon < 0$ ) results that each node does not forward packets whenever its utility has not exceeded its cost and goes against the cooperation among nodes. Therefore it is critically important that the parameter $\varepsilon$ is positive and nodes should always be slightly generous for the nodes to achieve the pareto optimal utilities.

For verify that NEPT constitutes a *Nash Equilibrium*, we set a nodes in class 2 to be the parasite who deviates from NEPT and always rejects relay request. The utilities vs. time in this case can be observed in Fig. 3. It shows the performance of type 2, 3,4 and 5 sessions degrades. This means that the parasite in class 2 who deviates from NEPT will drop the utilities associated with session type 2, 3, 4 and 5. Any node who deviates NEPT will degrade the utilities of all related nodes. However, since the parasite is chosen to be the relay node in a smaller probability in the session of higher type, the parasite's influence will become weaker as the session type rises, which is also shown in Fig. 3. All these above verify that the NEPT is a *Nash Equilibrium.*

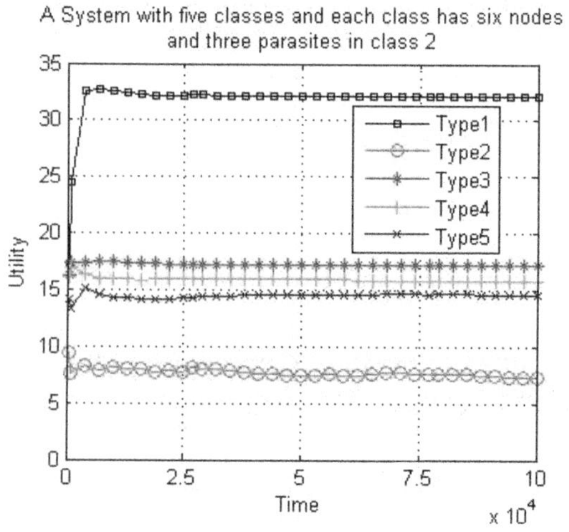


**Fig. 3.** Utilities versus time when a node deviates NEPT

## 6 Summary

This paper focuses on the packet transferring problem in MANET. A NEPT algorithm is proposed for propelling the nodes to operate at the pareto optimal utility of *Nash Equilibrium* with the Game theory. The simulation results verify that NEPT assures the nodes to achieve the pareto optimal utility by stimulating cooperation; deviation at any one node will decrease utilities of all related nodes.

## References

1. Akyildiz, F., Su, W., Sankarasubramaniam, Y., Cayirci, E.: Wireless Sensor Networks: A survey. Comput. Networks 38(4), 393–422 (2002)
2. Marbach, P., Qiu, Y.: Cooperation in Wireless Ad Hoc Networks: A Market-Based Approach. IEEE/ACM Transactions on Networking 13(6), 1325–1338 (2005)

3. Felegyhazi, M., Hubaux, J.P., Buttyan, L.: Nash Equilibria of Packet Forwarding Strategies in Wireless Ad Hoc Networks. IEEE Transactions on Mobile Computing 5, 463–476 (2005)
4. MacKenzie, A.B., Wicker, S.B.: Game Theory and the Design of Self-configuring, Adaptive Wireless Networks. IEEE Communications Magazine 39(11), 126–131 (2001)
5. Axelrod, R.: The Evolution of Cooperation. Basic Books, New York (1984)
6. Fudenburg, D., Tirole, J.: Game Theory. MIT Press, Cambridge (1991)
7. Bertsekas, D.P., Tsitsiklis, J.N.: Neuro-Dynamic Programming. Athena Scientific, Belmant (1996)

# Application of Three-Dimensional Finite Element Simulation Analysis for Homogeneous Earth Dam Structure

DongYu Ji

Department of Civil Engineering, Hunan Urban Construction College,
Xiangtan, Hunan, 411101, China
hnjdy@126.com

**Abstract.** Homogeneous earth dam is a common structure in hydraulic engineering, this paper adopt universal finite element calculation software to carry out three-dimensional finite element simulation analysis for homogeneous earth dam of Huaxi reservoir. Computer is fully used in calculation process, researching variation law of the dam's stress and displacement in construction process and operational process. The research results offer some reference for design and construction of homogeneous earth dam.

**Keywords:** Huaxi reservoir; homogeneous earth dam; finite element method; simulation analysis.

## 1 Project Summary

Huaxi reservoir is located Guiyang city in Guizhou province, the basin area that is controlled by reservoir is 325 km2, total reservoir storage capacity is 2000 million m3. It is a medium reservoir that function is mainly power generation, main dam is concrete gravity dam, dam high is 48 m, power station at dam toe, installed capacity is 2120 kW. There is assistant dam in the narrow mountain passes of reservoir, assistant dam is homogeneous earth dam, this paper will proceed three-dimensional finite element analysis on assistant dam. assistant dam high is 17.5 m, assistant dam length is 68 m, the dam builds on limestone foundation. Two cutoff trench are make which can prevent foundation leakage, the first cutoff trench is located the dam axis upstream 16.75m, trench depth is 2 m, trench bottom width is 1 m. The second cutoff trench is located the dam axis, trench bottom embeds bedrock depth is 2 m. Homogeneous earth dam's gradient of upstream face is 1:2.2, gradient of downstream face is 1:2, dam crest width is 4 m, dam bottom width is 77.5 m. Reservoir's normal storage level is 13 m, design level is 15.8 m, flood level is 17 m.

## 2 Analysis Model

### 2.1 Model Parameters

Homogeneous earth dam of Huaxi reservoir adopts clay, compression modulus $E_1$=52 MPa, Poisson ratio $\mu_1$=0.31[1], dry density $\gamma_1$=14.5 kN/m$^3$. Strata of dam site is limestone, rock mass elastic modulus $E_2$=16 GPa, Poisson ratio $\mu_2$=0.28[2].

S. Lin and X. Huang (Eds.): CSEE 2011, Part II, CCIS 215, pp. 560–565, 2011.

### 2.2 Model Selection

Three-dimensional finite element simulation analysis is proceeded for homogeneous earth dam of Huaxi reservoir. The dam and bedrock structure model is divided by eight nodes isoparametric block element. The element is often applied to three-dimensional model of entity structure, it has plasticity, creep, expansion, stress rigidization, large deformation and large strain characteristics. The element has eight nodes, each node has three translational degree of freedoms[3].

### 2.3 Simulation Range

When three-dimensional finite element simulative analysis for homogeneous earth dam of Huaxi reservoir is proceeded, the earth dam structure is simplified plane strain problem. Calculation model simulation range is listed below, calculation model along the river is 277.5 m, transverse of the river is 10 m, vertical direction is 77.5 m. The simulation range of whole calculation model is 277.5m×10m×77.5m [4]. Element division of dam and bedrock is shown in Fig. 1.

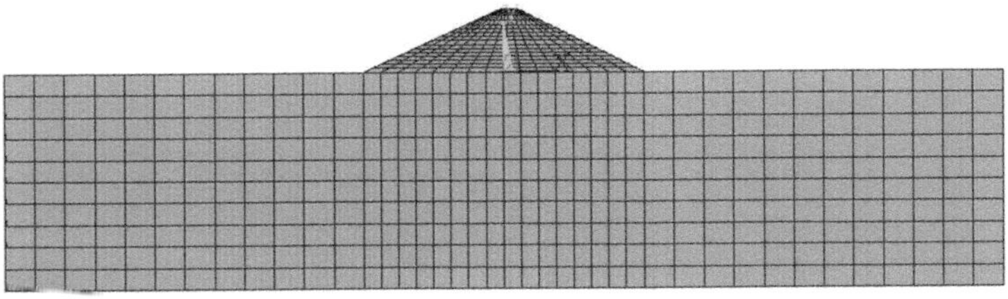

**Fig. 1.** Element division of dam and bedrock

### 2.4 Calculation Cases

Considering mechanical characteristics of homogeneous earth dam structure in construction and operating process[5], researching five cases as follows. Case 1, dam weight(construction condition); case 2, dam weight, normal storage level and tail water level(operating condition); case 3, dam weight, design level and tail water level(operating condition); case 4, dam weight, flood level and tail water level (operating condition); case 3, dam weight, design level, tail water level and earthquake effect(operating condition).

## 3 Calculation Results Analysis

### 3.1 Analysis Paths

In order to analyzing homogeneous earth dam of Huaxi reservoir, two calculation paths are selected in the middle section of homogeneous earth dam. Path 1, choosing one calculation point every 7.1 m from dam bottom to dam crest along homogeneous earth dam's upstream face, there are seven calculation points totally. Path 2, choosing one calculation point every 6.5 m from dam bottom to dam crest along homogeneous earth dam's downstream face, there are also seven calculation points totally.

### 3.2 Stress Analysis

In order to research stress distribution law of homogeneous earth dam, calculation point's first principal stress value on the analysis path of homogeneous earth dam's middle section under various cases are listed in the table 1.

**Table 1.** Calculation point's first principal stress value on the homogeneous earth dam's analysis path under various cases/kPa

| Calculation point | | 1 | 2 | 3 | 4 | 5 | 6 | 7 |
|---|---|---|---|---|---|---|---|---|
| Case 1 | Path 1 | -1.07 | -8.50 | -7.37 | -5.28 | -3.32 | -1.88 | -0.59 |
| | Path 2 | -0.76 | -8.98 | -7.73 | -5.50 | -3.51 | -2.00 | -0.62 |
| Case 2 | Path 1 | -37.49 | -47.84 | -36.53 | -21.22 | -4.25 | -1.88 | -0.59 |
| | Path 2 | -0.68 | -9.01 | -7.77 | -5.60 | -3.57 | -2.01 | -0.61 |
| Case 3 | Path 1 | -45.07 | -58.98 | -47.88 | -33.80 | -18.98 | -3.46 | -0.65 |
| | Path 2 | -0.52 | -9.04 | -7.81 | -5.71 | -3.64 | -2.00 | -0.61 |
| Case 4 | Path 1 | -48.30 | -63.73 | -52.68 | -38.87 | -24.78 | -12.87 | -0.68 |
| | Path 2 | -0.39 | -9.05 | -7.84 | -5.77 | -3.67 | -2.01 | -0.60 |
| Case 5 | Path 1 | -44.22 | -51.41 | -43.96 | -30.04 | -16.23 | -3.53 | -0.71 |
| | Path 2 | 2.41 | -8.51 | -7.46 | -5.51 | -3.49 | -1.89 | -0.53 |

We can see from table 1, under various cases, first principal stress of homogeneous earth dam's upstream and downstream face are basically compressive stress. But homogeneous earth dam's upstream face's compressive stress values gradually become smaller from dam bottom to dam crest, this is mainly because dam weight and water pressure effect together. Compressive stress values of homogeneous earth dam's upstream face are increase gradually along with increase of water pressure, this is mainly because the gradient of homogeneous earth dam's upstream face is gentle, so the first principal stress that is produced by water pressure is compressive stress on the dam's upstream face, and compressive stress of the dam is larger along with increase of water pressure. Compressive stress values of dam's downstream face are smaller than upstream face, this is mainly because dam's downstream face will produce tensile stress under water pressure, which offsets a part compressive stress. Earthquake effect along the river produces certain tensile stress on the homogeneous earth dam's downstream face, it offsets a part compressive stress, it causes the dam's downstream face's compressive stress reduced.

Because case 1 is construction condition, case 3 is common condition of dam in operating process, case 5 is serious condition of dam in operating process. We obtain homogeneous earth dam's contour maps of first and third principal stress under case 1, 3 and 5, contour maps are shown in from Fig. 2 to Fig. 7.

We can see from Fig. 2 to Fig. 7, homogeneous earth dam's first and third principal stress is compressive stress under case 1, 3 and 5, and principal stress is layered distribution along the direction of dam height, compressive stress value is larger at dam

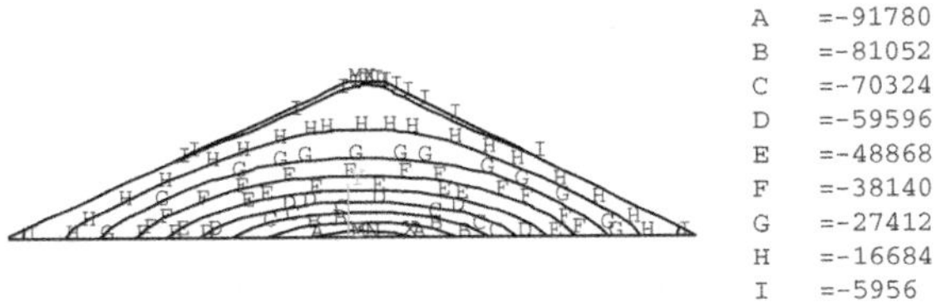


**Fig. 2.** Contour map of first principal stress under case 1/Pa

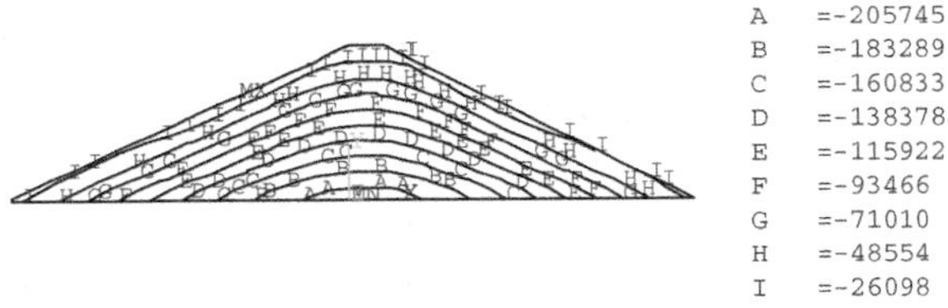


**Fig. 3.** Contour map of third principal stress under case 1/Pa

**Fig. 4.** Contour map of first principal stress under case 3/Pa

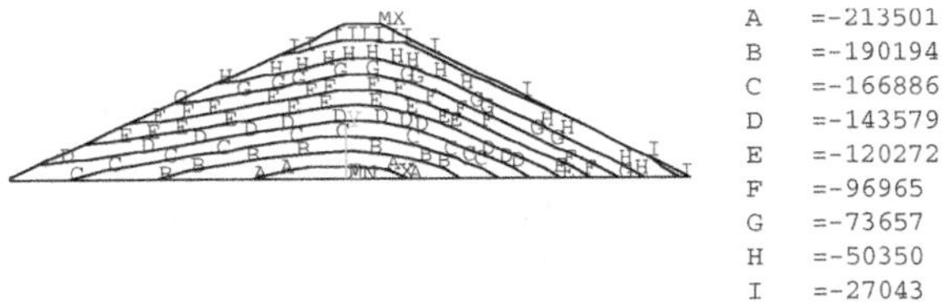


**Fig. 5.** Contour map of third principal stress under case 3/Pa

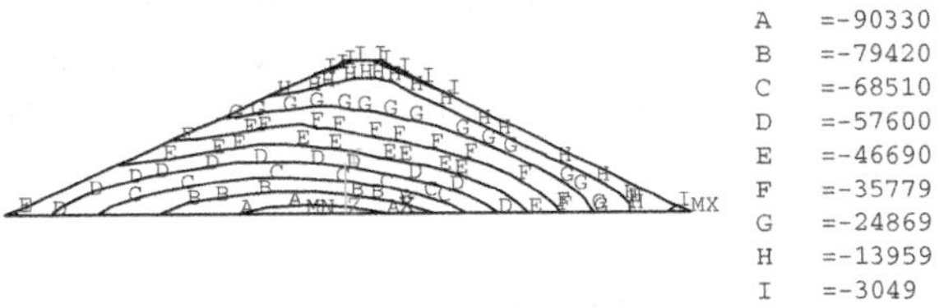


**Fig. 6.** Contour map of first principal stress under case 5/Pa

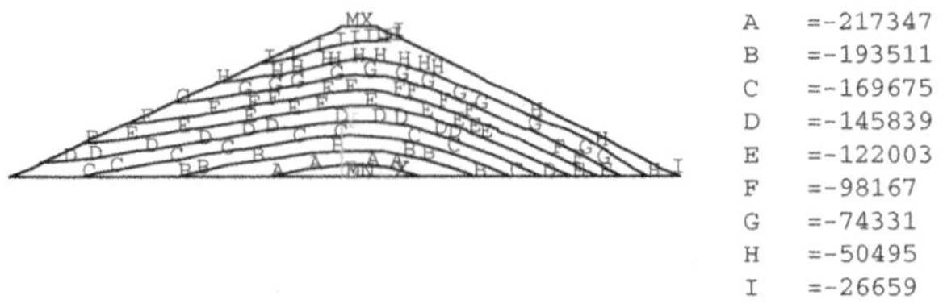


**Fig. 7.** Contour map of third principal stress under case 5/Pa

bottom, but tensile stress value is smaller at dam crest. Maximum compressive stress of the first principal stress is -0.094 MPa, which is located junction of dam bottom and bedrock under case 3. Maximum compressive stress of the third principal stress is -0.217 MPa, which is located junction of dam bottom and bedrock under case 5. This is mainly because dam weight and water pressure effect together.

### 3.3 Deformation Analysis

Through deformation analysis of homogeneous earth dam of Huaxi reservoir, getting calculation point's horizontal and vertical displacement values on the analysis path of homogeneous earth dam's middle section under various cases. Now displacement values of analysis path 1 under various cases are listed in the table 2.

**Table 2.** Calculation point's horizontal and vertical displacement values on the analysis path 1 of dam under various cases/mm

| Calculation point | | 1 | 2 | 3 | 4 | 5 | 6 | 7 |
|---|---|---|---|---|---|---|---|---|
| Case 1 | Horizontal displacement | 0.02 | -1.10 | -2.37 | -2.14 | -0.27 | 0.91 | 0.56 |
| | Vertical displacement | -0.14 | -1.32 | -4.63 | -10.23 | -18.50 | -26.35 | -29.42 |
| Case 2 | Horizontal displacement | 0.17 | 3.11 | 2.85 | 1.64 | 1.49 | 2.41 | 2.04 |
| | Vertical displacement | -0.46 | -5.48 | -10.09 | -14.15 | -18.92 | -25.96 | -29.00 |
| Case 3 | Horizontal displacement | 0.22 | 4.40 | 5.31 | 5.02 | 4.52 | 3.85 | 3.36 |
| | Vertical displacement | -0.54 | -6.66 | -12.39 | -17.69 | -22.82 | -26.81 | -29.15 |
| Case 4 | Horizontal displacement | 0.24 | 4.95 | 6.39 | 6.57 | 6.27 | 5.18 | 4.18 |
| | Vertical displacement | -0.57 | -7.17 | -13.38 | -19.19 | -24.77 | -28.50 | -29.72 |
| Case 5 | Horizontal displacement | 0.28 | 5.40 | 8.12 | 9.78 | 10.96 | 11.23 | 10.97 |
| | Vertical displacement | -0.53 | -6.61 | -12.28 | -17.55 | -22.71 | -26.78 | -29.15 |

***Note:*** *Negative horizontal displacements mean the displacements that direct upstream, negative vertical displacements mean the displacements that direct below in the table 2.*

We can see from table 2, horizontal displacement value of homogeneous earth dam is smaller under various cases, and horizontal displacement value is far less than vertical displacement value, this is mainly because the gradients of homogeneous earth dam's upstream face and downstream face are gentle, the dam produce mainly vertical settlement displacement under weight and water pressure. Under case 2, 3, 4 and 5, horizontal displacements of homogeneous earth dam direct downstream face, vertical displacement is downward displacement, and horizontal displacement value gradually become larger from dam bottom to dam crest. this is mainly because homogeneous earth dam's displacement is mainly vertical compression displacement under weight and water pressure. So dam crest's displacement is slightly greater than dam bottom's displacement. Earthquake effect along the river has more influence on the dam's horizontal displacement, horizontal displacement doubled nearly, but vertical displacement change little.

# 4 Conclusion

Research shows that, stress value of homogeneous earth dam of Huaxi reservoir is smaller, and these are all compressive stress, stress values can meet strength requirements. The dam's displacement is very small, and these are mainly compression displacement, displacement values can meet rigidity requirements. So homogeneous earth dam structure of Huaxi reservoir is safe and reliable, the dam can meet requirements.

# References

1. Li, G.: Advanced Soil Mechanics. Tsinghua University Press, Beijing (2005)
2. Shen, M., Chen, J.: Rock mechanics. Tongji University Press, Shanghai (2006)
3. Zhu, B.: Finite element method principle and application. China Water Conservancy and Hydropower Press, Beijing (1998)
4. Mai, J.: Hydraulic structures. Tsinghua University Press, Beijing (2005)
5. Li, M., Wang, F.: Earth-rockfill dam design and construction. China Water Conservancy and Hydropower Press, Beijing (2011)

# Finite Element Simulative Analysis of Inclined Clay-Core Wall Rockfill Dam Structure

DongYu Ji

Department of Civil Engineering, Hunan Urban Construction College,
Xiangtan, Hunan, 411101, China
hnjdy@126.com

**Abstract.** Inclined clay-core wall rockfill dam is a common structure form in water conservancy engineering, proceeding simulative analysis for inclined clay-core wall rockfill dam has certain theoretical value and application value. This paper adopt finite element method to carry out simulation analysis for inclined clay-core wall rockfill dam of Kuandian reservoir, researching distribution law of the dam's stress and displacement in construction process and operational process. Research results provide certain reference basis for design and construction of inclined clay-core wall rockfill dam.

**Keywords:** Kuandian reservoir; Inclined clay-core wall rockfill dam; Finite element method; Simulation analysis.

## 1 Introduction

Kuandian reservoir is located Tangjia gulf of Liaoniu river's tributaries, which is located Chengde county in Hebei province. Catchment area of dam site above is 27.6 $km^2$, total reservoir storage capacity is 317 million $m^3$, the project mainly function is irrigation, benefit farmland is 5000 mu. The hub project is make by water retaining dam, overfall dam, penstock, water pipe, stilling basin and hydropower station, etc. Water retaining structure is inclined clay-core wall rockfill dam, maximum dam high is 23 m, dam crest width is 5 m, dam crest length is 189.5 m, gradient of upstream face is 1:3.5, gradient of downstream face face is 1:1.5. The dam's upstream face is make by slope protection with paved rock blocks, gravel layer, sandy pebble protective layer, clay layer. There is rockfill dam in clay layer below, gradient of rockfill dam is 1:1. Reservoir's normal storage level is 18.1 m, design level is 20.7 m, flood level is 21.5 m.

## 2 Calculation Model

### 2.1 Model Parameters

Concrete strength grade of Kuandian reservoir's inclined clay-core wall rockfill dam is C20, elastic modulus $E_1$=25.5 GPa, Poisson ratio $\mu_1$=0.167[1], density $\gamma_1$ =24

S. Lin and X. Huang (Eds.): CSEE 2011, Part II, CCIS 215, pp. 566–571, 2011.

$kN/m^3$. Clay layer compression modulus $E_2$=55 MPa, Poisson ratio $\mu_2$=0.31[2], density $\gamma_2$ =16 $kN/m^3$. Dam masonry's elastic modulus $E_3$=18 GPa, Poisson ratio $\mu_3$=0.28, dry density $\gamma_3$ =30 $kN/m^3$. Dam foundation rock is glutenite[3], glutenite elastic modulus $E_4$=20 GPa, Poisson ratio $\mu_4$=0.28.

## 2.2 Element Selection

Finite element simulative analysis is proceeded for inclined clay-core wall rockfill dam of Kuandian reservoir. Slope protection with paved rock blocks, gravel layer, sandy pebble protective layer, clay layer, rockfill dam dan bedrock structure model is divided by eight nodes isoparametric block element[4]. The element is often applied to three-dimensional model of entity structure, it has eight nodes, each node has three translational degree of freedoms.

## 2.3 Simulation Range

Universal finite element calculation software is adopted, finite element simulative analysis for inclined clay-core wall rockfill dam of Kuandian reservoir is proceeded. Calculation model simulation range is listed below, calculation model along the river is 320 m, transverse of the river is 237 m, vertical direction is 88 m. The simulation range of whole calculation model is 320 m×237 m×88 m[5]. Element division of dam and bedrock is shown in Fig. 1, Element division of dam's section is shown in Fig. 2.

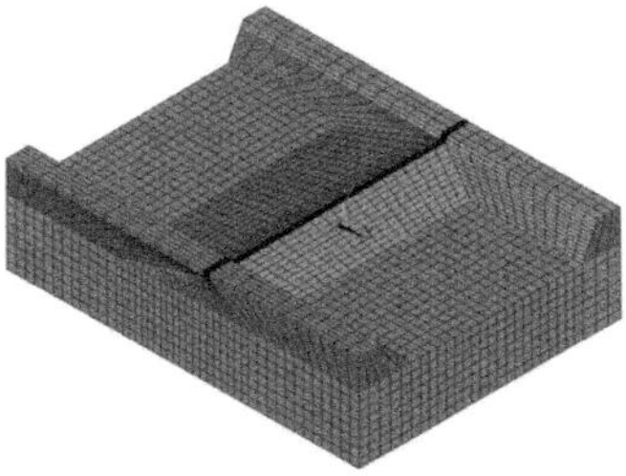

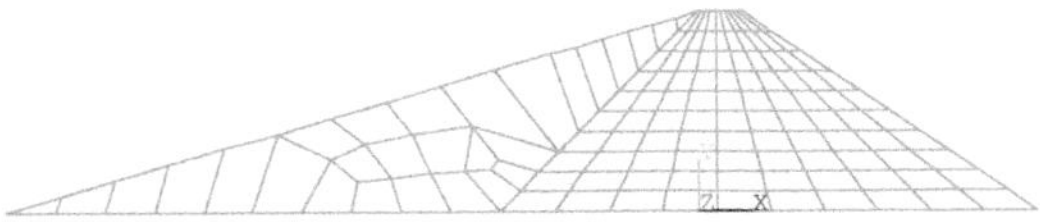

**Fig. 1.** Element division of dam and bedrock **Fig. 2.** Element division of dam's section

## 2.4 Calculation Cases

Considering mechanical characteristics of dam structure in construction and operating process[6], researching five cases as follows. Case 1, dam weight(construction condition); case 2, dam weight, normal storage level, tail water level and sediment pressure(operating condition); case 3, dam weight, design level, tail water level and sediment pressure(operating condition); case 4, dam weight, flood level, tail water level and sediment pressure(operating condition); case 5, dam weight, design level, tail water level, sediment pressure and earthquake effect(operating condition).

# 3 The Dam Structure Analysis

## 3.1 Calculation Paths

In order to analyzing inclined clay-core wall rockfill dam of Kuandian reservoir, three calculation paths are selected in the middle section of the dam. Path 1, choosing one calculation point every 12 m from dam bottom to dam crest along inclined clay-core wall rockfill dam's upstream face, there are eight calculation points totally. Path 2, choosing one calculation point every 6.6 m from dam bottom to dam crest along interface of clay layer and rockfill dam, there are six calculation points totally. Path 3, choosing one calculation point every 8.4 m from dam bottom to dam crest along inclined clay-core wall rockfill dam's downstream face, there are six calculation points totally.

## 3.2 Stress Analysis

Calculation point's first principal stress value on the calculation paths of inclined clay-core wall rockfill dam's middle section under various cases are listed in table 1.

**Table 1.** Calculation point's first principal stress value on the dam's calculation paths under various cases/kPa

| Calculation point | | 1 | 2 | 3 | 4 | 5 | 6 | 7 | 8 |
|---|---|---|---|---|---|---|---|---|---|
| | Path 1 | -3.4 | -12.9 | -22.6 | -14.7 | -24.3 | 3.5 | 29.3 | 0.7 |
| Case 1 | Path 2 | -111.1 | -118.3 | -67.9 | -31.4 | -7.5 | 0.7 | — | — |
| | Path 3 | -23.3 | -28.1 | -23.7 | -17.8 | -11.7 | -2.9 | — | — |
| | Path 1 | -23.1 | -77.8 | -73.8 | -50.5 | -40.4 | -3.8 | 27.5 | 0.6 |
| Case 2 | Path 2 | -133.5 | -135.0 | -81.5 | -33.5 | -8.1 | 0.6 | — | — |
| | Path 3 | -18.3 | -28.6 | -23.8 | -17.8 | -11.6 | -3.0 | — | — |
| | Path 1 | -25.0 | -89.0 | -85.2 | -62.5 | -52.5 | -13.3 | 28.4 | 0.8 |
| Case 3 | Path 2 | -139.9 | -143.8 | -100.8 | -44.9 | -7.3 | 0.8 | — | — |
| | Path 3 | -20.5 | -28.8 | -23.8 | -17.8 | -11.6 | -3.0 | — | — |
| | Path 1 | -26.3 | -92.4 | -88.8 | -66.3 | -56.4 | -17.2 | 27.1 | 1.1 |
| Case 4 | Path 2 | -145.0 | -146.9 | -106.8 | -50.5 | -9.7 | 1.1 | — | — |
| | Path 3 | -21.2 | -28.9 | -23.8 | -17.8 | -11.6 | 3.0 | — | — |
| | Path 1 | -4.7 | -84.4 | -80.9 | -62.5 | -53.1 | -21.9 | 17.2 | -2.2 |
| Case 5 | Path 2 | -122.9 | -137.9 | -109.5 | -57.8 | -17.1 | -2.2 | — | — |
| | Path 3 | -37.5 | -27.8 | -22.3 | -16.6 | -10.9 | -2.0 | — | — |

We can see from table 1, under various cases, first principal stress on the earth dam's calculation paths are basically compressive stress in the dam's mid-lower, but dam crest appears smaller tensile stress. The stress values of dam's upstream face and downstream face less than stress values of interface of clay layer and rockfill dam. First principal stress values of dam's calculation paths are increase gradually along

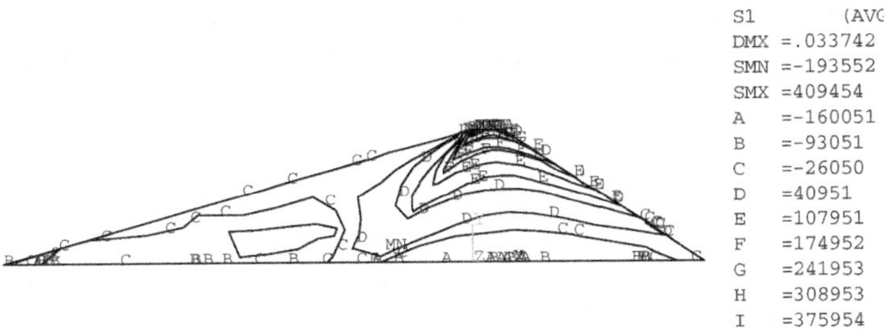


**Fig. 3.** Contour map of first principal stress under case 3/Pa

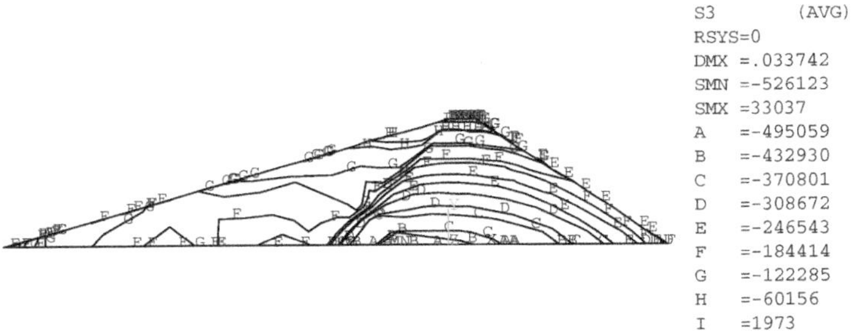


**Fig. 4.** Contour map of third principal stress under case 3/Pa

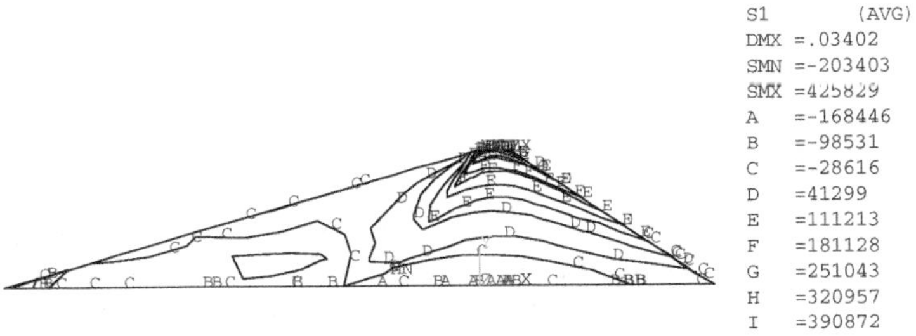


**Fig. 5.** Contour map of first principal stress under case 5/Pa

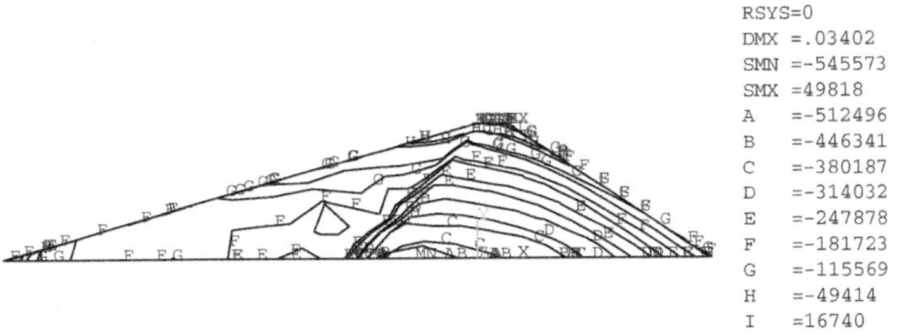


**Fig. 6.** Contour map of third principal stress under case 5/Pa

with increase of water pressure, this is mainly because the gradient of inclined clay-core wall rockfill dam's upstream face and downstream face is gentle, so dam weight and water pressure produces compressive stress.

Because case 3 is common condition of dam in operating process, case 5 is serious condition of dam in operating process. We obtain inclined clay-core wall rockfill dam's contour maps of first and third principal stress under case 3 and case 5, contour maps of first and third principal stress are shown in from Fig. 3 to Fig. 6.

We can see from Fig. 3 to Fig. 6, under case 3 and 5, first principal stress of the dam's upstream's clay layer are basically compressive stress, rockfill dam section appears tensile stress, maximum tension stress is 0.391 MPa, it is located junction of dam crest and bedrock under case 5. third principal stress of the dam are basically compressive stress, maximum compressive stress is -0.512 MPa, it is located junction of dam bottom and bedrock under case 5. And first and third principal stress of rockfill dam section is layered distribution along the direction of dam height. Stress distribution is complex in junction of dam bottom, crest and bedrock, this is mainly because these parts appears stress concentration phenomenon.

### 3.3 Deformation Analysis

In order to researching deformation distribution of inclined clay-core wall rockfill dam of Kuandian reservoir. We obtain inclined clay-core wall rockfill dam's contour maps of horizontal and vertical displacement values under case 1, 3 and 5, contour maps are shown in from Fig. 7 to Fig. 12.

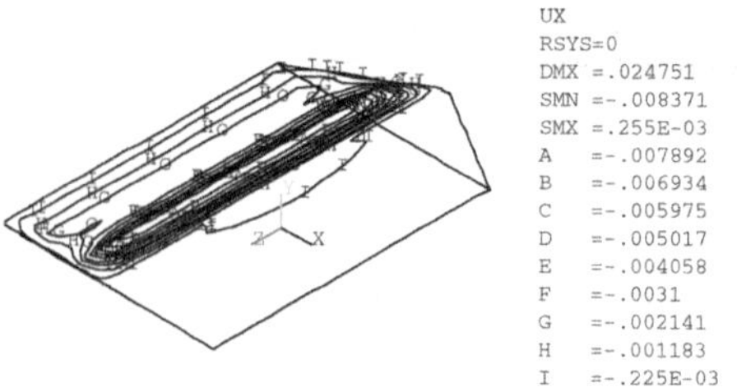


**Fig. 7.** Contour map of horizontal displacement under case 1/m

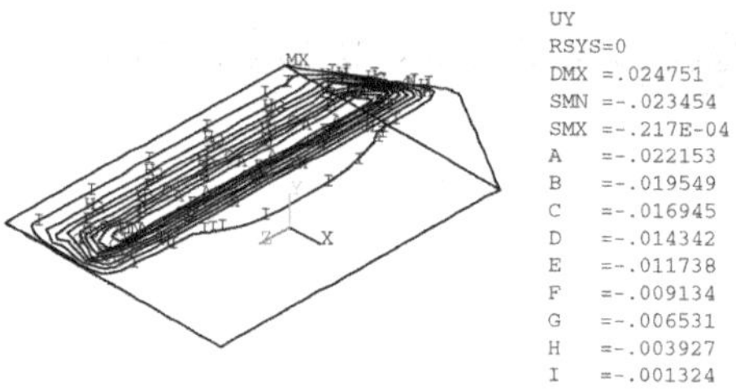


**Fig. 8.** Contour map of vertical displacement under case 1/m

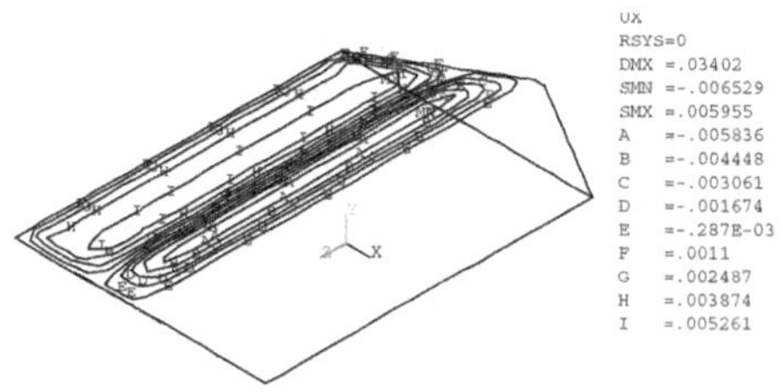


**Fig. 9.** Contour map of horizontal displacement under case 3/m

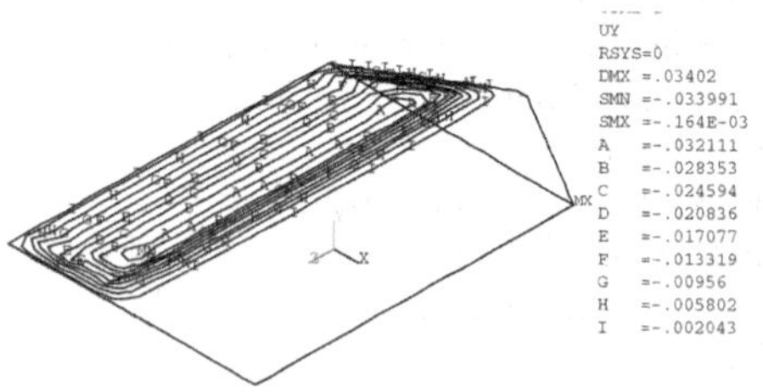


**Fig. 10.** Contour map of vertical displacement under case 3/m

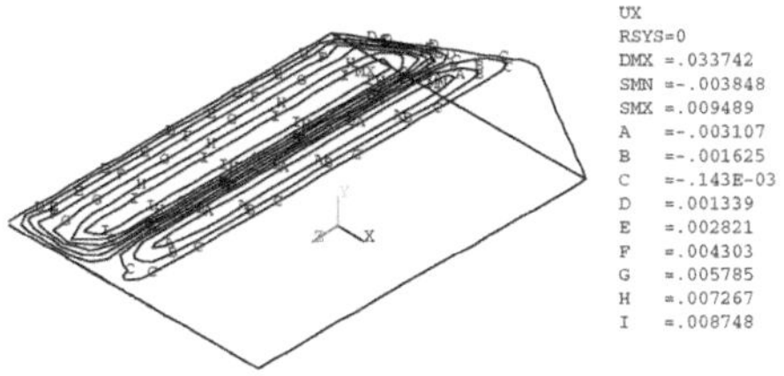


**Fig. 11.** Contour map of horizontal displacement under case 5/m

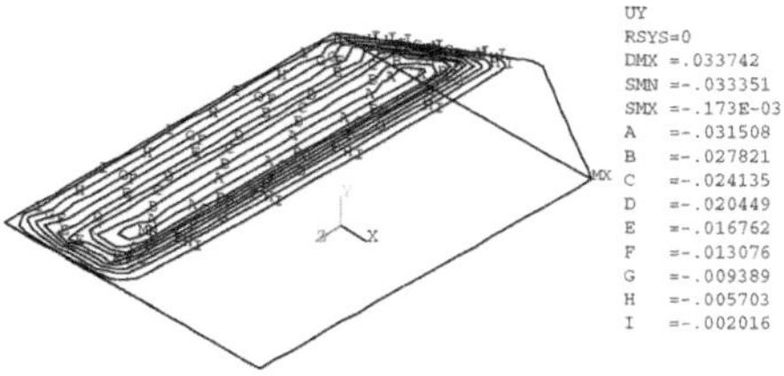


**Fig. 12.** Contour map of vertical displacement under case 5/m

We can see from Fig. 7 to Fig. 12, horizontal and vertical displacement value of inclined clay-core wall rockfill dam is smaller under various cases. horizontal displacements of inclined clay-core wall rockfill dam direct downstream face, vertical displacement is downward displacement, this is mainly because the gradients of inclined clay-core wall rockfill dam's upstream face and downstream face are gentle, the dam produce mainly vertical settlement displacement under weight and water pressure. The dam's displacement appears mainly upstream's clay layer, this is mainly because clay layer's compression modulus is smaller, it produce easily compression deformation.

## 4 Conclusion

From the above, stress and displacement values of inclined clay-core wall rockfill dam of Kuandian reservoir is smaller, the dam can meet design requirements. Analysis results show that, inclined clay-core wall rockfill dam of Kuandian reservoir is economic and reasonable, structure is safe and reliable.

## References

1. SL/T191-96 Design code for hydraulic concrete structures. China Water Conservancy and Hydropower Press, Beijing (1996)
2. Li, G.: Advanced Soil Mechanics. Tsinghua University Press, Beijing (2005)
3. Shen, M., Chen, J.: Rock mechanics. Tongji University Press, Shanghai (2006)
4. Wang, X.: Finite element method. Tsinghua University Press, Beijing (2003)
5. Mai, J.: Hydraulic structures. Tsinghua University Press, Beijing (2005)
6. Li, M., Wang, F.: Earth-rockfill dam design and construction. China Water Conservancy and Hydropower Press, Beijing (2011)

# An Improved Algorithm of Sensors Localization in Wireless Sensor Network

Xiaohui Chen[1,*], Jing He[1], and Bangjun Lei[2]

[1] College of Computer and Information Technology, China Three Gorges University, 443002 Yichang, Hubei, P.R. China
[2] Institute of Intelligent Vision and Image Information, China Three Gorges University, 443002 Yichang, Hubei, P.R. China
chui@ctgu.edu.cn

**Abstract.** To improve the location precision of nodes and simplify localization algorithm in WSN, this paper analyzed the reasons of localization error caused by the degraded state when using the least square algorithm, and then proposed an improved algorithm with a new participle given to choose benchmark anchor node. So, we compared the three localization algorithms which are the basic least square algorithm, the algorithm choosing the nearest node as the benchmark anchor node in LSM and the algorithm choosing the synthetic nearest node as the benchmark anchor node in LSM. The simulation results indicate that the proposed algorithm can improve the location precision.

**Keywords:** WSN node, localization precision, least square method.

## 1 Introduction

In recent years, there has been a number of tremendous applications of wireless sensor network in many domains, such as environmental monitoring fields, battlefields[1-5]. When the sensors send messages about the location of the sensors themselves, it is necessary to localize the sensors in the WSN. Generally, the sensor nodes are deployed randomly and their locations may not be acquired prematurely. The localization of the sensors has evoked a tremendous attention in these occasions. At the same time, the node energy consumption is one of the key factors which should be considered in WSN, so the algorithm of the localization must shorten its computing time.

To date, great deals of methods have already been developed to solve the localization of nodes in WNS. In order to reduce the influence of the measuring distance error and the distance estimation error, some existent localization algorithms[6-11] such as Savarese proposed two localization algorithms: cycle accuracy-Cooperative ranging[6] and Two-Phase localization[7] that can decrease the influence of distance error to localization; in 2002, Savvides[8] proposed n-hop multilateration primitive localization algorithm, where Kalman filtering technique was used to calculate the accurate coordinates circularly, it reduced the accumulation

* Corresponding author.

S. Lin and X. Huang (Eds.): CSEE 2011, Part II, CCIS 215, pp. 572–578, 2011.

error; Bergamo[9] averaged the measuring distance results to increase the localization accuracy with the attenuation of analogy signal; in 2005, in order to improve the accuracy of localization, Guha[10] used the method of non-convex constraints and time detecting to estimate the nodes' localization; in 2006, Cao[11] proposed a localization refinement algorithm based on Cayley-Menger determinant; in 2007, based on Second-order Cone Programming, Srirangarajan[12] solved the accuracy problem in the case that anchor position of the nodes is not accurate localization.

In this paper, firstly, the reasons why the location error exits during the degraded state when using the least squares algorithm was analyzed, and secondly, an improved algorithm was proposed with a principle of choosing the benchmark anchor node, and we compared the sumulator results by choosing different benchmark anchor node with the principle in the end. It is proved that the proposed algorithm can improve the localization precision.

## 2 The Model of Nodes' Localization

The localization in wireless sensor networks usually calculates the coordinates of unknown nodes through measuring the distances between the nodes to be measured and the three anchors around. It respectively uses the three anchor nodes to be the centre of three circles, similarly uses the distance between unknown nodes and anchor nodes to be the radius of three circles, and the coordinates of this point which the three circles have intersected is exactly that of the unknown node.

Assuming the coordinate of the unknown node to be (x, y); the coordinates of three anchor nodes are respectively set to be $(x_1, y_1)$, $(x_2, y_2)$, $(x_3, y_3)$; the distance between the unknown node and the three anchor nodes are ordinal set to be d1, d2, d3, the model of the localization figure is as shown in figure1.

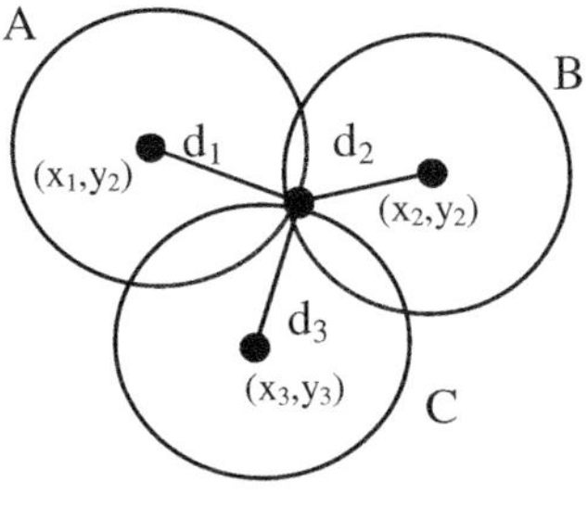


**Fig. 1.** The model of Nodes' localization

According to this model can get the equations as follow:

$$\begin{cases} (x_1 - x)^2 + (y_1 - y)^2 = d_1^2 \\ (x_2 - x)^2 + (y_2 - y)^2 = d_2^2 \\ (x_3 - x)^2 + (y_3 - y)^2 = d_3^2 \end{cases} . \quad (1)$$

Subtracting the third equation from the first and the second in expression(1),we get the following expression.

$$\begin{cases} 2(x_1 - x_3)x + 2(y_1 - y_3)y = (x_1^2 - x_3^2) + (y_1^2 - y_3^2) - (d_1^2 - d_3^2) \\ 2(x_2 - x_3)x + 2(y_2 - y_3)y = (x_2^2 - x_3^2) + (y_2^2 - y_3^2) - (d_2^2 - d_3^2) \end{cases} . \tag{2}$$

Solving the equations simultaneously, the coordinate of the unknown node D can be obtained.

## 3 Least-Square Algorithm for Localization Refinement

Because of the environmental influencing on distance measurement, such as, the signal of wireless are mainly influenced by transmission medium, the multipath transmission, signal reflections, antenna gain, etc, the error is produced.When the error appeared, these three circumferences will not have an intersection. So, we cannot get the coordinates of the nodes. In order to solve this problem, the least square algorithm has been used.

Assuming the coordinate of the unknown node to be (x, y), every node coordinate is respectively given to be $(x_1, y_1)$, $(x_2, y_2)$, $(x_3, y_3)$... $(x_n, y_n)$, so there are the expressions as following:

$$\begin{cases} (x_1 - x)^2 + (y_1 - y)^2 = d_1^2 \\ (x_2 - x)^2 + (y_2 - y)^2 = d_2^2 \\ \vdots \\ (x_n - x)^2 + (y_n - y)^2 = d_n^2 \end{cases} . \tag{3}$$

Respectively Subtracting the last equation from the others in expression(3),we get the following expression.

$$\begin{cases} 2(x_1 - x_n)x + 2(y_1 - y_n)y = (x_1^2 - x_n^2) + (y_1^2 - y_n^2) - (d_1^2 - d_n^2) \\ 2(x_2 - x_n)x + 2(y_2 - y_n)y = (x_2^2 - x_n^2) + (y_2^2 - y_n^2) - (d_2^2 - d_n^2) \\ \vdots \\ 2(x_{n-1} - x_n)x + 2(y_{n-1} - y_n)y = (x_{n-1}^2 - x_n^2) + (y_{n-1}^2 - y_n^2) - (d_{n-1}^2 - d_n^2) \end{cases} . \tag{4}$$

suppose:

$$A = 2 \times \begin{bmatrix} x_1 - x_n & y_1 - y_n \\ x_2 - x_n & y_n - y_n \\ \vdots & \vdots \\ x_{n-1} - x_n & y_{n-1} - y_n \end{bmatrix} .$$

$$B = \begin{bmatrix} (x_1^2 - x_n^2) + (y_1^2 - y_n^2) - (d_1^2 - d_n^2) \\ (x_2^2 - x_n^2) + (y_2^2 - y_n^2) - (d_2^2 - d_n^2) \\ \vdots \\ (x_{n-1}^2 - x_n^2) + (y_{n-1}^2 - y_n^2) - (d_{n-1}^2 - d_n^2) \end{bmatrix} .$$

$$X = \begin{bmatrix} x \\ y \end{bmatrix} .$$

Arranging the expression(4) in a matrix form is expression(5), we can solve the linear equation and get the corresponding value of X as expression(6).

$$AX = B\,. \tag{5}$$

$$X = (A^T A)^{-1} A^T B\,. \tag{6}$$

## 4 LSM_DS (Choosing the Benchmark Anchor Nodes of the Nearest Node)

It can effectively reduce the influence of measurement errors by using the least-square algorithm, and it can achieve the location estimation, but a part of the location information is lost in reducing factorial; meanwhile all the coordinate equations of the nodes are subtracted to be the benchmark node's distance equation, so the distance error between the benchmark node and ordinary node is influence heavily.

Therefore, it is hoped to choosing the benchmark node with the minimum distance error, as the error is usually relative to the distance, we can choose the node whose distance is the minimum to be the benchmark node.it can be presented by the expression(7) and (8) .

$$\begin{cases} (x_1 - x)^2 + (y_1 - y)^2 = d_1^2 + e_1 \\ (x_2 - x)^2 + (y_2 - y)^2 = d_2^2 + e_2 \\ \vdots \\ (x_n - x)^2 + (y_n - y)^2 = d_n^2 + e_n \end{cases}. \tag{7}$$

$$\begin{cases} 2(x_1 - x_n)x + 2(y_1 - y_n)y = (x_1^2 - x_n^2) + (y_1^2 - y_n^2) - (d_1^2 - d_n^2) - (e_1 - e_n) \\ 2(x_2 - x_n)x + 2(y_2 - y_n)y = (x_2^2 - x_n^2) + (y_2^2 - y_n^2) - (d_2^2 - d_n^2) - (e_2 - e_n) \\ \vdots \\ 2(x_{n-1} - x_n)x + 2(y_{n-1} - y_n)y = (x_{n-1}^2 - x_n^2) + (y_{n-1}^2 - y_n^2) - (d_{n-1}^2 - d_n^2) - (e_{n-1} - e_n) \end{cases}. \tag{8}$$

The simulation (with 100 ordinary nodes) results are shown by the figure2.

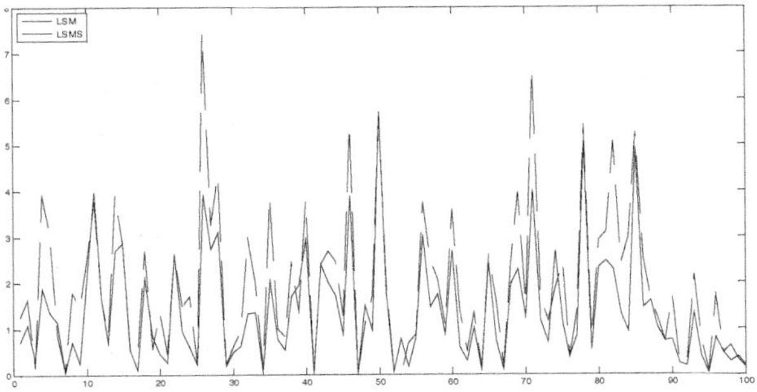


**Fig. 2.** The estimation error comparisons between using LSM and LSM_DS

To compare the errors between LSM and LSM_DS, the results are shown in fig.2 the LSM_DS algorithm only improves a part of the precision, the ensemble precision is not improved.

## 5 LSM_DR (Choosing the Benchmark Anchor Nodes of the Minimum Related Distance Node)

In order to expound clearly, we suppose the fellows:

$$J = \sum_{i=1}^{n-1} (e_i - e_n)^2, \text{ and } J^* = \sum_{i=1}^{n} (e_i)^2, \text{ and } J_e = \sum_{i=1}^{n-1} \left| e_i - e_n \right|.$$

As the error is $e_i$, we wish the $J^*$ rather than $J$ in expression (8) is minimum in the localizating by using LSM, So we have to reduce the extra error by $e_n$. we choose the anchor node with minimum related distance to be benchmark anchor node. In the algorithm, the anchor node which has the smallest $J_e$ is chosen to be the benchmark node, and the simulation is shown by the figure3.

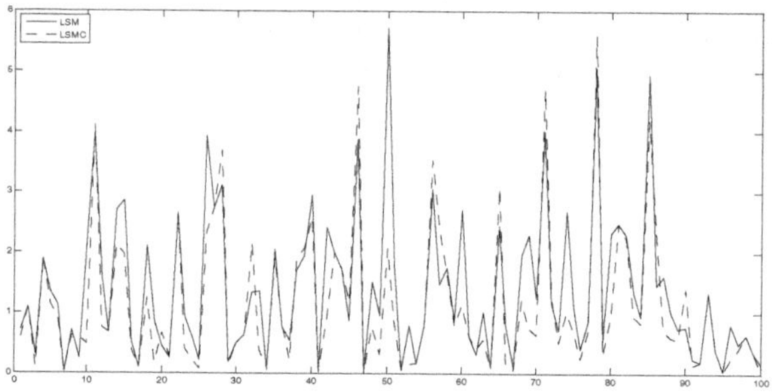


**Fig. 3.** Based on the minimum comprehensive distance to be the reference node

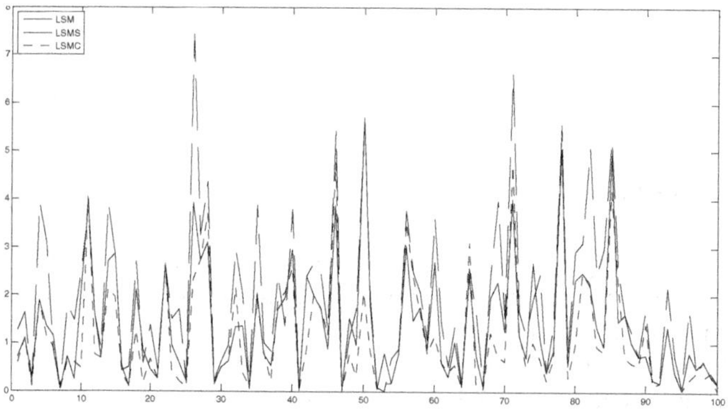


**Fig. 4.** The comparison of errors by the three algorithms

## 6 Simulation

By using random deployed 100 ordinary nodes with seven anchor nodes around them, the distances between the ordinary node and the anchor nodes are unequal and disproportionate, the distance error is a random positive number and it is less than the 30% of the distance. The simulation by using LSM, LSM_DS and LSM_DR about localization is shown as the figure4 and the table1. The simulation has proved that the LSM_DR can obtain more accurate localization while making the algorithm concisely.

**Table 1.** The errors of the three methods

| Algorithm | LSM | LSM_DS | LSM_DR |
|---|---|---|---|
| Average error(%) | 0.152 | 0.185 | 0.028 |

## 7 Conclusions

This paper proposes in improving localization precision with the application of the improved least squares algorithm, it analyses the reason why the extra error exists when subtracting the equation of distance between the anchor nodes and the ordinary nodes from the others in the LSM algorithm, and then proposes the principle of choosing the benchmark anchor nodes in reducing the errors with the distance' errors of the benchmark anchor nodes. The coordinate's errors of the nodes are decreased, thus, the goal to improve the localization accuracy has achieved.

**Acknowledgments.** Project supported by the National Natural Science Foundation of China (Grant No.60972162), 2010.1-2012.12.

## References

1. Akyildiz, I.F., Su, W., Sankarasubramaniam, Y., Cayirci, E.: A survey on sensor networks. IEEE Communications Magazine 40(8), 104–112 (2002)
2. Tilak, S., Abu-Ghazaleh, N.B., Heinzelman, W.W.: A taxonomy of wireless micro-sensor network models. Mobile Computing and Communications Review 1(2), 1–8 (2002)
3. Lorincz, K., Malan, D., Fulford-Jones, T.R.F., Nawoj, A., Clavel, A., Shnayder, V., Mainland, G., Welsh, M., Moulton, S.: Sensor networks for emergency response: challenges and opportunities. Pervasive Computing for First Response (Special Issue), IEEE Pervasive Computing (2004)
4. Gao, T., Greenspan, D., Welsh, M., Juang, R.R., Alm, A.: Vital signs monitoring and patient tracking over a wireless network. In: Proceedings of the 27th IEEE EMBS Annual International Conference (2005)
5. Wener-Allen, G., Lorincz, K., Ruiz, M., Marcillo, O., Johnson, J., Lees, J., Walsh, M.: Deploying a wireless sensor network on an active volcano. Data-Driven Applications in Sensor Networks (Special Issue). IEEE Internet Computing (2006)
6. Savarese, C., Rabaey, J.M., Beutel, J.: Localizationing in distributed ad-hoc wireless sensor network. In: Proceeding of the 2001 IEEE International Conference on Acoustics, Speech, and Signal Salt Lake (2001)
7. Savarese, C., Rabaey, J.M., Langendoen, K.: Robust Localization Algorithms for Distributed Ad Hoc Wireless Sensor Networks. In: Proceedings of the USENIX Technical Annual Conference, Monterey, USA (2002)

8. Avvides, A., Park, H., Srivastava, M.: The bits AND flops of the N-hop multilateration primitive for node localization problems. In: Proceeding of the 1st ACM International Workshop on Wireless Sensor Networks and Applications, Atlanta (2002)
9. Bergamo, P., Mazzini, G.: Localization in sensor networks with fading and mobility. In: Processing of the 13th IEEE International Symposium on Personal, Indoor and Mobile Radio Communications, NJ, USA (2002)
10. Guha, S., Murty, R.N., Sirer, E.G.: A unified node and event localization framework using non-convex constraints. In: Proceeding of the 6th ACM International Symposium on Mobile Ad Hoc Networking and Computing, IL, USA (2005)
11. Cao, M., Anderson, B.D.O., Morse, A.S.: Sensor network localization with imprecise distances. Systems and Control Letters (2006)
12. Srirangarajan, S., Tewfik, A.H., Luo, Z.Q.: Distributed sensor network localization with inaccurate anchor positions and noisy distance information. In: IEEE International Conference on Acoustics, Speech and Signal Processing, HI, USA (2007)
13. Chen, X., Du, L.: Adaptive Localization Algorithm for Wireless Sensor Networks Base on Back. In: Proceedings of 2010 Conference on Dependable Computing, China (2010)


## ABOUT THE AUTHOR(S)

Xiaohui Chen, he received his bachelor's degree from Wuhan University of Technology (WHUT) in 1988 and the master's degree from Huazhong University of Science and Technology (HUST) in 1996, He is currently a associate professor in the college of computer and information technology of China Three Gorges University. His research interests include wireless sensor networks, data mining, and intelligent control.

Jing He, he received his bachelor's degree from Huazhong University of Science and Technology (HUST) in 2010, he is seeking for his master's degree in the college of computer and information technology of China Three Gorges University now. His research fields include wireless sensor networks, embedded system.

# Identification of pH Neutralization Process Based on the T-S Fuzzy Model

Xiaohui Chen[1,*], Jinpeng Chen[1], and Bangjun Lei[2]

[1] College of Computer and Information Technology, China Three Gorges University, Yichang, China, 443002, P.R. China
chui@ctgu.edu.cn

[2] Institute of Intelligent Vision and Image Information, China Three Gorges University, Yichang, China 443002, P.R. China
qccjpgo@hotmail.com

**Abstract.** Considering the nonlinearity and gradient parameters of pH neutralization process, this paper concerned on the modeling and identification of pH neutralization process. As the approximate three sections linear characteristic of titration curve in the pH neutralization process, we discussed the use of T-S fuzzy model for modelling the pH neutralization process. Due to its gradient parameters, we identified the parameters of the system using its input-output data with the method of recursive least square with fading factor algorithm (RLS-RFF). Simulations including recursive least square (RLS) and RLS-RFF have shown the efficiency of the method, and RLS-RFF has better identification accuracy and adaptive ability in the reaction process of gradient parameters.

**Keywords:** T-S fuzzy model, fuzzy identification, pH neutralization process, least square method.

## 1 Introduction

PH process control has been widely used in chemical, light industry, wastewater treatment and environmental protection. Improving the identification accuracy is helpful to the effective control, and has a profound effect in improving the quality and output of the products, as well as the safety of production equipment and the environmental protection. Due to the influence of serious non-linear, time delay, and strong interference in pH neutralization process. The identification and control of the pH neutralization process has been one of the most difficult problems in relative fields[1]. Thus the research of the identification and control in pH neutralization process is very important.

For many years, extensive researches in the identification of pH neutralization process have been done by many relative experts. The mathematical model of pH neutralization process[2]. More complicated but more all-purpose pH neutralization model has been further proposed[3], which laid a research foundation for identification and control of the pH neutralization process. Modeling chemical

* Corresponding author.

S. Lin and X. Huang (Eds.): CSEE 2011, Part II, CCIS 215, pp. 579–585, 2011.

process systems via neural computation[4]. Modeling pH neutralization process using fuzzy neural approaches[5].Wiener model has been adopted to the study of pH neutralization process[6].Artificial neutral network of pH neutralization process[7]. The pH control process has been identified by universal learning network[8]. The approach of neutral network applied to pH neutralization process has the shortage that it is difficult to control the complexity of the network, and Wiener model applied to pH neutralization is complicated, what's more, the phenomenon of over-fitting is likely to appear; The T-S fuzzy model take linear equation as the consequent part[9], and it is possible to deal with nonlinear problems by using the theory of linear system. The universal approximation of T-S fuzzy model for the nonlinear system[10] provides the theory basis to identify the nonlinear system by using T-S fuzzy model.

In this paper, the identification of the pH neutralization process by using T-S fuzzy model has been discussed, and we analysed the characteristic of the titration curve in the pH neutralization process, which is very similar to the three sections linear line linked. It is feasible to take T-S fuzzy model as the model of pH neutralization process[11]. The parameters of the T-S fuzzy model are identified by the recursive least square with fading factor algorithm (RLS-RFF) in the reaction process of gradient parameters. The efficiency of the proposed approach is proved by MatLab simulation.

## 2 Identify the Model Parameters by Least Square

### 2.1 Establishment of the Model of pH Neutralization Process

T-S fuzzy model is an essential nonlinear model, and it can approach a nonlinear system infinitely. The pH neutralization process has the characteristic of three sections linear line linked. So the T-S fuzzy model is adopted as the model of pH neutralization process, the jth fuzzy rule can be described as follows:

$R^j$: IF $x_1$ is $A_1^j$ and $x_2$ is $A_2^j$ and ... $x_n$ is $A_n^j$, Then $y_j = p_0^j + p_1^j x_1 + ... + p_n^j x_n$ . (1)

The rule *j*th is represented by $R^j$, $x_j$ is the input variable, $A^j_n$ is the membership function of the input domain. $p^j_i$ is the consequent parameters, $y_j$ is the output of *j*th rule.

Aiming at the pH neutralization process, let *F(t)* be the acid flow rate, and *u(t)* be the alkali flow rate, then the input of the pH neutralization process *x=F(t)-u(t)*.

In this paper, Gaussian function is introduced, and it is shown as expression (2)

$$A_i^j(x_i) = \exp(-(\frac{x_i - \bar{x}_i^j}{\sigma_i^j})^2) \quad (2)$$

Set two fuzzy sets, so it can establish two fuzzy rules, and the premise parameters are $\bar{x}_i^j$ and $\sigma_i^j$ , *j*=1, 2.

The fuzzy logic system has the following form:

$$Y = f(X) = [\sum_{l=1}^{M} h^l \times (p_0^l + p_1^l x_1 + p_2^l x_2 + ... + p_n^l x_n)] \quad (3)$$

$$W=(h^1,h^1x_1,..h^1x_n,h^2,h^2x_1,...,h^2x_n,..h^M,h^Mx_1,..h^Mx_n)$$
$$\theta=(P_0^1,P_1^1,...,P_n^1,P_0^2,P_1^2,...,P_n^2,...,P_0^M,P_1^M,...,P_n^M)$$

$$Y = f(X) = W \bullet \theta^T . \tag{4}$$

Aiming at the representative model of pH neutralization process, then

$$Y = f(X) = [\sum_{l=1}^{2} h^l \times (p_0^l + p_1^l x)] = W \bullet \theta^T \tag{5}$$

$f(X)$ and its parameters in equation (5) are in a linear relationship, and $W$ is known as basis function, so the independent parameter $\theta$ is needed to be identified. So it is suitable for control design[12], and the linear parameter estimation method can be adopted. In this paper, Conventional recursive least square (RLS) and RLS-RFF algorithm are used to identify parameters.

## 2.2 Model Identification Based on Least Square

The titration curve of the pH neutralization process is very similar to the three sections linear line linked, so it can be modelled by the T-S fuzzy model. It is easy to identify premise part, and it can be obtained through several simple experiments. This paper mainly identifies the consequent part. Supposing the object output is $y$, identification output is $y_n$, then identification error $e = y-y_n$, identification Indicator $J = \sum e^2$.

### 2.2.1 Identification of Conventional RLS

Object model is represented by equation (5), according to the principle of RLS[13], with the observational data, then the least squares estimate $\theta_k$ of parameter vector $\theta$ is

$$\theta_k = (W_k^T W_k)^{-1} W_k^T Y_k . \tag{6}$$

$$Y_k = [y(1), y(2),... y(k)]^T , W_k = [w(0), w(1) ... w(k-1)]^T$$

Suppose
$$S_k = (W_k^T W_k)^{-1}. \tag{7}$$

According to the inversion formula of Matrix

$$S_k^{-1} = S_{k+1}^{-1} - w_{k+1} w_{k+1}^T . \tag{8}$$

Then
$$\theta_{k+1} = S_{k+1} W_{k+1}^T Y_{k+1} . \tag{9}$$

$$F_{k+1} = S_{k+1} w_{k+1} . \tag{10}$$

So
$$\begin{cases} \theta_{k+1} = \theta_k + F_{k+1}[y_{k+1} - w_{k+1}^T \theta_k] \\ F_{k+1} = \dfrac{S_k w_{k+1}}{1 + w_{k+1}^T S_k w_{k+1}} \\ S_{k+1} = [I - F_{k+1} w_{k+1}^T] S_k \end{cases} . \tag{11}$$

### 2.2.2 The RLS-RFF Algorithm

Actually the pH neutralization process is a time-varying dynamic process. With the growth of data, the conventional least square algorithm will appear a "data saturation" phenomenon, new data has been inundated by number of old data. It is impossible to track the varying parameters. In order to overcome the data saturation, the RLS-RFF algorithm is adopted. Specific approach is described as follows:

When get a new measured data, all previous data will be multiplied by a weighted factor $\rho(0 < \rho < 1)$. The form of the corresponding weight matrix is

$$M_k = \begin{bmatrix} \rho^{k-1} & & & & 0 \\ & \rho^{k-2} & & & \\ & & \ddots & & \\ & & & \rho & \\ 0 & & & & \rho^0 \end{bmatrix}.$$

According to the principle of recursive least square[13]

$$\theta_k = (W_k^T M_k W_k)^{-1} W_k M_k Y_k = (W_k^T W_k)^{-1} W_k Y_k = S_k W_k Y_k \cdot \quad (12)$$

Take the above expressions into the RLS parameter estimation algorithm

$$S_{k+1} = (W_{k+1}^T W_{k+1})^{-1} = (\rho^2 W_k^T W_k^T + w_{k+1} w_{k+1}^T)^{-1} = (\rho^2 S_k^{-1} + w_{k+1} w_{k+1}^T)^{-1} . \quad (13)$$

According to the matrix inversion formula

$$S_{k+1} = \frac{1}{\rho^2}[I - S_k w_{k+1}(\rho^2 + w_{k+1}^T S_k w_{k+1})^{-1} w_{k+1}^T] S_k . \quad (14)$$

Then

$$F_{k+1} = S_{k+1} w_{k+1} = S_k w_{k+1}[\rho^2 + w_{k+1}^T S_k w_{l+1}]^{-1} \cdot \quad (15)$$

So

$$\begin{cases} \theta_{k+1} = \theta_k + F_{k+1}[y_{k+1} - w_{k+1}^T \theta_k]. \\ F_{k+1} = \dfrac{S_k w_{k+1}}{\rho^2 + w_{k+1}^T S_k w_{k+1}} \\ S_{k+1} = \dfrac{1}{\rho^2}[I - F_{k+1} w_{k+1}^T] S_k \end{cases} \quad (16)$$

# 3 Mechanism of pH Neutralization Process

PH =- log [H +], and the representative model of the pH is described as follows[3]:

$$\begin{cases} V\dfrac{dy}{dt} = F(t)(a - y(t)) - u(t)(b + y(t)) \cdot \\ y(t) = 10^{-pH(t)} - 10^{pH(t)} K_w \end{cases} \quad (17)$$

The discretization of equation (17)

$$\frac{dy}{dt} = \frac{y(k) - y(k-1)}{T} . \quad (18)$$

$$y(k) = \frac{\frac{V}{T} y(k-1) - bu(k) + aF(k) + w(k+1)}{\frac{V}{T} + u(k) + F(k)} = \frac{y(k-1) - b\frac{T}{V}u(k) + a\frac{T}{V}F(k) + w(k+1)}{1 + \frac{T}{V}u(k) + \frac{T}{V}F(k)} . \quad (19)$$

$$pH(t) = \lg \frac{-y(t) + (y(t)^2 + 4K_w)^{1/2}}{2K_w} . \tag{20}$$

*Kw*: water equilibrium constant, $10^{-14}$ *V*: reactor volume, L
*F (t)*: the acid flow rate, L/min *u (t)*: the alkali flow rate, L/min
*a*: the concentration of acid, mol/L *b*: the concentration of alkali, mol/L
*T*: sampling time, min *w*: system noise signal
*y (t)*: the distance from the neutral point, [H+] - [OH-]

## 4 Simulation

Input the control signal $U = 21(1+\sin(2\pi t/f))$ and choose the model of formula (19) and (20), the parameters of simulation: $f = 200$, $a = 0.001$mol/L, $V = 2$L, $T = 1$ min, $F(t) = 0.1125$L/min. Initial $b_0$ of $b$ equals 0.001mol/L and increase slowly by $\nabla b$=0.15/400* $b_0$. The fuzzy center $x_1 = 7$, $x_2 = 30$; the fuzzy radius $\sigma_1 = 5$, $\sigma_2 = 5$.

The error curve is shown in Fig.1 and the titration curve is shown in Fig.2.

The Clustering method is adopted to identify the same objections, after 400 times calculations.

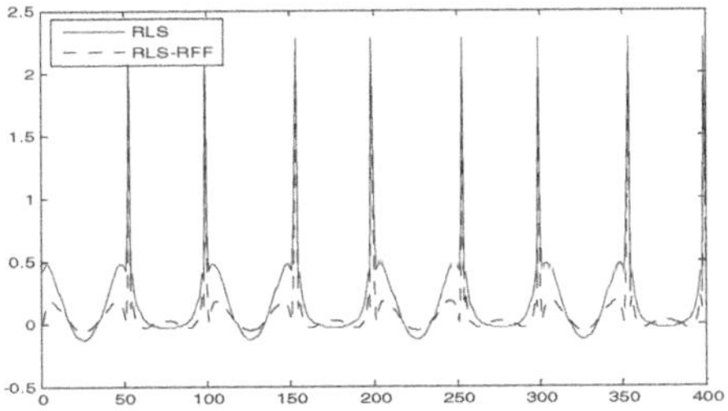


**Fig. 1.** The errors comparison between the RLS and RLS-RFF

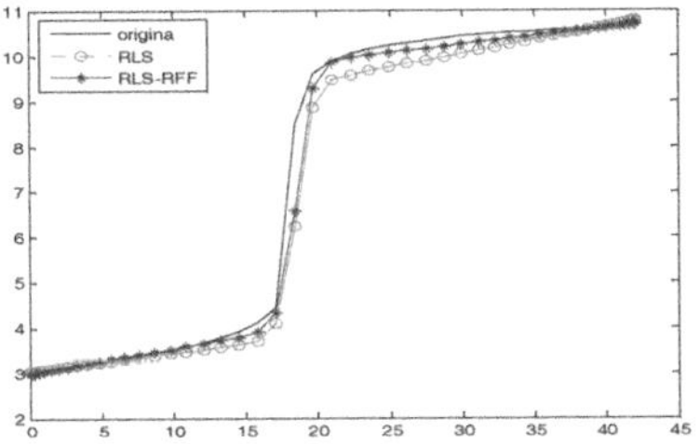


**Fig. 2.** The comparison of the pH titration curve among the original, RLS and RLS-RFF, where the red line is identified by the RLS-RFF algorithm, the green line is identified by the RLS algorithm, and the blue pH titration curve represents the original pH titration curve

**Table 1.** The errors of the two methods

| Approach | RLS | RLS-RFF |
|---|---|---|
| Error | 0.0203 | 0.0145 |

## 5 Conclusion

This paper is based on the characteristic of pH value neutralization titration curve, and the model of pH neutralization process is constructed by the T-S fuzzy model. The parameters of T-S fuzzy model are identified by the RLS algorithm, and the online identification of the model is realized. In consideration of the variation of parameters during reaction process, the real-time identification is taken by the RLS-RFF algorithm. The simulation results show that the RLS-RFF algorithm can fit the titration curve of pH neutralization process quite well, and the precision of identification has been greatly improved, as well good effect has been achieved. The use of T-S fuzzy model to describe the rationality and efficiency of pH neutralization process can realize the online identification of related parameters better.

**Acknowledgments.** Project supported by the National Natural Science Foundation of China (Grant No.60972162), 2010.1-2012.12.

## References

1. Du, S.: Set range intelligent controller for pH neutralization of industrial waste water. Chinese Journal of Scientific Instrument 25(2), 202–204 (2004)
2. Mcavoyt, T.J., Hsu, E., Lowenthal, S.: Dynamics of pH in controlled stirred tank reactor. J. Ind. Eng. Chem. Process Des. 11(1), 68–70 (1972)
3. Goodwith, G.C., Sin, K.S.: Adaptive filtering prediction and control. Prentice Hall, Inc., Englewood Cliffs (1984)
4. Bhat, N.V., Minderman, P.A., Macvoy Jr., T., et al.: Modeling chemical process systems via neural computation. IEEE Control Systems Magazine 10(3), 24–30 (1990)
5. Nie, J.H., Lonap, P., Hang, C.C.: Modeling pH neutralization processes using fuzzy neural approaches. Fuzzy Sets and Syst. 78(1), 5–22 (1996)
6. Gomez, J.C., Jutan, A., Baeyens, E.: Wiener model identification and predictive control of a pH neutralization process. IEEE Proc-Control Theory App. 151(3), 329–338 (2004)
7. Liu, T., Chen, Z., Zhou, Y., et al.: Application of artificial neutral for pH neutralization system identification. J. Computers and Applied Chemistry 23(12), 1280–1282 (2006)
8. Han, M., Jia, X.-m.: Identification of neutralization control process based on universal learning network. Journal of Dalian University of Technology 43(1), 119–123 (2003)
9. Takagi, T., Sugeno, M.: Fuzzy identification of systems and its application to modeling and control. IEEE Syst., Man, Cybern. SMC-15(1) (1985)
10. Wang, L.: Adaptive fuzzy system and control. National Defence Industry Press, Beijing (1995)

11. Chen, X.: Identification Method of pH characteristic Based on T-S model. In: The Proceedings of the 23rd Chinese Control Conference, pp. 980–983 (2004)
12. Chen, X., et al.: Application of Fuzzy Control to Production of Urea-Formaldehyde Resin Adhesive (UF). Journal of China Three Gorges University (Natural Sciences) 24(4), 314–316 (2002)
13. Yang, C., et al.: System Identification and Adaptive Control. Chongqing University Press, Chongqing (2003)

## ABOUT THE AUTHOR(S)

Xiaohui Chen, he received his bachelor's degree from Wuhan University of Technology (WHUT) in 1988 and the master's degree from Huazhong University of Science and Technology (HUST) in 1996, He is currently a associate professor in the college of computer and information technology of China Three Gorges University. His research interests include wireless sensor networks, data mining, and intelligent control.

Jinpeng Chen, he received his bachelor's degree from Huazhong University of Science and Technology (HUST) in 2010, he is seeking for his master's degree in the college of computer and information technology of China Three Gorges University now. His research fields include wireless sensor networks, intelligent control, and embedded system.

# The Reseach on Distributed Data Storage Method Based on Composite Threshold

Yue Fuqiang

School of automotive and electronic engineering,
Xichang College, Xichang, China

**Abstract.** This paper presents a distributed data storage method based on sleep scheduling to resolve the problems of network imbalance and storage hot spots problems. Compared with existing methods, our distributed data storage method has the following advantages: First, take node's residual energy into account when decides the order of storage node. This can avoid choosing the lower residual energy node blindly which closer to the hash location as frequently data storage node. Second, set a scheduling factor p for each node, which determined by its storage space and residual energy. As the scheduling factor p changes, sensor node switching the state between suspended state and active state, this can turn off the redundant node and reduce unnecessary energy consumption. Finally, multi group analysis of simulate experiments results show that compared to other data storage method the distributed data storage method based on composite threshold have obviously advantages on the sides of overall energy consumption, data storage capacity, the number of failure node and data quality, thus have a significant effect on reducing energy consumption and extending network life cycle.

**Keywords:** Distributed Data; Storage Method; Composite Threshold.

## 1 Introduction

In response to these questions Wen-Hwa Liao et al in 2009 proposed a grid-based dynamic load balancing of data storage method DLB [22]. It does this by covering the grid (cover-up scheme) and multi-threshold (the multi-threshold) and other programs to improve the GHT were present in the storage nodes fixed and stored in hot spots. One multi-threshold (the multi-threshold) threshold is based on the storage capacity of the node set. If the blind will be a storage capacity of nodes as storage nodes, regardless of its battery power, then, from the perspective of the network life cycle is unreasonable.

In this paper, the shortcomings of DLB, the threshold will be stored with the nodes form a comprehensive energy factor binding threshold, the threshold level under the Comprehensive threshold calculation node scheduling method of scheduling factor p, scheduling factor to determine the status of the node, so node job rotation and to dynamically adjust the data storage nodes, storage nodes within the grid balance the effect of the load to save energy and prolong network lifetime.

S. Lin and X. Huang (Eds.): CSEE 2011, Part II, CCIS 215, pp. 586–591, 2011.

## 2 Network Model

In a wireless sensor network, in order to be able to perceive their own position, each sensor node is equipped with a positioning device with the GPS or other positioning methods can get your current location. This article will cover the area of wireless sensor networks as shown in Figure 1. is divided into a number of area d × d dimensional grid of small squares, each grid number is based on horizontal and vertical coordinates (x, y), this number unique for each grid ID (similar to the IP address and MAC address). Sink in the sink node to set the coordinate origin. Any position of sensor nodes can be represented by formula (1) calculated in the grid mesh ID:

$$y = \lfloor (y_i - y_0) / d \rfloor \tag{1}$$

Where: $(x_i, y_i)$ ----- sensor nodes horizontal and vertical coordinates of current position

$(x_0, y_0)$ ----- Coordinate origin

d ----- grid side length, m

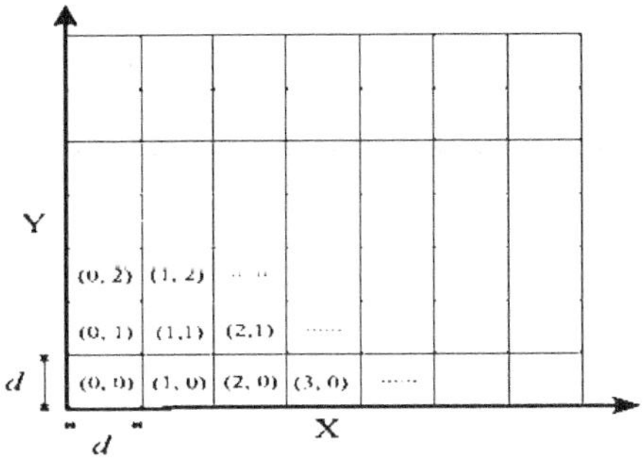


**Fig. 1.** The area is divided into two-dimensional perception of the logic grid

The size of the grid which is a very important parameter, in order not to affect the implementation of network routing protocols, the need for the grid side length d to a reasonable setting. Let r is the distance signal transmission of sensor nodes, d, 6 the choice of possibilities:

(1) d → + ∞. Broadcast signal within the grid is difficult to reach locations outside the grid, so grid of communication between the grid will not be successful. Therefore, d can not be too long (see Figure 2a, d = 2r the case).

(2) d = r. Two adjacent grid nodes in the grid center of communication situations can be the maximum d (see Figure 2b).

(3) $d = 2r/\sqrt{10}$ . A central location in the grid nodes to the grid and four neighbors of any node can communicate in case of the maximum value of d (see Figure 2c).

(4) $d = \sqrt{2}r/3$. Located in the center of a node in the grid and adjacent to the 8 grid nodes can communicate in any case the maximum value of d (see Figure 2d).

(5) $d \le r/(2\sqrt{2})$. At any location within the grid nodes and adjacent to the 8 grid nodes can communicate in any case the maximum value of d (see Figure 2e).

(6) d → - ∞. A grid of little or no sensor nodes. Means that the extreme case, d infinitely small, there are an infinite number of the grid. In fact, this extreme case does not tend to the case of the grid concept (see Figure 2 f, d = r/10 case).

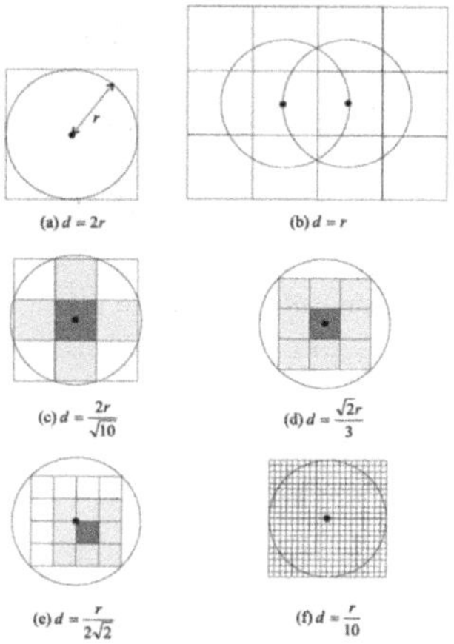


**Fig. 2.** d and the relationship between r

Not difficult to find, d the smaller the value, will lead to greater contact between the grid neighbors. Also means that the number of nodes within a grid will be less. But d is greater than a certain range, there will be "blind spot" problem. Therefore, the value in the choice of d, when there are some trade-offs. D Based on the above analysis paper will be set to.

Located in the network model, there are 5 packet:

When a node detects an event,λ it sends a packet to put the event storage node. A put packet includes the following three content points:

Event type ET
ED event data
Gvid each sensor node ID in the virtual grid

When a user issues a queryλ request, the sink node sends a packet to get the event storage node, a get packet includes the following two elements:

Event type ET
Gvid each sensor node ID in the virtual grid

When a storage node within theλ grid when the threshold is reached, a Storage_Full packet will be sent to inform neighbors within the same grid nodes. A Storage_Full packet includes the following two elements:

Grid in the reality of each sensor node ID grid
Nid identity of each sensor node ID

When a grid of all storageλ nodes are reached with a threshold level, the grid reaches a threshold value of the last node sends a data packet to the same Change_Threshold a grid of all the neighbor nodes. All the same grid nodes at the same time they change the two thresholds to the next level. A Change_Threshold 1 packet includes the following elements:

Grid in the reality of each sensor node ID grid

When a grid of all storageλ nodes have reached the threshold of the highest threshold level, the grid to reach the highest level of the last node sends a packet to all Grid_Extend neighbor grid. A Grid_Extend packet includes the following two elements:

Gvid each sensor node ID in the virtual grid
Event type ET

## 3 Distributed Data Storage Method Based on Composite Threshold

### 3.1 Method Description

Introducing the specific data storage method prior to introduction within the wireless sensor network neighbor discovery phase.

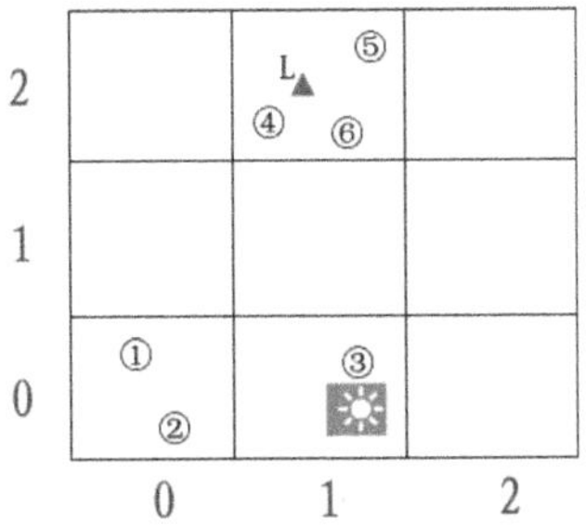


**Fig. 3.** Data Storage

For example, as shown in Figure 3, node 3 detects the temperature information, find the event according to event type storage location L, L from (1) into the grid ID (1,2). Grid (1,2) of nodes in the node 3, 6 received data packets sent after put in the grid (1,2) put the packet flood. Nodes within the grid mesh nodes first check their own form of Gvid Grid_Node table and whether the ET packet Gvid and put the same ET. If not identical, to continue to forward the packet to the neighbor grid. If the same, each node receives the packet first put its own residual energy calculation and storage space Si Ei product, comparison between the neighbor nodes after completing the maximum value from the Ei * Si node 5 as the current round of activity nodes, Save put packets, other nodes into the suspended state. Node 5 continues the process of storing data to calculate the scheduling factor p, as found in p <0, when the data packet to send a Storage_Full grid (1,2) in all nodes, enter the suspended state. Node receives data packets Storage_Full to calculate their remaining energy and storage space Si Ei product, between the neighbor nodes within the grid after completing comparison, the maximum value of Ei * Si node 4 as the active node to continue this round store data, other nodes continue to be suspended.

### 3.2 Method Characteristics

In summary, according to the threshold node integrated scheduling algorithm, the rotation through the work of the node to store data, to the saving power consumption, load balancing purposes. First, since only one node at a time to store data, avoiding duplication of multiple nodes send data to the emergence of information redundancy situation and the resulting energy consumption of network bandwidth and overhead; second, to solve all the queries are concentrated in the individual nodes, the query often occurs when hot issues; the third to reach the threshold of the node using the virtual coordinates, reducing the number of packets passed.

## 4 Simulation and Analysis of Results

Network at two different experimental conditions: (1) N = 800, X * Y = 400m * 400m. (2) N = 1800, X * Y = 600m * 600m. In both cases, the node density is the same.

Figure 4 shows the number of events given to changing the DBSS, DLB, and GHT's overall energy consumption. The overall energy consumption of DBSS which these three methods is the lowest, while the DLB's overall energy consumption is also lower than GHT. The results also show that: the node density is constant, the network time scale, GHT's energy consumption compared to DBSS and DLB is even more apparent. DBSS and DLB in the storage node is not fixed, DBSS nodes in the grid rotation under the Comprehensive threshold as a storage node; DLB near the storage node in the mesh nodes; GHT storage node close to the hash location. Overall energy consumption by the data storage and data retrieval of two parts. In the data store, DBSS, and DLB in a time event data generated is stored in a node only, while the GHT events generated by a certain time replication of data stored in the structure, which is stored in many nodes, thus DBSS stored energy than GHT and DLB. In data retrieval, DBSS, and DLB send a query to the same grid node, only the requested data stored in line with the sensor node to send data back together, and GHT sends a query request to all nodes in the mirror, each corresponding sensor data to be sent back to the sink node. Therefore, GHT's energy consumption is greater than DBSS data retrieval and DLB. In DLB, the storage node by node in order to determine the distance between grid points, with the number of incidents and increase in number of queries, the grid near the transmission of data packets frequently lead to high energy near the nodes, affect the DLB's overall energy consumption. Therefore, the contrast and DLB and GHT, DBSS's overall energy consumption is the lowest.

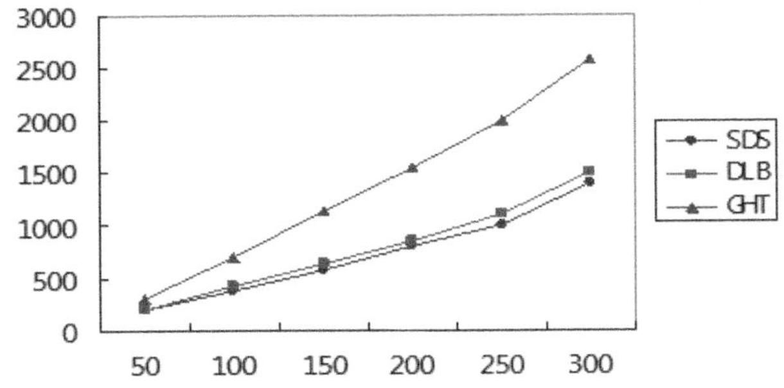


(A) X * Y = 400m * 400m

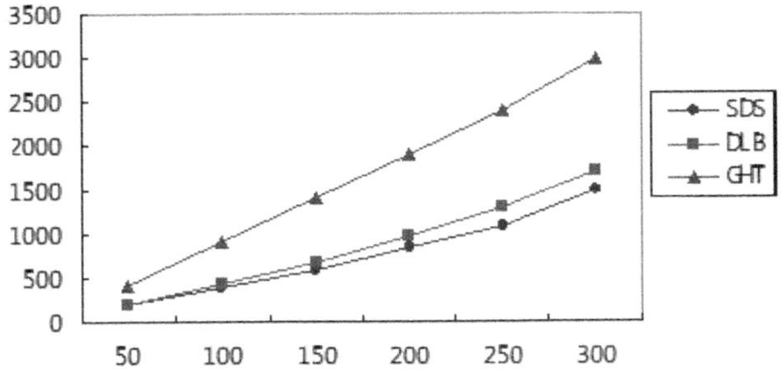


(B) X * Y = 600m * 600m

**Fig. 4.** Overall energy consumption

## References


1. Albano, M., Chessa, S.: Publish/subscribe in wireless sensor networks based on data centric storage. In: Proceedings of the 1st International Workshop on Context-Aware Middleware and Services, CAMS 2009, pp. 37–42 (2009)
2. Yu, Z., Xiao, B., Zhou, S.: Achieving 0ptimal data storage position in wireless sensor networks. Computer Communications 33(1), 92–102 (2010)
3. Ratnasamy, S., Karp, B., Yin, L., Yu, F., Estrin, D., Govindan, R., Shenker, S.: GHT: A geographic hash table for data-centric storage. In: Proceedings of the 1st ACM International Workshop on Wireless Sensor Networks and Applications, pp. 78–87 (2002)
4. Mathur, G., Desnoyers, P., Ganesan, D.: Ultra-low power data storage for sensor networks. In: Proc. of the 5th Interntional Conference on Information Processing in Sensor Networks, pp. 374–381. ACM Press, New York (2006)
5. Ren, Y., Shchuk, V., Li, F.Y.: A Distributed data storage and retrieval scheme in unattended WSNs using homomorphic encryption and secret sharing. In: 2009 2nd IFIP Wireless Days (WD), vol. 15, pp. 1–6 (2009)

# Open-Closed-Loop PD-Type Iterative Learning Control with Variable Gains

JianMing Wei[1] and YunAn Hu[2]

[1] Graduate Students' Brigade, Naval Aeronautical Engineering Institute, Yantai, China
[2] Department of Control Engineering, Naval Aeronautical Engineering Institute, Yantai, China
wjm604@163.com, hya507@sina.com

**Abstract.** In this paper an open-closed-loop PD-type Iterative Learning Control (ILC) algorithm with variable learning gains is proposed. The learning gains are varying with the system errors or the iteration times. Thus it can eliminate the errors fast and reduce the overshoot. Therefore the capability of the target tracking is greatly improved. A rigorous proof of the sufficient condition for the algorithm is given, which shows the convergence of the tracking error.

**Keywords:** open-closed-loop ILC; variable gains.

## 1 Introduction

In practice many tasks are repetitive over a finite time. Then can we improve the control performance by using the information collected from previous executions? Inspirited by this idea, Arimoto[1] proposed Iterative Learning Control (ILC) in 1984. Whereafter ILC attracted a lot of attention because of its simple control structure and good tracking performances. According to the learning action type, ILC can be classified as P-type, D-type, PD-type, and PID-type. However the learning gains of these ILC schemes are invariant. Though the tracking performances are improved to a certain extent compared with PID controller, the convergence speed is slow. Aim at solving this problem in [2] an ILC law with variable gain was proposed. In [3] P.R. Ouang introduced an adaptive switching learning control method to deal with trajectory tracking of robot manipulators and in 2011 he proposed switching gain PD-type ILC for trajectory tracking control of time-varying nonlinear systems with uncertainty and disturbance[4].

To improve the speed of convergence an open-closed-loop PD-type ILC with variable learning gains is proposed in this paper. Varying gains for the open-loop can be adjusted according to the system's errors and varying gains according the iteration times are introduced for closed-loop. And in general the gains of the system errors are large relatively at the initial iteration times and decrease with trial because the errors are large and can be eliminated quickly by large system error information. As the system errors are small the error differential gains become larger, thus to reduce the overshoot.

S. Lin and X. Huang (Eds.): CSEE 2011, Part II, CCIS 215, pp. 592–597, 2011.

## 2 Problem Formulation

Let's consider the class of time-varying nonlinear system which runs repeatedly over $[0,T]$:

$$\begin{aligned} \dot{x}_k(t) &= f\left(x_k,t\right) + B(t)u_k(t) \\ y(t) &= C(t)x_k(t) \end{aligned} . \tag{1}$$

where $k$ denotes the iteration index. $x \in \Re^n$, $y \in \Re^m$ and $u \in \Re^r$ are the state vector, output vector and control vector of the system, respectively. $f, B, C$ are the uncertain matrices with approximate dimension. The system is assumed to satisfy the following assumptions.

*Assumption 1*. Nonlinear function $f\left(x(t),t\right)$ is globally Lipschitz continuous with respect to $x$, i.e.

$$\left\|f(x_1,t) - f(x_2,t)\right\| \le M\left\|x_1 - x_2\right\|, \tag{2}$$

where $M$ is an unknown Lipschitz constant.

*Assumption 2*. For the given trajectory $y_d(t)$, there exists an unique $u_d(t)$ such that

$$\dot{x}_d(t) = f\left(x_d,t\right) + B(t)u_d(t), y_d(t) = C(t)x_d(t) \quad \forall t \in [0,T]. \tag{3}$$

*Remark 1*. Since $u_d(t)$ exists uniquely, the uniform convergence of the control $u(t)$ to $u_d(t)$ implies that the state and output tracking errors will vanish.

*Assumption 3*. $B(t), C(t)$ and $\dot{C}(t)$ are continuous and $\dot{C}(t)$ is bounded.

*Assumption 4*. The sequence of the original state error $\left\{\delta x_k(0)\right\}$ converges to zero.

The control target is to find an approximate learning scheme, such $\lim\limits_{k\to\infty}\left\|\delta u_k(t)\right\|_\lambda = 0$ and $\lim\limits_{k\to\infty}\left\|e_k(t)\right\|_\lambda = 0$, where $\delta u_k(t) = u_d(t) - u_k(t)$, $e_k(t) = \delta y_k(t) = y_d(t) - y_k(t)$ is the tracking error, $\left\|\cdot\right\|_\lambda = \max\limits_{t\in[0,T]} e^{-\lambda t}\left\|\cdot\right\|$ $(\lambda > 0)$ is the $\lambda$-norm.

## 3 Control Design and Convergence Analysis

Define $\delta x_k(t)$ and $f_1\left(\delta x_k(t),t\right)$ as follows:

$$\begin{cases} \delta x_k(t) = x_d(t) - x_k(t) \\ f_1\left(\delta x(t),t\right) = f\left(x_d(t),t\right) - f\left(x_d - \delta x(t),t\right) = f\left(x_d(t),t\right) - f\left(x(t),t\right) \end{cases} . \tag{4}$$

Then, we have

$$\begin{cases} \delta\dot{x}_k(t)=\dot{x}_d(t)-\dot{x}_k(t)=f_1(\delta x_k(t),t)+B(t)\delta u_k(t) \\ e_k(t)=\delta y_k(t)=C(t)\delta x_k(t) \end{cases}, \tag{5}$$

$$\dot{e}_k(t)=\dot{C}(t)\delta x_k(t)+C(t)\delta\dot{x}_k(t)=\dot{C}(t)\delta x_k(t)+C(t)f_1(\delta x_k(t),t)+C(t)B(t)\delta u(t). \tag{6}$$

Considering the following open-closed-loop PD-type learning law

$$u_{k+1}(t)=u_k(t)+K_{p1}e_k(t)+K_{d1}\dot{e}_k(t)+K_{p2}(k+1)e_{k+1}(t)+K_{d2}(k+1)\dot{e}_{k+1}(t), \tag{7}$$

with $K_{p1}=K_{p0}\left[\alpha\left(1-\frac{2}{\pi}arc\cot\left(\left\|e_k(t)\right\|\right)\right)\right]$, $K_{d1}=K_{d0}\left[1-\alpha\left(1-\frac{2}{\pi}arc\cot\left(\left\|e_k(t)\right\|\right)\right)\right]$, $K_{p2}(k+1)=g(k+1)K_{p2}(0)$, $K_{d2}(k+1)=\frac{1}{g(k+1)}K_{d2}(0)$, where $0<\alpha<1$, $g(k+1)=\begin{cases}\beta^{k+1} & ,\beta^{k+1}\ge g_{\min} \\ g_{\min} & ,otherwise\end{cases}$, $0.8<\beta<0.95$, $K_{p0},K_{d0},K_{p2}(0)$ and $K_{d2}(0)$ are set to be initial diagonal gain matrices. We obtain the following conclusion.

**Theorem 1.** For nonlinear system (1) with arbitrary original input $u_k(0)$, if the iterative learning algorithm is given as (9), then the sequences $\{x_k(t)\},\{y_k(t)\}$ and $\{u_k(t)\}$ will converge to their respective expected signals $x_d(t),y_d(t),u_d(t)$ under the following condition:

$$\rho\left(\left[I+K_{d2}(0)C(t)B(t)\right]^{-1}\left[I-K_{d1}C(t)B(t)\right]\right)<1, \tag{8}$$

where $\rho(A)$ denotes the spectral radius of matrix $A$.

*Proof*.

From (7) we know $\delta u_{k+1}=\delta u_k-K_{p1}e_k(t)-K_{d1}\dot{e}_k(t)-g(k+1)K_{p2}(0)e_{k+1}(t)$ $-\frac{1}{g(k+1)}K_{d2}(0)\dot{e}_{k+1}(t)$.

Substitute (4) and (5) into the above equation, we can get

$$\begin{aligned} &\delta u_{k+1}(t)+\left[I+\frac{1}{g(k+1)}K_{d2}(0)C(t)B(t)\right]^{-1}\left[\frac{1}{g(k+1)}K_{d2}(0)\dot{C}(t)\delta x_{k+1}(t)\right. \\ &\left.+\frac{1}{g(k+1)}K_{d2}(0)C(t)f_1(\delta x_{k+1}(t),t)+g(k+1)K_{p2}(0)C(t)\delta x_{k+1}(t)\right] \\ &=\left[I+\frac{1}{g(k+1)}K_{d2}(0)C(t)B(t)\right]^{-1}\left[I-K_{d1}C(t)B(t)\right]\delta u_k(t)- \\ &\left[I+\frac{1}{g(k+1)}K_{d2}(0)C(t)B(t)\right]^{-1}\left[K_{p1}C(t)\delta x_k(t)-K_{d1}\dot{C}(t)\delta x_k(t)-K_{d1}C(t)f_1(\delta x_k(t),t)\right]. \end{aligned} \tag{9}$$

Define $G_{k+1},\hat{G}_k : C_r[0,T] \to C_r[0,T]$ as follows:

$$G_{k+1}(u)(t) = \frac{1}{g(k+1)} K_{d2}(0)\dot{C}(t)x_{k+1}(t) + \frac{1}{g(k+1)} K_{d2}(0)C(t)f_1(x_{k+1}(t),t) + g(k+1)K_{p2}(0)C(t)x_{k+1}(t), \tag{10}$$

$$\hat{G}_k(u)(t) = K_{p1}C(t)x_k(t) - K_{d1}\dot{C}(t)x_k(t) - K_{d1}C(t)f_1(x_k(t),t), \tag{11}$$

where $x_k(t)$ and $x_{k+1}(t)$ are the solutions of the following equations: $\dot{x}(t) = f_1(x(t),t) + B(t)u(t), x(0) = x_d(0) - x_k(0)$.

Define the operators $Q_{k+1},\hat{Q}_k,P : C_r[0,T] \to C_r[0,T]$ as follows:

$$\begin{cases} Q_{k+1}(u)(t) = \left[I + \frac{1}{g(k+1)} K_{d2}(0)C(t)B(t)\right]^{-1} G_{k+1}(u)(t) \\ \hat{Q}_k(u)(t) = \left[I + \frac{1}{g(k+1)} K_{d2}(0)C(t)B(t)\right]^{-1} \hat{G}_k(u)(t) \\ P_k(u)(t) = \left[I + \frac{1}{g(k+1)} K_{d2}(0)C(t)B(t)\right]^{-1} \left[I - K_{d1}C(t)B(t)\right]u(t) \end{cases}, \tag{12}$$

Then the function $W_k$ can be defined as $W_k : C_r[0,T] \to C_r[0,T]$

$$W_k(u)(t) = (\hat{Q}_k + P_k)(u)(t). \tag{13}$$

Substituting (10)-(12) into (9), then we have

$$\delta u_{k+1}(t) + Q_{k+1}(\delta u_{k+1}(t))(t) = W_k(\delta u_k(t))(t). \tag{14}$$

According to the Lemma of Bellman-Gronwell, Assumption 1 and (5), we know that there exists $M_1 > 0$ which satisfy

$$\|\delta x(t)\| \le \|\delta x(0)\| + M\int_0^t \|\delta x(s)\| ds + \int_0^t \|B(s)\delta u(s)\| ds \le M_1\left(\|\delta x(0)\| + \int_0^t \|B(s)\delta u(s)\| ds\right). \tag{15}$$

Substituting (15) into (10) and (11), we have

$$\begin{cases} \|G_{k+1}(\delta u_{k+1}(t))\| \le M_2\left(\|\delta x_{k+1}(0)\| + \int_0^t \|\delta u_{k+1}(s)\| ds\right) \\ \|\hat{G}_k(\delta u_k(t))\| \le M_3\left(\|\delta x_k(0)\| + \int_0^t \|\delta u_k(s)\| ds\right) \end{cases}, \tag{16}$$

In the sequence we obtain

$$\begin{cases} \|Q_{k+1}(\delta u_{k+1}(t))\| \le M_4\left(\|\delta x_{k+1}(0)\| + \int_0^t \|\delta u_{k+1}(s)\| ds\right) \\ \|\hat{Q}_k(\delta u_k(t))\| \le M_5\left(\|\delta x_k(0)\| + \int_0^t \|\delta u_k(s)\| ds\right) \end{cases}, \tag{17}$$

Since $\frac{1}{g(k+1)} > 1$, then we can get

$$\|P_k(u)(t)\| \le \left\|\left[I + K_{d2}(0)C(t)B(t)\right]^{-1}\left[I - K_{d1}C(t)B(t)\right]u(t)\right\| = \|P(u)(t)\|. \tag{18}$$

From (13), (17) and (18) we have

$$\begin{aligned}
&\int_0^t \|W(\delta u_k(s))\| ds = \int_0^t \left\|\left(\hat{Q}_k + P_k\right)(\delta u_k(s))\right\| ds \le \int_0^t \left\|\hat{Q}_k(\delta u_k(s))\right\| ds + \int_0^t \left\|P_k(\delta u_k(s))\right\| ds \\
&\le \int_0^t \left[M_5\left(\|\delta x_k(0)\| + \int_0^s \|\delta u_k(\tau)\| d\tau\right)\right] ds + \sup_{t\in[0,T]} \left\|\left[I + K_{d2}(0)C(t)B(t)\right]^{-1}\right. \\
&\left.\times\left[I - K_{d1}C(t)B(t)\right]\right\| \int_0^t \|\delta u_k(s)\| ds \\
&\le M_5 T \|\delta x_k(0)\| + M_5 T \int_0^t \|\delta u_k(s)\| ds + M_6 \int_0^t \|\delta u_k(s)\| ds \\
&\le (M_5 T + M_6)\left(\|\delta x_k(0)\| + \int_0^t \|\delta u_k(s)\| ds\right) = M_7\left(\|\delta x_k(0)\| + \int_0^t \|\delta u_k(s)\| ds\right).
\end{aligned} \tag{19}$$

According to the Theorem 2.4 in [5], define operator $V_{k+1} : C_r[0,T] \to C_r[0,T]$ as

$$V_{k+1}(y)(t) = Q_{k+1}(u)(t),\ \forall y(t) \in C_r[0,T], \tag{20}$$

where $u(t)$ is the sole solution to $u(t) + Q_{k+1}(u)(t) = y(t)$, by comparing the above equation with (14) we have $V_{k+1}(W_k(\delta u_k))(t) = Q_{k+1}(\delta u_{k+1})(t)$. Then we obtain $\|V_{k+1}(W_k(\delta u_k))(t)\| \le M_8\left(\|\delta x_{k+1}(0)\| + \int_0^t \|W_k(\delta u_k(s))\| ds\right)$.

Define $\bar{V}_{k+1} : C_r[0,T] \to C_r[0,T]$ as

$$\bar{V}_{k+1}(u)(t) = -V_{k+1}(W_k(u))(t). \tag{21}$$

According to (17),(18) and (19) we have

$$\begin{aligned}
\bar{V}_{k+1}(u)(t) &\le M_8\left(\|\delta x_{k+1}(0)\| + \int_0^t \|W_k(\delta u_k(s))\| ds\right) \\
&\le M_9\left(\|\delta x_{k+1}(0)\| + \|\delta x_k(0)\| + \int_0^t \|\delta u_k(s)\| ds\right),
\end{aligned} \tag{22}$$

where $M_9 = M_8 \max\{1, M_8\}$.

Define $U_k : C_r[0,T] \to C_r[0,T]$ as follow:

$$U_k(u)(t) = \left(Q_k + \bar{V}_{k+1}\right)(u)(t). \tag{23}$$

According to (18) and (23), we have

$$\begin{aligned}
&\|U_k(\delta u_k)(t)\| \le \|Q_k(\delta u_k)(t)\| + \left\|\bar{V}_{k+1}(\delta u_k)(t)\right\| \\
&\qquad \le M_5\left(\|\delta x_k(0)\| + \int_0^t \|\delta u_k(s)\| ds\right) + M_9\left(\|\delta x_{k+1}(0)\| + \|\delta x_k(0)\| + \int_0^t \|\delta u_k(s)\| ds\right) \\
&\qquad \le M_{10}\left(\|\delta x_{k+1}(0)\| + \|\delta x_k(0)\| + \int_0^t \|\delta u_k(s)\| ds\right).
\end{aligned} \tag{24}$$

where $M_{10} = M_5 + M_9$.

According to (12) we know $\delta u_{k+1}(t)=W_k(\delta u_k)(t)+\bar{V}_{k+1}(\delta u_k)(t)=(P_k+U_k)(\delta u_k)(t)$. From (18), we know $\|P_k(u)(t)\|\le\|P(u)(t)\|$. Then we have

$$\begin{aligned}\|\delta u_{k+1}(t)\| &= \|(P_k+U_k)(\delta u_k)(t)\| \le \|(P+U_k)(\delta u_k)(t)\| \\ &= \|(P+U_k)(P+U_0)\cdots(P+U_0)(\delta u_0)(t)\|.\end{aligned} \tag{25}$$

According to the theorem 2.3 in [5], we know that if $\rho\left(\left[I+K_{d2}(0)C(t)B(t)\right]^{-1}\left[I-K_{d1}C(t)B(t)\right]\right)<1$, then we have $\lim_{k\to\infty}\|(P+U_k)(P+U_0)\cdots(P+U_0)(\delta u_0)(t)\|=0$. Thus we can conclude that the sequence $\{\delta u_k(t)\}$ converges to zero, if condition (8) comes into existence.

According to Lemma 2.1 in [5], from (4), (5) and Assumption 2 we have $\delta x_k(t)\le\int_0^t e^{M(t-\tau)}B(\tau)\delta u_k(\tau)d\tau$ and $e_k(t)=C(t)\delta x_k(t)\le\int_0^t e^{M(t-\tau)}C(\tau)B(\tau)\delta u_k(\tau)d\tau$.

According to the definition of $\lambda$-norm , we have $\|\delta x_k\|_\lambda\le b\frac{1-e^{(m-\lambda)T}}{\lambda-m}\|\delta u_k\|_\lambda$, $\|e_k\|_\lambda\le cb\frac{1-e^{(m-\lambda)T}}{\lambda-m}\|\delta u_k\|_\lambda$, where $m=\|M\|$, $b=\|B\|$, $c=\|C\|$, and $\lambda>m$.

In the above section, $M_i>0, i=1,2,\cdots,10$.

Then the convergence of $\{\delta u_k(t)\}$ implies that the sequences $\{\delta x_k(t)\}$ and $\{e_k(t)\}$ will both converge to zero. The proof is completed.

## 4 Conclusion

In this paper an open-closed-loop PD-type Iterative Learning Control (ILC) algorithm with variable learning gains is proposed. The learning gains are varying with the system errors or the iteration times. Rigorous analysis shows the convergence of the system error.

## References

1. Arimoto, S., Kawamrua, S.: Miyazaki: Bettering operation of robots by learning. Joural of Robotic Systems 1(2), 123–140 (1984)
2. Xu, M., Lin, H., Liu, Z.: Iterative learning control law with variable learning gain. Control Theory & Applications 24(5), 856–860 (2007)
3. Ouang, P.R., Zhang, W.J., Gupta, M.M.: An adaptive switching learning control method for trajectory tracking of robot manipulators. Mchatronics 16, 51–61 (2006)
4. Ouyang, P.R., Petz, B.A., Xi, F.F.: Iterative Learning Control With Switching Gain Feedback for Nonlinear Systems. Journal of Computational and Nonlinear Dynamics 6, 011020-1–7 (2011)
5. Yu, S., Qi, X., Wu, J.: Iterative Learning Control Theory and Its Applications. China Machine Press, Peking (2005)

# Matching Web Services Based on Semantic Similarity

Xiang ji, Yongqing Li, Lei Liu, and Rui Zhang

CCST, Jilin University. Street Qianjin No. 2699, Changchun, China 130012
Ji.x@husky.neu.edu, yqli08@mails.jlu.edu.cn,
{liulei,rui}@jlu.edu.cn

**Abstract.** Classical matching on web services provides only coarse granularity methods, which classify the matching into several levels. Such level restricted methods prohibit the evaluation of matching within the same level which is necessary for modern web service compositions. This paper offers a fine-grained matching method based on the semantic similarity of the services. By the semantic distances and the depth of concepts of ontology, a novel precise matching method is proposed to match the web services between levels and within the same level. Experiments show that the approach leads to better precision and recall.

**Keywords:** Ontology, Semantic Distance, Similarity, Web Service Matching.

## 1 Introduction

One of the main challenges of Web service composition is to find proper services according to different requests [1]. Classical service matching algorithms such as UDDI[2] are based on keywords matching. With the enormous growth of syntactical different but semantically similar services, the research on Web service composition shifts to semantics gradually.

Many service management solutions employ semantic service discovery such as [3][4][4][6]. Existing semantic service matching algorithms based on semantics are coarse grained. Paolucci et al. proposed the classical level based matching [7]. A hybrid semantic web service retrieval method was proposed in [8] with a matcher OWLS-MX based on finer grained levels. The matcher OWLS-MX3 in [9] was a refined version of that in [8] which considered matching on the structures. Even though OWLS-MX3refined the classical algorithm, the levels are not defined grained enough.

In this paper, we present a matching method based on the semantic distances between ontology concepts. Different types of semantic distances exists such as in [10][11][12][14] and [15]. The definitions in [10] did not take into account of the direction. In [12], the distance was calculated with the factor on the times of appearance in the whole concept set. We propose a new definition of semantic distance, based on the path between the concepts and also the depth of the concepts into the ontology hierarchy. Based on the definition, we offer a new method to calculate the semantic similarity of the input and output of candidate services and discussed the trends of the value with respect to the factors such as length of path

S. Lin and X. Huang (Eds.): CSEE 2011, Part II, CCIS 215, pp. 598–604, 2011.

between the concepts and depth of the concepts. Evaluation shows the better performance of our method on classical level based approaches.

The paper is organized as follows. Section 2 analyzes the classical level based approaches; Section 3 defines the semantic similarities and discussed the related properties. Section 4 shows the experiments and we conclude in Section 5.

## 2 Analysis of Level Base Matching

OWL-S [11] is a language for the description of web services based on the OWL standard by W3C. The 3 high level ontology components consist of the Profile, the Process Model and the Grounding. As is shown in Figure 1, the Service Profile is used mainly in web service advertising and discovering to present the service. Typical functional parameters of an OWL-S service profile include the input and output, the precondition(s) and effect(s); while the nonfunctional parameter captures name, type and textual description etc. The Service Model gives a detailed description of the service operation. And the Service Grounding lists the interoperability of series via messages.

The input and output of web services are represented as concepts in one ontology by the service requester and provider. Level based service matching results are calculated by the subsumption relations between the concepts. OWL-S/UDDI is a classical degree based algorithm [7] which classifies the degree of matching into 4: *exact, plugin, subsume* and *fail*. Suppose the input and output of the service requester are InR and OutR, and those of the service provider are InS and OutS. For the input, if the in the concept tree hierarchy exists InR=InS or InS is the direct child concept of InR, then the matching result is *exact*; if InS is the descendant concept but not direct child of InR, the result is *plugin*; on the contrary, if InR is the descendant concept but not direct child of InS, it is *subsume*; and if no subsumption relation exists between InS and InR, a *fail* returns. A naïve algorithm would be like Figure 1.

```
SemanticMatching(InS,InR){
    if (InR==InS||ChildClassOf(InS,InR))    {
        return exact;
    }else if (SubClassOf(InR,InS))  {
        return plugin;
    }else if (SubClassOf(InS,InR))  {
        return subsume;
    }else return fail;
}
```

**Fig. 1.** A Degree Based Matching Algorithm [7]

The matching algorithm on the outputs is similar with the opposite in the definition of *exact, plugin* and *subsume*. Suppose in the sample ontology as in the hierarchy of Figure 2, if the InR by the requester is 'Person' and the InS by the provider is 'Master' or 'College Student', then the algorithm in Figure 1 will both return *plugin*. From the service matching point of view the latter is a better match as master students are only a small portion of the college students which is a portion of the person in turn. However, the classical algorithm of the above is not able to distinguish the two cases but only returns *plugin*.

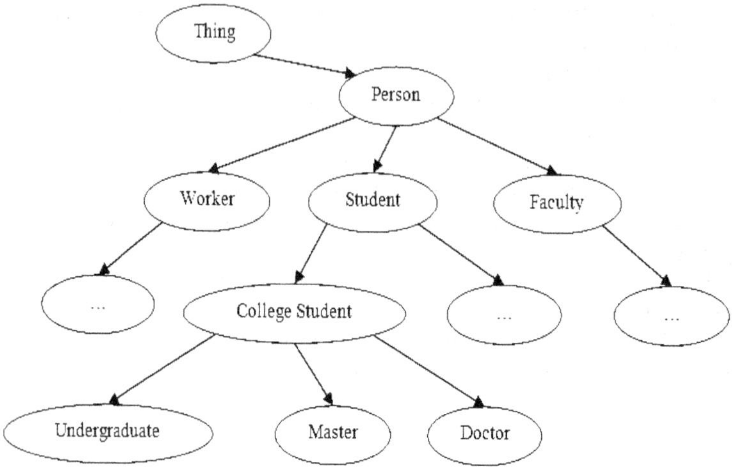


**Fig. 2.** An example of an ontology of classification

## 3 Matching via Semantic Similarity

Suppose N1 and N2 are two nodes in a directed tree T, as is shown in Figure 2, to introduce the matching algorithm via semantic similarity, the following definitions are provided.

**Definition 1.** *N2 is* ***reachable*** *by N1, which is also called N1 is* ***inverse reachable*** *by N2 iff there exists at least one path from N1 to N2, noted as N2<N1 or N1>N2; specially, if N2<N1 and N1>N2 are both satisfied, then N1 and N2 are* ***equal*** *noted as N1=N2; otherwise, N2 is* ***unreachable*** *by N1 and N1 is unreachable by N2, noted as N2≠N1 or N1≠N2.*

Note that reachable and inverse reachable and equal are all transitive relations.

**Definition 2.** *The* ***distance*** *between two notes N1 and N2, Dis(N1,N2) is calculated as*

$$Dis(N1, N2) = \begin{cases} 0, & if\ N1 = N2; \\ k, & if\ N2 < N1; \\ -Dis(N2, N1), & if\ N1 < N2 \\ \infty, & if\ N2 \neq N1. \end{cases}$$

*where k is an integer which represents the length of the shortest path from N1 to N2.*

Here the distances between any two nodes in Figure 2 can be calculated. For example, *Dis*(Master,Master)=0; *Dis*(Undergraduate,Person)=-*Dis*(Person,Undergraduate)=-3; and *Dis*(Master,Doctor)= ∞.

**Definition 3.** *Suppose the height of T is h, and the nearest ancestor node of N1 and N2 is in level m, then the* ***semantic similarity*** *between N1 and N2 is calculated as*

$$Sim(N1,N2)=\begin{cases}1, \quad if\ N1=N2\ i.e. Dis(N1,N2)=0\ or\ if\ Dis(N1,N2)=-1;\\ \dfrac{1}{2}+\dfrac{1}{1+e^{-Dis(N1,N2)}}\times\dfrac{m}{h}, \quad if\ Dis(N1,N2)<-1;\\ \dfrac{1}{1+e^{Dis(N1,N2)}}\times\dfrac{m}{h}, \quad if\ Dis(N1,N2)>0;\\ 0, \quad if\ Dis(N1,N2)=\infty\end{cases}$$

As described in Definition 3, the semantic similarities between any of the two nodes in Figure 2 can be calculated. For instance, *Sim*(Master,Master)=1; *Sim*(Person,Student)= $\frac{1}{1+e}\times\frac{2}{5}$ =0.1076; *Sim*(Master,Student)= $\frac{1}{2}+\frac{1}{1+e^2}\times\frac{3}{5}$ =0.5715; *Sim*(Master,Doctor)=0.

Based on the definitions, the following theorems can be concluded based on the assumption that the ontology tree structure is a top-down fashion directed tree like the one in Figure 2 (for ontolgy in the opposite direction, similar theory remains). The proofs of the theorems are straight forward and neglected for short of space in this paper.

**Theorem 1.** *For any two nodes with the same distance in between, deeper position into the tree results in more similarity. That is to say, for any nodes N11, N12 and N21 and N22, if Dis(N11,N12)=Dis(N21,N22), and the level of N11 is deeper than that of N21 (which means the level of N12 is deeper than N22 at the same time), then Sim(N11,N12)>Sim(N21,N22).*

This is coherent with the intuition that the deeper into the tree, the classification of the categories are more detailed and less distinguished. For example in Figure 2, it is common to classify 'Person' into 'worker', 'student' etc. near the root of the tree. But when it comes to the deeper part of the tree (far from the root), 'undergraduate' and 'master' students are usually not distinguished in many cases such as for identification, for medical insurance etc. In such cases, the semantics of the nodes are much closer than that of those in the upper part of the tree.

**Theorem 2.** *For any two nodes reachable in between, deeper position of the descendant node into the tree results in less similarity. That is to say, for any nodes N12is reachable by N11 and N22 is reachable by N21, if the level of N11 is the same as that of N21, then if the level of N12 is greater than N22, then Sim(N11,N12)<Sim(N21,N22).Correspondingly, For any two nodes reachable in between, higher position of the ancestor node into the tree results in less similarity.*

The theorem describes the influence of the distance on the similarity between nodes. As described in Theorem 1, the depth hints the similarity so here in Theorem 2 the depth factor is fixed if the higher node is at the same level in the tree which restricts the '*m*' in the formula in Definition 3 to be the same. It is easy to see the two sections of the function are both monotonously increasing with the *Dis* between the nodes.

**Theorem 3.** *For any two nodes reachable in between, N2 is reachable by N1 i.e. N1 is inverse reachable by N2, iff* $0<Sim(N1,N2)<\frac{1}{2}$*;otherwise iff* $\frac{1}{2}<Sim(N1,N2)<1$.

It is obvious that the semantic similarity is more precise in representation of the semantic relations between the concepts than the classical degree based matching by Paolucci et al. [4] Following the terminology in Section 1, if *Sim*(InR,InS) falls in (0,1/2), from Theorem 3, it can be inference that the concept at InR is the ancestor concept of InS, which is *subsume*; if *Sim*(InR,InS) falls in (1/2, 1), the relation is *plugin*; and if the value is 1 or 0, it is *exact* or *fail*. However, our method provides more information, a quantified value to represent the semantic similarity, which is the longer distance of the two concepts, the less similarity between, as inference from Theorem 2.

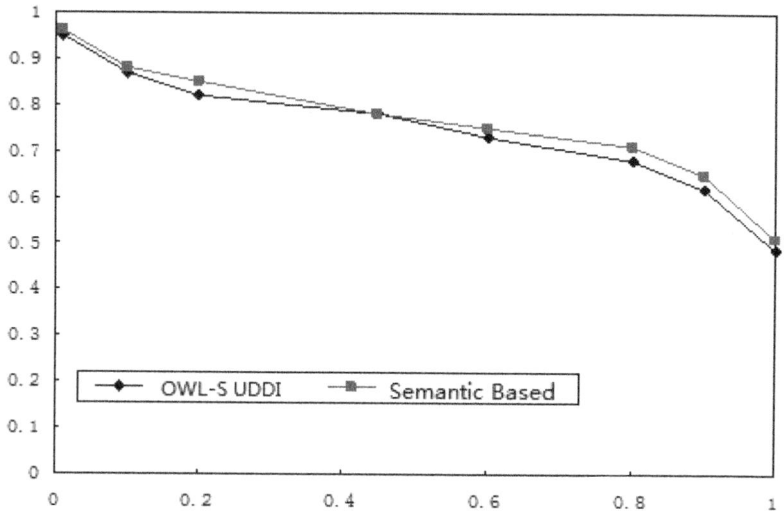


**Fig. 3.** Experiment Result: vertical axis for precision and horizontal for recall

## 4 Evaluation

Based on the method in Section 2, evaluation is made on a lab desktop with CPU Pentium 5.3GHz, RAM 4GB Operating System Windows XP with JDK 1.6.0. The test is based on 10 query files and 200 registered services from OWLS-TC3 [13], and for each test we use the average result of 100 attempts. The results in Figure 3 show that considering the same recall, denoted as the values on the vertical axis, the precision value, on the horizontal axis of our method wins the classical UDDI.

## 5 Conclusion

A method based on semantic similarity of the ontology concepts is proposed in this paper to calculate the matching between the service requester and the service providers. It exploits a new definition of the semantic distances considering not only the length of the path between the concepts in the ontology, but also the depth of the concepts into the ontology hierarchy. It allows finer grained levels than the classical semantics based service matching methods. The experiments show that the method performs better than the UDDI approach. The future work will be more experiments comparing with other approaches and refine the method for better performances.

**Acknowledgment.** Supported by the National Natural Science Foundation of China under Grant No. 60873044, the Opening Fund of Top Key Discipline of Computer Software and Theory in Zhejiang Provincial Colleges at Zhejiang Normal University of China under Grant No. ZSDZZZZXK11 and the Fundamental Research Funds for the Central Universities of China under Grant No. 200903174, 200903183, 200903188, 200903193, 201103124 and 201103133.

## References

[1] Cardoso, J.: Semantic Web Services: Theory, Tools, and Applications, pp. 191–216. Information Science Reference, USA (2007)

[2] Universal Description, Discovery and Integration (UDDI), `http://www.uddi.org/specification.html` (May 20, 2005)

[3] Gunay, A., Yolum, P.: Structure and Semantic Similarity Metrics for Web Service Match-making. In: Psaila, G., Wagner, R. (eds.) EC-Web 2007. LNCS, vol. 4655, pp. 129–138. Springer, Heidelberg (2007)

[4] Ziegler, P., Kiefer, C., Sturm, C., Dittrich, K., Bernstein, A.: Detecting Similarities in Ontologies with the SOQA-SimPack Toolkit. In: Ioannidis, Y., Scholl, M.H., Schmidt, J.W., Matthes, F., Hatzopoulos, M., Böhm, K., Kemper, A., Grust, T., Böhm, C. (eds.) EDBT 2006. LNCS, vol. 3896, pp. 59–76. Springer, Heidelberg (2006)

[5] Gomez-Pérez, González-Cabero, R., Lama, M.: A framework for designing and composing semantic web services. In: First International Semantic Web Services Symposium. Proceedings of 2004 AAAI Spring Symposium Series (March 2004)

[6] Bleul, S., Weise, T., Geihs, K.: An ontology for quality-aware service discovery. Computer Systems Science Engineering 20(4) (July 2006)

[7] Paolucci, M., Kawamura, T., Payne, T.R., et al.: Semantic Matching of Web Services Capabilities. In: International Semantic Web Conference, pp. 333–347 (2002)

[8] Klusch, M., Fries, B., Khalid, M., Sycara, K.: OWLS-MX: Hybrid Semantic Web Service Retrieval. In: Proceedings of the 1st International AAAI Fall Symposium on Agents and the Semantic Web. AAAI Press, Arlington VA (2005)

[9] Klusch, M., Kapahnke, P.: OWLS-MX3: An Adaptive Hybrid Semantic Service Matchmaker for OWL-S. In: Proceedings of the 3rd International SMR2 2009 Workshop on Service Matchmaking and Resource Retrieval in the Semantic Web, collocated with the 8th International Semantic Web Conference (ISWC 2009), Washington DC, USA (October 2009)

[10] Li, Y., Bandar, A., McLean, D.: An approach for measuring semantic similarity between words using multiple information sources. Transactions on Knowledge and Data Engineering 15 (2003)

[11] Klusch, M.: CASCOM:Intelligent Service Coordination in the Semantic Web, pp. 36–44. Birkhäuser, Berlin (2008)

[12] Peng, H., Niu, W., Huang, R.: Similarity Based Semantic Web Service Match. Web Information Systems and Mining, 252–260 (2009)
[13] Fries, B., Khalid, M.A., Kapahnke, P.: OWLS-TC version 3.0 revision 1 (November 2009), http://www.semwebcentral.org/projects/owls-tc/
[14] Ganesan, P., Garcia-Molina, H., Widom, J.: Exploiting hierarchical domain structure to compute similarity. ACM Transactions on Information Systems 21(1), 64–93 (2003)
[15] Bernstein, A., Kiefer, C.: iRDQL Imprecise RDQL Queries Using Similarity Joins. In: K-CAP 2005 Workshop on: Ontology Management: Searching, Selection, Ranking, and Segmentation (2005)

# Heilongjiang Ice-Snow Tourism Industry Competitiveness Evaluation Index System Design

Liang Zhang and XiaoMei Zhang

Sports Department
Northeast Agricultural University
No.59, Mucai Street, Xiangfang District,
150030, Haerbin, China

**Abstract.** Practice in Heilongjiang province is the earliest provinces of ice-snow tourism industry, In Heilongjiang province of ice-snow tourism industry as the most distinctive industries, in its economic development plays a leading and radiation. Based on comprehensive a large number of documents, professional consulting and reference related evaluation index system, constructing the contains five basic factors criterion layer and nineteen concrete influence factor index layer of Heilongjiang ice-snow tourism industry competitiveness evaluation index System, for the development of ice-snow tourism industry in Heilongjiang province has important reference value.

**Keywords:** Ice-snow tourism industry; Competitiveness; Index system.

## 1 Preface

In recent years, ice-snow tourism industry is one of the emerging tourist form, has broad prospects for development. The ice-snow tourism industry of our country relatively starts late, but its develops fast, now it is in rapid growth period. Ice-snow tourism is a distinctive characteristic in the North of travel products, It includes sightseeing, fitness and entertainment, has attracted many tourists.

Snow resources are advantageous resources in Heilongjiang province, Heilongjiang is "ice world and Snow Sea". As the highest latitude provinces in China, one year has 4-5 months snowfall period. Large amount of snow, snow lasted longer, the snow quality is good. And most of mountain has moderate high, slopes and slope is appropriate, very suitable for developing ice-snow tourism industry. Favorable natural conditions make Heilongjiang province become China's the richest ice-snow resources provinces. Heilongjiang relying on rich snow resources, created a world-renowned snow art, developed a world-class ski travel, making Heilongjiang province of ice-snow tourism become a "gold card" in China.

But, with a large number of ice-snow tourism destinations in China, Heilongjiang province is the earliest develop ice-snow tourism industry in China, but the competition is very fierce. In recent years, jilin, liaoning, Beijing, even sichuan, xinjiang, Inner Mongolia were also developed ice-snow tourism, preempted ice-snow tourism market share, Shared a lot of profit. Ice and snow is no longer "patent" in

S. Lin and X. Huang (Eds.): CSEE 2011, Part II, CCIS 215, pp. 605–610, 2011.

Heilongjiang province. How to meet the enormous market challenges, keep the core competitiveness of ice-snow tourism industry, creating a "longjiang snow is the world first " brand, make the snow into a "platinum", benefit the people and promote longjiang economic development is very important.

Therefore, establishing a scientific evaluation index system is particularly urgent, this also is this paper expatiates the original intention. Based on comprehensive a large number of documents, professional consulting and reference related evaluation index system, according to the purpose of Heilongjiang ice-snow tourism industry competitiveness evaluation, follow the systematic, independence, applicability and hierarchy principle, Put forward for Heilongjiang ice-snow tourism industry competitiveness evaluation index system, It can help to ascertain its own advantages and disadvantages, formulate realistic tourism development strategy, and better participates in the international and domestic tourism competition.

## 2 The Research Methods

### 2.1 Literature Reviewing

According to the research content and research mission needs, reading a lot of books, and through Chinese database retrieval to about ice-snow tourism, industrial competitiveness, evaluation index system and competitiveness evaluation index system of the related literature. On the basis of predecessors' achievements, put forward a set of evaluation index system of ice-snow tourism industry competitiveness in Heilongjiang province.

### 2.2 Expert Consultation and Expert Questionnaire

Consult experts about the factors of ice-snow tourism industry competitive in Heilongjiang province, understanding experts in opinion of index system and through the form of sending questionnaires for expert advice, put the preliminary results are comprehensive consolidation, then feedback to each expert, please they rethink again until experts advice trend consistent.

### 2.3 Logic Analysis

In the comprehensive numerous theoretical material and mathematical material basis, through the logic reasoning and summary, establishing ice-snow tourism industry competitiveness evaluation index system in Heilongjiang province, further improve the implementation of evaluation index system.

## 3 Heilongjiang Ice-Snow Tourism Industry Competitiveness Evaluation Index System Design

University students, reflecting the state's mental outlook and future direction are the pillars of the country in the contemporary era. Setting skiing course are not only benefit to the health of college students but also conducive to the development of skinning industry.

## 3.1 The Principle of Index Selection and the Idea of Build Index System

### 3.1.1 Systematic Principle

The Heilongjiang ice-snow tourism industry competitiveness evaluation index system as a system. Design index should be representative and coverage, and can comprehensively reflect the competitiveness of ice-snow tourism industry all aspects of content.

### 3.1.2 Applicability Principle

Index system shall apply to evaluation method, suitable for users to index level of acceptance and judgment, suitable for evaluation information base. Concrete can be divided into the following several aspects:

(1) Index number must be simplified properly
(2) The data easily acquisition
(3) Index should be comparable

### 3.1.3 Scientific Principle

Index system can be objectively reflecting the condition of ice-snow industrial competitiveness, avoiding the different aspects of the conflicts of values and cause index orientation one-sided. On the basis of an in-depth study on ice-snow tourism industry in Heilongjiang province, construction that can fully reflect the overall system characteristics, but also a structured scientific evaluation index system.

### 3.1.4 Qualitative Analysis and Quantitative Analysis Principle

When the index selection for unable to quantify and very important indexes can be adopted the form of qualitative indexes into index system. Index system designed to combine qualitative and quantitative, quantitative indicators provide strict control of the rigid control standards, descriptive indicators as a guiding principle of competitive evaluation, a combination of both strengthened the operability of the evaluation index system.

## 3.2 The Idea of Build Index System

The program of set up the ice-snow tourism industry competitiveness evaluation index system as shown in Figure 1. First, a lot of reading and reference literature related to evaluation index system to collect information relevant to establish the beginning of the proposed index system, then questionnaire investigation and consult the opinion about expert, using logic analysis method, and finally establish screening index evaluation index system, make the index system to meet scientific, applicability and rationality.

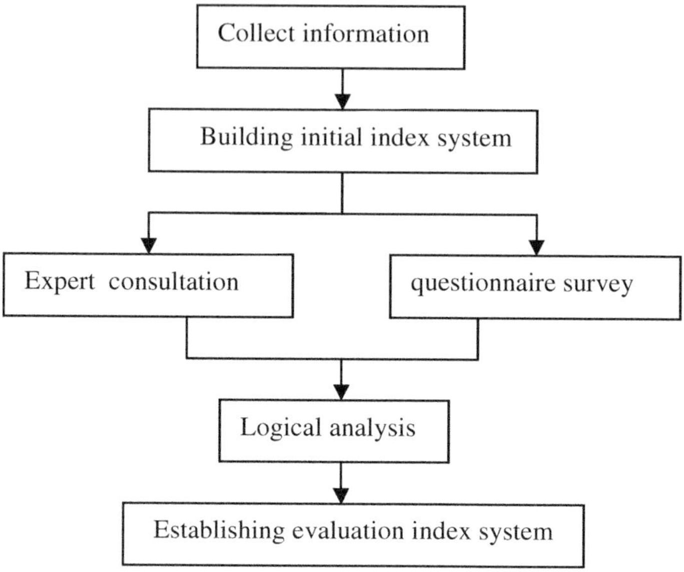


**Fig. 1.**

### 3.3 Establishment of Index System

Ice-snow tourism industry competitiveness evaluation system in Heilongjiang province is made up of several different subsystems, each subsystem is made up of several indicators. Therefore, the level of ice-snow tourism industry competitiveness affected by many different factors and constraints, which have a direct factors, but also indirect factors, then select the evaluation index should strive to be comprehensive, systematic, and must meet the systematic applicability, scientific, qualitative analysis and quantitative analysis principle, and to consider the reliability of data and indicators can be collected. Based on comprehensive a large number of documents, professional consulting and reference related evaluation index system, taking into account the availability of information and characteristics of the winter tourism industry development in Heilongjiang province, built a ice-snow tourism industry competitiveness evaluation index system in Heilongjiang province. This index includes three levels; from the top down is a hierarchical structure. The first level (target level) is the highest level of the structure, it is the overall objective of the study, but also the driving force behind the study, here to the competitiveness of the tourism industry as a general goal, but the goal is vague, comprehensive, operation is not easy, so there has a second level (criterion level) and the third level (index layer) index. The second level (criterion level) is the second high-level structure, it is the principle required to achieve the overall goal, including elements of tourism resources, tourism market conditions, tourism enterprises conditions, related industry conditions and external conditions. The third level (index layer) is the lowest level of the structure; it is a specific requirement to achieve the second level. Therefore, lower than the upper indicators more clear, specific and easy-to-computing, they can be used as some means to an upper-level indices.

After selecting of many indicators, the indicator system includes a level indicator layer, five secondary indicators, and 19 third indicator, the indicator system described as follows: Ice-snow tourism industry competitiveness in Heilongjiang province A=(B1, B2, B3, B4, B5), Hierarchical framework of specific targets, see Table 1:

**Table 1.** Heilongjiang ice-snow tourism industry competitiveness evaluation index system

| target level ( A ) | criterion level ( B ) | index layer ( C ) |
|---|---|---|
| Heilongjiang ice-snow tourism industry competitiveness | Elements of tourism resources ( B1 ) | Resource quality ( C1 ) Scale of resources ( C2 ) The uniqueness (C3)Historical, scientific and cultural values(C4) |
| | tourism market (B2) | Tourist market size(C5) Attractiveness of the tourism market (C6) Tourism industry revenues (C7) |
| | tourism enterprises (B3) | The quantity of tourist attractions(C8) The quantity of tourism service points (C9) Competition in tourist attractions (C10) management capabilities of tourism enterprises(C11) |
| | related industry conditions (B4) | Transportation Industry (C12) The number of tourist destination shopping center(C13) Catering and service facilities (C14) Level of fitness and recreation facilities (C15) |
| | external conditions (B5) | Degree of economic development (C16) Transportation convenient degree(C17) Government support(C18) Management and service level (C19) |

## 4 The Conclusion

With fast development of our economy and deepening of opening up, China's ice-snow tourism industry line with international standards gradually. Based on comprehensive a large number of documents, professional consulting and reference related evaluation index system, through expert advice, expert surveys, drawing on experts comments and suggestions , the indicator system was modified and improved continuously, and eventually established a set of evaluation index system. In the article, constructed of the ice-snow tourism industry competitiveness evaluation index system, the indicators are interrelated and complement each other, to reflect the reality of the situation and improve it Scientific, practical and operability. The establishment of index system, so as to fully understand the development situation of ice-snow tourism industry competitiveness in Heilongjiang province, to provide theoretical basis for it further development. Establish a realistic set of ice-snow tourism industry competitiveness evaluation system in Heilongjiang province, which is beneficial to the formation of ice-snow tourism industry development long-acting mechanism. It can provide valuable reference for the transportation industry, service industry and the harmonious development of economy in Heilongjiang province and provide the decision-making basis for the government tomacro-management of the ice -snow tourism industry. It has important application value and practical significance.

## References

1. Liu, Y.-f., Sun, H., Li, N., Yuan, Z.-h.: A System of Evaluation Indicators for the Provinicial Tourism Competition. Journal of Arid Land Resources and Environment (8) (2008)
2. Cao, H., Chen, Q.-h.: Evaluation Of Tourism Competitiveness In Rural Area Based On Entropy Technology And Ahp. Yunnan Geographic Environment Research (6) (2008)
3. Li, M.: The Study on the Index System of the Tourist Area Regulatory. Northwestern university master graduate thesis (2006)
4. Wang, J., Wang, Q.-Y.: China regional and above city tourism competitiveness evaluation research. Inquiry into Economic Issues (2), 132–133 (2010)
5. Yang, L., Wang, L.: Research on Establishment of Evaluation Index System of Highway Landscape. Forest Engineering (1) (2008)

# A Study on the Application of Computer-Assisted Teaching of English Writing to Vocational College Students

Sujuan Xiong[1], Wanwu Huang[1], and Yaqi Chen[2]

[1] Department of Foreign Languages, Hubei University of Technology, Wuhan, Hubei Prov. P.R.C 430068
[2] Department of Foreign Languages, Hubei Vocational College of Communications, Wuhan, Hubei Prov. P.R.C 430068
{79519240,121182680,59141948}@qq.com

**Abstract.** In this paper, theories of language learning motivation and language learning strategies were used to explore the influence of computer-assisted language teaching methods on vocational college students' learning motivation and learning strategies in English writing classes. The research lasted for one term. In the research, the English writing classes were designed to be computer-assisted. Before and after the research, a questionnaire was used respectively to measure students' language learning motivation and language learning strategies. Results of the research show that computer-assisted teaching of English writing can significantly improve vocational college students' language learning motivation. It is also shown from the results of the research that computer-assisted teaching of English writing can also help vocational college students to improve their language learning strategies.

**Keywords:** computer-assisted teaching; English writing; vocational colleges.

## 1 Introduction

Learning motivation is considered as an important factor that influences a language learner's learning achievements. Ellis [1] defined it as the effort that learners put into learning a second language as a result of their need or desire to learn it. Gardner [2] emphasized the important influence of learning motivation on foreign language learning.

Learning strategies is another important factor that has been widely studied in the field of foreign language teaching and learning. There is no consensus idea in learning strategies in today's academic circles. According to the existing literature reviews, it can be summed up as follows: regarding the learning strategies as specific learning methods or skills [3] (Mayer, 1978); regard the learning strategies as a student's learning process [4] (Nisbet, 1986). But the importance of it is never neglected.

Computer-assisted teaching of foreign languages is now a hot topic in the field of applied linguistics in China. Computer technologies are widely used in foreign language teaching classrooms, and many scholars have done a great number of

S. Lin and X. Huang (Eds.): CSEE 2011, Part II, CCIS 215, pp. 611–616, 2011.

researches to study the methods of designing computer-assisted foreign language classes, the influence of computer-assisted foreign language classes on students' achievements on foreign learning and so on. However, as to the influence of computer-assisted teaching of English writing on vocational students' learning motivation and learning strategies, not many people have studied yet.

This paper uses a questionnaire to measure vocational college students' learning motivation and learning strategies before and after a term's computer-assisted English writing class. It aims at finding out whether the changes before and after the research are positively significant or not. If they are, it means that the computer-assisted English writing class can help vocational college students improve their learning motivation and learning strategies.

## 2 Research Design and Methodology

**Hypothesis.** The hypothesis of the research is: the application of computer-assisted teaching method in English writing class can significantly improve vocational college students' learning motivation and learning strategy.

**Subjects.** The subjects of this study were the students in Hubei Vocational College of Communications. They were 59 freshmen of Grade 2008, coming from two classes in the Department of Logistics Management in this college. They were the students in China-Australia joint programs. Of the two classes, one is for the experiment class; one for the control class.

**Experiment Design.** This research began in Hubei Vocational College of Communications. This college is a full-time vocational technical college. The experimental time was 4 months, from February, 2008 to the end of that semester. During the process of the experiment, the experiment class and the control class have the same teacher, the same textbook, the same teaching hours. The only difference is that computer-assisted teaching method was used in the experimental class while in the control class the traditional method was used.

Before the experiment, a questionnaire to investigate the subjects' English writing motivation and strategies was carried. According to the independent-samples t-test analysis, there is no significant difference between the control class and the experiment class. (see Table 1 and Table 2).

**Table 1.** English writing motivation of the experimental class and the control class before the experiment

| | Number of subjects | mean | Std Dev | Independent-samples t-test | |
|---|---|---|---|---|---|
| | | | | T Value | P Value |
| Experiment class | 31 | 2.6.6 | 1.531 | 1.968 | 0.778 |
| Control class | 28 | 2.557 | 1.455 | | |

**Table 2.** English writing strategies of the experimental class and the control class before the experiment

| | Number of subjects | mean | Std Dev | Independent-samples t-test | |
|---|---|---|---|---|---|
| | | | | T Value | P Value |
| Experiment class | 31 | 2.627 | 1.301 | 1.964 | 0.250 |
| Control class | 28 | 2.500 | 1.238 | | |

**Design of Computer-assisted English Writing Class.** At the beginning of the semester, the teacher gave students of the experimental class a detailed description of contents, purpose and method of the experiment thus obtained the trust and cooperation of students. In the experiment, the teacher adopted computer assisted teaching technology to the new English writing teaching mode. Multimedia technology can deliver audio-visual information, such as word, voice, image, and animation, etc. It can provide the function of realistic symbol of beauty and visual images and sound. In order to make students accept the multimedia information automatically, and have strong sense of writing interests, the teacher encouraged them to participate in the teaching activities actively and complete the writing task on time. It can change the traditional single writing teaching mode which is a teacher-based teaching into a student-oriented teaching. And in the process of writing, they can choose topics by themselves and have discussion in the group. For example, when teaching how to describe a travel in one class, the teacher gave them a flash about traveling, and then the teacher displayed some pictures about landscape of a beautiful spot. The teacher then gave students a brief introduction of his own traveling to the other cities. On the basis of this, the teacher inspired the students to have a discussion about the awareness of traveling. After that, students wrote compositions about something interesting in their travel, or had description of beauty scenes in their travel.

The vivid interactive teaching activities, such as the online sharing writing materials, the introduction of writing strategies, and online teacher-students peer-assessment, can give full play to the communicative function in the writing, and can help students use subjective initiative and creativity in the students' writing process. It can also inspire the students' interest in writing, and make the students master the basic strategy of English writing. For example, planning strategy can help students plan and arrange the goal, processes, and procedures in writing activities; Monitoring strategy can make students monitor the method and process of writing consciously, etc; Evaluation strategy adopt various ways to help students to evaluate its effect, and according to the actual situation, adjust the writing process and strengthen the self-awareness and reflection of students. For another example, in the multimedia network environment, our preparation stage in writing conception, let the students list the relevant vocabularies and concepts, or let the students do online-searching of relevant information to the topic. And then we can use the projector to show the class. The teacher adopted the way of description before writing, let the students have the mutual inspiration and extent the train of thought through classroom discussion. Therefore, the interaction between teachers and students can not only give students more knowledge and inspiration, arouse their enthusiasm and strengthen the interest of their creation in writing, but also can active the atmosphere in classroom, establish the students' writing confidence and improve the students' oral ability. They are getting ready for the draft stage.

What's more, in multimedia network environment, the teachert directly took the outlines and final compositions from different groups on the screen using the projector. And then the whole analysed their respective advantages and disadvantages.

On one hand, in multimedia network environment, in the stage of the writing evaluation, students can use E-mail to have student-student peer assessment or teacher-student assessment. In this way, students would have double identities of authors and readers in the whole process of writing. It can raise the student's consciousness of readers, so as to get better grasp of writing strategy skills and knowledge. On the other hand, the samples and the best essays of students would be made into web pages, and dub the compositions in order that they can listen to it from the web. It can not only set up the author's confidence and pride, but also can inspire other students' writing desire and passion. Reading and analyzing the models online can not only let them know the differences between Chinese and English thinking, but also make them identify weaknesses in their writing essays, thereby further master writing skills and knowledge. Last but not least, E-mail and the BBS provide a broad space for students to develop on-line discussion in writing. The students would discuss their writing problems and give their own opinions using E-mail and BBS. From the reflection of students, most of the students think multimedia network teaching environment creates a real language communicative atmosphere for their English writing. It makes them have the chance of participating and evaluating the writing task actively. And it stimulates their writing passion.

**Instrument of the research.** In this research, the writer referred to the questionnaire designed by Wen [5] and designed one of his own. The questionnaire consists of two parts. Part I is mainly for the students background information. Part II is composed of 14 questions. Questions 1,2,3,4,5 are related to students' learning motivation, and questions 6,7,8,9,10,11,12,13,14 are related to students' learning strategies.

## 3 Findings and Discussions

**General change of students learning motivation.** From the results of the questionnaire, the percentage of students who like English writing increased from 32.3% before the experiment to 61.3% after the experiment. About 58.1% of students want to enhance their English test scores through improving English writing before the experiment, and after the experiment, 67.7 % of students feel that a lot of writing training is very important in learning foreign language, an increase of 29.0% before the experiment. Thus, the internal motivation of students enhanced after the computer-assisted English writing teaching. In addition, 32.3% of the students want to go abroad for further study before the experiment, and the percentage increased to 38.7% after the experiment.

**Paired-samples t-test of the learning motivation of the experimental class pre- and post-experiment.** Table 3 shows the results of the paired-samples t-test of the learning motivation of the experimental class pre- and post-experiment. From the table, it can be seen that after the experiment, learning motivation of students in the experimental class has been significantly improved.

**Table 3.** Paired-samples t-test of the learning motivation of the experimental class pre- and post-experiment

| | Number of subjects | mean | Std Dev | Paired-samples t-test | |
|---|---|---|---|---|---|
| | | | | T Value | P Value |
| Pre-experiment | 31 | 2.606 | 1.531 | 1.967 | <0.05 |
| Post-experiment | 31 | 3.290 | 1.468 | | |

**Paired-samples t-test of the learning strategies of the experimental class pre- and post-experiment.** Table 4 shows the results of the paired-samples t-test of the learning strategies of the experimental class pre- and post-experiment. It can be seen from the results that the results of pre- and post-experiment have changed significantly. Therefore it can be concluded that after the experiment, learning strategies of students in the experimental class has been significantly improved.

**Table 4.** Paired-samples t-test of the learning strategies of the experimental class pre- and post-experiment

| | Number of subjects | mean | Std Dev | Paired-samples t test | |
|---|---|---|---|---|---|
| | | | | T Value | P Value |
| Pre-experiment | 31 | 2.627 | 1.302 | 1.964 | <0.05 |
| Post-experiment | 31 | 3.293 | 1.424 | | |

**Independent-samples t-test of the learning strategies of the experimental class and the control class post-experiment.** Table 5 shows the results of the independent-samples t-test of the learning strategies of the experimental class and the control class post-experiment. It can be seen from the results that the change between the experimental class and the control class after the experiment is significant.

**Table 5.** The independent-samples t-test of the learning strategies of the experimental class and the control class post-experiment

| | Number of subjects | mean | Std Dev | Paired-samples t-test | |
|---|---|---|---|---|---|
| | | | | T Value | P Value |
| Experimenal class | 31 | 3.293 | 1.424 | 1.964 | <0.016 |
| Control class | 28 | 3.011 | 1.248 | | |

## 4 Summary

The following conclusions can be made from the results of the experiment:

1) Computer-assisted teaching of English writing can help vocational college students improve their learning motivation.

2) Computer-assisted teaching of English writing can help vocational college students improve their learning strategies.

## References


1. Ellis, R.: The Study of Second Language Acquisition. Oxford University Press, Oxford (1994)
2. Gardner, R.C.: Social Psychology and Second Language Learning. In: The Role of Attitudes and Motivation. Edward Arnold, London (1985)
3. Mayer, R.E.: Educational Psychology: A Cognitive Approach. Little Borwn, Boston (1987)
4. Nisbert, J., Shucksmith, J.: Learning Strategies. Routledge & Kegan parlplc (1986)
5. Wen, Q.F.: Applied Linguistics: Research Methods and Thesis Writing. Foreign Language Teaching & Researching Press, Beijing (2004)

# On Autonomous Learning Mode in Translation Practice Based on Web

Zhongyan Duan

Wuhan University of Science and Technology Zhongnan Branch,
Wuhan 430223

**Abstract.** In this paper, the author advanced a web-based autonomous learning mode in translation teaching according to the practicality of translation course. Specifically, the author explored new ways of making full use of on-line learning resources in such well-known websites as Globaltimes to conduct translation practice.

**Keywords:** autonomous learning, well-known websites, Translation Practice.

## 1 Introduction

Translation course is a typical skill-oriented course characterized by practicality. During the past decades, a number of teachers and scholars have explored various practice teaching as in the Table 1.

**Table 1.** Traditional Practice Teaching of Translation

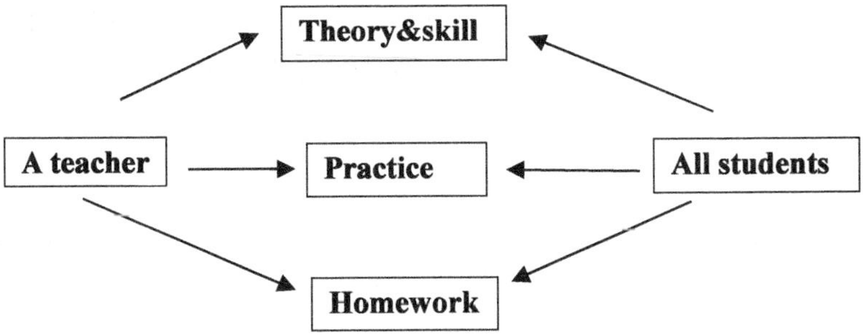


Table 1 indicates that the traditional translation teaching is still teacher-oriented. In the lecture, the beginning main step is to show a certain translation theory or skill with one or two examples to students by the teacher. The next step is to provide several sentences for all students to imply the instant theory or skill. After that, several sentences of greater difficulty, or sometimes a short passage are given to students to translate in their spare time. Almost all the learning materials are from the textbooks.

S. Lin and X. Huang (Eds.): CSEE 2011, Part II, CCIS 215, pp. 617–622, 2011.

This kind of teaching seems great. However, the result is not so satisfactory. One main cause is that the students are still passive in a teacher-oriented class, simple receivers of translation knowledge. With their teacher's instruction, they seem to know a certain theory or skill. Not active in the practice, how could they make use of the knowledge? Without application, how could they really master the skill? Another factor is the learning materials themselves. The present translation textbooks, with no more than two types of given exercises of a long or short history, fail to meet such requirement and have lost their attraction in students who are curious and fashionable (Zhongyan Duan, 2011: 377). In this paper, the author presents her new methods of autonomous learning mode apart from the traditional translation teaching.

## 2 Theoretical Foundation

In 1896, John Dewey advocated that education should be student-oriented and should satisfy the need of students. He insisted that the teacher was not the only source of knowledge, students could also be teachers. The teacher should play the role of assistants and supporters. Piaject, a Swiss psychologist, advanced constructivism, which claimed that knowledge was not simply from the teacher but through students' meaning construction in a certain environment with the help of others. In the following decades, scholars have put forward various student-oriented teaching theory and mode, including Caliber-oriented Education to Success, Task-based Language Teaching, Cooperative Language Learning Approach, and Autonomous Learning Mode.

### 2.1 Caliber-Oriented Education to Success

It is hold in Caliber-oriented Education to Success that caliber is basically innate and that its improvement is determined by education. In addition to the innate ones, caliber cannot be transferred directly; internalization is the fundamental law of its formation and development. Therefore, college education should be innovative in education models, mechanisms, content and methods, through the organization of autonomous learning, nurturing, training, experience, practice sublimation, contributing to the internalization and enhancement of students' success caliber(Zuobin Zhao, 2009: 104). Translation is one of the four skills in language learning. Obviously, internalization is an effective way for students to acquire the skill.

### 2.2 Autonomous Learning Mode

Autonomous Learning Mode is the latest one which has been researched by several scholars. Professor Tian Xianzhi has put forward three autonomous learning modes in self-study center and has applied them in her teaching practice (Xianzhi Tian, 2011: 19). In her modes, the class is student-oriented. Students are studying in groups in the study center. They are both competitive and cooperative within each groups. This mode could be well used in the teaching and learning of translation.

With a slight modification, the traditional teaching mode becomes an autonomous teaching mode as in Table 2.

**Table 2.** Autonomous Teaching Mode in Translation

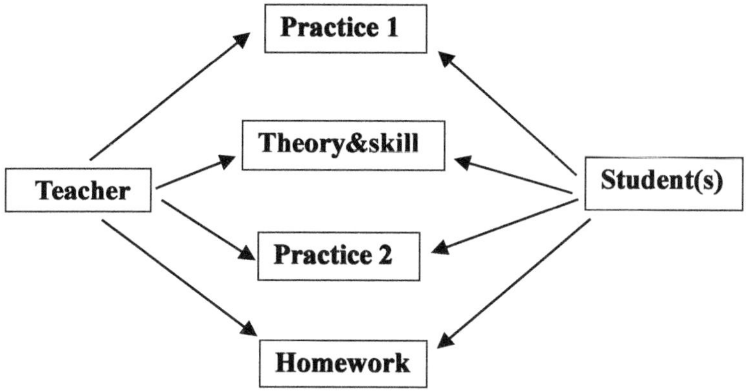


As Table 2 indicates, in autonomous teaching mode, practice comes first; therefore, this kind of mode is student-oriented. The students translate some several simple sentences individually and conclude the skill needed in the translation in groups or with the help of their teacher if necessary. At this time, the teacher illustrate the skill theoretically. After that, students will practice translating sentences of greater difficulties where the skill should be used. And there are much more exercises after class.

## 3 Autonomous Learning Mode in Translation Practice

A lot of well-known websites, such as www. Putclub.com, www.cuyoo. com, www. Globaltimes.cn, are quite practical English learning resources. In both Putclub and Globaltimes There are translation editions which include Bilingual Studio and Translation Tips, quite useful in the improvement of students' translation ability. With the help of such web-based learning materials, the teacher can maintain students' interest and expand the classroom teaching; the students themselves will improve their autonomous learning ability and be accustomed to the society.

The autonomous teaching mode in Table 2 has been applied in standard translation class for specific skill learning. Translation course is practicality-prominent. With limited class hours and immediate materials, it is difficult to internalize students' translation ability. As is known to all, translation tasks in society are mostly in the form of passage. So, it is necessary for students to translate passages regularly. Based on Professor Tian's autonomous learning modes in self-study center, the author has advanced an autonomous learning mode in translation practice which is shown in Table 3.

**Table 3.** Autonomous Teaching Mode in Translation Practice

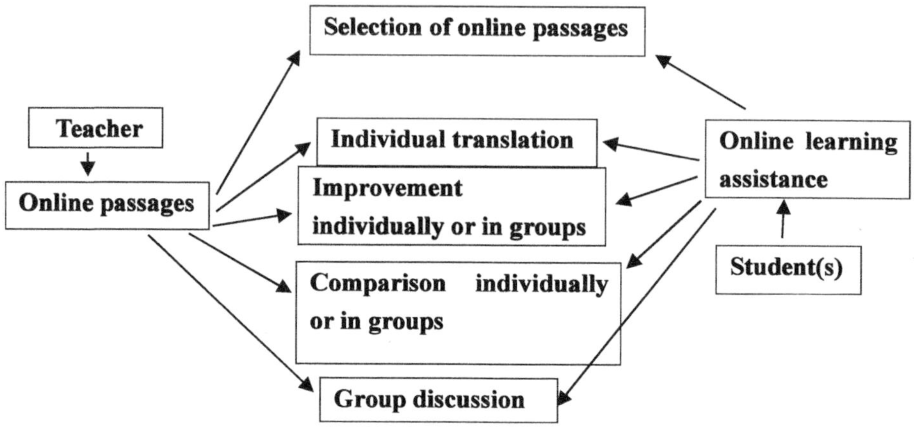


### 3.1 Selection of a Bilingual Passage of about 400 Words from Well-Known Websites

Translation practices are conducted monthly. Each time, the teacher will tell the students relevant topics in advance. Usually, the students choose one passage of about 400 words from the recommended websites or from one of the three passages selected by the teacher. With a bilingual passage, the students can compare their own drafts with the given version, which helps the development of their autonomous learning ability. Their own selection of the passage indicates that this mode is really student-oriented. It considers individual differences and encourages their participation in the task. And it can avoid their dissatisfaction from the teacher's only given passage.

### 3.2 Individual Translation Based on His Own Knowledge and the Online Learning Assistances

The first step in translation is to comprehend the original text, and here the choice of words are fundamental. It will greatly increase one's speed an accuracy of word choice when one use the online tools such as Iciba translation dictionary. In the traditional translation practice, misuses of word meaning were frequent with printed dictionaries and each time the translation was rather time-consuming. The voluminous works as Oxford Dictionary are quite heavy, and a considerable part of students were unwilling to carry. At the same time,many students can not afford Oxford Besta electronic dictionary or expensive phones with such e-dictionaries to download. Fortunately, the network is well developed now, and the education facilities are quite advanced. Students can take advantage of online translation tools to understand the original text quickly and accurately, choose the appropriate meaning, and change the original language into the target language to form the first draft, which is the most critical step in translation.

### 3.3 Improvement Individually or in Groups: Re-adjust the Printed Draft Manually in Red Ink

Readjust translations, also known as restructuring, is to organize the phrases, sentences and paragraphs in the draft, so that the target readers can understand and appreciate the translated version maximally. This is one of the important process of translation. However, the students most likely to ignore this step. That is the reason why the author makes it Step Three in translation practice. In this step, students should carefully check their translation, with special emphasis for the accuracy and consistency. They should remove unnecessary words and to add necessary words which are oversighted in their first draft. They should pay particular attention to consistency of key concepts in the translation.

Students can read the first draft on their own, or they can ask their partners to read it, or turn to the teacher if necessary. During the repeated reading, they will discover the poor expression and make appropriate modification. They are required to modify their printed draft in red ink, which shows their their seriousness during the process and which facilitates the teacher's supervision and inspection.

### 3.4 To Compare the Printed Final Draft with the Reference Translation and Write an Essay of the Practice Experience

The fourth step is to compare their printed final draft with the reference translation and write an essay of the practice experience. During the comparison, students will discover difficulties in their study and identify gaps between their own draft and the reference. This will, on the one hand, encourage them to further study and practice, on the other hand, inspire the teacher to develop appropriate teaching programs in the following teaching. Their writing of the whole process will definitely urge their serious performance and improve their writing ability which helps lay the foundation for their thesis writing. In the initial pilot period, they can write the essay in their mother tongue. In this mode, the essay should include the following three main elements:

A) A brief analysis of the original text, such as source, subject matter, genre, difficulty and length

B) Translation process, in particular the process of modification

C) comparison with the reference translation

### 3.5 Group Discussion of the Teacher's Feedback

Now the students should print out their essays and hand in all materials in the following order: the essay - the original text - the first draft - the final draft - the reference translation. The teacher should carefully review all the works, rating or giving comments. He or she should carefully examine whether the essay is consistent with the modification, and give special recognition to the excellent works. These students will be leaders in different groups and they can help the relatively poor ones in the future study. And the teacher can also spot difficulties arising in the practice and make them the focus in the future classroom teaching.

In this step, another important task is group discussion of the teacher's feedback. In each group, the members will be interested in reading and discussing the feedback, and they will be clear of their own performances with the feedback. In reading and discussion, they will see their own strength and shortages, learn from the reference translation, from each other, and from the teacher.

## 4 Conclusion

Autonomous Learning Mode meets the requirements of translation practice, and it is a useful supplement to classroom teaching. The after-class interviews and surveys indicate that the mode is generally welcomed by the students for it is scientific and rational, and the practice is necessary and feasible. It is student-oriented. It spurs students to be active in translation learning and practice. Therefore, translation ability can be internalized and become their own caliber.

Anyway, it could be further improved. The teachers are more skilled in the combination of translation theory and practice. They can consider actual teaching to set the practice hours and specific passages for students to choose. It hopefully that the mode will be of greater operability with such improvement.

## References

1. Duan, Z.: Task-based Teaching of English-Chinese Translation under "Caliber-oriented Education to Success" Based on Web. In: Advanced Research on Electronic Commerce, Web Application, and Communication, pp. 377–382 (2011)
2. Zhao, Z.: Practice and Theory of Caliber-oriented Education to Success, p. 104. Wuhan University Press, Wuhan (2009)
3. Tian, X.: A Brief Study on Autonomous Learning Mode in Self-study Center Based on Web. In: Advanced Research on Electronic Commerce, Web Application, and Communication, pp. 19–25 (2011)

# Author Index